Polyadic Algebraic Structures

Polyadic Algebraic Structures

Steven Duplij
Center for Information Processing WWU IT University of Münster,
48149 Münster, Germany

IOP Publishing, Bristol, UK

ISBN 978-0-7503-2648-3 (ebook)
ISBN 978-0-7503-2646-9 (print)
ISBN 978-0-7503-2649-0 (myPrint)
ISBN 978-0-7503-2647-6 (mobi)

DOI 10.1088/978-0-7503-2648-3

Version: 20220601

IOP ebooks

British Library Cataloguing-in-Publication Data: A catalogue record for this book is available from the British Library.

Published by IOP Publishing, wholly owned by The Institute of Physics, London

IOP Publishing, Temple Circus, Temple Way, Bristol, BS1 6HG, UK

US Office: IOP Publishing, Inc., 190 North Independence Mall West, Suite 601, Philadelphia, PA 19106, USA

To my parents Lidia and Anatoliy Duplij who made many vital efforts in my formation and development for the book to be born, and to my wife Pierpaola whose everyday assistance, inspiration and general discussions allowed me to crystallize ideas and not be afraid of introducing new concepts.

Contents

Part III Polyadic quantum groups

Preface

Nowadays, the arity of algebraic structures tends to be investigated from either a too general (e.g., in abstract algebra, as sets with operations of arbitrary arity satisfying axioms), or a too narrow (e.g., in physics, as group-like, ring-like and algebra-like concrete structures with mostly binary operations) point of view.

This book can be treated as a 'bridge' between these approaches. We propose using a full 'polyadization' procedure: take a concrete structure with binary operations and 'free-up' all arities while saving the 'interaction' between them. The latter gives equations relating arities, and since these are integers, we obtain 'quantization' rules for algebraic structures, restricting their arity shape. Another feature of this thorough 'polyadization' is the appearance of exotic properties which are not possible for binary arities, e.g., polyadic groups without identity and polyadic fields without zero. Another unusual possibility is mappings between structures of different arities, or heteromorphisms. This leads to a new kind of representation theory, in which the arity of an algebraic structure need not coincide with the arity of a representation. Further, new quantum groups, higher braidings, M-matrices and medialed categories are introduced.

The book contains original constructions and ideas developed by the author, while known material is given in brief for notational purposes only. It should attract the attention of researchers in modern theoretical physics and mathematics who are interested in new directions and developments.

Steven Duplij
Münster, Germany
December 2021

Acknowledgements

The author is deeply grateful to Vladimir Akulov, Bernd Billhardt, Andrzej Borowiec, Andrew Bruce, Vesna Celakoska, Miroslav Ćirić, Joachim Cuntz, Costantino Delizia, Christopher Deninger, Branko Dragovich, Wieslaw Dudek, Siegfried Echterhoff, Murray Geshtenhaber, Nick Gilbert, Gerald Goldin, Mike Hewitt, Lacrimioara Iancu, Vladimir Kalashnikov, Richard Kerner, Maurice Kibler, Mikhail Krivoruchenko, Grigorij Kurinnoj, Michel Lapidus, Dimitrij Leites, Daniel Lenz, Jerzy Lukierski, Yuri Manin, Riivo Must, Chi-Keung Ng, Peter Nonnenmann, Thomas Nordahl, Wolfram Von Oertzen, Valentin Ovsienko, Walther Paravicini, Christopher Phillips, Norbert Poncin, Maria Pop, Vasile Pop, Sergey Prokushkin, Anatoliy Prykarpatsky, Ioan Purdea, Mathias Pütz, Dominik Rudolph, Jim Stasheff, Vladimir Tkach, Ron Umble, Raimund Vogl, Thomas Timmermann, Alexander Voronov, Raimar Wulkenhaar, and Wend Werner for fruitful discussions, important help and valuable support.

Author biography

Steven Duplij

Steven Duplij (Stepan Douplii) is a theoretical and mathematical physicist from the University of Münster, Germany. He was born in Chernyshevsk–Zabaikalsky, Russia, and studied at Kharkiv University, Ukraine where he gained his PhD in 1983. While working at Kharkiv he received the title Doctor of Physical and Mathematical Sciences by Habilitation in 1999.

Dr Duplij is the editor-compiler of 'Concise Encyclopedia of Supersymmetry' (2005, Springer), and is the author of more than a hundred scientific publications and several books. He is listed in the World Directory Of Mathematicians, Marques Who Is Who In America, the Encyclopedia of Modern Ukraine, the Academic Genealogy of Theoretical Physicists and the Mathematics Genealogy Project. His scientific directions include supersymmetry and quantum groups, advanced algebraic structures, gravity and nonlinear electrodynamics, constrained systems and quantum computing.

Symbols

A list of symbols and notations used in the text.

S, T, G	— sets
$\mu^{(n)}$, $\nu^{(n)}$, $\rho^{(n)}$	— n-ary multiplications, additions, n-place actions ('elementwise' products)
$\boldsymbol{\mu}^{(n)}$, $\boldsymbol{\nu}^{(n)}$, $\boldsymbol{\rho}^{(n)}$	— n-ary multiplications, additions, n-place actions ('diagrammatic' linear maps)
$\Delta^{(n)}$	— n-ary comultiplications ('elementwise')
$\boldsymbol{\Delta}^{(n)}$, $\boldsymbol{\sigma}^{(n)}$	— n-ary comultiplications, coactions ('diagrammatic' linear maps)
$(\mu^{(n)})^{\circ\ell}$	— composition of ℓ n-ary multiplications (additions, actions)
s, **g**, **h**, **E**	— $(s_1, \ldots, s_n)$, n-tuples, polyads, sequences
$\mathcal{S}^{(n)}$, $\mathcal{G}^{(n)}$, $\mathcal{R}^{(n)}$, $\mathcal{A}^{(n)}$	—semigroups, groups, rings, algebras of arity n
$\mathbb{R}$, $\mathbb{C}$, $\mathbb{k}$	— ground fields
`Alg`, `Grp`	— category names
$\mathbf{\Phi}$, $\mathbf{\Psi}$, $\mathbf{\Pi}$	— homomorphisms, heteromorphisms, representations
$\mathsf{0}$, z_K, $\mathsf{1}$	— zero and unity of ground fields
z, z_V	— zero of semigroups and algebras, vector spaces
e	—identity of groups, semigroups and algebras
B, A	— braidings, associators
$\mathfrak{U}$, $\mathfrak{T}$	— unitors, natural transformations
F, C	— functors, categories
L, M	— Lie brackets, medial brackets
$A^{\times n}$	— direct product of n copies of A
$A^{\otimes n}$	— tensor product of n copies of A
$\mathcal{G}'$, a'	— grading groups, elements
f	— membership functions
$\mathfrak{A}^{(\mathsf{f})}$	— obscure (fuzzy) set
H	— cohomology classes
Italic	— definitions
San serif	— stressing
Quotes	— figurative meaning

Introduction

The study of polyadic (or higher arity) algebraic structures has a two-century long history, commencing with works by Cayley, Sylvester, Kasner, Prüfer, Dörnte, Lehmer, Post, etc. They took a single set, closed under one (main) binary operation having special properties (the so-called 'group-like' structure), and 'generalized' it by increasing the arity of that operation, which can then be called a *polyadic operation* and the corresponding algebraic structure *polyadic*. We will use the term 'polyadic' in this sense only, while other uses are extant in the literature (see, e.g., Halmos 1962). An 'abstract way' to study polyadic algebraic structures is via the use of universal algebras defined as sets with various axioms (equational laws) for the polyadic operations (Cohn 1965, Grätser 1968, Bergman 2012). However, some important algebraic structures cannot be described in this language, e.g., ordered groups, fields, etc (Denecke and Wismath 2009). It is therefore also worthwhile to pursue a 'concrete approach' by studying examples of binary algebraic structures and then to 'polyadize' them in the proper way. This method has initiated the development of a corresponding theory of n-ary quasigroups (Belousov 1972), n-ary semigroups (Monk and Sioson 1966, Zupnik 1967) and n-ary groups (Gal'mak 2003, Rusakov 1998) (for a more recent review, see, e.g., Duplij 2012 and a comprehensive list of references therein). Binary algebraic structures with two operations (addition and multiplication) on one set (the so-called 'ring-like' structures) were later generalized to (m, n)-rings (Celakoski 1977, Crombez 1972, Leeson and Butson 1980) and (m, n)-fields (Iancu and Pop 1996) (for a recent study, see Duplij 2017), while these were mostly investigated in a more restrictive manner by considering particular cases: ternary rings (or (2, 3)-rings) (Lister 1971), $(m, 2)$-rings (Boccioni 1965, Pop and Pop 2002), as well as (3, 2)-fields (Duplij and Werner 2021).

In chapter 1 one-set polyadic algebraic structures are reviewed and their classification is given. The general ways of changing arity and special elements are considered. Polyadic direct products of (semi)groups which can change arity are then introduced. A new form of the Hosszú–Gluskin theorem is presented in terms of polyadic powers and using the language of diagrams. It is shown that the Hosszú–Gluskin chain formula is not unique and can be 'deformed' using a parameter q which takes special integer values, which leads to the concept of a sandwich nonunital endomorphism. The 'q-deformed' analog of the Hosszú–Gluskin theorem is then formulated, and a 'q-deformed' homomorphism theorem is given. Then we generalize the Grothendieck construction of the completion group for a monoid (being the starting point of the algebraic K-theory) to the polyadic case, when an initial semigroup in m-ary and the corresponding final class group $K_0^{(m,n)}$ can be n-ary.

In chapter 2 a new multiplace generalization of associativity preserving homomorphisms, a 'heteromorphism' which connects polyadic systems having unequal arities, is introduced via an explicit formula, together with related definitions for multiplace representations and multiactions. The classification of heteromorphisms is given, and it is seen that their arity are not arbitrary, but 'quantized'. Then regular multiactions are introduced and the corresponding regular multiplace representations are defined.

In chapter 3 the regularity concept for semigroups is generalized in two ways simultaneously: to higher regularity and to higher arity. We show that in this approach the one-relational and multi-relational formulations of higher regularity do not coincide, and each element can have several inverses. For polyadic semigroups we introduce several types of higher regularity which satisfy the arity invariance principle: the expressions should not depend of the numerical arity values. We introduce sandwich higher polyadic regularity with generalized idempotents. We use the polyadic-binary correspondence in which the product of a special kind of matrices reproduces the higher regular semigroups and higher braid groups, and the latter are connected with the generalized polyadic braid equation. The higher degree (in our sense) Coxeter group and symmetry groups are then defined.

In chapter 4 polyadic rings and fields are introduced. Their direct products are unusual and can have different arity. The polyadic integer numbers, which form a polyadic ring, are defined. The basics of the new polyadic arithmetic thus introduced is presented: prime polyadic numbers, the polyadic Euler function, polyadic division with a remainder, etc. Secondary congruence classes of polyadic integer numbers and the corresponding finite polyadic rings are defined. A polyadic version of (prime) finite fields is introduced. These can be zeroless, zeroless and nonunital, or have several units; it is even possible for all of their elements to be units. None of the above situations is possible in the binary case. It is conjectured that a finite polyadic field should contain a certain canonical prime polyadic field, defined here as a minimal finite subfield, which can be considered as a polyadic analog of $GF(p)$. The Diophantine equations over the polyadic rings are then considered. Polyadic analogs of the Lander–Parkin–Selfridge conjecture and Fermat's last theorem are formulated. For polyadic numbers neither of the standard statements holds. Polyadic versions of the Frolov's theorem and the Tarry–Escott problem are presented.

In chapter 5 concrete two-set (module-like and algebra-like) algebraic structures are investigated from the viewpoint that the initial arities of all operations are arbitrary. Relations between operations arising from the structure definitions, however, lead to restrictions which determine their possible arity shapes (which are 'quantized') and allow us to formulate the partial arity freedom principle. Polyadic vector spaces and dual vector spaces, direct sums, associative algebras and tensor products are reconsidered, and a classification in terms of the units and zeroes of an algebra and its underlying field is provided. Medial polyadic associative algebras are introduced, their 'quantized' arity shape is given, and polyadic structure constants are defined.

In chapter 6 elements of polyadic operator theory are outlined: inner pairing spaces, multistars and polyadic analogs of adjoints, operator norms, isometries and projections are introduced. The arity signature and arity shape of polyadic algebraic structures is presented as a general table. Then polyadic C^*-algebras, Toeplitz algebras and Cuntz algebras are defined and represented by polyadic operators.

In chapter 7 algebraic structures in which the property of commutativity is replaced by the mediality property are introduced. We consider (associative) graded algebras and in place of almost commutativity, we introduce almost mediality: the commutativity-to-mediality ansatz. Higher graded twisted products and 'deforming' brackets (being the medial analog of Lie brackets) are defined. Toyoda's theorem which connects (universal) medial algebras with abelian algebras is proven for the almost medial graded algebras introduced here.

In chapter 8 a new general mechanism for 'breaking' commutativity in algebras is proposed: if the underlying set is taken to be not a crisp set, but rather an obscure/

fuzzy set, the membership function, reflecting the degree of truth that an element belongs to the set, can be incorporated into the commutation relations. Special 'deformations' of commutativity and almost commutativity are introduced in such a way that equal degrees of truth result in the 'nondeformed' case. We also sketch how to 'deform' ε-Lie algebras and Weyl algebras and introduce the double almost-Lie algebras. Further, the above constructions are extended to n-ary algebras for which the analogs of projective representations and ε-commutativity are studied.

In chapter 9 generalized polyadic coassociative coalgebras, bialgebras and Hopf algebras are defined and investigated. They have many unusual features in comparison with the binary case. For instance, both the algebra and its underlying field can be zeroless and nonunital, the existence of the unit and counit is not obligatory, while the dimension of the algebra is not arbitrary, but 'quantized'. The polyadic convolution product and bialgebra can be defined, and when the algebra and coalgebra have unequal arities, the polyadic version of the antipode, the querantipode, has different properties. We introduce the polyadic version of braidings, almost co-commutativity, quasitriangularity and the equations for the R-matrix (which can be treated as a polyadic analog of the Yang–Baxter equation). We propose another concept of deformation which is governed not by the twist map, but by the medial map. We present the corresponding braidings, almost co-mediality and M-matrix (the medial analog of the R-matrix), for which compatibility equations are found.

In chapter 10 constant matrix solutions of the polyadic (higher) braid equations are introduced and studied. They play a role as various quantum gates in quantum computation. We define two classes of 8-vertex matrices, star and circle, such that the magic matrices (connected with the Cartan decomposition) belong to the star class. We also define partial identity and unitarity. The general algebraic structure of the classes is described in terms of new semigroups, ternary and 5-ary groups and modules. We present the invertible and noninvertible 8-, 9- and 10-vertex Yang–Baxter maps. We then introduce polyadic braid operators for higher braid equations, find constant matrix solutions for the ternary 8- and 16-vertex cases and give their Pauli matrix presentation.

In chapter 11 we generalize tensor categories and braided tensor categories using the commutativity-to-mediality ansatz, as in chapter 7 for algebras. A polyadic (non-strict) tensor category has an n-ary tensor product as an additional multiplication with $n-1$ associators of the arity $2n-1$ satisfying a (n^2+1)-gon relation, which is a polyadic analog of the pentagon axiom. Polyadic monoidal categories may contain several unit objects, and it is also possible that all objects are units. A new kind of polyadic categories (called groupal) is defined: they are close to monoidal categories, but may not contain units: instead (by analogy with the querelements in n-ary groups) the querfunctor and (natural) functorial isomorphisms, the quertors, are considered. The arity-nonreducible n-ary braiding is introduced and the equation for it is derived, which for $n=2$ coincides with the Yang–Baxter equation. Then, analogously to the approach in chapter 7, we introduce 'medialing' instead of braiding and construct 'medialed' polyadic tensor categories.

New constructions and ideas

Here we list new constructions and ideas presented in the book:

- **Polyadization**—Exchanging all binary operations in an algebraic structure with the polyadic ones preserving 'interrelations' between them.
- **Sandwitch quasi-endomorphism**—A nonunital endomorphism of the sandwitch form (1.4.75).
- **Full polyadic direct product**—An external direct product of polyadic algebraic structures of the same arity (1.3.5), and for polyadic rings (4.2.5)–(4.2.18).
- **Iterated direct product**—A direct product of polyadic algebraic structures of different arities (1.3.12), see for (m, n)-rings (4.2.17)–(4.2.18).
- **Hetero ('entangled') k-power**—An external power of polyadic algebraic structures which changes arity, but preserves associativity (1.3.18).
- **Deformed Hosszú–Gluskin chain formula and theorem**—see (1.4.98) and theorem 1.4.12.
- **Heteromorphism**—A mapping between algebraic structures of different arities (2.2.4).
- **Multiplace representation**—A representation (2.4.5) of a on-set polyadic algebraic structure given by a multiplace function (2.4.4).
- **Arity changing formula**—Connection between arities of initial and final algebraic structures in heteromorphism (2.2.16).
- **Associative polyadic quiver**—A graph preserving associativity with respect to heteromorphism (2.2.30)–(2.2.31).
- ***k*-place action/multiaction**—see (2.5.2).
- **Higher regularity**—Regularity with multi-element inverses (3.1.3).
- **Arity invariance principle**—An expression of algebraic structure in term of operations does not depend on their arities (definition 3.1.4).
- **Sandwich polyadic regularity**—Polyadic analog of higher regularity in terms of idempotents (3.3.29).
- ***n*-ary matrices**—Set of matrices which is closed with respect to (not less than) n-ary multiplication (3.4.10).
- **Polyadic-binary correspondence**—Reproduction of higher regularity by product of n-ary matrices (3.4.15)–(3.4.21).
- **Higher *n*-degree braid (binary) group**—Generalization of the Artin braid group for n multipliers (3.4.81)–(3.4.88).
- **Higher *n*-degree Coxeter group**—see (3.4.115)–(3.4.116).
- **Quer-symmetric polyadic field**—A (m, n)-field in which unary queroperations commute, see definition 4.1.16.
- **Iterated direct product polyadic ring**—A direct product of polyadic rings of different arities (4.2.17)–(4.2.18).
- **Polyadic direct product field**—An external product of polyadic zeroless fields of the same arity shape, see definition 4.2.8.
- **Polyadic integer number**—A (m, n)-ring of representatives of a congruence (residue) class (4.3.9).

- **Secondary congruence class**—A double class of polyadic integers which gives finite polyadic ring (4.4.3).
- **Non-extended finite polyadic field**—An abstract non-extended prime finite (m, n)-field of polyadic integers (4.5.1).
- **Diophantine equation over polyadic integers**—see (4.6.1).
- **Polyadic Lander–Parkin–Selfridge conjecture**—see conjecture 4.6.7.
- **Polyadic Pythagorean theorem**—see definition 4.6.5.
- **Polyadic Fermat's Last Theorem**—see conjecture 4.6.6.
- **Polyadic vector space**—The 2-set 4-operation algebraic structure satisfying 7 axioms, see definition 5.1.6.
- **Polyadic associative algebra**—The 2-set 5-operation algebraic structure satisfying 4 compatibility conditions (5.3.4)–(5.3.12), see definition 5.3.1.
- **Polyadic convolution**—see definition 5.3.15.
- **Arity partial freedom principle**—theorem 5.3.12.
- **Polyadic unit map**—see (5.3.39).
- **Polyadic structure constants**—see (5.3.50).
- **Arity shape of polyadic algebraic structures**—see table 6.1.
- **Polyadic multistar adjoint**—see (6.2.21).
- **Polyadic isometry**—A polyadic operator which preserves the polyadic inner pairing, see definition 6.2.15.
- **Polyadic operator C^*-algebra**—The operator polyadic Banach algebra satisfying the multi-C^*-relations (6.2.58).
- **Membership deformation**—Deformation of commutativity using the membership function (8.3.2).
- **Higher level Lie bracket**—see (7.1.18)–(7.1.20).
- **Medial brackets**—Deviation from almost mediality (7.2.19).
- **Higher level mediality bracket**—see (7.4.23)–(7.4.25).
- **Ternary Toyoda theorem**—A shape of multiplication for almost medial algebras, see theorem 7.5.2.
- **Obscure algebra**—An algebra with the multiplication deformed by the membership function (8.3.1).
- **Double almost commutative algebra**—An ε-commutative algebra additionally deformed by the membership function, see definition 8.3.13.
- **Double Lie bracket**—Deviation from the double commutativity (8.3.23).
- **Polyadic projective representation**—A multiplace generalization of the projective representation (8.4.12).
- **Polyadic π-commutative algebra**—A polyadic graded algebra deformed by n-ary π-factor (8.5.3).
- **Polyadic counit**—see (9.1.22).
- **Polyadic duality**—see definition 9.1.33.
- **n-ary convolution product**—A product consisting of polyadic multiplication an comultiplication having different arity (9.1.64).
- **Polyadic querantipod**—A coquerelement for the identity with respect to the n-ary convolution product (9.3.10).
- **n-ary braid equation**—The equation for the morphism of n modules (9.5.23).

- *n*-**ary** *R*-**matrix**—An element of polyadic bialgebra which governs 'interaction' of product and coproduct (9.5.29).
- **Polyadic** *M*-**matrix**—An element of polyadic almost co-medial bialgebra which governs 'interaction' of product and coproduct (9.5.48).
- **Medial quasipolyangularity**—see (9.5.70)–(9.5.72).
- **Partial unitary matrix**—A matrix having a product with its conjugate-transposed as a partial identity (10.1.20).
- **Star and circle matrices**—Matrices of special form to which belong some solutions of the Yang–Baxter and higher braid equations (10.1.76)–(10.1.77) and (10.3.53)–(10.3.54).
- *n*-**ary associator**—Polyadic generalization of the associator (11.2.7).
- **Polyadic unitor**—An *n*-ary unit morphism (11.3.1).
- **Querfunctor, querobject, quermorphism**—Analogs of querelement in polyadic groupal categories, see definition 11.3.4.
- *n*-**ary braiding**—Natural transformation which permutes *n*-ary tensor product (11.4.9).
- *n*-**ary braid equation**—The equation for *n*-ary braiding (11.4.15).
- *n*-**ary medial braiding**—A functorial n^2-isomorphism (11.5.1).
- **Medialed polyadic monoidal category**—see definition 11.5.4.
- **Polyadic 'gauge' and 'twisted' shifts**—Polyadic analogs of componentwise and twisted shifts to define the equaivalence relations on a polyadic semigroup (1.5.44) and (1.5.45).
- *n*-**ary group completion of** *m*-**ary semigroup** $K_0^{(m,n)}$—A polyadic group of equivalence classes for a polyadic semigroup (1.5.70).

References

Belousov V D 1972 *N-ary Quasigroups* (Kishinev: Shtiintsa)

Bergman C 2012 *Universal Algebra: Fundamentals and Selected Topics* (New York: CRC Press)

Boccioni D 1965 Caratterizzazione di una classe di anelli generalizzati *Rend. Semin. Mat. Univ. Padova* **35** 116–27

Celakoski N 1977 On (F, G)-rings *God. Zb., Mat. Fak. Univ. Kiril Metodij Skopje* **28** 5–15

Cohn P M 1965 *Universal Algebra* (New York: Harper and Row)

Crombez G 1972 On (n, m)-rings *Abh. Math. Semin. Univ. Hamb.* **37** 180–99

Denecke K and Wismath S L 2009 *Universal Algebra and Coalgebra* (Singapore: World Scientific)

Duplij S 2012 Polyadic systems, representations and quantum groups *J. Kharkiv Natl Univ. Ser. Nuclei Part. Fields* **1017** 28–59 Expanded version arXiv:math.RT/1308.4060

Duplij S 2017 Polyadic integer numbers and finite (m, n)-fields *p-Adic Numbers Ultrametric Anal. Appl.* **9** 267–91

Duplij S and Werner W 2021 Structure of unital 3-fields *Math. Semesterber.* **68** 27–53

Gal'mak A M 2003 n-*Ary Groups, Part 1* (Gomel: Gomel University)

Grätser G 1968 *Universal Algebra* (New York: Van Nostrand)

Halmos P 1962 *Algebraic Logic* (New York: Chelsea)

Iancu L and Pop M S 1996 A Post type theorem for (*m*,*n*) fields *Proc. of the Scientific Communications Meeting of 'Aurel Vlaicu' University (Arad, Romania, 16–17 May 1996) Edition III* vol 14A, ed G Halic and G Cristescu (Arad: 'Aurel Vlaicu' Univ. of Arad Publishing Centre) pp 13–8
Leeson J J and Butson A T 1980 On the general theory of (m, n) rings *Algebra Univ.* **11** 42–76
Lister W G 1971 Ternary rings *Trans. Am. Math. Soc.* **154** 37–55
Monk J D and Sioson F 1966 *m*-Semigroups, semigroups and function representations *Fundam. Math.* **59** 233–41
Pop A and Pop M S 2002 Some embeddings theorems for $(n, 2)$-rings *Bul. Ştiinţ. Univ. Baia Mare Ser.* B **18** 311–6
Rusakov S A 1998 *Some Applications of* n-Ary Group Theory (Minsk: Belaruskaya Navuka)
Zupnik D 1967 Polyadic semigroups *Publ. Math. (Debrecen)* **14** 273–9

Part I

One-set polyadic algebraic structures

Chapter 1

One-set algebraic structures and Hosszú–Gluskin theorem

The goal of this chapter is to give a review of some polyadic one-set algebraic structures. First, we establish notations and classify their general properties, then introduce n-ary semigroups, quasigroups and groups, and briefly recall their properties.

1.1 General properties of one-set one-operation polyadic structures

We consider a nonempty set G (the underlying set), and its elements we denote by lower-case Latin letters $g_i \in G$. The *n-tuple* (or *polyad*) $g_1, \dots, g_n$ of elements from G is denoted by $(g_1, \dots, g_n)$. The Cartesian product $\overbrace{G \times \dots \times G}^{n} \equiv G^{\times n}$ consists of all n-tuples $(g_1, \dots, g_n)$, such that $g_i \in G$, $i = 1, \dots, n$. For all equal elements $g \in G$, we denote n-tuple (polyad) by power (g^n). If the number of elements in the n-tuple is clear from the context or is not important, we denote it with one bold letter $(\mathbf{g})$, in other cases we use the power in brackets $(\mathbf{g}^{(n)})$. The *i-projection* of the Cartesian product $G^{\times n}$ on its ith 'axis' is the map $\mathsf{Pr}_i^{(n)}: G^{\times n} \to G$ such that $(g_1, \dots, g_i, \dots, g_n) \longmapsto g_i$. The *i-diagonal* $\mathsf{Diag}_n: G \to G^{\times n}$ sends one element to the equal element n-tuple $g \longmapsto (g^n)$. Note that the one-point set $\{\bullet\}$ can be treated as a unit for the Cartesian product, since there are bijections between G and $G \times \{\bullet\}^{\times n}$, where G can be on any place. We define on the Cartesian product $G^{\times n}$ a polyadic (or n-ary) operation $\mu^{(n)}: G^{\times n} \to G$. We use small Greek letters and place arguments in square brackets and arity as the upper index in brackets $\mu^{(n)}[\mathbf{g}]$. Usually the operations with $n = 1, 2, 3$ are called *unary, binary and ternary*. The case of zero arity $n = 0$ is special and corresponds to fixing a distinguished element of G, a 'constant' $c \in G$, and it is called a *0-ary operation* $\mu_c^{(0)}$, which maps the one-point set $\{\bullet\}$ to G, such that $\mu_c^{(0)}: \{\bullet\} \to G$, and (formally) has the value $\mu_c^{(0)}[\{\bullet\}] = c \in G$. The 0-ary operation 'kills' arity, which can be seen from the following (Bergman 1995): the composition of n-ary and m-ary operations $\mu^{(n)} \circ \mu^{(m)}$ gives $(n + m - 1)$-ary operation by

$$\mu^{(n+m-1)}[\mathbf{g}, \mathbf{h}] = \mu^{(n)}[\mathbf{g}, \mu^{(m)}[\mathbf{h}]]. \tag{1.1.1}$$

If to compose $\mu^{(n)}$ with the 0-ary operation $\mu_c^{(0)}$, we obtain

$$\mu_c^{(n-1)}[\mathbf{g}] = \mu^{(n)}[\mathbf{g}, c], \tag{1.1.2}$$

since $\mathbf{g}$ is a polyad of length $(n - 1)$. Thus, it is necessary to make a clear distinction between the 0-ary operation $\mu_c^{(0)}$ and its value c in G, as will be seen and will become important below.

Definition 1.1.1. A *polyadic algebraic structure* $\mathcal{G}$ is a set G closed with respect to polyadic operations.

So, we write $\mathcal{G} = \langle \text{set} | \text{operations} \rangle$ or $\mathcal{G} = \langle \text{set} | \text{operations} | \text{relations} \rangle$, where 'relations' (axioms) are some additional properties of operations (e.g., associativity for semigroups or cancellativity). It is not necessary to list the images of 0-ary operations (e.g., the unit or zero in groups), as is done in various other definitions. We mostly consider concrete polyadic algebraic structures with one fundamental n-ary operation $\mu^{(n)}\colon G^{\times n} \to G$, which is called *polyadic multiplication* (or *n-ary multiplication*).

Definition 1.1.2. A one-set *n-ary algebraic structure* $\mathcal{M}^{(n)} = \langle M | \mu^{(n)} \rangle$ or *polyadic magma* (*n-ary magma*) is a set M closed under one n-ary operation $\mu^{(n)}$ (without any other additional structure).

Remark **1.1.3.** A set with one closed binary operation without any other relations was called a groupoid by Hausmann and Ore (1937) (see, also Clifford and Preston 1961). However, nowadays the term 'groupoid' is widely used in category theory and homotopy theory for a different construction (category with only invertible morphisms), the so-called Brandt groupoid (Brandt 1927) (see, also, Bruck 1966). Alternatively, and much later on, Bourbaki (1998) introduced the term 'magma' for binary algebraic structures. Then, the above terms were also extended to the case of one fundamental n-ary operation. We will also use some neutral notations 'polyadic algebraic structure' and 'n-ary algebraic structure' (when arity n is fixed/important), which adequately indicates all of their main properties.

1.1.1 Changing arity of one-set polyadic algebraic structures

Let us consider the ways to change arity of polyadic algebraic structures.

Problem 1.1.4. [Changing arity problem] For a given n-ary algebraic structure $\langle G | \mu^{(n)} \rangle$ to construct another polyadic algebraic structure $\langle G | \mu'^{(n')} \rangle$ over the same set G, which has a multiplication with a different arity n'.

The simplest examples of how to change the arity of a polyadic algebraic structure are given by (1.1.1) and (1.1.2). There are three general ways:

(i) *Iteration.* Using composition of the operation $\mu^{(n)}$ with itself, one can increase the arity from n to n'_{iter} (as in (1.1.1)) without changing the signature of the algebraic structure. Denote the number of iterating multiplications by ℓ_μ, and call the resulting composition an *iterated product* $(\mu^{(n)})^{\circ\ell_\mu}$ (or $(\mu^{(n)})^{\circ r}$ if the number of multiplications ℓ_μ is obvious or not important), such that

$$\mu'^{(n')} = (\mu^{(n)})^{\circ\ell_\mu} \stackrel{\mathrm{def}}{=} \overbrace{\mu^{(n)}\circ(\mu^{(n)}\circ\ldots(\mu^{(n)}\times\mathrm{id}^{\times(n-1)})\ldots\times\mathrm{id}^{\times(n-1)})}^{\ell_\mu}, \tag{1.1.3}$$

where the arities are connected by

$$n' = n_{\mathrm{iter}} = \ell_\mu(n-1)+1, \tag{1.1.4}$$

which gives the length of a polyad $(\mathbf{g})$ in the notation $(\mu^{(n)})^{\circ\ell_\mu}[\mathbf{g}]$. Without assuming associativity there are many variants for placing $\mu^{(n)}$ among id in the rhs of (1.1.3). The operation $(\mu^{(n)})^{\circ\ell_\mu}$ (1.1.3) is also named a *long product* (Dörnte 1929). An example of the iterated product can be given for a ternary operation $\mu^{(3)}$ ($n = 3$), where we can construct a 7-ary operation ($n' = 7$) by $\ell_\mu = 3$ compositions

$$\mu'^{(7)}[g_1, \ldots, g_7] = (\mu^{(3)})^{\circ 3}[g_1, \ldots, g_7] = \mu^{(3)}[\mu^{(3)}[\mu^{(3)}[g_1, g_2, g_3], g_4, g_5], g_6, g_7]. \tag{1.1.5}$$

(ii) *Reduction (Collapsing)*. Using n_c distinguished elements or constants (or n_c additional 0-ary operations $\mu^{(0)}_{c_i}$, $i = 1, \ldots, n_c$), one can decrease arity from n to n'_{red} (as in (1.1.2)), such that[1]

$$\mu'^{(n')} = \mu^{(n')}_{c_1\ldots c_{n_c}} \stackrel{\mathrm{def}}{=} \mu^{(n)}\circ\left(\overbrace{\mu^{(0)}_{c_1}\times\ldots\times\mu^{(0)}_{c_{n_c}}}^{n_c}\times\mathrm{id}^{\times(n-n_c)}\right), \tag{1.1.6}$$

where for arities

$$n' = n_{\mathrm{red}} = n - n_c, \tag{1.1.7}$$

and the 0-ary operations $\mu^{(0)}_{c_i}$ can be on any (allowed) places in (1.1.6). If we compose $\mu^{(n)}$ with the 0-ary operation $\mu^{(0)}_c$, then (see (1.1.2))

$$\mu^{(n-1)}_c[\mathbf{g}] = \mu^{(n)}[\mathbf{g}, c], \tag{1.1.8}$$

and this can be treated as a definition of a new $(n-1)$-ary operation $\mu^{(n-1)}_c = \mu^{(n)}\circ\mu^{(0)}_c$.

[1] In Dudek and Michalski (1984) $\mu^{(n')}_{c_1\ldots c_{n_c}}$ is named a retract (which term is already busy and widely used in category theory for another construction).

(iii) *Mixing.* Changing (increasing or decreasing) arity may be done by combining iterating and reducing (maybe with additional operations of different arity). If we do not use additional operations, the final arity can be presented in a general form using (1.1.4) and (1.1.7). So we have two subcases:

a. *Iteration → Reduction.* We have

$$n' = n_{\text{iter}\to\text{red}} = \ell_\mu(n-1) - n_c + 1. \tag{1.1.9}$$

The maximal number of constants (when $n'_{\text{iter}\to\text{red}} = 2$) is equal to

$$n_c^{\max} = \ell_\mu(n-1) - 1 \tag{1.1.10}$$

and can be increased by increasing the number of multiplications ℓ_μ.

b. *Reduction → Iteration.* We obtain

$$n' = n_{\text{red}\to\text{iter}} = \ell_\mu(n - 1 - n_c) + 1. \tag{1.1.11}$$

The maximal number of constants is

$$n_c^{\max} = n - 2 \tag{1.1.12}$$

which is achieved when $\ell_\mu = 1$.

***Example* 1.1.5.** If the initial multiplication is binary $\mu^{(2)} \equiv (\cdot)$, and there is one 0-ary operation $\mu_c^{(0)}$, we can construct the following mixing operation

$$\mu_c^{(n)}[g_1, \ldots, g_n] = g_1 \cdot g_2 \cdot \ldots g_n \cdot c, \tag{1.1.13}$$

which in our notation can be called a *c-iterated multiplication*[2].

It is important to find conditions where iterating and reducing compensate each other, i.e., they do not change arity overall. Indeed, let the number of the iterating multiplications ℓ_μ be fixed, then we can find such a number of reducing constants $n_c^{(0)}$, such that the final arity will coincide with the initial arity n. The result will depend on the order of operations. There are two cases:

(i) *Iteration →Reduction.* For the number of reducing constants $n_c^{(0)}$ we obtain from (1.1.4) and (1.1.7)

$$n_c^{(0)} = (n-1)(\ell_\mu - 1), \tag{1.1.14}$$

such that there is no restriction on ℓ_μ.

(ii) *Reduction →Iteration.* For $n_c^{(0)}$ we get

$$n_c^{(0)} = \frac{(n-1)(\ell_\mu - 1)}{\ell_\mu}, \tag{1.1.15}$$

[2] According to Dudek (2007) the operation (1.1.13) is called c-derived.

and now $\ell_\mu \leqslant n-1$. The requirement that $n_c^{(0)}$ should be an integer gives two further possibilities

$$n_c^{(0)} = \begin{cases} \frac{n-1}{2}, & \ell_\mu = 2, \\ n-2, & \ell_\mu = n-1. \end{cases} \tag{1.1.16}$$

These relations are important for the study of various polyadic multiplications.

1.1.2 Special elements in one-set polyadic structures

Recall the definitions of some standard algebraic structures and their special elements.

A *polyadic zero* of a polyadic algebraic structure $\mathcal{G}^{(n)}\langle G|\mu^{(n)}\rangle$ is a distinguished element $z \in G$ (and the corresponding 0-ary operation $\mu_z^{(0)}$) such that for any $(n-1)$-tuple (polyad) $\mathbf{g}^{(n-1)} \in G^{\times(n-1)}$ we have

$$\mu^{(n)}[\mathbf{g}^{(n-1)}, z] = z, \tag{1.1.17}$$

where z can be on any (allowed) place in the lhs of (1.1.17). If its place is not fixed it can be only one zero. As in the binary case, an analog of positive powers of an element (Post 1940) should coincide with the number of multiplications ℓ_μ in the iterating (1.1.3).

A (positive) *polyadic power* of an element is

$$g^{\langle \ell_\mu \rangle} = \left(\mu^{(n)}\right)^{\circ \ell_\mu}\left[g^{\ell_\mu(n-1)+1}\right]. \tag{1.1.18}$$

***Example* 1.1.6.** Let us consider a polyadic version of the binary q-addition which appears in study of nonextensive statistics (see, e.g., Tsallis 1994, Nivanen 2003)

$$\mu^{(n)}[\mathbf{g}] = \sum_{i=1}^{n} g_i + \hbar \prod_{i=1}^{n} g_i, \tag{1.1.19}$$

where $g_i \in \mathbb{C}$ and $\hbar = 1 - q_0$, q_0 is a real constant (we put here $q_0 \neq 1$ or $\hbar \neq 0$). It is obvious that $g^{\langle 0 \rangle} = g$, and

$$g^{\langle 1 \rangle} = \mu^{(n)}[g^{n-1}, g^{\langle 0 \rangle}] = ng + \hbar g^n. \tag{1.1.20}$$

So we have the following recurrence formula

$$g^{\langle k \rangle} = \mu^{(n)}[g^{n-1}, g^{\langle k-1 \rangle}] = (n-1)g + (1 + \hbar g^{n-1})g^{\langle k-1 \rangle}. \tag{1.1.21}$$

Solving this for an arbitrary polyadic power we get

$$g^{\langle k \rangle} = g\left(1 + \frac{n-1}{\hbar} g^{1-n}\right)(1 + \hbar g^{n-1})^k - \frac{n-1}{\hbar} g^{2-n}. \tag{1.1.22}$$

An element of a polyadic algebraic structure g is called ℓ_μ-*nilpotent* (or simply *nilpotent* for $\ell_\mu = 1$), if there exist such ℓ_μ that

$$g^{\langle \ell_\mu \rangle} = z. \tag{1.1.23}$$

Definition 1.1.7. A polyadic algebraic structure with zero z is called ℓ_μ-*nilpotent*, if there exists ℓ_μ such that for any $\left(\ell_\mu(n-1)+1\right)$-tuple (polyad) $\mathbf{g}$ we have

$$\left(\mu^{(n)}\right)^{\circ \ell_\mu}[\mathbf{g}] = z. \tag{1.1.24}$$

Therefore, the *index of nilpotency* (number of elements whose product is zero) of an ℓ_μ-nilpotent n-ary algebraic structure is $\left(\ell_\mu(n-1)+1\right)$, while its polyadic power is ℓ_μ.

A *polyadic (n-ary) identity* (or neutral element) of a polyadic algebraic structure is a distinguished element e (and the corresponding 0-ary operation $\mu_e^{(0)}$) such that for any element $g \in G$ we have

$$\mu^{(n)}[g, e^{n-1}] = g, \tag{1.1.25}$$

where g can be on any place in the lhs of (1.1.25).

In polyadic algebraic structures, there exist *neutral polyads* $\mathbf{n}$ consisting of elements of G satisfying

$$\mu^{(n)}[g, \mathbf{n}] = g, \tag{1.1.26}$$

where g can be on any of n-places in the lhs. The neutral polyads are not determined uniquely. It follows from (1.1.25) that the sequence of polyadic identities e^{n-1} is a neutral polyad.

Definition 1.1.8. An element of a polyadic algebraic structure g is called ℓ_μ-*idempotent* (or simply *idempotent* for $\ell_\mu = 1$), if there exist such ℓ_μ that

$$g^{\langle \ell_\mu \rangle} = g. \tag{1.1.27}$$

We define associativity as the invariance of the composition of two n-ary multiplications. A one-set polyadic algebraic structure $\langle G \mid \mu^{(n)} \rangle$ is called *totally associative*, if

$$\left(\mu^{(n)}\right)^{\circ 2}[\mathbf{g}, \mathbf{h}, \mathbf{u}] = \mu^{(n)}[\mathbf{g}, \mu^{(n)}[\mathbf{h}], \mathbf{u}] = \text{invariant}, \tag{1.1.28}$$

with respect to placement of the internal multiplication $\mu^{(n)}[\mathbf{h}]$ in the rhs on any of n-places, with a fixed order of elements in the any fixed polyad of $(2n-1)$ elements

$\mathbf{t}^{(2n-1)} = (\mathbf{g}, \mathbf{h}, \mathbf{u}) \in G^{\times(2n-1)}$. Thus, the internal multiplication in (1.1.28) can be 'moved' on any of n places $i = 1, \ldots, n$, which gives n relations

$$\mu^{(n)} \circ \left(\overset{i=1}{\mu^{(n)}} \times \mathrm{id}^{\times(n-1)} \right) = \mu^{(n)} \circ \left(\mathrm{id} \times \overset{i=2}{\mu^{(n)}} \times \mathrm{id}^{\times(n-2)} \right) = \cdots = \mu^{(n)} \circ \left(\mathrm{id}^{\times(n-1)} \times \overset{i=n}{\mu^{(n)}} \right). \quad (1.1.29)$$

Remark **1.1.9.** There exist other kinds of associativity which were introduced in Thurston (1949) and studied in Belousov (1972) and Sokhatsky (1997). We confine ourselves to the most general, the total associativity (1.1.28). In this case, the iteration (1.1.29) does not depend on the placement of internal multiplications (as in the rhs of (1.1.3)).

In a polyadic algebraic structure with zero (1.1.17) one can have *trivial associativity*, when all n terms in (1.1.28) are equal to zero, i.e.

$$\left(\mu^{(n)}\right)^{\circ 2}[\mathbf{g}] = z \quad (1.1.30)$$

for any $(2n - 1)$-tuple $\mathbf{g}^{(2n-1)} \in G^{\times(2n-1)}$.

1.2 Polyadic semigroups, quasigroups and groups

Semigroups with multiplication of higher arity ($n > 2$) were introduced in Sioson (1963) as algebraic systems, in Gluskin (1964) as positional operatives, in Monk and Sioson (1966) as m-semigroups, and in Zupnik (1967) as polyadic semigroups (we will use the latter term to be close in terminology to polyadic groups, Post 1940).

Definition 1.2.1. A *polyadic semigroup* (*n-ary semigroup*) $\mathcal{S}^{(n)}_{\text{semigrp}}$ is a one-set one-operation algebraic structure in which the n-ary operation is associative, $\mathcal{S}^{(n)}_{\text{semigrp}} = \langle S \mid \mu^{(n)} \mid \text{associativity (1.1.28)} \rangle$.

Corollary 1.2.2. Any 2-nilpotent n-ary algebraic structure (having index of nilpotency $(2n - 1)$) is a polyadic semigroup.

In the case of changing arity one should use in (1.1.30) not the changed final arity n', but the 'real' arity which is n for the reducing case and $\ell_\mu(n - 1) + 1$ for all other cases. Let us give some examples.

Example **1.2.3.** In the mixed (interaction–reduction) case with $n = 2$, $\ell_\mu = 3$, $n_c = 1$, we have a ternary algebraic structure $\langle G \mid \mu^{(3)} \rangle$ iterated from a binary algebraic structure $\left\langle G \middle| \mu^{(2)}, \mu_c^{(0)} \right\rangle$ with one distinguished element c (or an additional 0-ary operation)[3]

[3] This construction is named the b-derived groupoid in Dudek and Michalski (1984).

$$\mu_c^{(3)}[g, h, u] = (g \cdot (h \cdot (u \cdot c))), \tag{1.2.1}$$

where for binary multiplication we denote $g \cdot h = \mu^{(2)}[g, h]$. Thus, if the ternary algebraic structure $\left\langle G \middle| \mu_c^{(3)} \right\rangle$ is nilpotent of index 7 (see 1.1.7), then it is a ternary semigroup (because $\mu_c^{(3)}$ is trivially associative) independently of the associativity of $\mu^{(2)}$ (see, e.g., Borowiec *et al* 2006).

Important to find the *associativity preserving* conditions, where an associative initial operation $\mu^{(n)}$ leads to an associative final operation $\mu'^{(n')}$ during the change of arity.

***Example* 1.2.4.** An associativity preserving reduction can be given by the construction of a binary associative operation $\mu_{\mathbf{c}}^{(2)}$ using $(n-2)$-tuple $\mathbf{c}$ consisting of $n_c = n - 2$ different constants

$$\mu_{\mathbf{c}}^{(2)}[g, h] = \mu^{(n)}[g, \mathbf{c}, h], \quad \mathbf{c} \in G^{\times(n-2)}, \; g, h \in G. \tag{1.2.2}$$

Various associativity preserving mixing constructions with different arities and places were considered in Dudek and Michalski (1984), Michalski (1981) and Sokhatsky (1997).

An associative polyadic algebraic structure with the identity (1.1.25) is called a *polyadic monoid*. The structure of any polyadic monoid is fixed (Pop and Pop 2010): it can be obtained by iterating a binary operation (Čupona and Trpenovski 1961) (for polyadic groups this was shown in Dörnte 1929).

There are several analogs of binary commutativity in polyadic algebraic structures. The most straightforward one comes from commutation of the multiplication with permutations. A polyadic algebraic structure $\mathcal{G}^{(n)} = \langle G | \mu^{(n)} \rangle$ is *σ-commutative*, if $\mu^{(n)} = \mu^{(n)} \circ \sigma$, or

$$\mu^{(n)}[\mathbf{g}] = \mu^{(n)}[\sigma \circ \mathbf{g}], \quad \mathbf{g} \in G^{\times n}, \tag{1.2.3}$$

where $\sigma \circ \mathbf{g} = (g_{\sigma(1)}, \ldots, g_{\sigma(n)})$ is a permutated polyad and σ is a fixed element of S_n, the permutation group on n elements. If (1.2.3) holds for all $\sigma \in S_n$, then a polyadic algebraic structure is *commutative*. A special type of the σ-commutativity

$$\mu^{(n)}[g, \mathbf{t}^{(n-2)}, h] = \mu^{(n)}[h, \mathbf{t}^{(n-2)}, g], \tag{1.2.4}$$

where $\mathbf{t}^{(n-2)} \in G^{\times(n-2)}$ is any fixed $(n-2)$-polyad, is called *semicommutativity*. So for an n-ary semicommutative algebraic structure we have

$$\mu^{(n)}[g, h^{n-1}] = \mu^{(n)}[h^{n-1}, g]. \tag{1.2.5}$$

If an n-ary semigroup $\mathcal{S}^{(n)}$ is iterated from a commutative binary semigroup with identity, then $\mathcal{S}^{(n)}$ is semicommutative.

***Example* 1.2.5.** Let G be the set of natural numbers $\mathbb{N}$, and the 5-ary multiplication is defined by

$$\mu^{(5)}[\mathbf{g}] = g_1 - g_2 + g_3 - g_4 + g_5, \quad g_i \in \mathbb{N}, \tag{1.2.6}$$

then $\mathcal{G}_{\mathbb{N}}^{(5)} = \langle \mathbb{N} | \mu^{(5)} \rangle$ is a semicommutative 5-ary monoid having the identity $e_g = \mu^{(5)}[g^5] = g$ for each $g \in \mathbb{N}$. Therefore, $\mathcal{G}_{\mathbb{N}}^{(5)}$ is the idempotent monoid.

A possibility to get an analog of commutativity is to generalize the *mediality* (Murdoch 1939)

$$(g_{11} \cdot g_{12}) \cdot (g_{21} \cdot g_{22}) = (g_{11} \cdot g_{21}) \cdot (g_{12} \cdot g_{22}), \tag{1.2.7}$$

which in binary semigroups follows from (binary) commutativity. But for polyadic algebraic structures they are different. The mediality should contain $(n + 1)$ multiplications, it is a relation between $n \times n$ elements, and therefore can be presented in a matrix from (which can be achieved by placing the arguments of the external multiplication in a column).

A polyadic algebraic structure $\mathcal{G}^{(n)} = \langle G | \mu^{(n)} \rangle$ is *medial* (or *entropic*), if (Evans 1963, Belousov 1972)

$$\mu^{(n)} \begin{bmatrix} \mu^{(n)}[g_{11}, \ldots, g_{1n}] \\ \vdots \\ \mu^{(n)}[g_{n1}, \ldots, g_{nn}] \end{bmatrix} = \mu^{(n)} \begin{bmatrix} \mu^{(n)}[g_{11}, \ldots, g_{n1}] \\ \vdots \\ \mu^{(n)}[g_{1n}, \ldots, g_{nn}] \end{bmatrix}. \tag{1.2.8}$$

For polyadic semigroups we use the notation (1.1.3) and can present the mediality as follows

$$(\mu^{(n)})^{\circ n}[\mathbf{G}] = (\mu^{(n)})^{\circ n}[\mathbf{G}^T], \tag{1.2.9}$$

where $\mathbf{G} = \left\| g_{ij} \right\|$ is the $n \times n$ matrix of elements and $\mathbf{G}^T$ is its transpose. The semicommutative polyadic semigroups are medial, as in the binary case, but, in general (except $n = 3$) not vice versa (Głazek and Gleichgewicht 1982).

A polyadic algebraic structure $\mathcal{G}^{(n)} = \langle G | \mu^{(n)} \rangle$ is *cancellative*, if

$$\mu^{(n)}[g, \mathbf{t}] = \mu^{(n)}[h, \mathbf{t}] \Longrightarrow g = h, \tag{1.2.10}$$

where g, h can be on any place. This means that the mapping $\mu^{(n)}$ is one-to-one in each variable. If g, h are on the same ith place on both sides, the polyadic algebraic structure is called *i-cancellative*.

The *left* and *right* cancellativity are 1-cancellativity and n-cancellativity respectively. A right and left cancellative n-ary semigroup is *cancellative* (with respect to the same subset).

A polyadic algebraic structure is called (uniquely) *i-solvable*, if for all polyads $\mathbf{t}$, $\mathbf{u}$ and element h, one can (uniquely) resolve the equation (with respect to h) for the fundamental operation

$$\mu^{(n)}[\mathbf{u}, h, \mathbf{t}] = g, \tag{1.2.11}$$

where h can be on the ith place, and **u**, **t** are polyads of needed length.

Definition 1.2.6. A polyadic algebraic structure which is uniquely i-solvable for all places $i = 1, \ldots, n$ is called an *n-ary* (or *polyadic*) *quasigroup* $\mathcal{Q}^{(n)} = \langle Q | \mu^{(n)} | \text{solvability} \rangle$.

It follows, that, if (1.2.11) uniquely i-solvable for all places, than

$$(\mu^{(n)})^{\circ \ell_\mu}[\mathbf{u}, h, \mathbf{t}] = g, \quad g, h \in Q, \tag{1.2.12}$$

can be (uniquely) resolved with respect to h on any place, and **u**, **t** are polyads of needed lengths.

Definition 1.2.7. An associative polyadic quasigroup is called an *n-ary* (or *polyadic*) *group*.

***Remark* 1.2.8.** It was shown (Rusakov 1998), how to minimize the set of axioms to define polyadic groups (solvability not in all places, Post 1940, Celakoski 1977), decreasing or increasing the number of unknowns in determining equations (Gal'mak 2003). It was constructions in terms of additionally defined objects (various analogs of the identity and sequences; Ûsan 2003), as well as not using total associativity, but instead various partial ones (Sokolov 1976, Sokhatsky 1997, Yurevych 2001).

Definition 1.2.9. In a polyadic group $\mathcal{G}^{(n)} = \langle G | \mu^{(n)} \rangle$ the only solution of (1.2.11) is called a *querelement* of g and denoted by $\bar{g}$ (Dörnte 1929), such that

$$\mu^{(n)}[\mathbf{h}, \bar{g}] = g, \quad g, \bar{g} \in G, \tag{1.2.13}$$

where $\bar{g}$ can be on any place. So, any idempotent g coincides with its querelement $\bar{g} = g$.

***Example* 1.2.10.** For the q-addition (1.1.19) from example 1.1.6, using (1.2.13) with $\mathbf{h} = g^{n-1}$ we obtain

$$\bar{g} = -\frac{(n-2)g}{1 + \hbar g^{n-1}}. \tag{1.2.14}$$

We observe from (1.2.13) and (1.1.26) that the polyad

$$\mathbf{n}_g = (g^{n-2}\bar{g}) \tag{1.2.15}$$

is neutral for any element of a polyadic group, where $\bar{g}$ can be on any place. If this ith place is important, then we write $\mathbf{n}_{g;i}$.

Because any neutral polyad plays a role of identity (see (1.1.26)), for any element $g \in G$ we define its *polyadic inverse* (the sequence of length $(n-2)$ denoted by the same letter $\mathbf{g}^{-1}$ in bold) as (see Post 1940 and by modified analogy with Petrescu 2005, Gal'mak 2003)

$$\mathbf{n}_g = (\mathbf{g}^{-1}, g) = (g, \mathbf{g}^{-1}), \tag{1.2.16}$$

which can be written in terms of the multiplication as

$$\mu^{(n)}[g, \mathbf{g}^{-1}, h] = \mu^{(n)}[h, \mathbf{g}^{-1}, g] = h \tag{1.2.17}$$

for all $h \in G$. It is obvious that the polyads

$$\mathbf{n}_{g^k} = ((\mathbf{g}^{-1})^k, g^k) = (g^k, (\mathbf{g}^{-1})^k) \tag{1.2.18}$$

are neutral as well for any $k \geqslant 1$. It follows from (1.2.15) that the polyadic inverse of g is $(g^{n-3}, \bar{g})$, and one of $\bar{g}$ is (g^{n-2}), and in this case g is called *querable*. In a polyadic group all elements are querable (Dudek *et al* 1977, Celakoski 1977, Post 1940).

The number of relations in (1.2.13) can be reduced from n (the number of possible places) to only 2 (when g is on the first and last places (Dörnte 1929, Timm 1972), or on some other 2 places). In a polyadic group the *Dörnte relations*

$$\mu^{(n)}[g, \mathbf{n}_{h;i}] = \mu^{(n)}[\mathbf{n}_{h;j}, g] = g \tag{1.2.19}$$

hold true for any allowable i, j. In the case of a binary group the relations (1.2.19) become $g \cdot h \cdot h^{-1} = h \cdot h^{-1} \cdot g = g$.

The unique solvability relation (1.2.13) in an n-ary group can be treated as a definition of the unary (multiplicative) *queroperation*

$$\bar{\mu}^{(1)}[g] = \bar{g}. \tag{1.2.20}$$

Using the queroperation (1.2.20) one can give a *diagrammatic definition* of a polyadic group (see Gleichgewicht and Głazek 1967).

Definition 1.2.11. A *polyadic group* is a one-set algebraic structure (universal algebra)

$$\mathcal{G}^{(n)}_{\text{grp}} = \langle G \mid \mu^{(n)}, \bar{\mu}^{(1)} \mid \text{associativity, Dörnte relations} \rangle, \tag{1.2.21}$$

where $\mu^{(n)}$ is an n-ary associative operation and $\bar{\mu}^{(1)}$ is the *queroperation*, such that the following diagram

$$\begin{array}{ccccc}
G^{\times(n)} & \xrightarrow{\mathrm{id}^{\times(n-1)} \times \bar{\mu}^{(1)}} & G^{\times n} & \xleftarrow{\bar{\mu}^{(1)} \times \mathrm{id}^{\times(n-1)}} & G^{\times n} \\
\uparrow {\scriptstyle \mathrm{id} \times \mathrm{Diag}_{(n-1)}} & & \downarrow {\scriptstyle \mu^{(n)}} & & \uparrow {\scriptstyle \mathrm{Diag}_{(n-1)} \times \mathrm{id}} \\
G \times G & \xrightarrow{\mathrm{Pr}_1} & G & \xleftarrow{\mathrm{Pr}_2} & G \times G
\end{array} \tag{1.2.22}$$

commutes, where $\bar{\mu}^{(1)}$ can be only on the first and second places from the right (resp. left) on the left (resp. right) part of the diagram.

***Remark* 1.2.12.** A generalization of the queroperation and corresponding definitions can be made by substituting in the above formulas (1.2.13)–(1.2.20) the n-ary multiplication $\mu^{(n)}$ by iterating the multiplication $(\mu^{(n)})^{\circ\ell_\mu}$ (1.1.3) (see Dudek 1980 for $\ell_\mu = 2$).

Define the *querpower* k of g recursively

$$\bar{g}^{\langle\langle k\rangle\rangle} = \overline{(\bar{g}^{\langle\langle k-1\rangle\rangle})}, \tag{1.2.23}$$

where $\bar{g}^{\langle\langle 0\rangle\rangle} = g$, $\bar{g}^{\langle\langle 1\rangle\rangle} = \bar{g}$, or as the k composition $(\bar{\mu}^{(1)})^{\circ k} = \overbrace{\bar{\mu}^{(1)}\circ\bar{\mu}^{(1)}\circ\ldots\circ\bar{\mu}^{(1)}}^{k}$ of the queroperation (1.2.20).

For instance (Gal'mak 2003), $(\bar{\mu}^{(1)})^{\circ k} = (\mu^{(n)})^{\circ(n-3)}$, such that for any ternary group $(\bar{\mu}^{(1)})^{\circ 2} = \mathrm{id}$, i.e., one has $\bar{\bar{g}} = g$. Using the queroperation in polyadic groups we can define the *negative polyadic power* of an element g by the following recursive relation

$$\mu^{(n)}[g^{\langle \ell_\mu - 1\rangle}, g^{n-2}, g^{\langle -\ell_\mu\rangle}] = g, \tag{1.2.24}$$

or (after use of (1.1.18)) as a solution of the equation

$$(\mu^{(n)})^{\circ\ell_\mu}[g^{\ell_\mu(n-1)}, g^{\langle -\ell_\mu\rangle}] = g. \tag{1.2.25}$$

The querpower and the polyadic power are mutually connected (Dudek 1993). We reformulate this connection using the so-called Heine numbers (Heine 1878) or q-deformed numbers (Kac and Cheung 2002)

$$[[k]]_q = \frac{q^k - 1}{q - 1}, \tag{1.2.26}$$

which have the 'nondeformed' limit $q \to 1$ as $[k]_q \to k$. Then

$$\bar{g}^{\langle\langle k\rangle\rangle} = g^{\langle -[[k]]_{2-n}\rangle}, \tag{1.2.27}$$

which can be treated as the following 'deformation' statement:

Assertion 1.2.13. The querpower coincides with the negative polyadic deformed power with the 'deformation' parameter q which is equal to the 'deviation' $(2 - n)$ from the binary group.

***Example* 1.2.14.** Let us consider a binary group $\mathcal{G}^{(2)} = \langle G|\mu^{(2)}\rangle$, we denote $\mu^{(2)} = (\cdot)$, and construct (using (1.1.3) and (1.1.2)) the reduced 4-ary product by $\mu_4'[\mathbf{g}] = g_1 \cdot g_2 \cdot g_3 \cdot g_4 \cdot c$, where $g_i \in G$ and c is in the center of the group $\mathcal{G}^{(2)}$. In the 4-ary group $\mathcal{G}_4' = \left\langle G, \mu_4'\right\rangle$ we derive the following positive and negative polyadic powers (obviously $g^{\langle 0\rangle} = \bar{g}^{\langle\langle 0\rangle\rangle} = g$)

$$g^{\langle 1\rangle} = g^4 \cdot c,\ \ g^{\langle 2\rangle} = g^7 \cdot c^2, \ldots, g^{\langle k\rangle} = g^{3k+1} \cdot c^k, \tag{1.2.28}$$

$$g^{\langle -1\rangle} = g^{-2} \cdot c^{-1},\ \ g^{\langle -2\rangle} = g^{-5} \cdot c^{-2}, \ldots, g^{\langle -k\rangle} = g^{-3k+1} \cdot c^{-k}, \tag{1.2.29}$$

and the querpowers

$$\bar{g}^{\langle\langle 1\rangle\rangle} = g^{-2} \cdot c^{-1},\ \ \bar{g}^{\langle\langle 2\rangle\rangle} = g^{-4} \cdot c, \ldots,\ \ \bar{g}^{\langle\langle k\rangle\rangle} = g^{(-2)^k} \cdot c^{[[k]]_{-2}}. \tag{1.2.30}$$

1.3 Polyadic direct products and changed arity powers

The concept of direct product plays a crucial role for algebraic structures in the study of their internal constitution and their representation in terms of better known/ simpler structures (see, e.g., Lang 1965, Lambek 1966). The general method of the external direct product construction is to take the Cartesian product of the underlying sets and endow it with the operations from the algebraic structures under consideration. Usually this is an identical repetition of the initial multipliers' operations componentwise (Hungerford 1974). In the case of polyadic algebraic structures their arity comes into the game, such that endowing the product with operations becomes nontrivial in two aspects: the arities of all structures can be different (but 'quantized' and not unique) and the elements from different multipliers can be 'entangled' making the product not componentwise. The direct (componentwise) product of n-ary groups was considered in Michalski (1984a) and Shchuchkin (2014). We propose two corresponding polyadic analogs (changing arity and 'entangling') of the external direct product which preserve associativity, and therefore allow us to work out polyadic semigroups, groups, rings and fields (Duplij 2022).

The direct product is important, especially because it plays the role of a product in a corresponding category (see, e.g., Borceux 1994, Mac Lane 1971). For instance, the class of all polyadic groups for objects and polyadic group homomorphisms for morphisms form a category which is well-defined, because it has the polyadic direct product (Michalski 1984b, Iancu 1991) as a product.

1.3.1 Polyadic products of semigroups and groups

We start from the standard external direct product construction for semigroups. Then we show that consistent 'polyadization' of the semigroup direct product, which preserves associativity, can lead to additional properties (Duplij 2022):

(1) The arities of the polyadic direct product and power can differ from that of the initial semigroups. (1.3.1)

(2) The components of the polyadic power can contain elements from different multipliers. (1.3.2)

We use here a vector-like notation for clarity and convenience in passing to higher arity generalizations. Begin from the direct product of two (binary) semigroups $\mathcal{G}_{1,2} \equiv \mathcal{G}_{1,2}^{(2)} = \left\langle G_{1,2} \middle| \mu_{1,2}^{(2)} \equiv (\cdot_{1,2}) \middle| \text{assoc} \right\rangle$, where $G_{1,2}$ are underlying sets, while $\mu_{1,2}^{(2)}$ are multiplications in $\mathcal{G}_{1,2}$. On the Cartesian product of the underlying sets $G' = G_1 \times G_2$ we define a *direct product* $\mathcal{G}_1 \times \mathcal{G}_2 = \mathcal{G}' = \langle G' | \boldsymbol{\mu}'^{(2)} \equiv (\bullet') \rangle$ of the semigroups $\mathcal{G}_{1,2}$ by the componentwise multiplication of the doubles $\mathbf{G} = \begin{pmatrix} g_1 \\ g_2 \end{pmatrix} \in G_1 \times G_2$ (being the Kronecker product of doubles in our notation), as

$$\mathbf{G}^{(1)} \bullet' \mathbf{G}^{(2)} = \begin{pmatrix} g_1 \\ g_2 \end{pmatrix}^{(1)} \bullet' \begin{pmatrix} g_1 \\ g_2 \end{pmatrix}^{(2)} = \begin{pmatrix} g_1^{(1)} \cdot_1 g_1^{(2)} \\ g_2^{(1)} \cdot_2 g_2^{(2)} \end{pmatrix}, \tag{1.3.3}$$

and in the 'polyadic' notation

$$\boldsymbol{\mu}'^{(2)}[\mathbf{G}^{(1)}, \mathbf{G}^{(2)}] = \begin{pmatrix} \mu_1^{(2)}\left[g_1^{(1)}, g_1^{(2)}\right] \\ \mu_2^{(2)}\left[g_2^{(1)}, g_2^{(2)}\right] \end{pmatrix}. \tag{1.3.4}$$

Obviously, the associativity of $\boldsymbol{\mu}'^{(2)}$ follows immediately from that of $\mu_{1,2}^{(2)}$, because of the componentwise multiplication in (1.3.4). If $\mathcal{G}_{1,2}$ are groups with the identities $e_{1,2} \in G_{1,2}$, then the identity of the direct product is the double $\mathbf{E} = \begin{pmatrix} e_1 \\ e_2 \end{pmatrix}$, such that $\boldsymbol{\mu}'^{(2)}[\mathbf{E}, \mathbf{G}] = \boldsymbol{\mu}'^{(2)}[\mathbf{G}, \mathbf{E}] = \mathbf{G} \in \mathcal{G}$.

1.3.2 Full polyadic external product

The 'polyadization' of (1.3.4) is straightforward (Duplij 2022).

Definition 1.3.1. An n'-ary *full direct product* semigroup $\mathcal{G}'^{(n')} = \mathcal{G}_1^{(n)} \times \mathcal{G}_2^{(n)}$ consists of (two or k) n-ary semigroups (of the same arity $n' = n$)

$$\boldsymbol{\mu}'^{(n)}[\mathbf{G}^{(1)}, \mathbf{G}^{(2)}, \ldots, \mathbf{G}^{(n)}] = \begin{pmatrix} \mu_1^{(n)}\left[g_1^{(1)}, g_1^{(2)}, \ldots, g_1^{(n)}\right] \\ \mu_2^{(n)}\left[g_2^{(1)}, g_2^{(2)}, \ldots, g_2^{(n)}\right] \end{pmatrix}, \tag{1.3.5}$$

where the (total) polyadic associativity (1.1.28) of $\mu'^{(n')}$ is governed by those of the constituent semigroups $\mathcal{G}_1^{(n)}$ and $\mathcal{G}_2^{(n)}$ (or $\mathcal{G}_1^{(n)} \ldots \mathcal{G}_k^{(n)}$) and the componentwise construction (1.3.5).

If $\mathcal{G}_{1,2}^{(n)} = \left\langle G_{1,2} \middle| \mu_{1,2}^{(n)}, \bar{\mu}_{1,2}^{(1)} \right\rangle$ are n-ary groups (where $\bar{\mu}_{1,2}^{(1)}$ are the unary multiplicative queroperations (1.2.20)), then the queroperation $\bar{\boldsymbol{\mu}}'^{(1)}$ of the full direct product group $\mathcal{G}'^{(n')} = \langle G' \equiv G_1 \times G_2 | \boldsymbol{\mu}'^{(n')}, \bar{\boldsymbol{\mu}}'^{(1)} \rangle$ $(n' = n)$ is defined componentwise as follows

$$\bar{\mathbf{G}} \equiv \bar{\boldsymbol{\mu}}^{\prime(1)}[\mathbf{G}] = \begin{pmatrix} \bar{\mu}_1^{(1)}[g_1] \\ \bar{\mu}_2^{(1)}[g_2] \end{pmatrix}, \quad \text{or } \bar{\mathbf{G}} = \begin{pmatrix} \bar{g}_1 \\ \bar{g}_2 \end{pmatrix}, \tag{1.3.6}$$

which satisfies $\mu^{\prime(n)}[\mathbf{G}, \mathbf{G}, \ldots, \bar{\mathbf{G}}] = \mathbf{G}$ with $\bar{\mathbf{G}}$ on any place (see (1.2.13)).

Definition 1.3.2. A full polyadic direct product $\mathcal{G}^{\prime(n)} = \mathcal{G}_1^{(n)} \times \mathcal{G}_2^{(n)}$ is called *derived*, if its constituents $\mathcal{G}_1^{(n)}$ and $\mathcal{G}_2^{(n)}$ are derived, such that the operations $\mu_{1,2}^{(n)}$ are compositions of the binary operations $\mu_{1,2}^{(2)}$, correspondingly.

In the derived case all the operations in (1.3.5) have the form (see (1.1.3)–(1.1.4))

$$\mu_{1,2}^{(n)} = \left(\mu_{1,2}^{(2)}\right)^{\circ(n-1)}, \qquad \boldsymbol{\mu}^{(n)} = \left(\boldsymbol{\mu}^{(2)}\right)^{\circ(n-1)}. \tag{1.3.7}$$

The operations of the derived polyadic semigroup can be written as (see the binary direct product (1.3.3)–(1.3.4))

$$\boldsymbol{\mu}^{\prime(n)}[\mathbf{G}^{(1)}, \mathbf{G}^{(2)}, \ldots, \mathbf{G}^{(n)}] = \mathbf{G}^{(1)}\bullet'\mathbf{G}^{(2)}\bullet'\ldots\bullet'\mathbf{G}^{(n)} = \begin{pmatrix} g_1^{(1)} \cdot_1 g_1^{(2)} \cdot_1 \ldots \cdot_1 g_1^{(n)} \\ g_2^{(1)} \cdot_2 g_2^{(2)} \cdot_2 \ldots \cdot_2 g_2^{(n)} \end{pmatrix}. \tag{1.3.8}$$

We will be more interested in nonderived polyadic analogs of the direct product.

***Example* 1.3.3.** Let us have two ternary groups: the unitless nonderived group $\mathcal{G}_1^{(3)} = \left\langle \mathrm{i}\mathbb{R} \mid \mu_1^{(3)} \right\rangle$, where $\mathrm{i}^2 = -1$, $\mu_1^{(3)}\left[g_1^{(1)}, g_1^{(2)}, g_1^{(3)}\right] = g_1^{(1)} g_1^{(2)} g_1^{(3)}$ is a triple product in $\mathbb{C}$, the querelement is $\bar{\mu}_1^{(1)}[g_1] = 1/g_1$, and $\mathcal{G}_2^{(3)} = \left\langle \mathbb{R} \mid \mu_2^{(3)} \right\rangle$ with $\mu_2^{(3)}\left[g_2^{(1)}, g_2^{(2)}, g_2^{(3)}\right] = g_2^{(1)} \left(g_2^{(2)}\right)^{-1} g_2^{(3)}$, the querelement $\bar{\mu}_2^{(1)}[g_2] = g_2$. Then the ternary nonderived full direct product group becomes $\mathcal{G}^{\prime(3)} = \langle \mathrm{i}\mathbb{R} \times \mathbb{R} \mid \boldsymbol{\mu}^{\prime(3)}, \bar{\boldsymbol{\mu}}^{\prime(1)} \rangle$, where

$$\boldsymbol{\mu}^{\prime(3)}[\mathbf{G}^{(1)}, \mathbf{G}^{(2)}, \mathbf{G}^{(3)}] = \begin{pmatrix} g_1^{(1)} g_1^{(2)} g_1^{(3)} \\ g_2^{(1)} \left(g_2^{(2)}\right)^{-1} g_2^{(3)} \end{pmatrix}, \qquad \bar{\mathbf{G}} \equiv \bar{\boldsymbol{\mu}}^{\prime(1)}[\mathbf{G}] = \begin{pmatrix} 1/g_1 \\ g_2 \end{pmatrix}, \tag{1.3.9}$$

which contains no identity, because $\mathcal{G}_1^{(3)}$ is unitless and nonderived.

1.3.3 Mixed arity iterated product

In the polyadic case, the following question arises, which cannot even be stated in the binary case: is it possible to build a version of the associative direct product such that it can be nonderived and have different arity than the constituent semigroup arities? The answer is yes, which leads to two arity changing constructions: componentwise and noncomponentwise (Duplij 2022).

(i) *Iterated direct product* ($\circledast$). In each of the constituent polyadic semigroups we use the iterating (1.1.3) componentwise, but with different numbers of compositions, because the same number of compositions evidently leads to the iterated polyadic direct product. In this case the arity of the direct product is greater than or equal to the arities of the constituents $n' \geqslant n_1, n_2$.

(ii) *Hetero product* ($\boxtimes$). The polyadic product of k copies of the same n-ary semigroup is constructed using the associativity quiver technique, which mixes ('entangles') elements from different multipliers, it is noncomponentwise (by analogy with heteromorphisms in Duplij 2012), and so it can be called a *hetero product* or *hetero power* (for coinciding multipliers, i.e., constituent polyadic semigroups or groups). This gives the arity of the hetero product which is less than or equal to the arities of the equal multipliers $n' \leqslant n$.

In the first componentwise case (**1**), the constituent multiplications (1.3.5) are composed from the lower arity ones in the componentwise way, but the initial arities of up and down components can be different (as opposed to the binary derived case (1.3.7))

$$\mu_1^{(n)} = \left(\mu_1^{(n_1)}\right)^{\circ \ell_{\mu 1}}, \qquad \mu_2^{(n)} = \left(\mu_2^{(n_2)}\right)^{\circ \ell_{\mu 2}}, \qquad 3 \leqslant n_{1,2} \leqslant n - 1, \tag{1.3.10}$$

where we exclude the limits: the derived case $n_{1,2} = 2$ (1.3.7) and the undecomposed case $n_{1,2} = n$ (1.3.5). Since the total size of the up and down polyads is the same and coincides with the arity of the double $\mathbf{G}$ multiplication n', using (1.1.4) we obtain the *arity compatibility* relations

$$n' = \ell_{\mu 1}(n_1 - 1) + 1 = \ell_{\mu 2}(n_2 - 1) + 1. \tag{1.3.11}$$

Definition 1.3.4. A *mixed arity polyadic iterated direct product* semigroup $\mathcal{G}'^{(n')} = \mathcal{G}_1^{(n_1)} \circledast \mathcal{G}_2^{(n_2)}$ consists of (two) polyadic semigroups $\mathcal{G}_1^{(n_1)}$ and $\mathcal{G}_2^{(n_2)}$ of the different arity shapes n_1 and n_2

$$\boldsymbol{\mu}'^{(n')}\left[\mathbf{G}^{(1)}, \mathbf{G}^{(2)}, \ldots, \mathbf{G}^{(n')}\right] = \begin{pmatrix} \left(\mu_1^{(n_1)}\right)^{\circ \ell_{\mu 1}}\left[g_1^{(1)}, g_1^{(2)}, \ldots, g_1^{(n)}\right] \\ \left(\mu_2^{(n_2)}\right)^{\circ \ell_{\mu 2}}\left[g_2^{(1)}, g_2^{(2)}, \ldots, g_2^{(n)}\right] \end{pmatrix}, \tag{1.3.12}$$

and the arity compatibility relations (1.3.11) hold.

Observe that it is not the case that any two polyadic semigroups can be composed in the mixed arity polyadic direct product.

Assertion 1.3.5. *If the arity shapes of two polyadic semigroups $\mathcal{G}_1^{(n_1)}$ and $\mathcal{G}_2^{(n_2)}$ satisfy the compatibility condition*

$$a(n_1 - 1) = b(n_2 - 1) = c, \qquad a, b, c \in \mathbb{N}, \tag{1.3.13}$$

then they can form a mixed arity direct product $\mathcal{G}'^{(n')} = \mathcal{G}_1^{(n_1)} \circledast \mathcal{G}_2^{(n_2)}$, where $n' = c + 1$ (1.3.11).

Example 1.3.6. In the case of 4-ary and 5-ary semigroups $\mathcal{G}_1^{(4)}$ and $\mathcal{G}_2^{(5)}$ the direct product arity of $\mathcal{G}'^{(n')}$ is 'quantized' $n' = 3\ell_{\mu 1} + 1 = 4\ell_{\mu 2} + 1$, such that

$$n' = 12k + 1 = 13, 25, 37, \dots , \tag{1.3.14}$$

$$\ell_{\mu 1} = 4k = 4, 8, 12, \dots , \tag{1.3.15}$$

$$\ell_{\mu 2} = 3k = 3, 6, 9, \dots , \quad k \in \mathbb{N}, \tag{1.3.16}$$

and only the first mixed arity 13-ary direct product semigroup $\mathcal{G}'^{(13)}$ is nonderived. If $\mathcal{G}_1^{(4)}$ and $\mathcal{G}_2^{(5)}$ are polyadic groups with the queroperations $\bar{\mu}_1^{(1)}$ and $\bar{\mu}_2^{(1)}$ correspondingly, then the iterated direct $\mathcal{G}'^{(n')}$ is a polyadic group with the queroperation $\bar{\boldsymbol{\mu}}'^{(1)}$ given in (1.3.6).

In the same way one can consider the iterated direct product of any number of polyadic semigroups.

1.3.4 Polyadic hetero product

In the second noncomponentwise case (**2**) we allow multiplying elements from different components, and therefore we should consider the Cartesian k-power of sets $G' = G^{\times k}$ and endow the corresponding k-tuple with a polyadic operation in such a way that associativity of $\mathcal{G}^{(n)}$ will govern the associativity of the product $\mathcal{G}'^{(n)}$. In other words we construct a k-power of the polyadic semigroup $\mathcal{G}^{(n)}$ such that the result $\mathcal{G}'^{(n')}$ is an n'-ary semigroup.

The general structure of the hetero product formally coincides 'reversely' with the main heteromorphism equation (Duplij 2012). The additional parameter which determines the arity n' of the hetero power of the initial n-ary semigroup is the number of intact elements ℓ_{id}. Thus, we arrive at

Definition 1.3.7. The *hetero ('entangled') k-power* of the n-ary semigroup $\mathcal{G}^{(n)} = \langle G | \mu^{(n)} \rangle$ is the n'-ary semigroup defined on the kth Cartesian power $G' = G^{\times k}$, such that $\mathcal{G}'^{(n')} = \langle G' | \boldsymbol{\mu}'^{(n')} \rangle$,

$$\mathcal{G}'^{(n')} = \left(\mathcal{G}^{(n)}\right)^{\boxtimes k} \equiv \overbrace{\mathcal{G}^{(n)} \boxtimes \dots \boxtimes \mathcal{G}^{(n)}}^{k}, \tag{1.3.17}$$

and the n'-ary multiplication of k-tuples $\mathbf{G}^T = (g_1, g_2, \dots , g_k) \in G^{\times k}$ is given (informally) by

$$\boldsymbol{\mu}'^{(n')}\left[\begin{pmatrix} g_1 \\ \vdots \\ g_k \end{pmatrix}, \dots , \begin{pmatrix} g_{k(n'-1)} \\ \vdots \\ g_{kn'} \end{pmatrix}\right] = \begin{pmatrix} \left.\begin{matrix} \mu^{(n)}[g_1, \dots , g_n], \\ \vdots \\ \mu^{(n)}\left[g_{n(\ell_\mu - 1)}, \dots , g_{n\ell_\mu}\right] \end{matrix}\right\}\ell_\mu & \left.\begin{matrix} g_{n\ell_\mu + 1}, \\ \vdots \\ g_{n\ell_\mu + \ell_{\text{id}}} \end{matrix}\right\}\ell_{\text{id}} \end{pmatrix}, \quad g_i \in G, \tag{1.3.18}$$

where ℓ_{id} is the number of *intact elements* in the rhs, and $\ell_\mu = k - \ell_{\text{id}}$ is the number of multiplications in the resulting k-tuple of the direct product. The hetero power parameters are connected by the *arity changing formula* (Duplij 2012)

$$n' = n - \frac{n-1}{k}\ell_{\text{id}}, \tag{1.3.19}$$

with the integer $\frac{n-1}{k}\ell_{\text{id}} \geqslant 1$.

The concrete placement of elements and multiplications in (1.3.18) to obtain the associative $\boldsymbol{\mu}'^{(n')}$ is governed by the associativity quiver technique (Duplij 2012).

There exist important general numerical relations between the parameters of the twisted direct power n', n, k, ℓ_{id}, which follow from (1.3.18)–(1.3.19). First, there are non-strict inequalities for them

$$0 \leqslant \ell_{\text{id}} \leqslant k - 1, \tag{1.3.20}$$

$$\ell_\mu \leqslant k \leqslant (n-1)\ell_\mu, \tag{1.3.21}$$

$$2 \leqslant n' \leqslant n. \tag{1.3.22}$$

Second, the initial and final arities n and n' are not arbitrary, but 'quantized' such that the fraction in (1.3.19) has to be an integer (see table 1.1).

Assertion 1.3.8. *The hetero power is not unique in both directions, if we do not fix the initial* n *and final* n' *arities of* $\mathcal{G}^{(n)}$ *and* $\mathcal{G}'^{(n')}$.

Proof. This follows from (1.3.18) and the hetero power 'quantization' table 1.1. □

Table 1.1. Hetero power 'quantization'.

k	ℓ_μ	ℓ_{id}	n/n'
2	1	1	$n = 3, 5, 7, \ldots$ $n' = 2, 3, 4, \ldots$
3	1	2	$n = 4, 7, 10, \ldots$ $n' = 2, 3, 4, \ldots$
3	2	1	$n = 4, 7, 10, \ldots$ $n' = 3, 5, 7, \ldots$
4	1	3	$n = 5, 9, 13, \ldots$ $n' = 2, 3, 4, \ldots$
4	2	2	$n = 3, 5, 7, \ldots$ $n' = 2, 3, 4, \ldots$
4	3	1	$n = 5, 9, 13, \ldots$ $n' = 4, 7, 10, \ldots$

The classification of the hetero powers consists of two limiting cases.

(i) *Intactless power*: there are no intact elements $\ell_{\text{id}} = 0$. The arity of the hetero power reaches its maximum and coincides with the arity of the initial semigroup $n' = n$ (see example 1.3.12).

(ii) *Binary power*: the final semigroup is of lowest arity, i.e., binary $n' = 2$. The number of intact elements is (see example 1.3.11)

$$\ell_{\text{id}} = k\frac{n-2}{n-1}. \tag{1.3.23}$$

Example 1.3.9. Consider the cubic power of a 4-ary semigroup $\mathcal{G}^{'(3)} = (\mathcal{G}^{(4)})^{\boxtimes 3}$ with the identity e, then the ternary identity triple in $\mathcal{G}^{'(3)}$ is $\mathbf{E}^T = (e, e, e)$, and therefore this cubic power is a ternary semigroup with identity.

Proposition 1.3.10. *If the initial* n*-ary semigroup* $\mathcal{G}^{(n)}$ *contains an identity, then the hetero power* $\mathcal{G}^{'(n')} = (\mathcal{G}^{(n)})^{\boxtimes k}$ *can contain an identity in the intactless case and the Post-like quiver (Duplij 2012). For the binary power* k = 2 *only the one-sided identity is possible.*

Let us consider some concrete examples.

Example 1.3.11. Let $\mathcal{G}^{(3)} = \langle G \mid \mu^{(3)} \rangle$ be a ternary semigroup, then we can construct its power $k = 2$ (square) of the doubles $\mathbf{G} = \begin{pmatrix} g_1 \\ g_2 \end{pmatrix} \in G \times G = G'$ in two ways to obtain the associative hetero power

$$\boldsymbol{\mu}^{'(2)}[\mathbf{G}^{(1)}, \mathbf{G}^{(2)}] = \begin{cases} \begin{pmatrix} \mu^{(3)}\left[g_1^{(1)}, g_2^{(1)}, g_1^{(2)}\right] \\ g_2^{(2)} \end{pmatrix}, \\ \begin{pmatrix} \mu^{(3)}\left[g_1^{(1)}, g_2^{(2)}, g_1^{(2)}\right] \\ g_2^{(1)} \end{pmatrix}, \end{cases} \quad g_i^{(j)} \in G. \tag{1.3.24}$$

This means that the Cartesian square can be endowed with the associative multiplication $\boldsymbol{\mu}^{'(2)}$, and therefore $\mathcal{G}^{'(2)} = \langle G' \mid \boldsymbol{\mu}^{'(2)} \rangle$ is a binary semigroup being the hetero product $\mathcal{G}^{'(2)} = \mathcal{G}^{(3)} \boxtimes \mathcal{G}^{(3)}$. If $\mathcal{G}^{(3)}$ has a ternary identity $e \in G$, then $\mathcal{G}^{'(2)}$ has only the left (right) identity $\mathbf{E} = \begin{pmatrix} e \\ e \end{pmatrix} \in G'$, since $\boldsymbol{\mu}^{'(2)}[\mathbf{E}, \mathbf{G}] = \mathbf{G}$ ($\boldsymbol{\mu}^{'(2)}[\mathbf{G}, \mathbf{E}] = \mathbf{G}$), but not the right (left) identity. Thus, $\mathcal{G}^{'(2)}$ can be a semigroup only, even if $\mathcal{G}^{(3)}$ is a ternary group.

Example 1.3.12. Take $\mathcal{G}^{(3)} = \langle G \mid \mu^{(3)} \rangle$ a ternary semigroup, then the multiplication on the double $\mathbf{G} = \begin{pmatrix} g_1 \\ g_2 \end{pmatrix} \in G \times G = G'$ is ternary and noncomponentwise

$$\boldsymbol{\mu}'^{(3)}[\mathbf{G}^{(1)}, \mathbf{G}^{(2)}, \mathbf{G}^{(3)}] = \begin{pmatrix} \mu^{(3)}\left[g_1^{(1)}, g_2^{(2)}, g_1^{(3)}\right] \\ \mu^{(3)}\left[g_2^{(1)}, g_1^{(2)}, g_2^{(3)}\right] \end{pmatrix}, \quad g_i^{(j)} \in G, \tag{1.3.25}$$

and $\mu'^{(3)}$ is associative (and described by the Post-like associative quiver, Duplij 2012), and therefore the cubic hetero power is the ternary semigroup $\mathcal{G}'^{(3)} = \langle G \times G | \boldsymbol{\mu}'^{(3)} \rangle$, such that $\mathcal{G}'^{(3)} = \mathcal{G}^{(3)} \boxtimes \mathcal{G}^{(3)}$. In this case, as opposed to the previous example, the existence of a ternary identity in $\mathcal{G}^{(3)}$ implies the ternary identity in the direct cube $\mathcal{G}'^{(3)}$ by $\mathbf{E} = \binom{e}{e}$. If $\mathcal{G}^{(3)}$ is a ternary group with the unary queroperation $\bar{\mu}^{(1)}$, then the cubic hetero power $\mathcal{G}'^{(3)}$ is also a ternary group of the special class (Dudek 1990): all querelements coincide (see (1.3.6)), such that $\bar{\mathbf{G}}^T = (g_{\text{quer}}, g_{\text{quer}})$, where $\bar{\mu}^{(1)}[g] = g_{\text{quer}}$, $\forall g \in G$. This is because in (1.2.13) the querelement can be on any place.

Theorem 1.3.13. *If $\mathcal{G}^{(n)}$ is an* n*-ary group, then the hetero* k*-power $\mathcal{G}'^{(n')} = (\mathcal{G}^{(n)})^{\boxtimes k}$ can contain queroperations in the intactless case only.*

Corollary 1.3.14. If the power multiplication (1.3.18) contains no intact elements $\ell_{\text{id}} = 0$ and does not change arity $n' = n$, a hetero power can be a polyadic group which has only one querelement.

Next we consider more complicated hetero power ('entangled') constructions with and without intact elements, as well as Post-like and non-Post associative quivers (Duplij 2012).

Example 1.3.16. Let $\mathcal{G}^{(4)} = \langle G | \mu^{(4)} \rangle$ be a 4-ary semigroup, then we can construct its 4-ary associative cubic hetero power $\mathcal{G}'^{(4)} = \langle G' | \boldsymbol{\mu}'^{(4)} \rangle$ using the Post-like and non-Post associative quivers without intact elements. Taking in (1.3.18) $n' = n$, $k = 3$, $\ell_{\text{id}} = 0$, we get two possibilities for the multiplication of the triples $\mathbf{G}^T = (g_1, g_2, g_3) \in G \times G \times G = G'$.

(i) Post-like associative quiver. The multiplication of the hetero cubic power case takes the form

$$\boldsymbol{\mu}'^{(4)}[\mathbf{G}^{(1)}, \mathbf{G}^{(2)}, \mathbf{G}^{(3)}, \mathbf{G}^{(4)}] = \begin{pmatrix} \mu^{(4)}\left[g_1^{(1)}, g_2^{(2)}, g_3^{(3)}, g_1^{(4)}\right] \\ \mu^{(4)}\left[g_2^{(1)}, g_3^{(2)}, g_1^{(3)}, g_2^{(4)}\right] \\ \mu^{(4)}\left[g_3^{(1)}, g_1^{(2)}, g_2^{(3)}, g_3^{(4)}\right] \end{pmatrix}, \quad g_i^{(j)} \in G, \tag{1.3.26}$$

and it can be shown that $\boldsymbol{\mu}'^{(4)}$ is totally associative, therefore $\mathcal{G}'^{(4)} = \langle G' | \boldsymbol{\mu}'^{(4)} \rangle$ is a 4-ary semigroup.

(ii) Non-Post associative quiver. The multiplication of the hetero cubic power differs from (1.3.26)

$$\boldsymbol{\mu}'^{(4)}[\mathbf{G}^{(1)}, \mathbf{G}^{(2)}, \mathbf{G}^{(3)}, \mathbf{G}^{(4)}] = \begin{pmatrix} \mu^{(4)}\left[g_1^{(1)}, g_3^{(2)}, g_2^{(3)}, g_1^{(4)}\right] \\ \mu^{(4)}\left[g_2^{(1)}, g_1^{(2)}, g_3^{(3)}, g_2^{(4)}\right] \\ \mu^{(4)}\left[g_3^{(1)}, g_2^{(2)}, g_1^{(3)}, g_3^{(4)}\right] \end{pmatrix}, \quad g_i^{(j)} \in G, \tag{1.3.27}$$

and it can be shown that $\boldsymbol{\mu}'^{(4)}$ is totally associative, therefore $\mathcal{G}'^{(4)} = \langle G' | \boldsymbol{\mu}'^{(4)} \rangle$ is a 4-ary semigroup.

The following is valid for both the above cases. If $\mathcal{G}^{(4)}$ has the 4-ary identity satisfying

$$\mu^{(4)}[e, e, e, g] = \mu^{(4)}[e, e, g, e] = \mu^{(4)}[e, g, e, e] = \mu^{(4)}[g, e, e, e] = g, \quad \forall g \in G, \tag{1.3.28}$$

then the hetero power $\mathcal{G}'^{(4)}$ has the 4-ary identity

$$\mathbf{E} = \begin{pmatrix} e \\ e \\ e \end{pmatrix}, \quad e \in G. \tag{1.3.29}$$

In the case where $\mathcal{G}^{(3)}$ is a ternary group with the unary queroperation $\bar{\mu}^{(1)}$, then the cubic hetero power $\mathcal{G}'^{(4)}$ is also a ternary group with one querelement (see example 1.3.12)

$$\bar{\mathbf{G}} = \overline{\begin{pmatrix} g_1 \\ g_2 \\ g_3 \end{pmatrix}} = \begin{pmatrix} g_{\text{quer}} \\ g_{\text{quer}} \\ g_{\text{quer}} \end{pmatrix}, \quad g_{\text{quer}} \in G, \; g_i \in G, \tag{1.3.30}$$

where $g_{\text{quer}} = \bar{\mu}^{(1)}[g]$, $\forall g \in G$.

A more nontrivial example is a cubic hetero power which has different arity to the initial semigroup.

Example 1.3.16. Let $\mathcal{G}^{(4)} = \langle G | \mu^{(4)} \rangle$ be a 4-ary semigroup, then we can construct its ternary associative cubic hetero power $\mathcal{G}'^{(3)} = \langle G' | \boldsymbol{\mu}'^{(3)} \rangle$ using the associative quivers with one intact element and two multiplications (Duplij 2012). Taking in (1.3.18) the parameters $n' = 3$, $n = 4$, $k = 3$, $\ell_{\text{id}} = 1$ (see third line of table 1.1), we get for the ternary multiplication $\boldsymbol{\mu}'^{(3)}$ for the triples $\mathbf{G}^T = (g_1, g_2, g_3) \in G \times G \times G = G'$ of the hetero cubic power case the form

$$\boldsymbol{\mu}'^{(3)}[\mathbf{G}^{(1)}, \mathbf{G}^{(2)}, \mathbf{G}^{(3)}] = \begin{pmatrix} \mu^{(4)}\left[g_1^{(1)}, g_2^{(2)}, g_3^{(3)}, g_1^{(3)}\right] \\ \mu^{(4)}\left[g_2^{(1)}, g_3^{(2)}, g_1^{(2)}, g_2^{(3)}\right] \\ g_3^{(1)} \end{pmatrix}, \quad g_i^{(j)} \in G, \tag{1.3.31}$$

which is totally associative, and therefore the hetero cubic power of 4-ary semigroup $\mathcal{G}^{(4)} = \langle G|\mu^{(4)}\rangle$ is a ternary semigroup $\mathcal{G}'^{(3)} = \langle G'|\boldsymbol{\mu}'^{(3)}\rangle$, such that $\mathcal{G}'^{(3)} = (\mathcal{G}^{(4)})^{\boxtimes 3}$. If the initial 4-ary semigroup $\mathcal{G}^{(4)}$ has the identity satisfying (1.3.28), then the ternary hetero power $\mathcal{G}'^{(3)}$ has only the right ternary identity (1.3.29) satisfying one relation

$$\boldsymbol{\mu}'^{(3)}[\mathbf{G}, \mathbf{E}, \mathbf{E}] = \mathbf{G}, \quad \forall \mathbf{G} \in G^{\times 3}, \tag{1.3.32}$$

and therefore $\mathcal{G}'^{(3)}$ is a ternary semigroup with a right identity. If $\mathcal{G}^{(4)}$ is a 4-ary group with the queroperation $\bar{\mu}^{(1)}$, then the hetero power $\mathcal{G}'^{(3)}$ can only be a ternary semigroup, because in $\langle G'|\boldsymbol{\mu}'^{(3)}\rangle$ we cannot define the standard queroperation (Post 1940).

1.4 The deformed Hosszú–Gluskin theorem

Since the early days of 'polyadic history' (Kasner 1904, Prüfer 1924, Dörnte 1929), the interconnection between polyadic algebraic structures and binary ones has been one of the main areas of interest (Post 1940, Thurston 1954). Early constructions were confined to building some special polyadic (mostly ternary, Certaine 1943, Lehmer 1932) operations on elements of binary groups (Miller 1935, Lehmer 1936, Chunikhin 1946). A very special form of n-ary multiplication in terms of binary multiplication and a special mapping as a chain formula was found in Hosszú (1963) and Gluskin (1964, 1965). The theorem that any n-ary multiplication can be presented in this form is called the Hosszú–Gluskin theorem (for a review see Dudek and Głazek 2008, Gal'mak 2003). A concise and clear proof of the Hosszú–Gluskin chain formula was presented in Sokolov (1976), nevertheless a counter-example was presented in Chronowski (1991).

In this section we give a new form of the Hosszú–Gluskin theorem in terms of polyadic powers. Then we show that the Hosszú–Gluskin chain formula is not unique and can be generalized ('deformed') using a parameter q which takes special integer values. We present the 'q-deformed' analog of the Hosszú–Gluskin theorem in the form of an invariance and consider some examples. The 'q-deformed' homomorphism theorem is also given.

1.4.1 The Hosszú–Gluskin theorem

Let us consider a polyadic group $\mathcal{G}'^{(n)} = \langle G|\mu'^{(n)}\rangle$ and investigate possible concrete forms of polyadic multiplication $\mu'^{(n)}$ in terms of lesser arity operations on the same set G (see (1.1.3)). Obviously, the simplest way of constructing an n-ary product $\mu'^{(n)}$ from the binary one $\mu^{(2)} = (*)$ is $\ell_\mu = n$ iteration (1.1.3) (Suschkewitsch 1935, Miller 1935)

$$\mu'^{(n)}[\mathbf{g}] = g_1 * g_2 * \ldots * g_n, \quad g_i \in G. \tag{1.4.1}$$

In Dörnte (1929) it was noted that not all n-ary groups have a product of this special form. The binary group $\mathcal{G}_*^{(2)} = \langle G|\mu^{(2)} = *, e\rangle$ was called a *covering group* of the n-ary group $\mathcal{G}'^{(n)} = \langle G|\mu'^{(n)}\rangle$ in Post (1940) (see, also, Tvermoes 1953), where a theorem establishing a more general (than (1.4.1)) structure of $\mu'^{(n)}[\mathbf{g}]$ in terms of subgroup structure of the covering group was given. A manifest form of the n-ary group product $\mu'^{(n)}[\mathbf{g}]$ in terms of the binary one and a special mapping was found in Hosszú (1963) and Gluskin (1965), and is called the Hosszú–Gluskin theorem, despite the same formulas having appeared much earlier in Turing (1938) and Post (1940) (for the relationship between the formulations, see Gal'mak and Vorobiev 2013). A simple construction of $\mu'^{(n)}[\mathbf{g}]$ which is present in the Hosszú–Gluskin theorem was given in Sokolov (1976). Here we follow this scheme in the opposite direction, by just deriving the final formula step by step (without writing it immediately) with clear examples. Then we introduce a 'deformation' to it in such a way that a generalized 'q-deformed' Hosszú–Gluskin theorem can be formulated.

First, let us rewrite (1.4.1) in its equivalent form

$$\mu'^{(n)}[\mathbf{g}] = g_1 * g_2 * \ldots * g_n * e, \quad g_i, e \in G, \tag{1.4.2}$$

where e is a distinguished element of the binary group $\langle G \mid *, e \rangle$, that is the identity. Now we apply to (1.4.2) an 'extended' version of the homotopy relation (2.1.1) with $\varphi_i^{GG'} = \psi_i$, $i = 1, \ldots, n$, and the lhs mapping $\varphi_{n+1}^{GG'} = \mathrm{id}$, but add an action ψ_{n+1} on the identity e of the binary group $\langle G \mid *, e \rangle$. Then we get (see (1.1.8) and (1.1.13))

$$\mu^{(n)}[\mathbf{g}] = \mu_e^{(n)}[\mathbf{g}] = \psi_1(g_1) * \psi_2(g_2) * \ldots * \psi_n(g_n) * \psi_{n+1}(e) = \left(* \prod_{i=1}^{n} \psi_i(g_i) \right) * \psi_{n+1}(e). \tag{1.4.3}$$

In this way we have obtained the most general form of polyadic multiplication in terms of $(n+1)$ 'extended' homotopy maps ψ_i, $i = 1, \ldots, n+1$, such that the diagram

$$\begin{array}{ccccc} G^{\times(n)} \times \{\bullet\} & \xrightarrow{\mathrm{id}^{\times n} \times \mu_e^{(0)}} & G^{\times(n+1)} & \xrightarrow{\psi_1 \times \ldots \times \psi_{n+1}} & G^{\times(n+1)} \\ \uparrow \epsilon & & & & \downarrow \left(\mu^{(2)}\right)^{\times n} \\ G^{\times(n)} & & \xrightarrow{\quad \mu_e^{(n)} \quad} & & G \end{array} \tag{1.4.4}$$

commutes. A natural question arises, whether all associative polyadic algebraic structures have this form of multiplication or do we have others? In general, we can correspondingly classify polyadic algebraic structures as:

(1) *Homotopic* polyadic algebraic structures which can be presented in the form (1.4.3). (1.4.5)

(2) *Nonhomotopic* polyadic algebraic structures with multiplication of other than (1.4.3) shapes. (1.4.6)

If the second class is nonempty, it would be interesting to find examples of nonhomotopic polyadic algebraic structures. The Hosszú–Gluskin theorem considers the homotopic polyadic algebraic structures and gives one of the possible choices for the 'extended' homotopy maps ψ_i in (1.4.3). We will show that this choice can be extended ('deformed') to the infinite 'q-series'.

The main idea in constructing the 'automatically' associative n-ary operation $\mu^{(n)}$ in (1.4.3) is to express the binary multiplication ($*$) and the 'extended' homotopy maps ψ_i in terms of $\mu^{(n)}$ itself (Sokolov 1976). A simplest binary multiplication which can be built from $\mu^{(n)}$ is (see (1.2.2))

$$g *_{\mathbf{t}} h = \mu^{(n)}[g, \mathbf{t}^{(n-2)}, h], \tag{1.4.7}$$

where $\mathbf{t} = \mathbf{t}^{(n-2)} \in G^{\times(n-2)}$ is any fixed polyad of length $(n - 2)$. If we apply here the equations for the identity e in a binary group

$$g *_{\mathbf{t}} e = g, \quad e *_{\mathbf{t}} h = h, \tag{1.4.8}$$

then we obtain

$$\mu^{(n)}[g, \mathbf{t}, e] = g, \quad \mu^{(n)}[e, \mathbf{t}, h] = h. \tag{1.4.9}$$

We observe from (1.4.9) that $(\mathbf{t}, e)$ and $(e, \mathbf{t})$ are neutral sequences of length $(n - 1)$, and therefore using (1.2.16) we can take $\mathbf{t}$ as a polyadic inverse of e (the identity of the binary group) considered as an element (but not an identity) of the polyadic algebraic structure $\langle G|\mu^{(n)}\rangle$, that is $\mathbf{t} = \mathbf{e}^{-1}$. Then, the binary multiplication constructed from $\mu^{(n)}$ and which has the standard identity properties (1.4.8) can be chosen as

$$g * h = g *_e h = \mu^{(n)}[g, \mathbf{e}^{-1}, h]. \tag{1.4.10}$$

Using this construction any element of the polyadic algebraic structure $\langle G|\mu^{(n)}\rangle$ can be distinguished and may serve as the identity of the binary group, and is then denoted by e (for clarity and convenience).

We recognize in (1.4.10) a version of the Maltsev term (see, e.g., Bergman 2012), which can be called a *polyadic Maltsev term* and is defined as

$$p(g, e, h) \stackrel{\text{def}}{=} \mu^{(n)}[g, \mathbf{e}^{-1}, h] \tag{1.4.11}$$

having the standard term properties (Bergman 2012)

$$p(g, e, e) = g, \quad p(e, e, h) = h, \tag{1.4.12}$$

which now follow from (1.4.9), i.e., the polyads $(e, \mathbf{e}^{-1})$ and $(\mathbf{e}^{-1}, e)$ are neutral, as they should be (1.2.16). Denote by g^{-1} the inverse element of g in the binary group ($g * g^{-1} = g^{-1} * g = e$) and $\mathbf{g}^{-1} \in G^{\times(n-2)}$ its polyadic inverse in an n-ary group (1.2.16), then it follows from (1.4.10) that $\mu^{(n)}[g, \mathbf{e}^{-1}, g^{-1}] = e$. Thus, we get

$$g^{-1} = \mu^{(n)}[e, \mathbf{g}^{-1}, e], \tag{1.4.13}$$

which can be considered as a connection between the inverse g^{-1} in the binary group and the polyadic inverse in the polyadic algebraic structure related to the same element g. For an n-ary group we can write for the polyadic inverse sequence $\mathbf{g}^{-1} = (g^{n-3}, \bar{g})$ and the binary group inverse g^{-1} becomes

$$g^{-1} = \mu^{(n)}[e, g^{n-3}, \bar{g}, e]. \tag{1.4.14}$$

If $\langle G|\mu^{(n)}\rangle$ is an n-ary group, then the element e is querable (1.2.17), for the polyadic inverse $\mathbf{e}^{-1}$ one can choose $(e^{n-3}, \bar{e})$ with $\bar{e}$ being on any (allowed) place, and the polyadic Maltsev term becomes (Shchuchkin 2003) $p(g, e, h) = \mu^{(n)}[g, e^{n-3}, \bar{e}, h]$ (together with the multiplication (1.4.10)). For instance, if $n = 3$, we have

$$g * h = \mu^{(3)}[g, \bar{e}, h], \quad g^{-1} = \mu^{(3)}[e, \bar{g}, e], \tag{1.4.15}$$

and the neutral polyads are $(e, \bar{e})$ and $(\bar{e}, e)$.

Now let us turn to build the main construction, that of the 'extended' homotopy maps ψ_i (1.4.3) in terms of $\mu^{(n)}$, which will lead to the Hosszú–Gluskin theorem. We start with a simple example of a ternary algebraic structure (1.4.15), derive the Hosszú–Gluskin 'chain formula', and then it will be clear how to proceed for generic n. Instead of (1.4.3) we write

$$\mu^{(3)}[g, h, u] = \psi_1(g) * \psi_2(h) * \psi_3(u) * \psi_4(e) \tag{1.4.16}$$

and try to construct ψ_i in terms of the ternary product $\mu^{(3)}$ and the binary identity e. We already know the structure of the binary multiplication (1.4.15): it contains $\bar{e}$, and therefore we can insert between g, h and u in the lhs of (1.4.16) a neutral ternary polyad $(\bar{e}, e)$ or its powers $(\bar{e}^k, e^k)$. Thus, taking for all insertions the *minimal number* of neutral polyads, we get

$$\begin{aligned}\mu^{(3)}[g, h, u] &= \left(\mu^{(3)}\right)^{\circ 2}\left[g, \overset{*}{\overset{\downarrow}{\bar{e}}}, e, h, u\right] = \left(\mu^{(3)}\right)^{\circ 4}\left[g, \overset{*}{\overset{\downarrow}{\bar{e}}}, e, h, \bar{e}, \overset{*}{\overset{\downarrow}{\bar{e}}}, e, e, u\right] \\ &= \left(\mu^{(3)}\right)^{\circ 4}\left[g, \overset{*}{\overset{\downarrow}{\bar{e}}}, e, h, \bar{e}, \overset{*}{\overset{\downarrow}{\bar{e}}}, e, e, u, \bar{e}, \bar{e}, \overset{*}{\overset{\downarrow}{\bar{e}}}, e, e, e\right].\end{aligned} \tag{1.4.17}$$

We show by arrows the binary products in special places: there should be 1, 3, 5, … , $(2k - 1)$ elements inbetween them to form inner ternary products. Then we rewrite (1.4.17) as

$$\mu^{(3)}[g, h, u] = \left(\mu^{(3)}\right)^{\circ 3}\left[g, \overset{*}{\overset{\downarrow}{\bar{e}}}, \mu^{(3)}[e, h, \bar{e}], \overset{*}{\overset{\downarrow}{\bar{e}}}, \left(\mu^{(3)}\right)^{\circ 2}[e, e, u, \bar{e}, \bar{e}], \overset{*}{\overset{\downarrow}{\bar{e}}}, \mu^{(3)}[e, e, e]\right]. \tag{1.4.18}$$

Comparing this with (1.4.16), we can exactly identify the 'extended' homotopy maps ψ_i as

$$\psi_1(g) = g, \tag{1.4.19}$$

$$\psi_2(g) = \varphi(g), \tag{1.4.20}$$

$$\psi_3(g) = \varphi(\varphi(g)) \equiv \varphi^{\circ 2}(g), \tag{1.4.21}$$

$$\psi_4(e) = \mu^{(3)}[e, e, e], \tag{1.4.22}$$

where

$$\varphi(g) = \mu^{(3)}[e, g, \bar{e}], \tag{1.4.23}$$

which can be described by the commutative diagram

$$\begin{array}{ccccc} \{\bullet\} \times G \times \{\bullet\} & \xrightarrow{\mu_e^{(0)} \times \mathrm{id} \times \mu_e^{(0)}} & G^{\times 3} & \xrightarrow{\mathrm{id}^{\times 2} \times \bar{\mu}^{(1)}} & G^{\times 3} \\ \uparrow \epsilon & & & & \downarrow \mu^{(3)} \\ G & & \xrightarrow{\quad\quad\varphi\quad\quad} & & G \end{array} \tag{1.4.24}$$

The mapping ψ_4 is the first polyadic power (1.1.18) of the binary identity e in the ternary algebraic structure

$$\psi_4(e) = e^{\langle 1 \rangle}. \tag{1.4.25}$$

Thus, combining (1.4.18)–(1.4.25) we obtain the Hosszú–Gluskin 'chain formula' for $n = 3$

$$\mu^{(3)}[g, h, u] = g * \varphi(h) * \varphi^{\circ 2}(u) * b, \tag{1.4.26}$$

$$b = e^{\langle 1 \rangle}, \tag{1.4.27}$$

which depends on one mapping φ (taken in the chain of powers) only, and the first polyadic power $e^{\langle 1 \rangle}$ of the binary identity e. The corresponding Hosszú–Gluskin diagram

$$\begin{array}{ccccc} G^{\times 3} \times \{\bullet\}^{\times 3} & \xrightarrow{\mathrm{id} \times \varphi \times \varphi^{\circ 2} \times \left(\mu_e^{(0)}\right)^{\times 3}} & G^{\times 6} & \xrightarrow{\mathrm{id}^{\times 3} \times \mu^{(3)}} & G^{\times 4} \\ \uparrow \epsilon & & & & \downarrow \left(\mu^{(2)}\right)^{\times 3} \\ G \times G \times G & & \xrightarrow{\quad\quad\mu^{(3)}\quad\quad} & & G \end{array} \tag{1.4.28}$$

commutes.

The mapping φ is an automorphism of the binary group $\langle G | *, e \rangle$, because it follows from (1.4.15) and (1.4.23) that

$$\begin{aligned} \varphi(g) * \varphi(h) &= \mu^{(3)}[\mu^{(3)}[e, g, \bar{e}], \bar{e}, \mu^{(3)}[e, h, \bar{e}]] = \left(\mu^{(3)}\right)^{\circ 3}\left[e, g, \bar{e}, \overset{\text{neutral}}{(\bar{e}, e)}, h, \bar{e}\right] \\ &= \left(\mu^{(3)}\right)^{\circ 2}[e, g, \bar{e}, h, \bar{e}] = \mu^{(3)}[e, g * h, \bar{e}] = \varphi(g * h), \end{aligned} \tag{1.4.29}$$

$$\varphi(e) = \mu^{(3)}[e, e, \bar{e}] = \mu^{(3)}\left[e, \overset{\text{neutral}}{(e, \bar{e})}\right] = e. \tag{1.4.30}$$

It is important to note that not only the binary identity e, but also its first polyadic power $e^{\langle 1\rangle}$ is a fixed point of the automorphism φ, because

$$\varphi(e^{\langle 1\rangle}) = \mu^{(3)}[e, e^{\langle 1\rangle}, \bar{e}] = (\mu^{(3)})^{\circ 2}\left[e, e, e, \overbrace{(e, \bar{e})}^{\text{neutral}}\right] = \mu^{(3)}[e, e, e] = e^{\langle 1\rangle}. \quad (1.4.31)$$

Moreover, taking into account that in the binary group (see (1.4.15))

$$(e^{\langle 1\rangle})^{-1} = \mu^{(3)}[e, \overline{e^{\langle 1\rangle}}, e] = (\mu^{(3)})^{\circ 2}[e, \bar{e}, \bar{e}, \bar{e}, e] = \bar{e}, \quad (1.4.32)$$

we get

$$\varphi^2(g) = (\mu^{(3)})^{\circ 2}[e, e, g, \bar{e}, \bar{e}] = (\mu^{(3)})^{\circ 2}\left[e, e, \overbrace{(e, \bar{e})}^{\text{neutral}}g, \bar{e}, \bar{e}\right] = e^{\langle 1\rangle} * g * (e^{\langle 1\rangle})^{-1}. \quad (1.4.33)$$

The higher polyadic powers $e^{\langle k\rangle} = (\mu^{(3)})^{\circ k}[e^{2k+1}]$ of the binary identity e are obviously also fixed points

$$\varphi(e^{\langle k\rangle}) = e^{\langle k\rangle}. \quad (1.4.34)$$

The elements $e^{\langle k\rangle}$ form a subgroup $\mathcal{H}$ of the binary group $\langle G|*, e\rangle$, because

$$e^{\langle k\rangle} * e^{\langle l\rangle} = e^{\langle k+l\rangle}, \quad (1.4.35)$$

$$e^{\langle k\rangle} * e = e * e^{\langle k\rangle} = e^{\langle k\rangle}. \quad (1.4.36)$$

We can express the even powers of the automorphism φ through the polyadic powers $e^{\langle k\rangle}$ in the following way

$$\varphi^{2k}(g) = e^{\langle k\rangle} * g * (e^{\langle k\rangle})^{-1}. \quad (1.4.37)$$

This gives a manifest connection between the Hosszú–Gluskin 'chain formula' and the sequence of cosets (see Post 1940) for the particular case $n = 3$.

Example 1.4.1. Let us consider the ternary copula associative multiplication (Mesiar and Sarkoci 2010, Stupňanová and Kolesárová 2011)

$$\mu^{(3)}[g, h, u] = \frac{g(1-h)u}{g(1-h)u + (1-g)h(1-u)}, \quad (1.4.38)$$

where $g_i \in G = [0, 1]$ and $0/0 = 0$ is assumed[4]. It is associative and cannot be iterated from any binary group. Obviously, $\mu^{(3)}[g^3] = g$, and therefore this polyadic algebraic structure is ℓ_μ-idempotent (1.1.27) $g^{\langle \ell_\mu\rangle} = g$. The querelement is $\bar{g} = \bar{\mu}_1[g] = g$. Because each element is querable, then $\langle G|\mu^{(3)}, \bar{\mu}_1\rangle$ is a ternary group.

[4] In this example all denominators are supposed nonzero.

Take a fixed element $e \in [0, 1]$. We define the binary multiplication as $g * h = \mu^{(3)}[g, e, h]$ and the automorphism

$$\varphi(g) = \mu^{(3)}[e, g, e] = e^2 \frac{1 - g}{e^2 - 2ge + g}, \tag{1.4.39}$$

which has the property $\varphi^{2k} = \mathrm{id}$ and $\varphi^{2k+1} = \varphi$, where $k \in \mathbb{N}$. Obviously, in (1.4.39) g can be on any (allowed) place in the product $\mu^{(3)}[e, g, e] = \mu^{(3)}[e, e, g] = \mu^{(3)}[e, e, g]$. Now we can check the Hosszú–Gluskin 'chain formula' (1.4.26) for the ternary copula

$$\begin{aligned}
\mu^{(3)}[g, h, u] &= (((g * \varphi(h)) * u) * e) \\
&= (\mu^{(3)})^{\circ 2}\left[g, e, e^2 \frac{1 - h}{e^2 - 2he + g}, e, (u, e, e)\right] \\
&= (\mu^{(3)})^{\circ 2}\left[g, \left(e, e^2 \frac{1 - h}{e^2 - 2he + g}, e\right), u\right] \\
&= \mu^{(3)}[g, \varphi^2(h), u] = \mu^{(3)}[g, h, u].
\end{aligned} \tag{1.4.40}$$

The language of polyadic inverses allows us to generalize the Hosszú–Gluskin 'chain formula' from $n = 3$ (1.4.26) to arbitrary n in a clear way. The derivation coincides with (1.4.18) using the multiplication (1.4.10) (with substitution $\bar{e} \to \mathbf{e}^{-1}$), neutral polyads $(\mathbf{e}^{-1}, e)$ or their powers $((\mathbf{e}^{-1})^k, e^k)$, but contains n terms

$$\begin{aligned}
\mu^{(n)}[g_1, \ldots, g_n] &= \left(\mu^{(n)}\right)^{\circ 2}\left[g_1, \overset{*}{\overset{\downarrow}{\mathbf{e}^{-1}}}, e, g_2, \ldots, g_n\right] = \left(\mu^{(n)}\right)^{\circ 3}\left[g_1, \overset{*}{\overset{\downarrow}{\mathbf{e}^{-1}}}, e, g_2, \mathbf{e}^{-1}, \overset{*}{\overset{\downarrow}{\mathbf{e}^{-1}}}, e, e, g_3, \ldots, g_n\right] = \ldots \\
&= \left(\mu^{(n)}\right)^{\circ(n-1)}\left[g_1, \overset{*}{\overset{\downarrow}{\mathbf{e}^{-1}}}, e, g_2, \mathbf{e}^{-1}, \overset{*}{\overset{\downarrow}{\mathbf{e}^{-1}}}, e, e, g_3, \ldots, \overset{*}{\overset{\downarrow}{\mathbf{e}^{-1}}}, \overbrace{e, \ldots, e}^{n-1}, g_n, \overbrace{\mathbf{e}^{-1}, \ldots, \mathbf{e}^{-1}}^{n-1}, \overset{*}{\overset{\downarrow}{\mathbf{e}^{-1}}}, \overbrace{e, \ldots, e}^{n}\right].
\end{aligned} \tag{1.4.41}$$

We observe from (1.4.41) that the mapping φ in the n-ary case is

$$\varphi(g) = \mu^{(n)}[e, g, \mathbf{e}^{-1}], \tag{1.4.42}$$

and the last product of the binary identities $\mu^{(n)}[e, \ldots, e]$ is also the first n-ary power $e^{\langle 1 \rangle}$ (1.1.18). It follows from (1.4.42) and (1.4.10), that

$$\varphi^{n-1}(g) = e^{\langle 1 \rangle} * g * (e^{\langle 1 \rangle})^{-1}. \tag{1.4.43}$$

In this way, we obtain the Hosszú–Gluskin 'chain formula' for arbitrary n

$$\mu^{(n)}[g_1, \dots, g_n] = g_1 * \varphi(g_2) * \varphi^{\circ 2}(g_3) * \dots * \varphi^{\circ(n-2)}(g_{n-1}) * \varphi^{\circ(n-1)}(g_n) * e^{\langle 1 \rangle}$$
$$= \left(* \prod_{i=1}^{n} \varphi^{\circ(i-1)}(g_i) \right) * e^{\langle 1 \rangle}. \tag{1.4.44}$$

Thus, we have found the 'extended' homotopy maps ψ_i from (1.4.3) as

$$\psi_i(g) = \varphi^{\circ(i-1)}(g), \quad i = 1, \dots, n, \tag{1.4.45}$$

$$\psi_{n+1}(g) = g^{\langle 1 \rangle}, \tag{1.4.46}$$

where we put by definition $\varphi^{\circ 0}(g) = g$. Using (1.4.31) and (1.4.44) we can formulate the Hosszú–Gluskin theorem in the language of polyadic powers.

Theorem 1.4.2. *On a polyadic group $\mathcal{G}^{(n)} = \langle G | \mu^{(n)}, \bar{\mu}^{(1)} \rangle$ one can define a binary group $\mathcal{G}_*^{(2)} = \langle G | \mu^{(2)} = *, e \rangle$and its automorphism φ such that the Hosszú–Gluskin 'chain formula' (1.4.44) is valid, where the polyadic powers of the identity* e *are fixed points of φ (1.4.34), form a subgroup $\mathcal{H}$ of $\mathcal{G}_*^{(2)}$, and the $(n - 1)$ power of φ is a conjugation (1.4.43) with respect to $\mathcal{H}$.*

The following reverse Hosszú–Gluskin theorem holds.

Theorem 1.4.3. If in a binary group $\mathcal{G}_*^{(2)} = \langle G | \mu^{(2)} = *, e \rangle$ one can define an automorphism φ such that

$$\varphi^{\circ(n-1)}(g) = b * g * b^{-1}, \tag{1.4.47}$$

$$\varphi(b) = b, \tag{1.4.48}$$

where $b \in G$ is a distinguished element, then the 'chain formula'

$$\mu^{(n)}[g_1, \dots, g_n] = \left(* \prod_{i=1}^{n} \varphi^{\circ(i-1)}(g_i) \right) * b \tag{1.4.49}$$

determines an n-ary group, in which the distinguished element is the first polyadic power of the binary identity

$$b = e^{\langle 1 \rangle}. \tag{1.4.50}$$

1.4.2 'Deformation' of Hosszú–Gluskin chain formula

Let us raise the question: can the choice (1.4.45)–(1.4.46) of the 'extended' homotopy maps (1.4.3) be generalized? Before answering this question *positively* we consider some preliminary statements.

First, we note that we keep the general idea of inserting neutral sequences into a polyadic product (see (1.4.17) and (1.4.41)), because this is the only way to obtain 'automatic' associativity. Second, the number of the inserted neutral polyads can be chosen *arbitrarily*, not only minimally, as in (1.4.17) and (1.4.41) (as they are neutral). Nevertheless, we can show that this arbitrariness is somewhat restricted.

Ternary case. Indeed, let us consider a polyadic group $\langle G \mid \mu^{(n)}, \bar{\mu}_1 \rangle$ in the particular case $n = 3$, where for any $e_0 \in G$ and natural k the sequence $(\bar{e}_0^k, e_0^k)$ is neutral, then we can write

$$\mu^{(3)}[g, h, u] = (\mu^{(3)})^{\circ r}[g, \bar{e}_0^k, e_0^k, h, \bar{e}_0^{lk}, e_0^{lk}, u, \bar{e}_0^{mk}, e_0^{mk}], \tag{1.4.51}$$

where $r = k(l + m + 1) + 1$. If we make the change of variables $e_0^k = e$, then we obtain

$$\mu^{(3)}[g, h, u] = (\mu^{(3)})^{\circ r}[g, \bar{e}, e, h, \bar{e}^l, e^l, u, \bar{e}^m, e^m], \tag{1.4.52}$$

where $r = l + m + 2$. Because this should reproduce the formula (1.4.16), we immediately conclude that $\psi_1(g) = \mathrm{id}$, and the multiplication is the same as in (1.4.15), and e is again the identity of the binary group $\mathcal{G}_*^{(2)} = \langle G, *, e \rangle$. Moreover, if we put $\psi_2(g) = \varphi(g)$, as in the standard case, then we have a first 'half' of the mapping φ, that is $\varphi(g) = \mu^{(3)}[e, h, \text{something}]$. Now we are in a position to find this 'something' and other 'extended' homotopy maps ψ_i from (1.4.16), but without the requirement of a minimal number of inserted neutral polyads, as it was in (1.4.17). By analogy, we rewrite (1.4.52) as

$$\mu^{(3)}[g, h, u] = (\mu^{(3)})^{\circ r}[g, \bar{e}, (e, h, \bar{e}^q), \bar{e}, e^{q+1}, u, \bar{e}^m, e^m], \tag{1.4.53}$$

where $r = q + m + 3$ and we put $l = q + 1$. So we have found the 'something', and the map φ is

$$\varphi_q(g) = (\mu^{(3)})^{\circ 2\ell_\varphi(q)}[e, g, \bar{e}^q], \tag{1.4.54}$$

where the number of multiplications

$$\ell_\varphi(q) = \frac{q + 1}{2} \tag{1.4.55}$$

is an integer $\ell_\varphi(q) = 1, 2, 3, \ldots$, while $q = 1, 3, 5, 7, \ldots$. The diagram defined φ_q (e.g., for $q = 3$ and $\ell_\varphi(q) = 2$)

$$\begin{array}{ccccc}
\{\bullet\} \times G \times \{\bullet\}^{\times 3} & \xrightarrow{\mu_e^{(0)} \times \mathrm{id} \times (\mu_e^{(0)})^{\times 3}} & G^{\times 5} & \xrightarrow{\mathrm{id}^{\times 2} \times (\bar{\mu}^{(1)})^{\times 3}} & G^{\times 5} \\
\uparrow \epsilon & & & & \downarrow \mu^{(3)} \times \mu^{(3)} \\
G & & \xrightarrow{\quad \varphi_q \quad} & & G
\end{array} \tag{1.4.56}$$

commutes (see (1.4.24)). Then, we can find power m in (1.4.53)

$$\mu^{(3)}[g, h, u] = \left(\mu^{(3)}\right)^{\circ r}[g, \bar{e}, (e, h, \bar{e}^q), \bar{e}, (e, u, \bar{e}^q)^{q+1}, \bar{e}, e^{q(q+1)+1}], \tag{1.4.57}$$

and therefore $m = q(q + 1) + 1$. Thus, we have obtained the 'q-deformed' maps ψ_i (see (1.4.19)–(1.4.22))

$$\psi_1(g) = \varphi_q^{\circ[[0]]_q}(g) = \varphi_q^{\circ 0}(g) \equiv g, \tag{1.4.58}$$

$$\psi_2(g) = \varphi_q(g) = \varphi_q^{\circ[[1]]_q}(g), \tag{1.4.59}$$

$$\psi_3(g) = \varphi_q^{\circ(q+1)}(g) = \varphi_q^{\circ[[2]]_q}(g), \tag{1.4.60}$$

$$\psi_4(g) = \left(\mu^{(3)}\right)^{\circ r}[g^{q(q+1)+1}] = \left(\mu^{(3)}\right)^{\circ r}[g^{[[3]]_q}], \tag{1.4.61}$$

where φ is defined by (1.4.54) and $[[k]]_q$ is the q-deformed number (1.2.26), and we put $\varphi_q^{\circ 0} = \mathrm{id}$. The corresponding '$q$-deformed' chain formula (for $n = 3$) can be written as (see (1.4.26)–(1.4.27) for 'nondeformed' case)

$$\mu^{(3)}[g, h, u] = g * \varphi_q^{\circ[[1]]_q}(h) * \varphi_q^{\circ[[2]]_q}(u) * b_q, \tag{1.4.62}$$

$$b_q = e^{\langle \ell_e(q) \rangle}, \tag{1.4.63}$$

where the degree of the binary identity polyadic power

$$\ell_e(q) = q\frac{[[2]]_q}{2} = \ell_\varphi(q)\left(2\ell_\varphi(q) + 1\right) \tag{1.4.64}$$

is an integer. The corresponding 'deformed' chain diagram (e.g., for $q = 3$)

$$\begin{array}{ccccc} G^{\times 3} \times \{\bullet\}^{\times 13} & \xrightarrow{\mathrm{id} \times \varphi_q \times \varphi_q^{\circ 4} \times \left(\mu_e^{(0)}\right)^{\times 13}} & G^{\times 16} & \xrightarrow{\mathrm{id}^{\times 3} \times \left(\mu^{(3)}\right)^{\times 6}} & G^{\times 4} \\ \uparrow \epsilon & & & & \downarrow \left(\mu^{(2)}\right)^{\times 3} \\ G \times G \times G & & \xrightarrow{\mu_3} & & G \end{array} \tag{1.4.65}$$

commutes (see the Hosszú–Gluskin diagram (1.4.28)). In the 'deformed' case the polyadic power $e^{\langle \ell_e(q) \rangle}$ is not a fixed point of φ_q and satisfies

$$\varphi_q(e^{\langle \ell_e(q) \rangle}) = \varphi_q(\left(\mu^{(3)}\right)^{\circ \ell_e(q)}[e^{q^2+q+1}]) = e^{\langle \ell_e(q) \rangle} * \varphi_q(e) \tag{1.4.66}$$

or

$$\varphi_q\left(b_q\right) = b_q * \varphi_q(e). \tag{1.4.67}$$

Instead of (1.4.33) we have

$$\varphi_q^{\circ(q+1)}(g) * e^{\langle \ell_e(q) \rangle} = \left(\mu^{(3)}\right)^{\circ r}[e^{q+1}, g] = e^{\langle \ell_e(q) \rangle} * \varphi_q^{\circ(q+1)}(e) * g \tag{1.4.68}$$

or

$$\varphi_q^{\circ(q+1)}(g) * b_q = b_q * \varphi_q^{\circ(q+1)}(e) * g. \tag{1.4.69}$$

The 'nondeformed' limit $q \to 1$ of (1.4.62) gives the Hosszú–Gluskin chain formula (1.4.26) for $n = 3$.

Arbitrary arity case. Now let us turn to arbitrary n and write the n-ary multiplication using neutral polyads analogously to (1.4.53). By the same arguments, as in (1.4.52), we insert only one neutral polyad $(\mathbf{e}^{-1}, e)$ between the first and second elements in the multiplication, but in other places we insert powers $((\mathbf{e}^{-1})^k, e^k)$ (*allowed* by the chain properties), and obtain

$$\begin{aligned}
&\mu^{(n)}[g_1, \ldots, g_n] = \left(\mu^{(n)}\right)^{\circ r_1}\left[g_1, \mathbf{e}^{-1}, e, g_2, \ldots, g_n\right] \\
&= \left(\mu^{(n)}\right)^{\circ r_2}\left[g_1, \mathbf{e}^{-1}, \left(e, g_2, (\mathbf{e}^{-1})^q\right), \mathbf{e}^{-1}, e^{q+1}, g_3, \ldots, g_n\right] = \ldots \\
&= \left(\mu^{(n)}\right)^{\circ r_3}\left[g_1, \mathbf{e}^{-1}, \left(e, g_2, (\mathbf{e}^{-1})^q\right), \mathbf{e}^{-1}, \left(e^{q+1}, g_3, \overbrace{\mathbf{e}^{-1}, \ldots, \mathbf{e}^{-1}}^{q(q+1)}\right)\mathbf{e}^{-1}, e^{q(q+1)+1}, g_3, \ldots\right. \\
&\left.\ldots, \left(\overbrace{e, \ldots, e}^{q^{n-2}+\cdots+q+1}, g_{n-1}, \overbrace{\mathbf{e}^{-1}, \ldots, \mathbf{e}^{-1}}^{q(q^{n-2}+\cdots+q+1)}\right), \mathbf{e}^{-1}, \left(\overbrace{e, \ldots, e}^{q^{n-1}+\cdots+q+1}, g_n, \overbrace{\mathbf{e}^{-1}, \ldots, \mathbf{e}^{-1}}^{q(q^{n-1}+\cdots+q+1)}\right), \mathbf{e}^{-1}, \overbrace{e, \ldots, e}^{q^n+\cdots+q+1}\right].
\end{aligned} \tag{1.4.70}$$

So we observe that the binary product is now the same as in the 'nondeformed' case (1.4.10), while the map φ is

$$\varphi_q(g) = \left(\mu^{(n)}\right)^{\circ \ell_\varphi(q)}[e, g, (\mathbf{e}^{-1})^q], \tag{1.4.71}$$

where the number of multiplications

$$\ell_\varphi(q) = \frac{q(n-2)+1}{n-1} \tag{1.4.72}$$

is an integer and $\ell_\varphi(q) \to q$, as $n \to \infty$, in the nondeformed case $\ell_\varphi(1) = 1$, as in (1.4.42). Note that the 'deformed' map φ_q is the a-quasi-endomorphism (Gluskin and Shvarts 1972) of the binary group $\mathcal{G}_*^{(2)}$, because from (1.4.71) we get

$$\begin{aligned}
\varphi_q(g) * \varphi_q(h) &= \left(\mu^{(n)}\right)^{\circ r_1}[e, g, (\mathbf{e}^{-1})^q, \mathbf{e}^{-1}, e, h, (\mathbf{e}^{-1})^q] \\
&= \left(\mu^{(n)}\right)^{\circ r_2}[e, g, \mathbf{e}^{-1}, (e, e, (\mathbf{e}^{-1})^q), \mathbf{e}^{-1}, h, (\mathbf{e}^{-1})^q] \\
&= \varphi_q(g * a * h),
\end{aligned} \tag{1.4.73}$$

where

$$a = \left(\mu^{(n)}\right)^{\circ \ell_\varphi(q)}[e, e, (\mathbf{e}^{-1})^q] = \varphi_q(e). \tag{1.4.74}$$

Nevertheless, φ_q of the form (1.4.74) is a particular case of a more general morphism as in

Definition 1.4.4. A *sandwich quasi-endomorphism* $\tilde{\varphi}: G \to G$ is defined by

$$\tilde{\varphi}(g) * \tilde{\varphi}(h) = \tilde{\varphi}(g * \tilde{\varphi}(e) * h), \tag{1.4.75}$$

such that the corresponding diagram

$$\begin{array}{ccccc} G \times G & \xrightarrow{\mu^{(2)}} & G & \xleftarrow{\quad\varphi_q\quad} & G \\ {\scriptstyle \varphi_q \times \varphi_q}\uparrow & & & & \uparrow{\scriptstyle \mu^{(2)} \times \mu^{(2)}} \\ G \times G & \xrightarrow{\epsilon} & G \times \{\bullet\} \times G & \xrightarrow{\mathrm{id} \times \mu_e^{(0)} \times \mathrm{id}} & G \times G \times G \end{array} \tag{1.4.76}$$

commutes.

The general sandwich quasi-endomorphism $\tilde{\varphi}$ defined by (1.4.75) only (without any connection with φ_q and (1.4.73)–(1.4.74)), has interesting properties.

Proposition 1.4.5. *The image of identity $\tilde{\varphi}(e)$ is 'linear' with respect to the sandwich quasi-endomorphism*

$$\tilde{\varphi}(g) * \tilde{\varphi}(e) = \tilde{\varphi}(g * \tilde{\varphi}(e)), \tag{1.4.77}$$

$$\tilde{\varphi}(e) * \tilde{\varphi}(g) = \tilde{\varphi}(\tilde{\varphi}(e) * g). \tag{1.4.78}$$

Proof. It follows from (1.4.75), when one element is identity e. □

Proposition 1.4.6. *The sandwich quasi-endomorphism acts on the powers of identity $\tilde{\varphi}(e)$ as multiplication by $\tilde{\varphi}(e)$*

$$\tilde{\varphi}(\tilde{\varphi}^{\circ k}(e)) = (\tilde{\varphi}(e))^{\circ(k+1)}, \quad k = 1, 2, 3, 4, \ldots \tag{1.4.79}$$

Proof. First, we observe, it follows from (1.4.75) with $g = h = e$, that

$$\tilde{\varphi}(e) * \tilde{\varphi}(e) = \tilde{\varphi}(\tilde{\varphi}(e)) \equiv \tilde{\varphi}^{\circ 2}(e), \tag{1.4.80}$$

which gives (1.4.79) with $k = 1$. Then using (1.4.77) or (1.4.78) with $g = \tilde{\varphi}^{k-1}(e)$, we obtain (1.4.79). □

Another important property follows from (1.4.80)

$$\tilde{\varphi}^{\circ(-2)}(\tilde{\varphi}^{\circ 2}(e)) = e. \tag{1.4.81}$$

If the image of identity $\tilde{\varphi}(e)$ is invertible, we obtain some additional properties of the sandwich quasi-endomorphism.

Proposition 1.4.7. If $\tilde{\varphi}(e)^{-1}$ *exists, then*

$$\tilde{\varphi}(\tilde{\varphi}^{\circ(-1)}(e)) = e, \tag{1.4.82}$$

$$\tilde{\varphi}(g) = \tilde{\varphi}(e) * \tilde{\varphi}(\tilde{\varphi}^{\circ(-1)}(e) * g) = \tilde{\varphi}(g * \tilde{\varphi}^{\circ(-1)}(e)) * \tilde{\varphi}(e), \quad \forall g \in G, \tag{1.4.83}$$

and also

$$\tilde{\varphi}(g * h) = \tilde{\varphi}(g) * \tilde{\varphi}^{\circ(-1)}(e) * \tilde{\varphi}(h), \quad \forall g, h \in G, \tag{1.4.84}$$

Proof. It follows from (1.4.75), (1.4.77)–(1.4.78) and (1.4.80). □

Let us consider some examples.

Example 1.4.8. Let for any fixed $u \in G$ and $u^2 = \tilde{\varphi}(e)$ we define

$$\tilde{\varphi}(g) = u * g * u, \tag{1.4.85}$$

then (1.4.75) satisfies, because

$$\begin{aligned} \tilde{\varphi}(g) * \tilde{\varphi}(h) = u * g * u * u * h * u = u * g * \tilde{\varphi}(e) * h \\ * u = \tilde{\varphi}(g * \tilde{\varphi}(e) * h). \end{aligned} \tag{1.4.86}$$

This example inspires the following isomorphism. Let us consider the 'sandwich' semigroup S_{u^2} with the multiplication defined by

$$g *_{u^2} h = g * u^2 * h, \quad g, u, h \in G. \tag{1.4.87}$$

Lemma 1.4.9. *The 'sandwich' semigroup S_{u^2} is isomorphic to* G, *if* u *is invertible, and the isomorphism ψ: $G \to S_{u^2}$ is given by $g \to u^{-1} * g * u^{-1}$.*

Proof. It follows from (1.4.85) and (1.4.86). □

Now we consider an example with noninvertible image of identity $\tilde{\varphi}(e)$.

Example 1.4.10. Let $G = M_2(\mathbb{R})$ and $u = \begin{pmatrix} 1 & 1 \\ 0 & 0 \end{pmatrix}$, and e is the identity matrix. We define the sandwich quasi-endomorphism by

$$\tilde{\varphi}(g) = u * g * u, \tag{1.4.88}$$

$$\tilde{\varphi}\begin{pmatrix} a & b \\ c & d \end{pmatrix} = \begin{pmatrix} a + c & a + c \\ 0 & 0 \end{pmatrix}, \tag{1.4.89}$$

which satisfies (1.4.75), while the image of the identity $\tilde{\varphi}(e) = u^2 = u$ is clearly noninvertible.

In the special case, when $\tilde{\varphi} = \varphi_q$ and $q = 1$, then $\varphi_q(e) = e$, and the distinguished element a turns to the binary identity $a = e$, such that the a-quasi-endomorphism φ_q becomes invertible, i.e., an automorphism of $\mathcal{G}_*^{(2)}$.

***Remark* 1.4.11.** The choice (1.4.71) of the a-quasi-endomorphism φ_q is different from Gluskin and Shvarts (1972), the latter (in our notation) is $\varphi_k(g) = \mu^{(n)}[a^{k-1}, g, a^{n-k}]$, $k = 1, \dots, n-1$, it has only one multiplication and leads to the 'nondeformed' chain formula (1.4.44) (for semigroup case).

It follows from (1.4.70), that the 'extended' homotopy maps ψ_i (1.4.3) are (see (1.4.58)–(1.4.61))

$$\psi_1(g) = \varphi_q^{\circ[[0]]_q}(g) = \varphi_q^{\circ 0}(g) \equiv g, \tag{1.4.90}$$

$$\psi_2(g) = \varphi_q(g) = \varphi_q^{\circ[[1]]_q}(g), \tag{1.4.91}$$

$$\psi_3(g) = \varphi_q^{\circ(q+1)}(g) = \varphi_q^{\circ[[2]]_q}(g), \tag{1.4.92}$$
$$\vdots$$

$$\psi_{n-1}(g) = \varphi_q^{\circ(q^{n-3}+\dots+q+1)}(g) = \varphi_q^{\circ[[n-2]]_q}(g), \tag{1.4.93}$$

$$\psi_n(g) = \varphi_q^{\circ(q^{n-2}+\dots+q+1)}(g) = \varphi_q^{\circ[[n-1]]_q}(g), \tag{1.4.94}$$

$$\psi_{n+1}(g) = (\mu^{(n)})^{\circ r_1}[g^{q^{n-1}+\dots+q+1}] = (\mu^{(n)})^{\circ r_2}[g^{[[n]]_q}]. \tag{1.4.95}$$

In terms of the polyadic power (1.1.18), the last map is

$$\psi_{n+1}(g) = g^{\langle \ell_e \rangle}, \tag{1.4.96}$$

where (see (1.4.72))

$$\ell_e(q) = q\frac{[[n-1]]_q}{n-1} \tag{1.4.97}$$

is an integer. Thus the 'q-deformed' n-ary chain formula is (see (1.4.44))

$$\begin{aligned} \mu^{(n)}[g_1, \dots, g_n] = g_1 * \varphi_q^{\circ[[1]]_q}(g_2) * \varphi_q^{\circ[[2]]_q}(g_3) * \dots * \varphi_q^{\circ[[n-2]]_q}(g_{n-1}) \\ * \varphi_q^{\circ[[n-1]]_q}(g_n) * e^{\langle \ell_e(q) \rangle}. \end{aligned} \tag{1.4.98}$$

In the 'nondeformed' limit $q \to 1$ (1.4.98) reproduces the Hosszú–Gluskin chain formula (1.4.44). Let us obtain the 'deformed' analogs of the distinguished element relations (1.4.47)–(1.4.48) for arbitrary n (the case $n = 3$ is in (1.4.66)–(1.4.68)). Instead of the fixed point relation (1.4.48) we now have from (1.4.71), (1.4.97) and (1.4.95) the *quasi-fixed point*

$$\varphi_q(b_q) = b_q * \varphi_q(e), \tag{1.4.99}$$

where the 'deformed' distinguished element b_q is (see (1.4.50))

$$b_q = (\mu^{(n)})^{\circ r}[e^{[[n]]_q}] = e^{\langle \ell_e(q) \rangle}. \tag{1.4.100}$$

The conjugation relation (1.4.47) in the 'deformed' case becomes the *quasi-conjugation*

$$\varphi_q^{\circ[[n-1]]_q}(g) * b_q = b_q * \varphi_q^{\circ[[n-1]]_q}(e) * g. \tag{1.4.101}$$

This allows us to rewrite the 'deformed' chain formula (1.4.98) as

$$\begin{aligned} \mu^{(n)}[g_1, \dots, g_n] = g_1 * \varphi_q^{\circ[[1]]_q}(g_2) * \varphi_q^{\circ[[2]]_q}(g_3) * \dots * \varphi_q^{\circ[[n-2]]_q}(g_{n-1}) * b_q \\ * \varphi_q^{\circ[[n-1]]_q}(e) * g_n. \end{aligned} \tag{1.4.102}$$

Using the above proof sketch, we formulate the following 'q-deformed' analog of the Hosszú–Gluskin theorem:

Theorem 1.4.12. *On a polyadic group $\mathcal{G}^{(n)} = \langle G \mid \mu^{(n)}, \bar{\mu}^{(1)} \rangle$ one can define a binary group $\mathcal{G}_*^{(2)} = \langle G \mid \mu^{(2)} = *, e \rangle$ and (the infinite 'q-series' of) its automorphism φ_q such that the 'deformed' chain formula (1.4.98) is valid*

$$\mu^{(n)}[g_1, \dots, g_n] = \left(\overset{*}{\prod_{i=1}^{n}} \varphi^{\circ[[i-1]]_q}(g_i) \right) * b_q, \tag{1.4.103}$$

where (the infinite 'q-series' of) the 'deformed' distinguished element $\mathrm{b_q}$ *(being a polyadic power of the binary identity (1.4.100)) is the quasi-fixed point of φ_q (1.4.99) and satisfies the quasi-conjugation (1.4.101) in the form*

$$\varphi_q^{\circ[[n-1]]_q}(g) = b_q * \varphi_q^{\circ[[n-1]]_q}(e) * g * b_q^{-1}. \tag{1.4.104}$$

In the 'nondeformed' case $q = 1$ we obtain the Hosszú–Gluskin chain formula (1.4.44) and the corresponding theorem 1.4.2.

Example 1.4.13. Let us have a binary group $\langle G | (\cdot), 1 \rangle$ and a distinguished element $e \in G$, $e \neq 1$, then we can define a binary group $\mathcal{G}_*^{(2)} = \langle G | (*), e \rangle$ by the product

$$g * h = g \cdot e^{-1} \cdot h. \tag{1.4.105}$$

The quasi-endomorphism

$$\varphi_q(g) = e \cdot g \cdot e^{-q} \tag{1.4.106}$$

satisfies (1.4.75) with $\varphi_q(e) = e^{2-q}$, and we take

$$b_q = e^{[[n]]_q}. \tag{1.4.107}$$

Then we can obtain the 'q-deformed' chain formula (1.4.103) (for $q = 1$ see, e.g., Gal'mak and Vorobiev 2013).

We observe that the chain formula is the 'q-series' of equivalence relations (1.4.103), which can be formulated as an invariance. Indeed, let us denote the rhs of (1.4.103) by $\mathcal{M}_q(g_1, \dots, g_n)$, and the lhs as $\mathcal{M}_0(g_1, \dots, g_n)$, then the chain formula can be written as some invariance (cf associativity as an invariance (1.1.28)).

Theorem 1.4.14. *On a polyadic group $\mathcal{G}^{(n)} = \langle G | \mu^{(n)}, \bar{\mu}^{(1)} \rangle$ we can define a binary group $\mathcal{G}_*^{(2)} = \langle G | \mu^{(2)} = *, e \rangle$ such that the following invariance is valid*

$$\mathcal{M}_q(g_1, \dots, g_n) = invariant, \qquad q = 0, 1, \dots \tag{1.4.108}$$

where

$$\mathcal{M}_q(g_1, \dots, g_n) = \begin{cases} \mu^{(n)}[g_1, \dots, g_n], & q = 0, \\ \left(* \prod_{i=1}^{n} \varphi^{\circ [[i-1]]_q}(g_i) \right) * b_q, & q > 0, \end{cases} \tag{1.4.109}$$

and the distinguished element $b_q \in G$ and the quasi-endomorphism φ_q of $\mathcal{G}_*^{(2)}$ are defined in (1.4.100) and (1.4.71) respectively.

Example 1.4.15. Let us consider the ternary q-product used in the nonextensive statistics (Nivanen 2003)

$$\mu^{(3)}[g, t, u] = (g^{\hbar} + t^{\hbar} + u^{\hbar} - 3)^{\frac{1}{\hbar}}, \tag{1.4.110}$$

where $\hbar = 1 - q_0$, and $g, t, u \in G = \mathbb{R}_+$, $0 < q_0 < 1$, and also $g^\hbar + t^\hbar + u^\hbar - 3 > 0$ (as for other terms inside brackets with power $\frac{1}{\hbar}$ below). In case $\hbar \to 0$ the q-product becomes an iterated product in $\mathbb{R}_+$ as $\mu^{(3)}[g, t, u] \to gtu$. The quermap $\bar{\mu}_1$ is given by

$$\bar{g} = (3 - g^\hbar)^{\frac{1}{\hbar}}. \tag{1.4.111}$$

The polyadic algebraic structure $\mathcal{G}^{(n)} = \langle G | \mu^{(3)}, \bar{\mu}^{(1)} \rangle$ is a ternary group, because each element is querable. Take a distinguished element $e \in G$ and use (1.4.15), (1.4.110) and (1.4.111) to define the product

$$g * t = (g^\hbar - e^\hbar + t^\hbar)^{\frac{1}{\hbar}} \tag{1.4.112}$$

of the binary group $\mathcal{G}_*^{(2)} = \langle G | \mu^{(2)} = (*), e \rangle$.

(1) *The Hosszú–Gluskin chain formula ($q = 1$).* The automorphism (1.4.23) of $\mathcal{G}_*^{(2)}$ is now the identity map $\varphi = \mathrm{id}$. The first polyadic power of the distinguished element e is

$$b = e^{\langle 1 \rangle} = \mu^{(3)}[e^3] = (3e^\hbar - 3)^{\frac{1}{\hbar}}. \tag{1.4.113}$$

The chain formula (1.4.26) can be checked as follows

$$\begin{aligned} \mu^{(3)}[g, t, u] &= (((g * t) * u) * b) = (((g^\hbar - e^\hbar + t^\hbar) - e^\hbar + u^\hbar) - e^\hbar + b^\hbar)^{\frac{1}{\hbar}} \\ &= (g^\hbar - e^\hbar + t^\hbar - e^\hbar + u^\hbar - e^\hbar + 3e^\hbar - 3)^{\frac{1}{\hbar}} = (g^\hbar + t^\hbar + u^\hbar - 3)^{\frac{1}{\hbar}}. \end{aligned} \tag{1.4.114}$$

(2) *The 'q-deformed' chain formula* (for conciseness we consider only the case $q = 3$). Now the quasi-endomorphism φ_q (1.4.54) is not the identity, but is

$$\varphi_{q=3}(g) = (g^\hbar - 2e^\hbar + 3)^{\frac{1}{\hbar}}. \tag{1.4.115}$$

In case $q = 3$ we need its 4th ($=q + 1$) power (1.4.62)

$$\varphi_{q=3}^{\circ 4}(g) = (g^\hbar - 8e^\hbar + 12)^{\frac{1}{\hbar}}. \tag{1.4.116}$$

The deformed polyadic power $e^{\langle \ell_e \rangle}$ from (1.4.62) is (see, also, (1.4.61))

$$b_{q=3} = e^{\langle 5 \rangle} = (\mu^{(3)})^{\circ 5}[e^{13}] = (13e^\hbar - 18)^{\frac{1}{\hbar}}. \tag{1.4.117}$$

Now we check the 'q-deformed' chain formula (1.4.62) as

$$\begin{aligned} \mu^{(3)}[g, t, u] &= g * \varphi_{q=3}(t) * \varphi_{q=3}^{4}(u) \\ &* b_{q=3} = \Big(\Big(\Big(g * \varphi_{q=3}(t)\Big) * \varphi_{q=3}^{4}(u)\Big) * b_{q=3}\Big) \end{aligned} \tag{1.4.118}$$

$$= (g^{\hbar} - e^{\hbar} + (t^{\hbar} - 2e^{\hbar} + 3) - e^{\hbar} + (u^{\hbar} - 8e^{\hbar} + 12) - e^{\hbar} + (13e^{\hbar} - 18))^{\frac{1}{\hbar}} \tag{1.4.119}$$

$$= (g^{\hbar} + t^{\hbar} + u^{\hbar} - 3)^{\frac{1}{\hbar}}. \tag{1.4.120}$$

In a similar way, one can check the 'q-deformed' chain formula for any allowed q (determined by (1.4.72) and (1.4.97)) to obtain an infinite q-series of the chain representation of the same n-ary multiplication.

1.4.3 Generalized 'deformed' version of the homomorphism theorem

Let us consider a homomorphism of the binary groups entering into the 'deformed' chain formula (1.4.103) as $\Phi^*: \mathcal{G}_*^{(2)} \to \mathcal{G}_*^{\prime(2)}$, where $\mathcal{G}_*^{\prime(2)} = \langle G' | *', e' \rangle$. We observe that, because Φ^* commutes with the binary multiplication, we need its commutation also with the automorphisms φ_q in each term of (1.4.103) (which fixes equality of the 'deformation' parameters $q = q'$) and its homomorphic action on b_q. Indeed, if

$$\Phi^*\left(\varphi_q(g)\right) = \varphi_q'(\Phi^*(g)), \tag{1.4.121}$$

$$\Phi^*\left(b_q\right) = b_q', \tag{1.4.122}$$

then we get from (1.4.103)

$$\begin{aligned} \Phi^*(\mu^{(n)}[g_1, \ldots, g_n]) &= \Phi^*(g_1) *' \Phi^*\left(\varphi_q^{\circ[[1]]_q}(g_2)\right) *' \ldots *' \Phi^*\left(\varphi_q^{\circ[[n-1]]_q}(g_n)\right) *' \Phi^*\left(b_q\right) \\ &= \Phi^*(g_1) *' \varphi_q^{\prime\circ[[1]]_q}(\Phi^*(g_2)) *' \ldots *' \varphi_q^{\prime\circ[[n-1]]_q}(\Phi^*(g_n)) *' b_q' \\ &= \mu^{\prime(n)}[\Phi^*(g_1), \ldots, \Phi^*(g_n)], \end{aligned} \tag{1.4.123}$$

where

$$\begin{aligned} g' *' h' = \mu^{\prime(n)}[g', \mathbf{e}^{\prime -1}, h'],\ \varphi_q'(g') = \left(\mu^{\prime(n')}\right)^{\circ \ell_\varphi(q)}[e', g', (\mathbf{e}^{\prime -1})^q],\ b_q' \\ = \left(\mu^{\prime(n)}\right)^{\circ r}[e^{\prime [[n]]_q}]. \end{aligned} \tag{1.4.124}$$

Thus we have

Theorem 1.4.16. *A homomorphism Φ^* of the binary group $\mathcal{G}_*^{(2)}$ gives rise to a homomorphism Φ of the corresponding* n*-ary group $\mathcal{G}^{(n)}$, if Φ^* satisfies the 'deformed' compatibility conditions* (1.4.121)–(1.4.122).

The 'nondeformed' version ($q = 1$) of this theorem and the case of Φ^* being an isomorphism was considered in Dudek and Michalski (1984).

1.5 Polyadic analog of Grothendieck group

The Grothendieck construction of the completion group for a monoid is a ground and staring point of the algebraic K-theory (see, e.g. Karoubi 1978, Rosenberg 1994). Here we generalize this construction to the polyadic case, when an initial monoid (or semigroup) and a final class group are m-ary and n-ary, correspondingly, and we denote such a class polyadic group $\mathsf{K}_0^{(m,n)}$. As concrete examples, we provide several computations for $\mathsf{K}_0^{(m,n)}$, including the case when the arities do not coincide $m \neq n$.

1.5.1 Grothendieck group of commutative monoid

First, we describe the standard Grothendieck construction (see, e.g. Karoubi 1978, Rosenberg 1994, Weibel 1986) of a commutative group from a commutative semigroup with identity (monoid). We will use multiplicative notation, which will allow us to provide a straightforward "polyadization" according to the arity invariance principle (Duplij 2021).

Let us have a (binary, arity $m = 2$) commutative monoid $\mathcal{S} \equiv \mathcal{S}^{(2)} = \langle S | \mu = \mu^{(2)} \cong (\cdot) | assoc \rangle$, where S is the underlying set, and $\mu = \mu^{(2)}: S \times S \to S$ is the (associative) multiplication in $\mathcal{S}$. The Cartesian product of two underlying sets $S' = S \times S$ can be endowed with the componentwise multiplication $(a_1, b_1) \bullet' (a_2, b_2) = (a_1 \cdot a_2, b_1 \cdot b_2)$, $a_i, b_i \in S$, to define the binary direct product $\mathcal{S}' = \mathcal{S}'^{(2)} = \langle S' | \boldsymbol{\mu}' = \boldsymbol{\mu}'^{(2)} \equiv (\bullet') \rangle$ (which coincides with the direct sum, because of the finite number of sets in the product).

For convenience and conciseness, we introduce the doubles

$$\mathbf{S} = \begin{pmatrix} a \\ b \end{pmatrix} \in S \times S \tag{1.5.1}$$

and use vector-like notation for the multiplication (being the Kronecker product of the doubles)

$$\mathbf{S}_1 \bullet' \mathbf{S}_2 = \begin{pmatrix} a_1 \\ b_1 \end{pmatrix} \bullet' \begin{pmatrix} a_2 \\ b_2 \end{pmatrix} = \begin{pmatrix} a_1 \cdot a_2 \\ b_1 \cdot b_2 \end{pmatrix}, \quad \mathbf{S}_i \in S', \quad a_i, b_i \in S, \tag{1.5.2}$$

or in "polyadic" notation

$$\boldsymbol{\mu}'[\mathbf{S}_1, \mathbf{S}_2] = \begin{pmatrix} \mu[a_1, a_2] \\ \mu[b_1, b_2] \end{pmatrix}. \tag{1.5.3}$$

The associativity of the direct product $\boldsymbol{\mu}'$ follows immediately from that of μ, because of the componentwise multiplication in (1.5.3). Since $\mathcal{S}$ is a monoid with the neutral element (identity) $e \in S$, satisfying $\mu[e, a] = \mu[a, e] = a$, $a \in S$, then the identity of the direct product $\mathcal{S}'$ is the double

$$\mathbf{E} = \begin{pmatrix} e \\ e \end{pmatrix}, \tag{1.5.4}$$

such that

$$\mu'[\mathbf{E}, \mathbf{G}] = \mu'[\mathbf{G}, \mathbf{E}] = \mathbf{G} \in S \times S. \tag{1.5.5}$$

Therefore, $\mathcal{S}' = \mathcal{S} \times \mathcal{S}$ is a commutative monoid, as is $\mathcal{S}$.

Another associative direct product $\mathcal{S}'' = \mathcal{S}''^{(2)} = \left\langle S' \middle| \boldsymbol{\mu}'' = \boldsymbol{\mu}_2'' \equiv (\bullet'') \right\rangle$ can be obtained using the "twisted" multiplication of the doubles $\boldsymbol{\mu}''$ defined by

$$\mathbf{S}_1 \bullet'' \mathbf{S}_2 = \begin{pmatrix} a_1 \\ b_1 \end{pmatrix} \bullet'' \begin{pmatrix} a_2 \\ b_2 \end{pmatrix} = \begin{pmatrix} a_1 \cdot b_2 \\ a_2 \cdot b_1 \end{pmatrix}, \tag{1.5.6}$$

or

$$\mu''[\mathbf{S}_1, \mathbf{S}_2] = \begin{pmatrix} \mu_2[a_1, b_2] \\ \mu_2[a_2, b_1] \end{pmatrix}. \tag{1.5.7}$$

The neutral element (identity) **E** in the "twisted" direct product $\mathcal{S}''$ coincides with (1.5.4), and therefore $\mathcal{S}'' = \mathcal{S} \times \mathcal{S}$ is a commutative monoid as well.

The question arises: how to construct a (binary) group corresponding to the binary monoid $\mathcal{S} = \mathcal{S}_2$ which would reflect its substantial and important properties? The answer was provided by Grothendieck: to consider the equivalence relations and corresponding classes in the direct product $\mathcal{S} \times \mathcal{S}$. Because, indeed on classes one can define the inverse elements which are needed to build a group (in addition to associativity and existence of neutral elements which are sufficient for monoids). Here we briefly reproduce the construction of the Grothendieck group corresponding to the commutative monoid $\mathcal{S}$ (sometimes this is called the symmetrization of $\mathcal{S}$, Karoubi 1978, or the group completion of $\mathcal{S}$, Rosenberg 1994, Weibel 1986) in multiplicative notation, which will allow us to provide its "polyadization" in a straightforward way.

Let us consider two kinds of equivalence relations on the direct product $\mathcal{S}' = \mathcal{S} \times \mathcal{S}$. The first one ($\sim_1$) is reminiscent of "gauge invariance" (in physical language), because it identifies the doubles with equal "shifts", such that

$$\begin{pmatrix} a_1 \\ b_1 \end{pmatrix} \sim_1 \begin{pmatrix} a_2 \\ b_2 \end{pmatrix} \Longleftrightarrow \exists_{x,\, y \in S} \begin{pmatrix} \mu[a_1, x] \\ \mu[b_1, x] \end{pmatrix} = \begin{pmatrix} \mu[a_2, y] \\ \mu[b_2, y] \end{pmatrix}, \quad a_i, b_i \in S, \tag{1.5.8}$$

and we call it the *"gauge" shifts*. The second equivalence relation ($\sim_2$) uses only one "shift" as follows

$$\begin{pmatrix} a_1 \\ b_1 \end{pmatrix} \sim_2 \begin{pmatrix} a_2 \\ b_2 \end{pmatrix} \Longleftrightarrow \exists_{z \in S}\ (\mu)^{\circ 2}[a_1, b_2, z] = (\mu)^{\circ 2}[a_2, b_1, z], \quad a_i, b_i \in S, \tag{1.5.9}$$

so we call this the *"twisted" shift* (which was used originally by Grothendieck).

Assertion 1.5.1. Two equivalence relations above coincide $\sim_1 = \sim_2 \equiv \sim$.

Proof. Let (1.5.9) holds, then putting $x = \mu[z, b_2]$, $y = \mu[a_1, z]$, we obtain (1.5.8). Conversely, from (1.5.8) with

$$z = \mu[x, y] \tag{1.5.10}$$

it follows that

$$\begin{aligned}(\mu)^{\circ 2}[a_1, b_2, z] = \mu[\mu[a_1, x], \mu[y, b_2]] = \mu[\mu[y, b_1], \mu[a_2, x]] \\ = (\mu)^{\circ 2}[a_2, b_1, z].\end{aligned} \tag{1.5.11}$$

The group completion of the commutative monoid $\mathcal{S}$ is defined as the (binary $n = 2$) group $\widetilde{\mathcal{G}}^{(2)}$ (of isomorphism classes) being the factorization of $\mathcal{S} \times \mathcal{S}$ by the equivalence relation ($\sim$) given in (1.5.8) or (1.5.9). It is called the Grothendieck group, and is usually denoted by $\mathsf{K}_0(\mathcal{S}) = \widetilde{\mathcal{G}}^{(2)}$ (for this concrete case of monoid, Rosenberg 1994, Weibel 1986)

$$\mathsf{K}_0(\mathcal{S}) = \mathcal{S} \times \mathcal{S} \sim . \tag{1.5.12}$$

The representatives, "observables" (in physical language), are "gauge invariant" doubles (1.5.8)

$$\widetilde{\mathfrak{G}} = \begin{bmatrix} a \\ b \end{bmatrix}. \tag{1.5.13}$$

The multiplication of the representatives $\widetilde{\mathfrak{G}}$ in $\mathsf{K}_0(\mathcal{S})$ inherits the product of doubles in $\mathcal{S} \times \mathcal{S}$ (1.5.3)

$$\widetilde{\boldsymbol{\mu}}[\widetilde{\mathfrak{G}}_1, \widetilde{\mathfrak{G}}_2] = \begin{bmatrix} \mu[a_1, a_2] \\ \mu[b_1, b_2] \end{bmatrix}, \tag{1.5.14}$$

where the elements of the resulting binary factor group $\mathsf{K}_0(\mathcal{S}) = \widetilde{\mathcal{G}}^{(2)} = \widetilde{\mathcal{G}} = \langle \{\widetilde{\mathfrak{G}}\} \mid \widetilde{\boldsymbol{\mu}} \rangle$ (the representative doubles $\widetilde{\mathfrak{G}}$ (1.5.13) and the binary operation $\widetilde{\boldsymbol{\mu}} = \widetilde{\boldsymbol{\mu}}^{(2)}$ on the classes are marked by waves). The structure of the Grothendieck group for monoids $\mathsf{K}_0(\mathcal{S})$ was considered as the staring example in, e.g., Karoubi (1978), Rosenberg (1994), Weibel (1986).

The first equivalence relation ($\sim_1$) gives the form of the neutral element in $\mathsf{K}_0(\mathcal{S})$

$$\widetilde{\mathfrak{E}} = \begin{bmatrix} e \\ e \end{bmatrix} \sim \begin{bmatrix} a \\ a \end{bmatrix}, \quad \forall a \in S, \tag{1.5.15}$$

where $e \in S$ is the identity of the monoid $\mathcal{S}$. Indeed the shape (1.5.15) of the neutral element and commutativity of the initial monoid $\mathcal{S}$ allows us to obtain the inverse element in $\mathsf{K}_0(\mathcal{S})$ using the multiplication (1.5.14) in the following way

$$\widetilde{\boldsymbol{\mu}}\left[\begin{bmatrix} a \\ b \end{bmatrix}, \begin{bmatrix} b \\ a \end{bmatrix}\right] = \begin{bmatrix} \mu[a, b] \\ \mu[b, a] \end{bmatrix} = \begin{bmatrix} \mu[a, b] \\ \mu[a, b] \end{bmatrix} \sim \begin{bmatrix} e \\ e \end{bmatrix}, \tag{1.5.16}$$

or

$$\tilde{\mu}\left[\tilde{\mathfrak{G}}, \tilde{\mathfrak{G}}^{-1}\right] = \tilde{\mathfrak{E}}, \tag{1.5.17}$$

where the unique inverse is

$$\tilde{\mathfrak{G}}^{-1} = \begin{bmatrix} a \\ b \end{bmatrix}^{-1} = \begin{bmatrix} b \\ a \end{bmatrix}. \tag{1.5.18}$$

Thus, $\tilde{\mathcal{G}} = \mathsf{K}_0(\mathcal{S})$ is indeed a (binary) commutative group (of classes) corresponding to the (binary) monoid $\mathcal{S} = \mathcal{S}_2$. Using (1.5.8) and (1.5.15), the homomorphism (of monoids) $\mathbf{\Phi}_{SG}: \mathcal{S} \to \mathsf{K}_0(\mathcal{S})$ can be written as

$$\mathbf{\Phi}_{SG}(a) = \begin{bmatrix} \mu[a, a] \\ a \end{bmatrix}, \quad \forall a \in S, \tag{1.5.19}$$

where we did not use the identity $e \in S$ (this can be important in the "polyadization" below). It follows from (1.5.15), that $\mathbf{\Phi}_{SG}$ can be written with the identity in the form

$$\mathbf{\Phi}_{SG}(a) = \begin{bmatrix} a \\ e \end{bmatrix}, \quad \forall a \in S, e \in S. \tag{1.5.20}$$

Using (1.5.18) we observe that the image of $\mathbf{\Phi}_{SG}$ actually generates a group (being $\mathsf{K}_0(\mathcal{S})$), because

$$\begin{bmatrix} a \\ b \end{bmatrix} = \tilde{\mu}[\mathbf{\Phi}_{SG}(a), (\mathbf{\Phi}_{SG}(b))^{-1}], \quad \forall a, b \in S. \tag{1.5.21}$$

The universal property can be shown in the following way (see, e.g., Rosenberg 1994). Let us consider any commutative (binary) group $\tilde{\mathcal{G}}' = \langle G' | \tilde{\mu}' \rangle$ and the group homomorphism $\mathbf{\Phi}_{GG'}: \tilde{\mathcal{G}} \to \tilde{\mathcal{G}}'$, then there exists the unique homomorphism $\mathbf{\Phi}_{SG'}: \mathcal{S} \to \tilde{\mathcal{G}}'$, such that

$$\mathbf{\Phi}_{SG'} = \mathbf{\Phi}_{GG'} \circ \mathbf{\Phi}_{SG}. \tag{1.5.22}$$

Indeed, we derive, using (1.5.20) and (1.5.21)

$$\begin{aligned} \mathbf{\Phi}_{GG'}\left(\begin{bmatrix} a \\ b \end{bmatrix}\right) &= \mathbf{\Phi}_{GG'}\left(\tilde{\mu}\left[\begin{bmatrix} a \\ e \end{bmatrix}, \left(\begin{bmatrix} b \\ e \end{bmatrix}\right)^{-1}\right]\right) = \tilde{\mu}'\left[\mathbf{\Phi}_{GG'}\left(\begin{bmatrix} a \\ e \end{bmatrix}\right), \left(\mathbf{\Phi}_{GG'}\left(\begin{bmatrix} b \\ e \end{bmatrix}\right)\right)^{-1}\right] \\ &= \tilde{\mu}'\left[\mathbf{\Phi}_{GG'}(\mathbf{\Phi}_{SG}(a)), \left(\mathbf{\Phi}_{GG'}(\mathbf{\Phi}_{SG}(b))\right)^{-1}\right] = \tilde{\mu}'\left[\mathbf{\Phi}_{SG'}(a), \left(\mathbf{\Phi}_{SG'}(b)\right)^{-1}\right] \end{aligned} \tag{1.5.23}$$

Conversely, for a given homomorphism $\mathbf{\Phi}_{SG'}: \mathcal{S} \to \tilde{\mathcal{G}}'$ the group homomorphism $\mathbf{\Phi}_{GG'}: \tilde{\mathcal{G}} \to \tilde{\mathcal{G}}'$ is uniquely defined by (1.5.23) with $\mathbf{\Phi}_{SG'} = \mathbf{\Phi}_{GG'} \circ \mathbf{\Phi}_{SG}$ (1.5.22).

Example 1.5.2. The simplest example is the commutative monoid of nonnegative integers (natural numbers with the zero $\mathbb{N}_0$) under addition $\mathcal{S} = \langle \mathbb{N}_0 | \mu = (+) \rangle$, and $e = 0$.

The elements of the $\mathsf{K}_0(\mathcal{S}) = \langle \mathbb{N}_0 \times \mathbb{N}_0 | \widetilde{\boldsymbol{\mu}} = (\widetilde{\mp}) \rangle$ are doubles of natural numbers $\binom{n}{m} \in \mathbb{N}_0 \times \mathbb{N}_0$, and the neutral double is $\binom{0}{0}$. The equivalence relation (1.5.9) becomes

$$\binom{n_1}{m_1} \sim \binom{n_2}{m_2} \Longleftrightarrow n_1 + m_2 = n_2 + m_1, \quad n_i, m_i \in \mathbb{N}_0, \tag{1.5.24}$$

because the monoid $\mathcal{S}$ is cancellative. It follows from (1.5.24) by scaling, that in $\mathsf{K}_0(\mathcal{S})$ there exist two minimal representatives $\begin{bmatrix} n \\ 0 \end{bmatrix}$ and $\begin{bmatrix} 0 \\ m \end{bmatrix}$, and from (1.5.16) we have

$$\begin{bmatrix} n \\ 0 \end{bmatrix} \widetilde{\mp} \begin{bmatrix} 0 \\ n \end{bmatrix} = \begin{bmatrix} n \\ n \end{bmatrix} = \begin{bmatrix} 0 \\ 0 \end{bmatrix}. \tag{1.5.25}$$

Therefore, the representatives $\begin{bmatrix} n \\ 0 \end{bmatrix}$ and $\begin{bmatrix} 0 \\ m \end{bmatrix}$ can be treated as positive and negative integers in $\mathbb{Z}$, such that there exists the homomorphism $\mathsf{K}_0(\mathbb{N}_0) \to \mathbb{Z}$ defined by

$$\begin{bmatrix} n \\ m \end{bmatrix} \mapsto n - m \in \mathbb{Z}, \tag{1.5.26}$$

which is a bijection due to (1.5.24). The universal property (1.5.23) becomes

$$\begin{aligned} \boldsymbol{\Phi}_{GG'}\left(\begin{bmatrix} n \\ m \end{bmatrix}\right) = \boldsymbol{\Phi}_{GG'}\left(\begin{bmatrix} n \\ 0 \end{bmatrix} - \begin{bmatrix} m \\ 0 \end{bmatrix}\right) &= \boldsymbol{\Phi}_{GG'}\left(\begin{bmatrix} n \\ 0 \end{bmatrix}\right) - \boldsymbol{\Phi}_{GG'}\left(\begin{bmatrix} m \\ 0 \end{bmatrix}\right) \\ = \boldsymbol{\Phi}_{GG'}(\boldsymbol{\Phi}_{SG}(n)) - \boldsymbol{\Phi}_{GG'}(\boldsymbol{\Phi}_{SG}(m)) &= \boldsymbol{\Phi}_{SG'}(n) - \boldsymbol{\Phi}_{SG'}(m). \end{aligned} \tag{1.5.27}$$

Thus, the Grothendieck group of the monoid $\mathbb{N}_0$ of natural numbers with zero is the group of integers $\mathbb{Z}$ under addition $\mathsf{K}_0(\mathbb{N}_0) = \mathbb{Z}$.

1.5.2 The *n*-ary group completion of *m*-ary semigroup

Here we propose the "polyadization" (along the ideas of Duplij 2019) of the group completion concept using the arity invariance principle (Duplij 2021). By considering the polyadic algebraic structures, the main differences with the binary case will be the following Duplij (2022), (and refs there in):

1. There can exist several associative polyadic direct products (powers) that can be built from one polyadic semigroup.
2. If initial m-ary semigroup is commutative, the polyadic direct product can be noncommutative.
3. The arity n of the direct product of m-ary semigroups can be not equal to m.
4. The neutral element (identity) is not necessary for the m-ary semigroup and its power, because the polyadic analog of the binary inverse element is the querelement of an n-ary group, which is defined without the usage of a neutral element, and moreover some n-ary groups do not contain an identity at all.

Polyadic direct power construction. Consider an m-ary semigroup $\mathcal{S}^{(m)} = \langle S | \mu^{(m)} | total\ assoc \rangle$, where S is the underlying set, and $\mu^{(m)}: S^{\times m} \to S$ is the (totally associative) m-ary multiplication in $\mathcal{S}$. The Cartesian product of two underlying sets $S' = S^{\times m}$ can be endowed with an associative n-ary multiplication in various different ways, and also it can be that $n \neq m$ (Duplij 2022). Again, we introduce the doubles $\mathbf{S} = \begin{pmatrix} a \\ b \end{pmatrix} \in S \times S$, and the polyadic direct product which is the n-ary semigroup of doubles $\mathcal{S}'^{(n)} = \langle S' | \boldsymbol{\mu}'^{(n)} | \text{total assoc} \rangle$.

There are two possibilities to build the polyadic associative direct product (Duplij 2022):

1. The componentwise construction which corresponds to the standard binary direct product (1.5.3).
2. The noncomponentwise one, which corresponds to the twisted direct product (1.5.7).

In the first case, the n-ary semigroup of doubles $\mathcal{S}'^{(n)}$ corresponds to the full polyadic external product of Duplij (2022).

Definition 1.5.3. Ann-ary *componentwise direct product* (power) semigroup of doubles consists of two n-ary semigroups $\mathcal{S}'^{(n)} = \mathcal{S}^{(n)} \times \mathcal{S}^{(n)}$ (of the same arity)

$$\boldsymbol{\mu}'^{(n)}[\mathbf{S}_1, \mathbf{S}_2, \ldots, \mathbf{S}_n] = \begin{pmatrix} \mu^{(n)}[a_1, a_2, \ldots, a_n] \\ \mu^{(n)}[b_1, b_2, \ldots, b_n] \end{pmatrix}, \quad \mathbf{S}_i = \begin{pmatrix} a_i \\ b_i \end{pmatrix}, \quad a_i, b_i \in S, \tag{1.5.28}$$

where the (total) polyadic associativity (1.1.28) of $\boldsymbol{\mu}'^{(n)}$ is governed by the associativity of the constituent semigroup $\mathcal{S}^{(n)}$.

The simplest case is give by

Definition 1.5.4. A *full polyadic direct product (power)* $\mathcal{S}'^{(n)} = \mathcal{S}^{(n)} \times \mathcal{S}^{(n)}$ is called *derived*, if the constituent n-ary semigroup $\mathcal{S}^{(n)}$ is derived, such that its operation $\mu^{(n)}$ is composition of the binary operations $\mu^{(2)}$.

In the derived case all the operations in (1.5.28) have the form (see (1.1.3)–(1.1.4))

$$\mu^{(n)} = \left(\mu^{(2)}\right)^{\circ(n-1)}, \mu^{(2)} = (\cdot), \quad \boldsymbol{\mu}'^{(n)} = \left(\boldsymbol{\mu}'^{(2)}\right)^{\circ(n-1)}, \boldsymbol{\mu}^{(2)} = (\bullet'). \tag{1.5.29}$$

The operations of the derived polyadic direct product can be written as (cf. the binary case (1.5.3))

$$\boldsymbol{\mu}'^{(n)}_{\text{der}}[\mathbf{S}_1, \mathbf{S}_2, \ldots, \mathbf{S}_n] = \mathbf{S}_1 \bullet' \mathbf{S}_2 \bullet' \ldots \bullet' \mathbf{S}_n = \begin{pmatrix} a_1 \cdot a_2 \cdot \ldots \cdot a_n \\ b_1 \cdot b_2 \cdot \ldots \cdot b_n \end{pmatrix}, \tag{1.5.30}$$

and so it is simply a repitition of the binary products (1.5.3). Therefore, it would be more interesting to consider nonderived polyadic analogs of the direct product which do not come down to the binary ones.

The nonderived version of the polyadic direct product, the hetero ("entangled") product, was introduced in Duplij (2022) for an arbitrary number of constituents. Here we apply it for two m-ary semigroups to get the nonderived associative direct product, which can have a different arity $n \neq m$.

The general structure of the hetero product formally coincides ("reversely") with the main heteromorphism equation (Duplij 2012). The additional parameter which determines the arity n of the hetero power of the initial m-ary semigroup is the number of intact elements ℓ_{id} and number of constituents k. In our case of $k = 2$ multipliers, we have only two possibilities $\ell_{id} = 0, 1$. Thus, we arrive at

Definition 1.5.5. The *hetero ("entangled") power* $\boxtimes$ *(square)* of the m-ary semigroup $\mathcal{S}^{(m)} = \langle S | \mu^{(m)} \rangle$ is the n-ary semigroup defined on the Cartesian power $S' = S \times S$, such that $\mathcal{S}'^{(n)} = \langle S' | \boldsymbol{\mu}'^{(n)} \rangle$,

$$\mathcal{S}'^{(n)} = \mathcal{S}^{(m)} \boxtimes \mathcal{S}^{(m)}, \tag{1.5.31}$$

and the n-ary multiplication of doubles $\mathbf{S}_{ij} = \begin{pmatrix} a_i \\ a_j \end{pmatrix} \in S \times S$, $a_i, a_j \in S$, is given (informally) by

$$\boldsymbol{\mu}'^{(n)}\left[\begin{pmatrix} a_1 \\ a_2 \end{pmatrix}, \dots, \begin{pmatrix} a_{2n-1} \\ a_{2n} \end{pmatrix}\right] = \begin{cases} \begin{pmatrix} \mu^{(m)}[a_1, \dots, a_m], \\ \mu^{(m)}[a_{m+1}, \dots, a_{2m}] \end{pmatrix}, & \ell_{id} = 0, \quad n = m, \\ \begin{pmatrix} \mu^{(m)}[a_1, \dots, a_m], \\ a_{m+1} \end{pmatrix}, & \ell_{id} = 1, \quad n = \dfrac{m+1}{2}, \end{cases} \tag{1.5.32}$$

where $\ell_{id} = 0, 1$ is the number of *intact elements* in the r.h.s. of the polyadic direct product. The hetero power arities are connected by the *arity changing formula* (with $k = 2$) (Duplij 2012)

$$n = m - \frac{m-1}{2}\ell_{id}, \tag{1.5.33}$$

with the integer $\frac{n-1}{2}\ell_{id} \geqslant 1$.

In the case $\ell_{id} = 1$, the initial and final arities m and n are not arbitrary, but "quantized" such that the fraction in (1.5.33) has to be an integer (see the first row in table 1 of Duplij 2022)

$$\begin{aligned} m &= 3, 5, 7, \dots \\ n &= 2, 3, 4, \dots \end{aligned} \tag{1.5.34}$$

The concrete placement of elements and multiplications in (1.5.32) to obtain the associative $\boldsymbol{\mu}'^{(n')}$ is governed by the associativity quiver technique (Duplij 2012).

Thus, the classification of the hetero powers for a nonbinary initial semigroup $m \geqslant 3$ consists of two limiting cases:

1. *Intactless power*: there are no intact elements $\ell_{id} = 0$. The arity of the hetero power reaches its maximum and coincides with the arity of the initial semigroup $n = m$. Since $m \geqslant 3$, there is no binary hetero product.
2. *Binary power*: the final semigroup is of lowest arity, i.e. binary $n = 2$, and it follows from (1.5.34), that the only possibility is $m = 3$.

Let us consider some concrete examples of the hetero powers.

Example 1.5.6. Let $\mathcal{S}^{(3)} = \langle S | \mu^{(3)} \rangle$ be a commutative ternary semigroup, then we can construct its square of the doubles $\mathbf{S} = \begin{pmatrix} a \\ b \end{pmatrix} \in S \times S$ in two ways to obtain the associative hetero power

$$\boldsymbol{\mu}'^{(2)}[\mathbf{S}_1, \mathbf{S}_2] = \boldsymbol{\mu}'^{(2)}\left[\begin{pmatrix} a_1 \\ b_1 \end{pmatrix}, \begin{pmatrix} a_2 \\ b_2 \end{pmatrix}\right] \begin{cases} \begin{pmatrix} \mu^{(3)}[a_1, b_1, a_2] \\ b_2 \end{pmatrix}, \\ \begin{pmatrix} \mu^{(3)}[a_1, b_2, a_2] \\ b_1 \end{pmatrix}, \end{cases} \quad a_i, b_i \in S. \tag{1.5.35}$$

This means that the Cartesian square $S \times S$ can be endowed with an associative multiplication $\boldsymbol{\mu}'^{(2)}$ in two ways, and therefore $\mathcal{S}'^{(2)} = \langle S' | \boldsymbol{\mu}'^{(2)} \rangle$ is a noncommutative semigroup, being the hetero power $\mathcal{S}'^{(2)} = \mathcal{S}^{(3)} \boxtimes \mathcal{S}^{(3)}$. If the ternary semigroup $\mathcal{S}^{(3)}$ has a ternary identity $e \in S$, then $\mathcal{S}'^{(2)}$ has only the left (right) identity $\mathbf{E} = \begin{pmatrix} e \\ e \end{pmatrix} \in S \times S$, since $\boldsymbol{\mu}'^{(2)}[\mathbf{E}, \mathbf{S}] = \mathbf{S}$ ($\boldsymbol{\mu}'^{(2)}[\mathbf{S}, \mathbf{E}] = \mathbf{S}$), but not the right (left) identity. Thus, $\mathcal{S}'^{(2)}$ can be a semigroup only, even in the case when $\mathcal{S}^{(3)}$ is a ternary group.

Example 1.5.7. Take $\mathcal{S}^{(3)} = \langle S | \mu^{(3)} \rangle$ to be a commutative ternary semigroup, then the multiplication of the doubles is ternary and is noncomponentwise (as opposed to the componentwise product (1.5.28))

$$\boldsymbol{\mu}'^{(3)}[\mathbf{S}_1, \mathbf{S}_2, \mathbf{S}_3] = \boldsymbol{\mu}'^{(3)}\left[\begin{pmatrix} a_1 \\ b_1 \end{pmatrix}, \begin{pmatrix} a_2 \\ b_2 \end{pmatrix}, \begin{pmatrix} a_3 \\ b_3 \end{pmatrix}\right] = \begin{pmatrix} \mu^{(3)}[a_1, b_2, a_3] \\ \mu^{(3)}[b_1, a_2, b_3] \end{pmatrix}, \quad a_i, b_i \in S. \tag{1.5.36}$$

Nevertheless $\boldsymbol{\mu}'^{(3)}$ is associative (and is described by the Post-like associative quiver, Duplij 2012), therefore the hetero power $\mathcal{S}'^{(3)} = \mathcal{S}^{(3)} \boxtimes \mathcal{S}^{(3)}$ is indeed the noncommutative ternary semigroup $\mathcal{S}'^{(3)} = \langle G \times G | \boldsymbol{\mu}'^{(3)} \rangle$. In this case, as opposed to the previous example, the existence of the ternary identity e in $\mathcal{S}^{(3)}$ implies the ternary identity in the direct product $\mathcal{S}'^{(3)}$ by $\mathbf{E} = \begin{pmatrix} e \\ e \end{pmatrix}$, such that

$$\boldsymbol{\mu}'^{(3)}[\mathbf{S}, \mathbf{E}, \mathbf{E}] = \boldsymbol{\mu}'^{(3)}[\mathbf{E}, \mathbf{S}, \mathbf{E}] = \boldsymbol{\mu}'^{(3)}[\mathbf{E}, \mathbf{E}, \mathbf{S}] = \mathbf{S}. \tag{1.5.37}$$

Proposition 1.5.8. If the initial m-ary semigroup $\mathcal{S}^{(m)}$ contains an identity, then the hetero square n-ary semigroup $\mathcal{S}'^{(n)} = \mathcal{S}^{(m)} \boxtimes \mathcal{S}^{(m)}$ can contain an identity only in the intactless case and the Post-like quiver (Duplij 2012). For the binary power $n = 2$ only the one-sided identity is possible.

Next we consider more complicated hetero power ("entangled") constructions with and without intact elements, which is possible for $m \geqslant 4$ only (Duplij 2012).

Example 1.5.9. Let $\mathcal{S}^{(5)} = \langle S \mid \mu^{(5)} \rangle$ be a 5-ary semigroup, then we construct its 5-ary totally associative hetero square $\mathcal{G}'^{(5)} = \langle G' \mid \boldsymbol{\mu}'^{(5)} \rangle$ using the Post-like associative quiver without intact elements. We define the 5-ary multiplication of the doubles by

$$\boldsymbol{\mu}'^{(5)}[\mathbf{S}_1, \mathbf{S}_2, \mathbf{S}_3, \mathbf{S}_4, \mathbf{S}_5] = \begin{pmatrix} \mu^{(5)}[a_1, b_2, a_3, b_4, a_5] \\ \mu^{(5)}[b_1, a_2, b_3, a_4, b_5] \end{pmatrix}, \quad a_i, b_i \in S. \tag{1.5.38}$$

It can be shown that $\boldsymbol{\mu}'^{(5)}$ is totally associative, and therefore $\mathcal{S}'^{(5)} = \langle S' \mid \boldsymbol{\mu}'^{(5)} \rangle$ is a 5-ary commutative semigroup. If $\mathcal{S}^{(5)}$ has the 5-ary identity e satisfying

$$\mu^{(5)}[e, e, e, e, a] = a, \quad \forall a \in S, \ e \in S, \tag{1.5.39}$$

then the hetero power $\mathcal{S}'^{(5)}$ has the 5-ary identity

$$\mathbf{E} = \begin{pmatrix} e \\ e \end{pmatrix}, \quad e \in S. \tag{1.5.40}$$

A more nontrivial example is a hetero product (power) which has a different arity than that of the initial semigroup.

Example 1.5.10. Let $\mathcal{S}^{(5)} = \langle S \mid \mu^{(5)} \rangle$ be a commutative 5-ary semigroup, then we can construct its ternary associative hetero power $\mathcal{S}'^{(3)} = \langle S' \mid \boldsymbol{\mu}'^{(3)} \rangle$ using the associative quivers with one intact element (see (1.5.34) for allowed "quantized" arities). We propose for the triples $\mathbf{S}_i$ the following ternary multiplication

$$\boldsymbol{\mu}'^{(3)}[\mathbf{S}_1, \mathbf{S}_2, \mathbf{S}_3] = \boldsymbol{\mu}'^{(3)}\left[\begin{pmatrix} a_1 \\ b_1 \end{pmatrix}, \begin{pmatrix} a_2 \\ b_2 \end{pmatrix}, \begin{pmatrix} a_3 \\ b_3 \end{pmatrix}\right] = \begin{pmatrix} \mu^{(5)}[a_1, b_2, a_3, b_1, a_2] \\ b_3 \end{pmatrix}, \tag{1.5.41}$$

$$a_i, b_i \in S.$$

It can be seen that $\boldsymbol{\mu}'^{(3)}$ is totally associative, and therefore the hetero power of 5-ary semigroup $\mathcal{S}^{(5)} = \langle S \mid \mu^{(5)} \rangle$ is a noncommutative ternary semigroup $\mathcal{S}'^{(3)} = \langle S' \mid \boldsymbol{\mu}'^{(3)} \rangle$, such that $\mathcal{S}'^{(3)} = \mathcal{S}^{(5)} \boxtimes \mathcal{S}^{(5)}$. If the initial 5-ary semigroup $\mathcal{S}^{(5)}$ has the identity satisfying

$$\mu^{(5)}[e, e, e, e, a] = a, \quad \forall a \in S, \ e \in S, \tag{1.5.42}$$

then the ternary hetero power $\mathcal{S}'^{(3)}$ has only the left ternary identity (1.5.40) satisfying one relation

$$\boldsymbol{\mu}'^{(3)}[\mathbf{E}, \mathbf{E}, \mathbf{S}] = \mathbf{S}, \quad \forall \mathbf{S} \in S \times S, \tag{1.5.43}$$

and therefore $\mathcal{S}'^{(3)}$ is a ternary semigroup with the left identity.

Equivalence relations for m-ary semigroups. Consider of the extension of the equivalence relations (1.5.8) and (1.5.9) for m-ary semigroups. On the polyadic direct product (square) $\mathcal{S}^{(m)} \times \mathcal{S}^{(m)}$ we define two kinds of corresponding binary relations.

Definition 1.5.11. The first relation ($\sim_{\text{gauge}}$) is described by the *polyadic "gauge" shifts* of the doubles $\mathbf{S} = \begin{pmatrix} a \\ b \end{pmatrix} \in S \times S$ in the form

$$\begin{pmatrix} a_1 \\ b_1 \end{pmatrix} \sim_{\text{gauge}} \begin{pmatrix} a_2 \\ b_2 \end{pmatrix} \Longleftrightarrow \exists_{x,\, y \in S} \begin{pmatrix} \mu^{(m)}[a_1^{m-1}, x] \\ \mu^{(m)}[b_1^{m-1}, x] \end{pmatrix} = \begin{pmatrix} \mu^{(m)}[a_2^{m-1}, y] \\ \mu^{(m)}[b_2^{m-1}, y] \end{pmatrix}, \quad \forall\, a_i, b_i \in S. \tag{1.5.44}$$

The second relation ($\sim_2$) is given by the *polyadic "twisted"* shift

$$\begin{pmatrix} a_1 \\ b_1 \end{pmatrix} \sim_{\text{twist}} \begin{pmatrix} a_2 \\ b_2 \end{pmatrix} \Longleftrightarrow \exists_{z \in S}\ \left(\mu^{(m)}\right)^{\circ 2}[a_1^{m-1}, b_2^{m-1}, z] = \left(\mu^{(m)}\right)^{\circ 2}[a_2^{m-1}, b_1^{m-1}, z], \quad \forall\, a_i, b_i \in S. \tag{1.5.45}$$

Proposition 1.5.12. The relations described by the polyadic "gauge" shifts and "twisted" shift coincide $\sim_{\text{gauge}} = \sim_{\text{twist}} \equiv \sim_m$.

Proof. ($\sim_{\text{twist}} \Rightarrow \sim_{\text{gauge}}$). Let (1.5.45) hold, then putting $x = \mu^{(m)}[b_2^{m-1}, z]$, $y = \mu^{(m)}[a_1^{m-1}, z]$, we get (1.5.44).

$\sim_{\text{gauge}} \Rightarrow \sim_{\text{twist}}$). Conversely, if (1.5.44) takes place, then we can always find $t_1, \ldots, t_{m-2} \in S$, such that $z = \mu^{(m)}[x, y, t_1, \ldots t_{m-2}]$ and

$$\begin{aligned} \left(\mu^{(m)}\right)^{\circ 2}[a_1^{m-1}, b_2^{m-1}, z] &= \mu^{(m)}[\mu^{(m)}[a_1^{m-1}, x], \mu^{(m)}[b_2^{m-1}, y], t_1, \ldots t_{m-2}] \\ &= \mu^{(m)}[\mu^{(m)}[a_2^{m-1}, x], \mu^{(m)}[b_1^{m-1}, y], t_1, \ldots t_{m-2}] = \left(\mu^{(m)}\right)^{\circ 2}[a_2^{m-1}, b_1^{m-1}, z]. \end{aligned} \tag{1.5.46}$$

Proposition 1.5.13. The relation $\sim_m$ is the equivalence relation on the set of doubles $\{\mathbf{S}\}$, $\mathbf{S} \in S \times S$.

Proof. **(1)** Reflexivity ($1 = 2$) of (1.5.44) is obvious, and for $\sim_{\text{twist}}$ it follows from (1.5.45), becoming the identity $\left(\mu^{(m)}\right)^{\circ 2}[a^{m-1}, b^{m-1}, z] = \left(\mu^{(m)}\right)^{\circ 2}[a^{m-1}, b^{m-1}, z]$.

(2) Symmetry ($1 \longleftrightarrow 2$) is evident for $\sim_{\text{gauge}}$, while (1.5.45) is symmetric with respect to ($1 \longleftrightarrow 2$).

(3) Transitivity. Let $\mathbf{S}_1 \sim_m \mathbf{S}_2$ and $\mathbf{S}_2 \sim_m \mathbf{S}_3$, and we will prove that $\mathbf{S}_1 \sim_m \mathbf{S}_3$, $\mathbf{S}_i \in S \times S$.

We start from the "gauge"-like relations $\mathbf{S}_1 \sim_{\text{gauge}} \mathbf{S}_2$ and $\mathbf{S}_2 \sim_{\text{gauge}} \mathbf{S}_3$ for the first components of the doubles

$$\mu^{(m)}[a_1^{m-1}, x_1] = \mu^{(m)}\left[a_2^{m-1}, y_1\right], \tag{1.5.47}$$

$$\mu^{(m)}[a_2^{m-1}, x_2] = \mu^{(m)}\left[a_3^{m-1}, y_2\right], \quad \exists\, x_i, y_i \in S. \tag{1.5.48}$$

Then we multiply separately the left hand sides and right hand sides of (1.5.47), (1.5.48) together with $(m-2)$ identities $t_1 = t_1$, $t_2 = t_2, \ldots, t_{m-2} = t_{m-2}$, $t_1, \ldots, t_{m-2} \in S$, and derive

$$(\mu^{(m)})^{\circ 3}[a_1^{m-1}, x_1, a_2^{m-1}, x_2, t_1, \ldots, t_{m-2}] = (\mu^{(m)})^{\circ 3}\left[a_2^{m-1}, y_1, a_3^{m-1}, y_2, t_1, \ldots, t_{m-2}\right]. \tag{1.5.49}$$

Denoting $x_3 = (\mu^{(m)})^{\circ 2}[a_2^{m-1}, x_1, x_2, t_1, \ldots, t_{m-2}]$ and $y_3 = (\mu^{(m)})^{\circ 2}[a_2^{m-1}, y_1, y_2, t_1, \ldots, t_{m-2}]$, we obtain in the form

$$\mu^{(m)}[a_1^{m-1}, x_3] = \mu^{(m)}\left[a_3^{m-1}, y_3\right]. \tag{1.5.50}$$

The second components of the doubles can be treated similarly

$$\mu^{(m)}[b_1^{m-1}, x_3] = \mu^{(m)}\left[b_3^{m-1}, y_3\right], \tag{1.5.51}$$

and so we have $\mathbf{S}_1 \sim_{\text{gauge}} \mathbf{S}_3$.

The "twisted"-like relations $\mathbf{S}_1 \sim_{\text{twist}} \mathbf{S}_2$ and $\mathbf{S}_2 \sim_{\text{twist}} \mathbf{S}_3$ have the form

$$(\mu^{(m)})^{\circ 2}[a_1^{m-1}, b_2^{m-1}, z_1] = (\mu^{(m)})^{\circ 2}[a_2^{m-1}, b_1^{m-1}, z_1], \tag{1.5.52}$$

$$(\mu^{(m)})^{\circ 2}[a_2^{m-1}, b_3^{m-1}, z_2] = (\mu^{(m)})^{\circ 2}[a_3^{m-1}, b_2^{m-1}, z_2]. \tag{1.5.53}$$

We multiply separately the left hand sides and right hand sides of (1.5.52), (1.5.53) together with $(m-2)$ identities $t_1 = t_1$, $t_2 = t_2, \ldots, t_{m-2} = t_{m-2}$, $t_1, \ldots, t_{m-2} \in S$, to get

$$\begin{aligned}(\mu^{(m)})^{\circ 5}&[a_1^{m-1}, b_2^{m-1}, z_1, a_2^{m-1}, b_3^{m-1}, z_2, t_1, \ldots, t_{m-2}] \\ &= (\mu^{(m)})^{\circ 5}[a_2^{m-1}, b_1^{m-1}, z_1, a_3^{m-1}, b_2^{m-1}, z_2, t_1, \ldots, t_{m-2}],\end{aligned} \tag{1.5.54}$$

which can be written as

$$(\mu^{(m)})^{\circ 2}[a_1^{m-1}, b_3^{m-1}, z_3] = (\mu^{(m)})^{\circ 2}[a_3^{m-1}, b_1^{m-1}, z_3], \tag{1.5.55}$$

where $z_3 = (\mu^{(m)})^{\circ 3}[a_2^{m-1}, b_2^{m-1}, z_1, z_2, t_1, \ldots, t_{m-2}]$, $z_3 \in S$. Therefore, it follows from (1.5.55), that $\mathbf{S}_1 \sim_{\text{twist}} \mathbf{S}_3$.

We next introduce the corresponding equivalence classes.

Definition 1.5.14. The *equivalence class* of doubles corresponding to the equivalence relation $\sim_m$ on the m-ary semigroup $\mathcal{S}^{(m)}$ is the factor

$$\widetilde{\mathfrak{S}} = \begin{bmatrix} a \\ b \end{bmatrix}_m = \{\mathbf{S}\}/\sim_m, \quad \mathbf{S} = \begin{pmatrix} a \\ b \end{pmatrix} \in S \times S, \tag{1.5.56}$$

where $\sim_m$ is defined in (1.5.44) or (1.5.45).

Next we consider the structure of the equivalence classes (1.5.56) in some examples.

Example 1.5.15. [Negative numbers]Let $S = \mathbb{N}_- \subset \mathbb{Z}$ be the set of negative (integer) numbers (without zero), which do not form a binary semigroup, obviously. However, we can introduce the ternary multiplication $\mu^{(3)}[a, b, c] = abc$ (ordinary product in $\mathbb{Z}$), such that $\mathcal{S}^{(3)}_{\text{neg}} = \langle \mathbb{N}_- | \mu^{(3)} \rangle$ becomes the ternary semigroup (of negative numbers). The doubles have the form

$$\mathbf{S}_{\text{neg}} = \begin{pmatrix} -p \\ -q \end{pmatrix} \in \mathbb{N}_- \times \mathbb{N}_-, \quad p, q \in \mathbb{N}. \tag{1.5.57}$$

The equivalence relations (1.5.44) and (1.5.45) become ($m = 3$)

$$\begin{pmatrix} -p_1 \\ -q_1 \end{pmatrix} \sim_{\text{gauge}} \begin{pmatrix} -p_2 \\ -q_2 \end{pmatrix} \Longleftrightarrow \exists_{x,\, y \in \mathbb{N}} \begin{pmatrix} -p_1^2 x \\ -q_1^2 x \end{pmatrix} = \begin{pmatrix} -p_2^2 y \\ -q_2^2 y \end{pmatrix}, \quad p_i, q_i \in \mathbb{N}, \tag{1.5.58}$$

and

$$\begin{pmatrix} -p_1 \\ -q_1 \end{pmatrix} \sim_{\text{twist}} \begin{pmatrix} -p_2 \\ -q_2 \end{pmatrix} \Longleftrightarrow \exists_{x,\, y \in \mathbb{N}}\ p_1^2 q_2^2 z = p_2^2 q_1^2 z. \tag{1.5.59}$$

For instance, we compute

$$\begin{pmatrix} -1 \\ -2 \end{pmatrix} \sim \begin{pmatrix} -2 \\ -4 \end{pmatrix} \sim \begin{pmatrix} -3 \\ -6 \end{pmatrix} \ldots, \tag{1.5.60}$$

$$\begin{pmatrix} -4 \\ -3 \end{pmatrix} \sim \begin{pmatrix} -8 \\ -6 \end{pmatrix} \sim \begin{pmatrix} -12 \\ -9 \end{pmatrix} \ldots. \tag{1.5.61}$$

In general, the two elements

$$\begin{pmatrix} -p \\ -q \end{pmatrix}, \begin{pmatrix} -kp \\ -kq \end{pmatrix} \tag{1.5.62}$$

are in the same class, $k \in \mathbb{N}$. Each equivalence class $\widetilde{\mathfrak{S}}_{\text{neg}}$ contains the maximal representative $\begin{bmatrix} -p_{\min} \\ -q_{\min} \end{bmatrix}_3$ corresponding to mininal $p = p_{\min}$ and $q = q_{\min}$, together with other elements (1.5.62) of the class. Thus, for the ternary semigroup of negative integers the structure of the binary equivalence classes in $\mathcal{S}^{(3)}_{\text{neg}} \times \mathcal{S}^{(3)}_{\text{neg}} \sim_3$ is given by the following maximal representatives

$$\widetilde{\mathfrak{S}}_{\mathrm{neg},\, p>q} = \left\{ \begin{bmatrix} -p_{\min} \\ -q_{\min} \end{bmatrix}_3 \right\}, \quad p_{\min} > q_{\min}, \tag{1.5.63}$$

$$\widetilde{\mathfrak{S}}_{\mathrm{neg},\, p<q} = \left\{ \begin{bmatrix} -p_{\min} \\ -q_{\min} \end{bmatrix}_3 \right\}, \quad p_{\min} < q_{\min}, \quad p_{\min},\, q_{\min} \in \mathbb{N}, \tag{1.5.64}$$

where $p_{\min} \neq q_{\min}$, and they are mutually prime $\gcd(p_{\min}, q_{\min}) = 1$.

Polyadic group completion. In general, the procedure of group completion consists of two steps:

1. Endow the set of equivalence classes $\{\widetilde{\mathfrak{S}}\}$ with the multiplication $\widetilde{\boldsymbol{\mu}}_S^{(\tilde{n})}$ (in the binary case (1.5.14)).
2. Find (a polyadic analog of) an inverse and construct a group (as in (1.5.18)).

It is commonly accepted that the multiplication of classes inherits the product of representatives. So the informal formula of class multiplication coincides with the hetero ("entangled") power (1.5.32), indeed

$$\widetilde{\boldsymbol{\mu}}_S^{(\tilde{n})}\left[\begin{bmatrix} a_1 \\ a_2 \end{bmatrix}_{\tilde{n}}, \ldots, \begin{bmatrix} a_{2n-1} \\ a_{2n} \end{bmatrix}_{\tilde{n}} \right] = \begin{cases} \begin{bmatrix} \mu^{(m)}[a_1, \ldots, a_m], \\ \mu^{(m)}[a_{m+1}, \ldots, a_{2m}] \end{bmatrix}_m, & \ell_{\mathrm{id}} = 0, \ \tilde{n} = m, \\ \begin{bmatrix} \mu^{(m)}[a_1, \ldots, a_m], \\ a_{m+1} \end{bmatrix}_m, & \ell_{\mathrm{id}} = 1, \ \tilde{n} = \dfrac{m+1}{2}, \end{cases} \tag{1.5.65}$$

where $\ell_{\mathrm{id}} = 0, 1$ is the number of *intact elements* in the r.h.s. of (1.5.65).

Definition 1.5.16. The *n-ary semigroup of the binary equivalence classes* is defined as

$$\widetilde{\mathcal{S}}^{(\tilde{n})} = \left\langle \{\widetilde{\mathfrak{S}}\} \,\middle|\, \widetilde{\boldsymbol{\mu}}_S^{(\tilde{n})} \right\rangle. \tag{1.5.66}$$

Remark 1.5.17. In general, for a fixed initial semigroup $\mathcal{S}^{(m)}$, the formulas of the doubles multiplication (1.5.32) and classes multiplication (1.5.65) can be different (as two choices of (1.5.35)), because in one case we multiply concrete elements, while in another case–their representatives. Moreover, the final arities n and $\tilde{n}$ need not coincide, if we use different choices in r.h.s. of (1.5.32) and (1.5.65). For instance, compare the 5-ary (1.5.38) and ternary (1.5.41) multiplications of doubles obtained from the same 5-ary semigroup. This could lead to diverse possible polyadic analogs of the group completions of the same semigroup. Nevertheless, in what follows we will assume that

$$\tilde{n} = n \tag{1.5.67}$$

and choose the same product for elements and their representatives. Therefore, all examples of polyadic direct products in subsection 1.5.2 can be applied for the equivalence class multiplication (1.5.65) as well.

The polyadic analog of the inverse is the querelement (1.2.13) (Dörnte 1929), which does not need an identity at all. Therefore, the binary formulas (1.5.16)–(1.5.18) do not work for $m \geqslant 3$. So we have

Assertion 1.5.18. The presence of an identity is not necessary for constructing polyadic groups, it can be optional and plays no role, and therefore we consider polyadic group completion not of a polyadic monoid, but of a polyadic semigroup $\mathcal{S}^{(m)}$.

Let us suppose that we can construct the querelement for any double $\widetilde{\mathfrak{S}}$ in the n-ary semigroup $\widetilde{\mathcal{S}}^{(n)}$ of equivalence classes, which means that we define the unary queroperation $\widetilde{\boldsymbol{\mu}}^{(1)}$ (see (1.2.20))

$$\overline{\widetilde{\mathfrak{S}}} = \widetilde{\boldsymbol{\mu}}^{(1)}[\widetilde{\mathfrak{S}}] = \widetilde{\boldsymbol{\mu}}^{(1)}\left[\begin{bmatrix} a \\ b \end{bmatrix}_n\right] = \begin{bmatrix} \bar{a} \\ \bar{b} \end{bmatrix}_n, \tag{1.5.68}$$

which satisfies (1.2.13)

$$\widetilde{\boldsymbol{\mu}}^{(n)}\left[\widetilde{\mathfrak{S}}^{n-1}, \overline{\widetilde{\mathfrak{S}}}\right] = \widetilde{\mathfrak{S}}. \tag{1.5.69}$$

Then the n-ary semigroup $\widetilde{\mathcal{S}}^{(n)}$ becomes a polyadic group.

Definition 1.5.19. The *n-ary group completion* of the m-ary semigroup $\mathcal{S}^{(m)}$ is

$$\mathsf{K}_0^{(m,n)}(\mathcal{S}^{(m)}) = \langle \{\widetilde{\mathfrak{S}}\} \mid \widetilde{\boldsymbol{\mu}}^{(n)}, \widetilde{\boldsymbol{\mu}}^{(1)} \rangle, \tag{1.5.70}$$

where $\widetilde{\boldsymbol{\mu}}^{(n)}$ and $\widetilde{\boldsymbol{\mu}}^{(1)}$ are the n-ary multiplication of the doubles (1.5.65) and the queroperation (1.5.68).

In this notation the standard binary case (1.5.12) is

$$\mathsf{K}_0(\mathcal{S}) = \mathsf{K}_0^{(2,2)}(\mathcal{S}^{(2)}). \tag{1.5.71}$$

Remark 1.5.20. The computation of the group completion $\mathsf{K}_0^{(m,n)}$ with abitrary m and n is possible, if we know both multiplications: of the initial m-ary semigroup $\mu^{(m)}$ and of the n-ary group of equivalent classes $\widetilde{\boldsymbol{\mu}}^{(n)}$ together with its queroperation $\widetilde{\boldsymbol{\mu}}^{(1)}$.

Remark 1.5.21. As opposed to the binary case (1.5.71), for a given m-ary semigroup there can exist several associative products of doubles $\boldsymbol{\mu}'^{(n)}$ and corresponding products of classes $\widetilde{\boldsymbol{\mu}}^{(n)}$ (see, e.g. (1.5.28) and (1.5.36) or (1.5.38) and (1.5.41)), and therefore the polyadic group completion $\mathsf{K}_0^{(m,n)}$ is not unique, in general. The number of different $\mathsf{K}_0^{(m,n)}$ coincides with the number of distinct associative quivers in the doubles multiplication (1.5.32) and class products (1.5.65) having the same arity shape (Duplij 2012).

Remark 1.5.22. If we consider polyadic groups with the same arity shape of the multiplication of classes $\widetilde{\boldsymbol{\mu}}^{(n)}$ and the same queroperation $\widetilde{\boldsymbol{\mu}}^{(1)}$, then we can have a polyadic analog of universality (1.5.22) for this fixed completion group arity n.

Example 1.5.23. Let us compute $\mathsf{K}_0^{(m,n)}$ for the n-ary componentwise direct power (1.5.28) of the m-ary semigroup $\mathcal{S}^{(m)} = \langle S \mid \mu^{(m)} \rangle$. Taking to account *Remark* 1.5.17 and (1.5.67) we obtain for the equivalence classes doubles $\widetilde{\mathfrak{S}}_i$ the following m-ary multiplication

$$\widetilde{\boldsymbol{\mu}}^{(m)}_{\text{compw}}[\widetilde{\mathfrak{S}}_1, \widetilde{\mathfrak{S}}_2, \ldots, \widetilde{\mathfrak{S}}_m] = \begin{pmatrix} \mu^{(m)}[a_1, a_2, \ldots, a_m] \\ \mu^{(m)}[b_1, b_2, \ldots, b_m] \end{pmatrix}, \quad \widetilde{\mathfrak{S}}_i = \begin{bmatrix} a_i \\ b_i \end{bmatrix}_m, \quad a_i, b_i \in S, \tag{1.5.72}$$

because $n = m$ for the componentwise multiplication (1.5.28). We resolve the equation for the querelement (1.5.69) as

$$\overline{\widetilde{\mathfrak{S}}}_{\text{compw}} = \begin{bmatrix} \bar{a} \\ \bar{b} \end{bmatrix}_m \sim \begin{bmatrix} \mu^{(m)}[a^{\alpha}, b^{m-\alpha}] \\ \mu^{(m)}[a^{m-2+\alpha}, b^{2-\alpha}] \end{bmatrix}_m, \tag{1.5.73}$$

where $\alpha = 0, 1, 2$. Because all three solutions are in the same equivalence class, we choose as the representative the symmetric choice $\alpha = 1$

$$\overline{\widetilde{\mathfrak{S}}}_{\text{compw}} = \widetilde{\boldsymbol{\mu}}^{(1)}_{\text{compw}}[\widetilde{\mathfrak{S}}] = \begin{bmatrix} \bar{a} \\ \bar{b} \end{bmatrix}_m = \begin{bmatrix} \mu^{(m)}[a, b^{m-1}] \\ \mu^{(m)}[a^{m-1}, b] \end{bmatrix}_m. \tag{1.5.74}$$

Therefore, for any m-ary semigroup with m-ary componentwise direct power we obtain

$$\mathsf{K}_0^{(m,m)}(\mathcal{S}^{(m)}) = \left\langle \{\widetilde{\mathfrak{S}}\} \middle| \widetilde{\boldsymbol{\mu}}^{(m)}_{\text{compw}}, \widetilde{\boldsymbol{\mu}}^{(1)}_{\text{compw}} \right\rangle. \tag{1.5.75}$$

Example 1.5.24. Negative numbers (continued). Let us introduce for the equivalence classes $\widetilde{\mathfrak{S}}_{\text{neg}} = \begin{bmatrix} -p \\ -q \end{bmatrix}_3$, $p, q \in \mathbb{N}$, $\gcd(p, q) = 1$ (see *Example* 1.5.15), their ternary componentwise multiplication (1.5.72)

$$\widetilde{\boldsymbol{\mu}}^{(3)}_{\text{neg}}\left[\begin{bmatrix} -p_1 \\ -q_1 \end{bmatrix}_3 \begin{bmatrix} -p_2 \\ -q_2 \end{bmatrix}_3 \begin{bmatrix} -p_3 \\ -q_3 \end{bmatrix}_3\right] = \begin{bmatrix} -p_1p_2p_3 \\ -q_1q_2q_3 \end{bmatrix}_3, \quad p_i, q_i \in \mathbb{N}. \tag{1.5.76}$$

We derive from (1.5.69) the querelement

$$\bar{\bar{\mathfrak{S}}}_{\mathrm{neg}} = \widetilde{\boldsymbol{\mu}}_{\mathrm{neg}}^{(1)}[\widetilde{\mathfrak{S}}] = \widetilde{\boldsymbol{\mu}}_{\mathrm{neg}}^{(1)}\left[\begin{bmatrix} -p \\ -q \end{bmatrix}_3\right] = \begin{bmatrix} -pq^2 \\ -p^2q \end{bmatrix}_3, \quad \forall p, q \in \mathbb{N}, \ \gcd(p, q) = 1. \tag{1.5.77}$$

Thus, we obtain the group completion of the negative numbers (with the componentwise multiplication of classes $\widetilde{\boldsymbol{\mu}}_{\mathrm{neg}}^{(3)}$)

$$\mathsf{K}_0^{(3,3)}(\mathbb{N}_-)_{\mathrm{comp}} = \left\langle \left\{ \widetilde{\mathfrak{S}}_{\mathrm{neg}} \right\} \middle| \widetilde{\boldsymbol{\mu}}_{\mathrm{neg}}^{(3)}, \widetilde{\boldsymbol{\mu}}_{\mathrm{neg}}^{(1)} \right\rangle. \tag{1.5.78}$$

If we choose the ternary noncomponentwise multiplication (1.5.36), then

$$\widetilde{\boldsymbol{\mu}}_{\mathrm{neg2}}^{(3)}\left[\begin{bmatrix} -p_1 \\ -q_1 \end{bmatrix}_3 \begin{bmatrix} -p_2 \\ -q_2 \end{bmatrix}_3 \begin{bmatrix} -p_3 \\ -q_3 \end{bmatrix}_3\right] = \begin{bmatrix} -p_1q_2p_3 \\ -q_1p_2q_3 \end{bmatrix}_3, \quad p_i, q_i \in \mathbb{N}, \tag{1.5.79}$$

and the querelement will be different from (1.5.77)

$$\bar{\bar{\mathfrak{S}}}_{\mathrm{neg2}} = \widetilde{\boldsymbol{\mu}}_{\mathrm{neg2}}^{(1)}[\widetilde{\mathfrak{S}}] = \widetilde{\boldsymbol{\mu}}_{\mathrm{neg2}}^{(1)}\left[\begin{bmatrix} -p \\ -q \end{bmatrix}_3\right] = \begin{bmatrix} -pq^2 \\ -q^3 \end{bmatrix}_3, \quad \forall p, q \in \mathbb{N}, \ \gcd(p, q) = 1. \tag{1.5.80}$$

Therefore, the second polyadic group completion of the negative numbers (with the noncomponentwise multiplication of classes $\widetilde{\boldsymbol{\mu}}_{\mathrm{neg2}}^{(3)}$ (1.5.79) being different from $\widetilde{\boldsymbol{\mu}}_{\mathrm{neg}}^{(3)}$ (1.5.76)) becomes

$$\mathsf{K}_0^{(3,3)}(\mathbb{N}_-)_{\mathrm{noncomp}} = \left\langle \left\{ \widetilde{\mathfrak{S}}_{\mathrm{neg}} \right\} \middle| \widetilde{\boldsymbol{\mu}}_{\mathrm{neg2}}^{(3)}, \widetilde{\boldsymbol{\mu}}_{\mathrm{neg2}}^{(1)} \right\rangle. \tag{1.5.81}$$

Another simple example is the set of odd positive numbers.

Example 1.5.25. Let $\mathbb{N}_{\mathrm{odd}} = \{2k+1\}$, $k \in \mathbb{N}_0$, then it is the ternary semigroup $\mathcal{S}_{\mathrm{odd}}^{(3)} = \left\langle \mathbb{N}_{\mathrm{odd}} \middle| \mu_{\mathrm{odd}}^{(3)} \right\rangle$, where the multiplication is the ordinary addition

$$\mu_{\mathrm{odd}}^{(3)}[a, b, c] = a + b + c, \ a, b, c \in \mathbb{N}_{\mathrm{odd}}. \tag{1.5.82}$$

The doubles (1.5.1) have the form

$$\mathbf{S}_{\mathrm{odd}} = \begin{pmatrix} 2k_1 + 1 \\ 2k_2 + 1 \end{pmatrix} \in \mathbb{N}_{\mathrm{odd}} \times \mathbb{N}_{\mathrm{odd}}, \quad k_i \in \mathbb{N}_0. \tag{1.5.83}$$

The equivalence relations (1.5.44) and (1.5.45) become ($m = 3$)

$$\begin{pmatrix} 2k_1+1 \\ 2k_2+1 \end{pmatrix} \sim_{\text{gauge}} \begin{pmatrix} 2l_1+1 \\ 2l_2+1 \end{pmatrix}$$
$$\Leftrightarrow \exists_{x,\, y\in\mathbb{N}_{\text{odd}}} \begin{pmatrix} \mu_{\text{odd}}^{(3)}[(2k_1+1)^2, x] \\ \mu_{\text{odd}}^{(3)}[(2k_2+1)^2, x] \end{pmatrix} = \begin{pmatrix} \mu_{\text{odd}}^{(3)}[(2l_1+1)^2, y] \\ \mu_{\text{odd}}^{(3)}[(2l_2+1)^2, y] \end{pmatrix}, \quad k_i, l_i \in \mathbb{N}_0, \tag{1.5.84}$$

and

$$\begin{pmatrix} 2k_1+1 \\ 2k_2+1 \end{pmatrix} \sim_{\text{twist}} \begin{pmatrix} 2l_1+1 \\ 2l_2+1 \end{pmatrix}$$
$$\Leftrightarrow \exists_{x,\, y\in\mathbb{N}_{\text{odd}}} \left(\mu_{\text{odd}}^{(3)}\right)^{\circ 2}[(2k_1+1)^2, (2l_2+1)^2, z]$$
$$= \left(\mu_{\text{odd}}^{(3)}\right)^{\circ 2}[(2k_2+1)^2, (2l_1+1)^2, z]. \tag{1.5.85}$$

For instance,

$$\begin{pmatrix} 3 \\ 1 \end{pmatrix} \sim \begin{pmatrix} 5 \\ 3 \end{pmatrix} \sim \begin{pmatrix} 7 \\ 5 \end{pmatrix} \ldots, \begin{pmatrix} 5 \\ 1 \end{pmatrix} \sim \begin{pmatrix} 7 \\ 3 \end{pmatrix} \sim \begin{pmatrix} 9 \\ 5 \end{pmatrix} \ldots,$$
$$\begin{pmatrix} 7 \\ 1 \end{pmatrix} \sim \begin{pmatrix} 9 \\ 3 \end{pmatrix} \sim \begin{pmatrix} 11 \\ 5 \end{pmatrix} \ldots, \begin{pmatrix} 1 \\ 3 \end{pmatrix} \sim \begin{pmatrix} 3 \\ 5 \end{pmatrix} \sim \begin{pmatrix} 5 \\ 7 \end{pmatrix} \ldots, \tag{1.5.86}$$
$$\begin{pmatrix} 1 \\ 5 \end{pmatrix} \sim \begin{pmatrix} 3 \\ 7 \end{pmatrix} \sim \begin{pmatrix} 5 \\ 9 \end{pmatrix} \ldots, \begin{pmatrix} 1 \\ 7 \end{pmatrix} \sim \begin{pmatrix} 3 \\ 9 \end{pmatrix} \sim \begin{pmatrix} 5 \\ 11 \end{pmatrix} \ldots.$$

In general, two elements

$$\begin{pmatrix} a \\ b \end{pmatrix}, \begin{pmatrix} a+2 \\ b+2 \end{pmatrix}, \quad a, b \in \mathbb{N}_{\text{odd}}, \tag{1.5.87}$$

are in the same equivalence class $\begin{bmatrix} a \\ b \end{bmatrix} \in \mathbb{N}_{\text{odd}} \times \mathbb{N}_{\text{odd}} \sim$. The minimal representatives are the doubles

$$\widetilde{\mathfrak{S}}_{\text{up}} = \begin{bmatrix} 2k+1 \\ 1 \end{bmatrix}, \quad \widetilde{\mathfrak{S}}_{\text{down}} = \begin{bmatrix} 1 \\ 2k+1 \end{bmatrix}, \quad k \in \mathbb{N}. \tag{1.5.88}$$

Thus, the structure of the classes is the following

$$\{\widetilde{\mathfrak{S}}\} = \left\{\widetilde{\mathfrak{S}}_{\text{up}}\right\} \bigcup \left\{\widetilde{\mathfrak{S}}_{\text{down}}\right\}, \quad \left\{\widetilde{\mathfrak{S}}_{\text{up}}\right\} \bigcap \left\{\widetilde{\mathfrak{S}}_{\text{down}}\right\} = \varnothing. \tag{1.5.89}$$

We do not have an identity in the ternary semigroup $\mathcal{S}_{\text{odd}}^{(3)}$, and therefore the representatives (1.5.88) cannot be mapped to negative and positive numbers, as in the binary case (1.5.26), and moreover their multiplication is ternary (1.5.82).

To find the polyadic group structure (i.e. querelements (1.2.13)), for the ternary multiplication of the classes $\widetilde{\mathfrak{S}}$ we consider two cases:

1. The componentwise multiplication (1.5.28). We use the ternary multiplication of the classes as ordinary addition

$$\widetilde{\boldsymbol{\mu}}^{(3)}_{\text{odd1}}\left[\begin{bmatrix}2k_1+1\\1\end{bmatrix},\begin{bmatrix}2k_2+1\\1\end{bmatrix},\begin{bmatrix}2k_3+1\\1\end{bmatrix}\right] = \begin{bmatrix}2(k_1+k_2+k_3)+3\\3\end{bmatrix}\sim\begin{bmatrix}2(k_1+k_2+k_3)+1\\1\end{bmatrix},\quad k_i\in\mathbb{N}, \tag{1.5.90}$$

and

$$\widetilde{\boldsymbol{\mu}}^{(3)}_{\text{odd1}}\left[\begin{bmatrix}1\\2k_1+1\end{bmatrix},\begin{bmatrix}1\\2k_2+1\end{bmatrix},\begin{bmatrix}1\\2k_3+1\end{bmatrix}\right] = \begin{bmatrix}3\\2(k_1+k_2+k_3)+3\end{bmatrix}\sim\begin{bmatrix}1\\2(k_1+k_2+k_3)+1\end{bmatrix},\quad k_i\in\mathbb{N}. \tag{1.5.91}$$

Then, apply the general formula for the querelement (1.5.74) to our ternary case

$$\overline{\overline{\mathfrak{S}}}_{\text{compw}}=\begin{bmatrix}\bar{a}\\\bar{b}\end{bmatrix}_3=\begin{bmatrix}\mu^{(3)}[a,b^2]\\\mu^{(3)}[a^2,b]\end{bmatrix}_3, \tag{1.5.92}$$

which gives using the equivalence relations (1.5.87)

$$\overline{\overline{\mathfrak{S}}}_{\text{up}}=\widetilde{\boldsymbol{\mu}}^{(1)}_{\text{odd1}}\left[\widetilde{\mathfrak{S}}_{\text{up}}\right]=\overline{\begin{bmatrix}2k+1\\1\end{bmatrix}}\sim\begin{bmatrix}1\\2k+1\end{bmatrix}=\widetilde{\mathfrak{S}}_{\text{down}}, \tag{1.5.93}$$

$$\overline{\overline{\mathfrak{S}}}_{\text{down}}=\widetilde{\boldsymbol{\mu}}^{(1)}_{\text{odd1}}[\widetilde{\mathfrak{S}}_{\text{down}}]=\overline{\begin{bmatrix}1\\2k+1\end{bmatrix}}\sim\begin{bmatrix}2k+1\\1\end{bmatrix}=\widetilde{\mathfrak{S}}_{\text{up}},\quad k\in\mathbb{N}. \tag{1.5.94}$$

Thus, for the componentwise ternary group completion of the ternary semigroup of odd numbers $\mathbb{N}_{\text{odd}}$ we obtain

$$\mathsf{K}_0^{(3,3)}(\mathbb{N}_{\text{odd}})_{\text{compw}}=\left\langle\left\{\widetilde{\mathfrak{S}}_{\text{up}}\right\}\cup\left\{\widetilde{\mathfrak{S}}_{\text{down}}\right\}\middle|\widetilde{\boldsymbol{\mu}}^{(3)}_{\text{odd1}},\widetilde{\boldsymbol{\mu}}^{(1)}_{\text{odd1}}\right\rangle. \tag{1.5.95}$$

2. The noncomponentwise Post-like multiplication (1.5.36). In this case, we denote the multiplication of classes corresponding to the double product (1.5.36) by $\widetilde{\boldsymbol{\mu}}^{(3)}_{\text{odd2}}$, then the general formula for the querelement will be different from (1.5.92) and have the form

$$\overline{\overline{\mathfrak{S}}}_{n\text{compw}}=\begin{bmatrix}\bar{a}\\\bar{b}\end{bmatrix}_3=\begin{bmatrix}\mu^{(3)}[a^2,b]\\\mu^{(3)}[a,b^2]\end{bmatrix}_3, \tag{1.5.96}$$

which gives (using the equivalence relations (1.5.87))

$$\overline{\overline{\mathfrak{S}}}_{\text{up}}=\widetilde{\boldsymbol{\mu}}^{(1)}_{\text{odd2}}\left[\widetilde{\mathfrak{S}}_{\text{up}}\right]=\overline{\begin{bmatrix}2k+1\\1\end{bmatrix}}\sim\begin{bmatrix}2k+1\\1\end{bmatrix}=\widetilde{\mathfrak{S}}_{\text{up}}, \tag{1.5.97}$$

$$\overline{\overline{\mathfrak{S}}}_{\text{down}} = \tilde{\boldsymbol{\mu}}^{(1)}_{\text{odd2}}[\overline{\mathfrak{S}}_{\text{down}}] = \overline{\left[\frac{1}{2k+1}\right]} \sim \left[\frac{1}{2k+1}\right] = \overline{\mathfrak{S}}_{\text{down}}, \qquad k \in \mathbb{N}. \tag{1.5.98}$$

Now, each element is a kind of polyadic reflection obeying $\overline{\overline{\mathfrak{S}}} = \overline{\mathfrak{S}}$ (an analog of $g^{-1} = g$ in binary groups). Thus, the noncomponentwise ternary group completion of the ternary semigroup of odd numbers $\mathbb{N}_{\text{odd}}$ is

$$\mathsf{K}_0^{(3,3)}(\mathbb{N}_{\text{odd}})_{\text{ncompw}} = \left\langle \left\{\overline{\mathfrak{S}}_{\text{up}}\right\} \cup \left\{\overline{\mathfrak{S}}_{\text{down}}\right\} \middle| \tilde{\boldsymbol{\mu}}^{(3)}_{\text{odd2}}, \tilde{\boldsymbol{\mu}}^{(1)}_{\text{odd2}} \right\rangle. \tag{1.5.99}$$

Example 1.5.26. Consider the following 4-ary semigroup of complex matrices

$$\mathcal{S}^{(4)}_{\text{matr}} = \left\langle S_{\text{matr}} \middle| \mu^{(4)}_{\text{matr}} \right\rangle, \quad S_{\text{matr}} = \{a\}, \quad a = \begin{pmatrix} u & 0 \\ 0 & 0 \end{pmatrix}, \quad u \in \mathbb{C}, \tag{1.5.100}$$

$$\mu^{(4)}_{\text{matr}}[a_1, a_2, a_3, a_4] = a_1 + \varepsilon a_2 + \varepsilon^2 a_3 + a_4, \tag{1.5.101}$$

where $\varepsilon = e^{\frac{2\pi}{3}}$, $\varepsilon^3 = 1$.

We construct the doubles $\mathbf{S} = \begin{pmatrix} a \\ b \end{pmatrix} \in S_{\text{matr}} \times S_{\text{matr}}$ with the componentwise multiplication (1.5.28) which does not change the arity. Then we write the "twist-like" equivalence relation (1.5.45) for our case

$$\begin{pmatrix} a_1 \\ b_1 \end{pmatrix} \sim_{\text{twist}} \begin{pmatrix} a_2 \\ b_2 \end{pmatrix} \Longleftrightarrow \exists_{z \in S}\ (\mu^{(4)})^{\circ 2}[a_1^3, b_2^3, z] = (\mu^{(4)})^{\circ 2}[a_2^3, b_1^3, z], \quad a_i, b_i, z \in S_{\text{matr}}. \tag{1.5.102}$$

It follows from (1.5.100)–(1.5.101), that $\mathcal{S}^{(4)}_{\text{matr}}$ is an idempotent semigroup, because $\mu^{(4)}[a^4] = a$, $\forall a \in S_{\text{matr}}$, moreover, $\mu^{(4)}[a^3, b] = b$, $\forall a, b \in S_{\text{matr}}$, and so $(\mu^{(4)})^{\circ 2}[a_1^3, b_2^3, z] = (\mu^{(m)})^{\circ 2}[a_2^3, b_1^3, z] = z$ identically for $\forall a_i, b_i, z \in S_{\text{matr}}$. Thus, we do not have different equivalence classes at all, and therefore for (1.5.100)

$$\mathsf{K}_0^{(4,4)}(\mathcal{S}^{(4)}_{\text{matr}}) = 1, \tag{1.5.103}$$

in our multiplicative notation.

Let us consider the case, when the completion group of classes has different arity from that of the initial semigroup, i.e $\mathsf{K}_0^{(m,n)}(\mathcal{S}^{(m)})$ with $n \neq m$.

Example 1.5.27. The representatives of the residue (congruence) class $S_{\text{res}} = [[7]]_{10} = \{10k + 7\}$, $k \in \mathbb{N}_0$ form a 5-ary semigroup with respect to multiplication (see the arity shape for residue classes $[[a]]_b$ in Duplij 2017). We take

$$\mathcal{S}^{(5)}_{\text{res}} = \left\langle \{10k + 7\} \middle| \mu^{(5)}_{\text{res}} \right\rangle, \tag{1.5.104}$$

where 5-ary multiplication $\mu^{(5)}_{\text{res}}$ is the ordinary product (in $\mathbb{Z}_+$) $\mu^{(5)}_{\text{res}}[a_1, a_2, a_3, a_4, a_5] = a_1 a_2 a_3 a_4 a_5$, $a_i = 10k_i + 7$, $k_i \in \mathbb{N}_0$.

Now the doubles (1.5.1) are

$$\mathbf{S}_{\mathrm{res}} = \begin{pmatrix} 10k + 7 \\ 10l + 7 \end{pmatrix} \in [[7]]_{10} \times [[7]]_{10}, \quad k, l \in \mathbb{N}_0. \tag{1.5.105}$$

The equivalence relations (1.5.44) and (1.5.45) for $m = 5$ are

$$\begin{pmatrix} 10k_1 + 7 \\ 10l_1 + 7 \end{pmatrix} \sim_{\mathrm{gauge}} \begin{pmatrix} 10k_2 + 7 \\ 10l_2 + 7 \end{pmatrix} \Leftrightarrow \exists_{x,\, y \in S_{\mathrm{res}}} \begin{pmatrix} (10k_1 + 7)^4 x \\ (10l_1 + 7)^4 x \end{pmatrix} = \begin{pmatrix} (10k_2 + 7)^4 y \\ (10l_2 + 7)^4 y \end{pmatrix}, \quad k_i, l_i \in \mathbb{N}_0, \tag{1.5.106}$$

and

$$\begin{pmatrix} 10k_1 + 7 \\ 10l_1 + 7 \end{pmatrix} \sim_{\mathrm{twist}} \begin{pmatrix} 10k_2 + 7 \\ 10l_2 + 7 \end{pmatrix} \Leftrightarrow \exists_{x,\, y \in S_{\mathrm{res}}} (10k_1 + 7)^4 (10l_2 + 7)^4 z = (10k_2 + 7)^4 (10l_1 + 7)^2 z. \tag{1.5.107}$$

It follows from the "gauge" shifts (1.5.106) that the components of the double (1.5.105) are mutually prime. Some of the equivalence relations are (we present only those with $k_1 < k_2$)

$$\begin{gathered} \begin{pmatrix} 7 \\ 17 \end{pmatrix} \sim \begin{pmatrix} 77 \\ 187 \end{pmatrix} \sim \begin{pmatrix} 147 \\ 357 \end{pmatrix} \ldots, \begin{pmatrix} 7 \\ 77 \end{pmatrix} \sim \begin{pmatrix} 17 \\ 187 \end{pmatrix} \sim \begin{pmatrix} 27 \\ 297 \end{pmatrix} \ldots, \\ \begin{pmatrix} 17 \\ 47 \end{pmatrix} \sim \begin{pmatrix} 187 \\ 517 \end{pmatrix} \sim \begin{pmatrix} 357 \\ 987 \end{pmatrix} \ldots, \begin{pmatrix} 27 \\ 37 \end{pmatrix} \sim \begin{pmatrix} 297 \\ 407 \end{pmatrix} \sim \begin{pmatrix} 567 \\ 777 \end{pmatrix} \ldots, \\ \begin{pmatrix} 57 \\ 87 \end{pmatrix} \sim \begin{pmatrix} 247 \\ 377 \end{pmatrix} \sim \begin{pmatrix} 437 \\ 667 \end{pmatrix} \ldots, \begin{pmatrix} 87 \\ 377 \end{pmatrix} \sim \begin{pmatrix} 147 \\ 637 \end{pmatrix} \sim \begin{pmatrix} 177 \\ 767 \end{pmatrix} \ldots. \end{gathered} \tag{1.5.108}$$

We denote the equivalence class by

$$\widetilde{\mathfrak{S}} = \widetilde{\mathfrak{S}}_{\mathrm{res}} = \begin{bmatrix} a \\ b \end{bmatrix}_{\mathrm{res}} = \begin{bmatrix} 10k + 7 \\ 10l + 7 \end{bmatrix}_{\mathrm{res}} \in S_{\mathrm{res}} \times S_{\mathrm{res}} \sim, \quad k, l \in \mathbb{N}_0, \tag{1.5.109}$$

and we use the minimal representative, when needed and possible. The polyadic group of the equivalence classes can be obtained, if we define the classes multiplication and the querelement. If we choose the 5-ary componentwise multiplication (1.5.72), then we obtain the group completion of the same arity $\mathsf{K}_0^{(5,5)}(\mathcal{S}_{\mathrm{res}}^{(5)})$, as in the previous *Example* 1.5.25. The more exotic case is to use for the classes doubles $\widetilde{\mathfrak{S}}_{\mathrm{res}}$ the changing arity product (1.5.41), which gives us the ternary group of equivalence classes, which is built from the 5-ary semigroup. Indeed, let $\widetilde{\boldsymbol{\mu}}_{\mathrm{res}}^{(3)}$ be the ternary multiplication of the equivalence classes defined by (cf (1.5.41))

$$\widetilde{\boldsymbol{\mu}}_{\mathrm{res}}^{(3)}[\widetilde{\mathfrak{S}}_1, \widetilde{\mathfrak{S}}_2, \widetilde{\mathfrak{S}}_3] = \begin{bmatrix} (10k_1 + 7)(10k_2 + 7)(10k_3 + 7)(10l_1 + 7)(10l_2 + 7) \\ 10l_3 + 7 \end{bmatrix}_{\mathrm{res}}, \quad k_i, l_i \in \mathbb{N}_0. \tag{1.5.110}$$

The querelement $\overline{\overline{\mathfrak{S}}}$ can be determined from the manifest form of the ternary multiplication $\widetilde{\mu}_{\text{res}}^{(3)}$ of the equivalence classes, and it has the general form

$$\overline{\overline{\mathfrak{S}}} = \begin{bmatrix} \bar{a} \\ \bar{b} \end{bmatrix}_{\text{res}} \equiv \widetilde{\mu}_{\text{res}}^{(1)}[\widetilde{\mathfrak{S}}] \sim \begin{bmatrix} a^3b^2 \\ a^2b^3 \end{bmatrix}_{\text{res}} \sim \begin{bmatrix} (10k+7)^3(10l+7)^2 \\ (10k+7)^2(10l+7)^3 \end{bmatrix}_{\text{res}}, \; a, b \in S_{\text{res}}, \; k_i, l_i \in \mathbb{N}. \quad (1.5.111)$$

Thus, the ternary group of the classes (1.5.109)

$$\mathsf{K}_0^{(5,3)}(\mathcal{S}_{\text{res}}^{(5)}) = \left\langle \{\widetilde{\mathfrak{S}}_{\text{res}}\} \middle| \widetilde{\mu}_{\text{res}}^{(3)}, \widetilde{\mu}_{\text{res}}^{(1)} \right\rangle \quad (1.5.112)$$

is the completion group of the 5-ary semigroup $\mathcal{S}_{\text{res}}^{(5)}$ (1.5.104).

References

Belousov V D 1972 *N-ary Quasigroups* (Kishinev: Shtiintsa)

Bergman C 2012 *Universal Algebra: Fundamentals and Selected Topics* (New York: CRC Press)

Bergman G M 1995 *An Invitation to General Algebra and Universal Constructions* (Berkeley, CA: University of California)

Borceux F 1994 Handbook of categorical algebra 1. Basic category theory *Encyclopedia of Mathematics and its Applications* vol 50 (Cambridge: Cambridge University Press)

Borowiec A, Dudek W and Duplij S 2006 Bi-element representations of ternary groups *Commun. Algebra* **34** 1651–70

Bourbaki N 1998 *Elements of Mathematics: Algebra* I (Berlin: Springer)

Brandt H 1927 Über eine Verallgemeinerung des Gruppenbegriffes *Math. Annalen* **96** 360–7

Bruck R H 1966 *A Survey on Binary Systems* (New York: Springer)

Celakoski N 1977 On some axiom systems for *n*-groups *Mat. Bilt.* **1** 5–14

Certaine J 1943 The ternary operation $(abc) = ab^{-1}c$ of a group *Bull. Am. Math. Soc.* **49** 869–77

Chronowski A 1991 On extensions of homomorphisms homotopies of commutative *n*-adic groups *Arch. Math. (Brno)* **27** 139–48

Chunikhin S A 1946 On the theory of non-associative n-groups satisfying postulate K *Dokl. Akad. Nauk SSSR* **48** 7–10

Clifford A H and Preston G B 1961 *The Algebraic Theory of Semigroups* vol 1 (Providence, RI: American Mathematical Society)

Čupona G and Trpenovski B 1961 Finitary associative operations with neutral elements *Bull. Soc. Math. Phys. Macedoine* **12** 15–24

Dörnte W 1929 Unterschungen über einen verallgemeinerten Gruppenbegriff *Math. Z.* **29** 1–19

Dudek W 1990 On n-ary groups with only one skew elements *Radovi Mat. (Sarajevo)* **6** 171–5

Dudek W A 1980 Remarks on *n*-groups *Demonstr. Math.* **13** 165–81

Dudek W A 1993 Autodistributive *n*-groups *Ann. Sci. Math. Pol.* **23** 1–11

Dudek W A 2007 Remarks to Głazek's results on *n*-ary groups *Discuss. Math. Gen. Algebra Appl.* **27** 199–233

Dudek W A and Głazek K 2008 Around the Hosszú–Gluskin theorem for *n*-ary groups *Discrete Math.* **308** 4861–76

Dudek W A, Głazek K and Gleichgewicht B 1977 A note on the axioms of *n*-groups *Colloquia Math. Soc. J. Bolyai 29* vol 29 (Amsterdam: North-Holland) pp 195–202, Universal Algebra (Esztergom, 1977)

Dudek W A and Michalski J 1984 On retracts of polyadic groups *Demonstr. Math.* **17** 281–301
Duplij S 2012 Polyadic systems, representations and quantum groups *J. Kharkiv Natl Univ. Ser. Nuclei Part. Fields* **1017** 28–59 Expanded version arXiv:math.RT/1308.4060
Duplij S 2017 Polyadic integer numbers and finite (m, n)-fields *p-Adic Numbers, Ultrametric Analysis and Appl.* **9** 257–81 arXiv:math.RA/1707.00719
Duplij S 2019 Arity shape of polyadic algebraic structures *J. Math. Phys. Anal. Geom.* **15** 3–56
Duplij S 2021 Polyadic Hopf algebras and quantum groups East *Eur. J. Phys.* **2** 5–50
Duplij S 2022 *Polyadic analogs of direct product* Univ. Münster, Münster arXiv:math.RA/2201.08479
Evans T 1963 Abstract mean values *Duke Math. J.* **30** 331–47
Gal'mak A M 2003 *n-Ary Groups, Part 1* (Gomel University: Gomel)
Gal'mak A M and Vorobiev G N 2013 On Post–Gluskin–Hosszu theorem *Probl. Fiz. Mat. Tekh.* **2011** 55–60
Głazek K and Gleichgewicht B 1982 Abelian *n*-groups *Colloq. Math. Soc. János Bolyai* vol 29 (Amsterdam: North-Holland) pp 321–9, Universal Algebra (Esztergom, 1977)
Gleichgewicht B and Głazek K 1967 Remarks on *n*-groups as abstract algebras *Colloq. Math.* **17** 209–19
Gluskin L M 1964 On positional operatives *Sov. Math. Dokl.* **5** 1001–4
Gluskin L M 1965 Position operatives *Math. Sbornik* **68** 444–72
Gluskin L M and Shvarts V Y 1972 Towards a theory of associatives *Math. Notes* **11** 332–7
Hausmann B A and Ore Ø 1937 Theory of quasigroups *Am. J. Math.* **59** 983–1004
Heine E 1878 *Handbuch der Kugelfunktionan* (Reimer: Berlin)
Hosszú M 1963 On the explicit form of *n*-group operation *Publ. Math. Debrecen* **10** 88–92
Hungerford T W 1974 *Algebra* (Springer: New York)
Iancu L 1991 On the category of *n*-groups *Bul. Stiint. Univ. Baia Mare Ser.* B **7** 9–14
Kac V and Cheung P 2002 *Quantum Calculus* (New York: Springer)
Karoubi M 1978 *K-theory. An Introduction* (Berlin: Springer)
Kasner E 1904 An extension of the group concept *Bull. Am. Math. Soc* **10** 290–1
Lambek J 1966 *Lectures on Rings and Modules* (Providence, RI: Blaisdell)
Lang S 1965 *Algebra* (Reading, MA: Addison-Wesley)
Lehmer D H 1932 A ternary analogue of Abelian groups *Am. J. Math* **54** 329–38
Lehmer D H 1936 Polynomials for the n-ary composition of numerical functions *Am. J. Math.* **58** 563–72
Mac Lane S 1971 *Categories for the Working Mathematician* (Berlin: Springer)
Mesiar R and Sarkoci P 2010 Open problems posed at the Tenth international conference on fuzzy set theory and applications (FSTA 2010, Liptovský Ján, Slovakia) *Kybernetika* 46 585–99
Michalski J 1981 Covering *k*-groups of *n*-groups *Arch. Math. (Brno)* **17** 207–26
Michalski J 1984a Free products of *n*-groups *Fundam. Math.* **123** 11–20
Michalski J 1984b On the category of *n*-groups *Fundam. Math.* **122** 187–97
Miller G A 1935 Sets of group elements involving only products of more than *n Proc. Natl. Acad. Sci. USA* **21** 45–7
Monk J D and Sioson F 1966 *m*-Semigroups, semigroups and function representations *Fundam. Math.* **59** 233–41
Murdoch D C 1939 Quasi-groups which satisfy certain generalized associative laws *Am. J. Math.* **61** 509–22

Nivanen L, LeMéauté A and Wang Q A 2003 Generalized algebra within a nonextensive statistics *Rep. Math. Phys.* **52** 437–44

Petrescu A 2005 Skew elements in *n*-semigroups *Acta Univ. Apulensis Math. Inform.* **10** 39–48

Pop A and Pop M S 2010 On generalized algebraic structures *Creat. Math. Inf.* **19** 184–90

Post E L 1940 Polyadic groups *Polyadic groups Trans. Am. Math. Soc.* **48** 208–350

Prüfer H 1924 Theorie der abelshen Gruppen I. Grundeigenschaften *Math. Z.* **20** 165–87

Rosenberg J 1994 *Algebraic K-theory and its Applications* (New York: Springer)

Rusakov S A 1998 *Some Applications of* n-*Ary Group Theory* (Minsk: Belaruskaya Navuka)

Shchuchkin N A 2003 An interconnection between *n*-groups and groups *Chebysh. Sb.* **4** 125–41

Shchuchkin N A 2014 Direct product of *n*-ary groups *Chebysh. Sb.* **15** 101–21

Sioson F M 1963 On regular algebraic systems *Proc. Japan Acad* **39** 283–6

Sokhatsky F M 1997 On the associativity of multiplace operations *Quasigroups Relat. Syst.* **4** 51–66

Sokolov E I 1976 On the theorem of Gluskin–Hosszú on Dörnte groups *Mat. Issled.* **39** 187–9

Stupňanová A and Kolesárová A 2011 Associative *n*-dimensional copulas *Kybernetika* **47** 93–9

Suschkewitsch A 1935 Über die Erweiterung der Semigruppe bis zur ganzen Gruppe *Comm. Soc. Math. Kharkow* **12** 81–7

Thurston H A 1949 Partly associative operations *J. Lond. Math. Soc.* **24** 260–71

Thurston H A 1954 Reduction of finitary operations to binary and singulary operations *Proc. of the ICM* ed J C H Gerretsen and J De Groot (Amsterdam: North-Holland) pp 68–9

Timm J 1972 Verbandstheoretische Behandlung *n*-stelliger Gruppen *Abh. Math. Semin. Univ. Hamb.* **37** 218–24

Tsallis C 1994 Nonextensive physics: a possible connection between generalized statistical mechanics and quantum groups *Phys. Lett.* **A195** 329–34

Turing A 1938 The extensions of a group *Compos. Math.* **5** 357–67

Tvermoes H 1953 Über eine Verallgemeinerung des Gruppenbegriffs *Math. Scand.* **1** 18–30

Ûsan J 2003 *n*-Groups in the light of the neutral operations *Math. Morav.* **203**

Weibel C 1986 *The K-Book. An Introduction to Algebraic K-theory* (Providence, RI: AMS)

Yurevych O V 2001 Criteria for invertibility of elements in associates *Ukr. Math. J.* **53** 1895–905

Zupnik D 1967 Polyadic semigroups *Publ. Math. (Debrecen)* **14** 273–9

IOP Publishing

Polyadic Algebraic Structures

Steven Duplij

Chapter 2

Representations and heteromorphisms

The representation theory of (binary) groups (Weyl 1946, Fulton and Harris 1991) plays an important role in their physical applications (Cornwell 1997). It is initially based on a matrix realization of the group elements with the abstract group action realized as the usual matrix multiplication (Curtis and Reiner 1962, Collins 1990). The cubic and n-ary generalizations of matrices and determinants were made in Kapranov *et al* (1994) and Sokolov (1972), and their physical application appeared in Kawamura (2003) and Rausch de Traubenberg (2008). In general, particular questions of n-ary group representations were considered, and matrix representations derived in Borowiec *et al* (2006), and some general theorems connecting representations of binary and n-ary groups were presented in Dudek and Shahryari (2012). The intention here is to generalize the above constructions of n-ary group representations to more complicated and nontrivial cases. Then we consider the homomorphisms and their multiplace generalizations (heteromorphisms), paying attention to associativity. We define multiplace representations and multiactions, and give examples of matrix representations for some ternary groups.

2.1 Homomorphisms of one-set polyadic algebraic structures

Here we consider the arity 'interaction' of two one-set one-operation polyadic algebraic structures of any kind (quasigroup, semigroup, group, etc) $\mathcal{G}^{(n)} = \langle G|\mu^{(n)}\rangle$ and $\mathcal{G}'^{(n')} = \langle G'|\mu'^{(n')}\rangle$, in other words $\mathcal{G}^{(n)}$ and $\mathcal{G}'^{(n')}$ are two polyadic magmas or (Hausmann–Ore) groupoids, see remark (1.1.3). If they have the multiplications of the same arity $n = n'$, then one can define the standard (1-place) mappings from $\mathcal{G}^{(n)}$ to $\mathcal{G}'^{(n)}$. Usually such polyadic algebraic structures are called *similar* (but this is too much uncertain, in which sense), and therefore we call mappings between $\mathcal{G}^{(n)}$ and $\mathcal{G}'^{(n)}$ *the equiary mappings.*

Let us take $n + 1$ mappings $\varphi_i^{GG'}: G \to G'$, $i = 1, \ldots, n + 1$. An ordered system of mappings $\left\{\varphi_i^{GG'}\right\}$ is called a *homotopy* from $\mathcal{G}^{(n)}$ to $\mathcal{G}'^{(n)}$, if (Belousov 1972)

doi:10.1088/978-0-7503-2648-3ch2

$$\varphi_{n+1}^{GG'}(\mu^{(n)}[g_1, \ldots, g_n]) = \mu'^{(n)}\left[\varphi_1^{GG'}(g_1), \ldots, \varphi_n^{GG'}(g_n)\right], \quad g_i \in G. \tag{2.1.1}$$

One should add to this definition the 'mapping' of the multiplications

$$\mu^{(n)} \overset{\psi_{nn'}^{(\mu\mu')}}{\mapsto} \mu'^{(n')}. \tag{2.1.2}$$

In such a way, homotopy can be defined as the extended system of mappings $\left\{\varphi_i^{GG'}; \psi_{nn}^{(\mu\mu')}\right\}$.

The existence of the additional 'mapping' $\psi_{nn}^{(\mu\mu')}$ acting on the second component of $\langle G|\mu^{(n)}\rangle$ is tacitly implied. We will write/mention the 'mappings' $\psi_{nn'}^{(\mu\mu')}$ manifestly, e.g.,

$$\mathcal{G}^{(n)} \overset{\left\{\varphi_i^{GG'};\psi_{nn}^{(\mu\mu')}\right\}}{\rightrightarrows} \mathcal{G}'^{(n')}, \tag{2.1.3}$$

only as needed. If all the components $\varphi_i^{GG'}$ of a homotopy are bijections, it is called an *isotopy*. In the case of polyadic quasigroups (Belousov 1972) all mappings $\varphi_i^{GG'}$ are usually taken as permutations of the same underlying set $G = G'$. If the multiplications also coincide $\mu^{(n)} = \mu'^{(n)}$, then $\left\{\varphi_i^{GG}; \mathrm{id}\right\}$ is called an *autotopy* of the polyadic algebraic structure $\mathcal{G}^{(n)}$. Various properties of homotopy in universal algebras were studied, e.g., in Petrescu (1977) and Halaš (1994).

Homotopy, isotopy and autotopy are widely used equiary mappings in the study of polyadic quasigroups and loops, while their diagonal equiary counterparts (all $\varphi_i^{GG'}$ coincide), the homomorphism, isomorphism and automorphism, are more suitable in investigation of polyadic semigroups, groups and rings and their wide applications in physics. Usually, the latter is written about between similar (equiary) polyadic algebraic structures: they '...are so well known that we shall not bother to define them carefully' (Hobby 1988). Nevertheless, we give a diagrammatic definition of the standard homomorphism between similar polyadic algebraic structures in our notation, which will be convenient to explain the clear way of its generalization.

Definition 2.1.1. A *homomorphism* from $\mathcal{G}^{(n)}$ to $\mathcal{G}'^{(n)}$ is given, if there exists a mapping $\varphi^{GG'}: G \to G'$ satisfying

$$\varphi^{GG'}\left(\mu^{(n)}[g_1, \ldots, g_n]\right) = \mu'^{(n)}\left[\varphi^{GG'}(g_1), \ldots, \varphi^{GG'}(g_n)\right], \quad g_i \in G. \tag{2.1.4}$$

The homomorphism is denoted by the same one letter $\varphi^{GG'}$, while sometimes it would be more consistent to use for its notation the extended pair of mappings $\left\{\varphi^{GG'}; \psi_{nn}^{(\mu\mu')}\right\}$.

Let us mention a small subset of known generalizations of the homomorphism (for bibliography till 1982 see, e.g., Głazek and Gleichgewicht 1982) and then introduce a concrete construction for an analogous mapping which can change the arity of the multiplication (fundamental operation) without introducing additional (term) operations. A general approach to mappings between free algebraic structures was initiated in Fujiwara (1959), where the so-called basic mapping formulas for generators were introduced, and its generalization to many-sorted algebras was given in Vidal and Tur (2010). In Novotný (2002) it was shown that the construction of all homomorphisms between similar polyadic algebraic structures can be reduced to some homomorphisms between corresponding mono-unary algebras (Novotný 1990). The above constructions do not change the arity of polyadic algebraic structures, because they are based on the corresponding diagram which gives a definition of an *equiary* mapping.

To change arity during the mapping one has:

(i) To add another equiary diagram with additional operations using the same formula (2.1.4), where both do not change arity.
(ii) To use one modified (and not equiary) diagram and the underlying formula (2.1.4) by themselves, which will allow us to change arity without introducing additional operations.

The first way (i) leads to the concept of *weak homomorphism*, which was introduced in Goetz (1966), Marczewski (1966) and Głazek and Michalski (1974) for non-indexed algebras, and in Głazek (1980) for indexed algebras, then developed in Traczyk (1965) for Boolean and Post algebras, in Denecke and Wismath (2009) for coalgebras and *F*-algebras (Denecke and Saengsura 2008) (see also Chung and Smith 2008). To define the weak homomorphism in our notation we should incorporate into the polyadic algebraic structures $\langle G|\mu^{(n)}\rangle$ and $\langle G'|\mu'^{(n')}\rangle$ the following additional term operations of opposite arity $\nu_{n'}$: $G^{\times n'} \to G$ and ν'_n: $G'^{\times n} \to G'$ and consider two equiary mappings between $\langle G|\mu^{(n)}, \nu^{(n')}\rangle$ and $\langle G'|\mu'^{(n')}, \nu'^{(n)}\rangle$. If only one of these relations holds, such a mapping is called a *semi-weak homomorphism* (Kolibiar 1984). If $\varphi^{GG'}$ is bijective, then it defines a *weak isomorphism*. Any weak epimorphism can be decomposed into a homomorphism and a weak isomorphism (Głazek and Michalski 1977), and therefore the study of weak homomorphisms reduces to weak isomorphisms (see also Czákány 1962, Mal'tcev 1957, Mal'tsev 1958).

2.2 Heteromorphisms of one-set polyadic algebraic structures

2.2.1 Multiplace mappings and heteromorphisms

We now turn to the second way of changing the arity of the multiplication (ii) and use only one relation which we then modify in some natural manner. Recall that in any set G there always exists the additional distinguished mapping, viz. the identity id_G. We use the multiplication $\mu^{(n)}$ with its combination of id_G. Define an (ℓ_{id}-*intact*) id-*product* for the polyadic algebraic structure $\langle G|\mu^{(n)}\rangle$ as

$$\mu_n^{(\ell_{\mathrm{id}})} = \mu^{(n)} \times (\mathrm{id}_G)^{\times \ell_{\mathrm{id}}}, \tag{2.2.1}$$

$$\mu_n^{(\ell_{\mathrm{id}})}: G^{\times(n+\ell_{\mathrm{id}})} \to G^{\times(1+\ell_{\mathrm{id}})}. \tag{2.2.2}$$

We write $\mu_n^{(\ell_{\mathrm{id}})}(i)$ to indicate the exact ith place of $\mu^{(n)}$ in the rhs of (2.2.1). Here we use the id-product to generalize the homomorphism and consider mappings between polyadic algebraic structures of different arity. It follows from (2.2.2) that, if the image of the id-product is G alone, than $\ell_{\mathrm{id}} = 0$. Let us introduce a *multiplace mapping* $\mathbf{\Phi}_k^{(n,n')}$ acting as follows

$$\mathbf{\Phi}_k^{(n,n')}\colon G^{\times k} \to G'. \tag{2.2.3}$$

While constructing the corresponding diagram, we are allowed to take only one upper $\mathbf{\Phi}_k^{(n,n')}$, because of one G' in the upper right corner. Since we already know that the lower right corner is exactly $G'^{\times n'}$ (as a pre-image of one multiplication $\mu'^{(n')}$), the lower horizontal arrow should be a product of n' multiplace mappings $\mathbf{\Phi}_k^{(n,n')}$. So we can write a definition of a multiplace analog of homomorphisms which changes the arity of the multiplication using one relation.

Definition 2.2.1. A *k-place heteromorphism* from $\mathcal{G}^{(n)}$ to $\mathcal{G}'^{(n')}$ is given, if there exists a k-place mapping $\mathbf{\Phi}_k^{(n,n')}$ (2.2.3) such that the following (arity changing or *unequiary*) diagram is commutative

$$\begin{array}{ccc} G^{\times k} & \xrightarrow{\Phi_k} & G' \\ \Big\uparrow{\scriptstyle \mu_n^{(\ell_{\mathrm{id}})}} & & \Big\uparrow{\scriptstyle \mu'^{(n')}} \\ G^{\times kn'} & \xrightarrow{(\Phi_k)^{\times n'}} & (G')^{\times n'} \end{array} \tag{2.2.4}$$

and the corresponding defining equation (a modification of (2.1.4)) depends on the place i of $\mu^{(n)}$ in (2.2.1).

***Remark* 2.2.2.** The notion 'heteromorphism' is motivated by Ellerman (2006, 2007), where mappings between objects from different categories were considered and called 'chimera morphisms'. See, also, Pécsi (2012).

***Example* 2.2.3.** In the case $n = 3$, $n' = 2$, $k = 2$, $\ell_{\mathrm{id}} = 1$ we have

$$\mathbf{\Phi}_2^{(3,2)}\begin{pmatrix} \mu^{(3)}[g_1, g_2, g_3] \\ g_4 \end{pmatrix} = \mu'^{(2)}\left[\mathbf{\Phi}_2^{(3,2)}\begin{pmatrix} g_1 \\ g_2 \end{pmatrix}, \mathbf{\Phi}_2^{(3,2)}\begin{pmatrix} g_3 \\ g_4 \end{pmatrix}\right]. \tag{2.2.5}$$

This formula was used in the construction of the bi-element representations of ternary groups (Borowiec *et al* 2006).

***Example* 2.2.4.** Let $G = M_2^{\mathrm{adiag}}(\Bbbk)$, a set of antidiagonal 2×2 matrices over the field $\Bbbk$ and $G' = \Bbbk$, where $\Bbbk = \mathbb{R}, \mathbb{C}, \mathbb{Q}, \mathbb{H}$. The ternary multiplication $\mu^{(3)}$ is a product of 3 matrices. Obviously, $\mu^{(3)}$ is not derived from a binary multiplication. For the elements $g_i = \begin{pmatrix} 0 & a_i \\ b_i & 0 \end{pmatrix}$, $i = 1, 2$, we construct a 2-place mapping $G \times G \to G'$ as

$$\mathbf{\Phi}_2^{(3,\,2)}\begin{pmatrix} g_1 \\ g_2 \end{pmatrix} = a_1 a_2 b_1 b_2, \tag{2.2.6}$$

which is a heteromorphism, because it satisfies (2.2.5). Let us introduce a standard 1-place mapping by $\varphi(g_i) = a_i b_i$, $i = 1, 2$. It is important to note, that $\varphi(g_i)$ is not a homomorphism, because the product $g_1 g_2$ belongs to diagonal matrices. Consider the product of mappings

$$\varphi(g_1) \cdot \varphi(g_2) = a_1 b_1 a_2 b_2, \tag{2.2.7}$$

where the product $(\cdot)$ in lhs is taken in $\Bbbk$. We observe that (2.2.6) and (2.2.7) coincide for the commutative field $\Bbbk$ only ($=\mathbb{R}$, $\mathbb{C}$) only, and in this case we can have the relation between the heteromorphism $\mathbf{\Phi}_2^{(3,\,2)}$ and the 1-place mapping φ

$$\mathbf{\Phi}_2^{(3,\,2)}\begin{pmatrix} g_1 \\ g_2 \end{pmatrix} = \varphi(g_1) \cdot \varphi(g_2), \tag{2.2.8}$$

while for the noncommutative field $\Bbbk$ ($=\mathbb{Q}$ or $\mathbb{H}$) there is no such relation.

Definition 2.2.5. A heteromorphism is called *derived*, if it can be expressed through a 1-place mapping (not necessary a homomorphism).

In the above example 2.2.4 the heteromorphism is derived (by formula (2.2.8)) for the commutative field $\Bbbk$ and nonderived for the noncommutative $\Bbbk$.

A slightly modified construction (2.2.5) for arbitrary n with still binary final arity $n' = 2$, is defined by, $k = n - 1$, $\ell_{\mathrm{id}} = n - 2$, and

$$\mathbf{\Phi}_{n-1}^{(n,\,2)}\begin{pmatrix} \mu^{(n)}[g_1, \ldots, g_{n-1}, h_1] \\ h_2 \\ \vdots \\ h_{n-1} \end{pmatrix} = \mu'^{(2)}\left[\mathbf{\Phi}_{n-1}^{(n,\,2)}\begin{pmatrix} g_1 \\ \vdots \\ g_{n-1} \end{pmatrix}, \mathbf{\Phi}_{n-1}^{(n,\,2)}\begin{pmatrix} h_1 \\ \vdots \\ h_{n-1} \end{pmatrix}\right] \tag{2.2.9}$$

was used in Dudek (2007) to study representations of n-ary groups. However, no new results compared with Borowiec *et al* (2006) (other than changing 3 to n in some formulas) were obtained. This reflects the fact that a major role is played by the final arity n' and the number ℓ_μ of n-ary multiplications in the lhs of (2.2.5) and (2.2.9). Thus, in the above examples, the latter number was one $\ell_\mu = 1$, but can make it arbitrary, but below n, obviously, $\ell_\mu \leqslant n$.

Definition 2.2.6. A heteromorphism is called a *ℓ_μ-ple heteromorphism*, if it contains ℓ_μ multiplications in the argument of $\mathbf{\Phi}_k^{(n,n')}$ in its defining relation.

So by analogy with (2.2.1)–(2.2.2) we define a *ℓ_μ-ple ℓ_{id}-intact* id-*product* for the polyadic algebraic structure $\langle G \mid \mu^{(n)} \rangle$ as

$$\mu_n^{(\ell_\mu, \ell_{\mathrm{id}})} = (\mu^{(n)})^{\times \ell_\mu} \times (\mathrm{id}_G)^{\times \ell_{\mathrm{id}}}, \tag{2.2.10}$$

$$\mu_n^{(\ell_\mu, \ell_{\mathrm{id}})} : G^{\times(n\ell_\mu + \ell_{\mathrm{id}})} \to G^{\times(\ell_\mu + \ell_{\mathrm{id}})}. \tag{2.2.11}$$

Definition 2.2.7. A ℓ_μ*-ple k-place heteromorphism* from $\mathcal{G}^{(n)}$ to $\mathcal{G}^{\prime(n')}$ is given, if there exists a k-place mapping $\mathbf{\Phi}_k^{(n,n')}$ (2.2.3) such that the following *unequiary* diagram is commutative

$$\begin{array}{ccc} G^{\times k} & \xrightarrow{\Phi_k} & G' \\ \uparrow \mu_n^{(\ell_\mu, \ell_{\mathrm{id}})} & & \uparrow \mu'^{(n')} \\ G^{\times kn'} & \xrightarrow{(\Phi_k)^{\times n'}} & (G')^{\times n'} \end{array} \tag{2.2.12}$$

The corresponding *main heteromorphism equation* is

$$\mathbf{\Phi}_k^{(n,n')} \left(\begin{array}{c} \left. \begin{array}{c} \mu^{(n)}[g_1, \ldots, g_n], \\ \vdots \\ \mu^{(n)}\left[g_{n(\ell_\mu - 1)}, \ldots, g_{n\ell_\mu}\right] \end{array} \right\} \ell_\mu \\ \left. \begin{array}{c} g_{n\ell_\mu + 1}, \\ \vdots \\ g_{n\ell_\mu + \ell_{\mathrm{id}}} \end{array} \right\} \ell_{\mathrm{id}} \end{array} \right) = \mu^{\prime(n')} \left[\overbrace{\mathbf{\Phi}_k^{(n,n')} \begin{pmatrix} g_1 \\ \vdots \\ g_k \end{pmatrix}, \ldots, \mathbf{\Phi}_k^{(n,n')} \begin{pmatrix} g_{k(n'-1)} \\ \vdots \\ g_{kn'} \end{pmatrix}}^{n'} \right]. \tag{2.2.13}$$

Obviously, we can consider various permutations of the multiplications on both sides, as further additional demands (associativity, commutativity, etc), are introduced, which will be considered below. The commutativity of the diagram (2.2.12) leads to the system of equations connecting initial and final arities

$$kn' = \ell_\mu + \ell_{\mathrm{id}}, \tag{2.2.14}$$

$$k = \ell_\mu + \ell_{\mathrm{id}}. \tag{2.2.15}$$

Excluding ℓ_μ or ℓ_{id}, we obtain

Definition 2.2.8. For heteromorphisms there are two *arity changing formulas*

$$n' = n - \frac{n-1}{k}\ell_{\mathrm{id}}, \tag{2.2.16}$$

$$n' = \frac{n-1}{k}\ell_\mu + 1, \tag{2.2.17}$$

where both $\frac{n-1}{k}\ell_{\mathrm{id}} \geqslant 1$ and $\frac{n-1}{k}\ell_{\mu} \geqslant 1$ should be integer.

Therefore, the following inequalities hold valid

$$1 \leqslant \ell_{\mu} \leqslant k, \tag{2.2.18}$$

$$0 \leqslant \ell_{\mathrm{id}} \leqslant k - 1, \tag{2.2.19}$$

$$\ell_{\mu} \leqslant k \leqslant (n - 1)\ell_{\mu}, \tag{2.2.20}$$

$$2 \leqslant n' \leqslant n, \tag{2.2.21}$$

which are important for the further classification of heteromorphisms. The main statement follows from (2.2.21)

Theorem 2.2.9. The heteromorphism $\mathbf{\Phi}_k^{(n,n')}$ defined by the general relation (2.2.13) always decreases the arity of polyadic multiplication $n' \leqslant n$.

Another important observation is the fact that only the id-product (2.2.10) with $\ell_{\mathrm{id}} \neq 0$ leads to a change of the arity. In the extreme case, when k approaches its minimum, $k = k_{\mathrm{min}} = \ell_{\mu}$, the final arity approaches its maximum $n'_{\mathrm{max}} = n$, and the id-product becomes a product of ℓ_{μ} initial multiplications $\mu^{(n)}$ without id's, since now $\ell_{\mathrm{id}} = 0$ in (2.2.13). Therefore, we call a heteromorphism defined by (2.2.13) with $\ell_{\mathrm{id}} = 0$ *a* $k(= \ell_{\mu})$*-place homomorphism.*

***Remark* 2.2.10.** The ordinary homomorphism (2.1.1) corresponds to $k = \ell_{\mu} = 1$, and so it is actually a 1-place homomorphism.

An opposite extreme case, when the final arity approaches its minimum $n'_{\mathrm{min}} = 2$ (the final operation is binary), corresponds to the maximal value of k, that is $k = k_{\mathrm{max}} = (n - 1)\ell_{\mu}$. The number of id's now is $\ell_{\mathrm{id}} = (n - 2)\ell_{\mu} \geqslant 0$, which vanishes, when the initial operation is binary as well. This is the case of the ordinary homomorphism (2.1.1) for both binary operations $n' = n = 2$ and $k = \ell_{\mu} = 1$. We conclude that any one-set one-operation polyadic algebraic structure is connected with a binary algebraic structure. Our statement can be treated as:

Theorem 2.2.11. Any one-set n-ary algebraic structure can be mapped into a binary algebraic structure, using a suitable binarizing k-place heteromorphism (we call it a *binarizing heteromorphism*) $\mathbf{\Phi}_k^{(n,2)}: G^{\times k} \to G'$ (2.2.13), where $k = (n - 1)\ell_{\mu}$ and $\ell_{\mathrm{id}} = (n - 2)\ell_{\mu}$.

The case of 1-ple binarizing heteromorphism ($\ell_\mu = 1$) corresponds to the formula (2.2.9). Further requirements (associativity, commutativity, etc) will give additional relations between multiplications and $\mathbf{\Phi}_k^{(n,n')}$, and fix the exact structure of (2.2.13). Thus, we arrive to the following

Proposition 2.2.12. *Classification of ℓ_μ-ple* k*-place heteromorphisms:*

(i) $n' = n'_{\max} = n \Longrightarrow \mathbf{\Phi}_k^{(n,n)}$ *is the ℓ_μ-place or multiplace homomorphism, i.e.,*

$$k = k_{\min} = \ell_\mu, \quad \ell_{\text{id}} = 0. \tag{2.2.22}$$

(ii) $2 < n' < n \Longrightarrow \mathbf{\Phi}_k^{(n,n')}$ *is the intermediate heteromorphism with*

$$k = \frac{n-1}{n'-1}\ell_\mu, \tag{2.2.23}$$

*In this case the number of intact elements is proportional to the number of multiplications ℓ_μ in the lhs of (*2.2.13*)*

$$\ell_{\text{id}} = \frac{n-n'}{n'-1}\ell_\mu = \frac{n-n'}{n-1}k. \tag{2.2.24}$$

(iii) $n' = n'_{\min} = 2 \Longrightarrow \mathbf{\Phi}_k^{(n,2)}$ *is the $(n-1)\ell_\mu$-place (multiplace)* binarizing heteromorphism, *i.e.,*

$$k = k_{\max} = (n-1)\ell_\mu, \quad \ell_{\text{id}} = \frac{n-2}{n-1}k. \tag{2.2.25}$$

***Remark* 2.2.13.** In the first (2.2.22) and third (2.2.25) cases there are no restrictions on the initial arity n, while in the intermediate case n is 'quantized' due to the fact that fractions in (2.2.16) and (2.2.17) should be integer.

In the intermediate case, we obtain table 2.1 for the series of n and n' (we list only first ones, just for $2 \leqslant k \leqslant 4$ and include the binarizing case $n' = 2$ for completeness).

Thus, we established a general arity structure and classification of heteromorphisms of polyadic magmas defined by (2.2.13). The next important problem is considering the preservation of special properties (associativity, commutativity, solvability, etc) while passing from $\mu^{(n)}$ to $\mu'^{(n')}$, which means mappings of polyadics semigroups, groups, etc. This will further restrict the concrete artity shape of the main heteromorphism equation (2.2.13) for each choice of the heteromorphism parameters: arities n, n', number of places k, number of intacts ℓ_{id} and number of multiplications ℓ_μ in the lhs of (2.2.13).

Table 2.1. Heteromorphism 'quantization'.

k	ℓ_μ	ℓ_{id}	n/n'
2	1	1	$n = 3, 5, 7, \ldots$ $n' = 2, 3, 4, \ldots$
3	1	2	$n = 4, 7, 10, \ldots$ $n' = 2, 3, 4, \ldots$
3	2	1	$n = 4, 7, 10, \ldots$ $n' = 3, 5, 7, \ldots$
4	1	3	$n = 5, 9, 13, \ldots$ $n' = 2, 3, 4, \ldots$
4	2	2	$n = 3, 5, 7, \ldots$ $n' = 2, 3, 4, \ldots$
4	3	1	$n = 5, 9, 13, \ldots$ $n' = 4, 7, 10, \ldots$

2.2.2 Heteromorphisms and associativity quivers

We consider here the concrete form of semigroup heteromorphisms. In general, this is a complicated task, because it is not clear from (2.2.13), which permutation in the lhs should be taken to get an associative product in its rhs for each set of the heteromorphism parameters. Direct checking of the associativity of the final operation $\mu'^{(n')}$ for each permutation in the lhs of (2.2.13) is almost impossible, especially for higher n. To solve this difficulty we introduce the concept of the associative polyadic quiver and special rules to construct the associative final operation $\mu'^{(n')}$.

Definition 2.2.14. A *polyadic quiver of products* is the set of elements from $\mathcal{G}^{(n)}$ (presented as several copies of some matrix of the elements glued together) and arrows, such that the elements along arrows form n-ary products $\mu^{(n)}$.

For instance, for the 4-ary multiplication $\mu^{(4)}[g_1, h_2, g_2, u_1]$ (elements from $\mathcal{G}^{(n)}$ are arbitrary here) a corresponding 4-adic quiver will be denoted by $\{g_1 \to h_2 \to g_2 \to u_1\}$.

Definition 2.2.15. We define *polyadic quivers related to the main heteromorphism equation* (2.2.13):

(i) The matrix of elements is the transposed matrix from the rhs of (2.2.13), such that different letters correspond to their place in $\mathbf{\Phi}_k^{(n,n')}$ and the low index of an element is related to its position in the $\mu'^{(n')}$ product;
(ii) The number of polyadic quivers is ℓ_μ, which corresponds to ℓ_μ multiplications in the lhs of (2.2.13);
(iii) The heteromorphism parameters (n, n', k, ℓ_{id} and ℓ_μ) are not arbitrary, but satisfy the arity changing formulas (2.2.16)–(2.2.17);
(iv) The intact elements will be placed after a semicolon.

Obviously, that not all polyadic quivers preserve associativity.

Definition 2.2.16. An *associative polyadic quiver* is a polyadic quiver which ensures the final associativity of $\mu'^{(n')}$ in the main heteromorphism equation (2.2.13), when the initial multiplication $\mu^{(n)}$ is associative.

We propose a classification of associative polyadic quivers and the rules of construction of the corresponding heteromorphism equations, and then use the heteromorphism parameters for the classification of ℓ_μ-ple heteromorphisms (2.2.23). In other words, we describe a consistent procedure for building the *semigroup heteromorphisms*.

Consider the first class of heteromorphisms (without intact elements $\ell_{\text{id}} = 0$ or *intactless*), that is ℓ_μ-place (multiplace) homomorphisms. In the simplest case, associativity can be achieved, when all elements in a product are taken from the same row. The number of places k is not fixed by the arity relation (2.2.16) and can be arbitrary, while the arrows can have various directions. There are 2^k such combinations which preserve associativity.

If the arrows have the same vertical direction, this kind of mapping is also called a homomorphism. As an example, for $n = n' = 3$, $k = 2$, $\ell_\mu = 2$ we have

$$\mathbf{\Phi}_2^{(3,3)}\begin{pmatrix}\mu^{(3)}[g_1, g_2, g_3]\\ \mu^{(3)}[h_1, h_2, h_3]\end{pmatrix} = \mu'^{(3)}\left[\mathbf{\Phi}_2^{(3,3)}\begin{pmatrix}g_1\\ h_1\end{pmatrix}, \mathbf{\Phi}_2^{(3,3)}\begin{pmatrix}g_2\\ h_2\end{pmatrix}, \mathbf{\Phi}_2^{(3,3)}\begin{pmatrix}g_3\\ h_3\end{pmatrix}\right]. \tag{2.2.26}$$

Note that the analogous quiver with opposite arrow directions is

$$\mathbf{\Phi}_2^{(3,3)}\begin{pmatrix}\mu^{(3)}[g_1, g_2, g_3]\\ \mu^{(3)}[h_3, h_2, h_1]\end{pmatrix} = \mu'^{(3)}\left[\mathbf{\Phi}_2^{(3,3)}\begin{pmatrix}g_1\\ h_1\end{pmatrix}, \mathbf{\Phi}_2^{(3,3)}\begin{pmatrix}g_2\\ h_2\end{pmatrix}, \mathbf{\Phi}_2^{(3,3)}\begin{pmatrix}g_3\\ h_3\end{pmatrix}\right], \tag{2.2.27}$$

and it was used in constructing the middle representations of ternary groups (Borowiec *et al* 2006).

The main rule for nonvertical quivers is: all arrows of an associative quiver should have direction from left to right or vertical, and they should not intersect. We start always from the upper left corner, because of the permutation symmetry of (2.2.13), we can rearrange and rename variables in the necessary way.

Definition 2.2.17. The *Post-like associative quiver* is an important class of intactless heteromorphisms (with $\ell_{\text{id}} = 0$) preserving associativity can be constructed using an analogy with the Post substitutions (Post 1940).

The number of places k is now fixed by $k = n - 1$, while $n' = n$ and $\ell_\mu = k = n - 1$. An example of the Post-like associative quiver with the same heteromorphisms parameters as in (2.2.26)–(2.2.27) is

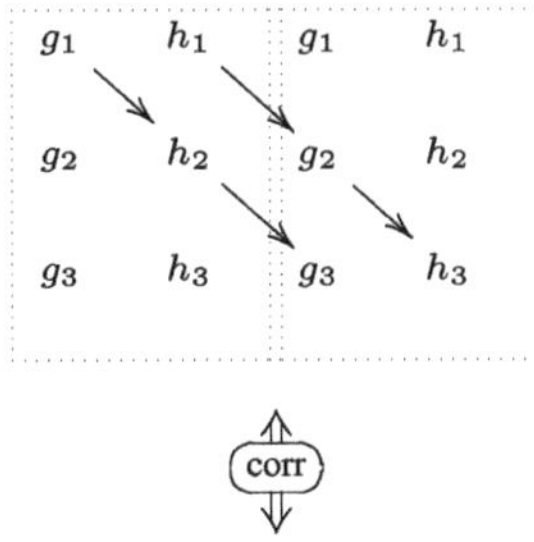

$$\mathbf{\Phi}_2^{(3,3)}\begin{pmatrix}\mu^{(3)}[g_1,h_2,g_3]\\ \mu^{(3)}[h_1,g_2,h_3]\end{pmatrix} = \mu'_3\left[\mathbf{\Phi}_2^{(3,3)}\begin{pmatrix}g_1\\ h_1\end{pmatrix}, \mathbf{\Phi}_2^{(3,3)}\begin{pmatrix}g_2\\ h_2\end{pmatrix}, \mathbf{\Phi}_2^{(3,3)}\begin{pmatrix}g_3\\ h_3\end{pmatrix}\right]. \quad (2.2.28)$$

***Remark* 2.2.18.** The equiary mapping (2.2.28) appeared in the study of ternary semigroups of morphisms (Chronowski 1994a, 1994b, Chronowski and Novotný 1995). Its n-ary generalization was used in the consideration of polyadic operations on Cartesian powers (Gal'mak 2008), polyadic analogs of the Cayley and Birkhoff theorems (Gal'mak 2001, 2007) and special representations of n-groups (Gleichgewicht 1983, Wanke-Jakubowska and Wanke-Jakubowska 1984) (where the n-group with the multiplication μ'_2 was called the *diagonal n-group*).

***Example* 2.2.19.** Let Λ be the Grassmann algebra consisting of even and odd parts $\Lambda = \Lambda_{\text{even}} \oplus \Lambda_{\text{odd}}$ (see e.g., Berezin 1987). The odd part can be considered as a ternary semigroup $\mathcal{G}_{\text{odd}}^{(3)} = \langle \Lambda_{\text{odd}} | \mu^{(3)} \rangle$, its multiplication $\mu^{(3)}\colon \Lambda_{\text{odd}} \times \Lambda_{\text{odd}} \times \Lambda_{\text{odd}} \to \Lambda_{\text{odd}}$ is defined by $\mu^{(3)}[\alpha, \beta, \gamma] = \alpha \cdot \beta \cdot \gamma$, where $(\cdot)$ is multiplication in Λ and $\alpha, \beta, \gamma \in \Lambda_{\text{odd}}$, so $\mathcal{G}_{\text{odd}}^{(3)}$ is nonderived and contains no unity. The even part can be treated as a ternary group $\mathcal{G}_{\text{even}}^{(3)} = \langle \Lambda_{\text{even}} | \mu'^{(3)} \rangle$ with the multiplication $\mu'^{(3)}\colon \Lambda_{\text{even}} \times \Lambda_{\text{even}} \times \Lambda_{\text{even}} \to \Lambda_{\text{even}}$, defined by $\mu^{(3)}[a, b, c] = a \cdot b \cdot c$, where $a, b, c \in \Lambda_{\text{even}}$, thus $\mathcal{G}_{\text{even}}^{(3)}$ is derived and contains unity. We introduce the heteromorphism $\mathcal{G}_{\text{odd}}^{(3)} \to \mathcal{G}_{\text{even}}^{(3)}$ as a mapping (2-place homomorphism) $\mathbf{\Phi}_2^{(3,3)}\colon \Lambda_{\text{odd}} \times \Lambda_{\text{odd}} \to \Lambda_{\text{even}}$ by the formula

$$\mathbf{\Phi}_2^{(3,3)}\begin{pmatrix}\alpha\\ \beta\end{pmatrix} = \alpha \cdot \beta, \quad (2.2.29)$$

where $\alpha, \beta \in \Lambda_{\text{odd}}$. It is seen that $\mathbf{\Phi}_2^{(3,3)}$ defined by (2.2.29) satisfies the Post-like heteromorphism equation (2.2.28), but not the 'vertical' one (2.2.26), due to the anticommutativity of odd elements from Λ_{odd}. In other words, $\mathcal{G}_{\text{even}}^{(3)}$ can be treated as a nontrivial example of the 'diagonal' semigroup of $\mathcal{G}_{\text{odd}}^{(3)}$ (according to the notation of Gleichgewicht 1983, Wanke-Jakubowska and Wanke-Jakubowska 1984). We conclude that the semigroup $\mathcal{G}_{\text{even}}^{(3)}$ can be the hetero-covering of the semigroup $\mathcal{G}_{\text{odd}}^{(3)}$ by definition 2.3.5.

For the number of places $k \geqslant 3$ there exist additional (to the above) associative quivers having the same heteromorphism parameters.

***Example* 2.2.20.** When $n' = n = 4$ and $k = 3$ we have the Post-like associative quiver

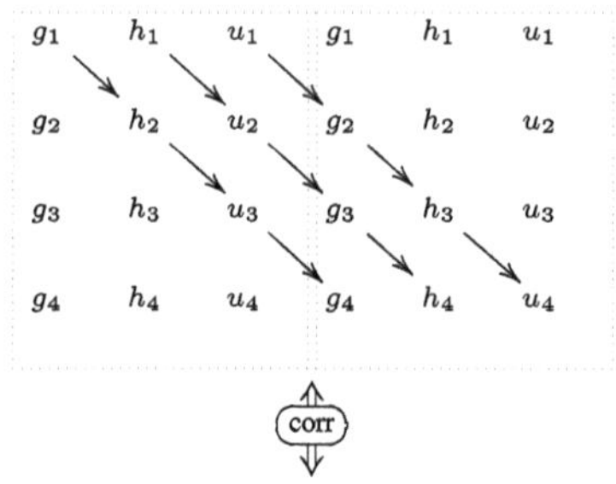

$$\Phi_3^{(4,4)}\begin{pmatrix}\mu^{(4)}\left[g_1,h_2,u_3,g_4\right]\\ \mu^{(4)}\left[h_1,u_2,g_3,h_4\right]\\ \mu^{(4)}\left[u_1,g_2,h_3,u_4\right]\end{pmatrix} = \mu'^{(4)}\left[\Phi_3^{(4,4)}\begin{pmatrix}g_1\\ h_1\\ u_1\end{pmatrix},\Phi_3^{(4,4)}\begin{pmatrix}g_2\\ h_2\\ u_2\end{pmatrix},\Phi_3^{(4,4)}\begin{pmatrix}g_3\\ h_3\\ u_3\end{pmatrix},\Phi_3^{(4,4)}\begin{pmatrix}g_4\\ h_4\\ u_4\end{pmatrix}\right]. \tag{2.2.30}$$

***Example* 2.2.21.** We have one intermediate non-Post associative quiver

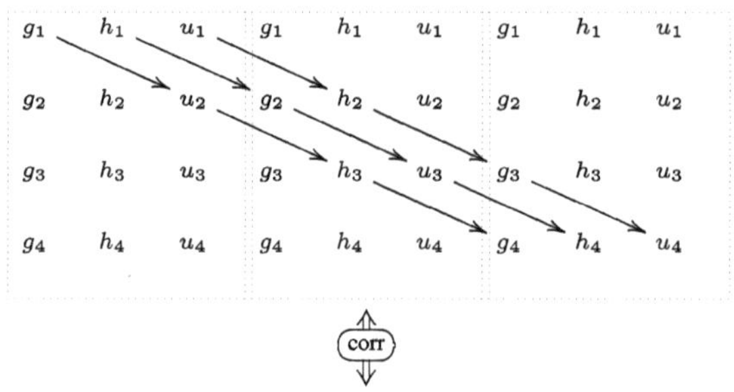

$$\Phi_3^{(4,4)}\begin{pmatrix}\mu^{(4)}\left[g_1,u_2,h_3,g_4\right]\\ \mu^{(4)}\left[h_1,g_2,u_3,h_4\right]\\ \mu^{(4)}\left[u_1,h_2,g_3,u_4\right]\end{pmatrix} = \mu'^{(4)}\left[\Phi_3^{(4,4)}\begin{pmatrix}g_1\\ h_1\\ u_1\end{pmatrix},\Phi_3^{(4,4)}\begin{pmatrix}g_2\\ h_2\\ u_2\end{pmatrix},\Phi_3^{(4,4)}\begin{pmatrix}g_3\\ h_3\\ u_3\end{pmatrix},\Phi_3^{(4,4)}\begin{pmatrix}g_4\\ h_4\\ u_4\end{pmatrix}\right]. \tag{2.2.31}$$

The next type of heteromorphisms (intermediate) is described by the equations (2.2.14)–(2.2.18), it contains intact elements ($\ell_{\mathrm{id}} \geqslant 1$) and changes (decreases) arity $n' < n$. For each fixed k the arities are not arbitrary and presented in table 2.1. The first general rule is: the associative quivers are nondecreasing in both, vertical (from up to down) and horizontal (from left to right), directions. Second, if there are several multiplications ($\ell_\mu \geqslant 2$), the corresponding associative quivers do not intersect.

***Example* 2.2.22.** For $k = 2$, the first (nonbinarizing $n' \geqslant 3$) case is $n = 5$, $n' = 3$, $\ell_{\mathrm{id}} = 1$ (see the first row and second n/n' pair of table 2.1). The corresponding associative quivers are

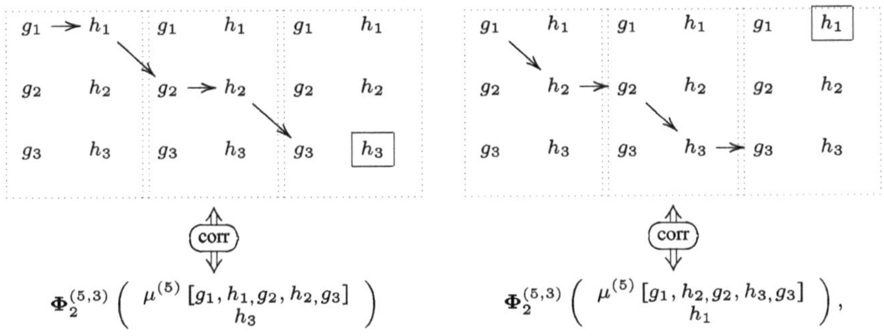

(2.2.32)

where we do not write the rhs of the heteromorphism equation (2.2.13), because it is simply related to the transposed quiver matrix of the size $n' \times k$.

The above examples are sufficient to understand the general rules of the associative quiver construction, such that the polyadic algebraic system associativity with respect to heteromorphisms preserves, which will allow to provide further investigation of semigroup heteromorphisms.

2.3 Hetero-covering of algebraic structures

The heteromorphism concept can be treated as an another approach which connects one-set one-operation polyadic algebraic structures of different arities, but already starting from the magma level. Previously, this connection was found for polyadic groups by direct artificial and cumbersome constructions of the polyadic product $\mu^{(n)}$ in terms of the binary multiplication $\mu'^{(2)}$ using ordinary (1-place) homomorphisms, that is the Hosszú–Gluskin theorem (Hosszú 1963, Gluskin 1965) (see section 4.1) and the Post theorems (Post 1940) (in Gal'mak and Vorobiev 2013, Gal'mak *et al* 2014 it was shown that they coincide). For semigroups the Hosszú–Gluskin theorem was considered in Zupnik (1967) and Pop (1979), and for n'-ary ($(k+1)$-ary) final multiplication $\mu'^{(n')}$ in Michalski (1981, 1987) and Dudek and Michalski (1982, 1984) (for polyadic semigroups, Pop 2002; and monoids, Dékov 1982 and Pop and Pop 2004) (see, also, Głazek and Gleichgewicht 1982, Rusakov 1998, Ûsan 2003).

Recall, that the standard covering n'-ary algebraic structure $\langle G' \mid \mu'^{(n')} \rangle$ for the n-ary structure $\langle G \mid \mu^{(n)} \rangle$ ($(k+1)$-ary and $(n+1)$-ary in Michalski 1981, 1987, respectively) is given by the 1-place mapping ψ: $G \to G'$ (in our notation)

$$\psi(\mu^{(n)}[g_1, \ldots, g_n]) = \left(\mu'^{(n')}\right)^{\circ \ell'_\mu}[\psi(g_1), \ldots, \psi(g_{n'})], \quad g_i \in G, \tag{2.3.1}$$

$$n - 1 = \ell'_\mu(n' - 1), \tag{2.3.2}$$

which differs from the homomorphism (2.1.4) by the long (iterated) ℓ'_μ-product in the rhs (ℓ'_μ is the number of the final n'-ary multiplications), such that $\psi(G)$ is the generating set of G', and (2.3.2) follows from the length of the iterated (long) product

(1.1.4). The noniterated product $\ell'_\mu = 1$ leads to $n' = n$ and the polyadic structure homomorphism formula (2.1.4). In the case $n' = 2$, (2.3.1) becomes the binary covering (1.4.1) (Dörnte 1929, Post 1940). If moreover $n = 2$, then $\ell'_\mu = 1$ and (2.3.1) is the ordinary binary homomorphism formula $\psi(g_1 \cdot g_2) = \psi(g_1) * \psi(g_2)$, where $\mu^{(2)} = (\cdot)$ and $\mu'^{(2)} = (*)$.

The direct generalization of the Hosszú–Gluskin–Post construction to the higher $n' \geqslant 3$ using the 1-place mapping ψ and (2.3.1) was considered in Michalski (1981, 1987) and Dudek and Michalski (1982, 1984). Here we propose another general concept of covering for polyadic magmas (Hausmann–Ore groupoids, see remark 1.1.3) and semigroups (using the associative quivers, see section 2.2.2), by introducing a 'hetero'-analog of (2.3.1).

Definition 2.3.1. A *long k-place heteromorphism* from $\mathcal{G}^{(n)}$ to $\mathcal{G}'^{(n')}$ is given, if there exists a k-place mapping $\mathbf{\Psi}^{(n,n')}_{k,\ell'_\mu}$ (2.2.3) such that the corresponding *long heteromorphism equation* is

$$\mathbf{\Psi}^{(n,n')}_{k,\ell'_\mu}\left(\begin{array}{l} \left.\begin{array}{c} \mu^{(n)}[g_1, \ldots, g_n], \\ \vdots \\ \mu^{(n)}\left[g_{n(\ell_\mu-1)}, \ldots, g_{n\ell_\mu}\right] \end{array}\right\}\ell_\mu \\ \left.\begin{array}{l} g_{n\ell_\mu+1}, \\ \vdots \\ g_{n\ell_\mu+\ell_{\mathrm{id}}} \end{array}\right\}\ell_{\mathrm{id}} \end{array}\right)$$
$$= \left(\mu'^{(n')}\right)^{\circ \ell'_\mu}\left[\overbrace{\mathbf{\Psi}^{(n,n')}_{k,\ell'_\mu}\begin{pmatrix} g_1 \\ \vdots \\ g_k \end{pmatrix}, \ldots, \mathbf{\Psi}^{(n,n')}_{k,\ell'_\mu}\begin{pmatrix} g_{k\ell'_\mu(n'-1)} \\ \vdots \\ g_{k(\ell'_\mu(n'-1)+1)} \end{pmatrix}}^{\ell'_\mu(n'-1)+1}\right]. \tag{2.3.3}$$

The system of equations connecting initial and final arities becomes (see (2.2.14)–(2.2.15))

$$k\left(\ell'_\mu(n' - 1) + 1\right) = n\ell_\mu + \ell_{\mathrm{id}}, \tag{2.3.4}$$

$$k = \ell_\mu + \ell_{\mathrm{id}}. \tag{2.3.5}$$

Excluding ℓ_{id} from (2.3.4), we obtain

Definition 2.3.2. The *arity changing formula* for the long heteromorphism is

$$n - 1 = \frac{k}{\ell_\mu} \ell'_\mu (n' - 1). \tag{2.3.6}$$

The inequalities for the parameters are the same as in (2.2.18)–(2.2.21). In table 2.2 we present the dependence of allowed initial and final arities from the long heteromorphism parameters k, ℓ_μ, ℓ_{id}, ℓ'_μ.

Definition 2.3.3. An n'-ary magma $\mathcal{G}'^{(n')} = \langle G' | \mu'^{(n')} \rangle$ is a *hetero-covering* of n-ary magma $\mathcal{G}^{(n)} = \langle G | \mu^{(n)} \rangle$, if there exists a k-place long heteromorphism $\Psi^{(n,n')}_{k,\ell'_\mu}: G^{\times k} \to G'$ which satisfies (2.3.3). In this case the n-ary magma $\mathcal{G}^{(n)}$ is called *hetero-derived* from $\mathcal{G}'^{(n')}$.

Table 2.2. Long heteromorphism 'quantization'.

k	ℓ_μ	ℓ_{id}	$\ell'_\mu / n / n'$
2	1	1	$\ell'_\mu =$ 1, 1, 2, 2, 3, 3, 4, 4, ...
			$n =$ 3, 5, 5, 9, 13, 19, 9 17, ...
			$n' =$ 2, 3, 2, 3, 3, 4, 2, 3, ...
3	1	2	$\ell'_\mu =$ 1, 1, 2, 2, 3, 3, 4, 5, ...
			$n =$ 4, 13, 7, 13, 10, 19, 13 16, ...
			$n' =$ 2, 5, 2, 3, 2, 3, 2, 2, ...
3	2	1	$\ell'_\mu =$ 1, 1, 2, 2, 3, 3, 4, 4, ...
			$n =$ 4, 10, 4, 10, 10, 19, 7 13, ...
			$n' =$ 3, 7, 2, 4, 3, 5, 2, 3, ...
4	1	3	$\ell'_\mu =$ 1, 1, 2, 2, 3, 3, 4, 4, ...
			$n =$ 5, 9, 9, 17, 13, 25, 17 33, ...
			$n' =$ 2, 3, 2, 3, 2, 3, 2, 3, ...
4	2	2	$\ell'_\mu =$ 1, 1, 2, 2, 3, 3, 4, 4, ...
			$n =$ 3, 5, 5, 9, 7, 13, 9 17, ...
			$n' =$ 2, 3, 2, 3, 2, 3, 2, 3, ...
4	3	1	$\ell'_\mu =$ 1, 1, 2, 2, 3, 3, 4, 4, ...
			$n =$ 5, 9, 9, 17, 5, 9, 17 33, ...
			$n' =$ 4, 7, 4, 7, 2, 3, 4, 7, ...

***Remark* 2.3.4.** The arity changing formula (2.3.6) contains the additional (as compared with the Hosszú–Gluskin–Post approach (2.3.2)) possibilities of non-integer multiplier $\frac{k}{\ell_\mu}\ell'_\mu$. This means that it becomes integer, when $\ell_\mu \big| k\ell'_\mu$, as well as for the limiting cases $k = 1$ (1-place homomorphism) and for the equiary case $n' = n$, $\ell'_\mu = 1$ and $\ell_\mu = k$.

By analogy with definition 2.3.3, we propose

Definition 2.3.5. An n'-ary semigroup $\mathcal{G}'^{(n')} = \langle G' | \mu'^{(n')} \rangle$ can be a *hetero-covering* of n-ary semigroup $\mathcal{G}^{(n)} = \langle G | \mu^{(n)} \rangle$, if there exists a k-place heteromorphism $\boldsymbol{\Psi}^{(n,n')}_{k,\ell'_\mu}\colon G^{\times k} \to G'$ (2.3.3) which is defined by the associative polyadic quiver. The n-ary semigroup $\mathcal{G}^{(n)}$ is called *hetero-derived* from $\mathcal{G}'^{(n')}$.

To obtain free (universal) hetero-covering semigroups one should introduce corresponding equivalence relations and factorize by them in the standard way (Michalski 1981).

2.4 Multiplace representations of polyadic algebraic structures

Representation theory deals with mappings from abstract algebraic structures into linear algebraic structures, such as, e.g., linear operators in vector spaces, or into general (semi)groups of transformations of some set (see, e.g., Kirillov 1976, Fulton and Harris 1991). In our notation, this means that in the mapping of polyadic algebraic structures (2.1.3) the final multiplication $\mu'^{(n')}$ is a linear map, which leads to some restrictions on the final polyadic structure $\mathcal{G}'^{(n')}$.

Let $\mathcal{V} = \left\langle V \middle| \mu_V \right\rangle$ be a vector space over a field $\Bbbk$ (usually algebraically closed) and $\operatorname{End} V$ be a set of linear endomorphisms of V, which is a binary group $\langle \operatorname{End} V | * \rangle$. In the standard way, a linear representation of a binary semigroup $\mathcal{G}^{(2)} = \langle G | \mu^{(2)} \rangle$ is a (1-place) map $\boldsymbol{\Pi}_1\colon \mathcal{G}^{(2)} \to \operatorname{End} V$, such that $\boldsymbol{\Pi}_1$ is a homomorphism

$$\boldsymbol{\Pi}_1(\mu^{(2)}[g, h]) = \boldsymbol{\Pi}_1(g) * \boldsymbol{\Pi}_1(h), \tag{2.4.1}$$

where $g, h \in G$ and $(*)$ is the binary multiplication in $\operatorname{End} V$ (usually, it is a (semi)group with multiplication as composition of operators or product of matrices, if a basis is chosen). If $\mathcal{G}^{(2)}$ is a binary group with the unity e, then we have the additional condition

$$\boldsymbol{\Pi}_1(e) = \mathrm{id}_V. \tag{2.4.2}$$

Let us generalize these well-known formulas to the corresponding polyadic algebraic structures along with the heteromorphism concept introduced in section 2.2.1. Our general idea is to use the heteromorphism equation (2.2.13) instead of the standard homomorphism equation (2.4.1), such that the arity of the representation will be different from the arity of the initial polyadic algebraic structure $n' \neq n$.

Consider the structure of the final n'-ary multiplication $\mu'^{(n')}$ in (2.2.13), taking into account that the final polyadic algebraic structure $\mathcal{G}'^{(n')}$ should be constructed from $\mathrm{End}V$. The most natural and physically applicable way is to consider the binary $\mathrm{End}V$ and to put the final algebraic structure to be derived $\mathcal{G}'^{(n')} = \mathrm{der}_{n'}(\mathrm{End}V)$, as it was proposed for the ternary case in Borowiec *et al* (2006). In this way $\mathcal{G}'^{(n')}$ becomes a derived n'-ary (semi)group of endomorphisms of V with the multiplication $\mu'^{(n')}$: $(\mathrm{End}V)^{\times n'} \to \mathrm{End}V$, where

$$\mu'^{(n')}[v_1, \ldots, v_{n'}] = v_1 * \ldots * v_{n'}, \quad v_i \in \mathrm{End}V. \tag{2.4.3}$$

The multiplication $\mu'^{(n')}$ (2.4.3) is obviously derived and is therefore associative by definition, we may consider the associative initial polyadic algebraic structures (semigroups and groups) and the associativity preserving mappings that are the special heteromorphisms constructed in section 2.2.1.

Let $\mathcal{G}^{(n)} = \langle G | \mu^{(n)} \rangle$ be an associative n-ary polyadic algebraic structure. By analogy with (2.2.3), we introduce the following k-place mapping

$$\mathbf{\Pi}_k^{(n,n')}: G^{\times k} \to \mathrm{End}V. \tag{2.4.4}$$

Definition 2.4.1. A *multiplace representation* of an associative one-set polyadic algebraic structure $\mathcal{G}^{(n)}$ in a vector space $\mathcal{V}$ is given, if there exists a k-place mapping (2.4.4) which satisfies the (associativity preserving) heteromorphism equation (2.2.13), that is

$$\mathbf{\Pi}_k^{(n,n')}\left(\begin{array}{c} \left.\begin{array}{c} \mu^{(n)}[g_1, \ldots, g_n], \\ \vdots \\ \mu^{(n)}\left[g_{n(\ell_\mu - 1)}, \ldots, g_{n\ell_\mu}\right] \end{array}\right\} \ell_\mu \\ \left.\begin{array}{c} g_{n\ell_\mu+1}, \\ \vdots \\ g_{n\ell_\mu+\ell_{\mathrm{id}}} \end{array}\right\} \ell_{\mathrm{id}} \end{array}\right) = \overbrace{\mathbf{\Pi}_k^{(n,n')}\begin{pmatrix} g_1 \\ \vdots \\ g_k \end{pmatrix} * \ldots * \mathbf{\Pi}_k^{(n,n')}\begin{pmatrix} g_{k(n'-1)} \\ \vdots \\ g_{kn'} \end{pmatrix}}^{n'}, \tag{2.4.5}$$

and the following diagram commutes

$$\begin{array}{ccc} G^{\times k} & \xrightarrow{\Pi_k} & \mathrm{End}\,V \\ \Big\uparrow {\scriptstyle \mu_n^{(\ell_\mu, \ell_{\mathrm{id}})}} & & \Big\uparrow {\scriptstyle (*)^{n'}} \\ G^{\times kn'} & \xrightarrow{(\Pi_k)^{\times n'}} & (\mathrm{End}\,V)^{\times n'} \end{array} \tag{2.4.6}$$

where $\mu_n^{(\ell_\mu, \ell_{\mathrm{id}})}$ is given by (2.2.10), ℓ_μ and ℓ_{id} are the numbers of multiplications and intact elements in the lhs of (2.4.5), respectively.

The exact permutation in the lhs of (2.4.5) is given by the associative quiver presented in the previous section. The representation parameters (n, n', k, ℓ_μ *and* ℓ_{id}) in (2.4.5) are the same as the heteromorphism parameters, and they satisfy the same arity changing formulas (2.2.16) and (2.2.17). Therefore, a general classification of multiplace representations can be done by analogy with that of the heteromorphisms (2.2.22)–(2.2.25) as follows:

(i) The *hom-like multiplace representation* which is a multiplace homomorphism with $n' = n'_{\max} = n$, without intact elements $l_{\mathrm{id}} = l_{\mathrm{id}}^{(\min)} = 0$, and minimal number of places

$$k = k_{\min} = \ell_\mu. \tag{2.4.7}$$

(ii) The *intact element multiplace representation* which is the intermediate heteromorphism with $2 < n' < n$ and the number of intact elements is

$$l_{\mathrm{id}} = \frac{n - n'}{n' - 1}\ell_\mu. \tag{2.4.8}$$

(iii) The *binary multiplace representation* which is a binarizing heteromorphism (2.2.25) with $n' = n'_{\min} = 2$, the maximal number of intact elements $l_{\mathrm{id}}^{(\max)} = (n - 2)\ell_\mu$ and maximal number of places

$$k = k_{\max} = (n - 1)\ell_\mu. \tag{2.4.9}$$

In case of n-ary semigroups the multiplace representations have no additional defining relations, as compared with (2.4.5). In case of n-ary groups, we need an analog of the 'normalizing' relation (2.4.2). If an n-ary group has the unity e, then one can put

$$\mathbf{\Pi}_k^{(n,n')}\left(\left.\begin{matrix} e \\ \vdots \\ e \end{matrix}\right\}k\right) = \mathrm{id}_V. \tag{2.4.10}$$

If there is no unity at all, one can 'normalize' the multiplace representation, using analogy with (2.4.2) in the form

$$\mathbf{\Pi}_1(h^{-1} * h) = \mathrm{id}_V, \tag{2.4.11}$$

as follows

$$\mathbf{\Pi}_k^{(n,n')}\left(\begin{matrix} \left.\begin{matrix} \bar{h} \\ \vdots \\ \bar{h} \end{matrix}\right\}\ell_\mu \\ \left.\begin{matrix} h \\ \vdots \\ h \end{matrix}\right\}\ell_{\mathrm{id}} \end{matrix}\right) = \mathrm{id}_V, \tag{2.4.12}$$

for all $h \in \mathcal{G}^{(n)}$, where $\bar{h}$ is the querelement of h. The latter ones can be placed on any (allowed) places in the lhs of (2.4.12) due to the Dörnte identities. Also, the multiplications in the lhs of (2.4.5) can change their place due to the same reason.

A general form of multiplace representations can be found by applying the Dörnte identities to each n-ary product in the lhs of (2.4.5). Then, using (2.4.12) we have schematically

$$\mathbf{\Pi}_k^{(n,n')}\begin{pmatrix} g_1 \\ \vdots \\ g_k \end{pmatrix} = \mathbf{\Pi}_k^{(n,n')}\begin{pmatrix} t_1 \\ \vdots \\ t_{\ell_\mu} \\ \left.\begin{matrix} g \\ \vdots \\ g \end{matrix}\right\} \ell_{\mathrm{id}} \end{pmatrix}, \tag{2.4.13}$$

where g is an arbitrary fixed element of the n-ary group and

$$t_a = \mu^{(n)}[g_{a1}, \ldots, g_{an-1}, \bar{g}], \quad a = 1, \ldots, \ell_\mu. \tag{2.4.14}$$

This is the general shape of multiplace representations, while the concrete formulas should be obtained in each case separately. Nevertheless, some conclusions can be drawn from (2.4.13). Firstly, the equivalence classes on which $\mathbf{\Pi}_k^{(n,n')}$ is constant are determined by fixing $\ell_\mu + 1$ elements, i.e., by the surface $t_a = \mathrm{const}$, $g = \mathrm{const}$. Secondly, some k-place representations of a n-ary group can be reduced to ℓ_μ-place representations of its retract. In the case $\ell_\mu = 1$, multiplace representations of a n-ary group derived from a binary group correspond to ordinary representations of the latter (see Borowiec *et al* 2006, Dudek 2007).

***Example* 2.4.2.** Let us consider the case of a binary multiplace representation $\mathbf{\Pi}_{2n-2}^{(n,2)}$ of n-ary group $\mathcal{G}^{(n)} = \langle G | \mu^{(n)} \rangle$ with two multiplications $\ell_\mu = 2$ defined by the associativity preserving equation

$$\mathbf{\Pi}_{2n-2}^{(n,2)}\begin{pmatrix} \mu^{(n)}[g_1, u_1, \ldots u_{n-2}, g_2] \\ u_1' \\ \vdots \\ u_{n-2}' \\ \mu^{(n)}[h_1, v_1, \ldots v_{n-2}, h_2] \\ v_1' \\ \vdots \\ v_{n-2}' \end{pmatrix} = \mathbf{\Pi}_{2n-2}^{(n,2)}\begin{pmatrix} g_1 \\ u_1 \\ \vdots \\ u_{n-2} \\ h_1 \\ v_1 \\ \vdots \\ v_{n-2} \end{pmatrix} * \mathbf{\Pi}_{2n-2}^{(n,2)}\begin{pmatrix} g_2 \\ u_1' \\ \vdots \\ u_{n-2}' \\ h_2 \\ v_1' \\ \vdots \\ v_{n-2}' \end{pmatrix} \tag{2.4.15}$$

and normalizing condition

$$\mathbf{\Pi}_{2n-2}^{(n,2)}\begin{pmatrix} \left.\begin{matrix} h \\ \vdots \\ h \end{matrix}\right\} n-2 \\ \bar{h} \\ \left.\begin{matrix} h \\ \vdots \\ h \end{matrix}\right\} n-2 \\ \bar{h} \end{pmatrix} = \mathrm{id}_V, \tag{2.4.16}$$

where $h \in \mathcal{G}^{(n)}$ is arbitrary. Using (2.4.15) and (2.4.16) for a general form of this $(2n-2)$-place representation we have

$$\mathbf{\Pi}^{(n,2)}_{2n-2} = \mathbf{\Pi}^{(n,2)}_{2n-2}\begin{pmatrix} g_1\\ u_1\\ \vdots\\ u_{n-2}\\ h_1\\ v_1\\ \vdots\\ v_{n-2}\end{pmatrix} = \mathbf{\Pi}^{(n,2)}_{2n-2}\begin{pmatrix} g_1\\ u_1\\ \vdots\\ u_{n-2}\\ h_1\\ v_1\\ \vdots\\ v_{n-2}\end{pmatrix} * \mathbf{\Pi}^{(n,2)}_{2n-2}\begin{pmatrix} \left.\begin{matrix}h\\ \vdots\\ h\end{matrix}\right\} n-2\\ \bar{h}\\ \left.\begin{matrix}h\\ \vdots\\ h\end{matrix}\right\} n-2\\ \bar{h}\end{pmatrix}$$

$$= \mathbf{\Pi}^{(n,2)}_{2n-2}\begin{pmatrix} \mu^{(n)}[g_1, u_1, \ldots u_{n-2}, h]\\ \left.\begin{matrix}h\\ \vdots\\ h\end{matrix}\right\} n-3\\ \bar{h}\\ \mu^{(n)}[h_1, v_1, \ldots v_{n-2}, h]\\ \left.\begin{matrix}h\\ \vdots\\ h\end{matrix}\right\} n-3\\ \bar{h}\end{pmatrix}. \tag{2.4.17}$$

The Dörnte identity applied to the first elements of the products in the rhs of (2.4.17) together with associativity of $\mu^{(n)}$ gives

$$\mathbf{\Pi}^{(n,2)}_{2n-2} = \mathbf{\Pi}^{(n,2)}_{2n-2}\begin{pmatrix} \mu^{(n)}\Big[\overbrace{g, \ldots, g}^{n-2}, t_g, h\Big]\\ \left.\begin{matrix}h\\ \vdots\\ h\end{matrix}\right\} n-3\\ \bar{h}\\ \mu^{(n)}\Big[\overbrace{g, \ldots, g}^{n-2}, t_h, h\Big]\\ \left.\begin{matrix}h\\ \vdots\\ h\end{matrix}\right\} n-3\\ \bar{h}\end{pmatrix} = \mathbf{\Pi}^{(n,2)}_{2n-2}\begin{pmatrix} \left.\begin{matrix}g\\ \vdots\\ g\end{matrix}\right\} n-2\\ t_g\\ \left.\begin{matrix}g\\ \vdots\\ g\end{matrix}\right\} n-2\\ t_h\end{pmatrix} * \mathbf{\Pi}^{(n,2)}_{2n-2}\begin{pmatrix} \left.\begin{matrix}h\\ \vdots\\ h\end{matrix}\right\} n-2\\ \bar{h}\\ \left.\begin{matrix}h\\ \vdots\\ h\end{matrix}\right\} n-2\\ \bar{h}\end{pmatrix}$$

$$= \mathbf{\Pi}^{(n,2)}_{2n-2}\begin{pmatrix} \left.\begin{matrix}g\\ \vdots\\ g\end{matrix}\right\} n-2\\ t_g\\ \left.\begin{matrix}g\\ \vdots\\ g\end{matrix}\right\} n-2\\ t_h\end{pmatrix}, \tag{2.4.18}$$

where

$$t_g = \mu^{(n)}[\bar{g}, g_1, u_1, \ldots u_{n-2}], \quad t_h = \mu^{(n)}[\bar{g}, h_1, v_1, \ldots v_{n-2}]. \tag{2.4.19}$$

Thus, the equivalent classes of the multiplace representation $\mathbf{\Pi}_{2n-2}^{(n,2)}$ (2.4.15) are determined by the $(\ell_\mu + 1 = 3)$-element surface

$$t_g = \text{const}, \; t_h = \text{const}, \; g = \text{const}. \tag{2.4.20}$$

Note that the 'ℓ_μ-place reduction' of a multiplace representation is possible not for all associativity preserving heteromorphism equations. For instance, if to exchange $v_i \leftrightarrow v_i'$ in lhs of (2.4.15), then the associativity remains, but the 'ℓ_μ-place reduction', analogous to (2.4.18) will not be possible.

In case, when it is possible, the corresponding ℓ_μ-place representation can be realized on the binary retract of $\mathcal{G}^{(n)}$ (for some special n-ary groups and $\ell_\mu = 1$ see see Borowiec *et al* 2006, Dudek 2007). Indeed, let $\mathcal{G}_{\text{ret}}^{(2)} = \langle G, \circledast \rangle = \text{ret}_g \langle G | \mu^{(n)} \rangle$, where $g_1 \circledast g_2 \stackrel{\text{def}}{=} \mu^{(n)}\left[g_1, \overbrace{g, \ldots, \bar{g}}^{n\ 2}, g_2\right]$, $g_1, g_2, g \in G$ and we define the reduced ℓ_μ-place representation (now $\ell_\mu = 2$) through (2.4.18) as follows

$$\mathbf{\Pi}_2^{\text{retg}}\begin{pmatrix} g_1 \\ g_2 \end{pmatrix} \stackrel{\text{def}}{=} \mathbf{\Pi}_{2n-2}^{(n,2)}\begin{pmatrix} \left.\begin{matrix} g \\ \vdots \\ g \end{matrix}\right\} n-2 \\ g_1 \\ \left.\begin{matrix} g \\ \vdots \\ g \end{matrix}\right\} n-2 \\ g_2 \end{pmatrix}. \tag{2.4.21}$$

Due to our classification $\mathbf{\Pi}_2^{\text{retg}}\begin{pmatrix} g_1 \\ g_2 \end{pmatrix}$ is a hom-like 2-place (binary) representation.

The above formulas describe various properties of multiplace representations, but they give no idea of how to build representations for concrete polyadic algebraic structures. The most common method of representation construction uses the concept of a group action on a set (see, e.g., Kirillov 1976). Below we extend this concept to the multiplace case and use corresponding heteromorphisms, as it was done above for homomorphisms and representations.

2.5 *k*-Place actions

2.5.1 *k*-Place actions and *G*-spaces

Let $\mathcal{G}^{(n)} = \langle G | \mu^{(n)} \rangle$ be a polyadic algebraic structure and X be a set. A (left) 1-place action of $\mathcal{G}^{(n)}$ on X is the external binary operation $\rho_1^{(n)}: G \times X \to X$ such that it is consistent with the multiplication $\mu^{(n)}$, i.e., composition of the binary operations $\rho_1\{g|\mathsf{x}\}$ gives the n-ary product, that is,

$$\rho_1^{(n)}\{\mu^{(n)}[g_1, \ldots g_n]|\mathsf{x}\} = \rho_1^{(n)}\left\{g_1\middle|\rho_1^{(n)}\left\{g_2\middle|\ldots\middle|\rho_1^{(n)}\{g_n|\mathsf{x}\}\right\}\ldots\right\}, \qquad g_1, \ldots, g_n \in G, \mathsf{x} \in X. \tag{2.5.1}$$

If the polyadic algebraic structure is an n-ary group, then in addition to (2.5.1) it is implied that there exists such $e_x \in G$ (which may or may not coincide with the unity of $\mathcal{G}^{(n)}$) that $\rho_1^{(n)}\{e_x|\mathsf{x}\} = \mathsf{x}$ for all $\mathsf{x} \in X$, and the mapping $\mathsf{x} \mapsto \rho_1^{(n)}\{e_x|\mathsf{x}\}$ is a bijection of X. The right 1-place actions of $\mathcal{G}^{(n)}$ on X are defined in a symmetric way, and therefore we will consider below only one of them. Obviously, we cannot compose $\rho_1^{(n)}$ and $\rho_1^{(n')}$ with $n \neq n'$. Usually X is called a G-set or G-space depending on its properties (see, e.g., Husemöller *et al* 2008).

The application of the 1-place action defined by (2.5.1) to the representation theory of n-ary groups gave mostly repetitions of the ordinary (binary) group representation results (except for trivial b-derived ternary groups) (Dudek and Shahryari 2012). Also, it is obviously seen that the construction (2.5.1) with the binary external operation ρ_1 cannot be applied for studying the most important regular representations of polyadic algebraic structures, when the X coincides with the underlying set G of the group $\mathcal{G}^{(n)}$ itself and the action arises from translations.

Now we introduce the multiplace concept of action for polyadic algebraic structures, which is consistent with heteromorphisms and multiplace representations. Then we will show how it naturally appears when $X = G$ of $\mathcal{G}^{(n)}$ and apply it to construct examples of representations including the regular ones.

Definition 2.5.1. For a polyadic algebraic structure $\mathcal{G}^{(n)} = \langle G|\mu^{(n)}\rangle$ and a set X we introduce an external polyadic operation

$$\rho_k\colon G^{\times k} \times X \to X, \tag{2.5.2}$$

which is called a (left) *k-place action* or *multiaction.*

To generalize the 1-action composition (2.5.1), we use the analogy with multiplication laws of the heteromorphisms (2.2.13) and the multiplace representations (2.4.5) and propose (schematically)

$$\rho_k^{(n)}\left\{\left.\begin{matrix}\left.\begin{matrix}\mu^{(n)}\left[g_1, \ldots, g_n\right], \\ \vdots \\ \mu^{(n)}\left[g_{n(\ell_\mu-1)}, \ldots, g_{n\ell_\mu}\right]\end{matrix}\right\}\ell_\mu \\ \left.\begin{matrix}g_{n\ell_\mu+1}, \\ \vdots \\ g_{n\ell_\mu+\ell_{\text{id}}}\end{matrix}\right\}\ell_{\text{id}}\end{matrix}\right|\mathsf{x}\right\} = \overbrace{\rho_k^{(n)}\left\{\begin{matrix}g_1 \\ \vdots \\ g_k\end{matrix}\middle|\ldots\middle|\rho_k^{(n)}\left\{\begin{matrix}g_{k(n'-1)} \\ \vdots \\ g_{kn'}\end{matrix}\middle|\mathsf{x}\right\}\ldots\right\}}^{n'}. \tag{2.5.3}$$

Note that the connection between all the parameters here is the same as in the arity changing formulas (2.2.16)–(2.2.17). Composition of mappings is associative, and therefore in concrete cases we can use the associative quiver technique, as it is described in the previous sections. If $\mathcal{G}^{(n)}$ is an n-ary group, then we should add to (2.5.3) the 'normalizing' relations analogous with (2.4.10) or (2.4.12). So, if there is a unity $e \in \mathcal{G}^{(n)}$, then

$$\rho_k^{(n)}\left\{\begin{matrix} e \\ \vdots \\ e \end{matrix} \middle| \mathsf{x} \right\} = \mathsf{x}, \quad \text{for all } \mathsf{x} \in X. \tag{2.5.4}$$

In terms of the querelement, the normalization has the form

$$\rho_k^{(n)}\left\{\begin{matrix} \left.\begin{matrix} \bar{h} \\ \vdots \\ \bar{h} \end{matrix}\right\} \ell_\mu \\ \left.\begin{matrix} h \\ \vdots \\ h \end{matrix}\right\} \ell_{\mathrm{id}} \end{matrix} \middle| \mathsf{x} \right\} = \mathsf{x}, \quad \text{for all } \mathsf{x} \in X \quad \text{and for all } h \in \mathcal{G}^{(n)}. \tag{2.5.5}$$

Definition 2.5.2. The multiaction $\rho_k^{(n)}$ is *transitive*, if any two points x and y in X can be 'connected' by $\rho_k^{(n)}$, i.e., there exist $g_1, \ldots, g_k \in G_n$ such that

$$\rho_k^{(n)}\left\{\begin{matrix} g_1 \\ \vdots \\ g_k \end{matrix} \middle| \mathsf{x} \right\} = \mathsf{y}. \tag{2.5.6}$$

If $g_1, \ldots, g_k$ are unique, then $\rho_k^{(n)}$ is *sharply transitive*. The subset of X, in which any points are connected by (2.5.6) with fixed $g_1, \ldots, g_k$ can be called the *multiorbit* of X. If there is only one multiorbit, then we call X the *heterogenous* G-space (by analogy with the homogeneous one). By analogy with the (ordinary) 1-place actions, we define a G-equivariant map $\mathbf{\Psi}$ between two G-sets X and Y by (in our notation)

$$\mathbf{\Psi}\left(\rho_k^{(n)}\left\{\begin{matrix} g_1 \\ \vdots \\ g_k \end{matrix} \middle| \mathsf{x} \right\}\right) = \rho_k^{(n)}\left\{\begin{matrix} g_1 \\ \vdots \\ g_k \end{matrix} \middle| \mathbf{\Psi}(\mathsf{x}) \right\} \in Y, \quad \mathsf{x} \in X \tag{2.5.7}$$

which makes G-space into a category (for details, see, e.g., Husemöller *et al* 2008). In the particular case, when X is a vector space over $\Bbbk$, the multiaction (2.5.2) can be called a *multi-G-module* which satisfies (2.5.4) and the additional (linearity) conditions

$$\rho_k^{(n)}\left\{\begin{matrix} g_1 \\ \vdots \\ g_k \end{matrix}\middle| a\mathsf{x} + b\mathsf{y}\right\} = a\rho_k^{(n)}\left\{\begin{matrix} g_1 \\ \vdots \\ g_k \end{matrix}\middle| \mathsf{x}\right\} + b\rho_k^{(n)}\left\{\begin{matrix} g_1 \\ \vdots \\ g_k \end{matrix}\middle| \mathsf{y}\right\}, \ \mathsf{x}, \mathsf{y} \in X, \tag{2.5.8}$$

where $a, b \in \Bbbk$. Comparing (2.4.5) and (2.5.3) we can define a multiplace representation as a multi-G-module by the following formula

$$\mathbf{\Pi}_k^{(n,n')}\begin{pmatrix} g_1 \\ \vdots \\ g_k \end{pmatrix}(\mathsf{x}) = \rho_k^{(n)}\left\{\begin{matrix} g_1 \\ \vdots \\ g_k \end{matrix}\middle| \mathsf{x}\right\}. \tag{2.5.9}$$

In a similar way, one can generalize to polyadic algebraic structures many other notions from group action theory (Kirillov 1976).

2.5.2 Regular k-place actions

The most important role in the study of polyadic algebraic structures is played by the case when $X = \mathcal{G}^{(n)}$, and the multiaction coincides with the n-ary analog of translations (Mal'tcev 1954), so-called i-translations (Belousov 1972). In the binary case, ordinary translations lead to regular representations (Kirillov 1976), and therefore we call such an action a *regular multiaction* (*k -place action*) $\rho_k^{\mathrm{reg}(n)}$. In this connection, the analog of the Cayley theorem for n-ary groups was obtained in Gal'mak (1986, 2001). Now we will show in examples how the regular multiactions can arise from i-translations.

***Example* 2.5.3.** Let $\mathcal{G}^{(3)}$ be a ternary semigroup, $k = 2$, and $X = G$ from $\mathcal{G}^{(3)}$, then 2-place (left) action can be defined as

$$\rho_2^{\mathrm{reg}(3)}\left\{\begin{matrix} g \\ h \end{matrix}\middle| u\right\} \stackrel{\mathrm{def}}{=} \mu^{(3)}[g, h, u]. \tag{2.5.10}$$

This gives the following composition law for two regular multiactions

$$\begin{aligned} \rho_2^{\mathrm{reg}(3)}\left\{\begin{matrix} g_1 \\ h_1 \end{matrix}\middle| \rho_2^{\mathrm{reg}(3)}\left\{\begin{matrix} g_2 \\ h_2 \end{matrix}\middle| u\right\}\right\} &= \mu^{(3)}[g_1, h_1, \mu^{(3)}[g_2, h_2, u]] \\ &= \mu^{(3)}[\mu^{(3)}[g_1, h_1, g_2], h_2, u] \\ &= \rho_2^{\mathrm{reg}(3)}\left\{\begin{matrix} \mu^{(3)}[g_1, h_1, g_2] \\ h_2 \end{matrix}\middle| u\right\}. \end{aligned} \tag{2.5.11}$$

Thus, using the regular 2-action (2.5.10) we have, in fact, derived the associative quiver corresponding to (2.2.5).

The formula (2.5.10) can be simultaneously treated as a 2-translation (Belousov 1972). In this way, the following *left regular multiaction*

$$\rho_k^{\mathrm{reg}(n)}\left\{\begin{matrix} g_1 \\ \vdots \\ g_k \end{matrix} \middle| h \right\} \stackrel{\mathrm{def}}{=} \mu^{(n)}[g_1, \ldots, g_k, h], \tag{2.5.12}$$

corresponds to (2.2.9), where in the rhs there is the i-translation with $i = n$. The *right regular multiaction* corresponds to the i-translation with $i = 1$. The binary composition of the left regular multiactions corresponds to (2.2.9). In general, the value of i fixes the minimal final arity n'_{reg}, which differs for even and odd values of the initial arity n.

Corollary 2.5.4. It follows from (2.5.12) that for regular multiactions the number of places is fixed

$$k_{\mathrm{reg}} = n - 1, \tag{2.5.13}$$

and the arity changing formulas (2.2.16)–(2.2.17) become

$$n'_{\mathrm{reg}} = n - \ell_{\mathrm{id}} \tag{2.5.14}$$

$$n'_{\mathrm{reg}} = \ell_\mu + 1. \tag{2.5.15}$$

From (2.5.14)–(2.5.15) we conclude that for any n a regular multiaction having one multiplication $\ell_\mu = 1$ is binarizing and has $n - 2$ intact elements. For $n = 3$ see (2.5.11). Also, it follows from (2.5.14) that for regular multiactions the number of intact elements gives exactly the difference between initial and final arities.

***Remark* 2.5.5.** In case the initial arity is odd, then there exists a special *middle regular multiaction* generated by the i-translation with $i = (n + 1)/2$. For $n = 3$ the corresponding associative quiver is (2.2.27) and such 2-actions were used in Borowiec *et al* (2006) to construct middle representations of ternary groups, which did not change arity ($n' = n$).

Now we give a more complicated example of a middle regular multiaction, which can contain intact elements and can therefore change arity.

***Example* 2.5.6.** Let us consider a 5-ary semigroup and the following middle 4-action

$$\rho_4^{\mathrm{reg}(5)}\left\{\begin{matrix} g \\ h \\ u \\ v \end{matrix} \middle| s \right\} = \mu^{(5)}\left[g, h, \overset{\overset{i=3}{\downarrow}}{s}, u, v\right]. \tag{2.5.16}$$

Using (2.5.15) we observe that there are two possibilities for the number of multiplications $\ell_\mu = 2, 4$. The last case $\ell_\mu = 4$ is similar to the vertical associative quiver (2.2.27), but with a more complicated lhs, that is

$$\rho_4^{\mathrm{reg}(5)}\left\{\begin{array}{c} \mu^{(5)}[g_1, h_1, g_2, h_2 g_3] \\ \mu^{(5)}[h_3, g_4, h_4, g_5, h_5] \\ \mu^{(5)}[u_5, v_5, u_4, v_4, u_3] \\ \mu^{(5)}[v_3, u_2, v_2, u_1, v_1] \end{array}\middle| s\right\} =$$
$$\rho_4^{\mathrm{reg}(5)}\left\{\begin{array}{c} g_1 \\ h_1 \\ u_1 \\ v_1 \end{array}\middle| \rho_4^{\mathrm{reg}(5)}\left\{\begin{array}{c} g_2 \\ h_2 \\ u_2 \\ v_2 \end{array}\middle| \rho_4^{\mathrm{reg}(5)}\left\{\begin{array}{c} g_3 \\ h_3 \\ u_3 \\ v_3 \end{array}\middle| \rho_4^{\mathrm{reg}(5)}\left\{\begin{array}{c} g_4 \\ h_4 \\ u_4 \\ v_4 \end{array}\middle| \rho_4^{\mathrm{reg}(5)}\left\{\begin{array}{c} g_5 \\ h_5 \\ u_5 \\ v_5 \end{array}\middle| s\right\}\right\}\right\}\right\}\right\}. \tag{2.5.17}$$

Now we have an additional case with two intact elements ℓ_{id} and two multiplications $\ell_\mu = 2$ as

$$\rho_4^{\mathrm{reg}(5)}\left\{\begin{array}{c} \mu^{(5)}[g_1, h_1, g_2, h_2 g_3] \\ h_3 \\ \mu^{(5)}[h_3, v_3, u_2, v_2, u_1] \\ v_1 \end{array}\middle| s\right\} = \rho_4^{\mathrm{reg}(5)}\left\{\begin{array}{c} g_1 \\ h_1 \\ u_1 \\ v_1 \end{array}\middle| \rho_4^{\mathrm{reg}(5)}\left\{\begin{array}{c} g_2 \\ h_2 \\ u_2 \\ v_2 \end{array}\middle| \rho_4^{\mathrm{reg}(5)}\left\{\begin{array}{c} g_3 \\ h_3 \\ u_3 \\ v_3 \end{array}\middle| s\right\}\right\}\right\}, \tag{2.5.18}$$

with arity changing from $n = 5$ to $n'_{\mathrm{reg}} = 3$. In addition to (2.5.18) we have 3 more possible regular multiactions due to the associativity of $\mu^{(5)}$, when the multiplication brackets in the sequences of 6 elements in the first two rows and the second two ones can be shifted independently.

***Remark* 2.5.7.** For $n > 3$, in addition to left, right and middle multiactions, there exist intermediate cases. First, observe that the i-translations with $i = 2$ and $i = n - 1$ immediately fix the final arity $n'_{\mathrm{reg}} = n$. Therefore, the composition of multiactions will be similar to (2.5.17), but with some permutations in the lhs.

Consider some multiplace analogs of regular representations of binary groups (Kirillov 1976). The straightforward generalization is to consider the previously introduced regular multiactions (2.5.12) in the rhs of (2.5.9). Let $\mathcal{G}^{(n)}$ be a finite polyadic associative algebraic structure and $\Bbbk\mathcal{G}^{(n)}$ be a vector space spanned by $\mathcal{G}^{(n)}$ (some properties of n-ary group rings were considered in Zeković and Artamonov 1999, 2002). This means that any element of $\Bbbk\mathcal{G}^{(n)}$ can be uniquely presented in the form $w = \sum_l a_l \cdot h_l$, $a_l \in \Bbbk$, $h_l \in G$. Then, using (2.5.12) and (2.5.9) we define the *i-regular k-place representation* by

$$\mathbf{\Pi}_k^{\mathrm{reg}(i)}\begin{pmatrix} g_1 \\ \vdots \\ g_k \end{pmatrix}(w) = \sum_l a_l \cdot \mu_{k+1}\left[g_1 \ldots g_{i-1} h_l g_{i+1} \ldots g_k\right]. \tag{2.5.19}$$

Comparing (2.5.12) and (2.5.19) one can conclude that all the general properties of multiplace regular representations are similar to those of the regular multiactions.

Definition 2.5.8. If $i = 1$ or $i = k$, the multiplace representation is called a *right* or *left regular representation* respectively. If k is even, the representation with $i = k/2 + 1$ is called a *middle regular representation.*

For ternary examples of right, left and middle regular representations, see Borowiec *et al* (2006).

References

Belousov V D 1972 *N-ary Quasigroups* (Kishinev: Shtiintsa)

Berezin F A 1987 *Introduction to Superanalysis* (Dordrecht: Reidel)

Borowiec A, Dudek W and Duplij S 2006 Bi-element representations of ternary groups *Commun. Algebra* **34** 1651–70

Chronowski A 1994a On ternary semigroups of homomorphisms of ordered sets *Arch. Math. (Brno)* **30** 85–95

Chronowski A 1994b A ternary semigroup of mappings *Demonstr. Math.* **27** 781–91

Chronowski A and Novotný M 1995 Ternary semigroups of morphisms of objects in categories *Arch. Math. (Brno)* **31** 147–53

Chung K O and Smith J D H 2008 Weak homomorphisms and graph coalgebras *Arab. J. Sci. Eng.* **33** 107–21

Collins M J 1990 *Representations and Characters of Finite Groups* (Cambridge: Cambridge University Press)

Cornwell J F 1997 *Group Theory in Physics: An Introduction* (London: Academic)

Curtis C W and Reiner I 1962 *Representation Theory of Finite Groups and Associative Algebras* (Providence, RI: American Mathematical Society)

Czákány B 1962 On the equivalence of certain classes of algebraic systems *Acta Sci. Math. (Szeged)* **23** 46–57

Dékov D V 1982 On the analog of Post theorem for polyadic monoids *Proc. of Int. Symp. on* n-*Ary Structures* (Skopje: Macedonian Academy of Sciences and Arts) pp 171–3

Denecke K and Saengsura K 2008 State-based systems and (F_1, F_2)-coalgebras *East-West J. Math.* **Special vol** 1–31

Denecke K and Wismath S L 2009 *Universal Algebra and Coalgebra* (Singapore: World Scientific)

Dörnte W 1929 Unterschungen über einen verallgemeinerten Gruppenbegriff *Math. Z.* **29** 1–19

Dudek W A 2007 Remarks to Głazek's results on n-ary groups *Discuss. Math. Gen. Algebra Appl.* **27** 199–233

Dudek W A and Michalski J 1982 On a generalization of Hosszú theorem *Demonstr. Math.* **15** 437–41

Dudek W A and Michalski J 1984 On retracts of polyadic groups *Demonstr. Math.* **17** 281–301

Dudek W A and Shahryari M 2012 Representation theory of polyadic groups *Algebr. Represent. Theory* **15** 29–51

Ellerman D 2006 A theory of adjoint functors—with some thoughts about their philosophical significance *What is Category Theory?* ed G Sica (Monza: Polimetrica) pp 127–83

Ellerman D 2007 Adjoints and emergence: applications of a new theory of adjoint functors *Axiomathes* **17** 19–39
Fujiwara T 1959 On mappings between algebraic systems *Osaka Math. J.* **11** 153–72
Fulton W and Harris J 1991 *Representation Theory: A First Course* (New York: Springer)
Gal'mak A M 1986 Translations of n-ary groups *Dokl. Akad. Nauk BSSR* **30** 677–80
Gal'mak A M 2001 Polyadic analogs of the Cayley and Birkhoff theorems *Russ. Math.* **45** 10–5
Gal'mak A M 2007 n-*Ary Groups, Part 2* (Minsk: Belarus State University)
Gal'mak A M 2008 Polyadic associative operations on Cartesian powers *Ser. Phys.-Math. Sci.* 3 28–34
Gal'mak A M, Balan V and Vorobiev G N 2014 On Post–Gluskin–Hosszu theorem *Appl. Sci.* **16** 11–22
Gal'mak A M and Vorobiev G N 2013 On Post-Gluskin-Hosszu theorem *Probl. Fiz. Mat. Tekh.* **2011** 55–60
Głazek K 1980 Weak homomorphisms of general algebras and related topics *Math. Semin. Notes Kobe Univ.* **8** 1–36
Głazek K and Gleichgewicht B 1982 Bibliography of n-groups (polyadic groups) and some group like n-ary systems *Proc. of the Symp. on* n-*Ary Structures* (Skopje: Macedonian Academy of Sciences and Arts) pp 253–89
Głazek K and Michalski J 1974 On weak homomorphisms of general non-indexed algebras *Bull. Acad. Pol. Sci., Sér. Sci. Math. Astron. Phys.* **22** 651–6
Głazek K and Michalski J 1977 Weak homomorphisms of general algebras *Comment. Math.* **19 (1976)** 211–28
Gleichgewicht B, Wanke-Jakubowska M B and Wanke-Jakubowska M E 1983 On representations of cyclic n-groups *Demonstr. Math.* **16** 357–65
Gluskin L M 1965 Position operatives *Math. Sbornik* **68(110)** 444–72
Goetz A 1966 On weak isomorphisms and weak homomorphisms of abstract algebras *Colloq. Math.* **14** 163–7
Halaš R 1994 A note on homotopy in universal algebra *Acta Univ. Palacki. Olomuc., Fac. Rerum Nat. Math.* **33** 39–42
Hobby D and McKenzie R 1988 *The Structure of Finite Algebras* (Providence, RI: American Mathematical Society)
Hosszú M 1963 On the explicit form of n-group operation *Publ. Math. Debrecen* **10** 88–92
Husemöller D, Joachim M, Jurčo B and Schottenloher M 2008 G-spaces, G-bundles, and G-vector bundles *Basic Bundle Theory and K-Cohomology Invariants* ed D Husemöller and S Echterhoff (Berlin: Springer) pp 149–61
Kapranov M, Gelfand I M and Zelevinskii A 1994 *Discriminants, Resultants and Multidimensional Determinants* (Berlin: Birkhäuser)
Kawamura Y 2003 Cubic matrices, generalized spin algebra and uncertainty relation *Progr. Theor. Phys.* **110** 579–87
Kirillov A A 1976 *Elements of the Theory of Representations* (Berlin: Springer)
Kolibiar M 1984 Weak homomorphisms in some classes of algebras *Stud. Sci. Math. Hung.* **19** 413–20
Mal'tcev A I 1954 On general theory of algebraic systems *Mat. Sb.* **35** 3–20
Mal'tcev A I 1957 Free topological algebras *Izv. Akad. Nauk SSSR. Ser. Mat.* **21** 171–98
Mal'tsev A I 1958 The structural characteristic of some classes of algebras *Dokl. Akad. Nauk SSSR* **120** 29–32

Marczewski E 1966 Independence in abstract algebras. Results and problems *Colloq. Math.* **14** 169–88
Michalski J 1981 Covering k-groups of n-groups *Arch. Math. (Brno)* **17** 207–26
Michalski J 1987 Associated *k*-groups of *n*-groups *Zb. Rad. Prir.-Mat. Fak. Univ. Novom Sadu Ser. Mat.* **17** 9–30
Novotný M 1990 Mono-unary algebras in the work of Czechoslovak mathematicians *Arch. Math. (Brno)* **26** 155–64
Novotný M 2002 Homomorphisms of algebras *Czech. Math. J.* **52** 345–64
Pécsi B 2012 On Morita contexts in bicategories *Appl. Categor. Struct.* **20** 415–32
Petrescu A 1977 On the homotopy of universal algebras. I *Rev. Roum. Math. Pures Appl.* **22** 541–51
Pop M S 1979 On the reduction of *n*-groups to groups *Stud. Univ. Babes-Bolyai Math.* **24** 38–40
Pop M S 2002 Two functorial *k*-reductions of *n*-semigroups *Proc. of the Algebra Symp.* (Cluj: Babes-Bolyai University) pp 203–18
Pop M S and Pop A 2004 On some relations on *n*-monoids *Carpathian J. Math.* **20** 87–94
Post E L 1940 Polyadic groups *Trans. Am. Math. Soc.* **48** 208–350
Rausch de Traubenberg M 2008 Cubic extentions of the Poincaré algebra *Phys. Atom. Nucl.* **71** 1102–8
Rusakov S A 1998 *Some Applications of* n-Ary Group Theory (Minsk: Belaruskaya Navuka)
Sokolov N P 1972 *Introduction to the Theory of Multidimensional Matrices* (Kyiv: Naukova Dumka)
Traczyk T 1965 Weak isomorphisms of Boolean and Post algebras *Colloq. Math.* **13** 159–64
Ûsan J 2003 *n*-Groups in the light of the neutral operations *Math. Morav.* **Special vol 203**
Vidal J C and Tur J S 2010 A 2-categorial generalization of the concept of institution *Stud. Log.* **95** 301–44
Wanke-Jakubowska M B and Wanke-Jakubowska M E 1984 On representations of *n*-groups *Ann. Sci. Math. Pol.* **24** 335–41
Weyl H 1946 *Classical Groups, Their Invariants and Representations* (Princeton, NJ: Princeton University Press)
Zeković B and Artamonov V A 1999 A connection between some properties of *n*-group rings and group rings *Math. Montisnigri* **15** 151–8
Zeković B and Artamonov V A 2002 On two problems for n-group rings *Math. Montisnigri* **15** 79–85
Zupnik D 1967 Polyadic semigroups *Publ. Math. (Debrecen)* **14** 273–9

Chapter 3

Polyadic semigroups and higher regularity

The concept of regularity was introduced in von Neumann (1936) and then widely used in the construction of regular and inverse semigroups (see, e.g., Clifford and Preston 1961, Ljapin 1968, Howie 1976, Grillet 1995 and Lawson 1998, Petrich 1984, also refs therein).

In this chapter we propose to generalize the concept of regularity for semigroups in two different aspects simultaneously: (1) higher regularity, which can be informally interpreted that each element has several inverse elements; (2) higher arity, which extends the binary multiplication to that of arbitrary arity, i.e., the consideration of polyadic semigroups.

The higher regularity concept was introduced in semisupermanifold theory (Duplij 1998) for generalized transition functions, which then gave rise to the development of a new kind of so-called regular obstructed category (Duplij and Marcinek 2000) and to their application to Topological Quantum Field Theory (Duplij and Marcinek 2002), the Yang–Baxter equation (Duplij and Marcinek 2001a) and statistics with a doubly regular R-matrix (Duplij and Marcinek 2003). Moreover, it was shown that the higher regular semigroups naturally appear in the framework of the polyadic-binary correspondence principle (Duplij 2021a). Categorical aspects of regularity in the language of arrow flows were considered in Gilbert (2003) and Wazzan (2019). Regular and inverse polyadic semigroups were investigated in Sioson (1963), Slipenko (1977) and Kolesnikov (1979).

3.1 Generalized q-regular elements in semigroups

Here we formulate some basic notions in terms of tuples and the 'quantity/number of multiplications', in such a way that they can be naturally generalized from the binary to polyadic case (Duplij 2012).

doi:10.1088/978-0-7503-2648-3ch3

3.1.1 Binary q-regular single elements

If S is a set, its ℓth Cartesian product $S^{\times\ell} = \overbrace{S \times S \times \ldots \times S}^{\ell}$ consists of all ℓ-tuples $\mathbf{x}^{(\ell)} = (x_1, x_2, \ldots, x_\ell)$, $x_i \in S$ (we omit powers (ℓ) in the obvious cases $\mathbf{x}^{(\ell)} \equiv \mathbf{x}$, $\ell \in \mathbb{N}$). Let $\mathcal{S}_2 = \langle S | \mu^{(2)}, \mathrm{assoc}_2 \rangle$ be a binary (having the arity $n = 2$) semigroup with the underlying set S, and the (binary $n = 2$) multiplication $\mu^{(2)}$: $S \times S \to S$ (sometimes we will denote $\mu^{(2)}[x_1, x_2] = x_1 x_2 \in SS$). The (binary) associativity assoc_2: $(x_1 x_2) x_3 = x_1 (x_2 x_3)$, $\forall x_i \in S$, allows us to omit brackets, and in the language of ℓ-tuples $\mathbf{x}^{(\ell)}$ the product of ℓ elements $x_1 x_2 \ldots x_\ell = \mu_2^{[\ell_\mu]}[\mathbf{x}^{(\ell)}] = \mathfrak{x}^{(\ell)} \in \overbrace{SS \ldots S}^{\ell}$, where ℓ_μ is the 'number of multiplications' (for the binary multiplication $\ell_\mu = \ell - 1$, i.e., the product of ℓ elements contains $\ell - 1$ (binary) multiplications, whereas for polyadic (n-ary) multiplications this is different, Post 1940). If the number of elements is not important or evident, we will write for the product $\mathfrak{x}^{(\ell)} \equiv \mathfrak{x} \in \overbrace{SS \ldots S}^{\ell}$, and so distinguish between the product of ℓ elements $\mathfrak{x}^{(\ell)} \in \overbrace{SS \ldots S}^{\ell}$ and the ℓ-tuples $\mathbf{x}^{(\ell)} \in S^{\times\ell} = \overbrace{S \times S \times \ldots \times S}^{\ell}$. If all elements in a ℓ-tuple coincide, we write $x^\ell = \mu_2^{[\ell_\mu]}[x^{(\ell)}]$ (this form will be important in the polyadic generalization below). An element $x \in S$ satisfying $x^\ell = x$ (or $\mu_2^{[\ell_\mu]}[x^{(\ell)}] = x$) is called (binary) ℓ-idempotent.

An element x of the semigroup $\mathcal{S}_2$ is called (von Neumann) regular, if the equation

$$xyx = x, \quad \text{or} \tag{3.1.1}$$

$$(\mu^{(2)})^{\circ 2}[x, y, x] = x, \quad x, y \in S, \tag{3.1.2}$$

has a solution y in $\mathcal{S}_2$, which need not be unique. An element y is called an inverse element (Clifford and Preston 1961) (the superseded terms for y are: a reciprocal element, Thierrin 1952; a generalized inverse, Vagner 1952; a regular element conjugated to x, Ljapin 1968). The set of the inverse elements of x is denoted by V_x (Howie 1976).

In search of generalizing the regularity condition (3.1.1) in $\mathcal{S}_2$ for the fixed element $x \in S$, we arrive at the following possibilities:

(i) *Higher regularity generalization*: instead of one element $y \in S$, use an q-tuple $\mathbf{y}^{(q)} \in S^q$ and the corresponding product $\mathfrak{y}^{(q)} \in \overbrace{SS \ldots S}^{q}$.

(ii) *Higher arity generalization*: instead of the binary product $\mu^{(2)}$, consider the polyadic (or n-ary semigroup) product with the multiplication $\mu^{(n)}$: $\overbrace{S \times S \times \ldots \times S}^{n} \to S$.

It follows from the higher regularity generalization (i) applied to for (3.1.1)–(3.1.2)

Definition 3.1.1. An element x of the (binary) semigroup $\mathcal{S}_2$ is called *higher q-regular*, if there exists (at least one, not necessarily unique) $(q-1)$-element solution as $(q-1)$-tuple $\mathbf{y}^{(q-1)}$ of

$$x\mathbf{y}^{(q-1)}x = x, \quad \text{or} \tag{3.1.3}$$

$$\mu_{n=2}^{[q]}[x, \mathbf{y}^{(q-1)}, x] = x, \quad x, y_1, \dots, y_{q-1} \in S. \tag{3.1.4}$$

Definition 3.1.2. The $(q-1)$-tuple $\mathbf{y}^{(q-1)}$ from (3.1.4) is called a *higher q -inverse to* x, or $\mathbf{y}^{(q-1)}$ is an *q-inverse (tuple)* for x.

Without any additional conditions, for instance splitting the $(q-1)$-tuple $\mathbf{y}^{(q-1)}$ for some reason (as below), the definition (3.1.3) reduces to the ordinary regularity (3.1.1), because the binary semigroup is closed with respect to the multiplication of any number of elements, and so there exists $z \in S$, such that $\mathbf{y}^{(q-1)} = z$ for any $y_1, \dots, y_{q-1} \in S$. Therefore, we have the following reduction

Assertion 3.1.3. *Any higher* q*-regular element in a binary semigroup is 2-regular (or regular in the ordinary sense* (3.1.1)).

If we consider the higher 3-regularity condition $x_1x_2x_3x_1 = x_1$ for a single element x_1, and $x_2, x_3 \in S$, it obviously coincides with the ordinary regularity condition for the single element x (3.1.1). However, if we cycle both the regularity conditions, they will not necessarily coincide, and new structures can appear. Therefore, using several different mutually consistent q-regularity conditions (3.1.3) we will be able to construct a corresponding (binary) semigroup which will not be reduced to an ordinary regular semigroup (see below).

3.1.2 Polyadic q-regular single elements

Following the higher arity generalization (ii), we introduce higher q-regular single elements in polyadic (n-ary) semigroups and show that the definition of regularity for higher arities cannot be done similarly to ordinary regularity, but requires a polyadic analog of q-regularity for a single element of a polyadic semigroup. Before proceeding we introduce

Definition 3.1.4. (Arity invariance principle). In an algebraic universal system with operations of different arities the general form of expressions should not depend on numerical arity values.

For instance, take a binary product of 3 elements $(xy)z \equiv \mu^{(2)}[\mu^{(2)}[x, y], z]$, its ternary analog will not be $\mu^{(3)}[x, y, z]$, but the expression of the same operation structure $\mu^{(3)}[\mu^{(3)}[x, y, t], z, u]$ which contains 5 elements.

This gives the following prescription for how to generalize expressions from the binary shape to a polyadic shape:

(i) Write down an expression using the operations manifestly.
(ii) Change arities from the binary values to the needed higher values.
(iii) Take into account the corresponding changes of tuple lengths according to the concrete argument numbers of operations.

***Example* 3.1.5.** If on a set S one defines a binary operation (multiplication) $\mu^{(2)}: S \times S \to S$, then one (left) neutral element $e \in S$ for $x \in S$ is defined by $\mu^{(2)}[e, x] = x$ (usually $ex = x$). However, for 4-ary operation $\mu^{(4)}: S^4 \to S$ it becomes $\mu^{(4)}[e_1, e_2, e_3, x] = x$, where $e_1, e_2, e_3 \in S$, so we can have a neutral sequence $\mathbf{e} = \mathbf{e}^{(3)} = (e_1, e_2, e_3)$ (a tuple $\mathbf{e}$ of the length 3, triple), and only in the particular case $e_1 = e_2 = e_3 = e$, do we obtain one neutral element e. Obviously, the composition of two ($\ell_\mu = 2$) 4-ary ($n = 4$) operations, denoted by $(\mu^{(4)})^{\circ 2}$ (as in (3.1.2)) should act on a word of a length $\ell = 7$, since $\mu^{(4)}[\mu^{(4)}[x_1, x_2, x_3, x_4], x_5, x_6, x_7]$, $x_i \in S$.

In general, if a polyadic system $\langle S | \mu^{(n)} \rangle$ has only one operation $\mu^{(n)}$ (n-ary multiplication), the length of words is 'quantized'

$$L_n^{(\ell_\mu)} = \ell_\mu (n - 1) + 1, \tag{3.1.5}$$

where ℓ_μ is a 'number of multiplications'.

Let $\mathcal{S}^{(n)} = \langle S | \mu^{(n)}, \mathrm{assoc}_n \rangle$ be a *polyadic* (*n-ary*) *semigroup* with the underlying set S, and the n-ary multiplication $\mu^{(n)}: S^n \to S$ (Post 1940). The n-ary (*total*) *associativity* assoc_n can be treated as invariance of two n-ary products composition $(\mu^{(n)})^{\circ 2}$ with respect to the brackets (Duplij 2012)

$$\mathrm{assoc}_n: (\mu^{(n)})^{\circ 2}[\mathbf{x}, \mathbf{y}, \mathbf{z}] = \mu^{(n)}[\mathbf{x}, \mu^{(n)}[\mathbf{y}], \mathbf{z}] = \text{invariant}, \quad \forall x_i, y_i, z_i \in S, \tag{3.1.6}$$

for all placements of the inner multiplication $\mu^{(n)}$ in the rhs of (3.1.6) with the fixed ordering of $(2n - 1)$ elements (following from (3.1.5)), and $\mathbf{x}$, $\mathbf{y}$, $\mathbf{z}$ are tuples of the allowed by (3.1.5) lengths, such that the sum of their lengths is $(2n - 1)$. The total associativity (3.1.6) allows us to write a product containing ℓ_μ multiplications using external brackets only $\mu_n^{[\ell_\mu]}[\mathbf{x}]$ (the *long product*, Dörnte 1929), where $\mathbf{x}$ is a tuple of length $\ell_\mu(n - 1) + 1$, because of (3.1.5). If all elements of the ℓ-tuple $\mathbf{x} = \mathbf{x}^{(\ell)}$ are the same we write it as $x^{(\ell)}$. A *polyadic power* (reflecting the 'number of multiplications', but not of the number of elements, as in the binary case) becomes

$$x^{\langle \ell_\mu \rangle} = (\mu^{(n)})^{\circ \ell_\mu}[x^{\ell_\mu(n-1)+1}], \tag{3.1.7}$$

because of (3.1.5). So the ordinary (binary) power p (as a number of elements in a product x^p) is $p = \ell_\mu + 1$.

Definition 3.1.6. An element x from a polyadic semigroup $\mathcal{S}^{(n)}$ is called *generalized polyadic* $\langle \ell_\mu \rangle$*-idempotent*, if

$$x^{\langle \ell_\mu \rangle} = x, \tag{3.1.8}$$

and it is *polyadic idempotent*, if (Post 1940)

$$x^{\langle 1 \rangle} \equiv x^{\langle \ell_\mu = 1 \rangle} = \mu^{(n)}[x^{(n)}] = x. \tag{3.1.9}$$

In the binary case, (3.1.9) and (3.1.8) correspond to the ordinary idempotent $x^2 = x$ and to the *generalized p -idempotent* $x^p = x$, where $p = \ell_\mu + 1$.

Recall also that a *left polyadic identity* $e \in S$ is defined by

$$\mu^{(n)}[e^{(n-1)}, x] = x, \quad \forall x \in S, \tag{3.1.10}$$

and e is a *polyadic identity*, if x can be on any place. If $e = e(x)$ depends on x, then we call it a *local polyadic identity* (see Pop and Pop 2010 for the binary case), and if e depends on several elements of S, we call it a *generalized local polyadic identity*. It follows from (3.1.10) that all polyadic identities (or neutral elements) are polyadic idempotents (3.1.9).

An element x of a n-ary semigroup $\mathcal{S}^{(n)}$ is called *regular* (Kolesnikov 1979), if the equation (see (3.1.2))

$$(\mu^{(n)})^{\circ 2}[x, \mathbf{y}^{(2n-3)}, x] = x, \ \ x \in S, \tag{3.1.11}$$

has $2n - 3$ solutions $y_1, \ldots, y_{2n-3} \in S$, which need not be unique (since $L_n^{(2)} - 2 = 2n - 3$, see (3.1.5)). If we have only the relation (3.1.11), then polyadic associativity allows us to apply one (internal) polyadic multiplication and remove $n - 1$ elements to obtain (Kolesnikov 1979)

$$\mu^{(n)}[x, \mathbf{z}^{(n-2)}, x] = x, \ \ x \in S, \tag{3.1.12}$$

and we call (3.1.12) the *reduced polyadic regularity* of a single element, while (3.1.11) will be called the *full regularity* of a single element. Thus, for a single element $x \in S$, by analogy with assertion 3.1.3, we have

Assertion 3.1.7. *In a polyadic semigroup $\mathcal{S}^{(n)}$, if any single element $x \in S$ satisfies the full regularity (3.1.11), it also satisfies the reduced regularity* (3.1.12).

From (3.1.12) it follows

Corollary 3.1.8. Reduced regularity exists, if and only if the arity of multiplication is strongly more than binary, i.e., $n \geqslant 3$.

***Example* 3.1.9.** In the minimal ternary case $n = 3$ we have the regularity $(\mu^{(3)})^{\circ 2}[x_1, x_2, x_3, x_4, x_1] = x_1$, $x_i \in S$, and putting $z = \mu^{(3)}[x_2, x_3, x_4]$ gives the reduced regularity

$$\mu^{(3)}[x_1, z, x_1] = x_1. \tag{3.1.13}$$

However, these regularities will give different cycles: one is of length 5, and the other is of length 3.

We now construct a polyadic analog of the higher q-regularity condition (3.1.4) for a single element.

Definition 3.1.10. An element x of a polyadic semigroup $\mathcal{S}^{(n)}$ is called *full higher q-regular*, if there exists (at least one, not necessarily unique) $(q(n-1)-1)$-element solution (tuple) $\mathbf{y}^{(q(n-1)-1)}$ of

$$(\mu^{(n)})^{\circ q}[x, \mathbf{y}^{(q(n-1)-1)}, x] = x, \quad x, y_1, \ldots, y_{q(n-1)-1} \in S, \; q \geqslant 2. \tag{3.1.14}$$

Definition 3.1.11. The tuple $\mathbf{y}^{(q(n-1)-1)}$ from (3.1.14) is called *higher q-regular n-ary inverse* to x, or $\mathbf{y}^{(q(n-1)-1)}$ is *q-inverse n-ary tuple* for x of the length $(q(n-1)-1)$.

***Example* 3.1.12.** In the full higher 3-regular ternary semigroup $\mathcal{S}^{(3)} = \langle S | \mu^{(3)} \rangle$ we have for each element $x \in S$ the condition

$$(\mu^{(3)})^{\circ 2}[x, y_1, y_2, y_3, y_4, y_5, x] = x, \tag{3.1.15}$$

and so each element $x \in S$ has a 3-inverse ternary 5-tuple $\mathbf{y}^{(5)} = (y_1, y_2, y_3, y_4, y_5)$, or the 5-tuple $\mathbf{y}^{(5)}$ is the higher 3-regular ternary inverse of x.

Comparing (3.1.14) with the full regularity condition (3.1.11) and its reduction (3.1.12), we observe that the full q-regularity condition can be reduced several times to get different versions of reduced regularity for a single element.

Definition 3.1.13. An element x of a polyadic semigroup $\mathcal{S}^{(n)}$ is called *m-reduced higher q-regular*, if there exists a (not necessarily unique) $((q-m)(n-1)-1)$-element solution (tuple) $\mathbf{y}^{((q-m)(n-1)-1)}$ of

$$(\mu^{(n)})^{\circ q}[x, \mathbf{y}^{((q-m)(n-1)-1)}, x] = x, \quad x, y_1, \ldots, y_{(q-m)(n-1)-1} \in S, \; 1 \leqslant m \leqslant q-1, \; q \geqslant 2. \tag{3.1.16}$$

***Example* 3.1.14.** In the ternary case $n = 3$, we have (for a single element $x_1 \in S$) the higher 3-regularity condition

$$(\mu^{(3)})^{\circ 3}[x_1, x_2, x_3, x_4, x_5, x_6, x_1] = x_1, \quad x_i \in S. \tag{3.1.17}$$

We can reduce this relation twice, for instance, as follows: put $t = \mu^{(3)}[x_2, x_3, x_4]$ to get

$$(\mu^{(3)})^{\circ 2}[x_1, t, x_5, x_6, x_1] = x_1, \tag{3.1.18}$$

and then $z = \mu^{(3)}[t, x_5, x_6]$ and obtain

$$\mu^{(3)}[x_1, z, x_1] = x_1, \quad t, z \in S, \tag{3.1.19}$$

which coincides with (3.1.13). Again we observe that (3.1.17), (3.1.18) and (3.1.19) give different cycles of the length 7, 5 and 3, respectively.

3.2 Higher q-inverse semigroups

Here we introduce and study the higher q-regular and q-inverse (binary) semigroups. We recall the standard definitions to establish notations (see, e.g., Clifford and Preston 1961, Ljapin 1968, Grillet 1995).

There are two different definitions of a regular semigroup:

$$\text{(i)}\, One - relation\ definition \text{ (Green 1951).} \tag{3.2.1}$$

$$\text{(ii)}\, Multi - relation\ definition \text{ (Liber 1953; Vagner 1952).} \tag{3.2.2}$$

According to the one-relation definition (i) (3.2.1): A semigroup $\mathcal{S} \equiv \mathcal{S}_2 = \langle S | \mu^{(2)}, \text{assoc}_2 \rangle$ is called regular if each element $x \in S$ has its inverse element $y \in S$ defined by $xyx = x$ (3.1.1), and y need not be unique (Green 1951).

Elements x_1, x_2 of S are called inverse to each other (Clifford and Preston 1961) (or regular conjugated, Ljapin 1968, mutually regular), if

$$x_1 x_2 x_1 = x_1, \quad x_2 x_1 x_2 = x_2, \tag{3.2.3}$$

$$(\mu^{(2)})^{\circ 2}[x_1, x_2, x_1] = x_1, \quad (\mu^{(2)})^{\circ 2}[x_2, x_1, x_2] = x_2, \quad x_1, x_2 \in S. \tag{3.2.4}$$

According to the multi-relation definition (ii) (3.2.2): A semigroup is called regular, if each element x_1 has its inverse x_2 (3.2.3), and x_2 need not be unique.

For a binary semigroup $\mathcal{S}_2$ and ordinary 2-regularity these definitions are equivalent (Thierrin 1952, Clifford and Preston 1961). Indeed, if x_1 has some inverse element $y \in S$, and $x_1 y x_1 = x_1$, then choosing $x_2 = y x_1 y$, we immediately obtain (3.2.3).

Definition 3.2.1. A 2-tuple $\mathbf{X}^{(2)} = (x_1, x_2) \in S \times S$ is called a *2-regular sequence of inverses*, if it satisfies (3.2.3).

Thus we can define regular semigroups in terms of regular sequences of inverses.

Definition 3.2.2. A binary semigroup is called a *regular* (or 2-*regular*) *semigroup*, if each element belongs to a (not necessary unique) 2-regular sequence of inverses $\mathbf{X}^{(2)} \in S \times S$.

3.2.1 Higher q-regular semigroups

Following Sioson (1963), in a regular semigroup $\mathcal{S}_2$, for each $x_1 \in S$ there exists $x_2 \in S$ such that $x_1 x_2 x_1 = x_1$, and for $x_2 \in S$ there exists $y \in S$ satisfying and $x_2 y x_2 = x_2$. Denoting $x_3 = y x_2$, then $x_2 = x_2(y x_2) = x_2 x_3$, and we observe that $x_1 x_2 x_3 x_1 = x_1$, i.e., in a regular semigroup, from 2-regularity there follows 3-regularity of $\mathcal{S}_2$ in the one-relation definition (i) (cf for a single element assertion 3.1.3). By analogy, in a regular semigroup, 2-regularity of an element implies its q-regularity in the one-relation definition (i) (3.2.1). Indeed, in this point there is a difference between the definitions (i) (3.2.1) and (ii) (3.2.2).

Let us introduce higher analogs of the regular sequences (3.2.3) (see definition 3.2.1).

Definition 3.2.3. A q-tuple $\mathbf{X}^{(q)} = (x_1, \ldots, x_q) \in S^{\times q}$ is called an (ordered) q-*regular sequence of inverses*, if it satisfies the q relations

$$x_1 x_2 x_3 \ldots x_q x_1 = x_1, \tag{3.2.5}$$

$$x_2 x_3 \ldots x_q x_1 x_2 = x_2, \tag{3.2.6}$$

$$\vdots$$

$$x_q x_1 x_2 \ldots x_{q-1} x_q = x_q, \tag{3.2.7}$$

or in the manifest form of the binary multiplication $\mu^{(2)}$ (which is needed for the higher arity generalization)

$$\left(\mu^{(2)}\right)^{\circ q}[x_1, x_2, x_3, \ldots, x_q, x_1] = x_1, \tag{3.2.8}$$

$$\left(\mu^{(2)}\right)^{\circ q}[x_2, x_3, \ldots, x_q, x_1, x_2] = x_2, \tag{3.2.9}$$

$$\vdots$$

$$\left(\mu^{(2)}\right)^{\circ q}[x_q, x_1, x_2, \ldots, x_{q-1}, x_q] = x_q. \tag{3.2.10}$$

Denote for each element $x_i \in S$, $i = 1, \ldots, q$, its q-inverse $(q-1)$-tuple (see definition 3.1.2) by $\bar{\mathbf{x}}_i \in S^{\times(q-1)}$, where

$$\bar{\mathbf{x}}_1 = (x_2, x_3, \ldots, x_{q-1}, x_q), \tag{3.2.11}$$

$$\bar{\mathbf{x}}_2 = (x_3, x_4 \ldots, x_{q-1}, x_q, x_1), \tag{3.2.12}$$

$$\vdots$$

$$\bar{\mathbf{x}}_q = (x_1, x_2, \ldots, x_{q-2}, x_{q-1}). \tag{3.2.13}$$

Then the definition of a q-regular sequence of inverses (3.2.5)–(3.2.10) will take the concise form (see (3.1.1) and (3.1.3))

$$x_i \bar{x}_i x_i = x_i, \tag{3.2.14}$$

$$(\mu^{(2)})^{\circ q}[x_i, \bar{\mathbf{x}}_i, x_i] = x_i, \quad x_i \in S, \;\; i = 1, \ldots, q, \tag{3.2.15}$$

where the products $\bar{x}_i \in \overbrace{SS\ldots S}^{q\;\;1}$ and the $(q-1)$-tuples $\bar{\mathbf{x}}_i \in \overbrace{S \times S \times \ldots \times S}^{q\;\;1}$ (3.2.11)–(3.2.13).

Now by analogy with definition 3.2.2 we have the following (multi-relation)

Definition 3.2.4. A binary semigroup $\mathcal{S}_2$ is called a *(cyclic) q-regular semigroup* $\mathcal{S}_2^{q-\mathrm{reg}}$, if each of its elements belongs to a (not necessary unique) q-regular sequence of inverses $\mathbf{X}^{(q)} \in S^{\times q}$ satisfying (3.2.5)–(3.2.7).

This definition incorporates the ordinary regular semigroups (3.2.3) by putting $q = 2$.

***Remark* 3.2.5.** The proposed concept of q-regularity is strongly multi-relational, in that all the relations in (3.2.5)–(3.2.7) or (3.2.14) should hold, which leads us to consider all the elements in the sequence of inverses on a par and cyclic, analogously to the ordinary case (3.2.3). This can be also expressed more traditionally as 'one element of a q-regular semigroup has $q - 1$ inverses'.

***Example* 3.2.6.** In the 3-regularity case we have three relations defining the regularity

$$x_1 x_2 x_3 x_1 = x_1, \tag{3.2.16}$$

$$x_2 x_3 x_1 x_2 = x_2, \tag{3.2.17}$$

$$x_3 x_1 x_2 x_3 = x_3. \tag{3.2.18}$$

Now the sequence of inverses is $\mathbf{X}^{(3)} = (x_1, x_2, x_3)$, and we can say in a symmetrical way: all of them are 'mutually inverses one to another', iff all three relations (3.2.16)–(3.2.18) hold simultaneously. Also, using (3.2.11)–(3.2.13) we see that each element x_1, x_2, x_3 has its 3-inverse 2-tuple (or, informally, the pair of inverses)

$$\bar{\mathbf{x}}_1 = (x_2, x_3), \tag{3.2.19}$$

$$\bar{\mathbf{x}}_2 = (x_3, x_1), \tag{3.2.20}$$

$$\bar{\mathbf{x}}_3 = (x_1, x_2). \tag{3.2.21}$$

***Remark* 3.2.7.** In what follows, we will use this 3-regularity example, when providing derivations and proofs, for clarity and conciseness. This is worthwhile, because the general q-regularity case does not differ from $q = 3$ in principle, and only forces us to consider cumbersome computations with many indices and variables without enhancing our understanding of the structures.

Let us formulate the Thierrin theorem (Thierrin 1952) in the q-regular setting, which connects one-element regularity and multi-element regularity.

Lemma 3.2.8. *Every* q*-regular element in a semigroup has its* q*-inverse tuple with* $q - 1$ *elements.*

Proof. Let x_1 be a 3-regular element x_1 of a semigroup $\mathcal{S}_2$, then we can write

$$x_1 = x_1 y_2 y_3 x_1, \qquad x_1, y_2, y_3 \in S. \tag{3.2.22}$$

We are to find elements x_2, x_3 which satisfy (3.2.16)–(3.2.18). Put

$$x_2 = y_2 y_3 x_1 y_2, \tag{3.2.23}$$

$$x_3 = y_3 x_1 y_2 y_3. \tag{3.2.24}$$

Then for the lhs of (3.2.16)–(3.2.18) we derive

$$\begin{aligned} x_1 x_2 x_3 x_1 &= x_1 y_2 y_3 (x_1 y_2 y_3 x_1) y_2 y_3 x_1 = x_1 y_2 y_3 (x_1 y_2 y_3 x_1) \\ &= x_1 y_2 y_3 x_1 = x_1, \end{aligned} \tag{3.2.25}$$

$$\begin{aligned} x_2 x_3 x_1 x_2 &= y_2 (y_3 x_1 y_2 y_3)(x_1 y_2 y_3 x_1) y_2 y_3 (x_1 y_2 y_3 x_1) y_2 \\ &= y_2 x_3 (x_1 x_2 y_3 x_1) y_2 = y_2 x_3 x_1 y_2 = x_2, \end{aligned} \tag{3.2.26}$$

$$\begin{aligned} x_3 x_1 x_2 x_3 &= y_3 x_1 y_2 y_3 (x_1 y_2 y_3 x_1) y_2 y_3 (x_1 y_2 y_3 x_1) y_2 y_3 \\ &= y_3 (x_1 y_2 y_3 x_1) y_2 y_3 = y_3 x_1 y_2 y_3 = x_3. \end{aligned} \tag{3.2.27}$$

Therefore, if we have any 3-regular element (3.2.22), the relations (3.2.16)–(3.2.21) hold. □

We now show that in semigroups the higher q-regularity (in the multi-relation formulation) is wider than for 2-regularity.

Lemma 3.2.9. *If a semigroup* $\mathcal{S}_2$ *is* q*-regular, such that the* q *relations (3.2.14) are valid, then* $\mathcal{S}_2$ *is 2-regular as well.*

Proof. In the case $q = 3$ we use (3.2.16)–(3.2.18) and denote $x_2 x_3 = y \in S$. Then $x_1 y x_1 = x_1$, and multiplying (3.2.17) and (3.2.18) by x_3 from the right and x_2 from the left, respectively, we obtain

$$(x_2x_3x_1x_2)x_3 = (x_2)x_3 \Longrightarrow (x_2x_3)x_1(x_2x_3) = (x_2x_3) \Longrightarrow yx_1y = y, \qquad (3.2.28)$$

$$x_2(x_3x_1x_2x_3) = x_2(x_3) \Longrightarrow (x_2x_3)x_1(x_2x_3) = (x_2x_3) \Longrightarrow yx_1y = y. \qquad (3.2.29)$$

This means that x_1 and y are mutually 2-inverse (ordinary inverse; Clifford and Preston 1961), and so $\mathcal{S}_2$ is 2-regular. □

Definition 3.2.10. A semigroup $\mathcal{S}_2$ is called a *pure q-regular semigroup*, if there are no additional relations beyond the q-regularity (3.2.5)–(3.2.7) imposed on q-regular sequence of inverses $\mathbf{X}^{(q)} = (x_1, \ldots, x_q)$. Oppositely, a q-regular semigroup $\mathcal{S}_2$ with extra relations on $\mathbf{X}^{(q)}$ is called *impure*, and these relations are called *impuring relations.*

For instance, an ordinary regular (2-regular) semigroup defined by (3.2.3) only is a pure regular semigroup. If we add the commutativity condition $x_1x_2 = x_2x_1$, then we obtain an impure regular semigroup, which is an inverse semigroup being actually a semilattice of abelian groups (see, e.g., Lawson 1998).

The converse is connected with imposing extra relations in the semigroup $\mathcal{S}_2$.

Theorem 3.2.11. *The following statements are equivalent for a pure regular (2-regular) semigroup* $\mathcal{S}_2$:

(i) The pure 2-regular semigroup $\mathcal{S}_2$ is pure q-regular.
(ii) The pure 2-regular semigroup $\mathcal{S}_2$ is cancellative.

Proof. (i)⇒(ii) Suppose $\mathcal{S}_2$ is 2-regular, which means that there are tuples satisfying (3.2.3), or in our notation here for each $x_1 \in S$ there exists $y \in S$ such that $x_1yx_1 = x_1$ and $yx_1y = y$. Let y be presented as a product of two arbitrary elements $y = x_2x_3$, $x_2, x_3 \in S$, which is possible since the underlying set S of the semigroup is closed with respect to multiplication. We now try to get the (3.2.16)–(3.2.18) without additional relations. After substitution y into the 2-regularity conditions, we obtain

$$x_1yx_1 = x_1 \Longrightarrow x_1(x_2x_3)x_1 = x_1 \Longrightarrow x_1x_2x_3x_1 = x_1, \qquad (3.2.30)$$

$$yx_1y = y \Longrightarrow (x_2x_3)x_1(x_2x_3) = (x_2x_3) \Longrightarrow (x_2x_3x_1x_2)x_3 = (x_2)x_3 \qquad (3.2.31)$$

$$\Longrightarrow x_2(x_3x_1x_2x_3) = x_2(x_3). \qquad (3.2.32)$$

We observe that the first line coincides with (3.2.16), but obtaining (3.2.17) and (3.2.18) from the second and third lines here requires right and left cancellativity (by the elements outside the brackets in the last equalities), respectively.

(ii) ⇒ (i) After applying cancellativity to (3.2.31) and (3.2.32), then equating expressions in brackets one obtains (3.2.17) and (3.2.18), correspondingly. The first line (3.2.30) coincides with (3.2.16) in any case. □

It is known that a cancellative regular semigroup is a group (Clifford and Preston 1961, Petrich 1984).

Corollary 3.2.12. If a pure 2-regular semigroup $\mathcal{S}_2$ is pure q-regular, it is a group.

If we consider impure q-regular semigroups, these can be constructed from pure 2-regular semigroups without the cancellativity requirement, but with special additional relations and extra idempotents. For instance, in the case of an impure 3-regular semigroup we have

Proposition 3.2.13. *If a semigroup is pure 2-regular we can construct an impure 3-regular semigroup with three impuring relations.*

Proof. Let $\mathcal{S}_2$ be a pure 2-regular semigroup (in multi-relation definition (3.2.3)), such that

$$x_1x_3x_1 = x_1, \quad x_3x_1x_3 = x_3, \quad x_1, x_3 \in S, \tag{3.2.33}$$

have at least one solution. Introduce an additional element $x_2 = x_3x_1 \in S$, which is idempotent (from (3.2.33)). It is seen that the triple $\mathbf{X}^{(3)} = \{x_1, x_2, x_3\}$ is a 3-inverse sequence, because (3.2.16)–(3.2.18) take place, indeed,

$$x_1 = x_1x_3x_1 = x_1(x_3x_1)x_3x_1 = x_1x_2x_3x_1, \tag{3.2.34}$$

$$x_2 = x_3x_1 = (x_3x_1)x_3x_1(x_3x_1) = x_2x_3x_1x_2, \tag{3.2.35}$$

$$x_3 = x_3x_1x_3 = x_3x_1(x_3x_1)x_3 = x_3x_1x_2x_3. \tag{3.2.36}$$

The impuring relations

$$x_1x_2 = x_1, \quad x_2x_3 = x_3, \quad x_3x_1 = x_2 \tag{3.2.37}$$

follow from (3.2.33) and the definition of x_2. □

In the same way one can construct recursively an impure $(q+1)$-regular semigroup from a pure q-regular semigroup. Taking into account lemma 3.2.9, we arrive at

Corollary 3.2.14. Impure q-regular semigroups are equivalent to pure 2-regular semigroups.

3.2.2 Idempotents and higher q-inverse semigroups

Lemma 3.2.15. *In an* q*-regular semigroup there are at least* q *(binary higher) idempotents (no summation)*

$$e_i = x_i \bar{x}_i, \quad i = 1, \ldots, q. \tag{3.2.38}$$

Proof. Multiply (3.2.14) by $\bar{x}_i$ from the right to get $x_i \bar{x}_i x_i \bar{x}_i = x_i \bar{x}_i$. □

In the case $q = 3$ the (higher) idempotents (3.2.38) become

$$e_1 = x_1 x_2 x_3,\ e_2 = x_2 x_3 x_1,\ e_3 = x_3 x_1 x_2, \tag{3.2.39}$$

having the following 'chain' commutation relations with elements

$$e_1 x_1 = x_1 e_2 = x_1, \tag{3.2.40}$$

$$e_2 x_2 = x_2 e_3 = x_2, \tag{3.2.41}$$

$$e_3 x_3 = x_3 e_1 = x_3. \tag{3.2.42}$$

Corollary 3.2.16. Each higher idempotent e_i is a left unit (neutral element) for x_i and a right unit (neutral element) for x_{i-1}.

Lemma 3.2.17. *If in a* q*-regular semigroup (higher) idempotents commute, then each element has a unique* q*-regular sequence of inverses.*

Proof. Suppose in a 3-regular semigroup defined by (3.2.16)–(3.2.18) the element x_1 has two pairs of inverses (3.2.11) $\bar{\mathbf{x}}_1 = (x_2, x_3)$ and $\bar{\mathbf{x}}'_1 = (x'_2, x_3)$. Then, in addition to (3.2.16)–(3.2.18) and the triple of idempotents (3.2.39) we have

$$x_1 x'_2 x_3 x_1 = x_1, \tag{3.2.43}$$

$$x'_2 x_3 x_1 x'_2 = x'_2, \tag{3.2.44}$$

$$x_3 x_1 x'_2 x_3 = x_3 \tag{3.2.45}$$

and

$$e'_1 = x_1 x'_2 x_3,\ e'_2 = x'_2 x_3 x_1,\ e'_3 = x_3 x_1 x'_2, \tag{3.2.46}$$

and also the idempotents (3.2.39) and (3.2.46) commute. We derive

$$\begin{aligned} x_2 &= x_2 x_3 x_1 x_2 = x_2 x_3 (x_1 x'_2 x_3 x_1) x_1 x_2 = x_2 x_3 (x_1 x'_2 x_3 (x_1 x'_2 x_3 x_1)) x_2 \\ &= x_2 x_3 x_1 x'_2 (x_3 x_1 x'_2)(x_3 x_1 x_2) = x_2 x_3 x_1 x'_2 (x_3 x_1 x_2)(x_3 x_1 x'_2) \\ &= (x_2 x_3 x_1)(x'_2 x_3 x_1) x_2 x_3 x_1 x'_2 = (x'_2 x_3 x_1)(x_2 x_3 x_1) x_2 x_3 x_1 x'_2 \\ &= x'_2 x_3 (x_1 x_2 x_3 x_1) x_2 x_3 x_1 x'_2 = x'_2 x_3 (x_1 x_2 x_3 x_1) x'_2 = x'_2 x_3 x_1 x'_2 = x'_2. \end{aligned} \tag{3.2.47}$$

A similar derivation can be done for $\bar{\mathbf{x}}_1'' = (x_2, x_3'')$ and other indices. □

Definition 3.2.18. A higher q-regular semigroup is called a *higher q-inverse semigroup*, if each element $x_i \in S$ has its unique q-inverse $(q-1)$-tuple $\bar{\mathbf{x}}_i \in S^{\times(q-1)}$ (3.2.11)–(3.2.13).

Thus, as in the regular (multi-relation 2-regular) semigroups (Liber 1954, Munn and Penrose 1955, Schein 1965), we arrive at a higher regular generalization of the characterization of the inverse semigroups by

Theorem 3.2.19. *A higher* q*-regular semigroup is an inverse semigroup, if their idempotents commute.*

It is known that an inverse (2-inverse) semigroup is a group, if it contains exactly one idempotent, and so groups are inverse semigroups (see, e.g., Clifford and Preston 1961, Lawson 1998). Nevertheless, for q-inverse semigroups this is not true. For instance, if the elements x, 1 are in a group, then 3-regularity conditions (3.2.16)–(3.2.18) are satisfied by $x_1 = 1$, $x_2 = x$, $x_3 = x^{-1}$ for all x, and therefore 2-inverses are not unique for groups. Thus, we have

Assertion 3.2.20. *The* q*-inverse semigroups do not contain groups.*

Further properties of q-inverse semigroups will be studied elsewhere using the semigroup theory methods (see, e.g., Clifford and Preston 1961, Lawson 1998, Petrich 1984, Higgins 1992).

3.3 Higher q-inverse polyadic semigroups

We now introduce a polyadic version of the higher q-regular and q-inverse semigroups defined in the previous section. For this we will use the relations with the manifest form of the binary multiplication to follow (ii) –(iii) of the arity invariance principle (see definition 3.1.4).

3.3.1 Higher q-regular polyadic semigroups

First, define the polyadic version of the higher q-regular sequences (3.2.8)–(3.2.10) in the framework of the arity invariance principle by substituting $(\mu^{(2)})^{\circ q} \mapsto (\mu^{(n)})^{\circ q}$ and then changing sizes of tuples, and generalizing to higher q (3.1.11).

Definition 3.3.1. In a polyadic (n-ary) semigroup $\mathcal{S}^{(n)} = \langle S \mid \mu^{(n)} \rangle$ (see section 3.1.2) a $q(n-1)$-tuple $\mathbf{X}^{(q(n-1))} = (x_1, \ldots, x_{q(n-1)}) \in S^{\times q(n-1)}$ is called a *polyadic q-regular sequence of inverses*, if it satisfies the $q(n-1)$ relations

$$(\mu^{(n)})^{\circ q}[x_1, x_2, x_3, \ldots, x_{q(n-1)}, x_1] = x_1, \tag{3.3.1}$$

$$(\mu^{(n)})^{\circ q}[x_2, x_3, \ldots, x_{q(n-1)}, x_1, x_2] = x_2, \tag{3.3.2}$$
$$\vdots$$

$$(\mu^{(n)})^{\circ q}[x_{q(n-1)}, x_1, x_2, \ldots, x_{q(n-1)-1}, x_{q(n-1)}] = x_{q(n-1)}. \tag{3.3.3}$$

Definition 3.3.2. A polyadic semigroup $\mathcal{S}^{(n)}$ is called a *q-regular semigroup*, if each element belongs to a (not necessary unique) q-regular sequence of inverses $\mathbf{X}^{(q(n-1))} \in S^{\times q(n-1)}$.

In a polyadic q-regular semigroup, each element instead of having one inverse element will have a tuple, and we now, instead of (3.1.1)–(3.1.2), have

Definition 3.3.3. In a polyadic (n-ary) q-regular semigroup each element $x_i \in S$, where $i = 1, \ldots, q(n-1)$, has a *polyadic q-inverse*, as the $(q(n-1)-1)$-tuple $\bar{\mathbf{x}}_i^{[n]} \in S^{\times(q(n-1)-1)}$ (see the binary case definition 3.1.2 and (3.2.11)–(3.2.13))

$$\bar{\mathbf{x}}_1^{[n]} = (x_2, x_3, \ldots, x_{q(n-1)-1}, x_{q(n-1)}), \tag{3.3.4}$$

$$\bar{\mathbf{x}}_2^{[n]} = (x_3, x_4 \ldots, x_{q(n-1)-1}, x_{q(n-1)}, x_1), \tag{3.3.5}$$
$$\vdots$$

$$\bar{\mathbf{x}}_{q(n-1)}^{[n]} = (x_1, x_2, \ldots, x_{q(n-1)-2}, x_{q(n-1)-1}). \tag{3.3.6}$$

In this notation the polyadic (n-ary) q-regular sequence of inverses (3.3.1)–(3.3.3) can be written in the customary concise form (see (3.2.15) for a binary semigroup)

$$(\mu^{(n)})^{\circ q}[x_i, \bar{\mathbf{x}}_i^{[n]}, x_i] = x_i, \quad x_i \in S, \;\; i = 1, \ldots, q(n-1). \tag{3.3.7}$$

Note that at first sight one could consider here also the procedure of reducing the number of multiplications from any number to one (Kolesnikov 1979), as in the single element regularity case (see assertion 3.1.7 and examples 3.1.9, 3.1.14). However, if we consider the whole polyadic q-regular sequence of inverses (3.3.1)–(3.3.3) consisting of $q(n-1)$ relations (3.3.1)–(3.3.3), it will not be possible in general to reduce arity in all of them simultaneously.

***Example* 3.3.4.** If we consider the ordinary regularity (2-regularity) for a ternary semigroup $\mathcal{S}^{(3)}$, we obtain the ternary regular sequence of $q(n-1) = 4$ elements and four regularity relations

$$(\mu^{(3)})^{\circ 2}[x_1, x_2, x_3, x_4, x_1] = x_1, \tag{3.3.8}$$

$$(\mu^{(3)})^{\circ 2}[x_2, x_3, x_4, x_1, x_2] = x_2, \tag{3.3.9}$$

$$(\mu^{(3)})^{\circ 2}[x_3, x_4, x_1, x_2, x_3] = x_3, \tag{3.3.10}$$

$$(\mu^{(3)})^{\circ 2}[x_4, x_1, x_2, x_3, x_4] = x_4, \quad x_i \in S. \tag{3.3.11}$$

Elements x_1, x_2, x_3, x_4 have these ternary inverses (see the binary 3-regularity (3.2.19)–(3.2.21))

$$\bar{\mathbf{x}}_1^{[3]} = (x_2, x_3, x_4), \tag{3.3.12}$$

$$\bar{\mathbf{x}}_2^{[3]} = (x_3, x_4, x_1), \tag{3.3.13}$$

$$\bar{\mathbf{x}}_3^{[3]} = (x_4, x_1, x_2), \tag{3.3.14}$$

$$\bar{\mathbf{x}}_4^{[3]} = (x_1, x_2, x_3). \tag{3.3.15}$$

***Remark* 3.3.5.** This definition of regularity in ternary semigroups is in full agreement with the arity invariance principle (definition 3.1.4): it has (the minimum) two (n-ary) multiplications as in the binary case (3.1.2). Although we can reduce the number of multiplications to one, as in (3.1.13) for a single relation, e.g., by the substitution $z = \mu^{(3)}[x_2, x_3, x_4]$, this cannot be done in all the cyclic relations (3.3.8)–(3.3.11): the third relations (3.3.10) cannot be presented in terms of z as well as the right hand sides.

***Remark* 3.3.6.** If we take the definition of regularity for a single relation, reduce it to one (n-ary) multiplication and then 'artificially' cycle it (see Kolesnikov 1979 and references citing it), we find a conflict with the arity invariance principle. Indeed, in the ternary case we obtain $\mu^{(3)}[x, y, x] = x$, $\mu^{(3)}[y, x, y] = y$. Despite the similarity to the standard binary regularity (3.2.3)–(3.2.4), it contains only one multiplication, and therefore it could be better treated as a symmetry property of the ternary product $\mu^{(3)}$, rather than as a relation between variables, e.g., as a regularity which needs at least two products. Also, the length of the sequence is two, as in the binary case, but it should be $q(n - 1) = 4$, as in the ternary 2-regular sequence (3.3.8)–(3.3.11) according to the arity invariance principle.

***Example* 3.3.7.** For a 3-regular ternary semigroup $\mathcal{S}^{(3)} = \langle S | \mu^{(3)} \rangle$ with $q = 3$ and $n = 3$ we have a 3-regular sequence of 6 (mutual) inverses $\mathbf{X}^{(6)} = (x_1, x_2, x_3, x_4, x_5, x_6) \in S^{\times 6}$, which satisfy the following six 3-regularity conditions

$$(\mu^{(3)})^{\circ 3}[x_1, x_2, x_3, x_4, x_5, x_6, x_1] = x_1, \tag{3.3.16}$$

$$(\mu^{(3)})^{\circ 3}[x_2, x_3, x_4, x_5, x_6, x_1, x_2] = x_2, \tag{3.3.17}$$

$$(\mu^{(3)})^{\circ 3}[x_3, x_4, x_5, x_6, x_1, x_2, x_3] = x_3, \tag{3.3.18}$$

$$(\mu^{(3)})^{\circ 3}[x_4, x_5, x_6, x_1, x_2, x_3, x_4] = x_4, \tag{3.3.19}$$

$$(\mu^{(3)})^{\circ 3}[x_5, x_6, x_1, x_2, x_3, x_4, x_5] = x_5, \tag{3.3.20}$$

$$(\mu^{(3)})^{\circ 3}[x_6, x_1, x_2, x_3, x_4, x_5, x_6] = x_6, \quad x_i \in S. \tag{3.3.21}$$

This is the first nontrivial case in both arity $n \neq 2$ and regularity $q \neq 2$ (see remark 3.2.7).

***Remark* 3.3.8.** In this example we can reduce the number of multiplications, as in example 3.1.14, to one single relation, but not to all of them. For instance, we can put in the first relations (3.3.16) $t = \mu^{(3)}[x_2, x_3, x_4]$ and further $z = \mu^{(3)}[t, x_5, x_6]$, but, for instance, the third relation (3.3.18) cannot be presented in terms of t, z, as in the first one in (3.3.16)–(3.3.21), because of the splitting of variables in t, z, as well as in the right hand sides.

The polyadic analog of the Thierrin theorem (Thierrin 1952) in q-regular setting is given by:

Lemma 3.3.9. *Every q-regular element in a polyadic (n-ary) semigroup has a polyadic q-inverse tuple with $(q(n-1)-1)$ elements.*

Its proof literally repeats that of lemma 3.2.8, but with different lengths of sequences. The same is true for lemma 3.2.9 and theorem 3.2.11 by exchanging $\mathcal{S}_2 \to \mathcal{S}^{(n)}$.

Definition 3.3.10. A higher q-regular polyadic semigroup $\mathcal{S}^{(n)}$ is called a *higher q-inverse polyadic semigroup*, if each element $x_i \in S$ has a unique q-inverse $(q(n-1)-1)$-tuple $\bar{\mathbf{x}}_i \in S^{\times(q(n-1)-1)}$ (3.3.4)–(3.3.6).

In searching for polyadic idempotents we observe that the binary regularity (3.2.15) and polyadic regularity (3.3.7) differ considerably in lengths of tuples. In the binary case any length is allowed, and one can define idempotents e_i by (3.2.38), as a left neutral element for x_i, such that (no summation) $e_i x_i = x_i$, $x_i \in \mathcal{S}_2$ (see (3.2.40)–(3.2.42)). At first sight, for polyadic q-regularity (3.3.7), we could proceed in a similar way. However we have

Proposition 3.3.11. *In the* q*-regular polyadic (*n*-ary) semigroup the length of the tuple $(x_i, \bar{\mathbf{x}}_i)$ is allowed to give an idempotent, only if $n = 2$, i.e., the semigroup is binary.*

Proof. The allowed length of the tuple $(x_i, \bar{\mathbf{x}}_i)$ (to give one element, an idempotent, see (3.1.5)), where $\bar{\mathbf{x}}_i$ are in (3.3.4)–(3.3.6), is $\ell_\mu(n-1)+1$ to apply ℓ_μ multiplications. While the tuple $(x_i, \bar{\mathbf{x}}_i, x_i)$ in (3.3.7) has the given length $q(n-1)+1$, and we obtain the equation for ℓ as

$$\ell_\mu(n-1)+1 = q(n-1) \Longrightarrow \ell_\mu = q - \frac{1}{n-1}. \tag{3.3.22}$$

The equation (3.3.22) has only one solution over $\mathbb{N}$ namely $\ell_\mu = q-1$, iff $n=2$. □

This means that to go beyond the binary semigroups $n>2$ and have idempotents, one needs to introduce a different regularity condition to (3.1.11) and (3.3.7), which we will do below.

3.3.2 Sandwich polyadic q-regularity

Here we go in the opposite direction to that above: we will define the idempotents and then construct the needed regularity conditions using them, which in the limit $n=2$ and $q=2$ will give the ordinary binary regularity (3.2.3)–(3.2.4).

Let us formulate the binary higher q-regularity (3.2.15) in terms of the local polyadic identities (3.1.10). We write the idempotents (3.2.38) in the form

$$e_i = \mu^{(2)}[x_i, \bar{\mathbf{x}}_i], \tag{3.3.23}$$

where $\bar{\mathbf{x}}_i$ are q-inverses being $(q-1)$-tuples (3.2.11)–(3.2.13).

In terms of the idempotents e_i (3.3.23) the higher q-regularity conditions (3.2.15) become

$$\mu^{(2)}[e_i, x_i] = x_i. \tag{3.3.24}$$

Assertion 3.3.12. *The binary* q*-regularity conditions coincide with the definition of the local left identities.*

Proof. Compare (3.3.24) and the definition (3.1.10) with $n=2$ (see Pop and Pop 2010). □

Thus, the main idea is to generalize (using the arity invariance principle) to the polyadic case the binary q-regularity in the form (3.3.24) and also idempotents (3.3.23), but not (3.2.14)–(3.2.15), as was done in (3.3.7).

By analogy with (3.3.23) let us introduce

$$e_i' = \left(\mu^{(n)}\right)^{\circ(q-1)}\left[x_i, \overbrace{\bar{\mathbf{x}}_i', \ldots, \bar{\mathbf{x}}_i'}^{q-1}\right], \quad x_i \in S, \quad i = 1, \ldots, n, \tag{3.3.25}$$

where $\bar{\mathbf{x}}_i'$ is the polyadic $(n-1)$-tuple for x_i (which differs from (3.3.4)–(3.3.6))

$$\bar{\mathbf{x}}_1' = (x_2, x_3, \dots, x_n), \tag{3.3.26}$$

$$\bar{\mathbf{x}}_2' = (x_3, x_4, \dots, x_n, x_1), \tag{3.3.27}$$
$$\vdots$$

$$\bar{\mathbf{x}}_n' = (x_1, x_2, \dots, x_{n-2}, x_{n-1}). \tag{3.3.28}$$

Definition 3.3.13. *Sandwich polyadic q-regularity* conditions are defined as

$$\mu^{(n)}\left[\overbrace{e_i', \dots, e_i'}^{n-1}, x_i\right] = x_i, \quad i = 1, \dots, n. \tag{3.3.29}$$

Observe that the conditions (3.3.29) coincide with the definition of the generalized local left polyadic identities (3.1.10) (see assertion 3.3.12), and therefore we can define the sandwich polyadic q-regularity (3.3.29) in an alternative (multi-relation (3.2.2)) way

Definition 3.3.14. A polyadic semigroup $\mathcal{S}^{(n)}$ is *sandwich q-regular*, if there exists n-tuple $\mathbf{X}^{(n)} = (x_1, x_2, x_3, \dots, x_n)$ in which each element x_i has its left generalized local polyadic identity e_i' of the form (3.3.25).

It follows from (3.3.29), that e_i' are n polyadic idempotents (3.1.9)

$$(e_i')^{\langle 1 \rangle} = \mu^{(n)}\left[\overbrace{e_i', \dots, e_i'}^{n}\right] = e_i', \quad i = 1, \dots, n. \tag{3.3.30}$$

In the binary limit $n = 2$ and $q = 2$ (3.3.25)–(3.3.29) give the ordinary regularity (3.2.3) in the multi-relation definition (3.2.2).

By analogy with definition 3.3.2 we have

Definition 3.3.15. An n-ary semigroup $\mathcal{S}^{(n)}$ is called a *sandwich polyadic q-regular semigroup* $\mathcal{S}^{(n)}_{\text{sandw}}$, if each element has a (not necessary unique) n-tuple of inverses satisfying (3.3.25)–(3.3.29).

Now instead of (3.3.4)–(3.3.6) we have for the inverses

Definition 3.3.16. Each element of a sandwich polyadic (n-ary) q-regular semigroup $x_i \in S$, $i = 1, \dots, n-1$ has its (*sandwich*) *polyadic n-inverse*, as the $(n-1)$-tuple $\bar{\mathbf{x}}_i^{'[n]} \in S^{\times(n-1)}$ (see (3.3.26)–(3.3.28)).

Example **3.3.17.** In the ternary case $n = 3$ and $q = 2$ we have the following 3 ternary idempotents

$$e_1' = \mu^{(3)}[x_1, x_2, x_3], \tag{3.3.31}$$

$$e_2' = \mu^{(3)}[x_2, x_3, x_1], \tag{3.3.32}$$

$$e_3' = \mu^{(3)}[x_3, x_1, x_2], \quad x_i \in S, \tag{3.3.33}$$

such that

$$\mu^{(3)}[e_i', e_i', e_i'] = e_i', \quad i = 1, 2, 3. \tag{3.3.34}$$

The sandwich ternary regularity conditions become (see (3.3.8)–(3.3.11))

$$(\mu^{(3)})^{\circ 3}[x_1, x_2, x_3, x_1, x_2, x_3, x_1] = x_1, \tag{3.3.35}$$

$$(\mu^{(3)})^{\circ 3}[x_2, x_3, x_1, x_2, x_3, x_1, x_2] = x_2, \tag{3.3.36}$$

$$(\mu^{(3)})^{\circ 3}[x_3, x_1, x_2, x_3, x_1, x_2, x_3] = x_3. \tag{3.3.37}$$

Each of the elements $x_1, x_2, x_3 \in S$ has a 2-tuple of the ternary inverses (see the inverses for the binary 3-regularity (3.2.19)–(3.2.21) and the ternary regularity (3.3.12)–(3.3.15)) as

$$\bar{\mathbf{x}}_1' = (x_2, x_3), \tag{3.3.38}$$

$$\bar{\mathbf{x}}_2' = (x_3, x_1), \tag{3.3.39}$$

$$\bar{\mathbf{x}}_3' = (x_1, x_2). \tag{3.3.40}$$

A ternary semigroup in which the sandwich ternary regularity (3.3.35)–(3.3.37) has a solution (starting from any of element x_i) is a sandwich regular ternary semigroup $\mathcal{S}^{(3)}_{\text{sandw}}$.

It can be seen from (3.3.35)–(3.3.37) why we call such regularity 'sandwich': each x_i appears in the lhs not only two times, on the first and the last places, as in the previous definitions, but also in the middle, which gives the possibility for us to define idempotents. In general, the ith condition of sandwich regularity (3.3.29) will contain $n - 2$ middle elements x_i.

Remark **3.3.18.** In (3.3.35) we can also reduce the number of multiplications to one (see Kolesnikov 1979), but only in a single relation (see example 3.1.14 and remark 3.3.8). However, in this case the whole system (3.3.35)–(3.3.37) will lose its self-consistency, since we are using the multi-relation definition of the sandwich regular semigroup (see (3.2.3) for the ordinary regularity).

We can obtain a matrix representation of a sandwich polyadic semigroup by using the method of antitriangle supermatrices (Duplij 1996) for regular semigroups and bands together with the general form of a polyadic matrix (Duplij 2021a).

***Example* 3.3.19.** Consider the sandwich regular ternary semigroup $\mathcal{S}^{(3)}_{\text{sandw}}$ from example 3.3.17 and its continuous 4-parameter representation in the set M_4^{adg} of 4×4 noninvertible matrices of the special antitriangle form (see Duplij 1996 for the binary case)

$$\mathrm{M}_\varepsilon(t, u, v, w) = \begin{pmatrix} 0 & 0 & 0 & \varepsilon t \\ 0 & 0 & \varepsilon u & 1 \\ 0 & \varepsilon v & 0 & 0 \\ \varepsilon w & 1 & 0 & 0 \end{pmatrix} \in M_4^{\text{adg}}, \tag{3.3.41}$$

where $t, u, v, w \in \mathbb{C}$ and ε is a nilpotent dual number $\varepsilon^2 = 0$ (see, e.g., Yaglom 1968). The set M_4^{adg} is closed with respect to the ordinary triple matrix product only, while the multiplication of two matrices (3.3.41) is obviously out of the set M_4^{adg}. This allows us to define the ternary product $\mu_{\mathrm{M}}^{(3)}$ on the set M_4^{adg} as

$$\begin{aligned} &\mu_{\mathrm{M}}^{(3)}[\mathrm{M}_\varepsilon(t_1, u_1, v_1, w_1), \mathrm{M}_\varepsilon(t_2, u_2, v_2, w_2), \mathrm{M}_\varepsilon(t_3, u_3, v_3, w_3)] \\ &= \mathrm{M}_\varepsilon(t_1, u_1, v_1, w_1)\mathrm{M}_\varepsilon(t_2, u_2, v_2, w_2)\mathrm{M}_\varepsilon(t_3, u_3, v_3, w_3) \\ &= \mathrm{M}_\varepsilon(t_1, u_3, v_1, w_3). \end{aligned} \tag{3.3.42}$$

Thus, $\mathcal{M}_3^{\text{adg}} = \left\langle M_4^{\text{adg}} \middle| \mu_{\mathrm{M}}^{(3)} \right\rangle$ is a ternary semigroup, because the total ternary associativity of $\mu_{\mathrm{M}}^{(3)}$ (3.1.6) is governed by the associativity of the ordinary matrix multiplication in the rhs of (3.3.42).

Then the matrix representation $\Phi_3^{\text{sandw}}: \mathcal{S}^{(3)}_{\text{sandw}} \to \mathcal{M}_3^{\text{adg}}$ is given by

$$x \mapsto \mathrm{M}_\varepsilon(t, u, v, w), \quad x \in \mathcal{S}^{(3)}_{\text{sandw}}, \quad \mathrm{M}_\varepsilon(t, u, v, w) \in \mathcal{M}_3^{\text{adg}}, \tag{3.3.43}$$

such that putting $x_i \mapsto \mathrm{M}_\varepsilon(t_i, u_i, v_i, w_i)$, $i = 1, 2, 3$, we obtain the sandwich ternary regularity (3.3.35)–(3.3.37).The semigroup $\mathcal{M}_3^{\text{adg}}$ is an example of the ternary band, because, as it follows from (3.3.41), each element of M_4^{adg} is a ternary idempotent (see (3.1.9))

$$(\mathrm{M}_\varepsilon(t, u, v, w))^3 = \mathrm{M}_\varepsilon(t, u, v, w). \tag{3.3.44}$$

Therefore, in the terminology of definition 3.2.1, $\mathcal{M}_3^{\text{adg}}$ is an impure regular semigroup, since it contains an additional (to regularity) condition (3.3.44) following from the concrete matrix structure (3.3.41), while $\mathcal{S}^{(3)}_{\text{sandw}}$ is a pure regular semigroup, if no other conditions except the regularity (3.3.35)–(3.3.37) are imposed. For the representation Φ_3^{sandw} (3.3.43) to be faithful, the semigroup $\mathcal{S}^{(3)}_{\text{sandw}}$ should be also impure regular and idempotent (as (3.3.44)), i.e., the additional impuring conditions are $\mu^{(3)}[x, x, x] = x$, $\forall x \in S$, which means that $\mathcal{S}^{(3)}_{\text{sandw}}$ becomes a ternary band as well.

***Remark* 3.3.20.** The introduced sandwich regularity (3.3.29) differs from Sioson (1963) considerably. For simplicity we consider the ternary case (3.3.35)–(3.3.37) and compare it with Sioson (1963) (in our notation)

$$(\mu^{(3)})^{\circ 4}[x_1, x_2, x_3, x_4, x_1, x_5, x_6, x_7, x_1] = x_1. \tag{3.3.45}$$

First, we note that only the one-relation definition (3.2.1) was used in Sioson (1963). By analogy with (3.1.12), we can provide in (3.3.45) the reduction as $y_1 = \mu^{(3)}[x_2, x_3, x_4]$, $y_2 = \mu^{(3)}[x_5, x_6, x_7]$ to get

$$(\mu^{(3)})^{\circ 2}[x_1, y_1, x_1, y_2, x_1] = x_1. \tag{3.3.46}$$

We cannot connect the sandwich ternary regularity (3.3.35) and (3.3.45)–(3.3.46), because the length of ternary words is 'quantized' (see (3.1.5)), and therefore it can be only odd $L_3^{(\ell_\mu)} = 2\ell_\mu + 1$, $\ell_\mu \in \mathbb{N}$, where ℓ_μ is the number of ternary multiplications. Second, for the same reason there is no natural way to introduce idempotents, as in (3.3.31)–(3.3.33), using (3.3.45)–(3.3.46) only.

***Example* 3.3.21.** The nontrivial case in both higher arity $n \neq 2$ and higher sandwich regularity $q \neq 2$ is the sandwich 3-regular 4-ary semigroup $\mathcal{S}_4$ in which there exist 4 elements satisfying the higher sandwich 3-regularity relations

$$\begin{aligned}(\mu^{(4)})^{\circ 4}[&x_1, x_2, x_3, x_4, x_2, x_3, x_4, x_1, x_2, x_3, x_4, x_2, x_3, x_4, x_1, x_2, x_3, x_4,\\ &x_2, x_3, x_4, x_1] = x_1,\end{aligned} \tag{3.3.47}$$

$$\begin{aligned}(\mu^{(4)})^{\circ 4}[&x_2, x_3, x_4, x_1, x_3, x_4, x_1, x_2, x_3, x_4, x_1, x_3, x_4, x_1, x_2, x_3, x_4, x_1,\\ &x_3, x_4, x_1, x_2] = x_2,\end{aligned} \tag{3.3.48}$$

$$\begin{aligned}(\mu^{(4)})^{\circ 4}[&x_3, x_4, x_1, x_2, x_4, x_1, x_2, x_3, x_4, x_1, x_2, x_4, x_1, x_2, x_3, x_4, x_1, x_2,\\ &x_4, x_1, x_2, x_3] = x_3,\end{aligned} \tag{3.3.49}$$

$$\begin{aligned}(\mu^{(4)})^{\circ 4}[&x_4, x_1, x_2, x_3, x_1, x_2, x_3, x_4, x_1, x_2, x_3, x_1, x_2, x_3, x_4, x_1, x_2, x_3,\\ &x_1, x_2, x_3, x_4] = x_4.\end{aligned} \tag{3.3.50}$$

The 4-ary idempotents are

$$e_1' = \mu^{(4)}[x_1, x_2, x_3, x_4, x_2, x_3, x_4], \quad e_2' = \mu^{(4)}[x_2, x_3, x_4, x_1, x_3, x_4, x_1], \tag{3.3.51}$$

$$e_3' = \mu^{(4)}[x_3, x_4, x_1, x_2, x_4, x_1, x_2], \quad e_4' = \mu^{(4)}[x_4, x_1, x_2, x_3, x_1, x_2, x_3], \tag{3.3.52}$$

and they obey the following commutation relations with elements (see (3.2.40)–(3.2.42))

$$\mu^{(4)}[e_1', e_1', e_1', x_1] = \mu^{(4)}[x_1, e_2', e_2', e_2'] = x_1, \tag{3.3.53}$$

$$\mu^{(4)}[e_2', e_2', e_2', x_2] = \mu^{(4)}[x_2, e_3', e_3', e_3'] = x_2, \tag{3.3.54}$$

$$\mu^{(4)}[e_3', e_3', e_3', x_3] = \mu^{(4)}[x_3, e_4', e_4', e_4'] = x_3, \tag{3.3.55}$$

$$\mu^{(4)}[e_4', e_4', e_4', x_4] = \mu^{(4)}[x_4, e_1', e_1', e_1'] = x_4. \tag{3.3.56}$$

Each of the elements x_1, x_2, x_3, x_4 has its triple of 4-ary inverses

$$\bar{\mathbf{x}}_1' = (x_2, x_3, x_4), \quad \bar{\mathbf{x}}_2' = (x_3, x_4, x_1), \tag{3.3.57}$$

$$\bar{\mathbf{x}}_3' = (x_4, x_1, x_2), \quad \bar{\mathbf{x}}_4' = (x_1, x_2, x_3). \tag{3.3.58}$$

By analogy with corollary 3.2.16 we now have

Corollary 3.3.22. Each polyadic idempotent e_i' of the sandwich higher q-regular n-ary semigroup $\mathcal{S}_{\text{sandw}}^{(n)}$ is a local left polyadic identity (neutral element) for x_i and a local right polyadic identity for x_{i-1}.

Definition 3.3.23. A sandwich higher q-regular polyadic semigroup $\mathcal{S}_{\text{sandw}}^{(n)}$ is called a *sandwich q-inverse polyadic semigroup*, if each element $x_i \in S$ has a unique q-inverse $(n-1)$-tuple $\bar{\mathbf{x}}_i' \in S^{\times(n-1)}$ (3.3.26)–(3.3.28).

In search of a polyadic analog of the lemma 3.2.17, we have found that the commutation of polyadic idempotents does not lead to uniqueness of the q-inverse elements (3.3.26)–(3.3.28). The problem appears because of the presence of the middle elements in (3.3.25) and, e.g., in (3.3.35) (see the discussion after (3.3.37)), while the middle elements do not exist in the previous formulations of regularity.

3.3.3 Sandwich regularity with generalized idempotents

Now we using the method of the previous section, we construct a sandwich regularity containing the generalized polyadic idempotents satisfying (3.1.8) instead of (3.3.30).

Definition 3.3.24. *Sandwich n-ary q -regularity* conditions for the sequence of inverses $\mathbf{X}^{(n)} = (x_1, x_2, x_3, \dots, x_n) \in S^{\times n}$ with generalized polyadic $\langle m\rangle$-idempotents e_i' are defined as

$$\left(\mu^{(n)}\right)^{\circ m}\left[\overbrace{e_i', \dots, e_i'}^{m(n-1)}, x_i\right] = x_i, \quad i = 1, \dots, n, \tag{3.3.59}$$

where e_i' are the same as in (3.3.25)–(3.3.28), but (instead of (3.3.30)) now satisfying

$$(e_i')^{\langle m \rangle} = (\mu^{(n)})^{\circ m} \left[\overbrace{e_i', \dots, e_i'}^{m(n-1)+1} \right] = e_i', \quad i = 1, \dots, n. \tag{3.3.60}$$

Definition 3.3.25. A n-ary semigroup $\mathcal{S}^{(n)}$ in which every element belongs to a (not necessary unique) sequence of inverses $\mathbf{X}^{(n)}$ satisfying (3.3.25)–(3.3.28) and (3.3.59)–(3.3.60) is called a *sandwich q-regular semigroup with* $\langle m \rangle$*-idempotents.*

***Example* 3.3.26.** In the lowest case, for the sandwich regular binary semigroup $\mathcal{S}_2^{\text{sandw}}$ with generalized $\langle 2 \rangle$-idempotents ($n = 2$, $q = 2$, $m = 2$) we have (instead of (3.2.3))

$$x_1 x_2 x_1 x_2 x_1 = x_1, \tag{3.3.61}$$

$$x_2 x_1 x_2 x_1 x_2 = x_2. \tag{3.3.62}$$

And $\langle 2 \rangle$-idempotents become

$$e_1 = x_1 x_2, \quad e_2 = x_2 x_1, \tag{3.3.63}$$

$$e_1^3 = e_1, \quad e_2^3 = e_2. \tag{3.3.64}$$

In the binary case ($n = 2$) e_1, e_2 are called tripotents, and for arbitrary m generalized p-idempotents, where $p = m + 1$.

It is obvious that (3.3.61)–(3.3.62) follow from (3.2.3), but not vice versa.

***Example* 3.3.27.** For the sandwich 3-regular binary semigroup with 3-idempotents we have ($n = 2$, $q = 3$, $m = 2$)

$$x_1 x_2 x_3 x_1 x_2 x_3 x_1 = x_1, \tag{3.3.65}$$

$$x_2 x_3 x_1 x_2 x_3 x_1 x_2 = x_2, \tag{3.3.66}$$

$$x_3 x_1 x_2 x_3 x_1 x_2 x_3 = x_3, \tag{3.3.67}$$

and

$$e_1 = x_1 x_2 x_3, \quad e_2 = x_2 x_3 x_1, \quad e_3 = x_3 x_1 x_2, \tag{3.3.68}$$

$$e_1^3 = e_1, \quad e_2^3 = e_2, \quad e_3^3 = e_3. \tag{3.3.69}$$

Note that (3.3.65)–(3.3.67) are different from (3.3.35)–(3.3.37), because in the latter the multiplication is ternary $\mu^{(3)}$. In opposition to example 3.3.27 we have

***Example* 3.3.28.** For the sandwich regular ternary semigroup $\mathcal{S}^{(3)}_{\text{sandw}}$ with $\langle 2\rangle$-idempotents we obtain ($n = 3$, $q = 2$, $m = 2$)

$$(\mu^{(3)})^{\circ 3}[x_1, x_2, x_3, x_1, x_2, x_3, x_1, x_2, x_3, x_1, x_2, x_3, x_1] = x_1, \tag{3.3.70}$$

$$(\mu^{(3)})^{\circ 3}[x_2, x_3, x_1, x_2, x_3, x_1, x_2, x_3, x_1, x_2, x_3, x_1, x_2] = x_2, \tag{3.3.71}$$

$$(\mu^{(3)})^{\circ 3}[x_3, x_1, x_2, x_3, x_1, x_2, x_3, x_1, x_2, x_3, x_1, x_2, x_3] = x_3, \quad x_i \in S. \tag{3.3.72}$$

The three polyadic $\langle 2\rangle$-idempotents are the same as those in (3.3.31)–(3.3.33)

$$e_1' = \mu^{(3)}[x_1, x_2, x_3], \tag{3.3.73}$$

$$e_2' = \mu^{(3)}[x_2, x_3, x_1], \tag{3.3.74}$$

$$e_3' = \mu^{(3)}[x_3, x_1, x_2], \tag{3.3.75}$$

but now they satisfy the $\langle 2\rangle$-idempotence condition (3.1.8)

$$(\mu^{(3)})^{\circ 2}[e_i', e_i', e_i', e_i', e_i'] = e_i', \quad i = 1, 2, 3 \tag{3.3.76}$$

instead of (3.3.34) being the $\langle 1\rangle$-idempotence (3.1.9).

In Duplij (2017) it was shown that the representatives of a congruence (residue) class $[[a]]_b$ form an n-ary semigroup (see section 4.3)

$$\begin{gathered}\mathcal{R}^{(n)}_{a,\,b} = \langle \{\mathsf{x}_{a,b}(l)\} \mid \mu^{(n)}\rangle,\ \mathsf{x}_{a,b}(l) = a + lb \in [[a]]_b,\\ l \in \mathbb{Z},\ a \in \mathbb{Z}_+,\ b \in \mathbb{N},\ 0 \leqslant a \leqslant b-1,\end{gathered} \tag{3.3.77}$$

with respect to the multiplication

$$\mu^{(n)}[\mathsf{x}_{a,b}(l_1), \mathsf{x}_{a,b}(l_2), \ldots, \mathsf{x}_{a,b}(l_n)] = \mathsf{x}_{a,b}(l_1)\mathsf{x}_{a,b}(l_2)\ldots\mathsf{x}_{a,b}(l_n) \bmod b, \tag{3.3.78}$$

if

$$a^n = a \bmod b. \tag{3.3.79}$$

The limiting cases $a = 0$ and $a = b - 1$ correspond to the binary semigroup $\mathcal{R}^{(2)}_{0,\,b}$ (residue class $[[0]]_b$) and the ternary semigroup $\mathcal{R}^{(3)}_{b-1,\,b}$ (residue class $[[b-1]]_b$), respectively (for details, see Duplij 2017).

***Example* 3.3.29.** To construct a concrete realization of the lowest sandwich regularity (3.3.70)–(3.3.72) we consider the ternary semigroup of congruence class representatives $\mathcal{R}^{(3)}_{b-1,\,b}$ (the second limiting case) with the multiplication $\mu^{(3)}[\mathsf{x}_{b-1,b}(l_1), \mathsf{x}_{b-1,b}(l_2), \mathsf{x}_{b-1,b}(l_n)] = \mathsf{x}_{b-1,b}(l_1)\mathsf{x}_{b-1,b}(l_2)\mathsf{x}_{b-1,b}(l_3) \bmod b$ which is totally commutative. In terms of the representative numbers l the ternary product is $l = l_1 + l_2 + l_3 + 2,\ l,\ l_i \in \mathbb{Z}$, which shows that the idempotents (3.3.73)–(3.3.75) coincide. The 3-idempotent sandwich regularity relations (3.3.61)–(3.3.62) become $l_1 + l_2 + l_3 + 3 = 0$.

Further analysis of the constructions introduced here and examples of regularity extensions would be interesting to investigate in more detail, which can lead to new classes of higher regular and inverse semigroups and polyadic semigroups.

3.4 Polyadic-binary correspondence, regular semigroups, braid groups

We begin by observing that the defining relations of the von Neumann regular semigroups (e.g., Grillet 1995, Howie 1976, Petrich 1984) and the Artin braid group (Kassel and Turaev 2008, Kauffman 1991) correspond to such properties of ternary matrices (over the same set) as idempotence and the orders of elements (period). We then generalize the correspondence thus introduced to the polyadic case and thereby obtain higher degree (in our definition) analogs of the former. The higher (degree) regular semigroups obtained in this way have appeared previously in semisupermanifold theory (Duplij 2000), and higher regular categories in TQFT (Duplij and Marcinek 2002). The representations of the higher braid relations in vector spaces coincide with the higher braid equation and corresponding generalized R-matrix obtained in Duplij (2021b), as do the ordinary braid group and the Yang–Baxter equation (Turaev 1988). The proposed constructions use polyadic group methods and differ from the tetrahedron equation (Zamolodchikov 1981) and n-simplex equations (Hietarinta 1997) connected with the braid group representations (Li and Hu 1995, Hu 1997), also from higher braid groups of Manin and Schechtman (1989). Finally, we define higher degree (in our sense) versions of the Coxeter group and the symmetric group and show that they are connected in the classical (i.e., non-higher) case only.

3.4.1 Polyadic-binary correspondence

There is a general observation (Nikitin 1984), that a block-matrix, forming a semisimple $(2, n)$-ring (Artinian ring with binary addition and n-ary multiplication) has the shape

$$M(n-1) \equiv M^{((n-1)\times(n-1))} = \begin{pmatrix} 0 & \mathsf{m}^{(i_1\times i_2)} & 0 & \dots & 0 \\ 0 & 0 & \mathsf{m}^{(i_2\times i_3)} & \dots & 0 \\ 0 & 0 & \vdots & \ddots & \vdots \\ \vdots & \vdots & 0 & \ddots & \mathsf{m}^{(i_{n-2}\times i_{n-1})} \\ \mathsf{m}^{(i_{n-1}\times i_1)} & 0 & 0 & \dots & 0 \end{pmatrix}. \tag{3.4.1}$$

In other words, it is given by the cyclic shift $(n-1) \times (n-1)$ matrix, in which identities are replaced by blocks of suitable sizes and with arbitrary entries.

The set $\mathsf{M}(n-1) = \{M(n-1)\}$ is closed with respect to the product of n matrices of the form (3.4.1), and we will therefore call them *n-ary matrices*. They form an n-ary semigroup, and when the blocks are over an associative binary ring $\mathcal{R}$, then total n-ary associativity follows from the associativity of the ordinary matrix multiplication of (3.4.1).

Our proposal is to use single arbitrary elements (from rings with associative multiplication) in place of the blocks $\mathsf{m}^{(i\times j)}$, supposing that the elements of the multiplicative part $\mathcal{G}$ of the rings form binary (semi)groups having some special properties. Then we investigate the similar correspondence between the (multiplicative) properties of the matrices $M^{(n-1)}$, related to idempotence and order, and the appearance of the relations in $\mathcal{G}$ leading to higher regular semigroups (see definition 3.2.4) and braid groups, respectively. We call this connection a polyadic matrix-binary (semi)group correspondence (or in short the *polyadic-binary correspondence*).

In the lowest arity (nonbinary) case $n = 3$, the ternary case, the 2×2 matrices $M^{(2)}$ are anti-diagonal ones. From $(M(2))^3 = M(2)$ and $(M(2))^3 \sim E(2)$ (where $E(2)$ is the ternary identity, see below), we obtain the correspondences of the above conditions on $M(2)$ with the ordinary regular semigroups and braid groups, respectively. For higher arity we have

Definition 3.4.1. The *higher polyadic-binary correspondence* (for arities $n \geqslant 4$) states that

$$\begin{aligned} &\text{Idempotence } (M(n-1))^n = M(n-1)\text{—corresponds to higher} \\ &\qquad\qquad n\text{−regular semigroups,} \\ &(M(n-1))^n = qE(n-1)\text{ —corresponds to higher } n\text{−degree} \\ &\qquad\qquad \text{braid groups,} \end{aligned} \tag{3.4.2}$$

where $E(n-1)$ is the n-ary identity (see below) and q is a fixed element of the braid group.

3.4.2 Ternary matrix group corresponding to the regular semigroup

Let $\mathcal{G} = \left\{ G \middle| \mu_g^{(2)} \right\}$ be a semigroup with the underlying set $G = \{g\}$ and the binary multiplication $\mu_g^{(2)}$: $G \times G \to G$. The anti-diagonal matrices over $\mathcal{G}$

$$M^g(2) = M^{(2\times 2)}(g^{(1)}, g^{(2)}) = \begin{pmatrix} 0 & g^{(1)} \\ g^{(2)} & 0 \end{pmatrix}, \quad g^{(1)}, g^{(2)} \in G, \tag{3.4.3}$$

form a ternary semigroup $\mathcal{M}_g^{(3)} \equiv \mathcal{M}_g^{(n=3)} = \left\{ \mathrm{M}^g(2) \middle| \mu_g^{(3)} \right\}$, where $\mathrm{M}^g(2) = \{M^g(2)\}$ is the set of ternary matrices (3.4.3) closed under the ternary multiplication

$$\mu_g^{(3)}[M_1^g(2), M_2^g(2), M_3^g(2)] = M_1^g(2)M_2^g(2)M_3^g(2), \qquad \forall\, M_1^g(2), M_2^g(2), M_3^g(2) \in \mathrm{M}^g(2), \tag{3.4.4}$$

being the ordinary matrix product. Recall that an element of the semigroup $\mathcal{M}_g^{(3)}$ is idempotent, if

$$\mu_g^{(3)}[M^g(2), M^g(2), M^g(2)] = M^g(2), \tag{3.4.5}$$

which in the matrix form (3.4.4) leads to

$$(M^g(2))^3 = M^g(2). \tag{3.4.6}$$

We denote the set of idempotent ternary matrices by $\mathrm{M}_{\mathrm{id}}^g(2) = \left\{ M_{\mathrm{id}}^g(2) \right\}$.

Definition 3.4.2. A ternary matrix semigroup in which every element is idempotent (3.4.6) is called an *idempotent ternary semigroup.*

Using (3.4.3) and (3.4.6) the idempotence expressed in components gives the regularity conditions

$$g^{(1)}g^{(2)}g^{(1)} = g^{(1)}, \tag{3.4.7}$$

$$g^{(2)}g^{(1)}g^{(2)} = g^{(2)}, \quad \forall\, g^{(1)}, g^{(2)} \in G. \tag{3.4.8}$$

It is well known (Clifford and Preston 1961) that a binary semigroup $\mathcal{G}$ in which any two elements are mutually regular (3.4.7)–(3.4.8) is called a *regular semigroup* $\mathcal{G}_{\mathrm{reg}}$ (see (3.2.3)).

Proposition 3.4.3. The set of idempotent ternary matrices (3.4.6) form a ternary semigroup $\mathcal{M}_{3,*\mathrm{id}}^g = \left\{ \mathrm{M}_{*\mathrm{id}}^g(2) | \mu_g^{(3)} \right\}$, if $\mathcal{G}_{\mathrm{reg}}$ is abelian.

Proof. It follows from (3.4.7)–(3.4.8), that idempotence (and following from it regularity) is preserved with respect to the ternary multiplication (3.4.4), only when any $g^{(1)}, g^{(2)} \in G$ commute. □

Definition 3.4.4. We say that the set of idempotent ternary matrices $\mathrm{M}_{\mathrm{id}}^g(2)$ (3.4.6) is in *ternary–binary correspondence* with the regular (binary) semigroup $\mathcal{G}_{\mathrm{reg}}$ and write this as

$$\mathrm{M}_{\mathrm{id}}^g(2) \rightleftharpoons \mathcal{G}_{\mathrm{reg}}. \tag{3.4.9}$$

This means that such property of the ternary matrices as their idempotence (3.4.6) leads to the regularity conditions (3.4.7)–(3.4.8) in the correspondent binary group $\mathcal{G}$.

Remark. The correspondence (3.4.9) is not a homomorphism and not a bi-element mapping (Borowiec *et al* 2006), and also not a heteromorphism in sense of Duplij (2012), because we do not demand that the set of idempotent matrices $M_{\mathrm{id}}^{g}(2)$ form a ternary semigroup (which is possible in commutative case of $\mathcal{G}$ only, see proposition 3.4.3).

3.4.3 Polyadic matrix semigroup corresponding to the higher regular semigroup

We next extend the ternary–binary correspondence (3.4.9) to the n-ary matrix case (3.4.1) and thereby obtain higher n-regular binary semigroups. Here we use the following notation:

Round brackets: (n) is size of matrix $n \times n$, also the sequential number of a matrix element.

Square brackets: $[n]$ is number of multipliers in the regularity and braid conditions.

Angle brackets: $\langle \ell \rangle_n$ is the polyadic power (number of n-ary multiplications).

Let us introduce the $(n-1) \times (n-1)$ matrix over a binary group $\mathcal{G}$ of the form (3.4.1)

$$M^{g}(n-1) \equiv M^{((n-1)\times(n-1))}\left(g^{(1)}, g^{(2)}, \dots, g^{(n-1)}\right) = \begin{pmatrix} 0 & g^{(1)} & 0 & \dots & 0 \\ 0 & 0 & g^{(1)} & \dots & 0 \\ 0 & 0 & \vdots & \ddots & \vdots \\ \vdots & \vdots & 0 & \ddots & g^{(n-2)} \\ g^{(n-1)} & 0 & 0 & \dots & 0 \end{pmatrix}, \tag{3.4.10}$$

where $g^{(i)} \in G$.

Definition 3.4.6. The set of n-ary matrices $\mathrm{M}^{g}(n-1)$ (3.4.10) over $\mathcal{G}$ is an *n-ary matrix semigroup* $\mathcal{M}_n^g = \left\{\mathrm{M}^{g}(n-1) \middle| \mu_g^{(n)}\right\}$, where the multiplication

$$\begin{aligned} &\mu_g^{(n)}\left[M_1^g(n-1), M_2^g(n-1), \dots, M_n^g(n-1)\right] \\ &= M_1^g(n-1) M_2^g(n-1) \dots M_n^g(n-1), \; M_i^g(n-1) \in \mathcal{M}_n^g \end{aligned} \tag{3.4.11}$$

is the ordinary product of n matrices $M_i^g(n-1) \equiv M^{((n-1)\times(n-1))}\left(g_i^{(1)}, g_i^{(2)}, \dots, g_i^{(n-1)}\right)$, see (3.4.10).

Recall that the polyadic power ℓ of an element M from a n-ary semigroup $\mathcal{M}^{(n)}$ is defined by (e.g., Post 1940)

$$M^{\langle \ell \rangle_n} = \left(\mu^{(n)}\right)^{\circ \ell} \left[\overbrace{M, \ldots, M}^{\ell(n-1)+1} \right], \tag{3.4.12}$$

such that ℓ coincides with the *number of n-ary multiplications*. In the binary case $n = 2$ the polyadic power is connected with the ordinary power p (*number of elements* in the product) as $p = \ell + 1$, i.e., $M^{\langle \ell \rangle_2} = M^{\ell+1} = M^p$. In the ternary case $n = 3$ we have $\langle \ell \rangle_3 = 2\ell + 1$, and so the lhs of (3.4.6) is of polyadic power $\ell = 1$.

Definition 3.4.7. An element of a n-ary semigroup $M \in \mathcal{M}_3$ is called *idempotent*, if its first polyadic power coincides with itself

$$M^{\langle 1 \rangle_n} = M, \tag{3.4.13}$$

and $\langle \ell \rangle$-*idempotent*, if

$$M^{\langle \ell \rangle_n} = M, \; M^{\langle \ell-1 \rangle_n} \neq M. \tag{3.4.14}$$

Definition 3.4.8. An n-ary semigroup $\mathcal{M}_n$ is called *idempotent* (ℓ*-idempotent*), if each of its elements $M \in \mathcal{M}_n$ is idempotent ($\langle \ell \rangle$*-idempotent*).

Assertion 3.4.9. From $M^{\langle 1 \rangle_n} = M$ it follows that $M^{\langle \ell \rangle_n} = M$, but not vice-versa, therefore all $\langle 1 \rangle$-idempotent elements are $\langle \ell \rangle$-idempotent, but an $\langle \ell \rangle$-idempotent element need not be $\langle 1 \rangle$-idempotent.

Therefore, the definition given in (3.4.14) makes sense.

Proposition 3.4.10. If an n-ary matrix $M^g(n-1) \in \mathcal{M}_n^g$ is idempotent (3.4.13), then its elements satisfy the $(n-1)$ relations

$$g^{(1)} g^{(2)}, \ldots g^{(n-2)} g^{(n-1)} g^{(1)} = g^{(1)}, \tag{3.4.15}$$

$$g^{(2)} g^{(3)}, \ldots g^{(n-1)} g^{(1)} g^{(2)} = g^{(2)}, \tag{3.4.16}$$

$$\vdots$$

$$g^{(n-1)} g^{(1)} g^{(2)}, \ldots g^{(n-2)} g^{(n-1)} = g^{(n-1)}, \;\; \forall g^{(1)}, \ldots, g^{(n-1)} \in G. \tag{3.4.17}$$

Proof This follows from (3.4.10), (3.4.11) and (3.4.13). □

Comparing (3.4.15)–(3.4.17) with the q-regularity (definition 3.2.4), we have

Assertion 3.4.11. *From idempotence of the* n*-ary matrix* $M^g(n-1)$ *it follows that the matrix elements satisfy* q*-regularity conditions with* $q = n - 1$.

The lowest (ternary) case $n = 3$ gives the ordinary regularity (2-regularity in our notation) (3.4.7)–(3.4.8).

Proposition 3.4.12. If an n-ary matrix $M^g(n-1) \in \mathcal{M}_n^g$ is $\langle \ell \rangle$-idempotent (3.4.14), then its elements satisfy the following $(n-1)$ relations

$$\overbrace{(g^{(1)}g^{(2)} \ldots g^{(n-2)}g^{(n-1)}) \ldots (g^{(1)}g^{(2)}, \ldots g^{(n-2)}g^{(n-1)})}^{\ell} g^{(1)} = g^{(1)}, \tag{3.4.18}$$

$$\overbrace{(g^{(2)}g^{(3)}, \ldots g^{(n-2)}g^{(n-1)}g^{(1)}) \ldots (g^{(2)}g^{(3)} \ldots g^{(n-2)}g^{(n-1)}g^{(1)})}^{\ell} g^{(2)} = g^{(2)}, \tag{3.4.19}$$

$$\vdots$$

$$\overbrace{(g^{(n-1)}g^{(1)}g^{(2)} \ldots g^{(n-3)}g^{(n-2)}) \ldots (g^{(n-1)}g^{(1)}g^{(2)} \ldots g^{(n-3)}g^{(n-2)})}^{\ell} g^{(n-1)} = g^{(n-1)}, \tag{3.4.20}$$

$$\forall g^{(1)}, \ldots, g^{(n-1)} \quad \in G.$$

Proof. This also follows from (3.4.10), (3.4.11) and (3.4.14). □

Comparing (3.4.18)–(3.4.20) with definition 3.3.24 we arrive to

Assertion 3.4.13. *From $\langle \ell \rangle$-idempotence of the* n*-ary matrix $M^g(n-1)$ it follows that the matrix elements satisfy the (binary) sandwich* q*-regularity conditions with* $q = n - 1$ and generalized $\langle m \rangle$-idempotents with $m = \ell$.

We here denote the binary semigroup $\mathcal{G}$, in which $n-1$ elements are $(n-1)$-regular (with generalized $\langle \ell \rangle$-idempotents) as $\mathcal{G}_{\text{reg}}[n-1]$ ($\mathcal{G}_{\langle \ell \rangle-\text{reg}}[n-1]$).

Similarly to assertion 3.4.9, it is seen that $[n]$-$\langle \ell \rangle$-regularity (3.4.18)–(3.4.20) follows from $[n]$-regularity (3.4.15)–(3.4.17), but not the other way around, and therefore we have

Assertion 3.4.14. *If a binary semigroup $\mathcal{G}_{\text{reg}}[n-1]$ is $(n-1)$-regular, then it is $\mathcal{G}_{\langle \ell \rangle-\text{reg}}[n-1]$ as well, but not vice-versa.*

Proposition 3.4.15. *The set of idempotent ($\langle \ell \rangle$-idempotent)* n*-ary matrices* $\mathrm{M}_{*id}^g(n-1)$ *form a* n*-ary semigroup* $\mathcal{M}_{3,\,*id}^g = \left\{ \mathrm{M}_{*id}^g(n-1) \middle| \mu_g^{(n)} \right\}$, *if and only if* $\mathcal{G}_{\text{reg}}[n-1]$ ($\mathcal{G}_{\langle \ell \rangle-\text{reg}}[n-1]$) *is abelian.*

Proof. It follows from (3.4.15)–(3.4.20) that the idempotence ($\langle \ell \rangle$-idempotence) and the following $(n-1)$-regularity ($(n-1)$-regularity with $\langle \ell \rangle$-idempotents) are preserved with respect the n-ary multiplication (3.4.11) only in the case, when all $g^{(1)}, \dots, g^{(n-1)} \in G$ mutually commute. □

By analogy with (3.4.9), we have

Definition 3.4.16. We will say that the set of n-ary $(n-1) \times (n-1)$ matrices $\mathrm{M}_{\mathrm{id}}^{g}(n-1)$ (3.4.10) over the underlying set G is in *polyadic-binary correspondence* with the binary $(n-1)$-regular semigroup $\mathcal{G}_{\mathrm{reg}}[n-1]$ and write this as

$$\mathrm{M}_{\mathrm{id}}^{g}(n-1) \rightleftharpoons \mathcal{G}_{\mathrm{reg}}[n]. \tag{3.4.21}$$

Thus, using the idempotence condition for n-ary matrices in components (being simultaneously elements of a binary semigroup $\mathcal{G}_{\mathrm{free}}$) and the polyadic-binary correspondence (3.4.21) we obtain the higher regularity conditions (3.4.15)–(3.4.20) generalizing the ordinary regularity (3.4.7)–(3.4.8), which allows us to define the higher $(n-1)$-regular binary semigroups $\mathcal{G}_{\mathrm{reg}}[n-1]$ ($\mathcal{G}_{\langle\ell\rangle-\mathrm{reg}}[n-1]$).

Example **3.4.17.** The lowest nontrivial ($n \geqslant 3$) case is $n = 4$, where the 3×3 matrices over $\mathcal{G}$ are of the shape

$$M(3) = M^{(3\times 3)} = \begin{pmatrix} 0 & a & 0 \\ 0 & 0 & b \\ c & 0 & 0 \end{pmatrix}, \quad a, b, c \in G, \tag{3.4.22}$$

and they form the 4-ary matrix semigroup $\mathcal{M}_{g}^{(4)}$. The idempotence $(M(3))^{\langle 1\rangle_4} = (M(3))^3 = M(3)$ gives three 3-regularity conditions

$$abca = a, \tag{3.4.23}$$

$$bcab = b, \tag{3.4.24}$$

$$cabc = c. \tag{3.4.25}$$

According to the polyadic-binary correspondence (3.4.21), the conditions (3.4.23)–(3.4.25) are 3-regularity relations for the binary semigroup $\mathcal{G}$, which defines to the higher 3-regular binary semigroup $\mathcal{G}_{\mathrm{reg}}[3]$ (see (3.2.16)–(3.2.18)).

In the case $\ell = 2$, we have $(M(3))^{\langle 2\rangle_4} = (M(3))^7 = M(3)$, which gives three 3-regularity conditions with generalized $\langle 2\rangle$-idempotents (they are different from 6-regularity)

$$abcabca = a, \tag{3.4.26}$$

$$bcabcab = b, \tag{3.4.27}$$

$$cabcabc = c, \tag{3.4.28}$$

and these define the semigroup $\mathcal{G}_{\langle 2\rangle-\mathrm{reg}}[3]$ in our notation. Obviously, (3.4.26)–(3.4.28) follow from (3.4.23)–(3.4.25), but not vice-versa.

The higher regularity conditions (3.4.23)–(3.4.25) obtained above from the idempotence of polyadic matrices using the polyadic-binary correspondence, appeared first in Duplij (1998) and were then used for transition functions in the investigation of semisupermanifolds (Duplij 2000) and higher regular categories in Topological Quantum Field Theory (Duplij and Marcinek 2001b, 2002).

Now we turn to the second line of (3.4.2), and in the same way as above introduce higher degree braid groups.

3.4.4 Ternary matrix group corresponding to the braid group

Recall the definition of the Artin braid group (Artin 1947) in terms of generators and relations (Kassel and Turaev 2008) (we follow the algebraic approach, see, e.g., Markov 1945).

The *Artin braid group* $\mathcal{B}_k$ (with k strands and the identity $e \in \mathcal{B}_k$) has the presentation by $k-1$ generators $\sigma_1, \dots, \sigma_{k-1}$ satisfying $k(k-1)/2$ relations

$$\sigma_i\sigma_{i+1}\sigma_i = \sigma_{i+1}\sigma_i\sigma_{i+1}, \quad 1 \leqslant i \leqslant k-2, \tag{3.4.29}$$

$$\sigma_i\sigma_j = \sigma_j\sigma_i, \quad \mid i-j \mid \geqslant 2, \tag{3.4.30}$$

where (3.4.29) are called the *braid relations*, and (3.4.30) are called *far commutativity*. A general element of $\mathcal{B}_k$ is a word of the form

$$w = \sigma_{i_1}^{p_1} \dots \sigma_{i_r}^{p_r} \dots \sigma_{i_m}^{p_m}, \quad i_m = 1, \dots, k, \tag{3.4.31}$$

where $p_r \in \mathbb{Z}$ are (positive or negative) powers of the generators σ_{i_r}, $r = 1, \dots, m$ and $m \in \mathbb{N}$.

For instance, $\mathcal{B}_3$ is generated by σ_1 and σ_2 satisfying one relation $\sigma_1\sigma_2\sigma_1 = \sigma_2\sigma_1\sigma_2$, and is isomorphic to the trefoil knot group. The group $\mathcal{B}_4$ has 3 generators $\sigma_1, \sigma_2, \sigma_3$ satisfying

$$\sigma_1\sigma_2\sigma_1 = \sigma_2\sigma_1\sigma_2, \tag{3.4.32}$$

$$\sigma_2\sigma_3\sigma_2 = \sigma_3\sigma_2\sigma_3, \tag{3.4.33}$$

$$\sigma_1\sigma_3 = \sigma_3\sigma_1. \tag{3.4.34}$$

The representation theory of $\mathcal{B}_k$ is well known and well established (Kassel and Turaev 2008, Kauffman 1991). The connections with the Yang–Baxter equation were investigated, e.g., in Turaev (1988).

Now we build a ternary group of matrices over $\mathcal{B}_k$ having generators satisfying relations which are connected with the braid relations (3.4.29)–(3.4.30). We then generalize our construction to an n-ary matrix group, which gives us the possibility to 'go back' and define some special higher analogs of the Artin braid group.

Let us consider the set of anti-diagonal 2×2 matrices over $\mathcal{B}_k$

$$M(2) = M^{(2\times 2)}(b^{(1)}, b^{(2)}) = \begin{pmatrix} 0 & b^{(1)} \\ b^{(2)} & 0 \end{pmatrix}, \quad b^{(1)}, b^{(2)} \in \mathcal{B}_k. \tag{3.4.35}$$

Definition 3.4.18. The set of matrices $\mathrm{M}(2) = \{M(2)\}$ (3.4.35) over $\mathcal{B}_k$ form a *ternary matrix semigroup* $\mathcal{M}_3 = \{\mathrm{M}(2)|\mu^{(3)}\}$, where $n = 3$ is the arity of the following multiplication

$$\mu_3[M_1(2), M_2(2), M_3(2)] \equiv M_1(2), M_2(2), M_3(2) = M(2), \tag{3.4.36}$$

$$b_1^{(1)} b_2^{(2)} b_3^{(1)} = b^{(1)}, \tag{3.4.37}$$

$$b_1^{(2)} b_2^{(1)} b_3^{(2)} = b^{(2)}, \quad b_i^{(1)}, b_i^{(2)} \in \mathcal{B}_k, \; M_i(2) = \begin{pmatrix} 0 & b_i^{(1)} \\ b_i^{(2)} & 0 \end{pmatrix} \tag{3.4.38}$$

and the associativity is governed by the associativity of both the ordinary matrix product in the rhs of (3.4.36) and $\mathcal{B}_k$.

Proposition 3.4.19. $\mathcal{M}^{(3)}$ is a ternary matrix group.

Proof. Each element of the ternary matrix semigroup $M(2) \in \mathcal{M}_3$ is invertible (in the ternary sense) and has a *querelement* $\bar{M}(2)$ (a polyadic analog of the group inverse; Dörnte 1929) defined by

$$\begin{aligned} \mu^{(3)}[M(2), M(2), \bar{M}(2)] &= \mu^{(3)}[M(2), \bar{M}(2), M(2)] \\ &= \mu^{(3)}[\bar{M}(2), M(2), M(2)] = M(2). \end{aligned} \tag{3.4.39}$$

It follows from (3.4.36)–(3.4.38), that

$$\bar{M}(2) = (M(2))^{-1} = \begin{pmatrix} 0 & (b^{(1)})^{-1} \\ (b^{(2)})^{-1} & 0 \end{pmatrix}, \quad b^{(1)}, b^{(2)} \in \mathcal{B}_k, \tag{3.4.40}$$

where $(M^{(2)})^{-1}$ denotes the ordinary matrix inverse (but not the binary group inverse which does not exist in the n-ary case, $n \geqslant 3$). Non-commutativity of $\mu^{(3)}$ is provided by (3.4.37)–(3.4.38). □

The ternary matrix group $\mathcal{M}_3$ has the *ternary identity*

$$E(2) = \begin{pmatrix} 0 & e \\ e & 0 \end{pmatrix}, \quad e \in \mathcal{B}_k, \tag{3.4.41}$$

where e is the identity of the binary group $\mathcal{B}_k$, and

$$\mu_3[M(2), E(2), E(2)] = \mu_3[E(2), M(2), E(2)] = \mu_3[E(2), E(2), M(2)] = M(2). \tag{3.4.42}$$

We observe that the ternary product $\mu^{(3)}$ in components is 'naturally braided' (3.4.37)–(3.4.38). This allows us to ask the question: which generators of the ternary group $\mathcal{M}_3$ can be constructed using the Artin braid group generators $\sigma_i \in \mathcal{B}_k$ and the relations (3.4.29)–(3.4.30)?

3.4.5 Ternary matrix generators

Let us introduce $(k-1)^2$ ternary 2×2 *matrix generators*

$$\Sigma_{ij}(2) = \Sigma_{ij}^{(2\times 2)}(\sigma_i, \sigma_j) = \begin{pmatrix} 0 & \sigma_i \\ \sigma_j & 0 \end{pmatrix}, \tag{3.4.43}$$

where $\sigma_i \in \mathcal{B}_k$, $i = 1\ldots, k-1$ are generators of the Artin braid group. The querelement of $\Sigma_{ij}(2)$ is defined by analogy with (3.4.40) as

$$\bar{\Sigma}_{ij}(2) = \left(\Sigma_{ij}(2)\right)^{-1} = \begin{pmatrix} 0 & \sigma_j^{-1} \\ \sigma_i^{-1} & 0 \end{pmatrix}. \tag{3.4.44}$$

Now we are in a position to present a ternary matrix group with multiplication μ_3 in terms of generators and relations in such a way that the braid group relations (3.4.29)–(3.4.30) will be reproduced.

Proposition 3.4.20. The relations for the matrix generators $\Sigma_{ij}(2)$ corresponding to the braid group relations for σ_i (3.4.29)–(3.4.30) have the form

$$\begin{aligned}&\mu^{(3)}\left[\Sigma_{i,j+1}(2), \Sigma_{i,j+1}(2), \Sigma_{i,j+1}(2)\right] = \mu^{(3)}\left[\Sigma_{j+1,i}(2), \Sigma_{j+1,i}(2), \Sigma_{j+1,i}(2)\right] \\ &= q_i^{[3]} E(2), \quad 1 \leqslant i \leqslant k-2,\end{aligned} \tag{3.4.45}$$

$$\mu^{(3)}\left[\Sigma_{ij}(2), \Sigma_{ij}(2), E(2)\right] = \mu^{(3)}\left[\Sigma_{ji}(2), \Sigma_{ji}(2), E(2)\right], \quad |\, i - j\,| \geqslant 2, \tag{3.4.46}$$

where $q_i^{[3]} = \sigma_i \sigma_{i+1} \sigma_i = \sigma_{i+1} \sigma_i \sigma_{i+1}$, and $E(2)$ is the ternary identity (3.4.41).

Proof. Use $\mu^{(3)}$ as the triple matrix product (3.4.36)–(3.4.38) and the braid relations (3.4.29)–(3.4.30). □

Definition 3.4.21. We say that the ternary matrix group $\mathcal{M}_3^{\text{gen}-\Sigma}$ generated by the matrix generators $\Sigma_{ij}(2)$ satisfying the relations (3.4.45)–(3.4.46) is in *ternary–binary correspondence* with the braid (binary) group $\mathcal{B}_k$, which is denoted as (see (3.4.9))

$$\mathcal{M}_3^{\text{gen}-\Sigma} \rightleftharpoons \mathcal{B}_k. \tag{3.4.47}$$

Indeed, in components the relations (3.4.45) give (3.4.29), and (3.4.46) leads to (3.4.30).

***Remark* 3.4.22.** Note that the above construction is totally different from the bi-element representations of ternary groups considered in Borowiec *et al* (2006) (for n-ary groups see Duplij 2012).

Definition 3.4.23 An element $M(2) \in \mathcal{M}_3$ is of *finite polyadic (ternary) order*, if there exists a finite ℓ such that

$$M(2)^{\langle \ell \rangle_3} = M(2)^{2\ell+1} = E(2), \tag{3.4.48}$$

where $E(2)$ is the ternary matrix identity (3.4.41).

Definition 3.4.24. An element $M(2) \in \mathcal{M}_3$ is of *finite q-polyadic (q-ternary) order*, if there exists a finite ℓ such that

$$M(2)^{\langle \ell \rangle_3} = M(2)^{2\ell+1} = qE(2), \quad q \in \mathcal{B}_k. \tag{3.4.49}$$

The relations (3.4.45) therefore say that the ternary matrix generators $\Sigma_{i,j+1}(2)$ are of finite q-ternary order. Each element of $\mathcal{M}_3^{\text{gen}-\Sigma}$ is a ternary matrix word (analogous to the binary word (3.4.31)), being the ternary product of the polyadic powers (3.4.12) of the 2 × 2 matrix generators $\Sigma_{ij}^{(2)}$ and their querelements $\bar{\Sigma}_{ij}^{(2)}$ (on choosing the first or second row)

$$\begin{aligned} W &= \begin{pmatrix} \Sigma_{ij_1}(2) \\ \bar{\Sigma}_{ij_1}(2) \end{pmatrix}^{\langle \ell_1 \rangle_3}, \dots, \begin{pmatrix} \Sigma_{ij_r}(2) \\ \bar{\Sigma}_{ij_r}(2) \end{pmatrix}^{\langle \ell_r \rangle_3}, \dots, \begin{pmatrix} \Sigma_{i_m j_m}(2) \\ \bar{\Sigma}_{i_m j_m}(2) \end{pmatrix}^{\langle \ell_m \rangle_3} \\ &= \begin{pmatrix} \Sigma_{ij_1}(2) \\ \bar{\Sigma}_{ij_1}(2) \end{pmatrix}^{2\ell_1+1}, \dots, \begin{pmatrix} \Sigma_{ij_r}(2) \\ \bar{\Sigma}_{ij_r}(2) \end{pmatrix}^{2\ell_r+1}, \dots, \begin{pmatrix} \Sigma_{i_m j_m}(2) \\ \bar{\Sigma}_{i_m j_m}(2) \end{pmatrix}^{2\ell_m+1}, \end{aligned} \tag{3.4.50}$$

where $r = 1, \dots, m$, $i_r, j_r = 1, \dots, k$ (from $\mathcal{B}_k$), $\ell_r, m \in \mathbb{N}$. In the ternary case the total number of multipliers in (3.4.50) should be compatible with (3.4.12), i.e., $(2\ell_1 + 1) + \cdots + (2\ell_r + 1) + \cdots + (2\ell_m + 1) = 2\ell_W + 1$, $\ell_W \in \mathbb{N}$, and m is therefore odd. Thus, we have

***Remark* 3.4.25.** The ternary words (3.4.50) in components give only a subset of the binary words (3.4.31), and so $\mathcal{M}_3^{\text{gen}-\Sigma}$ corresponds to $\mathcal{B}_k$, but does not present it.

***Example* 3.4.26.** For $\mathcal{B}_3$ we have only two ternary 2 × 2 matrix generators

$$\Sigma_{12}(2) = \begin{pmatrix} 0 & \sigma_1 \\ \sigma_2 & 0 \end{pmatrix}, \quad \Sigma_{21}(2) = \begin{pmatrix} 0 & \sigma_2 \\ \sigma_1 & 0 \end{pmatrix}, \tag{3.4.51}$$

satisfying

$$(\Sigma_{12}(2))^{\langle 1 \rangle_3} = (\Sigma_{12}(2))^3 = q_1^{[3]} E(2), \tag{3.4.52}$$

$$(\Sigma_{21}(2))^{\langle 1\rangle_3} = (\Sigma_{21}(2))^3 = q_1^{[3]}E(2), \tag{3.4.53}$$

where $q_1^{[3]} = \sigma_1\sigma_2\sigma_1 = \sigma_2\sigma_1\sigma_2$, and both matrix relations (3.4.52)–(3.4.53) coincide in components.

***Example* 3.4.27.** For $\mathcal{B}_4$, the ternary matrix group $\mathcal{M}_3^{\text{gen}-\Sigma}$ is generated by more generators satisfying the relations

$$(\Sigma_{12}(2))^3 = q_1^{(3)}E(2), \tag{3.4.54}$$

$$(\Sigma_{21}(2))^3 = q_1^{(3)}E(2), \tag{3.4.55}$$

$$(\Sigma_{23}(2))^3 = q_2^{(3)}E(2), \tag{3.4.56}$$

$$(\Sigma_{32}(2))^3 = q_2^{(3)}E(2), \tag{3.4.57}$$

$$\Sigma_{13}(2)\Sigma_{13}(2)E(2) = \Sigma_{31}(2)\Sigma_{31}(2)E(2), \tag{3.4.58}$$

where $q_1^{[3]} = \sigma_1\sigma_2\sigma_1 = \sigma_2\sigma_1\sigma_2$ and $q_2^{[3]} = \sigma_2\sigma_3\sigma_2 = \sigma_3\sigma_2\sigma_3$. The first two relations give the braid relations (3.4.32)–(3.4.33), while the last relation corresponds to far commutativity (3.4.34).

3.4.6 Generated n-ary matrix group corresponding the higher braid group

The above construction of the ternary matrix group $\mathcal{M}_3^{\text{gen}-\Sigma}$ corresponding to the braid group $\mathcal{B}_k$ can be naturally extended to the n-ary case, which will allow us to 'go in the opposite way' and build so-called higher degree analogs of $\mathcal{B}_k$ (in our sense: the number of factors in braid relations more than 3). We denote such a braid-like group with k generators by $\mathcal{B}_k[n-1]$, where n is the number of generator multipliers in the braid relations (as in the regularity relations (3.4.15)–(3.4.17)). Simultaneously n is the arity of the matrices (3.4.35), we therefore call $\mathcal{B}_k[n-1]$ a higher $(n-1)$-*degree analog* of the braid group $\mathcal{B}_k$. In this notation the Artin braid group $\mathcal{B}_k$ is $\mathcal{B}_k[3]$. Now we build $\mathcal{B}_k[n-1]$ for any degree n exploiting the 'reverse' procedure, as for $n=3$ and $\mathcal{B}_k$ in section 3.4.4. For that we need an n-ary generalization of the matrices over $\mathcal{B}_k$, which in the ternary case are the anti-diagonal matrices $M(2)$ (3.4.35), and the generator matrices $\Sigma_{ij}(2)$ (3.4.43). Then, using the n-ary analog of multiplication (3.4.37)–(3.4.38) we will obtain the higher degree (than (3.4.29)) braid relations which generate the so-called *higher n-degree braid group*. In distinction to the higher degree regular semigroup construction from section 3.4.3, where the n-ary matrices form a semigroup for the Abelian group $\mathcal{G}_{\text{free}}$, using the generator matrices, we construct an n-ary matrix semigroup (presented by generators and relations) for any (even non-commutative) matrix entries. In this way the polyadic-binary correspondence will connect n-ary matrix groups of finite order with higher binary braid groups (cf idempotent n-ary matrices and higher regular semigroups (3.4.21)).

Let us consider a binary group $\mathcal{B}$ and construct over it an n-ary matrix group along the lines of Nikitin (1984), similarly to the ternary matrix group $\mathcal{M}_3$ in (3.4.35)–(3.4.38).

Definition 3.4.28. A set $\mathrm{M}(n-1) = \{M(n-1)\}$ of n-ary $(n-1)\times(n-1)$ matrices

$$M(n-1) = M^{((n-1)\times(n-1))}(\mathsf{b}^{(1)}, \mathsf{b}^{(2)}, \dots, \mathsf{b}^{(n-1)}) = \begin{pmatrix} 0 & \mathsf{b}^{(1)} & 0 & \dots & 0 \\ 0 & 0 & \mathsf{b}^{(1)} & \dots & 0 \\ 0 & 0 & \vdots & \ddots & \vdots \\ \vdots & \vdots & 0 & \ddots & \mathsf{b}^{(n-2)} \\ \mathsf{b}^{(n-1)} & 0 & 0 & \dots & 0 \end{pmatrix}, \quad (3.4.59)$$

$$\mathsf{b}^{(j)} \in \mathcal{B},\ j = 1, \dots, n-1,$$

form an n-ary matrix semigroup $\mathcal{M}^{(n)} = \{\mathrm{M}(n-1) | \mu^{(n)}\}$, where $\mu^{(n)}$ is the following n-ary multiplication

$$\mu_n[M_1(n-1), M_2(n-1), \dots, M_n(n-1)] = M_1(n-1)M_2(n-1)\dots M_n(n-1) = M(n-1), \quad (3.4.60)$$

$$\mathsf{b}_1^{(1)}\mathsf{b}_2^{(2)}, \dots \mathsf{b}_{n-1}^{(n-1)}\mathsf{b}_n^{(1)} = \mathsf{b}^{(1)}, \quad (3.4.61)$$

$$\mathsf{b}_1^{(2)}\mathsf{b}_2^{(3)}, \dots \mathsf{b}_{n-1}^{(1)}\mathsf{b}_n^{(2)} = \mathsf{b}^{(2)}, \quad (3.4.62)$$

$$\vdots$$

$$\mathsf{b}_1^{(n-1)}\mathsf{b}_2^{(1)}, \dots \mathsf{b}_{n-1}^{(n-2)}\mathsf{b}_n^{(n-1)} = \mathsf{b}^{(n-1)}, \quad (3.4.63)$$

where the rhs of (3.4.60) is the ordinary matrix multiplication of n-ary matrices (3.4.59) $M_i(n-1) = M^{((n-1)\times(n-1))}(\mathsf{b}_i^{(1)}, \mathsf{b}_i^{(2)}, \dots, \mathsf{b}_i^{(n-1)})$, $i = 1, \dots, n$.

Proposition 3.4.29. $\mathcal{M}^{(n)}$ is an n-ary matrix group.

Proof. Because $\mathcal{B}$ is a (binary) group with the identity $\mathbf{e} \in \mathcal{B}$, each element of the n-ary matrix semigroup $M(n-1) \in \mathcal{M}_n$ is invertible (in the n-ary sense) and has a *querelement* $\bar{M}(n-1)$ (see Dörnte 1929) defined by (see (3.4.42))

$$\mu_n\Bigg[\overbrace{M(n-1), \dots, M(n-1)}^{n\ 1}, \bar{M}(n-1)\Bigg] = \dots = M(n-1), \quad (3.4.64)$$

where $\bar{M}(n-1)$ can be on any place, and so we have n conditions (see (3.4.39) for $n = 3$). □

The n-ary matrix group has the polyadic identity

$$E(n-1) = E^{((n-1)\times(n-1))} = \begin{pmatrix} 0 & \mathbf{e} & 0 & \ldots & 0 \\ 0 & 0 & \mathbf{e} & \ldots & 0 \\ 0 & 0 & \vdots & \ddots & \vdots \\ \vdots & \vdots & 0 & \ddots & \mathbf{e} \\ \mathbf{e} & 0 & 0 & \ldots & 0 \end{pmatrix}, \quad \mathbf{e} \in \mathcal{B}, \tag{3.4.65}$$

satisfying

$$\mu_n[M(n-1), E(n-1), \ldots, E(n-1)] = \ldots = M(n-1), \tag{3.4.66}$$

where $M(n-1)$ can be on any place, and so we have n conditions (see (3.4.42)).

Definition 3.4.30. An element of an n-ary group $M(n-1) \in \mathcal{M}^{(n)}$ has the *polyadic order* ℓ, if

$$(M(n-1))^{\langle \ell \rangle_n} \equiv (M(n-1))^{\ell(n-1)+1} = E(n-1), \tag{3.4.67}$$

where $E(n-1) \in \mathcal{M}^{(n)}$ is the polyadic identity (3.4.65), for $n = 3$ see (3.4.41).

Definition 3.4.31. An element ($(n-1) \times (n-1)$-matrix over $\mathcal{B}$) $M(n-1) \in \mathcal{M}^{\{n\}}$ is of *finite q-polyadic order*, if there exists a finite ℓ such that

$$(M(n-1))^{\langle \ell \rangle_n} \equiv (M(n-1))^{\ell(n-1)+1} = \mathsf{q}E(n-1), \quad \mathsf{q} \in \mathcal{B}_{\text{free}}. \tag{3.4.68}$$

Let us assume that the binary group $\mathcal{B}$ is presented by generators and relations (see the Artin braid group (3.4.29)–(3.4.30)), i.e., it is generated by $k-1$ generators $\boldsymbol{\sigma}_i$, $i = 1, \ldots, k-1$. An element of $\mathcal{B}_k^{\text{gen}-\sigma} \equiv \mathcal{B}(\mathbf{e}, \boldsymbol{\sigma}_i)$ is the word of the form (3.4.31). To find the relations between $\boldsymbol{\sigma}_i$ we construct the corresponding n-ary matrix generators analogous to the ternary ones (3.4.43). Then using an n-ary version of the relations (3.4.45)–(3.4.46) for the matrix generators, as the finite order conditions (3.4.68), we will obtain the corresponding higher degree braid relations for the binary generators $\boldsymbol{\sigma}_i$, and can therefore present a higher degree braid group $\mathcal{B}_k[n-1]$ in the form of generators and relations.

Using $k-1$ generators $\boldsymbol{\sigma}_i$ of $\mathcal{B}_k^{\text{gen}-\sigma}$ we build $(k-1)^n$ polyadic (or n-ary) $(n-1) \times (n-1)$-matrix generators having $n-1$ indices $i_1, \ldots, i_{n-1} = 1, \ldots, k-1$, as follows

$$\Sigma_{i_1,\ldots,i_{n-1}}(n-1) \equiv \Sigma_{i_1,\ldots,i_{n-1}}^{((n-1)\times(n-1))}(\boldsymbol{\sigma}_{i_1}, \ldots, \boldsymbol{\sigma}_{i_{n-1}}) = \begin{pmatrix} 0 & \boldsymbol{\sigma}_{i_1} & 0 & \ldots & 0 \\ 0 & 0 & \boldsymbol{\sigma}_{i_2} & \ldots & 0 \\ 0 & 0 & \vdots & \ddots & \vdots \\ \vdots & \vdots & 0 & \ddots & \boldsymbol{\sigma}_{i_{n-2}} \\ \boldsymbol{\sigma}_{i_{n-1}} & 0 & 0 & \ldots & 0 \end{pmatrix}. \tag{3.4.69}$$

For the matrix generator $\Sigma_{i_1,\ldots,i_{n-1}}(n-1)$ (3.4.69) its querelement $\bar{\Sigma}_{i_1,\ldots,i_{n-1}}(n-1)$ is defined by (3.4.64).

We now build an n-ary matrix analog of the braid relations (3.4.29), (3.4.45) and of far commutativity (3.4.30), (3.4.46). Using (3.4.69) we obtain $(n-1)$ conditions that the matrix generators are of finite polyadic order (analog of (3.4.45))

$$\mu^{(n)}[\Sigma_{i,i+1,\dots,i+n-2}(n-1), \Sigma_{i,i+1,\dots,i+n-2}(n-1), \dots, \Sigma_{i,i+1,\dots,i+n-2}(n-1)] \quad (3.4.70)$$

$$=\mu^{(n)}\left[\Sigma_{i+1,i+2,\dots,i+n-2,i}(n-1), \Sigma_{i+1,i+2,\dots,i+n-2,i}(n-1), \dots, \Sigma_{i+1,i+2,\dots,i+n-2,i}(n-1)\right] \quad (3.4.71)$$

$$\vdots$$

$$\mu^{(n)}\left[\Sigma_{i+n-2,i,i+1,\dots,i+n-3}(n-1), \Sigma_{i+n-2,i,i+1,\dots,i+n-3}(n-1), \dots, \Sigma_{i+n-2,i,i+1,\dots,i+n-3}(n-1)\right] \quad (3.4.72)$$

$$= q_i^{[n]} E(n-1),\ 1 \leqslant i \leqslant k-n+1, \quad (3.4.73)$$

where $E(n-1)$ are polyadic identities (3.4.65) and $q_i^{[n]} \in \mathcal{B}_k^{\text{gen}-\sigma}$.

We propose an n-ary version of the far commutativity relation (3.4.46) in the following form

$$\mu^{(n)}\left[\overbrace{\Sigma_{i_1,\dots,i_{n-1}}(n-1), \dots, \Sigma_{i_1,\dots,i_{n-1}}(n-1)}^{n-1}, E(n-1)\right] = \dots \quad (3.4.74)$$

$$= \mu^{(n)}\left[\overbrace{\Sigma_{\tau(i_1),\tau(i_2),\dots,\tau(i_{n-1})}(n-1), \dots, \Sigma_{\tau(i_1),\tau(i_2),\dots,\tau(i_{n-1})}(n-1)}^{n-1}, E(n-1)\right], \quad (3.4.75)$$

$$\text{if all } \left| i_p - i_s \right| \geqslant n-1, \quad p, s = 1, \dots, n-1, \quad (3.4.76)$$

where τ is an element the permutation symmetry group $\tau \in S_{n-1}$.

In matrix form we can define

Definition 3.4.32. An *n-ary (generated) matrix group* $\mathcal{M}_n^{\text{gen}-\Sigma}$ is presented by the $(n-1) \times (n-1)$ matrix generators $\Sigma_{i_1,\dots,i_{n-1}}(n-1)$ (3.4.69) and the relations (we use (3.4.60))

$$(\Sigma_{i,i+1,\dots,i+n-2}(n-1))^n \quad (3.4.77)$$

$$= (\Sigma_{i+1,i+2,\dots,i+n-2,i}(n-1))^n \quad (3.4.78)$$

$$\vdots$$

$$(\Sigma_{i+n-2,i,i+1,\dots,i+n-3}(n-1))^n$$
$$= q_i^{[n-1]} E(n-1),\ 1 \leqslant i \leqslant k-n+1, \quad (3.4.79)$$

and

$$\left(\Sigma^{(n-1)}_{i_1,\, i_2,\, \dots,\, i_{n-1}}\right)^{n-1} E(n-1) = \left(\Sigma^{(n-1)}_{\tau(i_1),\, \tau(i_2),\, \dots,\, \tau(i_{n-1})}\right)^{n-1} E(n-1),$$
$$\text{if all } \left| i_p - i_s \right| \geqslant n-1, \quad p, s = 1, \dots, n-1, \tag{3.4.80}$$

where $\tau \in S_{n-1}$ and $q_i^{[n-1]} \in \mathcal{B}_k^{\mathrm{gen}-\sigma}$.

Each element of $\mathcal{M}_n^{\mathrm{gen}-\Sigma}$ is an n-ary matrix word (analogous to the binary word (3.4.31)) being the n-ary product of the polyadic powers (3.4.12) of the matrix generators $\Sigma_{i_1,\dots,i_{n-1}}(n-1)$ and their querelements $\bar{\Sigma}_{i_1,\dots,i_{n-1}}(n-1)$ as in (3.4.50).

Similarly to the ternary case $n = 3$ (section 3.4.4) we now develop the n-ary 'reverse' procedure and build from $\mathcal{B}_k^{\mathrm{gen}-\sigma}$ the higher n-degree braid group $\mathcal{B}_k[n-1]$ using (3.4.69). Because the presentation of $\mathcal{M}_n^{\mathrm{gen}-\Sigma}$ by generators and relations has already been given in (3.4.77)–(3.4.80), we need to expand them into components and postulate that these new relations between the (binary) generators $\boldsymbol{\sigma}_i$ present a new higher degree analog of the braid group. This gives

Definition 3.4.33. A *higher n-degree braid (binary) group* $\mathcal{B}_k[n-1]$ is presented by $(k-1)$ generators $\boldsymbol{\sigma}_i \equiv \boldsymbol{\sigma}_i^{[n]}$ (and the identity $\mathbf{e}$) satisfying the following relations

- $(n-1)$ higher braid relations

$$\overbrace{\boldsymbol{\sigma}_i \boldsymbol{\sigma}_{i+1} \dots \boldsymbol{\sigma}_{i+n-3} \boldsymbol{\sigma}_{i+n-2} \boldsymbol{\sigma}_i}^{n} \tag{3.4.81}$$

$$= \boldsymbol{\sigma}_{i+1} \boldsymbol{\sigma}_{i+2} \dots \boldsymbol{\sigma}_{i+n-2} \boldsymbol{\sigma}_i \boldsymbol{\sigma}_{i+1} \tag{3.4.82}$$
$$\vdots$$

$$= \boldsymbol{\sigma}_{i+n-2} \boldsymbol{\sigma}_i \boldsymbol{\sigma}_{i+1} \boldsymbol{\sigma}_{i+2} \dots \boldsymbol{\sigma}_i \boldsymbol{\sigma}_{i+1} \boldsymbol{\sigma}_{i+n-2} \equiv q_i^{[n-1]}, \quad q_i^{[n-1]} \in \mathcal{B}_k[n-1], \tag{3.4.83}$$

$$i \in \mathfrak{I}_{\mathrm{braid}} = \{1, \dots, k-n+1\}, \tag{3.4.84}$$

- $(n-1)$-ary far commutativity

$$\overbrace{\boldsymbol{\sigma}_{i_1} \boldsymbol{\sigma}_{i_2} \dots \boldsymbol{\sigma}_{i_{n-3}} \boldsymbol{\sigma}_{i_{n-2}} \boldsymbol{\sigma}_{i_{n-1}}}^{n-1} \tag{3.4.85}$$
$$\vdots$$

$$= \boldsymbol{\sigma}_{\tau(i_1)} \boldsymbol{\sigma}_{\tau(i_2)} \dots \boldsymbol{\sigma}_{\tau(i_{n-3})} \boldsymbol{\sigma}_{\tau(i_{n-2})} \boldsymbol{\sigma}_{\tau(i_{n-1})}, \tag{3.4.86}$$

$$\text{if all } \left| i_p - i_s \right| \geqslant n-1, \quad p, s = 1, \dots, n-1, \tag{3.4.87}$$

$$\mathfrak{I}_{\mathrm{far}} = \{k-n, \dots, k-1\}, \tag{3.4.88}$$

where τ is an element of the permutation symmetry group $\tau \in S_{n-1}$.

A general element of the higher n-degree braid group $\mathcal{B}_k[n-1]$ is a word of the form

$$\mathfrak{w} = \boldsymbol{\sigma}_{i_1}^{p_1} \ldots \boldsymbol{\sigma}_{i_r}^{p_r} \ldots \boldsymbol{\sigma}_{i_m}^{p_m}, \quad i_m = 1, \ldots, k, \tag{3.4.89}$$

where $p_r \in \mathbb{Z}$ are (positive or negative) powers of the generators σ_{i_r}, $r = 1, \ldots, m$ and $m \in \mathbb{N}$.

***Remark* 3.4.34.** The ternary case $n = 3$ coincides with the Artin braid group $\mathcal{B}_k[2] = B_k$ (3.4.29)–(3.4.30).

***Remark* 3.4.35.** The representation of the higher n-degree braid relations in $\mathcal{B}_k[n-1]$ in the tensor product of vector spaces (similarly to B_k and the Yang–Baxter equation (Turaev 1988)) can be obtained using the k'-ary braid equation introduced in Duplij (2021b) (Proposition 7.2 and next there).

Definition 3.4.36. We say that the n-ary matrix group $\mathcal{M}_n^{\text{gen}-\Sigma}$ generated by the matrix generators $\Sigma_{i_1,i_2,\ldots,i_{n-1}}(n-1)$ satisfying the relations (3.4.77)–(3.4.80) is in *polyadic-binary correspondence* with the higher n-degree braid group $\mathcal{B}_k[n-1]$, which is denoted as (see (3.4.47))

$$\mathcal{M}_n^{\text{gen}-\Sigma} \rightleftharpoons \mathcal{B}_k[n-1]. \tag{3.4.90}$$

***Example* 3.4.37.** Let $n = 4$, then the 4-ary matrix group $\mathcal{M}_4^{\text{gen}-\Sigma}$ is generated by the matrix generators $\Sigma_{i_1,i_2,i_3}(3)$ satisfying (3.4.77)–(3.4.80)

• 4-ary relations of q-polyadic order (3.4.48)

$$(\Sigma_{i,i+1,i+2}(3))^4 = (\Sigma_{i+2,i,i+1}(3))^4 = (\Sigma_{i+1,i+2,i}(3))^4 = q_i^{[3]} E(3), \; 1 \leqslant i \leqslant k-3, \tag{3.4.91}$$

• far commutativity

$$\begin{aligned} &(\Sigma_{i_1,i_2,i_3}(3))^3 E(3) = (\Sigma_{i_3,i_1,i_2}(3))^3 E(3) = (\Sigma_{i_2,i_3,i_1}(3))^3 E(3) \\ &= (\Sigma_{i_1,i_3,i_2}(3))^3 E(3) = (\Sigma_{i_3,i_2,i_1}(3))^3 E(3) = (\Sigma_{i_2,i_1,i_3}(3))^3 E(3), \\ &\mid i_1 - i_2 \mid \geqslant 3, \mid i_1 - i_3 \mid \geqslant 3, \mid i_2 - i_3 \mid \geqslant 3. \end{aligned} \tag{3.4.92}$$

Let $\boldsymbol{\sigma}_i \equiv \boldsymbol{\sigma}_i^{[3]} \in \mathcal{B}_k[3]$, $i = 1, \ldots, k-1$, then we use the 4-ary 3×3 matrix presentation for the generators (see example 3.4.7)

$$\Sigma_{i_1,i_2,i_3}(3) \equiv \Sigma^{(3\times3)}(\boldsymbol{\sigma}_{i_1}, \boldsymbol{\sigma}_{i_2}, \boldsymbol{\sigma}_{i_3}) = \begin{pmatrix} 0 & \boldsymbol{\sigma}_{i_1} & 0 \\ 0 & 0 & \boldsymbol{\sigma}_{i_2} \\ \boldsymbol{\sigma}_{i_3} & 0 & 0 \end{pmatrix}, \quad i_1, i_2, i_3 = 1, \ldots, k-1. \tag{3.4.93}$$

The querelement $\bar{\Sigma}_{i_1,i_2,i_3}(3)$ satisfying

$$(\Sigma_{i_1,i_2,i_3}(3))^3\bar{\Sigma}_{i_1,i_2,i_3}(3) = \Sigma_{i_1,i_2,i_3}(3), \tag{3.4.94}$$

has the form

$$\bar{\Sigma}_{i_1,i_2,i_3}(3) = \begin{pmatrix} 0 & \sigma_{i_3}^{-1}\sigma_{i_2}^{-1} & 0 \\ 0 & 0 & \sigma_{i_1}^{-1}\sigma_{i_3}^{-1} \\ \sigma_{i_2}^{-1}\sigma_{i_1}^{-1} & 0 & 0 \end{pmatrix}. \tag{3.4.95}$$

Expanding (3.4.91)–(3.4.92) in components, we obtain the relations for the higher 4-degree braid group $\mathcal{B}_k[3]$ as follows

- higher 4-degree braid relations

$$\sigma_i\sigma_{i+1}\sigma_{i+2}\sigma_i = \sigma_{i+1}\sigma_{i+2}\sigma_i\sigma_{i+1} = \sigma_{i+2}\sigma_i\sigma_{i+1}\sigma_{i+2} \equiv q_i^{[3]},\ 1 \leqslant i \leqslant k-3, \tag{3.4.96}$$

- ternary far (total) commutativity

$$\sigma_{i_1}\sigma_{i_2}\sigma_{i_3} = \sigma_{i_2}\sigma_{i_3}\sigma_{i_1} = \sigma_{i_3}\sigma_{i_1}\sigma_{i_2} = \sigma_{i_1}\sigma_{i_3}\sigma_{i_2} = \sigma_{i_2}\sigma_{i_1}\sigma_{i_3} = \sigma_{i_3}\sigma_{i_2}\sigma_{i_1}, \tag{3.4.97}$$

$$\mid i_1 - i_2 \mid \geqslant 3,\ \mid i_1 - i_3 \mid \geqslant 3,\ \mid i_2 - i_3 \mid \geqslant 3. \tag{3.4.98}$$

In the higher 3-degree braid group the minimum number of generators is 4, which follows from (3.4.96). In this case we have a braid relation for $i = 1$ only and no far commutativity relations, because of (3.4.98). Then

***Example* 3.4.39.** The higher 4-degree braid group $\mathcal{B}_4[3]$ is generated by 3 generators σ_1, σ_2, σ_3, which satisfy only the braid relation

$$\sigma_1\sigma_2\sigma_3\sigma_1 = \sigma_2\sigma_3\sigma_1\sigma_2 = \sigma_3\sigma_1\sigma_2\sigma_3. \tag{3.4.99}$$

If $k \leqslant 7$, then there will be no far commutativity relations at all, which follows from (3.4.98), and so the first higher 4-degree braid group containing far commutativity should have $k = 8$ elements.

***Example* 3.4.39.** The higher 4-degree braid group $\mathcal{B}_8[3]$ is generated by 7 generators $\sigma_1, \ldots, \sigma_7$, which satisfy the braid relations with $i = 1, \ldots, 5$

$$\sigma_1\sigma_2\sigma_3\sigma_1 = \sigma_2\sigma_3\sigma_1\sigma_2 = \sigma_3\sigma_1\sigma_2\sigma_3, \tag{3.4.100}$$

$$\sigma_2\sigma_3\sigma_4\sigma_2 = \sigma_3\sigma_4\sigma_2\sigma_3 = \sigma_4\sigma_2\sigma_3\sigma_4, \tag{3.4.101}$$

$$\sigma_3\sigma_4\sigma_5\sigma_3 = \sigma_4\sigma_5\sigma_3\sigma_4 = \sigma_5\sigma_3\sigma_4\sigma_5, \tag{3.4.102}$$

$$\sigma_4\sigma_5\sigma_6\sigma_4 = \sigma_5\sigma_6\sigma_4\sigma_5 = \sigma_6\sigma_4\sigma_5\sigma_6, \tag{3.4.103}$$

$$\boldsymbol{\sigma}_5\boldsymbol{\sigma}_6\boldsymbol{\sigma}_7\boldsymbol{\sigma}_5 = \boldsymbol{\sigma}_6\boldsymbol{\sigma}_7\boldsymbol{\sigma}_5\boldsymbol{\sigma}_6 = \boldsymbol{\sigma}_7\boldsymbol{\sigma}_5\boldsymbol{\sigma}_6\boldsymbol{\sigma}_7, \tag{3.4.104}$$

together with the ternary far commutativity relation

$$\boldsymbol{\sigma}_1\boldsymbol{\sigma}_4\boldsymbol{\sigma}_7 = \boldsymbol{\sigma}_4\boldsymbol{\sigma}_7\boldsymbol{\sigma}_1 = \boldsymbol{\sigma}_7\boldsymbol{\sigma}_1\boldsymbol{\sigma}_4 = \boldsymbol{\sigma}_1\boldsymbol{\sigma}_7\boldsymbol{\sigma}_4 = \boldsymbol{\sigma}_4\boldsymbol{\sigma}_1\boldsymbol{\sigma}_7 = \boldsymbol{\sigma}_7\boldsymbol{\sigma}_4\boldsymbol{\sigma}_1. \tag{3.4.105}$$

***Remark* 3.4.40.** In polyadic group theory there are several possible modifications of the commutativity property, but nevertheless we assume here the *total commutativity* relations in the n-ary matrix generators and the corresponding far commutativity relations in the higher degree braid groups.

If $\mathcal{B}_k[n-1] \to \mathbb{Z}$ is the abelianization defined by $\boldsymbol{\sigma}_i^{\pm} \to \pm 1$, then $\boldsymbol{\sigma}_i^p = \mathbf{e}$, if and only if $p = 0$, and $\boldsymbol{\sigma}_i$ are of infinite order. Moreover, we can prove (as in the ordinary case $n = 3$; Dyer 1980).

Theorem 3.4.41. The higher n-degree braid group $\mathcal{B}_k[n-1]$ is torsion-free.

Recall (see, e.g., Kassel and Turaev 2008) that there exists a surjective homomorphism of the braid group onto the finite symmetry group $B_k \to S_k$ by $\sigma_i \to s_i = (i, i+1) \in S_k$. The generators s_i satisfy (3.4.29)–(3.4.30) together with the finite order demand

$$s_i s_{i+1} s_i = s_{i+1} s_i s_{i+1}, \quad 1 \leqslant i \leqslant k-2, \tag{3.4.106}$$

$$s_i s_j = s_j s_i, \quad |\, i - j \,| \geqslant 2, \tag{3.4.107}$$

$$s_i^2 = e, \quad i = 1, \ldots, k-1, \tag{3.4.108}$$

which is called the *Coxeter presentation* of the symmetry group S_k. Indeed, multiplying both sides of (3.4.106) from the right successively by s_{i+1}, s_i, and s_{i+1}, using (3.4.108), we obtain $(s_i s_{i+1})^3 = 1$, and (3.4.107) on s_i and s_j, we get $(s_i s_j)^2 = 1$. Therefore, a Coxeter group (Brieskorn and Saito 1972) corresponding (3.4.106)–(3.4.108) is presented by the same generators s_i and the relations

$$(s_i s_{i+1})^3 = 1, \quad 1 \leqslant i \leqslant k-2, \tag{3.4.109}$$

$$(s_i s_j)^2 = 1, \quad |\, i - j \,| \geqslant 2, \tag{3.4.110}$$

$$s_i^2 = e, \quad i = 1, \ldots, k-1. \tag{3.4.111}$$

A general Coxeter group $W_k = W_k(e, r_i)$ is presented by k generators r_i and the relations (Björner and Brenti 2005)

$$(r_i r_j)^{m_{ij}} = e, \quad m_{ij} = \begin{cases} 1, & i = j, \\ \geqslant 2, & i \neq j. \end{cases} \tag{3.4.112}$$

By analogy with (3.4.106)–(3.4.108), we make the following

Definition 3.4.42. A higher analog of S_k, the *n-degree symmetry group* $\mathcal{S}_k[n-1] = \mathcal{S}_k^{[n-1]}(\mathbf{e}, \mathbf{s}_i)$, is presented by generators $\mathbf{s}_i$, $i = 1, \ldots, k-1$ satisfying (3.4.81)–(3.4.86) together with the additional condition of finite $(n-1)$-order $\mathbf{s}_i^{(n-1)} = \mathbf{e}$, $i = 1, \ldots, k$.

***Example* 3.4.43.** The lowest higher degree case is $\mathcal{S}_4[3]$ which is presented by three generators $\mathbf{s}_1$, $\mathbf{s}_2$, $\mathbf{s}_3$ satisfying (see (3.4.99))

$$\mathbf{s}_1\mathbf{s}_2\mathbf{s}_3\mathbf{s}_1 = \mathbf{s}_2\mathbf{s}_3\mathbf{s}_1\mathbf{s}_2 = \mathbf{s}_3\mathbf{s}_1\mathbf{s}_2\mathbf{s}_3, \tag{3.4.113}$$

$$\mathbf{s}_1^3 = \mathbf{s}_2^3 = \mathbf{s}_3^3 = \mathbf{e}. \tag{3.4.114}$$

In a similar way we define a higher degree analog of the Coxeter group (3.4.112).

Definition 3.4.44. A higher *n-degree Coxeter group* $\mathcal{W}_k[n-1] = \mathcal{W}_k^{[n-1]}(\mathbf{e}, \mathbf{r}_i)$ is presented by k generators $\mathbf{r}_i$ obeying the relations

$$(\mathbf{r}_{i_1}, \mathbf{r}_{i_2}, \ldots, \mathbf{r}_{i_{n-1}})^{\mathbf{m}_{i_1 i_2, \ldots, i_{n-1}}} = \mathbf{e}, \tag{3.4.115}$$

$$\mathbf{m}_{i_1 i_2, \ldots, i_{n-1}} = \begin{cases} 1, & i_1 = i_2 = \ldots = i_{n-1}, \\ \geqslant n-1, & \left| i_p - i_s \right| \geqslant n-1, \quad p, s = 1, \ldots, n-1. \end{cases} \tag{3.4.116}$$

It follows from (3.4.116) that all generators are of $(n-1)$ order $\mathbf{r}_i^{n-1} = \mathbf{e}$. A *higher n -degree Coxeter matrix* is a hypermatrix $M_{k,\, Cox}^{[n-1]}\left(\overbrace{k \times k \times \ldots \times k}^{n-1}\right)$ having 1 on the main diagonal and other entries $\mathbf{m}_{i_1 i_2, \ldots, i_{n-1}}$.

***Example* 3.4.45.** In the lowest higher degree case $n = 4$ and all $\mathbf{m}_{i_1 i_2, \ldots, i_{n-1}} = 3$ we have (instead of commutativity in the ordinary case $n = 3$)

$$(\mathbf{r}_i\mathbf{r}_j)^2 = \mathbf{r}_j^2\mathbf{r}_i^2, \tag{3.4.117}$$

$$\mathbf{r}_i\mathbf{r}_j\mathbf{r}_i = \mathbf{r}_j^2\mathbf{r}_i^2\mathbf{r}_j^2. \tag{3.4.118}$$

***Example* 3.4.46.** A higher 3-degree analog of (3.4.109)–(3.4.111) is given by

$$(\mathbf{r}_i\mathbf{r}_{i+1}\mathbf{r}_{i+2})^4 = 1, \quad 1 \leqslant i \leqslant k-3, \tag{3.4.119}$$

$$(\mathbf{r}_{i_1}\mathbf{r}_{i_2}\mathbf{r}_{i_3})^3 = 1, \mid i_1 - i_2 \mid \geqslant 3, \mid i_1 - i_3 \mid \geqslant 3, \mid i_2 - i_3 \mid \geqslant 3, \tag{3.4.120}$$

$$\mathbf{r}_i^3 = \mathbf{e}, \quad i = 1, \dots, k - 1. \tag{3.4.121}$$

It follows from (3.4.120), that

$$(\mathbf{r}_{i_1}\mathbf{r}_{i_2}\mathbf{r}_{i_3})^2 = \mathbf{r}_{i_3}^2\mathbf{r}_{i_2}^2\mathbf{r}_{i_1}^2, \tag{3.4.122}$$

which cannot be reduced to total commutativity (3.4.97). From the first relation (3.4.119) we obtain

$$\mathbf{r}_i\mathbf{r}_{i+1}\mathbf{r}_{i+2}\mathbf{r}_i = \mathbf{r}_{i+2}^2\mathbf{r}_{i+1}^2, \tag{3.4.123}$$

which differs from the higher 4-degree braid relations (3.4.96).

***Example* 3.4.47.** In the simplest case the higher 4-degree Coxeter group $\mathcal{W}_4[3]$ has 3 generator $\mathbf{r}_1$, $\mathbf{r}_2$, $\mathbf{r}_3$ satisfying

$$(\mathbf{r}_1\mathbf{r}_2\mathbf{r}_3)^4 = \mathbf{r}_1^3 = \mathbf{r}_2^3 = \mathbf{r}_3^3 = \mathbf{e}. \tag{3.4.124}$$

***Example* 3.4.48.** The minimal case, when the conditions (3.4.120) appear is $\mathcal{W}_8[3]$

$$\mathbf{r}_1\mathbf{r}_2\mathbf{r}_3\mathbf{r}_1 = \mathbf{r}_3^2\mathbf{r}_2^2, \tag{3.4.125}$$

$$\mathbf{r}_2\mathbf{r}_3\mathbf{r}_4\mathbf{r}_2 = \mathbf{r}_4^2\mathbf{r}_3^2, \tag{3.4.126}$$

$$\mathbf{r}_3\mathbf{r}_4\mathbf{r}_5\mathbf{r}_3 = \mathbf{r}_5^2\mathbf{r}_4^2, \tag{3.4.127}$$

$$\mathbf{r}_4\mathbf{r}_5\mathbf{r}_6\mathbf{r}_4 = \mathbf{r}_6^2\mathbf{r}_5^2, \tag{3.4.128}$$

$$\mathbf{r}_5\mathbf{r}_6\mathbf{r}_7\mathbf{r}_5 = \mathbf{r}_7^2\mathbf{r}_6^2, \tag{3.4.129}$$

and an analog of commutativity

$$(\mathbf{r}_1\mathbf{r}_4\mathbf{r}_7)^2 = \mathbf{r}_7^2\mathbf{r}_4^2\mathbf{r}_1^2. \tag{3.4.130}$$

Thus, we arrive at

Theorem 3.4.49. *The higher* n*-degree Coxeter group can present the* n*-degree symmetry group in the lowest case only, if and only if $n = 3$.*

As a further development, it would be interesting to consider the higher degree (in our sense) groups constructed here from a geometric viewpoint (e.g., Birman 1976, Kauffman 1991).

References

Artin E 1947 Theory of braids *Ann. Math.* **48** 101–26

Birman J S 1976 *Braids, Links and Mapping Class Group* (Princeton, NJ: Princeton University Press)

Björner A and Brenti F 2005 *Combinatorics of Coxeter Groups* (New York: Springer)

Borowiec A, Dudek W and Duplij S 2006 Bi-element representations of ternary groups *Commun. Algebra* **34** 1651–70

Brieskorn E and Saito K 1972 Artin-Gruppen und Coxeter-Gruppen *Invent. Math.* **17** 245–71

Clifford A H and Preston G B 1961 *The Algebraic Theory of Semigroups* vol 1 (Providence, RI: American Mathematical Society)

Dörnte W 1929 Unterschungen über einen verallgemeinerten Gruppenbegriff *Math. Z.* **29** 1–19

Duplij S 1996 Supermatrix representations of semigroup bands *Pure Math. Appl.* **7** 235–61 (arXiv: funct-an/9609002)

Duplij S 1998 On semi-supermanifolds *Pure Math. Appl.* **9** 283–310

Duplij S 2000 *Semisupermanifolds and Semigroups* (Kharkiv: Krok)

Duplij S 2012 Polyadic systems, representations and quantum groups *J. Kharkiv Natl Univ. Ser. Nuclei Part. Fields* **1017** 28–59

Duplij S 2017 Polyadic integer numbers and finite (m, n)-fields *p-Adic Numbers Ultrametric Anal. Appl.* **9** 267–91 arXiv:math.RA/1707.00719

Duplij S 2021a Higher braid groups and regular semigroups from polyadic-binary correspondence *Mathematics* **9** 972

Duplij S 2021b Polyadic Hopf algebras and quantum groups East *Eur. J. Phys.* **2** 5–50

Duplij S and Marcinek W 2000 On higher regularity and monoidal categories *Kharkiv State Univ. J. (Vestnik KSU), Ser. Nuclei, Particles and Fields* **481** 27–30

Duplij S and Marcinek W 2001a Semisupermanifolds and regularization of categories, modules, algebras and Yang–Baxter equation *Supersymmetry and Quantum Field Theory* (Amsterdam: Elsevier) pp 110–5

Duplij S and Marcinek W 2001b Semisupermanifolds and regularization of categories, modules, algebras and Yang–Baxter equation (Int. Conf. on Supersymmetry and Quantum Field Theory: D.V. Volkov Memorial Conf., Kharkiv, Ukraine, July 2000) *Nucl. Phys. Proc. Suppl.* **102** 293–7

Duplij S and Marcinek W 2002 Regular obstructed categories and topological quantum field theory *J. Math. Phys* **43** 3329–41

Duplij S and Marcinek W 2003 Braid semistatistics and doubly regular R-matrix *J. Kharkiv Natl Univ. Ser. Nuclei Part. Fields* **601** 75–80

Dyer J L 1980 The algebraic braid groups are torsion-free: an algebraic proof *Math. Z.* **172** 157–60

Gilbert N D 2003 Flows on regular semigroups *Appl. Categ. Struct.* **11** 147–55

Green J A 1951 On the structure of semigroups *Ann. Math.* **54** 163–72

Grillet P-A 1995 *Semigroups. An Introduction to the Structure Theory* (New York: Dekker)

Hietarinta J 1997 Permutation-type solutions to the Yang–Baxter and other n-simplex equations *J. Phys.* **A30** 4757–71

Higgins P M 1992 *Techniques of Semigroup Theory* (Oxford: Oxford University Press)

Howie J M 1976 *An Introduction to Semigroup Theory* (London: Academic)

Hu Z-N 1997 Baxter-Bazhanov model, Frenkel-Moore equation and the braid group *Commun. Theor. Phys.* **27** 429–34

Kassel C and Turaev V 2008 *Braid Groups Graduate Texts in Mathematics* vol 247 (New York: Springer)
Kauffman L H 1991 *Knots and Physics* (Singapore: World Scientific)
Kolesnikov O V 1979 Inverse n-semigroups *Comment. Math. Prace Mat.* **21** 101–8
Lawson M V 1998 *Inverse Semigroups: The Theory of Partial Symmetries* (Singapore: World Scientific)
Li Y Q and Hu Z N 1995 Solutions of n-simplex equation from solutions of braid group representation *J. Phys. A. Math. Gen.* **28** L219–23
Liber A E 1953 On symmetric generalized groups *Mat. Sb. (N.S.)* **33** 531–44
Liber A E 1954 On the theory of generalized groups *Dokl. Akad. Nauk SSSR (N.S.)* **97** 25–8
Ljapin E S 1968 *Semigroups* (Providence, RI: American Mathematical Society)
Manin Y I and Schechtman V V 1989 Arrangements of hyperplanes, higher braid groups and higher Bruhat orders *Algebraic Number Theory* ed J Coates, R Greenberg, B Mazur and I Satake, 17th edn (Boston, MA: Academic) pp 289–308
Markov A A 1945 Fundamentals of algebraic theory of braid groups *Trudy Mat. Inst. Akad. Nauk SSSR* **16** 1–54
Munn W D and Penrose R 1955 A note on inverse semigroups *Proc. Camb. Phil. Soc.* **51** 396–9
Nikitin A N 1984 Semisimple Artinian $(2, n)$-rings *Vestnik Moskov. Univ. Ser. I Mat. Mekh.* **6** 3–7
Petrich M 1984 *Inverse Semigroups* (Wiley: New York)
Pop A and Pop M S 2010 On generalized algebraic structures *Creat. Math. Inf.* **19** 184–90
Post E L 1940 Polyadic groups *Trans. Am. Math. Soc.* **48** 208–350
Schein B M 1965 On the theory of generalized groups and generalized heaps *Theory of Semigroups and Appl.* vol 1 (Saratov: Izdat. Saratov. University) pp 286–324
Sioson F M 1963 On regular algebraic systems *Proc. Japan Acad.* **39** 283–6
Slipenko A K 1977 Regular operatives and ideal equivalences *Dokl. AN Ukr. SSR. Seriya A. Fiz-Mat i Tekhn Nauki* **3** 218–21
Thierrin G 1952 Sur les éléments inversifs et les éléments unitaires d'un demi-groupe inversif *C. R. Acad. Sci. Paris* **234** 33–4
Turaev V G 1988 The Yang-Baxter equation and invariants of links *Invent. Math.* **92** 527–53
Vagner V V 1952 Generalized groups *Dokl. Akad. Nauk SSSR (N.S.)* **84** 1119–22
von Neumann J 1936 On regular rings *Proc. Nat. Acad. Sci. USA* **22** 707–13
Wazzan S A 2019 Flows on classes of regular semigroups and Cauchy categories *J. Math. (Hindawi)* **8027391** 9
Yaglom I M 1968 *Complex Numbers in Geometry* (New York: Academic)
Zamolodchikov A B 1981 Tetrahedron equations and the relativistic S-matrix of straight-strings in 2 + 1-dimensions *Commun. Math. Phys.* **79** 489–505

Chapter 4

Polyadic rings, fields and integer numbers

Parts of this chapter have been reprinted with permission from Duplij (2019).

In this chapter we introduce a special—polyadic—version of the finite fields (Lidl and Niederreiter 1997, Mullen and Panario 2013, Menezes *et al* 1997) in such a way that, instead of the binary ring of integers $\mathbb{Z}$, we consider a polyadic ring. The concept of the polyadic integer numbers $\mathbb{Z}^{(m,n)}$ as representatives of a fixed congruence class, which form the (m, n)-ring (with m-ary addition and n-ary multiplication), was introduced in Duplij (2019). We analyze $\mathbb{Z}^{(m,n)}$ in more detail, by developing elements of a polyadic analog of binary arithmetic: polyadic prime numbers, polyadic division with a remainder, the polyadic Euler totient function, etc. We define new secondary congruence classes and the corresponding finite (m, n)-rings $\mathbb{Z}^{(m,n)}(q)$ of polyadic integer numbers, which give $\mathbb{Z}/q\mathbb{Z}$ in the 'binary limit'. The conditions under which these rings become fields are given, they have unusual properties, and can be zeroless, zeroless–nonunital or have several units, and it is even possible for all elements to be units. It is shown that there exist non-isomorphic finite polyadic fields of the same arity shape and order. None of the above situations are possible for binary fields.

Note that some general properties of polyadic rings and fields were given in Crombez (1972), Leeson and Butson (1980), Pop and Pop (2002), Duplij and Werner (2021), but their concrete examples using integers differ considerably from our construction here, and the latter leads to so-called nonderived (proper) versions which have not been considered before. We conjecture that any (m, n)-field with $m > n$ contains as a subfield one of the prime polyadic fields constructed here, which can be considered as a polyadic analog of $GF(p)$.

We then consider Diophantine equations over the polyadic rings in a straightforward manner: we change only the arities of the operations ('additions' and 'multiplications'), but save their mutual 'interaction'. In this way we try to 'polyadize' the equal sums of like powers equation and formulate polyadic analogs of the Lander–Parkin–Selfridge conjecture and of Fermat's last theorem (Lander *et al* 1967). It is shown that in the simplest case, when the polyadic 'addition' and 'multiplication' are

doi:10.1088/978-0-7503-2648-3ch4

nonderived (e.g., for polyadic numbers), neither conjecture is valid, and counter-examples are presented. Finally, we apply Frolov's theorem to the Tarry–Escott problem (Dorwart and Brown 1937, Nguyen 2016) over polyadic rings to obtain new solutions to the equal sums of like powers equation for fixed congruence classes.

4.1 One-set polyadic 'linear' structures

We use the notations and definitions from Duplij (2012, 2019) (see, also, references therein).

4.1.1 Polyadic distributivity

Let us consider a polyadic structure with two operations on the same set R: the 'chief' (*multiplication*) n-ary operation $\mu^{(n)}: R^{\times n} \to R$ and the additional m-ary operation $\nu^{(m)}: R^{\times m} \to R$, that is $\langle R|\mu^{(n)}, \nu^{(m)}\rangle$. If there are no relations between $\mu^{(n)}$ and $\nu^{(m)}$, then nothing new, as compared with the polyadic structures having a single operation $\langle R|\mu^{(n)}\rangle$ or $\langle R|\nu^{(m)}\rangle$, can be said. Informally, the 'interaction' between operations can be described using the important relation of distributivity (an analog of $x \cdot (y + t) = x \cdot y + x \cdot t$, $x, y, t \in R$ in the binary case).

Definition 4.1.1. The *polyadic distributivity* for the operations $\mu^{(n)}$ and $\nu^{(m)}$ (no additional properties are implied for now) consists of n relations

$$\begin{aligned}&\mu^{(n)}\left[\nu^{(m)}[x_1, \ldots x_m], y_2, y_3, \ldots y_n\right]\\ &= \nu^{(m)}\left[\mu^{(n)}\left[x_1, y_2, y_3, \ldots y_n\right],\right.\\ &\quad \left.\mu^{(n)}\left[x_2, y_2, y_3, \ldots y_n\right], \ldots \mu^{(n)}\left[x_m, y_2, y_3, \ldots y_n\right]\right]\end{aligned} \tag{4.1.1}$$

$$\begin{aligned}&\mu^{(n)}\left[y_1, \nu^{(m)}[x_1, \ldots x_m], y_3, \ldots y_n\right]\\ &= \nu^{(m)}\left[\mu^{(n)}\left[y_1, x_1, y_3, \ldots y_n\right],\right.\\ &\quad \left.\mu^{(n)}\left[y_1, x_2, y_3, \ldots y_n\right], \ldots \mu^{(n)}\left[y_1, x_m, y_3, \ldots y_n\right]\right]\end{aligned} \tag{4.1.2}$$

$$\vdots$$

$$\begin{aligned}&\mu^{(n)}\left[y_1, y_2, \ldots y_{n-1}, \nu^{(m)}[x_1, \ldots x_m]\right]\\ &= \nu^{(m)}\left[\mu^{(n)}\left[y_1, y_2, \ldots y_{n-1}, x_1\right],\right.\\ &\quad \left.\mu^{(n)}\left[y_1, y_2, \ldots y_{n-1}, x_2\right], \ldots \mu^{(n)}\left[y_1, y_2, \ldots y_{n-1}, x_m\right]\right],\end{aligned} \tag{4.1.3}$$

where $x_i, y_j \in R$.

It is seen that the operations $\mu^{(n)}$ and $\nu^{(m)}$ enter into (4.1.1)–(4.1.3) in a nonsymmetric way, which allows us to distinguish them: one of them ($\mu^{(n)}$, the n-ary multiplication) 'distributes' over the other one $\nu^{(m)}$, and therefore $\nu^{(m)}$ is called the *addition*. If only some of the relations (4.1.1)–(4.1.3) hold, then such distributivity is *partial* (the analog of left and right distributivity in the binary case). Obviously, the operations $\mu^{(n)}$ and $\nu^{(m)}$ need have nothing to do with ordinary multiplication (in the binary case denoted by $\mu_2 \Longrightarrow (\cdot)$) and addition (in the binary case denoted by $\nu_2 \Longrightarrow (+)$), as in the example below.

***Example* 4.1.2.** Let $R = \mathbb{R}$, $n = 2$, $m = 3$, and $\mu^{(2)}[y_1, y_2] = y_1^{y_2}$, $\nu^{(3)}[x_1, x_2, x_3] = x_1x_2x_3$ (product in $\mathbb{R}$). The partial distributivity now is $(x_1x_2x_3)^{y_2} = x_1^{y_2}x_2^{y_2}x_3^{y_2}$ (only the first relation (4.1.1) holds).

4.1.2 Polyadic rings and fields

Let us we briefly recall one-set (ring-like) polyadic structures (informally). Let both operations $\mu^{(n)}$ and $\nu^{(m)}$ be (totally) *associative*, which (in our definition, Duplij 2012) means independence of the composition of two operations under placement of the internal operations (there are n and m such placements and therefore $(n + m)$ corresponding relations)

$$\mu^{(n)}[\mathbf{x}, \mu^{(n)}[\mathbf{y}], \mathbf{t}] = \text{invariant}, \tag{4.1.4}$$

$$\nu^{(m)}[\mathbf{u}, \nu^{(m)}[\mathbf{v}], \mathbf{w}] = \text{invariant}, \tag{4.1.5}$$

where the polyads $\boldsymbol{x}, \boldsymbol{y}, \mathbf{t}, \mathbf{u}, \mathbf{v}, \mathbf{w}$ have corresponding allowed lengths, and then both $\langle R|\mu^{(n)}|\text{assoc}\rangle$ and $\langle R|\nu^{(m)}|\text{assoc}\rangle$ are *polyadic semigroups* $\mathcal{S}^{(n)}$ and $\mathcal{S}^{(m)}$. A (totally) *commutative semigroup* $\langle R|\nu^{(m)}|\text{assoc, comm}\rangle$ is defined by $\nu^{(m)}[\mathbf{x}] = \nu^{(m)}[\sigma\circ\mathbf{x}]$, for all $\sigma \in S_n$, where S_n is the symmetry group. If the equation $\nu^{(m)}[\mathbf{x}, u, \mathbf{y}] = t$ is solvable with respect to u for any place of u with $\boldsymbol{x}$ and $\boldsymbol{y}$ of the summing length $m - 1$, then $\langle R|\nu^{(m)}|\text{assoc, solv}\rangle$ is a *polyadic group* $\mathcal{G}^{(m)}$, and such $u = \tilde{t}$ is called a (*additive*) *querelement* for t, which defines the (*additive*) *unary queroperation* $\tilde{\nu}^{(1)}$ by $\tilde{\nu}^{(1)}[t] = \tilde{t}$ (see (1.2.13)).

Definition 4.1.3. A *polyadic* (m, n)*-ring* $\mathcal{R}^{(m,n)}$ is a set R with two operations $\mu^{(n)}\colon R^{\times n} \to R$ and $\nu^{(m)}\colon R^{\times m} \to R$, such that:

(i) they are distributive (4.1.1)–(4.1.3);
(ii) $\langle R|\mu^{(n)}|\text{ assoc}\rangle$ is a polyadic semigroup;
(iii) $\langle R|\nu^{(m)}|\text{assoc, comm, solv}\rangle$ is a commutative polyadic group.

It is obvious that a (2, 2)-ring $\mathcal{R}^{(2,2)}$ is an ordinary (binary) ring $\mathcal{R}$. Polyadic rings have much richer structure and can have unusual properties (Celakoski 1977, Crombez 1972, Čupona 1965, Leeson and Butson 1980). If the multiplicative semigroup $\langle R|\mu^{(n)}|\text{assoc}\rangle$ is commutative, $\mu^{(n)}[\mathbf{x}] = \mu^{(n)}[\sigma\circ\mathbf{x}]$, for all $\sigma \in S_n$, then $\mathcal{R}^{(m,n)}$ is called a *commutative polyadic ring*, and if it contains the identity, then $\mathcal{R}^{(m,n)}$ is a (polyadic) (m, n)*-semiring*. If the distributivity is only partial, then $\mathcal{R}^{(m,n)}$ is called a *polyadic near-ring*.

Recall that, in distinction to binary rings, an *n-admissible* 'length of word $(\mathbf{x})$' should be congruent to $1 \bmod (n-1)$, containing $\ell_\mu(n-1)+1$ elements (ℓ_μ is a 'number of multiplications') $(\mu^{(n)})^{\circ \ell_\mu}[\mathbf{x}]$ ($\mathbf{x} \in R^{\times(\ell_\mu(n-1)+1)}$), so-called $(\ell_\mu(n-1)+1)$-ads, or *polyads* (Post 1940). An *m-admissible* 'quantity of words $(\mathbf{y})$' in a polyadic 'sum' has to be congruent to 1 mod $(m-1)$, i.e., consisting of $\ell_\nu(m-1)+1$ summands (ℓ_ν is a 'number of additions') $(\nu^{(m)})^{\circ \ell_\nu}[\mathbf{y}]$ ($\mathbf{y} \in R^{\times(\ell_\nu(m-1)+1)}$). Therefore, a straightforward 'polyadization' of any binary expression ($m = n = 2$) can be introduced as follows: substitute the number of multipliers $\ell_\mu + 1 \to \ell_\mu(n-1)+1$ and number of summands $\ell_\nu + 1 \to \ell_\nu(m-1)+1$, respectively.

An example of 'trivial polyadization' is the simplest (m, n)-ring derived from the ring of integers $\mathbb{Z}$ as the set of $\ell_\nu(m-1)+1$ 'sums' of n-admissible $(\ell_\mu(n-1)+1)$-ads $(\mathbf{x})$, where $\mathbf{x} \in \mathbb{Z}^{\times(\ell_\mu(n-1)+1)}$ (Leeson and Butson 1980).

Definition 4.1.4. A polyadic ring is called *derived*, if $\nu^{(m)}$ and $\mu^{(n)}$ are equivalent to a repetition of the binary addition and multiplication, while $\langle R|+\rangle$ and $\langle R|\cdot\rangle$ are commutative (binary) group and semigroup respectively. If only one operation $\nu^{(m)}$ (or $\mu^{(n)}$) has this property, we call such a $\mathcal{R}^{(m,n)}$ *additively* (or *multiplicatively*) derived (*half-derived*).

The additive *m-ary polyadic power* and the multiplicative *n-ary polyadic power* are defined by (inside polyadic products we denote repeated entries by $\overbrace{x, \ldots, x}^{k}$ as x^k)

$$x^{\langle \ell_\nu \rangle_{+m}} = (\nu^{(n)})^{\circ \ell_\nu}[x^{\ell_\nu(m-1)+1}], \quad x^{\langle \ell_\mu \rangle_{\times n}} = (\mu^{(n)})^{\circ \ell_\mu}[x^{\ell_\mu(n-1)+1}], \; x \in R, \tag{4.1.6}$$

such that the polyadic powers and ordinary powers differ by one: $x^{\langle \ell_\nu \rangle_{+2}} = x^{\ell_\nu+1}$, $x^{\langle \ell_\mu \rangle_{\times 2}} = x^{\ell_\mu+1}$.

The *polyadic idempotents* in $\mathcal{R}^{(m,n)}$ satisfy

$$x^{\langle \ell_\nu \rangle_{+m}} = x, \quad x^{\langle \ell_\mu \rangle_{\times n}} = x, \tag{4.1.7}$$

and are called the *additive ℓ_ν-idempotent* and the *multiplicative ℓ_μ-idempotent*, respectively.

The additive 1-idempotent, the *zero* $z \in R$, is (if it exists) defined by

$$\nu^{(m)}[\mathbf{x}, z] = z, \quad \forall \mathbf{x} \in R^{\times(m-1)}. \tag{4.1.8}$$

An element $x \in R$ is called (polyadic) *nilpotent*, if $x^{\langle 1 \rangle_{+m}} = z$, and all higher powers of a nilpotent element are nilpotent, as follows from (4.1.8) and associativity. Evidently, a zero (if it exists) is a multiplicative idempotent and is unique, and, if a polyadic ring has an additive idempotent, it is a zero (Leeson and Butson 1980). Due to the distributivity (4.1.1)–(4.1.3), there can be at most one zero in a polyadic ring $\mathcal{R}^{(m,n)}$.

The *unit* e of $\mathcal{R}^{(m,n)}$ is a multiplicative 1-idempotent which is defined (if it exists) as

$$\mu^{(n)}[e^{n-1}, x] = x, \quad \forall x \in R, \tag{4.1.9}$$

where (in case of a noncommutative polyadic ring) x can be on any place.

Definition 4.1.5. An element $x \in R$ is called a (polyadic) ℓ_μ-*reflection*, if $x^{\langle \ell_\mu \rangle_{\times n}} = e$ (the multiplicative analog of a nilpotent element).

In binary rings the identity is the only neutral element, while in polyadic rings there can exist many *neutral* $(n-1)$-*polyads* $\mathbf{n}$ satisfying

$$\mu^{(n)}[x, \mathbf{n}] = x, \quad \mathbf{n} \in R^{\times(n-1)}, \tag{4.1.10}$$

for any $x \in R$ which can also be on any place. The neutral polyads $\mathbf{n}$ are not determined uniquely. Obviously, the polyad (e^{n-1}) is neutral.

Definition 4.1.6. If a polyadic ring $\mathcal{R}^{(m,n)}$ contains no unit and no zero, we call it a *zeroless–nonunital polyadic ring*.

It is obvious that zeroless–nonunital rings can contain other idempotents of higher polyadic powers.

Remark **4.1.7.** There exist exotic polyadic rings which have no zero, no identity, and no additive idempotents at all (see, e.g., Crombez 1972), but, if $m = 2$, then a zero always exists (Leeson and Butson 1980).

Example **4.1.8.** Let us consider a polyadic ring $\mathcal{R}^{(3,4)}$ generated by 2 elements x, y and the relations

$$\begin{gathered} \mu^{(4)}[x^4] = x, \quad \mu^{(4)}[x^3, y] = y, \quad \mu^{(4)}[x^2, y^2] = x, \\ \mu^{(4)}[x, y^3] = y, \quad \mu^{(4)}[y^4] = x, \end{gathered} \tag{4.1.11}$$

$$\begin{gathered} \nu^{(3)}[x^3] = y, \quad \nu^{(3)}[x^2, y] = x, \quad \nu^{(3)}[x, y^2] = y, \\ \nu^{(3)}[y^3] = x, \end{gathered} \tag{4.1.12}$$

which has a multiplicative idempotent x only, but it has no zero and no unit.

In polyadic rings (including the zeroless–nonunital ones) invertibility can be governed in a way which is not connected with unit and zero elements. For a fixed element $x \in R$ its *additive querelement* $\tilde{x}$ and *multiplicative querelement* $\bar{x}$ are defined by

$$\nu^{(m)}[x^{m-1}, \tilde{x}] = x, \tag{4.1.13}$$

$$\mu^{(n)}[x^{n-1}, \bar{x}] = x, \tag{4.1.14}$$

where in the second equation, if the n-ary multiplication $\mu^{(n)}$ is noncommutative, $\bar{x}$ can be on any place. Because $\langle R | \nu^{(m)} \rangle$ is a commutative group, each $x \in R$ has its additive querelement $\tilde{x}$ (and is *additively querable* or 'polyadically invertible'). The n-ary semigroup $\langle R | \mu^{(n)} \rangle$ can have no multiplicatively querable elements at all. However, if every $x \in R$ has its unique querelement, then $\langle R | \mu^{(n)} \rangle$ is an n-ary group (Post 1940). Obviously, that an n-ary group cannot have nilpotent elements, but it can have ℓ_μ-reflections. Denote $R^* = R \backslash \{z\}$, if the zero z exists. Observe that (in distinction to binary rings) $\langle R^* | \mu^{(n)} | \text{assoc} \rangle$ is not a polyadic group, in general. But if $\langle R^* | \mu^{(n)} | \text{assoc} \rangle$ is an n-ary group, we arrive to (see Leeson and Butson 1980, Iancu and Pop 1997)

Definition 4.1.9. If $\langle R^* | \mu^{(n)} \rangle$ is the n-ary group, then $\mathcal{R}^{(m,n)}$ is called a (m, n)-*division ring* $\mathcal{D}^{(m,n)}$.

Definition 4.1.10. A (total) commutative (m, n)-division ring $\mathcal{R}^{(m,n)}$ is called a (m, n)-*field* $\mathcal{F}^{(m,n)}$.

In n-ary groups there exists an 'intermediate' commutativity, the so-called semicommutativity (1.2.4).

Definition 4.1.11. A semicommutative (m, n)-division ring $\mathcal{R}^{(m,n)}$ is called a *semicommutative* (m, n)-*field* $\mathcal{F}^{(m,n)}$.

The simplest example of a (m, n)-field derived from $\mathbb{R}$ is the set of $\ell_\nu(m-1)+1$ 'sums' of admissible $\left(\ell_\mu(n-1)+1\right)$-ads $(\mathbf{x})$, where $\mathbf{x} \in \mathbb{R}^{\times(\ell_\mu(n-1)+1)}$. Some nonderived (m, n)-fields are in

***Example* 4.1.12.** The set $i\mathbb{R}$ with $i^2 = -1$ is a $(2, 3)$-field with a zero and no unit (operations are made in $\mathbb{C}$), but the multiplicative querelement of ix is $-i/x$ $(x \neq 0)$.

***Example* 4.1.13.** The set of antidiagonal 2×2 matrices over $\mathbb{R}$ is a semicommutative $(2, 3)$-field $\mathcal{F}^{(2,3)} = \langle \mathbb{R} | \nu^{(2)} \equiv (+), \mu^{(3)} \rangle$, where $(+)$ is the matrix addition in $\mathbb{R}$ and $\mu^{(3)}$ is the product of three matrices. $\mathcal{F}^{(2,3)}$ has the zero $z = \begin{pmatrix} 0 & 0 \\ 0 & 0 \end{pmatrix}$ and two units $e = \pm\begin{pmatrix} 0 & 1 \\ 1 & 0 \end{pmatrix}$, and the unique multiplicative querelement of $x = \begin{pmatrix} 0 & a \\ b & 0 \end{pmatrix}$ is $\bar{x} = \begin{pmatrix} 0 & 1/b \\ 1/a & 0 \end{pmatrix}$.

The definition of a polyadic field can be done in the diagrammatic form, analogous to definition 1.2.11 and (1.2.22). We introduce the *double Dörnte relations*: for n-ary multiplication $\mu^{(n)}$ (1.2.19) and for m-ary addition $\nu^{(m)}$, as follows

$$\nu^{(m)}[\mathbf{m}_y, x] = x, \tag{4.1.15}$$

where the (additive) neutral sequence is $\mathbf{m}_y = (y^{m-2}, \tilde{y})$, and $\tilde{y}$ is the additive querelement for $y \in R$ (see (1.2.15)). As distinct from (1.2.19) we have only one (additive) Dörnte relation (4.1.15) and one diagram from (1.2.22) only, because of commutativity of $\nu^{(m)}$.

By analogy with the multiplicative queroperation $\bar{\mu}^{(1)}$ (1.2.20), introduce the *additive unary queroperation* by

$$\tilde{\nu}^{(1)}(x) = \tilde{x}, \quad \forall x \in R, \tag{4.1.16}$$

where $\tilde{x}$ is the additive querelement (4.1.13). Thus, we introduce (see definition 1.2.11)

Definition 4.1.14 Diagrammatic definition of (m, n)**-field..** A *(polyadic) (m, n)-field* is a one-set algebraic structure with 4 operations and 3 relations

$$\langle R | \nu^{(m)}, \tilde{\nu}^{(1)}, \mu^{(n)}, \bar{\mu}^{(1)} \\ | \text{associativity, distributivity, double Dörnte relations} \rangle, \tag{4.1.17}$$

where $\nu^{(m)}$ and $\mu^{(n)}$ are commutative associative m-ary addition and n-ary associative multiplication connected by polyadic distributivity (4.1.1)–(4.1.3), $\tilde{\nu}^{(1)}$ and $\bar{\mu}^{(1)}$ are unary additive queroperation (4.1.13) and multiplicative queroperation (4.1.14).

***Remark* 4.1.15.** The relations between the operations can be treated as 'interactions' described by diagrams (e.g., (1.2.22)), such that

(i) Associativity of addition (1.1.28) is the 'interaction' of two m-ary additions $\nu^{(m)}$.

(ii) Associativity of multiplication (1.1.28) is the 'interaction' of two n-ary multiplications $\mu^{(n)}$.

(iii) Distributivity (4.1.1)–(4.1.3) is the 'interaction' of the addition $\nu^{(m)}$ and the multiplication $\mu^{(n)}$.

(iv) The additive Dörnte relation is the 'interaction' of the m-ary addition $\nu^{(m)}$ and the unary additive queroperation $\tilde{\nu}^{(1)}$, because we can write (4.1.15) as

$$\nu^{(m)}[y^{m-2}, \tilde{\nu}^{(1)}[y], x] = x. \tag{4.1.18}$$

(v) The multiplicative Dörnte relations are the 'interaction' of the n-ary multiplication $\mu^{(n)}$ and the unary multiplicative queroperation $\bar{\mu}^{(1)}$, since (1.2.19) can be written as

$$\mu^{(n)}[y^{n-2}, \bar{\mu}^{(1)}[y], x] = x, \tag{4.1.19}$$

$$\mu^{(n)}[x, y^{n-2}, \bar{\mu}^{(1)}[y]] = x, \tag{4.1.20}$$

where $\bar{\mu}^{(1)}$ can be on any place to act on one corresponding y only.

The only 'interaction' remains is between the unary queroperations. In general, there is no initial relation between $\tilde{\nu}^{(1)}$ and $\bar{\mu}^{(1)}$, nevertheless a possible their 'interaction' can lead to further thorough classification of polyadic fields.

Definition 4.1.16. A polyadic field $\mathcal{F}^{(m,n)}$ is called *quer-symmetric*, if its unary queroperations commute

$$\tilde{\nu}^{(1)} \circ \bar{\mu}^{(1)} = \bar{\mu}^{(1)} \circ \tilde{\nu}^{(1)}, \tag{4.1.21}$$

$$\bar{\tilde{x}} = \tilde{\bar{x}}, \quad \forall x \in R, \tag{4.1.22}$$

in other case $\mathcal{F}^{(m,n)}$ is called *quer-nonsymmetric*.

***Example* 4.1.17.** Consider the nonunital zeroless (denoted by hat $\hat{\mathcal{F}}$) polyadic field $\hat{\mathcal{F}}^{(3,3)} = \langle \{ia/b\} \mid \nu^{(3)}, \mu^{(3)} \rangle$, $\mathrm{i}^2 = -1$, $a, b \in \mathbb{Z}^{\mathrm{odd}}$. The ternary addition $\nu^{(3)}[x, y, t] = x + y + t$ and the ternary multiplication $\mu^{(3)}[x, y, t] = xyt$ are nonderived, ternary associative and distributive (operations are in $\mathbb{C}$). For each $x = ia/b$ $(a, b \in \mathbb{Z}^{\mathrm{odd}})$ the additive querelement is $\tilde{x} = -ia/b$, and the multiplicative querelement is $\bar{x} = -ib/a$ (see (1.2.13)). Therefore, both $\langle \{ia/b\} \mid \mu^{(3)} \rangle$ and $\langle \{ia/b\} \mid \nu^{(3)} \rangle$ are ternary groups, but they both contain no neutral elements (no unit, no zero). The nonunital zeroless (3, 3)-field $\hat{\mathcal{F}}^{(3,3)}$ is quer-symmetric, because (see (4.1.22))

$$\bar{\tilde{x}} = \tilde{\bar{x}} = i\frac{b}{a}. \tag{4.1.23}$$

To find quer-nonsymmetric polyadic fields is a not simple task.

***Example* 4.1.18.** Consider the set of real 4×4 matrices over the fractions $\frac{4k+3}{4l+3}$, $k, l \in \mathbb{Z}$, of the form

$$M = \begin{pmatrix} 0 & \dfrac{4k_1+3}{4l_1+3} & 0 & 0 \\ 0 & 0 & \dfrac{4k_2+3}{4l_2+3} & 0 \\ 0 & 0 & 0 & \dfrac{4k_3+3}{4l_3+3} \\ \dfrac{4k_4+3}{4l_4+3} & 0 & 0 & 0 \end{pmatrix}, \quad k_i, l_i \in \mathbb{Z}. \tag{4.1.24}$$

The set $\{M\}$ is closed with respect to the ordinary addition of $m \geqslant 5$ matrices, because the sum of lower number of the fractions $\frac{4k+3}{4l+3}$ does not give a fraction of the same form (which follows from (4.3.10)), and with respect to the ordinary multiplication of $n \geqslant 5$ matrices, since the product of the lower number of matrices (4.1.24) does not have the same shape (3.4.10)–(3.4.11). The polyadic associativity and polyadic distributivity follows from the binary ones of the ordinary matrices over $\mathbb{R}$, and the product of 5 matrices is semicommutative (see 1.2.4). Taking the minimal values $m = 5$, $n = 5$, we define the semicommutative zeroless (5, 5)-field (see (4.1.11))

$$\mathcal{F}_M^{(5,5)} = \langle \{M\} | \nu^{(5)}, \mu^{(5)}, \bar{\nu}^{(1)}, \bar{\mu}^{(1)} \rangle, \tag{4.1.25}$$

where $\nu^{(5)}$ and $\mu^{(5)}$ are the ordinary sum and product of 5 matrices, while $\bar{\nu}^{(1)}$ and $\bar{\mu}^{(1)}$ are additive and multiplicative queroperations

$$\bar{\nu}^{(1)}[M] \equiv \tilde{M} = -3M, \; \bar{\mu}^{(1)}[M] \equiv \bar{M} = \frac{4l_1 + 3}{4k_1 + 3}\frac{4l_2 + 3}{4k_2 + 3}\frac{4l_3 + 3}{4k_3 + 3}\frac{4l_4 + 3}{4k_4 + 3}M. \tag{4.1.26}$$

The division ring $\mathcal{D}_M^{(5,5)}$ is zeroless, because the fraction $\frac{4k+3}{4l+3}$, is never zero for $k, l \in \mathbb{Z}$, and it is unital with the unit

$$M_e = \begin{pmatrix} 0 & 1 & 0 & 0 \\ 0 & 0 & 1 & 0 \\ 0 & 0 & 0 & 1 \\ 1 & 0 & 0 & 0 \end{pmatrix}. \tag{4.1.27}$$

Using (4.1.24) and (4.1.26), we obtain

$$\bar{\nu}^{(1)}[\bar{\mu}^{(1)}[M]] = -3\frac{4l_1 + 3}{4k_1 + 3}\frac{4l_2 + 3}{4k_2 + 3}\frac{4l_3 + 3}{4k_3 + 3}\frac{4l_4 + 3}{4k_4 + 3}M, \tag{4.1.28}$$

$$\bar{\mu}^{(1)}[\bar{\nu}^{(1)}[M]] = -\frac{1}{27}\frac{4l_1 + 3}{4k_1 + 3}\frac{4l_2 + 3}{4k_2 + 3}\frac{4l_3 + 3}{4k_3 + 3}\frac{4l_4 + 3}{4k_4 + 3}M, \tag{4.1.29}$$

or

$$\widetilde{\bar{M}} = 81\overline{\tilde{M}}, \tag{4.1.30}$$

and therefore the additive and multiplicative queroperations do not commute independently of field parameters. Thus, the matrix (5, 5)-division ring $\mathcal{D}_M^{(5,5)}$ (4.1.25) is a quer-nonsymmetric division ring.

Proposition 4.1.19. *In the case of polyadic structures with two operations on one set there are no conditions between arities of operations which could follow from distributivity* (4.1.1)–(4.1.3) *or the other relations above, and therefore they have no arity shape.*

Such conditions will appear in chapter 5, where we consider more complicated universal algebraic structures with two or more sets with operations and relations.

4.2 Polyadic direct products of rings and fields

Direct product is an important concept, especially because it plays a role of a product in a corresponding category (see, e.g., Borceux 1994, Mac Lane 1971). The general method of the direct product construction is taking the Cartesian product of the underlying sets and endow it by the operations from the algebraic structure under consideration (for one-set polyadic structures, see section 1.3 and Duplij

2022). We here consider polyadic rings and fields in the same way. Since there exist zeroless polyadic fields (Duplij 2017), the well-known statement (see, e.g., Lambek 1966) of the absence of binary fields that are a direct product of fields does not hold in the polyadic case. We construct polyadic fields that are products of zeroless fields, which can lead to a new category of polyadic fields.

Now we show that the thorough 'polyadization' of operations can lead to some unexpected new properties of ring and field external direct products (Duplij 2022). The main new peculiarities of the polyadic case are:

(i) The arity shape of the external product ring and its constituent rings can be different.

(ii) The external product of polyadic fields can be a polyadic field.

4.2.1 External direct product of binary rings

First, we recall the ordinary (binary) direct product of rings in notation which would be convenient to generalize to higher arity structures (Duplij 2017). Let us have two binary rings $\mathcal{R}_{1,2} \equiv \mathcal{R}_{1,2}^{(2,2)} = \left\langle R_{1,2} \middle| \nu_{1,2}^{(2)} \equiv (+_{1,2}), \mu_{1,2}^{(2)} \equiv (\cdot_{1,2}) \right\rangle$, where $R_{1,2}$ are underlying sets, while $\nu_{1,2}^{(2)}$ and $\mu_{1,2}^{(2)}$ are additions and multiplications (satisfying distributivity) in $\mathcal{R}_{1,2}$, correspondingly. On the Cartesian product of the underlying sets $R' = R_1 \times R_2$ one defines the *external direct product ring* $\mathcal{R}_1 \times \mathcal{R}_2 = \mathcal{R}' = \langle R' | \boldsymbol{\nu}'^{(2)} \equiv (+'), \boldsymbol{\mu}'^{(2)} \equiv (\bullet') \rangle$ by the componentwise operations (addition and multiplication) on the doubles $\mathbf{X} = \begin{pmatrix} x_1 \\ x_2 \end{pmatrix} \in R_1 \times R_2$ as follows

$$\begin{aligned} \mathbf{X}^{(1)} +' \mathbf{X}^{(2)} &= \begin{pmatrix} x_1 \\ x_2 \end{pmatrix}^{(1)} + \begin{pmatrix} x_1 \\ x_2 \end{pmatrix}^{(2)} \equiv \begin{pmatrix} x_1^{(1)} \\ x_2^{(1)} \end{pmatrix} +' \begin{pmatrix} x_1^{(2)} \\ x_2^{(2)} \end{pmatrix} \\ &= \begin{pmatrix} x_1^{(1)} +_1 x_1^{(2)} \\ x_2^{(1)} +_2 x_2^{(2)} \end{pmatrix}, \end{aligned} \tag{4.2.1}$$

$$\mathbf{X}^{(1)} \bullet' \mathbf{X}^{(2)} = \begin{pmatrix} x_1 \\ x_2 \end{pmatrix}^{(1)} \bullet' \begin{pmatrix} x_1 \\ x_2 \end{pmatrix}^{(2)} = \begin{pmatrix} x_1^{(1)} \cdot_1 x_1^{(2)} \\ x_2^{(1)} \cdot_2 x_2^{(2)} \end{pmatrix}, \tag{4.2.2}$$

or in the polyadic notation (with manifest operations)

$$\boldsymbol{\nu}'^{(2)}[\mathbf{X}^{(1)}, \mathbf{X}^{(2)}] = \begin{pmatrix} \nu_1^{(2)}[x_1^{(1)}, x_1^{(2)}] \\ \nu_2^{(2)}[x_2^{(1)}, x_2^{(2)}] \end{pmatrix}, \tag{4.2.3}$$

$$\boldsymbol{\mu}'^{(2)}[\mathbf{X}^{(1)}, \mathbf{X}^{(2)}] = \begin{pmatrix} \mu_1^{(2)}[x_1^{(1)}, x_1^{(2)}] \\ \mu_2^{(2)}[x_2^{(1)}, x_2^{(2)}] \end{pmatrix}. \tag{4.2.4}$$

The associativity and distributivity of the binary direct product operations $\boldsymbol{\nu}'^{(2)}$ and $\boldsymbol{\mu}'^{(2)}$ are obviously governed by those of the constituent binary rings $\mathcal{R}_1$ and $\mathcal{R}_2$, because of the componentwise construction on the rhs of (4.2.3)–(4.2.4). In the polyadic case, the construction of the direct product is not so straightforward and can have additional unusual peculiarities.

4.2.2 Full polyadic external direct product of (m, n)-rings

Let us consider the following task: for a given polyadic (m, n)-ring $\mathcal{R}'^{(m,n)} = \langle R' | \boldsymbol{\nu}'^{(m)}, \boldsymbol{\mu}'^{(n)} \rangle$ to construct a product of all possible (in arity shape) constituent rings $\mathcal{R}_1^{(m_1,n_1)}$ and $\mathcal{R}_2^{(m_2,n_2)}$. The first-hand 'polyadization' of (4.2.3)–(4.2.4) leads to

Definition 4.2.1. A *full polyadic direct product ring* $\mathcal{R}'^{(m,n)} = \mathcal{R}_1^{(m,n)} \times \mathcal{R}_2^{(m,n)}$ consists of (two) polyadic rings of the same arity shape, such that

$$\boldsymbol{\nu}'^{(m)}[\mathbf{X}^{(1)}, \mathbf{X}^{(2)}, \dots, \mathbf{X}^{(m)}] = \begin{pmatrix} \nu_1^{(m)}[x_1^{(1)}, x_1^{(2)}, \dots, x_1^{(m)}] \\ \nu_2^{(m)}[x_2^{(1)}, x_2^{(2)}, \dots, x_2^{(m)}] \end{pmatrix}, \tag{4.2.5}$$

$$\boldsymbol{\mu}'^{(n)}[\mathbf{X}^{(1)}, \mathbf{X}^{(2)}, \dots, \mathbf{X}^{(n)}] = \begin{pmatrix} \mu_1^{(n)}[x_1^{(1)}, x_1^{(2)}, \dots, x_1^{(n)}] \\ \mu_2^{(n)}[x_2^{(1)}, x_2^{(2)}, \dots, x_2^{(n)}] \end{pmatrix}, \tag{4.2.6}$$

where the polyadic associativity (1.1.28) and polyadic distributivity (4.1.1)–(4.1.3) of the direct product operations $\nu^{(m)}$ and $\mu^{(n)}$ follow from those of the constituent rings and the componentwise operations in (4.2.5)–(4.2.6).

***Example* 4.2.2.** Consider two (2, 3)-rings $\mathcal{R}_1^{(2,3)} = \left\langle \{\mathrm{i}x\} | \nu_1^{(2)} = (+), \mu_1^{(3)} = (\cdot), x \in \mathbb{Z}, \mathrm{i}^2 = -1 \right\rangle$ and $\mathcal{R}_2^{(2,3)} = \left\langle \left\{ \begin{pmatrix} 0 & a \\ b & 0 \end{pmatrix} \right\} \middle| \nu_2^{(2)} = (+), \mu_2^{(3)} = (\cdot), a, b \in \mathbb{Z} \right\rangle$, where $(+)$ and $(\cdot)$ are operations in $\mathbb{Z}$, then their polyadic direct product on the doubles $\mathbf{X}^T = \left(\mathrm{i}x, \begin{pmatrix} 0 & a \\ b & 0 \end{pmatrix} \right) \in (\mathrm{i}\mathbb{Z}, GL^{\mathrm{adiag}}(2, \mathbb{Z}))$ is defined by

$$\boldsymbol{\nu}'^{(2)}[\mathbf{X}^{(1)}, \mathbf{X}^{(2)}] = \begin{pmatrix} \mathrm{i}x^{(1)} + \mathrm{i}x^{(2)} \\ \begin{pmatrix} 0 & a^{(1)} + a^{(2)} \\ b^{(1)} + b^{(2)} & 0 \end{pmatrix} \end{pmatrix}, \tag{4.2.7}$$

$$\boldsymbol{\mu}'^{(3)}[\mathbf{X}^{(1)}, \mathbf{X}^{(2)}, \mathbf{X}^{(3)}] = \begin{pmatrix} \mathrm{i}x^{(1)}x^{(2)}x^{(3)} \\ \begin{pmatrix} 0 & a^{(1)}b^{(2)}a^{(3)} \\ b^{(1)}a^{(2)}b^{(3)} & 0 \end{pmatrix} \end{pmatrix}. \tag{4.2.8}$$

The polyadic associativity and distributivity of the direct product operations $\boldsymbol{\nu}'^{(2)}$ and $\boldsymbol{\mu}'^{(3)}$ are evident, and therefore $\mathcal{R}^{(2,3)} = \left\langle \left\{ \left(ix, \begin{pmatrix} 0 & a \\ b & 0 \end{pmatrix} \right) \right\} \middle| \boldsymbol{\nu}'^{(2)}, \boldsymbol{\mu}'^{(3)} \right\rangle$ is a (2, 3)-ring $\mathcal{R}^{(2,3)} = \mathcal{R}_1^{(2,3)} \times \mathcal{R}_2^{(2,3)}$.

Definition 4.2.3. A polyadic direct product $\mathcal{R}^{(m,n)}$ is called *derived*, if both constituent rings $\mathcal{R}_1^{(m,n)}$ and $\mathcal{R}_2^{(m,n)}$ are derived, such that the operations $\nu_{1,2}^{(m)}$ and $\mu_{1,2}^{(n)}$ are compositions of the binary operations $\nu_{1,2}^{(2)}$ and $\mu_{1,2}^{(2)}$, correspondingly.

So, in the derived case (see (1.1.3)) all the operations in (4.2.5)–(4.2.6) have the form (see (1.3.7))

$$\nu_{1,2}^{(m)} = \left(\nu_{1,2}^{(2)}\right)^{\circ(m-1)}, \quad \mu_{1,2}^{(n)} = \left(\mu_{1,2}^{(2)}\right)^{\circ(n-1)}, \tag{4.2.9}$$

$$\nu^{(m)} = \left(\nu^{(2)}\right)^{\circ(m-1)}, \quad \mu^{(n)} = \left(\mu^{(2)}\right)^{\circ(n-1)}. \tag{4.2.10}$$

Thus, the operations of the derived polyadic ring can be written as (see the binary direct product (4.2.1)–(4.2.2))

$$\boldsymbol{\nu}'^{(m)}[\mathbf{X}^{(1)}, \mathbf{X}^{(2)}, \ldots, \mathbf{X}^{(m)}] = \begin{pmatrix} x_1^{(1)} +_1 x_1^{(2)} +_1 \ldots +_1 x_1^{(m)} \\ x_2^{(1)} +_2 x_2^{(2)} +_2 \ldots +_2 x_2^{(m)} \end{pmatrix}, \tag{4.2.11}$$

$$\boldsymbol{\mu}'^{(n)}[\mathbf{X}^{(1)}, \mathbf{X}^{(2)}, \ldots, \mathbf{X}^{(n)}] = \begin{pmatrix} x_1^{(1)} \cdot_1 x_1^{(2)} \cdot_1 \ldots \cdot_1 x_1^{(n)} \\ x_2^{(1)} \cdot_2 x_2^{(2)} \cdot_2 \ldots \cdot_2 x_2^{(n)} \end{pmatrix}. \tag{4.2.12}$$

The external direct product (2, 3)-ring $\mathcal{R}^{(2,3)}$ from example 4.2.2 is not derived, because both multiplications $\mu_1^{(3)}$ and $\mu_2^{(3)}$ there are nonderived.

4.2.3 Mixed arity iterated product of (m, n)-rings

Recall, that some polyadic multiplications can be iterated, i.e. composed (1.1.3) from those of lower arity (1.1.4), also larger than 2, and so being nonderived, in general. The nontrivial 'polyadization' of (4.2.3)–(4.2.4) can arise, when the composition of the separate (up and down) components in the rhs of (4.2.5)–(4.2.6) will be different, and therefore the external product operations on the doubles $\mathbf{X} \in R_1 \times R_2$ cannot be presented in the iterated form (1.1.3).

Let now the constituent operations in (4.2.5)–(4.2.6) be composed from lower arity corresponding operations, but in different ways for the up and down components, such that

$$\nu_1^{(m)} = \left(\nu_1^{(m_1)}\right)^{\circ \ell_{\nu 1}}, \quad \nu_2^{(m)} = \left(\nu_2^{(m_2)}\right)^{\circ \ell_{\nu 2}}, \quad 3 \leqslant m_{1,2} \leqslant m-1, \tag{4.2.13}$$

$$\mu_1^{(n)} = \left(\mu_1^{(n_1)}\right)^{\circ \ell_{\mu 1}}, \quad \mu_2^{(n)} = \left(\mu_2^{(n_2)}\right)^{\circ \ell_{\mu 2}}, \qquad 3 \leqslant n_{1,2} \leqslant n-1, \tag{4.2.14}$$

where we exclude the limits: the derived case $m_{1,2} = n_{1,2} = 2$ (4.2.9)–(4.2.10) and the uncomposed case $m_{1,2} = m$, $n_{1,2} = n$ (4.2.5)–(4.2.6). Since the total size of the up and down polyads is the same and coincides with the arities of the double addition m and multiplication n, using (1.1.4) we obtain the *arity compatibility* relations

$$m = \ell_{\nu 1}(m_1 - 1) + 1 = \ell_{\nu 2}(m_2 - 1) + 1, \tag{4.2.15}$$

$$n = \ell_{\mu 1}(n_1 - 1) + 1 = \ell_{\mu 2}(n_2 - 1) + 1. \tag{4.2.16}$$

Definition 4.2.4. A *mixed arity polyadic direct product ring* $\mathcal{R}^{(m,\, n)} = \mathcal{R}_1^{(m_1,n_1)} \circledast \mathcal{R}_2^{(m_2,n_2)}$ consists of two polyadic rings of the different arity shape, such that

$$\boldsymbol{\nu}'^{(m)}[\mathbf{X}^{(1)}, \mathbf{X}^{(2)}, \ldots, \mathbf{X}^{(m)}] = \begin{pmatrix} \left(\nu_1^{(m_1)}\right)^{\circ \ell_{\nu 1}}\left[x_1^{(1)}, x_1^{(2)}, \ldots, x_1^{(m)}\right] \\ \left(\nu_2^{(m_2)}\right)^{\circ \ell_{\nu 2}}\left[x_2^{(1)}, x_2^{(2)}, \ldots, x_2^{(m)}\right] \end{pmatrix}, \tag{4.2.17}$$

$$\boldsymbol{\mu}'^{(n)}[\mathbf{X}^{(1)}, \mathbf{X}^{(2)}, \ldots, \mathbf{X}^{(n)}] = \begin{pmatrix} \left(\mu_1^{(n_1)}\right)^{\circ \ell_{\mu 1}}\left[x_1^{(1)}, x_1^{(2)}, \ldots, x_1^{(n)}\right] \\ \left(\mu_2^{(n_2)}\right)^{\circ \ell_{\mu 2}}\left[x_2^{(1)}, x_2^{(2)}, \ldots, x_2^{(n)}\right] \end{pmatrix}, \tag{4.2.18}$$

and the arity compatibility relations (4.2.15)–(4.2.16) hold valid.

Thus, two polyadic rings cannot always be composed in the mixed arity polyadic direct product.

Assertion 4.2.5. *If the arity shapes of two polyadic rings* $\mathcal{R}_1^{(m_1,n_1)}$ *and* $\mathcal{R}_2^{(m_2,n_2)}$ *satisfy the compatibility conditions*

$$a(m_1 - 1) = b(m_2 - 1), \tag{4.2.19}$$

$$c(n_1 - 1) = d(n_2 - 1), \qquad a, b, c, d \in \mathbb{N}, \tag{4.2.20}$$

then they can form a mixed arity direct product.

The limiting cases, undecomposed (4.2.5)–(4.2.6) and derived (4.2.11)–(4.2.12), satisfy the compatibility conditions (4.2.19)–(4.2.20) as well.

***Example* 4.2.6.** Let us consider two (nonderived) polyadic rings

$$\mathcal{R}_1^{(9,\, 3)} = \left\langle \{8l + 7\} \middle| \nu_1^{(9)}, \mu_1^{(3)}, l \in \mathbb{Z} \right\rangle, \tag{4.2.21}$$

$$\mathcal{R}_2^{(5,\,5)} = \left\langle \{M\} \middle| \nu_2^{(5)}, \mu_2^{(5)} \right\rangle, \tag{4.2.22}$$

where

$$M = \begin{pmatrix} 0 & 4k_1+3 & 0 & 0 \\ 0 & 0 & 4k_2+3 & 0 \\ 0 & 0 & 0 & 4k_3+3 \\ 4k_4+3 & 0 & 0 & 0 \end{pmatrix}, \quad k_i \in \mathbb{Z}, \tag{4.2.23}$$

and $\nu_2^{(5)}$ and $\mu_2^{(5)}$ are the ordinary sum and product of 5 matrices. Using (4.2.15)–(4.2.16) we obtain $m = 9$, $n = 5$, if we choose the smallest 'numbers of multiplications' $\ell_{\nu 1} = 1$, $\ell_{\nu 2} = 2$, $\ell_{\mu 1} = 2$, $\ell_{\mu 2} = 1$, and therefore the mixed arity direct product nonderived (9, 5)-ring becomes

$$\mathcal{R}^{(9,5)} = \langle \{\mathbf{X}\} | \boldsymbol{\nu}'^{(9)}, \boldsymbol{\mu}'^{(5)} \rangle, \tag{4.2.24}$$

where the doubles are $\mathbf{X} = \begin{pmatrix} 8l+7 \\ M \end{pmatrix}$ and the nonderived direct product operations are

$$\boldsymbol{\nu}'^{(9)}[\mathbf{X}^{(1)}, \mathbf{X}^{(2)}, \ldots, \mathbf{X}^{(9)}] = \begin{pmatrix} 8(l^{(1)} + l^{(2)} + l^{(3)} + l^{(4)} + l^{(5)} + l^{(6)} + l^{(7)} + l^{(8)} + l^{(9)} + 7) + 7 \\ \begin{pmatrix} 0 & 4K_1+3 & 0 & 0 \\ 0 & 0 & 4K_2+3 & 0 \\ 0 & 0 & 0 & 4K_3+3 \\ 4K_4+3 & 0 & 0 & 0 \end{pmatrix} \end{pmatrix}, \tag{4.2.25}$$

$$\boldsymbol{\mu}'^{(5)}[\mathbf{X}^{(1)}, \mathbf{X}^{(2)}, \mathbf{X}^{(3)}, \mathbf{X}^{(4)}, \mathbf{X}^{(5)}] = \begin{pmatrix} (8l_\mu + 7) \\ \begin{pmatrix} 0 & 4K_{\mu,1}+3 & 0 & 0 \\ 0 & 0 & 4K_{\mu,2}+3 & 0 \\ 0 & 0 & 0 & 4K_{\mu,3}+3 \\ 4K_{\mu,4}+3 & 0 & 0 & 0 \end{pmatrix} \end{pmatrix}, \tag{4.2.26}$$

where in the first line $K_i = k_i^{(1)} + k_i^{(2)} + k_i^{(3)} + k_i^{(4)} + k_i^{(5)} + k_i^{(6)} + k_i^{(7)} + k_i^{(8)} + k_i^{(9)} + 6 \in \mathbb{Z}$, $l_\mu \in \mathbb{Z}$ is a cumbersome integer function of $l^{(j)} \in \mathbb{Z}$, $j = 1, \ldots, 9$, and in the second line $K_{\mu,i} \in \mathbb{Z}$ are cumbersome integer functions of $k_i^{(s)}$, $i = 1, \ldots, 4$, $s = 1, \ldots, 5$. Therefore the polyadic ring (4.2.24) is the nonderived mixed arity polyadic external product $\mathcal{R}^{(9,5)} = \mathcal{R}_1^{(9,\,3)} \circledast \mathcal{R}_2^{(5,\,5)}$ (see definition 4.2.4).

Theorem 4.2.7. *The category of polyadic rings* `PolRing` *can exist (having the class of all polyadic rings for objects and ring homomorphisms for morphisms) and can be well-defined, because it has a product as the polyadic external product of rings.*

In the same way one can construct the iterated full and mixed arity products of any number k of polyadic rings, just by passing from the doubles $\mathbf{X}$ to k-tuples $\mathbf{X}_k^T = (x_1, \dots, x_k)$.

4.2.4 Polyadic hetero product of (m, n)-fields

The most crucial difference between the binary direct products and the polyadic ones arises for fields, because a direct product two binary fields is not a field (Lambek 1966). The reason lies in the fact that each binary field $\mathcal{F}^{(2,2)}$ necessarily contains 0 and 1, by definition. As follows from (4.2.2), a binary direct product contains nonzero idempotent doubles $\begin{pmatrix}1\\0\end{pmatrix}$ and $\begin{pmatrix}0\\1\end{pmatrix}$ which are noninvertible, and therefore the external direct product of fields $\mathcal{F}_1^{(2,\,2)} \times \mathcal{F}_2^{(2,\,2)}$ can never be a field. As opposite, polyadic fields (see definition 4.1.10) can be zeroless (we denote them by hat $\hat{\mathcal{F}}$), and the above arguments do not hold valid for them.

Definition 4.2.8. The *polyadic zeroless direct product field* $\hat{\mathcal{F}}'^{(m,\,n)} = \langle R' | \boldsymbol{\nu}'^{(m)}, \boldsymbol{\mu}'^{(n)} \rangle$ consists of (two) zeroless polyadic fields $\hat{\mathcal{F}}_1^{(m,n)} = \left\langle R_1 \middle| \nu_1^{(m)}, \mu_1^{(n)} \right\rangle$ and $\hat{\mathcal{F}}_2^{(m,n)} = \left\langle R_2 \middle| \nu_2^{(m)}, \mu_2^{(n)} \right\rangle$ of the same arity shape, while the componentwise operations on the doubles $\mathbf{X} \in R_1 \times R_2$ in (4.2.5)–(4.2.6) still hold valid, and $\left\langle R_1 \middle| \mu_1^{(n)} \right\rangle$, $\left\langle R_2 \middle| \mu_2^{(n)} \right\rangle$, $\langle R' = \{\mathbf{X}\} | \boldsymbol{\mu}'^{(n)} \rangle$ are n-ary groups.

Following definition 4.1.11, we have

Corollary 4.2.9. If at least one of the constituent fields is semicommutative, and another one is totally commutative, then the polyadic product will be a semicommutative (m, n)-field.

The additive and multiplicative unary queroperations (1.2.20) and (4.1.16) for the direct product field $\hat{\mathcal{F}}^{(m,\,n)}$ are defined componentwise on the doubles $\mathbf{X}$ as follows

$$\tilde{\boldsymbol{\nu}}'^{(1)}[\mathbf{X}] = \begin{pmatrix} \tilde{\nu}_1^{(1)}[x_1] \\ \tilde{\nu}_2^{(1)}[x_2] \end{pmatrix}, \tag{4.2.27}$$

$$\bar{\boldsymbol{\mu}}'^{(1)}[\mathbf{X}] = \begin{pmatrix} \bar{\mu}_1^{(1)}[x_1] \\ \bar{\mu}_2^{(1)}[x_2] \end{pmatrix}, \quad x_1 \in R_1,\ x_2 \in R_2. \tag{4.2.28}$$

Definition 4.2.10. A polyadic direct product field $\hat{\mathcal{F}}^{'(m,n)} = \langle R' | \boldsymbol{\nu}^{'(m)}, \tilde{\boldsymbol{\nu}}^{'(1)}, \boldsymbol{\mu}^{'(n)}, \bar{\boldsymbol{\mu}}^{'(1)} \rangle$ is called *quer-symmetric*, if its unary queroperations (4.2.27)–(4.2.28) commute

$$\tilde{\boldsymbol{\nu}}^{'(1)} \circ \bar{\boldsymbol{\mu}}^{'(1)} = \bar{\boldsymbol{\mu}}^{'(1)} \circ \tilde{\boldsymbol{\nu}}^{'(1)}, \tag{4.2.29}$$

$$\bar{\tilde{\mathbf{X}}} = \tilde{\bar{\mathbf{X}}}, \quad \forall \mathbf{X} \in R', \tag{4.2.30}$$

in other case $\hat{\mathcal{F}}^{'(m,n)}$ is called a *quer-nonsymmetric direct product* (m, n)*-field.*

***Example* 4.2.11.** Consider two nonunital zeroless (3, 3)-fields

$$\hat{\mathcal{F}}_{1,2}^{(3,3)} = \left\langle \left\{ \mathrm{i}\frac{a_{1,2}}{b_{1,2}} \right\} \middle| \nu_{1,2}^{(3)}, \mu_{1,2}^{(3)}, \tilde{\nu}_{1,2}^{(1)}, \bar{\mu}_{1,2}^{(1)} \right\rangle, \quad \mathrm{i}^2 = -1, \quad a_{1,2}, b_{1,2} \in \mathbb{Z}^{\mathrm{odd}}, \tag{4.2.31}$$

where ternary additions $\nu_{1,2}^{(3)}$ and ternary multiplications $\mu_{1,2}^{(3)}$ are sum and product in $\mathbb{Z}^{\mathrm{odd}}$, correspondingly, and the unary additive and multiplicative queroperations are $\tilde{\nu}_{1,2}^{(1)}[\mathrm{i}a_{1,2}/b_{1,2}] = -\mathrm{i}a_{1,2}/b_{1,2}$ and $\bar{\mu}_{1,2}^{(1)}[\mathrm{i}a_{1,2}/b_{1,2}] = -\mathrm{i}b_{1,2}/a_{1,2}$ (see example 4.1.17). Using (4.2.5)–(4.2.6) we build the operations of the polyadic nonderived nonunital zeroless product (3, 3)-field $\hat{\mathcal{F}}^{'(3,3)} = \hat{\mathcal{F}}_1^{(3,3)} \times \hat{\mathcal{F}}_2^{(3,3)}$ on the doubles $\mathbf{X}^T = (\mathrm{i}a_1/b_1, \mathrm{i}a_2/b_2)$ as follows

$$\boldsymbol{\nu}^{'(3)}[\mathbf{X}^{(1)}, \mathbf{X}^{(2)}, \mathbf{X}^{(3)}] = \begin{pmatrix} \mathrm{i}\dfrac{a_1^{(1)}b_1^{(2)}b_1^{(3)} + b_1^{(1)}a_1^{(2)}b_1^{(3)} + b_1^{(1)}b_1^{(2)}a_1^{(3)}}{b_1^{(1)}b_1^{(2)}b_1^{(3)}} \\ \mathrm{i}\dfrac{a_2^{(1)}b_2^{(2)}b_2^{(3)} + b_2^{(1)}a_2^{(2)}b_2^{(3)} + b_2^{(1)}b_2^{(2)}a_2^{(3)}}{b_2^{(1)}b_2^{(2)}b_2^{(3)}} \end{pmatrix}, \tag{4.2.32}$$

$$\boldsymbol{\mu}^{'(3)}[\mathbf{X}^{(1)}, \mathbf{X}^{(2)}, \mathbf{X}^{(3)}] = \begin{pmatrix} -\mathrm{i}\dfrac{a_1^{(1)}a_1^{(2)}a_1^{(3)}}{b_1^{(1)}b_1^{(2)}b_1^{(3)}} \\ -\mathrm{i}\dfrac{a_2^{(1)}a_2^{(2)}a_2^{(3)}}{b_2^{(1)}b_2^{(2)}b_2^{(3)}} \end{pmatrix}, \quad a_i^{(j)}, b_i^{(j)} \in \mathbb{Z}^{\mathrm{odd}}, \tag{4.2.33}$$

and the unary additive and multiplicative queroperations (4.2.27)–(4.2.28) of the direct product $\hat{\mathcal{F}}^{'(3,3)}$ are

$$\tilde{\boldsymbol{\nu}}^{'(1)}[\mathbf{X}] = \begin{pmatrix} -\mathrm{i}\dfrac{a_1}{b_1} \\ -\mathrm{i}\dfrac{a_2}{b_2} \end{pmatrix}, \tag{4.2.34}$$

$$\bar{\boldsymbol{\mu}}^{'(1)}[\mathbf{X}] = \begin{pmatrix} -\mathrm{i}\dfrac{b_1}{a_1} \\ -\mathrm{i}\dfrac{b_2}{a_2} \end{pmatrix}, \qquad a_i, b_i \in \mathbb{Z}^{\mathrm{odd}}. \tag{4.2.35}$$

Therefore, both $\langle \{\mathbf{X}\} | \boldsymbol{\nu}^{'(3)}, \tilde{\boldsymbol{\nu}}^{'(1)} \rangle$ and $\langle \{\mathbf{X}\} | \boldsymbol{\mu}^{'(3)}, \bar{\boldsymbol{\mu}}^{'(1)} \rangle$ are commutative ternary groups, which means that the polyadic direct product $\hat{\mathcal{F}}^{'(3,3)} = \hat{\mathcal{F}}_1^{(3,3)} \times \hat{\mathcal{F}}_2^{(3,3)}$ is the nonunital zeroless polyadic field. Moreover, $\hat{\mathcal{F}}^{'(3,3)}$ is quer-symmetric, because (4.2.29)–(4.2.30) hold valid

$$\bar{\boldsymbol{\mu}}^{'(1)} \circ \tilde{\boldsymbol{\nu}}^{'(1)}[\mathbf{X}] = \tilde{\boldsymbol{\nu}}^{'(1)} \circ \bar{\boldsymbol{\mu}}^{'(1)}[\mathbf{X}] = \begin{pmatrix} \mathrm{i}\dfrac{b_1}{a_1} \\ \mathrm{i}\dfrac{b_2}{a_2} \end{pmatrix}, \qquad a_i, b_i \in \mathbb{Z}^{\mathrm{odd}}. \tag{4.2.36}$$

***Example* 4.2.12.** Let us consider the polyadic direct product of two zeroless fields, one of them the semicommutative (5, 5)-field $\hat{\mathcal{F}}_1^{(5,5)} = \mathcal{F}_M^{(5,5)}$ from (4.1.25), and the other one the nonderived nonunital zeroless (5, 5)-field of fractions $\hat{\mathcal{F}}_2^{(5,5)} = \left\langle \left\{ \sqrt{\mathrm{i}}\,\frac{4r+1}{4s+1} \right\} \middle| \nu_2^{(5)}, \mu_2^{(5)} \right\rangle$, $r, s \in \mathbb{Z}$, $\mathrm{i}^2 = -1$. The double is $\mathbf{X}^T = \left(\sqrt{\mathrm{i}}\,\frac{4r+1}{4s+1}, M \right)$, where M is in (4.1.24). The polyadic nonunital zeroless direct product field $\hat{\mathcal{F}}^{'(5,5)} = \hat{\mathcal{F}}_1^{(5,5)} \times \hat{\mathcal{F}}_2^{(5,5)}$ is nonderived and semicommutative, and is defined by $\hat{\mathcal{F}}^{(5,5)} = \langle \mathbf{X} | \boldsymbol{\nu}^{'(5)}, \boldsymbol{\mu}^{'(5)}, \tilde{\boldsymbol{\nu}}^{'(1)}, \bar{\boldsymbol{\mu}}^{(1)} \rangle$, where its addition and multiplication are

$$\boldsymbol{\nu}^{'(5)}[\mathbf{X}^{(1)}, \mathbf{X}^{(2)}, \mathbf{X}^{(3)}, \mathbf{X}^{(4)}, \mathbf{X}^{(5)}] = \begin{pmatrix} \sqrt{\mathrm{i}}\,\dfrac{4R_\nu + 1}{4S_\nu + 1} \\ \begin{pmatrix} 0 & \dfrac{4K_{\nu,1}+3}{4L_{\nu,1}+3} & 0 & 0 \\ 0 & 0 & \dfrac{4K_{\nu,2}+3}{4L_{\nu,2}+3} & 0 \\ 0 & 0 & 0 & \dfrac{4K_{\nu,3}+3}{4L_{\nu,3}+3} \\ \dfrac{4K_{\nu,4}+3}{4L_{\nu,4}+3} & 0 & 0 & 0 \end{pmatrix} \end{pmatrix}, \tag{4.2.37}$$

$$\boldsymbol{\mu}'^{(5)}[\mathbf{X}^{(1)}, \mathbf{X}^{(2)}, \mathbf{X}^{(3)}, \mathbf{X}^{(4)}, \mathbf{X}^{(5)}]$$
$$= \begin{pmatrix} \sqrt{\mathrm{i}}\dfrac{4R_\mu + 1}{4S_\mu + 1} \\ \begin{pmatrix} 0 & \dfrac{4K_{\mu,1} + 3}{4L_{\mu,1} + 3} & 0 & 0 \\ 0 & 0 & \dfrac{4K_{\mu,2} + 3}{4L_{\mu,2} + 3} & 0 \\ 0 & 0 & 0 & \dfrac{4K_{\mu,3} + 3}{4L_{\mu,3} + 3} \\ \dfrac{4K_{\mu,4} + 3}{4L_{\mu,4} + 3} & 0 & 0 & 0 \end{pmatrix} \end{pmatrix}, \tag{4.2.38}$$

where $R_{\nu,\mu}$, $S_{\nu,\mu} \in \mathbb{Z}$ are cumbersome integer functions of $r^{(i)}$, $s^{(i)} \in \mathbb{Z}$, $i = 1, \ldots, 5$, and $K_{\nu,i}$, $K_{\mu,i}$, $L_{\nu,i}$, $L_{\mu,i} \in \mathbb{Z}$ are cumbersome integer functions of $k_j^{(i)}$, $l_j^{(i)} \in \mathbb{Z}$, $j = 1, \ldots, 4$, $i = 1, \ldots, 5$ (see (4.1.24)). The unary queroperations (4.2.27)–(4.2.28) of the direct product $\hat{\mathcal{F}}^{(5,5)}$ are

$$\bar{\boldsymbol{\nu}}'^{(1)}[\mathbf{X}] = \begin{pmatrix} -3\sqrt{\mathrm{i}}\dfrac{4r + 1}{4s + 1} \\ -3M \end{pmatrix}, \tag{4.2.39}$$

$$\bar{\boldsymbol{\mu}}'^{(1)}[\mathbf{X}] = \begin{pmatrix} -\sqrt{\mathrm{i}}\left(\dfrac{4s + 1}{4r + 1}\right)^3 \\ \dfrac{4l_1 + 3}{4k_1 + 3}\dfrac{4l_2 + 3}{4k_2 + 3}\dfrac{4l_3 + 3}{4k_3 + 3}\dfrac{4l_4 + 3}{4k_4 + 3}M \end{pmatrix}, \quad r, s, k_i, l_i \in \mathbb{Z}, \tag{4.2.40}$$

where M is in (4.1.24). Therefore, $\langle\{\mathbf{X}\}|\boldsymbol{\nu}'^{(5)}, \bar{\boldsymbol{\nu}}'^{(1)}\rangle$ is a commutative 5-ary group, and $\langle\{\mathbf{X}\}|\boldsymbol{\mu}'^{(5)}, \bar{\boldsymbol{\mu}}'^{(1)}\rangle$ is a semicommutative 5-ary group, which means that the polyadic direct product $\hat{\mathcal{F}}'^{(5,5)} = \hat{\mathcal{F}}_1^{(5,\,5)} \times \hat{\mathcal{F}}_2^{(5,\,5)}$ is the nonunital zeroless polyadic semicommutative (5, 5)-field. Using (4.1.30) we obtain

$$\bar{\boldsymbol{\nu}}'^{(1)}\bar{\boldsymbol{\mu}}'^{(1)}[\mathbf{X}] = 81\bar{\boldsymbol{\mu}}'^{(1)}\bar{\boldsymbol{\nu}}'^{(1)}[\mathbf{X}], \tag{4.2.41}$$

and therefore the direct product (5, 5)-field $\hat{\mathcal{F}}'^{(5,5)}$ is quer-nonsymmetric (see (4.1.21)).

Thus, we arrive at

Theorem 4.2.13. *The category of* zeroless *polyadic fields* `zlessPolField` *can exist (having the class of all zeroless polyadic fields for objects and field homomorphisms for morphisms) and can be well-defined, because it has a product as the polyadic field product.*

Further analysis of the direct product constructions introduced here and their examples for polyadic rings and fields would be interesting to provide in detail, which can also lead to new kinds of categories.

4.3 Polyadic integer numbers

The notion of the ring of polyadic integer numbers $\mathbb{Z}^{(m,n)}$ which was introduced in Duplij (2019), where its difference from the (m, n)-ring of integers from Leeson and Butson (1980) was outlined.

Let us consider a standard congruence class (residue class) of an integer a modulo b

$$[[a]]_b = \{\{a + bk\} | k \in \mathbb{Z},\ a \in \mathbb{Z}_+, b \in \mathbb{N}, 0 \leqslant a \leqslant b - 1\}. \tag{4.3.1}$$

We denote a representative element by $x_k \equiv x_k^{[a,b]} = a + bk$, where obviously $\{x_k\}$ is an infinite set.

4.3.1 Congruence classes and operations

There are two ways to equip the set (4.3.1) with operations:

(i) The 'External' way is to define (binary) operations between the congruence classes. Let us define on the finite underlying set of b congruence classes $\{[[a]]_b\}$, $a = 0, 1, \dots, b - 1$ the following new binary operations (here, if b is fixed, and we denote the binary class representative by an integer with one prime $[[a]]_b \equiv a'$, as well as the corresponding binary operations $+'$, $\cdot'$ between classes)

$$a_1' +' a_2' = (a_1 + a_2)', \tag{4.3.2}$$

$$a_1' \cdot' a_2' = (a_1 a_2)'. \tag{4.3.3}$$

Then, the binary *residue class ring* is defined by

$$\mathbb{Z}/b\mathbb{Z} = \{\{a'\} | +', \cdot', 0', 1'\}. \tag{4.3.4}$$

In the case of prime $b = p$, the ring $\mathbb{Z}/p\mathbb{Z}$ becomes a binary finite field having p elements.

(ii) The 'Internal' way is to introduce (polyadic) operations inside a given class $[[a]]_b$ (with both a and b fixed). We introduce the commutative m-ary addition and commutative n-ary multiplication of representatives x_{k_i} of the fixed congruence class by

$$\nu^{(m)}[x_{k_1}, x_{k_2}, \dots, x_{k_m}] = x_{k_1} + x_{k_2} + \dots + x_{k_m}, \tag{4.3.5}$$

$$\mu^{(n)}[x_{k_1}, x_{k_2}, \dots, x_{k_n}] = x_{k_1} x_{k_2} \dots x_{k_n}, \quad x_{k_i} \in [[a]]_b, k_i \in \mathbb{Z}. \tag{4.3.6}$$

In general, the binary sums $x_{k_1} + x_{k_2}$ and products $x_{k_1} x_{k_2}$ are not in $[[a]]_b$.

Proposition 4.3.1 Duplij 2019) *The polyadic operations* $\nu^{(m)}$ and $\mu^{(n)}$ *become closed in* $[[a]]_b$, *if the arities* (m, n) *have the minimal values satisfying*

$$ma \equiv a(\bmod b), \tag{4.3.7}$$

$$a^n \equiv a(\bmod b). \tag{4.3.8}$$

Remark **4.3.2.** If $n = b = p$ is prime, then (4.3.8) is valid for any $a \in \mathbb{N}$, which is another formulation of Fermat's little theorem.

Polyadic distributivity is inherited from that of $\mathbb{Z}$, and therefore we have

Definition 4.3.3 (Duplij 2019). The congruence class $[[a]]_b$ equipped with a structure of nonderived infinite commutative polyadic ring and satisfying (4.3.7)–(4.3.8) is called a (m, n)*-ring of polyadic integer numbers*

$$\mathbb{Z}^{(m,\, n)} \equiv \mathbb{Z}_{a,\, b}^{(m,n)} = \{[[a]]_b | \nu^{(m)}, \mu^{(n)}\}. \tag{4.3.9}$$

Obviously, $\mathbb{Z}^{(m,\, n)}$ (as in the binary case) cannot become a polyadic field with any choice of parameters.

Remark **4.3.4.** At first sight it seems that one can obtain a polyadic field only in the 'external' way, i.e., using the 'trivial polyadization' of the binary finite field $\mathbb{Z}/p\mathbb{Z}$ (just a repetition of the binary group operations). This leads to the derived polyadic finite fields, which have a very simple structure, in which the admissible binary sums and binary products of the congruence classes are used (Leeson and Butson 1980).

Here we propose a new approach, and thereby construct the nonderived finite (m, n)-fields of polyadic integer numbers $\mathbb{Z}^{(m,\, n)}$.

Remark **4.3.5.** The polyadic operations (4.3.5)–(4.3.6) are not derived (see, e.g., Głazek and Michalski 1984, Michalski 1988), because on the set $\{x_{k_i}\}$ one cannot define the binary semigroup structure with respect to ordinary addition and multiplication, and therefore $\mathbb{Z}_{a,\, b}^{(m,n)}$ is a nonderived polyadic ring.

Example **4.3.6.** In the residue class

$$[[3]]_4 = \{\dots -25, -21, -17, -13, -9, -5, -1, 3, 7, 11, 15, 19, 23, 27, 31, 35, 39\dots\} \tag{4.3.10}$$

we can add only $4\ell_\nu + 1$ representatives and multiply $2\ell_\mu + 1$ representatives (ℓ_ν, ℓ_μ are 'numbers' of m-ary additions and n-ary multiplications respectively) to retain the same class, e.g., take $\ell_\nu = 2, \ell_\mu = 3$ to get $(7 + 11 + 15 + 19 + 23) - 5 - 9 - 13 - 1 = 47 \in [[3]]_4$, $((7 \cdot 3 \cdot 11) \cdot 19 \cdot 15) \cdot 31 \cdot 27 = 55\,103\,895 \in [[3]]_4$. Obviously, we cannot add and multiply arbitrary quantities of numbers in $[[3]]_4$, only the admissible ones. This means that $[[3]]_4$ is the polyadic $(5, 3)$-ring $\mathbb{Z}^{(5,3)} = \mathbb{Z}_{3,\, 4}^{(5,\, 3)}$.

***Remark* 4.3.7.** After imposing the operations (4.3.5)–(4.3.6) the representatives $x_k^{[a,b]}$ become abstract variables (elements of the corresponding (m, n)-ring or polyadic integer numbers) which are not ordinary integers (forming a (2, 2)-ring), but have the latter as their 'binary limit'. So in computations the integral numbers (denoting representatives) should carry their arity shape (m, n) as additional indices. Indeed, the representative, e.g., $3 = 3_{(5,3)} \in \mathbb{Z}_{3,4}^{(5,3)}$ is different from $3 = 3_{(3,2)} \in \mathbb{Z}_{1,2}^{(3,2)}$, i.e., properly speaking $3_{(5,3)} \neq 3_{(3,2)}$, since their operations (multiplication and addition) are different, because they belong to different polyadic rings, $\mathbb{Z}_{3,4}^{(5,3)}$ and $\mathbb{Z}_{1,2}^{(3,2)}$, respectively. For conciseness, we omit the indices (m, n), if their value is clear from the context.

Lemma 4.3.8. In case

$$(m-1)\frac{a}{b} = \text{integer} \tag{4.3.11}$$

the algebraic structure $\langle [[a]]_b | \nu^{(m)} \rangle$ is a commutative m-ary group.

Proof. The closure of the operation (4.3.5) can be written as $x_{k_1} + x_{k_2} + \cdots + x_{k_m} = x_{k_0}$, or $ma + b(k_1 + k_2 + \cdots + k_m) = a + bk_0$, and then $k_0 = (m-1)a/b + (k_1 + k_2 + \cdots + k_m)$, from (4.3.11). The (total) associativity and commutativity of $\nu^{(m)}$ follows from those of the addition in the binary $\mathbb{Z}$. Each element x_k has its unique querelement $\tilde{x} = x_{\tilde{k}}$ determined by the equation $(m-1)x_k + x_{\tilde{k}} = x_k$, which (uniquely, for any $k \in \mathbb{Z}$) gives

$$\tilde{k} = bk(2-m) - (m-1)\frac{a}{b}. \tag{4.3.12}$$

Thus, each element is 'querable' (polyadic invertible), and so $\langle [[a]]_b | \nu^{(m)} \rangle$ is an m-ary group. □

***Example* 4.3.9.** For $a = 2$, $b = 7$ we have 8-ary group $\langle [[2]]_7 | \nu^{(8)} \rangle$, and the additive querelement of x_k is $\tilde{x} = x_{(-2-12k)}$.

Lemma 4.3.10. If

$$\frac{a^n - a}{b} = \text{integer}, \tag{4.3.13}$$

then $\langle [[a]]_b | \mu^{(n)} \rangle$ is a commutative n-ary semigroup.

Proof. It follows from (4.3.6), that the closeness of the operation $\mu^{(n)}$ is $x_{k_1} x_{k_2} \ldots x_{k_n} = x_{k_0}$, which can be written as $a^n + b(\text{integer}) = a + bk_0$ leading to (4.3.13). The (total) associativity and commutativity of $\mu^{(n)}$ follows from those of the multiplication in $\mathbb{Z}$. □

It follows from (4.3.11) and (4.3.13) that the characterization of the fixed congruence class $[[a]]_b$ and the corresponding (m, n)-ring of polyadic integer numbers $\mathbb{Z}_{a,b}^{(m,n)}$ can be done in terms of Duplij (2019).

Definition 4.3.11. The *shape invariants* I, $J \in \mathbb{Z}_+$ of the congruence class $[[a]]_b$ can be defined uniquely by

$$I = I_m^{[a,b]} = (m - 1)\frac{a}{b}, \tag{4.3.14}$$

$$J = J_n^{[a,b]} = \frac{a^n - a}{b}. \tag{4.3.15}$$

In the binary case, when $m = n = 2$ ($a = 0$, $b = 1$) both shape invariants vanish, $I = J = 0$. Nevertheless, there exist 'partially' binary cases, when only $n = 2$ and $m \neq 2$, while J is nonzero, for instance in $\mathbb{Z}_{6,\,10}^{(6,\,2)}$ we have $I = J = 3$.

Theorem 4.3.12. *The fixed congruence class* $[[a]]_b$ *is a polyadic* (m, n)*-ring*

$$\mathcal{R}_{a,\,b}^{(m,n)} = \langle [[a]]_b | \nu^{(m)}, \mu^{(n)} \rangle, \tag{4.3.16}$$

where the arities m *and* n *are minimal positive integers (more than or equal to 2), for which the congruences*

$$ma \equiv a(\operatorname{mod} b), \tag{4.3.17}$$

$$a^n \equiv a(\operatorname{mod} b) \tag{4.3.18}$$

take place simultaneously, fixating its polyadic shape invariants (I, J).

Proof. By lemma 4.3.8 and 4.3.10 the set $[[a]]_b$ is an m-ary group with respect to 'm-ary addition' $\nu^{(m)}$ and an n-ary semigroup with respect to 'n-ary multiplication' $\mu^{(n)}$, while the polyadic distributivity (4.1.1)–(4.1.3) follows from (4.3.5) and (4.3.6) and the binary distributivity in $\mathbb{Z}$. □

Definition 4.3.13. The infinite set of representatives of the congruence (residue) class $[[a]]_b$ having fixed arities and form the (m, n)-ring $\mathcal{R}_{a,\,b}^{(m,n)}$ (4.3.16) is called the set of *(polyadic)* (m, n)*-integers (integer numbers)* and denoted by $\mathbb{Z}^{(m,\,n)}$.

Obviously, for ordinary integers $\mathbb{Z} = \mathbb{Z}^{(2,2)}$, and they form the binary ring $\mathcal{R}_{0,\,1}^{(2,\,2)}$.

Remark **4.3.14.** For a fixed $b \geqslant 2$ there are b congruence classes $[[a]]_b$, $0 \leqslant a \leqslant b - 1$, and therefore exactly b corresponding polyadic (m, n)-rings $\mathcal{R}_{a,\,b}^{(m,n)}$, each of them is infinite-dimensional.

Corollary 4.3.15. In case $\gcd(a, b) = 1$ and b is prime, there exists the solution $n = b$.

Proof. Follows from (4.3.18) and Fermat's little theorem. □

***Remark* 4.3.16.** We exclude from consideration the zero congruence class $[[0]]_b$, because the arities of operations $\nu^{(m)}$ and $\mu^{(n)}$ cannot be fixed up by (4.3.17)–(4.3.18) becoming identities for any m and n. Since the arities are uncertain, their minimal values can be chosen $m = n = 2$, and therefore, it follows from (4.3.5) and (4.3.6), $\mathcal{R}_{0,\,b}^{(2,\,2)} = \mathbb{Z}$. Thus, in what follows we always imply that $a \neq 0$ (without using a special notation, e.g., $\mathcal{R}^{*}$, etc).

In table 4.1 we present the allowed (by (4.3.17)–(4.3.18)) arities of the polyadic ring $\mathcal{R}_{a,\,b}^{(m,n)}$ and the corresponding polyadic shape invariants $(I,\, J)$ for $b \leqslant 10$.

Table 4.1. The polyadic ring $\mathcal{R}_{a,\,b}^{(m,n)}$ of the fixed residue class $[[a]]_b$: arity shape.

a/b	2	3	4	5	6	7	8	9	10
1	$m=\mathbf{3}$ $n=\mathbf{2}$ $I=1$ $J=0$	$m=\mathbf{4}$ $n=\mathbf{2}$ $I=1$ $J=0$	$m=\mathbf{5}$ $n=\mathbf{2}$ $I=1$ $J=0$	$m=\mathbf{6}$ $n=\mathbf{2}$ $I=1$ $J=0$	$m=\mathbf{7}$ $n=\mathbf{2}$ $I=1$ $J=0$	$m=\mathbf{8}$ $n=\mathbf{2}$ $I=1$ $J=0$	$m=\mathbf{9}$ $n=\mathbf{2}$ $I=1$ $J=0$	$m=\mathbf{10}$ $n=\mathbf{2}$ $I=1$ $J=0$	$m=\mathbf{11}$ $n=\mathbf{2}$ $I=1$ $J=0$
2		$m=\mathbf{4}$ $n=\mathbf{3}$ $I=2$ $J=2$		$m=\mathbf{6}$ $n=\mathbf{5}$ $I=2$ $J=6$	$m=\mathbf{4}$ $n=\mathbf{3}$ $I=1$ $J=1$	$m=\mathbf{8}$ $n=\mathbf{4}$ $I=2$ $J=2$		$m=\mathbf{10}$ $n=\mathbf{7}$ $I=2$ $J=14$	$m=\mathbf{6}$ $n=\mathbf{5}$ $I=1$ $J=3$
3			$m=\mathbf{5}$ $n=\mathbf{3}$ $I=3$ $J=6$	$m=\mathbf{6}$ $n=\mathbf{5}$ $I=3$ $J=48$	$m=\mathbf{3}$ $n=\mathbf{2}$ $I=1$ $J=1$	$m=\mathbf{8}$ $n=\mathbf{7}$ $I=3$ $J=312$	$m=\mathbf{9}$ $n=\mathbf{3}$ $I=3$ $J=3$		$m=\mathbf{11}$ $n=\mathbf{5}$ $I=3$ $J=24$
4				$m=\mathbf{6}$ $n=\mathbf{3}$ $I=4$ $J=12$	$m=\mathbf{4}$ $n=\mathbf{2}$ $I=2$ $J=2$	$m=\mathbf{8}$ $n=\mathbf{4}$ $I=4$ $J=36$		$m=\mathbf{10}$ $n=\mathbf{4}$ $I=4$ $J=28$	$m=\mathbf{6}$ $n=\mathbf{3}$ $I=2$ $J=6$
5					$m=\mathbf{7}$ $n=\mathbf{3}$ $I=5$ $J=20$	$m=\mathbf{8}$ $n=\mathbf{7}$ $I=5$ $J=11\,160$	$m=\mathbf{9}$ $n=\mathbf{3}$ $I=5$ $J=15$	$m=\mathbf{10}$ $n=\mathbf{7}$ $I=5$ $J=8680$	$m=\mathbf{3}$ $n=\mathbf{2}$ $I=1$ $J=2$
6						$m=\mathbf{8}$ $n=\mathbf{3}$ $I=6$ $J=30$			$m=\mathbf{6}$ $n=\mathbf{2}$ $I=3$ $J=3$
7							$m=\mathbf{9}$ $n=\mathbf{3}$ $I=7$ $J=42$	$m=\mathbf{10}$ $n=\mathbf{4}$ $I=7$ $J=266$	$m=\mathbf{11}$ $n=\mathbf{5}$ $I=7$ $J=1680$
8								$m=\mathbf{10}$ $n=\mathbf{3}$ $I=8$ $J=56$	$m=\mathbf{6}$ $n=\mathbf{5}$ $I=4$ $J=3276$
9									$m=\mathbf{11}$ $n=\mathbf{3}$ $I=9$ $J=72$

Let us investigate the properties of $\mathcal{R}_{a,\,b}^{(m,n)}$ in more detail. First, we consider equal arity polyadic rings and find the relation between the corresponding congruence classes.

Proposition 4.3.17. *The residue (congruence) classes* $[[a]]_b$ *and* $[[a']]_{b'}$ *which are described by the polyadic rings of the same arities* $\mathcal{R}_{a,\,b}^{(m,n)}$ *and* $\mathcal{R}_{a',\,b'}^{(m,n)}$ *are related by*

$$\frac{b'I'}{a'} = \frac{bI}{a}, \tag{4.3.19}$$

$$a' + b'J' = (a + bJ)^{\log_a a'}. \tag{4.3.20}$$

Proof. Follows from (4.3.11) and (4.3.13). □

For instance, in table 4.1 the congruence classes $[[2]]_5$, $[[3]]_5$, $[[2]]_{10}$, and $[[8]]_{10}$ are (6, 5)-rings. If, in addition, $a = a'$, then the polyadic shapes satisfy

$$\frac{I}{J} = \frac{I'}{J'}. \tag{4.3.21}$$

4.3.2 Polyadic rings in limiting cases

The limiting cases $a \equiv \pm 1 (\mathrm{mod}\, b)$ are described by

Corollary 4.3.18. The polyadic ring of the fixed congruence class $[[a]]_b$ is: (1) multiplicative binary, if $a = 1$; (2) multiplicative ternary, if $a = b - 1$; (3) additive $(b + 1)$-ary in both cases. That is, the limiting cases contain the rings $\mathcal{R}_{1,\,b}^{(b+1,\,2)}$ and $\mathcal{R}_{b\,-\,1,\,b}^{(b+1,\,3)}$, respectively. They correspond to the first row and the main diagonal of table 4.1. Their intersection consists of the (3, 2)-ring $\mathcal{R}_{1,\,2}^{(3,\,2)}$.

Definition 4.3.19. The congruence classes $[[1]]_b$ and $[[b - 1]]_b$ are called the *limiting classes*, and the corresponding polyadic rings are named the *limiting polyadic rings* of a fixed congruence class.

Proposition 4.3.20. *In the limiting cases* $a = 1$ and $a = b - 1$ *the n-ary semigroup* $\langle [[a]]_b \mid \mu^{(n)} \rangle$:

(i) *has the neutral sequences of the form* $\bar{\mathbf{n}}_{n\,-\,1} = x_{k_1} x_{k_2} \ldots x_{k_{n\,-\,1}} = 1$, *where* $x_{k_i} = \pm 1$;

(ii) *has (a) the unit* $e = x_{k\,=\,1} = 1$, *for the limiting class* $[[1]]_b$, *(b) the unit* $e^{-} = x_{k\,=\,-1} = -1$, *if n is odd, for* $[[b - 1]]_b$, *(c) the class* $[[1]]_2$ *contains both polyadic units e and* e^{-};

(iii) *has the set of 'querable' (polyadic invertible) elements which consist of* $\bar{x} = x_{\bar{k}} = \pm 1$;

(iv) *has in the 'intersecting' case* $a = 1$, $b = 2$ and $n = 2$ *the binary subgroup* $\mathbb{Z}_2 = \{1, -1\}$, *while other elements have no inverses.*

Proof.

(i) The (multiplicative) neutral sequence $\bar{\mathbf{n}}_{n-1}$ of length $(n-1)$ is defined by $\mu^{(n)}[\bar{\mathbf{n}}_{n-1}, x_k] = x_k$. It follows from (4.3.6) and cancellativity in $\mathbb{Z}$, that $\bar{\mathbf{n}}_{n-1} = x_{k_1}x_{k_2}\ldots x_{k_{n-1}} = 1$ which is

$$(a + bk_1)(a + bk_2)\ldots(a + bk_{n-1}) = 1. \tag{4.3.22}$$

The solution of this equation in integers is the following: (a) all multipliers are $a + bk_i = 1$, $i = 1, \ldots, n-1$; (b) an even number of multipliers can be $a + bk_i = -1$, while the others are 1.

(ii) If the polyadic unit $e = x_{k_1} = a + bk_1$ exists, it should satisfy $\mu^{(n)}\left[\overbrace{e, e, \ldots, e}^{n-1}, x_k\right] = x_k \ \forall x_k \in \langle [[a]]_b \mid \mu^{(n)} \rangle$, such that the neutral sequence $\bar{\mathbf{n}}_{n-1}$ consists of one element e only, and this leads to $(a + bk_1)^{n-1} = 1$. For any n this equation has the solution $a + bk_1 = 1$, which uniquely gives $a = 1$ and $k_1 = 0$, thus $e = x_{k_1=0} = 1$. If n is odd, then there exists a 'negative unit' $e^- = x_{k_1=-1} = -1$, such that $a + bk_1 = -1$, which can be uniquely solved by $k_1 = -1$ and $a = b - 1$. The neutral sequence becomes $\bar{\mathbf{n}}_{n-1} = \overbrace{e^-, e^-, \ldots, e^-}^{n-1} = 1$, as a product of an even number of $e^- = -1$. The intersection of limiting classes consists of a single class $[[1]]_2$, and therefore it contains both polyadic units e and e^-.

(iii) An element x_k in $\langle [[a]]_b \mid \mu^{(n)} \rangle$ is 'querable', if there exists its querelement $\bar{x} = x_{\bar{k}}$ such that $\mu^{(n)}\left[\overbrace{x_k, x_k, \ldots, x_k}^{n-1}, \bar{x}\right] = x_k$. Using (4.3.6) and the cancellativity in $\mathbb{Z}$, we obtain the equation $(a + bk)^{n-2}(a + b\bar{k}) = 1$, which in integers has 2 solutions: (a) $(a + bk)^{n-2} = 1$ and $(a + b\bar{k}) = 1$, the last relation fixes up the class $[[1]]_b$, and the arity of multiplication $n = 2$, and therefore the first relation is valid for all elements in the class, each of them has the same querelement $\bar{x} = 1$. This means that all elements in $[[1]]_b$ are 'querable', but only one element $x = 1$ has an inverse, which is also 1; (b)$(a + bk)^{n-2} = -1$ and $(a + b\bar{k}) = -1$. The second relation fixes the class $[[b-1]]_b$, and from the first relation we conclude that the arity n should be odd. In this case only one element -1 is 'querable', which has $\bar{x} = -1$, as a querelement.

(iv) The 'intersecting' class $[[1]]_2$ contains 2 'querable' elements ± 1 which coincide with their inverses, which means that $\{+1, -1\}$ is a binary subgroup (that is $\mathbb{Z}_2$) of the binary semigroup $\langle [[1]]_2 \mid \mu^{(2)} \rangle$. □

Corollary 4.3.21. In the non-limiting cases $a \neq 1, b - 1$, the n-ary semigroup $\langle [[a]]_b | \mu^{(n)} \rangle$ contains no 'querable' (polyadic invertible) elements at all.

Proof. It follows from $(a + bk) \neq \pm 1$ for any $k \in \mathbb{Z}$ or $a \neq \pm 1 (\mathrm{mod}\, b)$. □

Based on the above statements, consider the properties of the polyadic rings $\mathcal{R}_{a,b}^{(m,n)}$ $(a \neq 0)$ describing nonzero congruence classes (see *remark* 4.3.16).

Proposition 4.3.22. *The polyadic ring* $\mathcal{R}_{a,b}^{(m,n)}$ *is an* (m, n)*-integral domain.*

Proof. It follows from the definitions (4.3.5)–(4.3.6), the condition $a \neq 0$, and commutativity and cancellativity in $\mathbb{Z}$. □

Lemma 4.3.23. There are no such congruence classes which can be described by polyadic (m, n)-field.

Proof. Follows from proposition 4.3.20 and corollary 4.3.21.

This statement for the particular limiting case $[[1]]_2$ appeared in Duplij and Werner (2021), while studying the ideal structure of the corresponding (3, 2)-ring.

Proposition 4.3.24. *In the limiting case* $a = 1$ *the polyadic ring* $\mathcal{R}_{1,b}^{(b+1,2)}$ *can be embedded into a* $(b + 1, 2)$*-ary field.*

Proof. Because the polyadic ring $\mathcal{R}_{1,b}^{(b+1,2)}$ of the congruence class $[[1]]_b$ is a $(b + 1, 2)$-integral domain by proposition 4.3.22, we can construct in a standard way the corresponding $(b + 1, 2)$-quotient ring which is a $(b + 1, 2)$-ary field up to isomorphism, as was shown in Crombez and Timm (1972). By analogy, it can be called the field of polyadic rational numbers which have the form

$$x = \frac{1 + bk_1}{1 + bk_2}, \quad k_i \in \mathbb{Z}. \tag{4.3.23}$$

Indeed, they form a $(b + 1, 2)$-field, because each element has an inverse under multiplication (which is obvious) and is additively 'querable', such that the equation for the querelement $\bar{x}$ becomes $\nu^{(b+1)}\left[\overbrace{x, x, \ldots, x}^{b}, \bar{x}\right] = x$ which can be solved for any x, giving uniquely $\bar{x} = -(b - 1)\frac{1 + bk_1}{1 + bk_2}$. □

The introduced polyadic inner structure of the residue (congruence) classes allows us to extend various number theoretic problems by considering the polyadic (m, n)-integers $\mathbb{Z}^{(m,n)}$ instead of $\mathbb{Z}$.

4.3.3 Prime polyadic integers

Let us introduce a polyadic analog of prime numbers in $\mathbb{Z}^{(m,n)}$. First we need

Definition 4.3.25. A *polyadically composite (reducible) number* is $x_k \in \mathbb{Z}^{(m,n)}$, such that the expansion

$$x_k = (\mu^{(n)})^{\circ\ell}[x_{k_1}, x_{k_2}, , x_{k_{\ell(n-1)+1}}], \quad x_{k_i} \in \mathbb{Z}^{(m,n)}, \tag{4.3.24}$$

is unique, where ℓ is a number of n-ary multiplications, and there exist at least one $x_{k_i} \neq x_k$ and $x_{k_i} \neq e$ (i.e., is not equal to unit of $\mathbb{Z}^{(m,n)}$, if it exists). Denote the set of such numbers $\{x_{k_i}\} = \mathbb{D}(x_k)$ which is called the *composition set* of x_k.

Definition 4.3.26. An *irreducible polyadic number* is $x_k \in \mathbb{Z}^{(m,n)}$ which cannot be expressed as any (long) polyadic product (4.3.24).

Proposition 4.3.27. *In the polyadic ring $\mathbb{Z}_{a,b}^{(m,n)}$ without the unit the elements satisfying*

$$-| a - b |^n < x_k < | a - b |^n \tag{4.3.25}$$

are irreducible.

Proof. Since $0 \leqslant a \leqslant b - 1$, the minimal absolute value of an element $x_k \in \mathbb{Z}_{a,b}^{(m,n)}$ is $| a - b |$. The minimum of its n-ary product is $| a - b |^n$, and therefore smaller elements cannot be decomposed. □

Example **4.3.28.** In the (6, 5)-ring $\mathbb{Z}_{8,10}^{(6,5)}$ all polyadic integer numbers are even, and there is no unit, and so they are binary composite

$$\mathbb{Z}_{8,10}^{(6,5)} = \{\ldots -72, -62, -52, -42, -32, -22, -12, -2, 8, 18, 28, 38, 48, 58, \ldots\}. \tag{4.3.26}$$

Nevertheless, the lowest elements, e.g., $\{-22, -12, 8, 18, 28\}$, are irreducible, while the smallest (by absolute value) polyadically composite element is $(-32) = \mu_5[(-2)^5]$.

Definition 4.3.29. A range in which all elements are indecomposable is called a *polyadic irreducible gap*.

Remark **4.3.30.** We do not demand positivity, as in the binary case, because polyadic integer numbers $\mathbb{Z}_{a,b}^{(m,n)}$ (4.3.9) are 'symmetric' not with respect to $x = 0$, but under $x = x_{k=0} = a$.

The polyadic analog of binary prime numbers plays an intermediate role between composite and irreducible elements.

Definition 4.3.31. A *polyadic prime number* is $x_{k_p} \in \mathbb{Z}^{(m,n)}$, such that it obeys only the unique expansion

$$x_{k_p} = \left(\mu^{(n)}\right)^{\circ \ell}\left[x_{k_p}, e^{\ell(n-1)}\right], \tag{4.3.27}$$

where e is a polyadic unit of $\mathbb{Z}^{(m,n)}$ (if exists).

So, the polyadic prime numbers can appear only in those polyadic rings $\mathbb{Z}_{a,b}^{(m,n)}$ which contain units. In Duplij (2019, proposition 6.15) it was shown that such rings correspond to the *limiting congruence classes* $[[1]]_b$ and $[[b-1]]_b$, and indeed only for them can $a + bk = 1 \bmod b$, and $e^{\ell(n-1)}$ can be a neutral sequence (for $e = 1$ always, while for $e = -1$ only when $\ell(n-1)$ is even).

Proposition 4.3.32. *The prime polyadic numbers can exist only in the limiting polyadic rings* $\mathbb{Z}_{1,b}^{(b+1,2)}$ *and* $\mathbb{Z}_{b-1,b}^{(b+1,3)}$.

Proof. The equation $a + bk = 1 \bmod b$ (for $0 \leqslant a \leqslant b-1$) has two solutions: $a = 1$ and $a = b - 1$ corresponding for two limiting congruence classes $[[1]]_b$ and $[[b-1]]_b$, which correspond to

$$x_k^+ = bk + 1, \tag{4.3.28}$$

$$x_k^- = b(k+1) - 1, \quad k \in \mathbb{Z}. \tag{4.3.29}$$

The parameters-to-arity mapping (4.3.54) fixes their multiplication arity to $n = 2$ and $n = 3$ respectively, which gives manifestly

$$\mu_2\left[x_{k_1}^+, x_{k_2}^+\right] = x_k^+, \qquad k = bk_1k_2 + k_1 + k_2, \tag{4.3.30}$$

$$\begin{aligned} &\mu^{(3)}[x_{k_1}^-, x_{k_2}^-, x_{k_3}^-] = x_k^-, \quad k, k_i \in \mathbb{Z}, \quad i = 1, 2, 3, \qquad b \in \mathbb{N}, \\ &k = b^2k_1k_2k_3 + (b-1)[b(k_1k_2 + k_2k_3 + k_1k_3) + (b-1)(k_1 + k_2 + k_3) \\ &\quad + (b-2)]. \end{aligned} \tag{4.3.31}$$

Therefore, for $\mathbb{Z}_{1,b}^{(b+1,2)}$ we have the unit $e = x_{k=0}^+ = 1$ (which is obvious for the binary multiplication), while in $\mathbb{Z}_{b-1,b}^{(b+1,3)}$ the unit is $e = x_{k=-1}^- = -1$ ($b \geqslant 3$), and the sequence is (e^2) is evidently neutral. □

Denote the set of ordinary binary prime numbers in the interval $1 \leqslant k \leqslant k_{\max}$ by $\mathbb{P}(k_{\max})$, $k_{\max} \in \mathbb{N}$. The set of prime polyadic numbers for the polyadic ring $\mathbb{Z}_{a,b}^{(m,n)}$ in the interval $x_{-k_{\max}} \leqslant x_k \leqslant x_{k_{\max}}$ is denoted by $\mathbb{P}_{a,b}^{(m,n)}(k_{\max})$. Obviously, in the binary limit $\mathbb{P}_{0,1}^{(2,2)}(k_{\max}) = \mathbb{P}(k_{\max}) \cup \{-\mathbb{P}(k_{\max})\}$. Nevertheless, prime polyadic numbers can be composite as binary numbers.

Assertion 4.3.33. The set of prime polyadic numbers in the interval $x_{-k_{\max}} \leqslant x_k \leqslant x_{k_{\max}}$ for $\mathbb{Z}_{a,b}^{(m,n)}$ can contain composite binary numbers, i.e.

$$\Delta\mathbb{P}_{a,\,b}^{(m,n)}(k_{\max}) = \mathbb{P}_{a,\,b}^{(m,n)}(k_{\max})\backslash\{\{\mathbb{P}(x_{k_{\max}}) \bigcup \{\mathbb{P}(-x_{-k_{\max}})\}\} \bigcap [[a]]_b\} \neq \varnothing. \quad (4.3.32)$$

Definition 4.3.34. (1) The cardinality of the set of ordinary binary prime numbers $\mathbb{P}(k_{\max})$ is called a *prime-counting function* and denoted by $\pi(k_{\max}) = |\,\mathbb{P}(k_{\max})\,|$.

(2) The cardinality of the set of prime polyadic numbers $\mathbb{P}_{a,\,b}^{(m,n)}(k_{\max})$ is called a *polyadic prime-counting function* and denoted by

$$\pi_{a,\,b}^{(m,n)}(k_{\max}) = \left|\ \mathbb{P}_{a,\,b}^{(m,n)}(k_{\max})\ \right|. \quad (4.3.33)$$

Example **4.3.35.** (1) Consider $\mathbb{Z}_{43,\,44}^{(45,\,3)}$ and $k_{\max} = 2$, then

$$\mathbb{P}_{43,\,44}^{(45,\,3)}(2) = \{-45,\,-1,\,43,\,87,\,131\}, \quad (4.3.34)$$

$$\Delta\mathbb{P}_{43,\,44}^{(45,\,3)}(2) = \{-45\},\quad \pi_{43,\,44}^{(45,\,3)}(2) = 5. \quad (4.3.35)$$

(2) For $\mathbb{Z}_{50,\,51}^{(52,\,3)}$ and $k_{\max} = 5$ we have

$$\mathbb{P}_{50,\,51}^{(52,\,3)}(5) = \{-205,\,-154,\,-103,\,-52,\,-1,\,50,\,101,\,152,\,203,\,254,\,305\}, \quad (4.3.36)$$

$$\Delta\mathbb{P}_{50,\,51}^{(52,\,3)}(5) = \{-205,\,-154,\,-52,\,50,\,152,\,203,\,254,\,305\},\ \pi_{50,\,51}^{(52,\,3)}(5) = 11. \quad (4.3.37)$$

Remark **4.3.36.** This happens because in $\mathbb{Z}_{a,\,b}^{(m,n)}$ the role of 'building blocks' (prime polyadic numbers) is played by those x_k which cannot be presented as a (long) ternary product of other polyadic integer numbers from the same $\mathbb{Z}_{a,\,b}^{(m,n)}$ as in (4.3.24), but which satisfy (4.3.27) only. Nevertheless, such prime polyadic numbers can be composite binary prime numbers.

In general, for the limiting cases, in which polyadic prime numbers exist, we have

Proposition 4.3.37. *(1) In $\mathbb{Z}_{1,\,b}^{(b+1,\,2)}$ the 'smallest' polyadic integer numbers satisfying*

$$-b < k_p < b + 2, \quad (4.3.38)$$

$$1 - b^2 < x_{k_p} < (b + 1)^2, \quad (4.3.39)$$

are not decomposable, and therefore such $x_{k_p} \in \mathbb{Z}_{1,\,b}^{(b+1,\,2)}$ are all *polyadic prime numbers.*

(2) For another limiting case $\mathbb{Z}_{b-1,\,b}^{(b+1,\,3)}$ the polyadic integer numbers satisfying

$$1 - b < k_p < b - 1, \quad (4.3.40)$$

$$-(b - 1)^2 < x_{k_p} < b^2 - 1, \quad (4.3.41)$$

are not ternary decomposable and so all *such* $x_{k_p} \in \mathbb{Z}_{b-1,b}^{(b+1,2)}$ *are polyadic prime numbers.*

Proof. This follows from determining the maximum of the negative values and the minimum of the positive values of the functions x_k^+ and x_k^- in (4.3.30)–(4.3.31). □

Definition 4.3.38. The range in which all elements are polyadically prime numbers is called the *polyadic primes gap*, and for the two limiting cases it is given by (4.3.39) and (4.3.41), respectively.

***Example* 4.3.39.** In the polyadic ring $\mathbb{Z}_{50,51}^{(52,3)}$ for the polyadic primes gap we have $-2500 < x_{k_p} < 2600$: all such polyadic integer numbers are polyadically prime, but there are many composite binary numbers among them.

Introduce now a polyadic analog of the *Euler (totient) function* which in the binary case counts the number of coprimes to a given natural number. Denote the set of ordinary binary numbers $k > 1$ which are *coprime* to $k_{\max} \in \mathbb{N}$ by $\mathbb{S}(k_{\max})$ (named totatives of $k_{\max}$). Then, the cardinality of $\mathbb{S}(k_{\max})$ is defined as Euler function $\varphi(k_{\max}) = |\,\mathbb{S}(k_{\max})\,|$. Obviously, if $k_{\max} = p$ is prime, then $\varphi(p) = p - 1$. The notion of coprime numbers is based on the divisors: the coprime numbers k_1 andk_2 have the greatest common divisor $\gcd(k_1, k_2) = 1$. In the polyadic case it is not so straightforward, and we need to start from the basic definitions.

Observe that in a (commutative) polyadic ring $\mathcal{R}^{(m,n)} = \{R \mid \nu^{(m)}, \mu^{(n)}\}$ the analog of the division operation is usually not defined uniquely, which makes it useless for real applications. Indeed, y divides x, where $x, y \in R$, if there exists a sequence $\mathbf{z} \in R^{\times(n-1)}$ of length $(n-1)$, such that $x = \mu^{(n)}[y, \mathbf{z}]$. To be consistent with the ordinary integer numbers $\mathbb{Z}$, we demand in the polyadic number ring $\mathbb{Z}^{(m,n)}$: (1) Uniqueness of the result; (2) Only one polyadic number (not a sequence) as the result. This naturally leads to

Definition 4.3.40. A polyadic number (quotient)x_{k_2} *polyadically divides* a polyadic number (dividend)x_{k_1}, if there exists $x_{k_q} := x_{k_1} \div_p x_{k_2}$, called the (unique) *result of division*, such that

$$x_{k_1} = \mu^{(n)}\left[x_{k_2}, (x_{k_q})^{n-1}\right], \qquad x_{k_1}, x_{k_2}, x_{k_q} \in \mathbb{Z}^{(m,n)}. \tag{4.3.42}$$

***Remark* 4.3.41.** For polyadic prime numbers (4.3.27) the only possibility for the quotient is $x_{k_2} = x_{k_1}$ such that $x_{k_1} = \mu^{(n)}[x_{k_1}, (e)^{n-1}]$ or $x_{k_1} \div_p x_{k_1} = e$, where e is the unit of $\mathbb{Z}^{(m,n)}$.

Assertion 4.3.42. *Polyadic division is distributive from the left*

$$\nu^{(m)}[x_{k_1}, x_{k_2}, \dots, x_{k_m}] \div_p x_k = \nu^{(m)}\left[(x_{k_1} \div_p x_k), (x_{k_2} \div_p x_k), \dots, (x_{k_m} \div_p x_k)\right], \tag{4.3.43}$$

but not distributive from the right.

Proof. This follows from the polyadic distributivity in the (m, n)-ring $\mathbb{Z}^{(m,n)}$. □

***Example* 4.3.43.** (1) In the polyadic ring $\mathbb{Z}_{4,9}^{(10,4)}$ we have uniquely $x_{28} \div_p x_4 = x_4$ or $256 \div_p 4 = 4$. (2) For the limiting ring $\mathbb{Z}_{3,4}^{(5,3)}$, we find $x_{43} \div_p x_1 = x_{-2}$ or $175 \div_p 7 = -5$.

In the same way we define a polyadic analog of division with a remainder.

Definition 4.3.44. A *polyadic division with a remainder* is defined, if for a polyadic dividend x_{k_1} and divisor x_{k_2} there exists a *polyadic remainder* x_{k_r} such that

$$x_{k_1} = \nu^{(m)}\Big[\mu^{(n)}\Big[x_{k_2}, (x_{k_q})^{n-1}\Big], (x_{k_r})^{m-1}\Big], \qquad x_{k_1}, x_{k_2}, x_{k_q}, x_{k_r} \in \mathbb{Z}^{(m,n)}, \quad (4.3.44)$$

which is denoted by $x_{k_r} = x_{k_1} \operatorname{mod}_p x_{k_2}$, and (4.3.44) can be presented in the following binary form

$$x_{k_1} = \left(x_{k_2} \boxdot_p x_{k_q}\right) \boxplus_p x_{k_r}. \qquad (4.3.45)$$

The distributivity of these operations is governed by distributivity in the polyadic ring $\mathbb{Z}^{(m,n)}$.

***Example* 4.3.45.** In the polyadic ring $\mathbb{Z}_{8,10}^{(6,5)}$ we can have different divisions for the same dividend as $38 = \big((-22)\boxdot_p(-2)\big)\boxplus_p 78 = \big((-92)\boxdot_p(-2)\big)\boxplus_p 238$.

Because divisibility in the polyadic case is not symmetric with respect to dividend and divisor (4.3.42), we define polyadically coprime numbers using the definition of compositeness (4.3.24).

Definition 4.3.46. The s polyadic integer numbers $x_{k_1}, \dots, x_{k_s} \in \mathbb{Z}^{(m,n)}$ are *polyadically coprime,* if their composition sets do not intersect $\mathbb{D}(x_{k_1}) \cap \mathbb{D}(x_{k_1}) \cap \dots \cap \mathbb{D}(x_{k_s}) = \varnothing$.

***Remark* 4.3.47.** This definition does not imply the existence of a unit in $\mathbb{Z}^{(m,n)}$, as opposed to the definition of the polyadic prime numbers (4.3.27) in which the availability of a unit is crucial.

Assertion 4.3.48. *Polyadically coprime numbers can exist in any polyadic ring* $\mathbb{Z}^{(m,n)}$, *and* not *only in the limiting cases with unit (see proposition* 4.3.32).

Corollary 4.3.49. All elements in the polyadic irreducible gap are polyadically coprime.

***Example* 4.3.50.** In $\mathbb{Z}_{8,10}^{(6,5)}$ (4.3.26) the polyadic integer numbers $(-32) = \mu_5[(-2)^5]$ and $32\,768 = \mu_5[(8)^5]$ are both composed, but polyadically coprime, because the composition sets $\mathbb{D}(-32) = \{-2\}$ and $\mathbb{D}(32\,768) = \{8\}$ do not intersect. Alternatively, not polyadically coprime numbers here are, e.g., $(-3072) = \mu_5[(-12), (-2)^2, (8)^2]$ and $(-64\,512) = \mu_5[-2, (8)^2, 18, 28]$, because $\mathbb{D}(-3072) \cap \mathbb{D}(-64\,512) = \{-2, 8\} \neq \varnothing$. We cannot multiply these two numbers, because the arity of multiplication is 5.

***Remark* 4.3.51.** For polyadic integer numbers $\mathbb{Z}^{(m,n)}$ we cannot measure the property 'to be coprime' in terms of a single element, as in the binary case, by their gcd, because the n-ary multiplication is only allowed for admissible sequences. Therefore, we need to consider the intersection of the composition sets.

In the polyadic ring $\mathbb{Z}_{a,b}^{(m,n)}$ for a given $k_{\max} \in \mathbb{Z}_+$ we denote by $\mathbb{S}_{a,b}^{(m,n)}(k_{\max})$ the set of polyadic prime numbers $x_k \in \mathbb{Z}_{a,b}^{(m,n)}$ which are polyadically coprime to $x_{k_{\max}}$ and $x_{-k_{\max}}$ in the open interval $x_{-k_{\max}} < x_k < x_{k_{\max}}$ (we assume that, if two numbers are coprime, then their opposite numbers are also coprime). For the binary limit, obviously, $\mathbb{S}_{0,1}^{(2,2)}(k_{\max}) = \mathbb{S}(k_{\max}) \cup \{-\mathbb{S}(k_{\max})\}$.

Definition 4.3.52. The cardinality of $\mathbb{S}_{a,b}^{(m,n)}(k_{\max})$ is called a *polyadic Euler function* denoted by

$$\varphi_{a,b}^{(m,n)}(k_{\max}) = \left| \mathbb{S}_{a,b}^{(m,n)}(k_{\max}) \right|. \tag{4.3.46}$$

In the binary case $\varphi_{0,1}^{(2,2)}(k_{\max}) = 2\varphi(k_{\max})$. Because of *remark* 4.3.51, the computation of the polyadic Euler function requires for each element in the range $x_{-k_{\max}} < x_k < x_{k_{\max}}$ a thorough consideration of composition sets.

***Example* 4.3.53.**

(i) For the limiting polyadic ring $\mathbb{Z}_{1,29}^{(30,2)}$ and $k_{\max} = 10$ we have $x_{-10} = -289$, $x_{10} = 291$ and

$$\mathbb{S}_{1,29}^{(30,2)}(10) = \{-260, -202, -173, -115, -86, -28, 1, 59, 88, 146, 175, 233, 262\}, \tag{4.3.47}$$

$$\varphi_{1,29}^{(30,2)}(10) = 13. \tag{4.3.48}$$

(ii) In another limiting case with ternary multiplication $\mathbb{Z}_{31,32}^{(33,3)}$ we get $x_{-5} = -129$, $x_5 = 191$ and

$$\mathbb{S}_{31,32}^{(33,3)}(5) = \{-97, -65, -1, 31, 95, 127\}, \tag{4.3.49}$$

$$\varphi_{1,29}^{(30,2)}(5) = 6. \tag{4.3.50}$$

(iii) In the non-limiting case $\mathbb{Z}_{7,10}^{(11,5)}$ and $k_{\max} = 10$ we have $x_{-10} = -93$, $x_{10} = 107$ with

$$\mathbb{S}_{7,10}^{(11,5)}(10) = \{-83, -73, -53, -43, -23, -13, 7, 17, 37, 47, 67, 77, 97\}, \qquad (4.3.51)$$

$$\varphi_{7,10}^{(11,5)}(10) = 13. \qquad (4.3.52)$$

(iv) For the polyadic Euler function in some other non-limiting cases we have

$$\varphi_{27,49}^{(50,15)}(7) = 6, \quad \varphi_{17,38}^{(39,10)}(20) = 21, \quad \varphi_{16,28}^{(8,4)}(30) = \varphi_{46,50}^{(26,6)}(15) = 0. \qquad (4.3.53)$$

4.3.4 The mapping of parameters to arity

Let us consider the connection between congruence classes and arities in more detail.

***Remark* 4.3.54.** Note that:

(i) Solutions to (4.3.7) and (4.3.8) do not exist simultaneously for all a and b;
(ii) The pair a, b determines m, n uniquely;
(iii) It can occur that for several different pairs a, b there can be the same arities m, n.

Therefore, we have

Assertion 4.3.55. The *parameters-to-arity mapping*

$$\psi\colon (a, b) \longrightarrow (m, n) \qquad (4.3.54)$$

is a partial surjection.

We list the lowest arities which can be obtained with different choices of (a, b).

$$\left.\begin{matrix} m = 3 \\ n = 2 \end{matrix}\right\}: \begin{pmatrix} a = 1 \\ b = 2 \end{pmatrix}, \begin{pmatrix} a = 3 \\ b = 6 \end{pmatrix}, \begin{pmatrix} a = 5 \\ b = 10 \end{pmatrix}, \begin{pmatrix} a = 7 \\ b = 14 \end{pmatrix}, \begin{pmatrix} a = 9 \\ b = 18 \end{pmatrix}, \begin{pmatrix} a = 11 \\ b = 22 \end{pmatrix};$$
$$\left.\begin{matrix} m = 4 \\ n = 2 \end{matrix}\right\}: \begin{pmatrix} a = 1 \\ b = 3 \end{pmatrix}, \begin{pmatrix} a = 4 \\ b = 6 \end{pmatrix}, \begin{pmatrix} a = 4 \\ b = 12 \end{pmatrix}, \begin{pmatrix} a = 10 \\ b = 15 \end{pmatrix}, \begin{pmatrix} a = 7 \\ b = 21 \end{pmatrix}, \begin{pmatrix} a = 16 \\ b = 24 \end{pmatrix};$$
$$\left.\begin{matrix} m = 4 \\ n = 3 \end{matrix}\right\}: \begin{pmatrix} a = 2 \\ b = 3 \end{pmatrix}, \begin{pmatrix} a = 2 \\ b = 6 \end{pmatrix}, \begin{pmatrix} a = 8 \\ b = 12 \end{pmatrix}, \begin{pmatrix} a = 14 \\ b = 21 \end{pmatrix}, \begin{pmatrix} a = 8 \\ b = 24 \end{pmatrix}, \begin{pmatrix} a = 20 \\ b = 30 \end{pmatrix};$$
$$\left.\begin{matrix} m = 5 \\ n = 2 \end{matrix}\right\}: \begin{pmatrix} a = 1 \\ b = 4 \end{pmatrix}, \begin{pmatrix} a = 9 \\ b = 12 \end{pmatrix}, \begin{pmatrix} a = 5 \\ b = 20 \end{pmatrix}, \begin{pmatrix} a = 21 \\ b = 28 \end{pmatrix}, \begin{pmatrix} a = 9 \\ b = 36 \end{pmatrix}, \begin{pmatrix} a = 33 \\ b = 44 \end{pmatrix};$$
$$\left.\begin{matrix} m = 5 \\ n = 3 \end{matrix}\right\}: \begin{pmatrix} a = 3 \\ b = 4 \end{pmatrix}, \begin{pmatrix} a = 3 \\ b = 12 \end{pmatrix}, \begin{pmatrix} a = 15 \\ b = 20 \end{pmatrix}, \begin{pmatrix} a = 7 \\ b = 28 \end{pmatrix}, \begin{pmatrix} a = 27 \\ b = 36 \end{pmatrix}, \begin{pmatrix} a = 39 \\ b = 52 \end{pmatrix}.$$

It has not been possible to derive a general formula for $\psi(a, b)$, nevertheless it can be done in some particular cases.

Proposition 4.3.56. *(1) In the limiting cases $(a = 1, b - 1)$ we have*

$$\psi(1, b) = (b + 1, 2), \qquad \psi(b - 1, b) = (b + 1, 3). \tag{4.3.55}$$

(2) If $a|b$, then

$$\psi(a, ad) = \left(d + 1, \min_l \log_a(ld + 1) + 1\right), \tag{4.3.56}$$

where l is the smallest integer for which log *takes its minimal integer value. (3) If* $\gcd(a, b) = d$, *then*

$$\psi(a, b) = \psi(a_0 d, b_0 d) = \left(b_0 + 1, \min_l \log_a\left(l\frac{b_0}{a_0} + 1\right) + 1\right), \tag{4.3.57}$$

with the same l.

Proof. All the statements follow directly from (4.3.7)–(4.3.8). □

In our approach, the concrete choice of operations (4.3.5)–(4.3.6) inside a congruence class $[[a]]_b$ gives

Assertion 4.3.57. *The number of additions in the polyadic ring $\mathbb{Z}_{a,b}^{(m,n)}$ is greater that the number of multiplications $m > n$ with any choice of (a, b).*

Also, not all pairs (a, b) are allowed due to (4.3.7)–(4.3.8). We list the forbidden (a, b) for $b \leqslant 20$

$$\begin{aligned}
&b = 4\colon a = 2;\\
&b = 8\colon a = 2, 4, 6;\\
&b = 9\colon a = 3, 6;\\
&b = 12\colon a = 2, 6, 10;\\
&b = 16\colon a = 2, 4, 6, 8, 12, 14;\\
&b = 18\colon a = 3, 6, 12, 15;\\
&b = 20\colon a = 2, 6, 10, 14, 18.
\end{aligned} \tag{4.3.58}$$

In example 4.3.6 for $\mathbb{Z}_{3,4}^{(5,3)}$ we have $I = 3$, $J = 6$. In the limiting cases (4.3.28)–(4.3.29) we have, in general, for a fixed b

$$I_{b+1}^{[1,b]} = 1, \quad J_2^{[1,b]} = 0, \tag{4.3.59}$$

$$I_{b+1}^{[b-1,b]} = b - 1, \quad J_3^{[b-1,b]} = (b-1)(b-2). \tag{4.3.60}$$

Thus, one can classify and distinguish the limiting cases of the congruence classes in terms of the invariants and their manifest form (4.3.59)–(4.3.60).

4.4 Finite polyadic rings of integers

Here we propose a special method of constructing a finite nonderived polyadic ring by combining the 'external' and 'internal' methods. Let us 'apply' (1) to (2), such that instead of (4.3.4), we introduce the finite polyadic ring $\mathbb{Z}^{(m,n)}/c\mathbb{Z}$, where $\mathbb{Z}^{(m,n)}$ is defined in (4.3.9). However, if we directly consider the 'double' class $\{a + bk + cl\}$ and fix a and b, then the factorization by $c\mathbb{Z}$ will not give closed operations for arbitrary c.

Assertion 4.4.1. If the finite polyadic ring $\mathbb{Z}_{a,b}^{(m,n)}/c\mathbb{Z}$ has q elements, then

$$c = bq. \tag{4.4.1}$$

Proof. It follows from (4.4.1), that the 'double' class remains in $[[a]]_b$. □

***Remark* 4.4.2.** The representatives $x_k^{[a,b]} = a + bk$ belong to an (m, n)-ring with the polyadic operations (4.3.5)–(4.3.6), while the notions of subtraction, division, modulo and remainder are defined for binary operations. Therefore, we cannot apply the standard binary modular arithmetic to $x_k^{[a,b]}$ directly, but we can define the equivalence relations and corresponding class operations in terms of k.

4.4.1 Secondary congruence classes

On the set of the 'double' classes $\{a + bk + bql\}$, $k, l \in \mathbb{Z}$ and fixed $b \in \mathbb{N}$ and $a = 0, \ldots, b-1$ we define the equivalence relation $\overset{k}{\sim}$ by

$$\{a + bk_1 + bql_1\} \overset{k}{\sim} \{a + bk_2 + bql_2\} \Longrightarrow k_1 - k_2 = ql, \quad l, l_1, l_2 \in \mathbb{Z}. \tag{4.4.2}$$

Proposition 4.4.3. *The equivalence relation* $\overset{k}{\sim}$ *is a congruence.*

Proof. It follows from the binary congruency of k such that $k_1 \equiv k_2 (\mathrm{mod}\, q)$ which is just a rewritten form of the last condition of (4.4.2). □

So now we can factorize $\mathbb{Z}_{a,b}^{(m,n)}$ by the congruence $\overset{k}{\sim}$ and obtain

Definition 4.4.4. A *secondary (equivalence) class* of a polyadic integer $x_k^{[a,b]} = a + bk \in \mathbb{Z}_{a,b}^{(m,n)}$ 'modulo' bq (with q being the number of representatives $x_k^{[a,b]}$, for fixed $b \in \mathbb{N}$ and $0 \leqslant a \leqslant b - 1$) is

$$[[x_k^{[a,b]}]]_{bq} = \{\{(a + bk) + bql\} | l \in \mathbb{Z}, q \in \mathbb{N}, 0 \leqslant k \leqslant q - 1\}. \tag{4.4.3}$$

***Remark* 4.4.5.** If the binary limit is given by $a = 0$, $b = 1$ and $\mathbb{Z}_{0,1}^{(2,2)} = \mathbb{Z}$, then the secondary class becomes the ordinary class (4.3.1).

If the values of the parameters a, b, q are clear from the context, we denote the secondary class representatives by an integer with two primes, as follows $[[x_k^{[a,b]}]]_{bq} \equiv x_k'' \equiv x''$.

***Example* 4.4.6.** (a) Let $a = 3$, $b = 5$, then for $q = 4$ elements we have the secondary classes with $k = 0, 1, 2, 3$ (the corresponding binary limits are in brackets)

$$[[x_k^{[3,5]}]]_{20} = 3'', 8'', 13'', 18'' = \begin{cases} 3'' = \{\dots -17, 3, 23, 43, 63, \dots\}, \\ 8'' = \{\dots -12, 8, 28, 48, 68, \dots\}, \\ 13'' = \{\dots -7, 13, 33, 53, 73, \dots\}, \\ 18'' = \{\dots -2, 18, 38, 58, 78, \dots\}, \end{cases} \tag{4.4.4}$$

$$\left([[k]]_4 = 0', 1', 2', 3' = \begin{cases} 0' = \{\dots -4, 0, 4, 8, 12, \dots\}, \\ 1' = \{\dots -3, 1, 5, 9, 13, \dots\}, \\ 2' = \{\dots -2, 2, 6, 10, 14, \dots\}, \\ 3' = \{\dots -1, 3, 7, 11, 15, \dots\}. \end{cases} \right). \tag{4.4.5}$$

(b) For $a = 3$, $b = 6$ and for 4 elements and $k = 0, 1, 2, 3$

$$[[x_k^{[3,6]}]]_{24} = 3'', 9'', 15'', 21'', \qquad ([[k]]_4 = 0', 1', 2', 3'). \tag{4.4.6}$$

(c) If $a = 4$, $b = 5$, for 3 elements and $k = 0, 1, 2$ we get

$$[[x_k^{[4,5]}]]_{15} = 4'', 9'', 14'', \qquad ([[k]]_3 = 0', 1', 2'). \tag{4.4.7}$$

The crucial difference between these sets of classes are: (1) they are described by rings of different arities determined by (4.3.7) and (4.3.8); (2) some of them are fields.

4.4.2 Finite polyadic rings of secondary classes

Now we determine the operations between secondary classes. The most significant difference with the binary class operations (4.3.2)–(4.3.3) is the fact that secondary classes obey nonderived polyadic operations.

Proposition 4.4.7. *The set* $\{x_k''\}$ *of* q *secondary classes* $k = 0, \dots, q - 1$ *(with the fixed* a, b*) can be endowed with the following commutative* m*-ary addition*

$$x''_{k_{\text{add}}} = \nu''^{(m)}\left[x''_{k_1}, x''_{k_1}, \ldots, x''_{k_m}\right], \tag{4.4.8}$$

$$k_{\text{add}} \equiv ((k_1 + k_2 + \cdots + k_m) + I_m^{[a,b]})(\operatorname{mod} q) \tag{4.4.9}$$

and commutative n-ary multiplication

$$x''_{k_{\text{mult}}} = \mu''^{(n)}\left[x''_{k_1}, x''_{k_1}, \ldots, x''_{k_n}\right], \tag{4.4.10}$$

$$\begin{aligned} k_{\text{mult}} \equiv (a^{n-1}(k_1 + k_2 + \cdots + k_n) \\ + a^{n-2}b(k_1k_2 + k_2k_3 + \cdots + k_{n-1}k_n) + \ldots \\ + b^{n-1}k_1\ldots k_n + J_n^{[a,b]})(\operatorname{mod} q), \end{aligned} \tag{4.4.11}$$

which satisfy the polyadic distributivity, and the shape invariants $I_m^{[a,b]}$, $J_n^{[a,b]}$ are defined in (4.3.14).

Proof. This follows from the definition of the secondary class (4.4.3) and manifest form of the 'underlying' polyadic operations (4.3.5)–(4.3.6), which are commutative and distributive. □

***Remark* 4.4.8.** The binary limit is given by $a = 0, b = 1$ and $m = n = 2$, $I_m^{[a,b]} = J_n^{[a,b]} = 0$, such that the secondary class becomes the ordinary congruence class $x''_k \to k'$, obeying the standard binary class operations (4.3.2)–(4.3.3), which in terms of k are $k_{\text{add}} \equiv (k_1 + k_2)(\operatorname{mod} q)$, $k_{\text{mult}} \equiv (k_1k_2)(\operatorname{mod} q)$.

Definition 4.4.9. The set of secondary classes (4.4.3) equipped with operations (4.4.8), (4.4.10) is denoted by

$$\mathbb{Z}^{(m,\, n)}(q) \equiv \mathbb{Z}_{a,\, b}^{(m,n)}(q) = \mathbb{Z}_{a,\, b}^{(m,n)}/(bq)\mathbb{Z} = \{\{x''_k\} | \nu''^{(m)}, \mu''^{(n)}\}, \tag{4.4.12}$$

and is a *finite secondary class (m, n)-ring* of polyadic integer numbers $\mathbb{Z}^{(m,\, n)} \equiv \mathbb{Z}_{a,\, b}^{(m,n)}$. The value q (the number of elements) is called its *order*.

Informally, $\mathbb{Z}^{(m,\, n)} = \mathbb{Z}^{(m,\, n)}(\infty)$. First, note that the constructed finite (m, n)-rings (4.4.12) have a much richer structure and exotic properties which do not exist in the binary finite residue class rings (4.3.4), and, in general, they give many concrete examples for possible different kinds of polyadic rings. One of such 'non-binary' properties is the availability of several units (for odd multiplicative arity n), and moreover sometimes all ring elements are units (such rings are 'automatically' fields, see below).

***Example* 4.4.10.** (a) In (5, 3)-ring $\mathbb{Z}_{3,\,4}^{(4,\,3)}(2)$ with 2 secondary classes both elements are units (we mark units by subscript e) $e_1 = 3''_e = 3''$, $e_2 = 7''_e = 7''$, because they are both multiplicative idempotents and satisfy the following ternary multiplication (see (4.1.9))

$$\mu^{(3)}[3'', 3'', 3''] = 3'', \mu^{(3)}[3'', 3'', 7''] = 7'', \mu^{(3)}[3'', 7'', 7''] = 3'', \\ \mu^{(3)}[7'', 7'', 7''] = 7''. \tag{4.4.13}$$

(b) In the same way the ring $\mathbb{Z}_{5,6}^{(7,3)}(4)$ consists of only 4 units $e_1 = 5''_e$, $e_2 = 11''_e$, $e_3 = 17''_e$, $e_4 = 23''_e$, and no zero. (c) Equal arity rings of the same order may be not isomorphic. For instance, $\mathbb{Z}_{1,3}^{(4,2)}(2)$ consists of unit $e = 1''_e = 1''$ and zero $z = 4''_z = 4''$ only, satisfying

$$\mu_2[1'', 1''] = 1'', \mu_2[1'', 4''] = 4'', \mu_2[4'', 4''] = 4'', \tag{4.4.14}$$

and therefore $\mathbb{Z}_{1,3}^{(4,2)}(2)$ is a field, because $\{1'', 4''_z\}\backslash 4''_z$ is a (trivial) binary group, consisting of one element $1''_e$. However, $\mathbb{Z}_{4,6}^{(4,2)}(2)$ has the zero $z = 4''_z = 4''$, $10''$ and has no unit, because

$$\mu_2[4'', 4''] = 4'', \mu_2[4'', 10''] = 4'', \mu_2[10'', 10''] = 4'', \tag{4.4.15}$$

so that $\mathbb{Z}_{4,6}^{(4,2)}(2)$ is not a field, because of the last relation (nilpotency of $10''$). Their additive 4-ary groups are also not isomorphic (which is easy to show). However, $\mathbb{Z}_{1,3}^{(4,2)}(2)$ and $\mathbb{Z}_{4,6}^{(4,2)}(2)$ have the same arity and order.

Recalling assertion 4.3.55, we conclude more concretely:

Assertion 4.4.11. For a fixed arity shape (m, n), there can be non-isomorphic secondary class polyadic rings $\mathbb{Z}^{(m,n)}(q)$ of the same order q, which describe different binary residue classes $[[a]]_b$.

A polyadic analog of the characteristic can be introduced, when there exist both a unit and zero in a finite ring. Recall, that if $\mathcal{R}$ is a finite binary ring with unit1 and zero 0, then its characteristic is defined as a smallest integer χ, such that

$$\left(\overbrace{1 + 1 +, \ldots, +1}^{\chi}\right) = \chi \cdot 1 = 0. \tag{4.4.16}$$

This means that the 'number' of unit additions being $\chi - 1$ produces zero. The same is evident for any other element $x \in \mathcal{R}$, because $x = x \cdot 1$.

Definition 4.4.12. For the finite polyadic ring $\mathbb{Z}^{(m,n)}(q)$ which contains both the unit e and the zero z, a *polyadic characteristic* χ_p is defined as a smallest additive polyadic power (4.1.7) of e which is equal to zero

$$e^{\langle \chi_p \rangle_{+m}} = z. \tag{4.4.17}$$

In the binary limit, obviously, $\chi_p = \chi - 1$. A polyadic analog of the middle term in (4.4.16) can be obtained by using the polyadic distributivity and (4.3.5) as

$$e^{\langle \chi_p \rangle_{+m}} = e^{\langle \chi_p (m - 1)+1 \rangle_{\times n}}. \tag{4.4.18}$$

In table 4.2 we present the parameters-to-arity mapping $\psi_{a,\,b}^{(m,n)}$ (4.3.54) together with the polyadic characteristics of those finite secondary class rings $\mathbb{Z}_{a,\,b}^{(m,n)}(q)$ which contain both unit(s) and zero, and which have order less than or equal to 10 for $b \leqslant 6$.

Now we turn to the question of which secondary classes can be described by polyadic finite fields.

4.5 Finite polyadic fields of integer numbers

Let us consider the structure of the finite secondary class rings $\mathbb{Z}_{a,\,b}^{(m,n)}(q)$ in more detail and determine which of them are polyadic fields.

Proposition 4.5.1. *A finite polyadic ring* $\mathbb{Z}_{a,\,b}^{(m,n)}(q)$ *is a secondary class finite* (m, n)*-field* $\mathcal{F}_{a,\,b}^{\prime\prime(m,n)}(q)$ *if all its elements except* z *(if it exists) are polyadically multiplicative invertible having a unique querelement.*

Proof. In both cases $\{\{x_k^{\prime\prime}\}|\mu^{\prime\prime(n)}\}$ and $\left\{\{x_k^{\prime\prime}\backslash z\}\middle|\mu^{\prime\prime(n)}\right\}$ are commutative and cancellative n-ary groups, which follows from the concrete form of multiplication (4.4.10). Therefore, according to definition 4.1.10, in such a case $\mathbb{Z}_{a,\,b}^{(m,n)}(q)$ becomes a polyadic field $\mathcal{F}_{a,\,b}^{\prime\prime(m,n)}(q)$. □

4.5.1 Abstract finite polyadic fields

Recall, that in the binary field theory (Lidl and Niederreiter 1997) the residue (congruence) class ring (4.3.4) with q elements $\mathbb{Z}/q\mathbb{Z}$ is a congruence class (non-extended) field, if its order $q = p$ is a prime number, such that $\mathcal{F}'(p) = \left\{\{[[a]]_p\}\middle|+', \cdot', 0', 1'\right\}$, $a = 0, 1, \ldots, p - 1$. Because all non-extended binary fields of a fixed prime order p are isomorphic to each other and, in turn, isomorphic to the congruence class field $\mathcal{F}'(p)$, it is natural to study them in a more 'abstract' way, i.e., without connection to a specific congruence class structure. This can be achieved by consideration of the one-to-one onto mapping from the congruence class to its representative which preserves the field (ring) structure and provides operations (binary multiplication and addition with ordinary 0 and 1) by modulo p. In other words, the mapping $\Phi_p([[a]]_p) = a$ is an isomorphism of binary fields Φ_p: $\mathcal{F}'(p) \to \mathcal{F}(p)$, where $\mathcal{F}(p) = \{\{a\}|+, \cdot, 0{,}1\}_{\mathrm{mod}\, p}$ is an 'abstract' non-extended (prime) finite field of order p (or *Galois field* $GF(p)$).

Similarly, we introduce a polyadic analog of the 'abstract' binary non-extended (prime) finite fields. Let us consider the set of polyadic integer numbers

Table 4.2. Polyadic characteristics χ_p for the finite secondary class (m, n)-rings $\mathbb{Z}_{a,b}^{(m,n)}(q)$ of order $2 \leqslant q \leqslant 10$ for $2 \leqslant b \leqslant 6$. The orders q which do not give fields are slanted.

$a \backslash b$	2	3	4	5	6
1	$m = 3$ $n = 2$ $q = 3, \chi_p = 1$ $q = 5, \chi_p = 2$ $q = 7, \chi_p = 3$ $q = 9, \chi_p = 4$	$m = 4$ $n = 2$ $q = 2, \chi_p = 1$ $q = 4, \chi_p = 1$ $q = 5, \chi_p = 3$ $q = 7, \chi_p = 2$ $q = 8, \chi_p = 5$ $q = 10, \chi_p = 3$	$m = 5$ $n = 2$ $q = 3, \chi_p = 2$ $q = 5, \chi_p = 1$ $q = 7, \chi_p = 5$ $q = 9, \chi_p = 2$	$m = 6$ $n = 2$ $q = 2, \chi_p = 1$ $q = 3, \chi_p = 1$ $q = 4, \chi_p = 3$ $q = 6, \chi_p = 1$ $q = 7, \chi_p = 4$ $q = 8, \chi_p = 3$ $q = 9, \chi_p = 7$	$m = 7$ $n = 2$ $q = 5, \chi_p = 4$ $q = 7, \chi_p = 1$
2		$m = 4$ $n = 3$ $q = 2, \chi_p = 1$ $q = 4, \chi_p = 1$ $q = 5, \chi_p = 3$ $q = 7, \chi_p = 2$ $q = 8, \chi_p = 5$ $q = 10, \chi_p = 3$		$m = 6$ $n = 5$ $q = 2, \chi_p = 1$ $q = 3, \chi_p = 1$ $q = 4, \chi_p = 3$ $q = 6, \chi_p = 1$ $q = 7, \chi_p = 4$ $q = 8, \chi_p = 3$ $q = 9, \chi_p = 7$	$m = 4$ $n = 3$ $q = 5, \chi_p = 3$ $q = 7, \chi_p = 2$ $q = 10, \chi_p = 3$
3			$m = 5$ $n = 3$ $q = 3, \chi_p = 2$ $q = 5, \chi_p = 1$ $q = 7, \chi_p = 5$ $q = 9, \chi_p = 2$	$m = 6$ $n = 5$ $q = 2, \chi_p = 1$ $q = 3, \chi_p = 1$ $q = 4, \chi_p = 3$ $q = 6, \chi_p = 1$ $q = 7, \chi_p = 4$ $q = 8, \chi_p = 3$	$m = 3$ $n = 2$ $q = 5, \chi_p = 2$ $q = 7, \chi_p = 3$
4				$m = 6$ $n = 3$ $q = 2, \chi_p = 1$ $q = 3, \chi_p = 1$ $q = 4, \chi_p = 3$ $q = 6, \chi_p = 1$ $q = 7, \chi_p = 4$ $q = 8, \chi_p = 3$ $q = 9, \chi_p = 7$	$m = 4$ $n = 2$ $q = 5, \chi_p = 3$ $q = 7, \chi_p = 2$ $q = 10, \chi_p = 3$
5					$m = 7$ $n = 3$ $q = 5, \chi_p = 4$ $q = 7, \chi_p = 1$

$\{x_k\} \equiv \{x_k^{[a,b]}\} = \{a + bk\} \in \mathbb{Z}_{a,b}^{(m,n)}$, $b \in \mathbb{N}$ and $0 \leqslant a \leqslant b - 1$, $0 \leqslant k \leqslant q - 1$, $q \in \mathbb{N}$, which obey the operations (4.3.5)–(4.3.6). The polyadic version of the prime finite field $\mathcal{F}(p)$ of order p (or Galois field $GF(p)$) is given by

Definition 4.5.2. The 'abstract' *non-extended (prime) finite* (m, n)*-field of order* q is

$$\mathcal{F}^{(m,n)}(q) \equiv \mathcal{F}_{a,b}^{(m,n)}(q) = \{\{a + bk\} | \nu^{(m)}, \mu^{(n)}\}_{\mathrm{mod}\, bq}, \tag{4.5.1}$$

if $\{\{x_k\} | \nu^{(m)}\}_{\mathrm{mod}\, bq}$ is an additive m-ary group, and $\{\{x_k\} | \mu^{(n)}\}_{\mathrm{mod}\, bq}$ (or, when zero z exists, $\left\{\{x_k \backslash z\} \middle| \mu^{(n)}\right\}_{\mathrm{mod}\, bq}$) is a multiplicative n-ary group.

Then we define a one-to-one onto mapping from the secondary congruence class to its representative by $\Phi_q^{[a,b]}\left([[x_k^{[a,b]}]]_{bq}\right) = x_k^{[a,b]}$ and arrive at the following

Proposition 4.5.3. *The mapping* $\Phi_q^{[a,b]}$: $\mathcal{F}_{a,b}^{\prime\prime(m,n)}(q) \to \mathcal{F}_{a,b}^{(m,n)}(q)$ is a polyadic ring homomorphism (being, in fact, an isomorphism) and satisfies (here we use the 'prime' notations)

$$\Phi_q^{[a,b]}(\nu^{\prime\prime(m)}[x_1^{\prime\prime}, x_2^{\prime\prime}, \ldots, x_m^{\prime\prime}]) = \nu^{(m)}\left[\Phi_q^{[a,b]}(x_1), \Phi_q^{[a,b]}(x_2), \ldots, \Phi_q^{[a,b]}(x_m)\right], \tag{4.5.2}$$

$$\Phi_q^{[a,b]}(\mu^{\prime\prime(n)}[x_1^{\prime\prime}, x_2^{\prime\prime}, \ldots, x_n^{\prime\prime}]) = \mu^{(n)}\left[\Phi_q^{[a,b]}(x_1), \Phi_q^{[a,b]}(x_2), \ldots, \Phi_q^{[a,b]}(x_n)\right]. \tag{4.5.3}$$

Proof. This follows directly from (4.3.5)–(4.3.6) and (4.4.8)–(4.4.10). Obviously, a mapping defined in this way governs the polyadic distributivity, and therefore $\Phi_q^{[a,b]}$ is a ring homomorphism, or, more exactly, a 1-place heteromorphism for m-ary addition together with n-ary multiplication (see Duplij 2019). Because $x_k^{\prime\prime} \to x_k$ is one-to-one for any fixed $0 \leqslant k \leqslant q - 1$, $\Phi_q^{[a,b]}$ is an isomorphism. □

In table 4.3 we present the 'abstract' non-extended polyadic *finite* fields $\mathcal{F}_{a,b}^{(m,n)}(q)$ of lowest arity shape (m, n) and orders q. The forbidden pairs (a, b) for $\mathcal{F}_{a,b}^{(m,n)}(q)$ coincide with ones for polyadic rings $\mathbb{Z}_{a,b}^{(m,n)}$ listed in (4.3.58).

4.5.2 Multiplicative structure of finite polyadic fields

We outline the following crucial differences in the multiplicative structure between the binary finite fields and $\mathcal{F}^{(m,n)}(q)$.

***Remark* 4.5.4.** The order of a non-extended finite polyadic field may not be prime (e.g., $\mathcal{F}_{1,2}^{(3,2)}(4)$, $\mathcal{F}_{3,4}^{(5,3)}(8)$, $\mathcal{F}_{2,6}^{(4,3)}(9)$), and may not even be a power of a prime binary number (e.g., $\mathcal{F}_{5,6}^{(7,3)}(6)$, $\mathcal{F}_{3,10}^{(11,5)}(10)$), see table 4.4.

Table 4.3. Content and arities of the secondary class finite polyadic rings $\mathbb{Z}_{a,b}^{(m,n)}(q)$ and the corresponding simple finite polyadic (m, n)-fields $\mathcal{F}_{a,b}^{(m,n)}(q)$ of order $2 \leqslant q \leqslant 4$ (framed) for $2 \leqslant b \leqslant 6$. The subscripts e and z mark those secondary classes (two primes are omitted) which play the roles of polyadic unit and polyadic zero, respectively. The double frames denote the finite polyadic fields containing both a unit and zero. The last line in a cell (corresponding to a fixed congruence class $[[a]]_b$) gives the allowed orders of finite polyadic fields for $5 \leqslant q \leqslant 10$, and bold numbers mark the orders of such fields which contain both unit(s) and zero.

a\b	2	3	4	5	6
1	$m = \mathbf{3}$ $n = \mathbf{2}$ $1_e,3$ $1_e,3_z,5$ $1_e,3,5,7$ q=**5**,**7**,8	$m = \mathbf{4}$ $n = \mathbf{2}$ $1_e,4_z$ $1_e,4,7$ $1_e,4_z,7,10$ q=**5**,**7**,9	$m = \mathbf{5}$ $n = \mathbf{2}$ $1_e,5$ $1_e,5,9_z$ $1_e,5,9,13$ q=**5**,**7**,8	$m = \mathbf{6}$ $n = \mathbf{2}$ $1_e,6_z$ $1_e,6_z,11$ $1_e,6,11,16_z$ q=5,**7**	$m = \mathbf{7}$ $n = \mathbf{2}$ $1_e,7$ $1_e,7,13$ $1_e,7,13,19$ q=**5**,6,**7**,8,9
2		$m = \mathbf{4}$ $n = \mathbf{3}$ $2_z,5_e$ $2,5,8_e$ $2,5_e,8_z,11_e$ q=**5**,**7**,9		$m = \mathbf{6}$ $n = \mathbf{5}$ $2_z,7_e$ $2_e,7,12_z$ $2,7_e,12_z,17_e$ q=5,**7**	$m = \mathbf{4}$ $n = \mathbf{3}$ $2,8_z$ $2,8_e,14$ $2,8_z,14,20$ q=**5**,**7**,9
3			$m = \mathbf{5}$ $n = \mathbf{3}$ $3_e,7_e$ $3_z,7_e,11_e$ $3,7_e,11,15_e$ q=**5**,6,**7**,8	$m = \mathbf{6}$ $n = \mathbf{5}$ $3_e,8_z$ $3_z,8_e,13_e$ $3_e,8_z,13_e,18$ q=5,**7**	$m = \mathbf{3}$ $n = \mathbf{2}$ $3,9_e$ $3,9_z,15$ $3,9_e,15,21$ q=**5**,**7**,8
4				$m = \mathbf{6}$ $n = \mathbf{3}$ $4_z,9_e$ $4_e,9_z,14_e$ $4_z,9_e,14,19_e$ q=5,**7**	$m = \mathbf{4}$ $n = \mathbf{2}$ $4_z,10$ $4,10_e,16$ $4,10,16_z,22$ q=**5**,**7**,9
5					$m = \mathbf{7}$ $n = \mathbf{3}$ $5_e,11_e$ $5,11,17_e$ $5_e,11_e,17_e,23_e$ q=**5**,6,**7**,8,9

Table 4.4. Idempotence polyadic orders $\lambda_{p}^{[a,b]}$ for (allowed) finite polyadic fields $\mathcal{F}_{a,\,b}^{(m,n)}(q)$, with $2 \leqslant b \leqslant 10$ and field orders $2 \leqslant q \leqslant 10$. If q_{*} is n-admissible, they are underlined. The zeroless–nonunital cases are framed. Subscripts correspond to the number of units $\kappa_{e} \geqslant 2$, while nonframed entries have exactly one unit.

$[[a]]_b$		*Arities*	*Finite polyadic field order q*								
b	a	(m,n)	2	3	4	5	6	7	8	9	10
2	1	$(3,2)$	2	$\mathbf{2}$	2	$\mathbf{4}$	∅	$\mathbf{6}$	4	∅	∅
3	1	$(4,2)$	$\mathbf{1}$	3	∅	$\mathbf{4}$	∅	$\mathbf{6}$	∅	9	∅
	2	$(4,3)$	$\mathbf{1}$	$\underline{3}$	∅	$\mathbf{2}_{2e}$	∅	$\mathbf{3}_{2e}$	∅	$\underline{9}$	∅
4	1	$(5,2)$	2	$\mathbf{2}$	4	$\mathbf{4}$	∅	$\mathbf{6}$	8	∅	∅
	3	$(5,3)$	1_{2e}	$\mathbf{1}_{2e}$	2_{2e}	$\mathbf{2}_{2e}$	∅	$\mathbf{3}_{2e}$	4_{2e}	∅	∅
5	1	$(6,2)$	$\mathbf{1}$	$\mathbf{2}$	∅	5	∅	$\mathbf{6}$	∅	∅	∅
	2	$(6,5)$	$\mathbf{1}$	$\mathbf{1}_{2e}$	∅	$\underline{5}$	∅	$\mathbf{3}_{2e}$	∅	∅	∅
	3	$(6,5)$	$\mathbf{1}$	$\mathbf{1}_{2e}$	∅	$\underline{5}$	∅	$\mathbf{3}_{2e}$	∅	∅	∅
	4	$(6,3)$	$\mathbf{1}$	$\mathbf{1}_{2e}$	∅	$\underline{5}$	∅	$\mathbf{3}_{2e}$	∅	∅	∅
6	1	$(7,2)$	2	3	2	$\mathbf{4}$	6	$\mathbf{6}$	4	9	∅
	2	$(4,3)$	∅	$\underline{3}$	∅	$\mathbf{2}_{2e}$	∅	$\mathbf{3}_{2e}$	∅	$\underline{9}$	∅
	3	$(3,2)$	2	∅	2	$\mathbf{4}$	∅	$\mathbf{6}$	4	∅	∅
	4	$(4,2)$	∅	3	∅	$\mathbf{4}$	∅	$\mathbf{6}$	∅	9	∅
	5	$(7,3)$	1_{2e}	$\underline{3}$	1_{4e}	$\mathbf{2}_{2e}$	3_{2e}	$\mathbf{3}_{2e}$	2	$\underline{9}$	∅
7	1	$(8,2)$	$\mathbf{1}$	$\mathbf{2}$	∅	$\mathbf{4}$	∅	7	∅	∅	∅
	2	$(8,4)$	$\mathbf{1}$	$\mathbf{2}$	∅	$\underline{\mathbf{4}}$	∅	$\underline{7}$	∅	∅	∅
	3	$(8,7)$	$\mathbf{1}$	$\mathbf{1}_{2e}$	∅	$\mathbf{2}_{2e}$	∅	$\underline{7}$	∅	∅	∅
	4	$(8,4)$	$\mathbf{1}$	$\mathbf{2}$	∅	$\underline{\mathbf{4}}$	∅	$\underline{7}$	∅	∅	∅
	5	$(8,7)$	$\mathbf{1}$	$\mathbf{1}_{2e}$	∅	$\mathbf{2}_{2e}$	∅	$\underline{7}$	∅	∅	∅
	6	$(8,3)$	$\mathbf{1}$	$\mathbf{1}_{2e}$	∅	$\mathbf{2}_{2e}$	∅	$\underline{7}$	∅	∅	∅
8	1	$(9,2)$	2	$\mathbf{2}$	4	$\mathbf{4}$	∅	$\mathbf{6}$	8	∅	∅
	3	$(9,3)$	$\boxed{2}$	$\mathbf{1}_{2e}$	$\boxed{4}$	$\mathbf{2}_{2e}$	∅	$\mathbf{3}_{2e}$	$\boxed{8}$	∅	∅
	5	$(9,3)$	$\boxed{2}$	$\mathbf{1}_{2e}$	$\boxed{4}$	$\mathbf{2}_{2e}$	∅	$\mathbf{3}_{2e}$	$\boxed{8}$	∅	∅
	7	$(9,3)$	1_{2e}	$\mathbf{1}_{2e}$	2_{2e}	$\mathbf{2}_{2e}$	∅	$\mathbf{3}_{2e}$	4_{2e}	∅	∅
9	1	$(10,2)$	$\mathbf{1}$	3	∅	$\mathbf{4}$	∅	$\mathbf{6}$	∅	9	∅
	2	$(10,7)$	$\mathbf{1}$	$\boxed{3}$	∅	$\mathbf{2}_{2e}$	∅	$\mathbf{1}_{6e}$	∅	$\boxed{9}$	∅
	4	$(10,4)$	$\mathbf{1}$	$\boxed{3}$	∅	$\underline{\mathbf{4}}$	∅	$\mathbf{2}_{3e}$	∅	$\boxed{9}$	∅
	5	$(10,7)$	$\mathbf{1}$	$\boxed{3}$	∅	$\mathbf{2}_{2e}$	∅	$\mathbf{1}_{6e}$	∅	$\boxed{9}$	∅
	7	$(10,4)$	$\mathbf{1}$	$\boxed{3}$	∅	$\underline{\mathbf{4}}$	∅	$\mathbf{2}_{3e}$	∅	$\boxed{9}$	∅
	8	$(10,3)$	$\mathbf{1}$	3	∅	$\mathbf{2}_{2e}$	∅	$\mathbf{3}_{2e}$	∅	9	∅
10	1	$(11,2)$	2	$\mathbf{2}$	2	5	∅	$\mathbf{6}$	8	∅	10
	2	$(6,5)$	∅	$\mathbf{1}_{2e}$	∅	$\underline{5}$	∅	$\mathbf{3}_{2e}$	∅	∅	∅
	3	$(11,5)$	1_{2e}	$\mathbf{1}_{2e}$	1_{4e}	$\underline{5}$	∅	$\mathbf{3}_{2e}$	1_{8e}	∅	5_{2e}
	4	$(6,3)$	∅	$\mathbf{1}_{2e}$	∅	$\underline{5}$	∅	$\mathbf{3}_{2e}$	∅	∅	∅
	5	$(3,2)$	2	$\mathbf{2}$	2	∅	∅	$\mathbf{6}$	∅	∅	∅
	6	$(6,2)$	∅	$\mathbf{2}$	∅	5	∅	$\mathbf{6}$	∅	∅	∅
	7	$(11,5)$	1_{2e}	$\mathbf{1}_{2e}$	1_{4e}	$\underline{5}$	∅	$\mathbf{3}_{2e}$	1_{8e}	∅	5_{2e}
	8	$(6,5)$	∅	$\mathbf{1}_{2e}$	∅	$\underline{5}$	∅	$\mathbf{3}_{2e}$	∅	∅	∅
	9	$(11,3)$	1_{2e}	$\mathbf{1}_{2e}$	1_{4e}	$\underline{5}$	∅	$\mathbf{3}_{2e}$	2_{4e}	∅	5_{2e}

***Remark* 4.5.5.** The polyadic characteristic χ_p of a non-extended finite polyadic field can have values such that $\chi_p + 1$ (corresponding in the binary case to the ordinary characteristic χ) can be nonprime (table 4.2).

Assertion 4.5.6. If a secondary class $[[x_k^{[a,b]}]]_{bq}$ contains no zero, it can be isomorphic to the 'abstract' *zeroless* finite polyadic field $\mathcal{F}_{a,\,b}^{(m,n)}(q)$.

Zeroless fields are marked by one frame in table 4.3. There exist finite polyadic fields with more than one unit, and also all elements can be units. Such cases are marked in table 4.4 by subscripts which indicate the number of units.

Denote the Abelian finite multiplicative n-ary group of $\mathcal{F}_{a,\,b}^{(m,n)}(q)$ by $\mathcal{G}_{a,\,b}^{(n)}(q)$.

***Example* 4.5.7.** (1) The finite (6, 3)-field $\mathcal{F}_{4,\,5}^{(6,\,3)}(3)$ of order 3 has two units$\{4,\,14\} \equiv \{4_e,\,14_e\}$ and zero $9 \equiv 9_z$, and its multiplicative 3-ary group $\mathcal{G}_{4,\,5}^{(3)}(2)$ is

$$\mu^{(3)}[4,\,4,\,4] = 4,\ \mu^{(3)}[4,\,4,\,14] = 14,\ \mu^{(3)}[4,\,14,\,14] = 4,\ \mu^{(3)}[14,\,14,\,14] = 14.$$

(2) In $\mathcal{F}_{5,\,6}^{(7,\,3)}(4)$ of order 4 all the elements $\{5,\,11,\,17,\,23\} \equiv \{5_e,\,11_e,\,17_e,\,23_e\}$ are units (see (4.1.9)), because for its 3-ary group $\mathcal{G}_{5,\,6}^{(3)}(4)$ we have

$$\begin{aligned}
&\mu^{(3)}[5,\,5,\,5] = 5,\ \mu^{(3)}[11,\,11,\,11] = 11,\ \mu^{(3)}[17,\,17,\,17] = 17,\ \mu^{(3)}[23,\,23,\,23] = 23,\\
&\mu^{(3)}[5,\,5,\,11] = 11,\ \mu^{(3)}[5,\,5,\,17] = 17,\ \mu^{(3)}[5,\,5,\,23] = 23,\\
&\mu^{(3)}[11,\,11,\,5] = 5,\ \mu^{(3)}[11,\,11,\,17] = 17,\ \mu^{(3)}[11,\,11,\,23] = 23,\\
&\mu^{(3)}[17,\,17,\,5] = 5,\ \mu^{(3)}[17,\,17,\,11] = 11,\ \mu^{(3)}[17,\,17,\,23] = 23,\\
&\mu^{(3)}[23,\,23,\,5] = 5,\ \mu^{(3)}[23,\,23,\,11] = 11,\ \mu^{(3)}[23,\,23,\,17] = 17.
\end{aligned}$$

In general, n-ary groups may contain no units (and multiplicative idempotents) at all, and invertibility is controlled in another way, by querelements: each element of any n-ary group should be (uniquely) 'quereable' (4.1.14). In the case of $(m,\,n)$-fields both m-ary additive group and n-ary multiplicative group $\mathcal{G}_{a,\,b}^{(n)}(q)$ can be of this kind. By analogy with zeroless–nonunital rings we have

Definition 4.5.8. A polyadic field $\mathcal{F}^{(m,\,n)}$ is called *zeroless–nonunital*, if it contains no zero and no unit.

Assertion 4.5.9. The zeroless–nonunital polyadic fields are totally (additively and multiplicatively) nonderived.

Proposition 4.5.10. (1) If $\gcd(bq) \neq 1$, then $\mathcal{F}_{a,\,b}^{(m,n)}(q)$ is zeroless. (2) The zero can exist only, if $\gcd(bq) = 1$ and the field order $q = p$ is prime.

Proof. It follows directly from the definition of the polyadic zero (4.1.8) and (4.4.9). □

Let us consider examples of zeroless–nonunital finite fields of polyadic integer numbers $\mathcal{F}_{a,\,b}^{(m,n)}(q)$.

***Example* 4.5.11.** (1) The zeroless–nonunital polyadic finite fields having lowest $|\, a+b \,|$ are, e.g., $\mathcal{F}_{3,\,8}^{(9,\,3)}(2)$, $\mathcal{F}_{3,\,8}^{(9,\,3)}(4)$, $\mathcal{F}_{5,\,8}^{(9,\,3)}(4)$, $\mathcal{F}_{5,\,8}^{(9,\,3)}(8)$, also $\mathcal{F}_{4,\,9}^{(10,\,4)}(3)$, $\mathcal{F}_{4,\,9}^{(10,\,4)}(9)$, and $\mathcal{F}_{7,\,9}^{(10,\,4)}(3)$, $\mathcal{F}_{7,\,9}^{(10,\,4)}(9)$. (2) The multiplication of the zeroless–nonunital (9, 3)-field $\mathcal{F}_{5,\,8}^{(9,\,3)}(2)$ is

$$\mu^{(3)}[5,\,5,\,5] = 13,\; \mu^{(3)}[5,\,5,\,13] = 5,\; \mu^{(3)}[5,\,13,\,13] = 13,\; \mu^{(3)}[13,\,13,\,13] = 5.$$

Using (4.1.7) we find the (unique) multiplicative querelements $\bar{5} = 13, \bar{13} = 5$. The addition of $\mathcal{F}_{5,\,8}^{(9,\,3)}(2)$ is

$$\nu_9[5^9] = 13,\; \nu_9[5^8,\,13] = 5,\; \nu_9[5^7,\,13^2] = 13,\; \nu_9[5^6,\,13^3] = 5,\; \nu_9[5^5,\,13^4] = 13,$$
$$\nu_9[5^4,\,13^5] = 5,\; \nu_9[5^3,\,13^6] = 13,\; \nu_9[5^2,\,13^7] = 5,\; \nu_9[5,\,13^8] = 13,\; \nu_9[13^9] = 5.$$

The additive (unique) querelements are $\tilde{5} = 13, \tilde{13} = 5$. So all elements are additively and multiplicatively querable (polyadically invertible), and therefore ν_9 is 9-ary additive group operation and $\mu^{(3)}$ is 3-ary multiplicative group operation, as it should be for a field. Because it contains no unit and no zero, $\mathcal{F}_{5,\,8}^{(9,\,3)}(2)$ is actually a zeroless–nonunital finite (9, 3)-field of order 2.

Other zeroless–nonunital finite polyadic fields are marked by frames in table 4.4.

***Remark* 4.5.12.** The absence of zero does not guarantee that an $(m,\,n)$-ring $\mathcal{R}_{a,\,b}^{(m,n)}(q)$ is a field. For that, both $\langle [[a]]_b | \nu^{(m)} \rangle$ and $\langle [[a]]_b | \mu^{(n)} \rangle$ have to be polyadic groups.

***Example* 4.5.13.** The (4, 3)-ring $\mathcal{R}_{2,\,3}^{(4,\,3)}(6)$ is zeroless, and $\langle [[3]]_4 | \nu^{(4)} \rangle$ is its 4-ary additive group (each element has a unique additive querelement). Despite each element of $\langle [[2]]_3 | \mu^{(3)} \rangle$ having a querelement, it is not a multiplicative 3-ary group, because for the two elements 2 and 14 we have nonunique querelements

$$\mu^{(3)}[2,\,2,\,5] = 2,\; \mu^{(3)}[2,\,2,\,14] = 2,\; \mu^{(3)}[14,\,14,\,2] = 14,\; \mu^{(3)}[14,\,14,\,11] = 14. \quad (4.5.4)$$

The conditions on the congruence classes $[[a]]_b$ and the invariants I, J (4.3.14), which give the same arity structure are given in Duplij (2019). Note, that there exist polyadic fields of the same arities $(m,\,n)$ and the same order q which are not isomorphic (in contrast with what is possible in the binary case).

***Example* 4.5.14.** The polyadic (9, 3)-fields corresponding to the congruence classes $[[5]]_8$ and $[[7]]_8$ are not isomorphic for orders $q = 2,\,4,\,8$ (see table 4.4). Despite both being zeroless, the first $\mathcal{F}_{5,\,8}^{(9,\,3)}(q)$ are nonunital, while the second $\mathcal{F}_{7,\,8}^{(9,\,3)}(q)$ has two units, which makes an isomorphism impossible.

In follows from the binary field theory (Lidl and Niederreiter 1997), that in a (non-extended, prime) finite binary field $\mathcal{F}(p)$, the *order of an element* $x \in \mathcal{F}(p)$ is defined as a smallest integer λ such that $x^{\lambda} = 1$. Obviously, the set of fixed order elements forms a cyclic subgroup $\mathcal{G}_{\lambda}$ of the multiplicative binary group of $\mathcal{F}(p)$, and $\lambda|(p-1)$. If $\lambda = p-1$, such an element is called a *primitive (root)*, it generates all elements, and these exist in any finite binary field. Moreover, any element of $\mathcal{F}(p)$ is idempotent $x^{p} = x$, while all its nonzero elements satisfy $x^{p-1} = 1$ (Fermat's little theorem). A non-extended (prime) finite field is fully determined by its order p (up to isomorphism), and, moreover, any $\mathcal{F}(p)$ is isomorphic to $\mathbb{Z}/p\mathbb{Z}$.

In the polyadic case, the situation is more complicated. Because the related secondary class structure (4.5.1) contains parameters in addition to the number of elements q, the order of (non-extended) polyadic fields may not be prime, or nor even a power of a prime integer (e.g., $\mathcal{F}_{5,\,6}^{(7,\,3)}(6)$ or $\mathcal{F}_{3,\,10}^{(11,\,5)}(10)$). Also, as was shown above, finite polyadic fields can be zeroless, nonunital and have many (or even all) units (see table 4.4). Therefore, we cannot use units in the definition of the element order. Instead, we propose an alternative:

Definition 4.5.15. If an element of the finite polyadic field $x \in \mathcal{F}^{(m,\,n)}(q)$ satisfies

$$x^{\langle \lambda_p \rangle_{\times n}} = x, \tag{4.5.5}$$

then the smallest such λ_p is called the *idempotence polyadic order* and denoted $^{*}ord x = \lambda_p$.

Obviously, $\lambda_p = \lambda$ (see (4.1.6)).

Definition 4.5.16. The *idempotence polyadic order* $\lambda_p^{[a,b]}$ of a finite polyadic field $\mathcal{F}_{a,\,b}^{(m,n)}(q)$ is the maximum λ_p of all its elements, and we call such field $\lambda_p^{[a,b]}$*-idempotent* and denote $^{*}ord\mathcal{F}_{a,\,b}^{(m,n)}(q) = \lambda_p^{[a,b]}$.

In table 4.4 we present the idempotence polyadic order $\lambda_p^{[a,b]}$ for the (allowed) finite polyadic fields $\mathcal{F}_{a,\,b}^{(m,n)}(q)$ (4.5.1) with $2 \leqslant b \leqslant 10$ and order $q \leqslant 10$.

Definition 4.5.17. Denote by q_{*} the number of nonzero distinct elements in $\mathcal{F}^{(m,\,n)}(q)$

$$q_{*} = \begin{cases} q-1, & \text{if } \exists z \in \mathcal{F}^{(m,\,n)}(q) \\ q, & \text{if } \nexists z \in \mathcal{F}^{(m,\,n)}(q), \end{cases} \tag{4.5.6}$$

which is called a *reduced (field) order*.

The second choice of (4.5.6) in the binary case is absent, because any commutative binary group (as the additive group of a field) contains a zero (the identity of this group), and therefore any binary field has a zero, which does not always hold for the m-ary additive group of $\mathcal{F}^{(m,n)}$ (see example 4.5.11).

Theorem 4.5.18. If a finite polyadic field $\mathcal{F}^{(m,n)}(q)$ has an order q, such that $q_* = q_*^{\mathrm{adm}} = \ell(n-1)+1$ is n-admissible, then (for $n \geqslant 3$ and one unit):

(i) *A sequence of the length $q_*(n-1)$ built from any fixed element $y \in \mathcal{F}^{(m,n)}(q)$ is neutral*

$$\mu_n^{(q_*)}[x, y^{q_*(n-1)}] = x, \quad \forall x \in \mathcal{F}^{(m,n)}(q). \tag{4.5.7}$$

(ii) *Any element* y *satisfies the polyadic idempotency condition*

$$y^{\langle q_* \rangle_{\times n}} = y, \quad \forall y \in \mathcal{F}^{(m,n)}(q). \tag{4.5.8}$$

Proof. (1) Take a long n-ary product of the q_* distinct nonzero elements $x_0 = (\mu^{(n)})^{\circ \ell}[x_1, x_2, \dots, x_{q_*}]$, such that q_* can take only multiplicatively n-admissible values q_*^{adm}, where $\ell \in \mathbb{N}$ is a 'number' of n-ary multiplications. Then polyadically multiply each x_i by a fixed element $y \in \mathcal{F}^{(m,n)}(q)$ such that all q_* elements $\mu^{(n)}[x_i, y^{n-1}]$ will be distinct as well. Therefore, their product should be the same x_0. Using commutativity and associativity, we obtain

$$\begin{aligned} x_0 &= (\mu^{(n)})^{\circ \ell}[x_1, x_2, \dots, x_{q_*}] \\ &= (\mu^{(n)})^{\circ \ell}\left[\mu^{(n)}[x_1, y^{n-1}], \mu^{(n)}[x_2, y^{n-1}], \dots, \mu^{(n)}\left[x_{q_*}, y^{n-1}\right]\right] \\ &= \mu_n^{(q_*)}\left[(\mu^{(n)})^{\circ \ell}[x_1, x_2, \dots, x_{q_*}], y^{q_*(n-1)}\right] \\ &= \mu_n^{(q_*)}[x_0, y^{q_*(n-1)}]. \end{aligned} \tag{4.5.9}$$

(2) Insert into the formula obtained above $x_0 = y$, then use (4.1.7) to get (4.5.8). □

Finite polyadic fields $\mathcal{F}_{a,b}^{(m,n)}(q)$ having n-admissible reduced order $q_* = q_*^{\mathrm{adm}} = \ell(n-1)+1$ ($\ell \in \mathbb{N}$) (underlined in table 4.4) are closest to the binary finite fields $\mathcal{F}(p)$ in their general properties: they are half-derived, while if they contain a zero, they are fully derived. If $q_* \neq q_*^{\mathrm{adm}}$, then $\mathcal{F}_{a,b}^{(m,n)}(q)$ can be nonunital or contain more than one unit (subscripts in table 4.4).

Assertion 4.5.19. *The finite fields $\mathcal{F}_{a,b}^{(m,n)}(q)$ of* n*-admissible reduced order $q_* = q_*^{\mathrm{adm}}$ cannot have more than one unit and cannot be zeroless–nonunital.*

Assertion 4.5.20. *If $q_* \neq q_*^{\mathrm{adm}}$, and $\mathcal{F}_{a,b}^{(m,n)}(q)$ is unital zeroless, then the reduced order* q_* *is the product of the idempotence polyadic field order $\lambda_p^{[a,b]} = {}^*ord\mathcal{F}_{a,b}^{(m,n)}(q)$ and the number of units κ_e (if $a|b$ and $n \geqslant 3$)*

$$q_* = \lambda_p^{[a,b]} \kappa_e. \tag{4.5.10}$$

Let us consider the structure of the multiplicative group $\mathcal{G}_{a,\,b}^{(n)}(q_*)$ of $\mathcal{F}_{a,\,b}^{(m,n)}(q)$ in more detail. Some properties of commutative cyclic n-ary groups were considered for particular relations between orders and arity. Here we have: (1) more parameters and different relations between these, the arity and order; (2) the (m, n)-field under consideration, which leads to additional restrictions. In such a way exotic polyadic groups and fields arise which have unusual properties that have not been studied before.

Definition 4.5.21. An element $x_{\text{prim}} \in \mathcal{G}_{a,\,b}^{(n)}(q_*)$ is called *n-ary primitive*, if its idempotence order is

$$\lambda_p = ord\, x_{\text{prim}} = q_*. \tag{4.5.11}$$

Then, all λ_p polyadic powers $x_{\text{prim}}^{\langle 1\rangle_{\times n}}, x_{\text{prim}}^{\langle 2\rangle_{\times n}}, \ldots, x_{\text{prim}}^{\langle q_*\rangle_{\times n}} \equiv x_{\text{prim}}$ generate other distinct elements, and so $\mathcal{G}_{a,\,b}^{(n)}(q_*)$ is a finite cyclic n-ary group generated by x_{prim}, i.e., $\mathcal{G}_{a,\,b}^{(n)}(q_*) = \left\langle \left\{ x_{\text{prim}}^{\langle i\rangle_{\times n}} \right\} \middle| \mu^{(n)} \right\rangle$. We denote a number of primitive elements in $\mathcal{F}_{a,\,b}^{(m,n)}(q)$ by κ_{prim}.

Assertion 4.5.22. For zeroless $\mathcal{F}_{a,\,b}^{(m,n)}(q)$ and prime order $q = p$, we have $\lambda_p^{[a,b]} = q$, and $\mathcal{G}_{a,\,b}^{(n)}(q)$ is indecomposable $(n \geqslant 3)$.

***Example* 4.5.23.** The smallest 3-admissible zeroless polyadic field is $\mathcal{F}_{2,\,3}^{(4,\,3)}(3)$ with the unit $e = 8$ and two 3-ary primitive elements 2, 5 having 3-idempotence order $^*ord 2 = {}^*ord 5 = 3$, so $\kappa_{\text{prim}} = 2$, because

$$2^{\langle 1\rangle_{\times 3}} = 8,\ 2^{\langle 2\rangle_{\times 3}} = 5,\ 2^{\langle 3\rangle_{\times 3}} = 2, \quad 5^{\langle 1\rangle_{\times 3}} = 8,\ 5^{\langle 2\rangle_{\times 3}} = 2,\ 5^{\langle 3\rangle_{\times 3}} = 5, \tag{4.5.12}$$

and therefore $\mathcal{G}_{2,\,3}^{(3)}(3)$ is a cyclic indecomposable 3-ary group.

Assertion 4.5.24. If $\mathcal{F}_{a,\,b}^{(m,n)}(q)$ is zeroless–nonunital, then every element is n-ary primitive, $\kappa_{\text{prim}} = q$, also $\lambda_p^{[a,b]} = q$ (the order q can be not prime), and $\mathcal{G}_{a,\,b}^{(n)}(q)$ is a indecomposable commutative cyclic n-ary group without identity $(n \geqslant 3)$.

***Example* 4.5.25.** The (10, 7)-field $\mathcal{F}_{5,\,9}^{(10,\,7)}(9)$ is zeroless–nonunital, each element (has $\lambda_p = 9$) is primitive and generates the whole field, and therefore $\kappa_{\text{prim}} = 9$, thus the 7-ary multiplicative group $\mathcal{G}_{5,\,9}^{(7)}(9)$ is indecomposable and without identity.

The structure of $\mathcal{G}_{a,\,b}^{(n)}(q_*)$ can be extremely nontrivial and may have no analogs in the binary case.

Assertion 4.5.26. *If there exists more than one unit, then:*

(i) *If $\mathcal{G}_{a,\,b}^{(n)}(q_*)$ can be decomposed on its* n*-ary subgroups, the number of units κ_e coincides with the number of its cyclic* n*-ary subgroups $\mathcal{G}_{a,\,b}^{(n)}(q_*) = \mathcal{G}_1 \cup \mathcal{G}_2 \ldots \cup \mathcal{G}_{\kappa_e}$ which do not intersect $\mathcal{G}_i \cap \mathcal{G}_j = \varnothing, i, j = i = 1, \ldots, \kappa_e, i \neq j$.*

(ii) *If a zero exists, then each* $\mathcal{G}_i$ *has its own unit* e_i, $i = 1, \ldots, \kappa_e$.
(iii) *In the zeroless case* $\mathcal{G}^{(n)}_{a,b}(q) = \mathcal{G}_1 \cup \mathcal{G}_2 \ldots \cup \mathcal{G}_{\kappa_e} \cup E(\mathcal{G})$, *where* $E(\mathcal{G}) = \{e_i\}$ *is the split-off subgroup of units.*

***Example* 4.5.27.** (1) In the (9, 3)-field $\mathcal{F}^{(9,3)}_{5,8}(7)$ there is a single zero $z = 21 \equiv 21_z$ and two units $e_1 = 13 \equiv 13_e$, $e_2 = 29 \equiv 29_e$, and so its multiplicative 3-ary group $\mathcal{G}^{(3)}_{5,8}(6) = \{5, 13, 29, 37, 45, 53\}$ consists of two nonintersecting (which is not possible in the binary case)3-ary cyclic subgroups $\mathcal{G}^{(3)}_1 = \{5, 13_e, 45\}$ and $\mathcal{G}^{(3)}_2 = \{29_e, 37, 53\}$ (for both $\lambda_p = 3$

$$\mathcal{G}^{(3)}_1 = \left\{5^{\langle 1\rangle_{\times 3}} = 13_e,\ 5^{\langle 2\rangle_{\times 3}} = 45,\ 5^{\langle 3\rangle_{\times 3}} = 5\right\},\ \bar{5} = 45,\ \bar{45} = 5,$$
$$\mathcal{G}^{(3)}_2 = \left\{37^{\langle 1\rangle_{\times 3}} = 29_e,\ 37^{\langle 2\rangle_{\times 3}} = 53,\ 37^{\langle 3\rangle_{\times 3}} = 37\right\},\ \bar{37} = 53,\ \bar{53} = 37.$$

All nonunital elements in $\mathcal{G}^{(3)}_{5,8}(6)$ are (polyadic)1-reflections, because $5^{\langle 1\rangle_{\times 3}} = 45^{\langle 1\rangle_{\times 3}} = 13_e$ and $37^{\langle 1\rangle_{\times 3}} = 53^{\langle 1\rangle_{\times 3}} = 29_e$, and so the subgroup of units $E(\mathcal{G}) = \{13_e, 29_e\}$ is unsplit $E(\mathcal{G}) \cap \mathcal{G}^{(3)}_{1,2} \neq \varnothing$.

(2) For the zeroless $\mathcal{F}^{(9,3)}_{7,8}(8)$, its multiplicative 3-group $\mathcal{G}^{(3)}_{5,8}(6) = \{7, 15, 23, 31, 39, 47, 55, 63\}$ has two units $e_1 = 31 \equiv 31_e$, $e_2 = 63 \equiv 63_e$, and it splits into two nonintersecting nonunital cyclic 3-subgroups ($\lambda_p = 4$ and $\lambda_p = 2$) and the subgroup of units

$$\mathcal{G}^{(3)}_1 = \left\{7^{\langle 1\rangle_{\times 3}} = 23,\ 7^{\langle 2\rangle_{\times 3}} = 39,\ 7^{\langle 3\rangle_{\times 3}} = 55,\ 7^{\langle 4\rangle_{\times 3}} = 4\right\},$$
$$\bar{7} = 55,\ \overline{55} = 7,\ \overline{23} = 39,\ \overline{39} = 23,$$
$$\mathcal{G}^{(3)}_2 = \left\{15^{\langle 1\rangle_{\times 3}} = 47,\ 15^{\langle 2\rangle_{\times 3}} = 15\right\},\ \overline{15} = 47,\ \overline{47} = 15,$$
$$E(\mathcal{G}) = \{31_e, 63_e\}.$$

There are no ℓ_μ-reflections, and so $E(\mathcal{G})$ splits out $E(\mathcal{G}) \cap \mathcal{G}^{(3)}_{1,2} = \varnothing$.

If all elements are units $E(\mathcal{G}) = \mathcal{G}^{(n)}_{a,b}(q)$, then, obviously, this group is 1-idempotent, and $\lambda_p = 1$.

Assertion 4.5.28. If $\mathcal{F}^{(m,n)}_{a,b}(q)$ is zeroless–nonunital, then there no n-ary cyclic subgroups in $\mathcal{G}^{(n)}_{a,b}(q)$.

The subfield structure of $\mathcal{F}^{(m,n)}_{a,b}(q)$ can coincide with the corresponding subgroup structure of the multiplicative n-ary group $\mathcal{G}^{(n)}_{a,b}(q_*)$, only if its additive m-ary group has the same subgroup structure. However, additive m-ary groups of all polyadic fields $\mathcal{F}^{(m,n)}_{a,b}(q)$ have the same structure: they are cyclic and have no proper m-ary subgroups, each element generates all other elements, i.e., it is a primitive root. Therefore, we arrive at

Theorem 4.5.29. *The polyadic field $\mathcal{F}_{a,b}^{(m,n)}(q)$, being isomorphic to the (m, n)-field of polyadic integer numbers $\mathbb{Z}_{a,b}^{(m,n)}(q)$, has no any proper subfield.*

In this sense, $\mathcal{F}_{a,b}^{(m,n)}(q)$ can be named a *prime polyadic field*.

Recall that any binary finite field has an order which is a power of a prime number $q = p^r$ (its characteristic), and all such fields are isomorphic and contain a prime subfield $GF(p)$ of order p which is isomorphic to the congruence (residue) class field $\mathbb{Z}/p\mathbb{Z}$ (Lidl and Niederreiter 1997).

Conjecture 4.5.30. A finite (m, n)-field (with $m > n$) should contain a minimal subfield which is isomorphic to one of the prime polyadic fields constructed above, and therefore $\mathcal{F}_{a,b}^{(m,n)}(q)$ can be interpreted as a polyadic analog of $GF(p)$.

4.5.3 Examples of exotic finite polyadic fields

We give concrete examples of finite polyadic fields which have properties that are not possible in the binary case (see table 4.4). Only the multiplication of fields will be shown, because their additive part is huge (many pages) for higher arities, and does not carry so much distinctive information.

(1) The first exotic finite polyadic field that has a number of elements that is not a prime number, or prime power (as it should be for a finite binary field) is $\mathcal{F}_{5,6}^{(7,3)}(6)$, which consists of 6 elements $\{5, 11, 17, 23, 29, 35\}$, $q = 6$, It is zeroless and contains two units $\{17, 35\} \equiv \{17_e, 35_e\}$, $\kappa_e = 2$, and each element has the idempotence polyadic order $\lambda_p = 3$, i.e., $\mu^{(3)}[x^7] = x$, $\forall x \in \mathcal{F}_{5,6}^{(7,3)}(6)$. The multiplication is

$$\mu^{(3)}[5, 5, 5] = 17,\ \mu^{(3)}[5, 5, 11] = 23,\ \mu^{(3)}[5, 5, 17] = 29,\ \mu^{(3)}[5, 5, 23] = 35,$$
$$\mu^{(3)}[5, 5, 29] = 5,\ \mu^{(3)}[5, 5, 35] = 11,\ \mu^{(3)}[5, 11, 11] = 29,\ \mu^{(3)}[5, 11, 17] = 35,$$
$$\mu^{(3)}[5, 11, 23] = 5,\ \mu^{(3)}[5, 11, 29] = 11,\ \mu^{(3)}[5, 11, 35] = 17,\ \mu^{(3)}[5, 17, 17] = 5,$$
$$\mu^{(3)}[5, 17, 23] = 11,\ \mu^{(3)}[5, 17, 29] = 17,\ \mu^{(3)}[5, 17, 35] = 23,\ \mu^{(3)}[5, 23, 23] = 17,$$
$$\mu^{(3)}[5, 23, 29] = 23,\ \mu^{(3)}[5, 23, 35] = 29,\ \mu^{(3)}[5, 29, 29] = 29,\ \mu^{(3)}[5, 29, 35] = 35,$$
$$\mu^{(3)}[5, 35, 35] = 5,\ \mu^{(3)}[11, 11, 11] = 35,\ \mu^{(3)}[11, 11, 17] = 5,\ \mu^{(3)}[11, 11, 23] = 11,$$
$$\mu^{(3)}[11, 11, 29] = 17,\ \mu^{(3)}[11, 11, 35] = 23,\ \mu^{(3)}[11, 11, 17] = 5,\ \mu^{(3)}[11, 17, 17] = 11,$$
$$\mu^{(3)}[11, 17, 23] = 17,\ \mu^{(3)}[11, 17, 29] = 23,\ \mu^{(3)}[11, 17, 35] = 29,\ \mu^{(3)}[11, 17, 23] = 17,$$
$$\mu^{(3)}[11, 17, 29] = 23,\ \mu^{(3)}[11, 17, 35] = 29,\ \mu^{(3)}[11, 23, 23] = 23,\ \mu^{(3)}[11, 23, 29] = 29,$$
$$\mu^{(3)}[11, 23, 23] = 23,\ \mu^{(3)}[11, 23, 35] = 35,\ \mu^{(3)}[11, 29, 29] = 35,\ \mu^{(3)}[11, 29, 35] = 5,$$
$$\mu^{(3)}[11, 35, 35] = 11,\ \mu^{(3)}[17, 17, 17] = 17,\ \mu^{(3)}[17, 17, 23] = 23,\ \mu^{(3)}[17, 17, 29] = 29,$$
$$\mu^{(3)}[17, 17, 35] = 35,\ \mu^{(3)}[17, 23, 23] = 29,\ \mu^{(3)}[17, 23, 29] = 35,\ \mu^{(3)}[17, 29, 29] = 5,$$
$$\mu^{(3)}[17, 29, 35] = 11,\ \mu^{(3)}[17, 35, 35] = 17,\ \mu^{(3)}[23, 23, 23] = 35,\ \mu^{(3)}[23, 23, 29] = 5,$$
$$\mu^{(3)}[23, 23, 35] = 11,\ \mu^{(3)}[23, 29, 29] = 11,\ \mu^{(3)}[23, 29, 35] = 17,\ \mu^{(3)}[23, 35, 35] = 23,$$
$$\mu^{(3)}[29, 29, 29] = 17,\ \mu^{(3)}[29, 29, 35] = 23,\ \mu^{(3)}[29, 35, 35] = 29,\ \mu^{(3)}[35, 35, 35] = 35.$$

The multiplicative querelements are $\bar{5} = 29$, $\overline{29} = 5$, $\overline{11} = 23$, $\overline{23} = 11$. Because

$$5^{\langle 1\rangle_{\times 3}} = 17_e,\ 5^{\langle 2\rangle_{\times 3}} = 29,\ 5^{\langle 3\rangle_{\times 3}} = 5,\ 29^{\langle 1\rangle_{\times 3}} = 17_e,\ 29^{\langle 2\rangle_{\times 3}} = 5,\ 29^{\langle 3\rangle_{\times 3}} = 29, \quad (4.5.13)$$

$$11^{\langle 1\rangle_{\times 3}} = 35_e,\ 11^{\langle 2\rangle_{\times 3}} = 23,\ 11^{\langle 3\rangle_{\times 3}} = 11,\ 23^{\langle 1\rangle_{\times 3}} = 35_e,\ 23^{\langle 2\rangle_{\times 3}} = 11,\ 23^{\langle 3\rangle_{\times 3}} = 23, \quad (4.5.14)$$

the multiplicative 3-ary group $\mathcal{G}^{(7,\,3)}_{5,\,6}(6)$ consists of two nonintersecting cyclic 3-ary subgroups

$$\mathcal{G}^{(7,\,3)}_{5,\,6}(6) = \mathcal{G}_1 \bigcup \mathcal{G}_2,\ \ \mathcal{G}_1 \bigcap \mathcal{G}_2 = \varnothing, \quad (4.5.15)$$

$$\mathcal{G}_1 = \{5,\ 17_e,\ 29\}, \quad (4.5.16)$$

$$\mathcal{G}_2 = \{11,\ 23,\ 35_e\}, \quad (4.5.17)$$

which is impossible for binary subgroups, as these always intersect in the identity of the binary group.

(2) The finite polyadic field $\mathcal{F}^{(7,\,3)}_{5,\,6}(4) = \{\{5, 11, 17, 23\} \mid \nu^{(7)}, \mu^{(3)}\}$ which has the same arity shape as above, but with order 4, has the exotic property that all elements are units, which follows from its multiplication

$$\mu^{(3)}[5, 5, 5] = 5,\ \mu^{(3)}[5, 5, 11] = 11,\ \mu^{(3)}[5, 5, 17] = 17,\ \mu^{(3)}[5, 5, 23] = 23,$$
$$\mu^{(3)}[5, 11, 11] = 5,\ \mu^{(3)}[5, 11, 17] = 23,\ \mu^{(3)}[5, 11, 23] = 17,\ \mu^{(3)}[5, 17, 17] = 5,$$
$$\mu^{(3)}[5, 17, 23] = 11,\ \mu^{(3)}[5, 23, 23] = 5,\ \mu^{(3)}[11, 11, 11] = 11,\ \mu^{(3)}[11, 11, 17] = 17,$$
$$\mu^{(3)}[11, 11, 23] = 23,\ \mu^{(3)}[11, 17, 17] = 11,\ \mu^{(3)}[11, 17, 23] = 5,\ \mu^{(3)}[11, 23, 23] = 11,$$
$$\mu^{(3)}[17, 17, 17] = 17,\ \mu^{(3)}[17, 17, 23] = 23,\ \mu^{(3)}[17, 23, 23] = 17,\ \mu^{(3)}[23, 23, 23] = 23.$$

(3) Next we show by construction, that (as opposed to the case of binary finite fields) there exist non-isomorphic finite polyadic fields of the same order and arity shape. Indeed, consider these two (9, 3)-fields of order 2, that are $\mathcal{F}^{(9,\,3)}_{3,\,8}(2)$ and $\mathcal{F}^{(9,\,3)}_{7,\,8}(2)$. The first is zeroless–nonunital, while the second is zeroless with two units, i.e., all elements are units. The multiplication of $\mathcal{F}^{(9,\,3)}_{3,\,8}(2)$ is

$$\mu^{(3)}[3, 3, 3] = 11,\ \mu^{(3)}[3, 3, 11] = 3,\ \mu^{(3)}[3, 11, 11] = 11,\ \mu^{(3)}[11, 11, 11] = 3,$$

having the multiplicative querelements $\bar{3} = 11$, $\overline{11} = 3$. For $\mathcal{F}^{(9,\,3)}_{7,\,8}(2)$ we get the 3-group of units

$$\mu^{(3)}[7, 7, 7] = 7,\ \mu^{(3)}[7, 7, 15] = 15,\ \mu^{(3)}[7, 15, 15] = 7,\ \mu^{(3)}[15, 15, 15] = 15.$$

They have different idempotence polyadic orders $\mathrm{ord}\mathcal{F}^{(9,3)}_{3,8}(2) = 2$ and $\mathrm{ord}\mathcal{F}^{(9,3)}_{7,8}(2) = 1$. Despite their additive m-ary groups being isomorphic, it follows from the above multiplicative structure, that it is not possible to construct an isomorphism between the fields themselves.

(4) The smallest exotic finite polyadic field with more than one unit is $\mathcal{F}^{(4,3)}_{2,3}(5) = \{\{2, 5, 8, 11, 14\} \mid \nu^{(4)}, \mu^{(3)}\}$ of order 5 with two units $\{11, 14\} \equiv \{11_e, 14_e\}$ and the zero $5 \equiv 5_z$. The presence of zero allows us to define

the polyadic characteristic (4.4.17) which is 3 (see table 4.2), because the 3rd additive power of all elements is equal to zero

$$2^{\langle 3\rangle_{+4}} = 8^{\langle 3\rangle_{+4}} = 11_e^{\langle 3\rangle_{+4}} = 14_e^{\langle 3\rangle_{+4}} = 5_z. \tag{4.5.18}$$

The additive querelements are

$$\tilde{2} = 11_e, \; \tilde{8} = 14_e, \; \widetilde{11}_e = 8, \; \widetilde{14}_e = 2. \tag{4.5.19}$$

The idempotence polyadic order is $\mathrm{ord}\mathcal{F}_{2,3}^{(4,3)}(5) = 2$, because for nonunit and nonzero elements

$$2^{\langle 2\rangle_{\times 3}} = 2, \; 8^{\langle 2\rangle_{\times 3}} = 8, \tag{4.5.20}$$

and their multiplicative querelements are $\bar{2} = 8, \bar{8} = 2$. The multiplication is given by the cyclic 3-ary group $\mathcal{G}_{2,\,3}^{(3)}(4) = \{\{2,\, 8,\, 11,\, 14\} \mid \mu^{(3)}\}$ as:

$$\begin{aligned}
&\mu^{(3)}[2,\, 2,\, 2] = 8,\; \mu^{(3)}[2,\, 2,\, 8] = 2,\; \mu^{(3)}[2,\, 2,\, 11] = 14,\; \mu^{(3)}[2,\, 2,\, 14] \\
&= 11,\; \mu^{(3)}[2,\, 8,\, 8] = 8, \\
&\mu^{(3)}[2,\, 8,\, 11] = 11,\; \mu^{(3)}[2,\, 8,\, 14] = 14,\; \mu^{(3)}[2,\, 11,\, 11] = 2,\; \mu^{(3)}[2,\, 11,\, 14] \\
&= 8,\; \mu^{(3)}[2,\, 14,\, 14] = 2, \\
&\mu^{(3)}[8,\, 8,\, 8] = 2,\; \mu^{(3)}[8,\, 8,\, 11] = 14,\; \mu^{(3)}[8,\, 8,\, 14] = 11,\; \mu^{(3)}[8,\, 11,\, 11] \\
&= 8,\; \mu^{(3)}[8,\, 11,\, 14] = 2, \\
&\mu^{(3)}[8,\, 14,\, 14] = 8,\; \mu^{(3)}[11,\, 11,\, 11] = 11,\; \mu^{(3)}[11,\, 11,\, 14] = 14,\; \mu^{(3)}[11,\, 14,\, 14] \\
&= 11,\; \mu^{(3)}[14,\, 14,\, 14] = 14.
\end{aligned}$$

We observe that, despite having two units, the cyclic 3-ary group $\mathcal{G}_{2,\,3}^{(3)}(4)$ has no decomposition into nonintersecting cyclic 3-ary subgroups, as in (4.5.15). One of the reasons is that the polyadic field $\mathcal{F}_{5,\,6}^{(7,\,3)}(6)$ is zeroless, while $\mathcal{F}_{2,\,3}^{(4,\,3)}(5)$ has a zero (see assertion 4.5.26).

4.6 Diophantine equations over polyadic integers and Fermat's theorem

First, recall the standard binary version of the equal sums of like powers *Diophantine equation* (Lander *et al* 1967, Ekl 1998). Take the fixed non-negative integers $p, q, l \in \mathbb{N}^0$, $p \leqslant q$, and the positive integer unknowns $u_i, v_j \in \mathbb{Z}_+$, $i = 1, \ldots p + 1$, $j = 1, 1, \ldots q + 1$, then the Diophantine equation is

$$\sum_{i=1}^{p+1} u_i^{l+1} = \sum_{j=1}^{q+1} v_j^{l+1}. \tag{4.6.1}$$

The trivial case, when $u_i = 0$, $v_j = 0$, for all i, j is not considered. In the binary case, the solutions of (4.6.1) are usually denoted by $(l + 1|p + 1,\, q + 1)_r$, which shows the number of summands on both sides and the powers of elements (Lander *et al* 1967). But in the polyadic case (see below), the number of summands and powers do not coincide with $l + 1$, $p + 1$, $q + 1$. We mark the solutions of (4.6.1) by the triple $(l|p, q)_r$ showing quantity of operations, where r (if it is used) is the order of the solution (ranked by the value of the sum) and the unknowns u_i, v_j are placed in ascending order $u_i \leqslant u_{i+1}$, $v_j \leqslant v_{j+1}$.

Let us recall the *Tarry–Escott problem* (or multigrades problem) (Dorwart and Brown 1937): to find the solutions to (4.6.1) for an equal number of summands on both sides of $p = q$ and s equations simultaneously, such that $l = 0, \ldots, s$. Known solutions exist for powers until $s = 10$, which are bounded such that $s \leqslant p$ (in our notations), see, also, Nguyen (2016). The solutions with highest powers $s = p$ are the most interesting and called the ideal solutions (Borwein 2002).

Theorem 4.6.1 (Frolov 1889). If the set of s Diophantine equations (4.6.1) with $p = q$ for $l = 0, \ldots, s$ has a solution $\{u_i, v_i, i = 1, \ldots p + 1\}$, then it has the solution $\{a + bu_i, a + bv_i, i = 1, \ldots p + 1\}$, where $a, b \in \mathbb{Z}$ are arbitrary and fixed.

In the simplest case $(1|0, 1)$, one term in the lhs, one addition on the rhs and one multiplication, the (coprime) positive numbers satisfying (4.6.1) are called a (primitive) *Pythagorean triple*. For the *Fermat's triple* $(l|0, 1)$ with one addition on the rhs and more than one multiplication $l \geqslant 2$, there are no solutions of (4.6.1), which is known as *Fermat's last theorem* proved in Wiles (1995). There are many solutions known with more than one addition on both sides, where the highest number of multiplications till now is 31.

Before generalizing (4.6.1) to the polyadic case we note the following.

***Remark* 4.6.2.** The notations in (4.6.1) are chosen in such a way that p and q are the numbers of binary additions on both sides, while l is the number of binary multiplications in each term, which is natural for using polyadic powers (Duplij 2012).

4.6.1 Polyadic analog of the Lander–Parkin–Selfridge conjecture

In Lander *et al* (1967), a generalization of Fermat's last theorem was conjectured, that the solutions of (4.6.1) exist for small powers only, which can be formulated in terms of the numbers of operations as

Conjecture 4.6.3. Lander–Parkin–Selfridge (Lander *et al* 1967). There exist solutions of (4.6.1) in positive integers, if the number of multiplications is less than or equal than the total number of additions plus one

$$3 \leqslant l \leqslant l_{\mathrm{LSP}} = p + q + 1, \tag{4.6.2}$$

where $p + q \geqslant 2$.

***Remark* 4.6.4.** If the equation (4.6.1) is considered over the binary ring of integers $\mathbb{Z}$, such that $u_i, v_j \in \mathbb{Z}$, it leads to a straightforward reformulation: for even powers it is obvious, but for odd powers all negative terms can be rearranged and placed on the other side.

Let us consider the Diophantine equation (4.6.1) over polyadic integers $\mathbb{Z}^{(m,n)}$ (i.e., over the polyadic (m, n)-ary ring, see definition 4.3.13) such that $u_i, v_j \in \mathbb{Z}^{(m,n)}$. We use the 'long products' $(\mu^{(n)})^{\circ l}$ and $(\nu^{(m)})^{\circ l}$ containing l operations, and also the 'polyadic power' for an element $x \in \mathbb{Z}^{(m,n)}$ with respect to n-ary multiplication (Duplij 2012)

$$x^{\langle l \rangle_n} = (\mu^{(n)})^{\circ l}\left[\overbrace{x, x, \ldots, x}^{l(n-1)+1}\right]. \tag{4.6.3}$$

In the binary case, $n = 2$, the polyadic power coincides with the $(l + 1)$ power of an element $x^{\langle l \rangle_2} = x^{l+1}$, which explains remark 4.6.2. In this notation the polyadic analog of the equal sums of like powers Diophantine equation has the form

$$(\nu^{(m)})^{\circ p}\left[u_1^{\langle l \rangle_n}, u_2^{\langle l \rangle_n}, \ldots, u_{p(m-1)+1}^{\langle l \rangle_n}\right] = (\nu^{(m)})^{\circ q}\left[v_1^{\langle l \rangle_n}, v_2^{\langle l \rangle_n}, \ldots, v_{q(m-1)+1}^{\langle l \rangle_n}\right], \tag{4.6.4}$$

where p and q are the number of m-ary additions on the lhs and rhs correspondingly. The solutions of (4.6.4) will be denoted by $\{u_1, u_2, \ldots, u_{p(m-1)+1}; v_1, v_2, \ldots, v_{q(m-1)+1}\}$. In the binary case $m = 2$, $n = 2$, (4.6.4) reduces to (4.6.1). Analogously, we mark the solutions of (4.6.4) by the *polyadic triple* $(l|p, q)_r^{(m,n)}$. Now the *polyadic Pythagorean triple* $(1|0,1)^{(m,n)}$, having one term on the lhs, one m-ary addition on the rhs and one n-ary multiplication (elements are in the first polyadic power $\langle 1 \rangle_n$), becomes

$$u_1^{\langle 1 \rangle_n} = \nu^{(m)}[v_1^{\langle 1 \rangle_n}, v_2^{\langle 1 \rangle_n}, \ldots, v_m^{\langle 1 \rangle_n}]. \tag{4.6.5}$$

Definition 4.6.5. The equation (4.6.5) solved by minimal $u_1, v_i \in \mathbb{Z}$, $i = 1, \ldots, m$ can be named the *polyadic Pythagorean theorem.*

The *polyadic Fermat's triple* $(l|0,1)^{(m,n)}$ has one term in the lhs, one m-ary addition in the rhs and l (n-ary) multiplications

$$u_1^{\langle l \rangle_n} = \nu^{(m)}[v_1^{\langle l \rangle_n}, v_2^{\langle l \rangle_n}, \ldots, v_m^{\langle l \rangle_n}]. \tag{4.6.6}$$

One may be interested in whether the polyadic analog of Fermat's last theorem is valid, and if not, in which cases the analogy with the binary case can be sustained.

Conjecture 4.6.6. Polyadic analog of Fermat's Last Theorem. The polyadic Fermat triple (4.6.6) has no solutions over the polyadic (m, n)-ary ring $\mathbb{Z}^{(m,n)}$, if $l \geqslant 2$, i.e., there is more than one n-ary multiplication.

Its straightforward generalization leads to the polyadic version of the Lander–Parkin–Selfridge conjecture, as

Conjecture 4.6.7. There exist solutions of the polyadic analog of the equal sums of like powers Diophantine equation (4.6.4) in integers, if the number of n-ary multiplications is less than or equal than the total number of m-ary additions plus one

$$3 \leqslant l \leqslant l_{\mathrm{pLPS}} = p + q + 1. \tag{4.6.7}$$

Below we will see a counterexample to both of the above conjectures.

***Example* 4.6.8.** Let us consider the (3, 2)-ring $\mathbb{Z}^{(3,2)} = \langle \mathbb{Z} \mid \nu^{(3)}, \mu^{(2)} \rangle$, where

$$\nu^{(3)}[x, y, z] = x + y + z + 2, \tag{4.6.8}$$

$$\mu^{(2)}[x, y] = xy + x + y. \tag{4.6.9}$$

Note that this exotic polyadic ring is commutative and cancellative, having unit 0, no multiplicative inverses, and for any $x \in \mathbb{Z}^{(3,2)}$ its additive querelement $\tilde{x} = -x - 2$, therefore $\langle \mathbb{Z} \mid \nu^{(3)} \rangle$ is a ternary group (as it should be). The polyadic power of any element is

$$x^{\langle l \rangle_2} = (x + 1)^{l+1} - 1. \tag{4.6.10}$$

(1) For the (3, 2)-ring $\mathbb{Z}^{(3,2)}$ the polyadic Pythagorean triple $(1|0,1)^{(3,2)}$ in (4.6.5) now is

$$u^{\langle 1 \rangle_2} = \nu^{(3)}[x^{\langle 1 \rangle_2}, y^{\langle 1 \rangle_2}, z^{\langle 1 \rangle_2}], \tag{4.6.11}$$

which, using (4.6.3), (4.6.9) and (4.6.10), becomes the (shifted) Pythagorean quadruple (Spira 1962)

$$(u + 1)^2 = (x + 1)^2 + (y + 1)^2 + (z + 1)^2, \tag{4.6.12}$$

and it has infinite number of solutions, among which two minimal ones $\{u = 2; x = 0, y = z = 1\}$ and $\{u = 14; x = 1, y = 9, z = 10\}$ give $3^2 = 1^2 + 2^2 + 2^2$ and $15^2 = 2^2 + 10^2 + 11^2$, correspondingly.

(2) For $\mathbb{Z}^{(3,2)}$ the polyadic Fermat's triple $(l|0,1)^{(3,2)}$ becomes

$$(u + 1)^{l+1} = (x + 1)^{l+1} + (y + 1)^{l+1} + (z + 1)^{l+1}. \tag{4.6.13}$$

If the polyadic analog of Fermat's last theorem 4.6.6 holds, then there are no solutions to (4.6.13) for more than one n-ary multiplication $l \geqslant 2$. But this is the particular case, $p = 0$, $q = 2$, of the binary Lander–Parkin–Selfridge conjecture 4.6.3 which now takes the form: the solutions to (4.6.13) exist, if $l \leqslant 3$. Thus, as a counterexample to the polyadic analog of Fermat's last theorem, we have two possible solutions with numbers of multiplications: $l = 2, 3$. In the case of $l = 2$ there exist two solutions: one well-known solution $\{u = 5; x = 2, y = 3, z = 4\}$ giving

$6^3 = 3^3 + 4^3 + 5^3$ and another one giving $709^3 = 193^3 + 461^3 + 631^3$, while for $l = 3$ there exist an infinite number of solutions, and one of them (minimal) gives $42\,2481^4 = 95\,800^4 + 217\,519^4 + 414\,560^4$ (Elkies 1988).

(3) The general polyadic triple $(l|p, q)^{(3,2)}$, using (4.6.4), can be presented in the standard binary form (as (4.6.1))

$$\sum_{i=1}^{2p+1} (u_i + 1)^{l+1} = \sum_{j=1}^{2q+1} (v_i + 1)^{l+1}, \quad u_i, v_j \in \mathbb{Z}. \tag{4.6.14}$$

Let us apply the polyadic Lander–Parkin–Selfridge conjecture 4.6.7 for this case: the solutions to (4.6.14) exist, if $3 \leqslant l \leqslant l_{\text{pLSP}} = p + q + 1$. But the binary Lander–Parkin–Selfridge conjecture 4.6.3, applied directly, gives $3 \leqslant l \leqslant l_{\text{LSP}} = 2p + 2q + 1$. So we should have counterexamples to the polyadic Lander–Parkin–Selfridge conjecture, when $l_{\text{pLSP}} < l \leqslant l_{\text{LSP}}$. For instance, for $p = q = 1$, we have $l_{\text{pLSP}} = 3$, while the (minimal) counterexample with $l = 5$ is $\{u_1 = 3,\ u_2 = 18,\ u_3 = 21,\ v_1 = 9,\ v_2 = 14,\ v_3 = 22\}$ giving $3^6 + 19^6 + 22^6 = 10^6 + 15^6 + 23^6$ (SubbaRao 1934).

As can be observed from example 4.6.8, the arity shape of the polyadic ring $\mathbb{Z}^{(m,\,n)}$ is crucial in constructing polyadic analogs of the equal sums of like powers conjectures. We can make some general estimations assuming a special (more or less natural) form of its operations over integers.

Definition 4.6.9. We call $\mathbb{Z}^{(m,\,n)}$ the *standard polyadic ring*, if the 'leading terms' of its m-ary addition and n-ary multiplication are

$$\nu^{(m)}\left[\overbrace{x, x, \ldots, x}^{m}\right] \sim mx, \tag{4.6.15}$$

$$\mu^{(n)}\left[\overbrace{x, x, \ldots, x}^{n}\right] \sim x^n, \quad x \in \mathbb{Z}. \tag{4.6.16}$$

The polyadic ring $\mathbb{Z}^{(3,2)}$ from example 4.6.8, as well as the congruence class polyadic ring $\mathcal{R}_{a,\,b}^{(m,n)}$ (4.3.16) are both standard.

Using (4.6.3), we obtain approximate behavior of the polyadic power in the standard polyadic ring

$$x^{\langle l \rangle_n} \sim x^{l(n\,-\,1)+1}, \quad x \in \mathbb{Z}, l \in \mathbb{N}, n \geqslant 2. \tag{4.6.17}$$

So increasing the arity of multiplication leads to higher powers, while increasing arity of addition gives more terms in sums. Thus, the estimation for the polyadic analog of the equal sums of like powers Diophantine equation (4.6.4) becomes

$$(p(m-1)+1)x^{l(n-1)+1} \sim (q(m-1)+1)x^{l(n-1)+1}, \quad x \in \mathbb{Z}. \tag{4.6.18}$$

Now we can apply the binary Lander–Parkin–Selfridge conjecture 4.6.3 in the form: the solutions to (4.6.18) can exist if $3 \leqslant l \leqslant l_{\mathrm{LPS}}$, where l_{LPS} is an integer solution of

$$(n-1)l_{\mathrm{LPS}} = (p+q)(m-1)+1. \tag{4.6.19}$$

On the other hand the polyadic Lander–Parkin–Selfridge conjecture 4.6.7 gives: the solutions to (4.6.18) can exist if $3 \leqslant l \leqslant l_{\mathrm{pLPS}} = p+q+1$. Note that $(p+q) \geqslant 2$ now.

An interesting question arises: which arities give the same limit, that is, when $l_{\mathrm{pLPS}} = l_{\mathrm{LPS}}$?

Proposition 4.6.10. *For any fixed number of additions in both sizes of the polyadic analog of the equal sums of like powers Diophantine equation* (4.6.4)$p+q \geqslant 2$*, there exist limiting arities m_0 and n_0 (excluding the trivial binary case $m_0 = n_0 = 2$), for which the binary and polyadic Lander–Parkin–Selfridge conjectures coincide $l_{\mathrm{pLPS}} = l_{\mathrm{LPS}}$, such that*

$$m_0 = 3+p+q+(p+q+1)k, \tag{4.6.20}$$

$$n_0 = 2+p+q+(p+q)k, \quad k \in \mathbb{N}^0. \tag{4.6.21}$$

Proof. To equate $l_{\mathrm{LPS}} = l_{\mathrm{pLPS}} = p+q+1$ we use (4.6.19) and solve in integers the equation

$$(n_0-1)(p+q+1) = (p+q)(m_0-1)+1. \tag{4.6.22}$$

In the trivial case, $m_0 = n_0 = 2$, this is an identity, while the other solutions can be found from $n_0(p+q+1) = (p+q)m_0 + 2$, which gives (4.6.20)–(4.6.21). □

Corollary 4.6.11. In the limiting case $l_{\mathrm{pLPS}} = l_{\mathrm{LPS}}$ the arity of multiplication always exceeds the arity of addition

$$m_0 - n_0 = k+1, \quad k \in \mathbb{N}^0, \tag{4.6.23}$$

and they start from $m_0 \geqslant 5$, $n_0 \geqslant 4$.

The first allowed arities m_0 and n_0 are presented in table 4.5. Their meaning is the following.

Corollary 4.6.12. For the polyadic analog of the equal sums of like powers equation over the standard polyadic ring $\mathbb{Z}^{(m,n)}$ (with fixed $p+q \geqslant 2$) the polyadic Lander–Parkin–Selfridge conjecture becomes weaker than the binary one $l_{\mathrm{pLPS}} \geqslant l_{\mathrm{LPS}}$, if: (1) the arity of multiplication exceeds its limiting value n_0 with fixed arity of the addition; (2) the arity of addition is lower, than its limiting value m_0 with the fixed arity of multiplication.

Table 4.5. The limiting arities m_0 and n_0, which give $l_{pLPS} = l_{LPS}$ in (4.6.18).

$p + q = 2$		$p + q = 3$		$p + q = 4$	
m_0	n_0	m_0	n_0	m_0	n_0
5	4	6	5	7	6
8	6	10	8	12	10
11	8	14	11	17	14
14	10	18	14	22	18

***Example* 4.6.13.** Consider the standard polyadic ring $\mathbb{Z}^{(m,n)}$ and fix the arity of addition $m_0 = 12$, then take in (4.6.18) the total number of additions $p + q = 4$ (the last column in table 4.5). We observe that the arity of multiplication $n = 16$, which exceeds the limiting arity $n_0 = 10$ (corresponding to m_0). Thus, we obtain $l_{pLPS} = 5$ and $l_{LPS} = 3$ by solving (4.6.19) in integers, and therefore the polyadic Lander–Parkin–Selfridge conjecture becomes now weaker than the binary one, and we do not obtain counterexamples to it, as in example 4.6.8 (where the situation was opposite $l_{pLPS} = 3$ and $l_{LPS} = 5$, and they cannot be equal).

A concrete example of the standard polyadic ring (definition 4.6.9) is the polyadic ring of the fixed congruence class $\mathcal{R}_{a,b}^{(m,n)}$ considered in section 4.3, because its operations (4.3.5)–(4.3.6) have the same straightforward behavior (4.6.15)–(4.6.16). Let us formulate the polyadic analog of the equal sums of like powers Diophantine equation (4.6.4) over $\mathcal{R}_{a,b}^{(m,n)}$ in terms of operations in $\mathbb{Z}$. Using (4.3.5)–(4.3.6) and (4.6.17) for (4.6.4) we obtain

$$\sum_{i=1}^{p(m-1)+1} (a + bk_i)^{l(n-1)+1} = \sum_{j=1}^{q(m-1)+1} \left(a + bk_j\right)^{l(n-1)+1}, \quad a, b, k_i \in \mathbb{Z}. \tag{4.6.24}$$

It is seen that the leading power behavior of both sides in (4.6.24) coincides with the general estimation (4.6.18). But now the arity shape (m, n) is fixed by (4.3.17)–(4.3.18) and given in table 4.1. Nevertheless, we can consider for (4.6.24) the polyadic analog of Fermat's last theorem 4.6.6, the Lander–Parkin–Selfridge conjecture 4.6.3 (solutions exist for $l \leqslant l_{LPS}$) and its polyadic version (conjecture 4.6.7, solutions exist for $l \leqslant l_{pLPS}$), as in the estimations above. Let us consider some examples of solutions to (4.6.24).

***Example* 4.6.14.** Let $[[2]]_3$ be the congruence class, which is described by (4, 3)-ring $\mathcal{R}_{2,3}^{(4,3)}$ (see table 4.1), and we consider the polyadic Fermat's triple $(l|0,5)^{(4,3)}$ (4.6.6). Now the powers are $l_{LPS} = 8$, $l_{pLPS} = 6$, and for instance, if $l = 2$, we have solutions, because $l < l_{pLPS} < l_{LPS}$, and one of them is

$$14^5 = 4 \cdot (-1)^5 + 7 \cdot 5^5 + 8^5 + 2 \cdot 11^5. \tag{4.6.25}$$

4.6.2 Frolov's theorem and the Tarry–Escott problem

A special set of solutions to the polyadic Lander–Parkin–Selfridge conjecture 4.6.7 can be generated, if we put $p = q$ in (4.6.24), which we call *equal-summand solutions* (the term 'symmetric solution' is already taken and widely used, Borwein 2002), by exploiting the Tarry–Escott problem approach (Dorwart and Brown 1937) and Frolov's theorem 4.6.1.

Theorem 4.6.15. If the set of integers $k_i \in \mathbb{Z}$ solves the Tarry–Escott problem

$$\sum_{i=1}^{p(m-1)+1} k_i^r = \sum_{j=1}^{p(m-1)+1} k_j^r, \quad r = 1, \dots, s = l(n-1) + 1, \tag{4.6.26}$$

then the polyadic equal sums of like powers equation with equal summands (4.6.4) has a solution over the polyadic (m, n)-ring $\mathcal{R}_{a,b}^{(m,n)}$ having the arity shape given by the following relations:

(i) *Inequality*

$$l(n-1) + 1 \leqslant p(m-1); \tag{4.6.27}$$

(ii) *Equality*

$$p(m-1) = 2^{l(n-1)+1}. \tag{4.6.28}$$

Proof. Using Frolov's theorem 4.6.1 applied to (4.6.26) we state that

$$\sum_{i=1}^{p(m-1)+1} (a + bk_i)^r = \sum_{j=1}^{p(m-1)+1} \left(a + bk_j\right)^r, \quad r = 1, \dots, s = l(n-1) + 1, \tag{4.6.29}$$

for any fixed integers $a, b \in \mathbb{Z}$. This means, that (4.6.29), with k_i (satisfying (4.6.26)) corresponds to a solution to the polyadic equal sums of like powers equation (4.6.4) for any congruence class $[[a]]_b$. Nevertheless, the values a and b are fixed by the restrictions on the arity shape and the relations (4.3.11) and (4.3.13).

(i) It is known that the Tarry–Escott problem can have a solution only when the powers are strongly less than the number of summands (Borwein 2002, Dorwart and Brown 1937), that is $(l(n-1)+1)+1 \leqslant p(m-1)+1$, which gives (4.6.27).

(ii) A special kind of solution, when the number of summands is equal to 2 into the number of powers, was found using the Thue–Morse sequence (Allouche and Shallit 1999), which always satisfies the bound (4.6.27), and in our notation it is (4.6.28).

In both cases the relations (4.6.27) and (4.6.28) should be solved in positive integers and with $m \geqslant 2$ and $n \geqslant 2$, which can lead to nonunique solutions. □

Let us consider some examples which give solutions to the polyadic equal sums of like powers equation (4.6.4) with $p = q$ over the polyadic (m, n)-ring $\mathcal{R}_{a,b}^{(m,n)}$ of the fixed congruence class $[[a]]_b$.

***Example* 4.6.16.** (1) One of the first ideal (nonsymmetric) solutions to the Tarry–Escott problem has 6 summands and 5 powers

$$\begin{aligned} &0^r + 19^r + 25^r + 57^r + 62^r + 86^r \\ &\quad = 2^r + 11^r + 40^r + 42^r + 69^r + 85^r, \ r = 1, \dots, 5. \end{aligned} \tag{4.6.30}$$

We compare with (4.6.26) and obtain

$$p(m-1) = 5, \tag{4.6.31}$$

$$l(n-1) = 4. \tag{4.6.32}$$

After ignoring binary arities we get $m = 6$, $p = 1$ and $n = 3$, $l = 2$. From theorem 4.3.12 and table 4.1 we observe the minimal choice $a = 4$ and $b = 5$. It follows from Frolov's theorem 4.6.1, that all equations in (4.6.30) have symmetry $k_i \to a + bk_i = 4 + 5k_i$. Thus, we obtain the solution of the polyadic equal sums of like powers equation (4.6.4) for the fixed congruence class $[[4]]_5$ in the form

$$\begin{aligned} &4^5 + 99^5 + 129^5 + 289^5 + 314^5 + 434^5 \\ &\quad = 14^5 + 59^5 + 204^5 + 214^5 + 349^5 + 429^5. \end{aligned} \tag{4.6.33}$$

It is seen from table 4.1 that the arity shape $(m = 6, n = 3)$ corresponds, e.g., to the congruence class $[[4]]_{10}$ as well. Using Frolov's theorem, we substitute in (4.6.30)$k_i \to 4 + 10k_i$ to obtain the solution in the congruence class $[[4]]_{10}$

$$\begin{aligned} &4^5 + 194^5 + 254^5 + 574^5 + 624^5 + 864^5 \\ &\quad = 24^5 + 114^5 + 404^5 + 424^5 + 694^5 + 854^5. \end{aligned} \tag{4.6.34}$$

(2) To obtain the special kind of solutions to the Tarry–Escott problem we start with the known one with 8 summands and 3 powers (see, e.g., Lehmer 1947)

$$\begin{aligned} &0^r + 3^r + 5^r + 6^r + 9^r + 10^r + 12^r + 15^r \\ &\quad = 1^r + 2^r + 4^r + 7^r + 8^r + 11^r + 13^r + 14^r, \ r = 1, 2, 3. \end{aligned} \tag{4.6.35}$$

So we have the concrete solution to the system (4.6.26) with the condition (4.6.28) which now takes the form $8 = 2^3$, and therefore

$$p(m-1) = 7, \tag{4.6.36}$$

$$l(n-1) = 2. \tag{4.6.37}$$

Excluding the trivial case containing binary arities, we have $m = 8$, $p = 1$ and $n = 3$, $l = 1$. It follows from theorem 4.3.12 and table 4.1, that $a = 6$ and $b = 7$, and so the polyadic ring is $\mathcal{R}_{6,7}^{(8,3)}$. Using Frolov's theorem 4.6.1, we can substitute entries in (4.6.35) as $k_i \to a + bk_i = 6 + 7k_i$ in the equation with highest power $r = 3$ (which is relevant to our task) and obtain the solution of (4.6.4) for $[[6]]_7$ as follows

$$\begin{aligned} &6^3 + 27^3 + 41^3 + 48^3 + 69^3 + 76^3 + 90^3 + 111^3 \\ &= 13^3 + 20^3 + 34^3 + 55^3 + 62^3 + 83^3 + 97^3 + 104^3. \end{aligned} \tag{4.6.38}$$

We conclude that consideration of the Tarry–Escott problem and Frolov's theorem over polyadic rings gives the possibility of obtaining many nontrivial solutions to the polyadic equal sums of like powers equation for fixed congruence classes.

Parts of this chapter have been reprinted from Duplij (2017). Reproduced with permission (©) Springer.

References

Allouche J-P and Shallit J 1999 The ubiquitous Prouhet–Thue–Morse sequence *Sequences and Their Applications. Proc. of the Int. Conf., SETA '98, Singapore, December, 1998* (London: Springer) pp 1–16

Borceux F 1994 Handbook of categorical algebra 1 *Basic category theory Encyclopedia of Mathematics and its Applications* vol 50 (Cambridge: Cambridge University Press)

Borwein P 2002 The Rouhet–Tarry–Escott problem *Computational Excursions in Analysis and Number Theory (CMS Books in Mathematics vol 10)* ed P Borwein (New York: Springer) pp 85–95

Celakoski N 1977 On (F, G)-rings *God. Zb. Mat. Fak. Univ. Kiril Metodij Skopje* **28** 5–15

Crombez G 1972 On (n, m)-rings *Abh. Math. Semin. Univ. Hamb.* **37** 180–99

Crombez G and Timm J 1972 On (n, m)-quotient rings *Abh. Math. Semin. Univ. Hamb.* **37** 200–3

Čupona G 1965 On $[m, n]$-rings *Bull. Soc. Math. Phys. Macedoine* **16** 5–9

Dorwart H L and Brown O E 1937 The Tarry–Escott problem *Amer. Math. Monthly* **44** 613–26

Duplij S 2012 Polyadic systems, representations and quantum groups *J. Kharkov Natl Univ. Ser. Nuclei Part. Fields* **1017** 28–59 Expanded version arXiv:math.RT/1308.4060

Duplij S 2017 Polyadic integer numbers and finite (m, n)-fields *p-Adic Numbers Ultrametric Anal. Appl.* **9** 267–91

Duplij S 2019 Arity shape of polyadic algebraic structures *J. Math. Phys. Anal. Geom.* **15** 3–56

Duplij S 2022 Polyadic analogs of direct product *preprint Univ. Münster, Münster* 21 arXiv:math.RA/2201.08479

Duplij S and Werner W 2021 Structure of unital 3-fields *Math. Semesterber.* **68** 27–53

Ekl R L 1998 New results in equal sums of like powers *Math. Comput.* **67** 1309–15

Elkies N D 1988 On $A^4 + B^4 + C^4 = D^4$ *Math. Comp.* **51** 825–35

Frolov M 1889 Égalités à deux degrés *Bull. Soc. Math. Fr.* **17** 69–83

Głazek K and Michalski J 1984 On polyadic groups which are term-derived from groups *Stud. Sci. Math. Hung.* **19** 307–15

Iancu L and Pop M S 1997 A Post type theorem for (m, n) fields *Proc. of the Scientific Communications Meeting of 'Aurel Vlaicu' University, Arad, Romania, May 16–17, 1996,*

Edition III ed G Halic and G Cristescu vol 14A (Arad: Vlaicu' Univ. of Arad Publishing Centre) pp 13–8
Lambek J 1966 *Lectures on Rings and Modules* (Providence, RI: Blaisdell)
Lander L, Parkin T and Selfridge J 1967 A survey of equal sums of like powers *Math. Comput.* **21** 446–59
Leeson J J and Butson A T 1980 On the general theory of (m, n) rings *Algebra Univ.* **11** 42–76
Lehmer D H 1947 The Tarry-Escott problem *Scr. Math.* **13** 37–41
Lidl R and Niederreiter H 1997 *Finite Fields* (Cambridge: Cambridge University Press)
Mac Lane S 1971 *Categories for the Working Mathematician* (Berlin: Springer)
Menezes A J, van Oorschot P C and Vanstone S A 1997 *Handbook of Applied Cryptography* (Boca Raton, FL: CRC Press)
Michalski J 1988 On J-derived polyadic groups *Mathematica (Cluj)* **30** 149–55
Mullen G L and Panario D 2013 *Handbook of Finite Fields* (Boca Raton, FL: CRC Press)
Nguyen H D 2016 A new proof of the Prouhet–Tarry–Escott problem *Integers* **16** 1–9
Pop A and Pop M S 2002 Some embeddings theorems for $(n, 2)$-rings *Bul. Ştiinţ. Univ. Baia Mare Ser.* B **18** 311–6
Post E L 1940 Polyadic groups *Trans. Am. Math. Soc* **48** 208–350
Spira R 1962 The Diophantine equation $x^2 + y^2 + z^2 = m^2$ *Am. Math. Mon.* **69** 360–5
SubbaRao K 1934 On sums of sixth powers *J. Lond. Math. Soc.* **s1-9** 172–3
Wiles A 1995 Modular elliptic curves and Fermat's last theorem *Ann. Math.* **141** 443–551

Part II

Two-sets polyadic algebraic structures

IOP Publishing

Polyadic Algebraic Structures

Steven Duplij

Chapter 5

Polyadic algebras and deformations

Parts of this chapter have been reprinted with permission from Duplij (2019).

In this chapter we consider two-set algebraic structures in the 'concrete way' and provide the consequent 'polyadization' of binary operations on them for the so-called module-like structures (vector spaces) and algebra-like structures (algebras and inner product spaces). The 'polyadization' of binary scalar multiplication is defined in terms of the multiactions (k-place actions) introduced in Duplij (2012), having special arity shapes parametrized by the number of intact elements in the corresponding multiactions (see section 2.5).

5.1 Two-set polyadic structures

In the case of one set the 'polyadization' of two operations' 'interaction' is straightforward, giving only polyadic distributivity which does not connect or restrict their arities (see chapter 4). However, when the number of sets becomes greater than one, the 'polyadization' turns out to be nontrivial, leading to special relations between the operation arities, and also introduces additional (to the arities) parameters, allowing us to classify them. We call a selection of such relations an arity shape and formulate the arity partial freedom principle that not all arities of the operations that arise during 'polyadization' of binary operations are possible.

5.1.1 Polyadic vector spaces

Let us consider a polyadic field $\Bbbk^{(m_k, n_k)} = \left\langle K \middle| \nu_k^{(m_k)}, \mu_k^{(n_k)} \right\rangle$ ('polyadic scalars'), having m_k-ary addition $\nu_k^{(m_k)}: K^{m_k} \to K$ and n_k-ary multiplication $\mu_k^{(n_k)}: K^{n_k} \to K$, and the identity $e_K \in K$, a neutral element with respect to multiplication $\mu_k^{(n_k)}[e_K^{n_k-1}, \lambda] = \lambda$, for all $\lambda \in K$. In polyadic structures, one can introduce a neutral $(n_k - 1)$-polyad (*identity polyad* for 'scalars')$\mathbf{e}_K \in K^{n_k-1}$ by

$$\mu_k^{(n_k)}[\mathbf{e}_K, \lambda] = \lambda \tag{5.1.1}$$

where $\lambda \in K$ can be on any place.

Next, take an m_V-ary commutative (abelian) group $\langle V | \nu_V^{(m_V)} \rangle$, which can be treated as 'polyadic vectors' with m_V-ary addition $\nu_V^{(m_V)}: V^{m_V} \to V$. Define in $\langle V | \nu_V^{(m_V)} \rangle$ an additive neutral element (zero) $\mathsf{z}_V \in V$ by

$$\nu_V^{(m_V)}[\mathsf{z}_V^{m_V-1}, \mathsf{v}] = \mathsf{v} \tag{5.1.2}$$

for any $\mathsf{v} \in V$, and a 'negative vector' $\bar{\mathsf{v}} \in V$ as its querelement

$$\nu_V^{(m_V)}[\mathbf{a}_V, \bar{\mathsf{v}}, \mathbf{b}_V] = \mathsf{v}, \tag{5.1.3}$$

where $\bar{\mathsf{v}}$ can be on any place in the lhs, and $\mathbf{a}_V$, $\mathbf{b}_V$ are polyads in V. Here, instead of one neutral element we can also introduce the $(m_V - 1)$-polyad z_V (which may not be unique), and so, for a *zero polyad* (for 'vectors') we have

$$\nu_V^{(m_V)}[\mathsf{z}_V, \mathsf{v}] = \mathsf{v}, \quad \forall \mathsf{v} \in V, \tag{5.1.4}$$

where $\mathsf{v} \in V$ can be on any place. The 'interaction' between 'polyadic scalars' and 'polyadic vectors' (the analog of binary multiplication by a scalar $\lambda \mathsf{v}$) can be defined as a *multiaction* (k_ρ-place action) introduced in Duplij (2012)

$$\rho_V^{(k_\rho)}: K^{k_\rho} \times V \longrightarrow V. \tag{5.1.5}$$

The set of all multiactions form a n_ρ-ary semigroup $\mathcal{S}_\rho$ under composition. We can 'normalize' the multiactions in a similar way, as multiplace representations (Duplij 2012), by (an analog of $1\mathsf{v} = \mathsf{v}$, $\mathsf{v} \in V$, $1 \in K$)

$$\rho_V^{(k_\rho)} \left\{ \begin{matrix} e_K \\ \vdots \\ e_K \end{matrix} \middle| \, \mathsf{v} \right\} = \mathsf{v}, \tag{5.1.6}$$

for all $\mathsf{v} \in V$, where e_K is the identity of $\Bbbk^{(m_k, n_k)}$. In the case of an (ordinary) 1-place (left) action (as an external binary operation) $\rho_1: K \times V \to V$, its consistency with the polyadic field multiplication $\mu_k^{(n_k)}$ under composition of the binary operations $\rho_1\{\lambda | a\}$ gives a product of the same arity

$$n_\rho = n_k, \tag{5.1.7}$$

that is (a polyadic analog of $\lambda(\mu \mathsf{v}) = (\lambda\mu)\mathsf{v}$, $\mathsf{v} \in V$, $\lambda, \mu \in K$)

$$\begin{aligned} &\rho_1 \left\{ \lambda_1 \middle| \rho_1 \left\{ \lambda_2 \middle| \ldots \middle| \rho_1 \left\{ \lambda_{n_k} \middle| \mathsf{v} \right\} \right\} \ldots \right\} = \rho_1 \left\{ \mu_k^{(n_k)}[\lambda_1, \lambda_2, \ldots \lambda_{n_k}] \middle| \mathsf{v} \right\}, \\ &\lambda_1, \ldots, \lambda_n \in K, \, \mathsf{v} \in V. \end{aligned} \tag{5.1.8}$$

In the general case of k_ρ-place actions, the multiplication in the n_ρ-ary semigroup $\mathcal{S}_\rho$ can be defined by the *arity changing formula* (Duplij 2012) (schematically)

$$\rho_V^{(k_\rho)}\left\{\begin{matrix}\lambda_1\\ \vdots\\ \lambda_{k_\rho}\end{matrix}\right| \ldots \mid \rho_V^{(k_\rho)}\left\{\begin{matrix}\lambda_{k_\rho(n_\rho-1)+1}\\ \vdots\\ \lambda_{k_\rho n_\rho}\end{matrix}\right| \left.\mathrm{v}\right\}\ldots\Bigg\}$$

$$= \rho_{k_\rho}\left\{\left.\begin{matrix}\left.\begin{matrix}\mu_k^{(n_k)}\left[\lambda_1, \ldots\lambda_{n_k}\right],\\ \vdots\\ \mu_k^{(n_k)}\left[\lambda_{n_k(\ell_\mu-1)}, \ldots\lambda_{n_k\ell_\mu}\right]\end{matrix}\right\}\ell_\mu\\ \left.\begin{matrix}\lambda_{n_k\ell_\mu+1},\\ \vdots\\ \lambda_{n_k\ell_\mu+\ell_{\mathrm{id}}}\end{matrix}\right\}\ell_{\mathrm{id}}\end{matrix}\right| \mathrm{v}\right\}, \tag{5.1.9}$$

where ℓ_μ and ℓ_{id} are both integers. The associativity of (5.1.9) in each concrete case can be achieved by applying the *associativity quiver* concept from Duplij (2012).

Definition 5.1.1. The ℓ-*shape* is a pair $(\ell_\mu, \ell_{\mathrm{id}})$, where ℓ_μ is the number of multiplications and ℓ_{id} is the number of *intact elements* in the composition of operations.

It follows from (5.1.9),

Proposition 5.1.2. *The arities of the polyadic field $\Bbbk^{(m_k, n_k)}$, the arity n_ρ of the multiaction semigroup $\mathcal{S}_\rho$ and the ℓ-shape of the composition satisfy*

$$k_\rho n_\rho = n_k \ell_\mu + \ell_{\mathrm{id}}, \tag{5.1.10}$$

$$k_\rho = \ell_\mu + \ell_{\mathrm{id}}. \tag{5.1.11}$$

We can exclude ℓ_μ or ℓ_{id} and obtain

$$n_\rho = n_k - \frac{n_k - 1}{k_\rho}\ell_{\mathrm{id}}, \quad n_\rho = \frac{n_k - 1}{k_\rho}\ell_\mu + 1, \tag{5.1.12}$$

respectively, where $\frac{n_k-1}{k_\rho}\ell_{\mathrm{id}} \geqslant 1$ and $\frac{n_k-1}{k_\rho}\ell_\mu \geqslant 1$ are integers. The following inequalities hold

$$1 \leqslant \ell_\mu \leqslant k_\rho, \quad 0 \leqslant \ell_{\mathrm{id}} \leqslant k_\rho - 1, \quad \ell_\mu \leqslant k_\rho \leqslant (n_k - 1)\ell_\mu, \quad 2 \leqslant n_\rho \leqslant n_k. \tag{5.1.13}$$

Remark **5.1.3.** The formulas (5.1.12) coincide with the *arity changing formulas* for heteromorphisms (Duplij 2012) applied to (5.1.9).

It follows from (5.1.10), that the ℓ-shape is determined by the arities and number of places k_ρ by

$$\ell_\mu = \frac{k_\rho(n_\rho - 1)}{n_k - 1}, \qquad \ell_{\mathrm{id}} = \frac{k_\rho(n_k - n_\rho)}{n_k - 1}. \tag{5.1.14}$$

Because we have two polyadic 'additions' $\nu_V^{(m_V)}$ and $\nu_k^{(m_k)}$, we need to consider, how the multiaction $\rho_V^{(k_\rho)}$ 'distributes' between each of them. First, consider distributivity of the multiaction $\rho_V^{(k_\rho)}$ with respect to 'vector addition' $\nu_V^{(m_V)}$ (a polyadic analog of the binary $\lambda(\mathsf{v} + \mathsf{u}) = \lambda\mathsf{v} + \lambda\mathsf{u}$, $\mathsf{v}, \mathsf{u} \in V$, $\lambda, \mu \in K$)

$$\rho_V^{(k_\rho)}\left\{\begin{matrix}\lambda_1\\ \vdots\\ \lambda_{k_\rho}\end{matrix}\middle| \nu_V^{(m_V)}[\mathsf{v}_1, \ldots, \mathsf{v}_{m_V}]\right\}$$
$$= \nu_V^{(m_V)}\left[\rho_V^{(k_\rho)}\left\{\begin{matrix}\lambda_1\\ \vdots\\ \lambda_{k_\rho}\end{matrix}\middle| \mathsf{v}_1\right\}, \ldots, \rho_V^{(k_\rho)}\left\{\begin{matrix}\lambda_1\\ \vdots\\ \lambda_{k_\rho}\end{matrix}\middle| \mathsf{v}_{m_V}\right\}\right]. \tag{5.1.15}$$

Observe that here, in distinction to (5.1.9), there is no connection between the arities m_V and k_ρ.

Secondly, the distributivity of the multiaction $\rho_V^{(k_\rho)}$ ('multiplication by scalar') with respect to the 'field addition' (a polyadic analog of $\lambda\mathsf{v} + \mu\mathsf{v} = (\lambda + \mu)\mathsf{v}$, $\mathsf{v} \in A$, $\lambda, \mu \in K$) has a form similar to (5.1.9) (which can be obtained from the arity changing formula, Duplij 2012)

$$\nu_V^{(m_V)}\left[\rho_V^{(k_\rho)}\left\{\begin{matrix}\lambda_1\\ \vdots\\ \lambda_{k_\rho}\end{matrix}\middle| \mathrm{v}\right\}, \ldots, \rho_V^{(k_\rho)}\left\{\begin{matrix}\lambda_{k_\rho(m_V-1)+1}\\ \vdots\\ \lambda_{k_\rho m_V}\end{matrix}\middle| \mathrm{v}\right\}\right]$$
$$= \rho_V^{(k_\rho)}\left\{\begin{matrix}\left.\begin{matrix}\nu_k^{(m_k)}\left[\lambda_1, \ldots\lambda_{m_k}\right],\\ \vdots\\ \nu_k^{(m_k)}\left[\lambda_{m_k(\ell'_\mu-1)}, \ldots\lambda_{m_k\ell'_\mu}\right]\end{matrix}\right\}\ell'_\mu\\ \left.\begin{matrix}\lambda_{m_k\ell'_\mu+1},\\ \vdots\\ \lambda_{m_k\ell'_\mu+\ell'_{\mathrm{id}}}\end{matrix}\right\}\ell'_{\mathrm{id}}\end{matrix}\middle| \mathrm{v}\right\}, \tag{5.1.16}$$

where ℓ'_ρ and ℓ'_{id} are the numbers of multiplications and *intact elements* in the resulting multiaction, respectively. Here the arities are not independent as in (5.1.15), and so we have

Proposition 5.1.4. *The arities of the polyadic field* $\Bbbk^{(m_k, n_k)}$*, the arity* m_V *of the vector addition* $\nu_V^{(m_V)}$ *and the* ℓ*-shape of the distributivity satisfy*

$$k_\rho m_V = m_k\ell'_\mu + \ell'_{\mathrm{id}}, \tag{5.1.17}$$

$$k_\rho = \ell'_\mu + \ell'_{\text{id}}. \tag{5.1.18}$$

It follows from (5.1.17)–(5.1.18)

$$m_V = m_k - \frac{m_k - 1}{k_\rho}\ell'_{\text{id}}, \qquad m_V = \frac{m_k - 1}{k_\rho}\ell'_\mu + 1. \tag{5.1.19}$$

Here $\frac{m_k - 1}{k_\rho}\ell'_{\text{id}} \geqslant 1$ and $\frac{m_k - 1}{k_\rho}\ell'_\mu \geqslant 1$ are integers, and we have the inequalities

$$1 \leqslant \ell'_\mu \leqslant k_\rho, \quad 0 \leqslant \ell'_{\text{id}} \leqslant k_\rho - 1, \quad \ell'_\mu \leqslant k_\rho \leqslant (m_k - 1)\ell'_\mu, \quad 2 \leqslant m_V \leqslant m_k. \tag{5.1.20}$$

Now, the ℓ-shape of the distributivity is fully determined from the arities and number of places k_ρ by the arity shape formulas

$$\ell'_\rho = \frac{k_\rho(m_V - 1)}{m_k - 1}, \qquad \ell'_{\text{id}} = \frac{k_\rho(m_k - m_V)}{m_k - 1}. \tag{5.1.21}$$

It follows from (5.1.20) that:

Corollary 5.1.5. The arity m_V of the vector addition is less than or equal to the arity m_k of the field addition.

Definition 5.1.6. A polyadic ($\Bbbk$)-vector ('linear') space over a polyadic field is the 2-set 4-operation algebraic structure

$$\mathcal{V}^{(m_k,\, n_k; m_V; k_\rho)} = \left\langle K;\, V \middle| \nu_k^{(m_k)},\, \mu_k^{(n_k)};\, \nu_V^{(m_V)} \middle| \rho_V^{(k_\rho)} \right\rangle, \tag{5.1.22}$$

such that the following axioms hold:

(1) $\left\langle K \middle| \nu_k^{(m_k)},\, \mu_k^{(n_k)} \right\rangle$ is a polyadic (m_k, n_k)-field $\Bbbk^{(m_k,\, n_k)}$;

(2) $\langle V | \nu_V^{(m_V)} \rangle$ is a commutative m_V-ary group;

(3) $\left\langle \rho_V^{(k_\rho)} \middle| \text{composition} \right\rangle$ is an n_ρ-ary semigroup $\mathcal{S}_\rho$;

(4) Distributivity of the multiaction $\rho_V^{(k_\rho)}$ with respect to the 'vector addition' $\nu_V^{(m_V)}$ (5.1.15);

(5) Distributivity of $\rho_V^{(k_\rho)}$ with respect to the 'scalar addition' $\nu_k^{(m_k)}$ (5.1.16);

(6) Compatibility of $\rho_V^{(k_\rho)}$ with the 'scalar multiplication' $\mu_k^{(n_k)}$ (5.1.9);

(7) Normalization of the multiaction $\rho_V^{(k_\rho)}$ (5.1.6).

All of the arities in (5.1.22) are independent and can be chosen arbitrarily, but they fix the ℓ-shape of the multiaction composition (5.1.9) and the distributivity (5.1.16) by (5.1.14) and (5.1.21), respectively. Note that the main distinction from the binary case is the possibility for the arity n_ρ of the multiaction semigroup $\mathcal{S}_\rho$ to be arbitrary.

Definition 5.1.7. A polyadic $\Bbbk$-vector subspace is

$$\mathcal{V}_{\text{sub}}^{(m_k,n_k;m_V;k_\rho)} = \left\langle K;\ V_{\text{sub}} \middle| \nu_k^{(m_k)},\ \mu_k^{(n_k)};\ \nu_V^{(m_V)} \middle| \rho_V^{(k_\rho)} \right\rangle, \tag{5.1.23}$$

where the subset $V_{\text{sub}} \subset V$ is closed under all operations $\nu_k^{(m_k)}$, $\mu_k^{(n_k)}$, $\nu_V^{(m_V)}$, $\rho_V^{(k_\rho)}$ and the axioms (1)–(7).

Let us consider a subset $\mathsf{S} = \{\mathsf{v}_1, \dots, \mathsf{v}_{d_V}\} \subseteq V$ (of d_V 'vectors'), then a *polyadic span* of S is (a 'linear combination')

$$\text{Span}_{\text{pol}}^{\lambda}(\mathsf{v}_1, \dots, \mathsf{v}_{d_V}) = \{w\}, \tag{5.1.24}$$

$$w = \left(\nu_V^{(m_V)}\right)^{\circ \ell_v} \left[\rho_V^{(k_\rho)} \left\{ \begin{matrix} \lambda_1 \\ \vdots \\ \lambda_{k_\rho} \end{matrix} \middle| \mathsf{v}_1 \right\}, \dots, \rho_V^{(k_\rho)} \left\{ \begin{matrix} \lambda_{(d_V-1)k_\rho} \\ \vdots \\ \lambda_{d_V k_\rho} \end{matrix} \middle| \mathsf{v}_s \right\} \right], \tag{5.1.25}$$

where $(d_V \cdot k_\rho)$ 'scalars' play the role of coefficients (or coordinates as columns consisting of k_ρ elements from the polyadic field $\Bbbk^{(m_k,\, n_k)}$), and the number of 'vectors' s is connected with the 'number of m_V-ary additions' ℓ_v by

$$d_V = \ell_v(m_V - 1) + 1, \tag{5.1.26}$$

while $\text{Span}_{\text{pol}}^{\lambda}\ \mathsf{S}$ is the set of all 'vectors' of this form (5.1.24) (we consider here finite 'sums' only).

Definition 5.1.8. A polyadic span $\mathsf{S} = \{\mathsf{v}_1, \dots, \mathsf{v}_{d_V}\} \subseteq V$ is *nontrivial*, if at least one multiaction $\rho_V^{(k_\rho)}$ in (5.1.24) is nonzero.

Since polyadic fields and groups do not contain zeroes, we need to redefine the basic notions of equivalences. Let us take two different spans of the set S.

Definition 5.1.9. A set $\{\mathsf{v}_1, \dots, \mathsf{v}_{d_V}\}$ is called 'linear' *polyadic independent*, if from the equality of nontrivial spans, as $\text{Span}_{\text{pol}}^{\lambda}(\mathsf{v}_1, \dots, \mathsf{v}_{d_V}) = \text{Span}_{\text{pol}}^{\lambda'}(\mathsf{v}_1, \dots, \mathsf{v}_{d_V})$, it follows that all $\lambda_i = \lambda_i'$, $i = 1, \dots, d_V k_\rho$.

Definition 5.1.10. A set $\{\mathsf{v}_1, \dots, \mathsf{v}_{d_V}\}$ is called a *polyadic basis* of a polyadic vector space $\mathcal{V}^{(m_k,\, n_k;m_V;k_\rho)}$, if it spans the whole space $\text{Span}_{\text{pol}}^{\lambda}(\mathsf{v}_1, \dots, \mathsf{v}_{d_V}) = V$.

In other words, any element of V can be uniquely presented in the form of the polyadic 'linear combination' (5.1.24). If a polyadic vector space $\mathcal{V}^{(m_k,\, n_k;m_V;k_\rho)}$ has a finite basis $\{\mathsf{v}_1, \dots, \mathsf{v}_{d_V}\}$, then any other basis $\{\mathsf{v}_1', \dots, \mathsf{v}_{d_V}'\}$ has the same number of elements.

Definition 5.1.11. The number of elements in the polyadic basis $\{\mathsf{v}_1, \dots, \mathsf{v}_{d_V}\}$ is called the *polyadic dimension* of $\mathcal{V}^{(m_k,\, n_k; m_V; k_\rho)}$.

Remark **5.1.12.** The so-called 3-vector space introduced and studied in Duplij and Werner (2021), corresponds to $\mathcal{V}_{m_k=3, n_k=2, m_V=3, k_\rho=1}$.

5.1.2 One-set polyadic vector space

A particular polyadic vector space is important: consider $V = K$, $\nu_V^{(m_V)} = \nu_k^{(m_k)}$ and $m_V = m_k$, which gives the following one-set 'linear' algebraic structure (we call it a *one-set polyadic vector space*)

$$\mathcal{K}^{(m_k,\, n_k; k_\rho)} = \left\langle K \middle| \nu_k^{(m_k)},\, \mu_k^{(n_k)} \middle| \rho_k^{(k_\rho)} \right\rangle, \tag{5.1.27}$$

where now the multiaction

$$\rho_k^{(k_\rho)} \left\{ \begin{matrix} \lambda_1 \\ \vdots \\ \lambda_{k_\rho} \end{matrix} \middle| \, \lambda \right\}, \; \lambda,\, \lambda_i \in K, \tag{5.1.28}$$

acts on K itself (in some special way), and therefore can be called a *regular multiaction*. In the binary case $n_k = m_k = 2$, the only possibility for the regular action is the multiplication (by 'scalars') in the field $\rho_k^{\mathrm{reg}(1)}\{\lambda_1 \mid \lambda\} = \mu_k^{(2)}[\lambda_1 \lambda] (\equiv \lambda_1 \lambda)$, which obviously satisfies the axioms (4)–(7) of a vector space in definition 5.1.6. In this way we arrive at the definition of the binary field $\Bbbk \equiv \Bbbk^{(2,2)} = \left\langle K \middle| \nu_k^{(2)},\, \mu_k^{(2)} \right\rangle$, and so a one-set binary vector space coincides with the underlying field $\mathcal{K}^{(m_k=2,\, n_k=2,\, k_\rho=1)} = \Bbbk$, or as it is said 'a field is a (one-dimensional) vector space over itself'.

Remark **5.1.13.** In the polyadic case, the regular multiaction $\rho_k^{(k_\rho)}$ can be chosen, as any (additional to $\nu_k^{(m_k)}$, $\mu_k^{(n_k)}$) function satisfying axioms (4)–(7) of a polyadic vector space and the number of places k_ρ and the arity of the semigroup of multiactions $\mathcal{S}_\rho$ can be arbitrary, in general. Also, $\rho_k^{(k_\rho)}$ can be taken as a some nontrivial combination of $\nu_k^{(m_k)}$, $\mu_k^{(n_k)}$ satisfying axioms (4)–(7) (which admits a nontrivial 'multiplication by scalars').

In the simplest regular (similar to the binary) case,

$$\rho_k^{\mathrm{reg}(k_\rho)} \left\{ \begin{matrix} \lambda_1 \\ \vdots \\ \lambda_{k_\rho} \end{matrix} \middle| \, \lambda \right\} = \left(\mu_k^{(n_k)}\right)^{\circ \ell_k} \left[\lambda_1, \dots, \lambda_{k_\rho}, \lambda\right], \tag{5.1.29}$$

where ℓ_κ is the number of multiplications $\mu_k^{(n_k)}$, and the number of places k_ρ is now fixed by

$$k_\rho = \ell_\kappa(n_k - 1), \tag{5.1.30}$$

while λ in (5.1.29) can be on any place due the commutativity of the field multiplication $\mu_k^{(n_k)}$.

***Remark* 5.1.14.** In general, the one-set polyadic vector space need not coincide with the underlying polyadic field, $\mathcal{K}^{(m_k, n_k; k_\rho)} \neq \Bbbk^{(m_k, n_k)}$ (as opposed to the binary case), but can have a more complicated structure which is determined by an additional multiplace function, the multiaction $\rho_k^{(k_\rho)}$.

5.2 Mappings between polyadic vector spaces

Let us consider D_V different polyadic vector spaces over the same polyadic field $\Bbbk^{(m_k, n_k)}$, as

$$\mathcal{V}_i^{\left(m_k, n_k, m_V^{(i)}, k_\rho^{(i)}\right)} = \left\langle K; V_i \middle| \nu_k^{(m_k)}, \mu_k^{(n_k)}; \nu_{V,i}^{\left(m_V^{(i)}\right)} \middle| \rho_{V,i}^{\left(k_\rho^{(i)}\right)} \right\rangle, \qquad i = 1, \ldots, D_V < \infty. \tag{5.2.1}$$

Here we define a polyadic analog of a 'linear' mapping for polyadic vector spaces which 'commutes' with the 'vector addition' and the 'multiplication by scalar' (an analog of the additivity $\mathbf{F}(\mathsf{v} + \mathsf{u}) = \mathbf{F}(\mathsf{v}) + \mathbf{F}(\mathsf{u})$, and the homogeneity of degree one $\mathbf{F}(\lambda \mathsf{v}) = \lambda \mathbf{F}(\mathsf{v})$, $\mathsf{v}, \mathsf{u} \in V$, $\lambda \in K$).

Definition 5.2.1. A *1-place ('*$\Bbbk$*-linear') mapping* between two polyadic vector spaces $\mathcal{V}^{(m_k, n_k; m_V; k_\rho)} = \left\langle K; V \middle| \nu_k^{(m_k)}, \mu_k^{(n_k)}; \nu_V^{(m_V)} \middle| \rho_k^{(k_\rho)} \right\rangle$ and $\mathcal{V}'^{(m_k, n_k; m_V; k_\rho)} = \left\langle K; V' \middle| \nu_k^{(m_k)}, \mu_k^{(n_k)}; \nu_{V'}^{(m_V)} \middle| \rho_k'^{(k_\rho)} \right\rangle$ over the same polyadic field $\Bbbk^{(m_k, n_k)} = \left\langle K \middle| \nu_k^{(m_k)}, \mu_k^{(n_k)} \right\rangle$ is $\mathbf{F}_1 \colon V \to V'$, such that

$$\mathbf{F}_1\left(\nu^{(m_V)}[\mathsf{v}_1, \ldots, \mathsf{v}_{m_V}]\right) = \nu'^{(m_V)}[\mathbf{F}_1(\mathsf{v}_1), \ldots, \mathbf{F}_1(\mathsf{v}_{m_V})], \tag{5.2.2}$$

$$\mathbf{F}_1\left(\rho_V^{(k_\rho)}\left\{\begin{matrix}\lambda_1\\ \vdots\\ \lambda_{k_\rho}\end{matrix} \middle| \mathsf{v}\right\}\right) = \rho_V'^{(k_\rho)}\left\{\begin{matrix}\lambda_1\\ \vdots\\ \lambda_{k_\rho}\end{matrix} \middle| \mathbf{F}_1(\mathsf{v})\right\}, \tag{5.2.3}$$

where $\mathsf{v}_1, \ldots, \mathsf{v}_{m_V}, \mathsf{v} \in V$, $\lambda_1, \ldots, \lambda_{k_\rho} \in K$.

If z_V is a 'zero vector' in V and $\mathsf{z}_{V'}$ is a 'zero vector' in V' (see (5.1.2)), then it follows from (5.2.2)–(5.2.3), that $\mathbf{F}_1(\mathsf{z}_V) = \mathsf{z}_{V'}$.

The initial and final arities of $\nu_V^{(m_V)}$ ('vector addition') and the multiaction $\rho_k^{(k_\rho)}$ ('multiplication by scalar') coincide, because $\mathbf{F}_1$ is a 1-place mapping (a linear homomorphism). In Duplij (2012) *multiplace mappings* and corresponding *heteromorphisms* were introduced. The latter allow us to change arities ($m_V \to m'_V$, $k_\rho \to k'_\rho$), which is the main difference between binary and polyadic mappings.

Definition 5.2.2. A k_F*-place ('*$\Bbbk$*-linear') mapping* between two polyadic vector spaces $\mathcal{V}^{(m_k,\, n_k; m_V; k_\rho)} = \left\langle K;\, V \middle| \nu_k^{(m_k)},\, \mu_k^{(n_k)};\, \nu_V^{(m_V)} \middle| \rho_V^{(k_\rho)} \right\rangle$ and $\mathcal{V}'^{\left(m_k,\, n_k; m'_V; k_\rho\right)} = \left\langle K;\, V' \middle| \nu_k^{(m_k)},\, \mu_k^{(n_k)};\, \nu'^{\left(m'_V\right)}_V \middle| \rho'^{\left(k'_\rho\right)}_V \right\rangle$ over the same polyadic field $\Bbbk^{(m_k,\, n_k)} = \left\langle K \middle| \nu_k^{(m_k)},\, \mu_k^{(n_k)} \right\rangle$ is defined, if there exists $\mathbf{F}_{k_F}\colon V^{\times k_F} \to V'$, such that

$$\mathbf{F}_{k_F}\begin{pmatrix} \left.\begin{matrix} \nu_V^{(m_V)}[\mathsf{v}_1, \ldots, \mathsf{v}_{m_V}] \\ \vdots \\ \nu_V^{(m_V)}\left[\mathsf{v}_{m_V\left(\ell_\mu^k - 1\right)}, \ldots \mathsf{v}_{m_V \ell_\mu^k}\right] \end{matrix}\right\} \ell_\mu^k \\ \left.\begin{matrix} \mathsf{v}_{m_V \ell_\mu^k + 1}, \\ \vdots \\ \mathsf{v}_{m_V \ell_\mu^k + \ell_{\mathrm{id}}^k} \end{matrix}\right\} \ell_{\mathrm{id}}^k \end{pmatrix} \tag{5.2.4}$$
$$= \nu'^{\left(m'_V\right)}_V\left[\mathbf{F}_{k_F}\begin{pmatrix} \mathsf{v}_1 \\ \vdots \\ \mathsf{v}_{k_F} \end{pmatrix}, \ldots, \mathbf{F}_{k_F}\begin{pmatrix} \mathsf{v}_{k_F\left(m'_V - 1\right)} \\ \vdots \\ \mathsf{v}_{k_F m'_V} \end{pmatrix}\right],$$

$$\mathbf{F}_{k_F}\begin{pmatrix} \left.\begin{matrix} \rho_V^{(k_\rho)}\left\{\begin{matrix} \lambda_1 \\ \vdots \\ \lambda_{k_\rho} \end{matrix} \middle| \mathsf{v}_1 \right\} \\ \vdots \\ \rho_V^{(k_\rho)}\left\{\begin{matrix} \lambda_{k_\rho\left(\ell_\mu^f - 1\right)} \\ \vdots \\ \lambda_{k_\rho \ell_\mu^f} \end{matrix} \middle| \mathsf{v}_{\ell_\mu^f} \right\} \end{matrix}\right\} \ell_\mu^f \\ \left.\begin{matrix} \mathsf{v}_{\ell_\mu^f + 1} \\ \vdots \\ \mathsf{v}_{k_F} \end{matrix}\right\} \ell_{\mathrm{id}}^f \end{pmatrix} = \rho'^{\left(k'_\rho\right)}_V\left\{\begin{matrix} \lambda_1 \\ \vdots \\ \lambda_{k'_\rho} \end{matrix} \middle| \mathbf{F}_{k_F}\begin{pmatrix} \mathsf{v}_1 \\ \vdots \\ \mathsf{v}_{k_F} \end{pmatrix} \right\}, \tag{5.2.5}$$

where $\mathsf{v}_1, \ldots, \mathsf{v}_{m_V}, \mathsf{v} \in V, \lambda_1, \ldots, \lambda_{k_\rho} \in K$, and the four integers $\ell_\rho^k, \ell_{\mathrm{id}}^k, \ell_\rho^f, \ell_{\mathrm{id}}^f$ define the ℓ-shape of the mapping.

It follows from (5.2.4)–(5.2.5), that the arities satisfy

$$k_F m'_V = m_V \ell_\mu^k + \ell_{\mathrm{id}}^k, \quad k_F = \ell_\mu^k + \ell_{\mathrm{id}}^k, \quad k_F = \ell_\mu^f + \ell_{\mathrm{id}}^f, \quad k'_\mu = k_\rho \ell_\mu^f. \tag{5.2.6}$$

The following inequalities hold

$$1 \leqslant \ell_\mu^k \leqslant k_F,\ 0 \leqslant \ell_{id}^k \leqslant k_F - 1,\ \ell_\mu^k \leqslant k_F \leqslant (m_V - 1)\ell_\mu^k,\ 2 \leqslant m_V' \leqslant m_V,\ 2 \leqslant k_\rho \leqslant k_\rho'. \quad (5.2.7)$$

Thus, we arrive to

Theorem 5.2.3. The ℓ-shape of the k_F-place mapping between polyadic vector spaces is determined by

$$\ell_\mu^k = \frac{k_F(m_V - 1)}{m_V - 1},\quad \ell_{id}^k = \frac{k_F(m_V - m_V')}{m_V - 1},\quad \ell_\mu^f = \frac{k_\rho}{k_\rho'},\quad \ell_{id}^f = k_F - \frac{k_\rho}{k_\rho'}. \quad (5.2.8)$$

5.2.1 Polyadic functionals and dual polyadic vector spaces

An important particular case of the k_F-place mapping can be considered, where the final polyadic vector space coincides with the underlying field (an analog of a 'linear functional').

Definition 5.2.4. A 'linear'*polyadic functional* (or *polyadic dual vector*, *polyadic covector*) is a k_L-place mapping of a polyadic vector space $\mathcal{V}^{(m_k, n_k; m_V; k_\rho)} = \left\langle K;\ V \middle| \nu_k^{(m_k)}, \mu_k^{(n_k)};\ \nu_V^{(m_V)} \middle| \rho_V^{(k_\rho)} \right\rangle$ into its polyadic field $\Bbbk^{(m_k, n_k)} = \left\langle K \middle| \nu_k^{(m_k)}, \mu_k^{(n_k)} \right\rangle$, such that there exists $\boldsymbol{L}_{k_L}\colon V^{\times k_L} \to K$, and

$$\mathbf{L}_{k_L}\begin{pmatrix} \left.\begin{matrix} \nu_V^{(m_V)}[\mathsf{v}_1, \ldots, \mathsf{v}_{m_V}] \\ \vdots \\ \nu_V^{(m_V)}\left[\mathsf{v}_{m_V(\ell_\nu^k - 1)}, \ldots \mathsf{v}_{m_V \ell_\nu^k}\right] \end{matrix}\right\} \ell_\nu^k \\ \left.\begin{matrix} \mathsf{v}_{m_V \ell_\nu^k + 1}, \\ \vdots \\ \mathsf{v}_{n_k \ell_\nu^k + \ell_{id}^\nu} \end{matrix}\right\} \ell_{id}^\nu \end{pmatrix} = \nu_k^{(m_k)}\left[\mathbf{L}_{k_L}\begin{pmatrix} \mathsf{v}_1 \\ \vdots \\ \mathsf{v}_{k_L} \end{pmatrix}, \ldots, \mathbf{L}_{k_L}\begin{pmatrix} \mathsf{v}_{k_L(m_k - 1)} \\ \vdots \\ \mathsf{v}_{k_L m_k} \end{pmatrix}\right], \quad (5.2.9)$$

$$\mathbf{L}_{k_L}\begin{pmatrix} \left.\begin{matrix} \rho_V^{(k_\rho)}\left\{\begin{matrix} \lambda_1 \\ \vdots \\ \lambda_{k_\rho} \end{matrix}\middle| \mathsf{v}_1\right\} \\ \vdots \\ \rho_V^{(k_\rho)}\left\{\begin{matrix} \lambda_{k_\rho(\ell_\mu^L - 1)} \\ \vdots \\ \lambda_{k_\rho \ell_\mu^L} \end{matrix}\middle| \mathsf{v}_{\ell_\mu^L}\right\} \end{matrix}\right\} \ell_\mu^L \\ \left.\begin{matrix} \mathsf{v}_{\ell_\mu^L + 1} \\ \vdots \\ \mathsf{v}_{k_L} \end{matrix}\right\} \ell_{id}^L \end{pmatrix} = \mu_k^{(n_k)}\left[\lambda_1, \ldots, \lambda_{n_k - 1}, \mathbf{L}_{k_L}\begin{pmatrix} \mathsf{v}_1 \\ \vdots \\ \mathsf{v}_{k_L} \end{pmatrix}\right], \quad (5.2.10)$$

where $\mathsf{v}_1, \ldots, \mathsf{v}_{m_V}, \mathsf{v} \in V, \lambda_1, \ldots, \lambda_{n_k} \in K$, and the integers $\ell_\nu^k, \ell_{\rm id}^\nu, \ell_\mu^L, \ell_{\rm id}^L$ define the ℓ-shape of $\boldsymbol{L}_{k_L}$.

It follows from (5.2.4)–(5.2.5), that the arities satisfy

$$k_L m_k = m_V \ell_\nu^k + \ell_{\rm id}^\nu, \quad k_L = \ell_\nu^k + \ell_{\rm id}^\nu, \quad k_L = \ell_\mu^h + \ell_{\rm id}^h, \quad n_k - 1 = k_\rho \ell_\mu^h, \tag{5.2.11}$$

and for them

$$\begin{gathered} 1 \leqslant \ell_\nu^k \leqslant k_L, \; 0 \leqslant \ell_{\rm id}^\nu \leqslant k_L - 1, \; \ell_\nu^k \leqslant k_L \leqslant (m_V - 1)\ell_\nu^k, \\ 2 \leqslant m_k \leqslant m_V, \; 2 \leqslant k_\rho \leqslant n_k - 1. \end{gathered} \tag{5.2.12}$$

Thus, the ℓ-shape of the polyadic functional is determined by

$$\begin{gathered} \ell_\nu^k = \frac{k_L(m_k - 1)}{m_V - 1}, \; \ell_{\rm id}^\nu = \frac{k_L(m_V - m_k)}{m_V - 1}, \; \ell_\mu^h = \frac{k_\rho}{n_k - 1}, \\ \ell_{\rm id}^h = k_L - \frac{k_\rho}{n_k - 1}. \end{gathered} \tag{5.2.13}$$

In the binary case, because the dual vectors (linear functionals) take their values in the underlying field, new operations between them, such that the dual vector 'addition' ($+^*$) and the 'multiplication by a scalar' ($\bullet^*$) can be naturally introduced by $(L^{(1)} +^* L^{(2)})(\mathsf{v}) = L^{(1)}(\mathsf{v}) + L^{(2)}(\mathsf{v})$, $(\lambda \bullet^* L)(\mathsf{v}) = \lambda \bullet L(\mathsf{v})$, which leads to another vector space structure, called a dual vector space. Note that operations $+^*$ and $+$, $\bullet^*$ and $\bullet$ are different, because $+$ and $\bullet$ are the operations in the underlying field $\Bbbk$. In the polyadic case, we have more complicated arity changing formulas, and here we consider finite-dimensional spaces only. The arities of operations between dual vectors can be different from ones in the underlying polyadic field $\Bbbk^{(m_k, n_k)}$, in general. In this way, we arrive at the following

Definition 5.2.5. A *polyadic dual vector space* over a polyadic field $\Bbbk^{(m_k, n_k)}$ is

$$\mathcal{V}_*^{(m_k, n_k; m_V^*; k_\rho^*)} = \left\langle K; \left\{\mathbf{L}_{k_L}^{(i)}\right\} \middle| \nu_k^{(m_k)}, \mu_k^{(n_k)}; \boldsymbol{\nu}_{m_L}^* \middle| \rho_{k_L}^* \right\rangle, \tag{5.2.14}$$

and the axioms are:

(1) $\left\langle K \middle| \nu_k^{(m_k)}, \mu_k^{(n_k)} \right\rangle$ is a polyadic (m_k, n_k)-field $\Bbbk^{(m_k, n_k)}$;

(2) $\left\langle \left\{\mathbf{L}_{k_L}^{(i)}\right\} \middle| \boldsymbol{\nu}_{m_L}^*, i = 1, \ldots, D_L \right\rangle$ is a commutative m_L-ary group (which is finite, if $D_L < \infty$);

(3) The 'dual vector addition' $\boldsymbol{\nu}_{m_L}^*$ is compatible with the polyadic field addition $\nu_k^{(m_k)}$ by

$$\boldsymbol{\nu}_{m_L}^*\left[\mathbf{L}_{k_L}^{(1)}, \ldots, \mathbf{L}_{k_L}^{(m_L)}\right](\mathbf{a}^{(k_L)}) = \nu_k^{(m_k)}\left[\mathbf{L}_{k_L}^{(1)}(\mathbf{a}^{(k_L)}), \ldots, \mathbf{L}_{k_L}^{(m_k)}(\mathbf{v}^{(k_L)})\right], \tag{5.2.15}$$

where $\mathbf{v}^{(k_L)} = \begin{pmatrix} \mathsf{v}_1 \\ \vdots \\ \mathsf{v}_{k_L} \end{pmatrix}$, $\mathsf{v}_1, \ldots, \mathsf{v}_{k_L} \in V$, and it follows that

$$m_L = m_k. \tag{5.2.16}$$

(4) The compatibility of $\rho^*_{k_L}$ with the 'multiplication by a scalar' in the underlying polyadic field

$$\rho^*_{k_L}\left\{\begin{matrix} \lambda_1 \\ \vdots \\ \lambda_{k_L} \end{matrix} \middle| \, \mathbf{L}_{k_L} \right\}(\mathbf{v}^{(k_L)}) = \mu_k^{(n_k)}[\lambda_1, \ldots, \lambda_{n_k-1}, \mathbf{L}_{k_L}(\mathbf{v}^{(k_L)})], \tag{5.2.17}$$

and then

$$k_L = n_k - 1 \tag{5.2.18}$$

(5) $\left\langle \left\{\rho^*_{k_L}\right\} \middle| \text{composition} \right\rangle$ is an n_L-ary semigroup $\mathcal{S}_L$ (similar to (5.1.9))

$$\overbrace{\rho^*_{k_L}\left\{\begin{matrix} \lambda_1 \\ \vdots \\ \lambda_{k_L} \end{matrix} \middle| \cdots \middle| \rho^*_{k_L}\left\{\begin{matrix} \lambda_{k_L(n_L-1)} \\ \vdots \\ \lambda_{k_L n_L} \end{matrix} \middle| \, \mathbf{L}_{k_L} \right\} \ldots \right\}}^{n_L}(\mathbf{v}^{(k_L)}) \tag{5.2.19}$$

$$= \rho^*_{k_L}\left\{\begin{matrix} \left.\begin{matrix} \mu_k^{(n_k)}\left[\lambda_1, \ldots \lambda_{n_k}\right], \\ \vdots \\ \mu_k^{(n_k)}\left[\lambda_{n_k(\ell^L_\mu - 1)}, \ldots \lambda_{n_k \ell^L_\mu}\right] \end{matrix}\right\} \ell^L_\mu \\ \left.\begin{matrix} \lambda_{n_k \ell^L_\mu + 1}, \\ \vdots \\ \lambda_{n_k \ell^L_\mu + \ell^L_{\mathrm{id}}} \end{matrix}\right\} \ell^L_{\mathrm{id}} \end{matrix} \middle| \, \mathbf{L}_{k_L} \right\}(\mathbf{v}^{(k_L)}), \tag{5.2.20}$$

where the ℓ-shape is determined by the system

$$k_L n_L = n_k \ell^L_\mu + \ell^L_{\mathrm{id}}, \qquad k_L = \ell^L_\mu + \ell^L_{\mathrm{id}}. \tag{5.2.21}$$

Using (5.2.18) and (5.2.21), we obtain the ℓ-shape as

$$\ell^L_\mu = n_L - 1, \quad \ell^L_{\mathrm{id}} = n_k - n_L. \tag{5.2.22}$$

Corollary 5.2.6. The arity n_L of the semigroup $\mathcal{S}_L$ is less than or equal to the arity n_k of the underlying polyadic field $n_L \leqslant n_k$.

5.2.2 Polyadic direct sum and tensor product

The Cartesian product of D_V polyadic vector spaces $\times\Pi_{i=1}^{m_V}\mathcal{V}_i^{\left(m_k,n_k,m_V^{(i)},k_\rho^{(i)}\right)}$ (sometimes we use the concise notation $\times\Pi\mathcal{V}_i$), $i = 1, \ldots, D_V$ is given by the D_V-ples (an analog of the Cartesian pair (v, u), $\mathsf{v} \in V^{(1)}$, $\mathsf{u} \in V^{(2)}$)

$$\begin{pmatrix} \mathsf{v}^{(1)} \\ \vdots \\ \mathsf{v}^{(D_V)} \end{pmatrix} \equiv \left(\mathsf{v}^{(D_V)}\right) \in V^{\times D_V}. \tag{5.2.23}$$

We introduce polyadic generalizations of the direct sum and tensor product of vector spaces by considering 'linear' operations on the D_V-ples (5.2.23).

In the first case, to endow $\times\Pi\mathcal{V}_i$ with the structure of a vector space we need to define a new operation between the D_V-ples (5.2.23) (this is similar to vector addition, but between elements from different spaces) and a rule, specifying how they are 'multiplied by scalars' (analogs of $(\mathsf{v}_1, \mathsf{v}_2) + (\mathsf{u}_1, \mathsf{u}_2) = (\mathsf{v}_1 + \mathsf{u}_1, \mathsf{v}_2 + \mathsf{u}_2)$ and $\lambda(\mathsf{v}_1, \mathsf{v}_2) = (\lambda\mathsf{v}_1, \lambda\mathsf{v}_2)$). In the binary case, a formal summation is used, but it can be different from the addition in the initial vector spaces. Therefore, we can define on the set of the D_V-ples (5.2.23) new operations χ_{m_V} ('addition of vectors from different spaces') and 'multiplication by a scalar' τ_{k_ρ}, which does not need to coincide with the corresponding operations $\nu_{V,i}^{\left(m_V^{(i)}\right)}$ and $\rho_{V,i}^{\left(k_\rho^{(i)}\right)}$ of the initial polyadic vector spaces $\mathcal{V}_i^{\left(m_k,n_k;m_V^{(i)};k_\rho^{(i)}\right)}$.

If all D_V-ples (5.2.23) are of fixed length, then we can define their 'addition' χ_{m_V} in the standard way, if all the arities $m_V^{(i)}$ coincide and equal the arity of the resulting vector space

$$m_V = m_V^{(1)} = \ldots = m_V^{(D_V)}, \tag{5.2.24}$$

while the operations ('vector additions') themselves $\nu_{V,i}^{(m_V)}$ between vectors in different spaces can be still different. Thus, a new commutative m_V-ary operation ('addition') χ_{m_V} of the D_V-ples of the same length is defined by

$$\chi_{m_V}\left[\begin{pmatrix} \mathsf{v}_1^{(1)} \\ \vdots \\ \mathsf{v}_1^{(D_V)} \end{pmatrix}, \ldots, \begin{pmatrix} \mathsf{v}_{m_V}^{(1)} \\ \vdots \\ \mathsf{v}_{m_V}^{(D_V)} \end{pmatrix}\right] = \begin{pmatrix} \nu_{V,1}^{(m_V)}\left[\mathsf{v}_1^{(1)}, \ldots, \mathsf{v}_{m_V}^{(1)}\right] \\ \vdots \\ \nu_{V,D_V}^{(m_V)}\left[\mathsf{v}_1^{(D_V)}, \ldots, \mathsf{v}_{m_V}^{(D_V)}\right] \end{pmatrix}, \tag{5.2.25}$$

where $D_V \neq m_V$, in general. However, it is also possible to add D_V-ples of different length such that the operation (5.2.25) is compatible with all arities $m_V^{(i)}$, $i = 1, \ldots, m_V$. For instance, if $m_V = 3$, $m_V^{(1)} = m_V^{(2)} = 3$, $m_V^{(3)} = 2$, then

$$\chi_3\left[\begin{pmatrix} \mathsf{v}_1^{(1)} \\ \mathsf{v}_1^{(2)} \\ \mathsf{v}_1^{(3)} \end{pmatrix}, \begin{pmatrix} \mathsf{v}_2^{(1)} \\ \mathsf{v}_2^{(2)} \\ \mathsf{v}_2^{(3)} \end{pmatrix}, \begin{pmatrix} \mathsf{v}_3^{(1)} \\ \mathsf{v}_3^{(2)} \end{pmatrix}\right] = \begin{pmatrix} \nu_{V,1}^{(3)}[\mathsf{v}_1^{(1)}, \mathsf{v}_2^{(1)}, \mathsf{v}_3^{(1)}] \\ \nu_{V,2}^{(3)}[\mathsf{v}_1^{(2)}, \mathsf{v}_2^{(2)}, \mathsf{v}_3^{(2)}] \\ \nu_{V,3}^{(2)}[\mathsf{v}_1^{(3)}, \mathsf{v}_2^{(3)}] \end{pmatrix}. \tag{5.2.26}$$

Assertion 5.2.7. In the polyadic case, a direct sum of polyadic vector spaces having *different arities* of 'vector addition' $m_V^{(i)}$ can be defined.

Let us introduce the multiaction $\boldsymbol{\tau}_{k_\rho}$ ('multiplication by a scalar') acting on D_V-ple $(\mathsf{v}^{(m_V)})$, then

$$\boldsymbol{\tau}_{k_\rho}\left\{\begin{matrix}\lambda_1\\ \vdots\\ \lambda_{k_\rho}\end{matrix}\,\middle|\,\begin{pmatrix}\mathsf{v}^{(1)}\\ \vdots\\ \mathsf{v}^{(D_V)}\end{pmatrix}\right\} = \begin{pmatrix}\rho_1^{\left(k_\rho^{(1)}\right)}\left\{\begin{matrix}\lambda_1\\ \vdots\\ \lambda_{k_\rho^{(1)}}\end{matrix}\,\middle|\,\mathsf{v}^{(1)}\right\}\\ \vdots\\ \rho_{m_V}^{\left(k_\rho^{(D_V)}\right)}\left\{\begin{matrix}\lambda_{k_\rho^{(1)}+\cdots+k_\rho^{(D_V-1)}+1}\\ \vdots\\ \lambda_{k_\rho^{(1)}+\cdots+k_\rho^{(D_V)}}\end{matrix}\,\middle|\,\mathsf{v}^{(D_V)}\right\}\end{pmatrix}, \tag{5.2.27}$$

where

$$k_\rho^{(1)} + \cdots + k_\rho^{(D_V)} = k_\rho. \tag{5.2.28}$$

Definition 5.2.8. A *polyadic direct sum* of m_V polyadic vector spaces is their Cartesian product equipped with the m_V-ary addition χ_{m_V} and the k_ρ-place multiaction $\boldsymbol{\tau}_{k_\rho}$, satisfying (5.2.25) and (5.2.27) respectively

$$\oplus\Pi_{i=1}^{D_V}\mathcal{V}_i^{\left(m_k,n_k,m_V^{(i)},k_\rho^{(i)}\right)} = \left\{\times\Pi_{i=1}^{D_V}\mathcal{V}_i^{\left(m_k,n_k,m_V^{(i)},k_\rho^{(i)}\right)}\middle|\chi_{m_V},\,\boldsymbol{\tau}_{k_\rho}\right\}. \tag{5.2.29}$$

Let us consider another way to define a vector space structure on the D_V-ples from the Cartesian product $\times\Pi\mathcal{V}_i$. Remember that in the binary case, the concept of bilinearity is used, which means 'distributivity' and 'multiplicativity by scalars' on *each place* separately in the Cartesian pair $(\mathsf{v}_1, \mathsf{v}_2) \in V_1 \times V_2$ (*as opposed to* the direct sum, where these relations hold on all places simultaneously, see (5.2.25) and (5.2.27)) such that

$$(\mathsf{v}_1 + \mathsf{u}_1, \mathsf{v}_2) = (\mathsf{v}_1, \mathsf{v}_2) + (\mathsf{u}_1, \mathsf{v}_2), \qquad (\mathsf{v}_1, \mathsf{v}_2 + \mathsf{u}_2) = (\mathsf{v}_1, \mathsf{v}_2) + (\mathsf{v}_1, \mathsf{u}_2), \tag{5.2.30}$$

$$\lambda(\mathsf{v}_1, \mathsf{v}_2) = (\lambda\mathsf{v}_1, \mathsf{v}_2) = (\mathsf{v}_1, \lambda\mathsf{v}_2), \tag{5.2.31}$$

respectively. If we denote the ideal corresponding to the relations (5.2.30)–(5.2.31) by $\boldsymbol{B}_2$, then the binary tensor product of the vector spaces can be defined from their Cartesian product by factoring out this ideal, as $\mathcal{V}_1 \otimes \mathcal{V}_2 \overset{\text{def}}{=} \mathcal{V}_1 \times \mathcal{V}_2/\boldsymbol{B}_2$. Note first, that the additions and multiplications by a scalar on both sides of (5.2.30)–(5.2.31) 'work' in different spaces, which sometimes can be concealed by adding the word 'formal' to them. Second, all these operations have the *same* arity (binary ones), which need not be the case when considering polyadic structures.

As in the case of the polyadic direct sum, we first define a new operation $\tilde{\chi}_{m_V}$ ('addition') of the D_V-ples of fixed length (different from χ_{m_V} in (5.2.25)), when all the arities $m_V^{(i)}$ coincide and equal to m_V (5.2.24). Then, a straightforward generalization of (5.2.30) can be defined for m_V-ples similar to polyadic distributivity (4.1.1)–(4.1.3), as in the following m_V relations

$$\begin{pmatrix} \nu_{V,1}^{(m_V)}\left[\mathsf{v}_1^{(1)}, \ldots, \mathsf{v}_{m_V}^{(1)}\right] \\ \mathsf{u}_2 \\ \vdots \\ \mathsf{u}_{D_V} \end{pmatrix} = \tilde{\chi}_{m_V}\left[\begin{pmatrix} \mathsf{v}_1^{(1)} \\ \mathsf{u}_2 \\ \vdots \\ \mathsf{u}_{D_V} \end{pmatrix}, \ldots, \begin{pmatrix} \mathsf{v}_{m_V}^{(1)} \\ \mathsf{u}_2 \\ \vdots \\ \mathsf{u}_{D_V} \end{pmatrix}\right], \tag{5.2.32}$$

$$\begin{pmatrix} \mathsf{u}_1 \\ \nu_{V,2}^{(m_V)}\left[\mathsf{v}_1^{(2)}, \ldots, \mathsf{v}_{m_V}^{(2)}\right] \\ \vdots \\ \mathsf{u}_{m_V} \end{pmatrix} = \tilde{\chi}_{m_V}\left[\begin{pmatrix} \mathsf{u}_1 \\ \mathsf{v}_1^{(2)} \\ \vdots \\ \mathsf{u}_{m_V} \end{pmatrix}, \ldots, \begin{pmatrix} \mathsf{u}_1 \\ \mathsf{v}_{m_V}^{(2)} \\ \vdots \\ \mathsf{u}_{m_V} \end{pmatrix}\right], \tag{5.2.33}$$

$\vdots$

$$\begin{pmatrix} \mathsf{u}_1 \\ \mathsf{u}_2 \\ \vdots \\ \nu_{V,m_V}^{(m_V)}\left[\mathsf{v}_1^{(D_V)}, \ldots, \mathsf{v}_{m_V}^{(D_V)}\right] \end{pmatrix} = \tilde{\chi}_{m_V}\left[\begin{pmatrix} \mathsf{u}_1 \\ \mathsf{u}_2 \\ \vdots \\ \mathsf{v}_1^{(D_V)} \end{pmatrix}, \ldots, \begin{pmatrix} \mathsf{u}_1 \\ \mathsf{u}_2 \\ \vdots \\ \mathsf{v}_{m_V}^{(D_V)} \end{pmatrix}\right]. \tag{5.2.34}$$

By analogy, if all $k_\rho^{(i)}$ are equal we can define a new multiaction $\tilde{\boldsymbol{\tau}}_{k_\rho}$ (different from $\boldsymbol{\tau}_{k_\rho}$ (5.2.27)) but with the same number of places

$$k_\rho = k_\rho^{(1)} = \ldots = k_\rho^{(D_V)} \tag{5.2.35}$$

as the D_V relations (an analog of (5.2.31))

$$\tilde{\boldsymbol{\tau}}_{k_\rho}\left\{\begin{matrix} \lambda_1 \\ \vdots \\ \lambda_{k_\rho} \end{matrix} \middle| \begin{pmatrix} \mathsf{v}^{(1)} \\ \vdots \\ \mathsf{v}^{(D_V)} \end{pmatrix}\right\} = \begin{pmatrix} \rho_1^{(k_\rho)}\left\{\begin{matrix} \lambda_1 \\ \vdots \\ \lambda_{k_\rho} \end{matrix} \middle| \mathsf{v}^{(1)}\right\} \\ \mathsf{v}^{(2)} \\ \vdots \\ \mathsf{v}^{(D_V)} \end{pmatrix} \tag{5.2.36}$$

$$= \begin{pmatrix} \mathsf{v}^{(1)} \\ \rho_2^{(k_\rho)}\left\{\begin{matrix} \lambda_1 \\ \vdots \\ \lambda_{k_\rho} \end{matrix} \middle| \mathsf{v}^{(2)}\right\} \\ \vdots \\ \mathsf{v}^{(D_V)} \end{pmatrix} \tag{5.2.37}$$

$\vdots$

$$=\begin{pmatrix} \mathsf{v}^{(1)} \\ \mathsf{v}^{(2)} \\ \vdots \\ \rho_{D_V}^{(k_\rho)}\left\{\begin{matrix}\lambda_1 \\ \vdots \\ \lambda_{k_\rho}\end{matrix}\right| \mathsf{v}^{(D_V)}\Bigg\} \end{pmatrix}. \tag{5.2.38}$$

Let us denote the ideal corresponding to the relations (5.2.32)–(5.2.34), (5.2.36)–(5.2.38) by $\boldsymbol{B}_{D_V}$.

Definition 5.2.9. A *polyadic tensor product* of D_V polyadic vector spaces $\mathcal{V}_i^{\left(m_k,n_k;m_V^{(i)};k_\rho^{(i)}\right)}$ is obtained from their Cartesian product equipped with the m_V-ary addition $\tilde{\chi}_{m_V}$ (of D_V-ples) and the k_ρ-place multiaction $\tilde{\boldsymbol{\tau}}_{k_\rho}$, satisfying (5.2.32)–(5.2.34) and (5.2.36)–(5.2.38), respectively, by factoring out the ideal $\boldsymbol{B}_{D_V}$

$$\otimes\Pi_{i=1}^{m_V}\mathcal{V}_i^{\left(m_k,n_k;m_V^{(i)};k_\rho^{(i)}\right)} = \left\{\times\Pi_{i=1}^{m_V}\mathcal{V}_i^{\left(m_k,n_k;m_V^{(i)};k_\rho^{(i)}\right)}\Big|\tilde{\chi}_{m_V}, \tilde{\boldsymbol{\tau}}_{k_\rho}\right\}\Big/ \boldsymbol{B}_{D_V}. \tag{5.2.39}$$

As in the case of the polyadic direct sum, we can consider distributivity for D_V-ples of different length. In a similar example (5.2.26), if $m_V = 3$, $m_V^{(1)} = m_V^{(2)} = 3$, $m_V^{(3)} = 2$, we have

$$\begin{pmatrix} \nu_{V,1}^{(3)}[\mathsf{v}_1^{(1)}, \mathsf{v}_2^{(1)}, \mathsf{v}_3^{(1)}] \\ \mathsf{u}_2 \\ \mathsf{u}_3 \end{pmatrix} = \tilde{\chi}_3\left[\begin{pmatrix} \mathsf{v}_1^{(1)} \\ \mathsf{u}_2 \\ \mathsf{u}_3 \end{pmatrix}, \begin{pmatrix} \mathsf{v}_2^{(1)} \\ \mathsf{u}_2 \\ \mathsf{u}_3 \end{pmatrix}, \begin{pmatrix} \mathsf{v}_3^{(1)} \\ \mathsf{u}_2 \end{pmatrix}\right], \tag{5.2.40}$$

$$\begin{pmatrix} \mathsf{u}_1 \\ \nu_{V,2}^{(3)}[\mathsf{v}_1^{(2)}, \mathsf{v}_2^{(2)}, \mathsf{v}_3^{(2)}] \\ \mathsf{u}_3 \end{pmatrix} = \tilde{\chi}_3\left[\begin{pmatrix} \mathsf{u}_1 \\ \mathsf{v}_1^{(2)} \\ \mathsf{u}_3 \end{pmatrix}, \begin{pmatrix} \mathsf{u}_1 \\ \mathsf{v}_2^{(2)} \\ \mathsf{u}_3 \end{pmatrix}, \begin{pmatrix} \mathsf{u}_1 \\ \mathsf{v}_3^{(2)} \end{pmatrix}\right], \tag{5.2.41}$$

$$\begin{pmatrix} \mathsf{u}_1 \\ \mathsf{u}_2 \\ \nu_{V,3}^{(2)}[\mathsf{v}_1^{(3)}, \mathsf{v}_2^{(3)}] \end{pmatrix} = \tilde{\chi}_3\left[\begin{pmatrix} \mathsf{u}_1 \\ \mathsf{u}_2 \\ \mathsf{v}_1^{(3)} \end{pmatrix}, \begin{pmatrix} \mathsf{u}_1 \\ \mathsf{u}_2 \\ \mathsf{v}_2^{(3)} \end{pmatrix}, \begin{pmatrix} \mathsf{u}_1 \\ \mathsf{u}_2 \end{pmatrix}\right]. \tag{5.2.42}$$

Assertion 5.2.10. A tensor product of polyadic vector spaces having *different arities* of the 'vector addition' $m_V^{(i)}$ can be defined.

In the case of modules over a polyadic ring, the formulas connecting arities and ℓ-shapes similar to those above hold, while their concrete properties (noncommutativity, mediality, etc) should be taken into account.

5.3 Polyadic associative algebras

By analogy with the binary case, introducing an additional operation on vectors of vector spaces, a multiplication which is distributive and 'linear' with respect to 'scalars', leads to a polyadic generalization of the (associative) algebra notion (Carlsson 1980, Michor and Vinogradov 1996). Here, we denote the second (except for the 'scalars' K) set by A (instead of V above), on which we define two operations: m_a-ary 'addition' $\nu_A^{(m_a)}: A^{\times m_a} \to A$ and n_a-ary 'multiplication' $\mu_A^{(n_a)}: A^{\times n_a} \to A$.

5.3.1 'Elementwise' description

We formulate the polyadic associative algebras in terms of sets and operations written in concrete manifest form. The arities will be initially taken as arbitrary, but then relations between them will follow from the compatibility conditions (Duplij 2019).

Definition 5.3.1. A *polyadic (associative) algebra* (or $\Bbbk$-algebra) is a tuple consisting of 2 sets and 5 operations

$$\mathcal{A} = \mathcal{A}^{(m_a, n_a; m_k, n_k; k_\rho)} = \left\langle A; K \middle| \nu_A^{(m_a)}, \mu_A^{(n_a)}; \nu_k^{(m_k)}, \mu_k^{(n_k)}; \rho_A^{(k_\rho)} \right\rangle, \tag{5.3.1}$$

where:

(i) $\Bbbk^{(m_k, n_k)} = \left\langle K \middle| \nu_k^{(m_k)}, \mu_k^{(n_k)} \right\rangle$ is a polyadic field with the m_k-ary *field (scalar) addition* $\nu_k^{(m_k)}: K^{m_k} \to K$ and n_k-ary *field (scalar) multiplication* $\mu_k^{(n_k)}: K^{m_k} \to K$;

(ii)

$$\mathcal{A}_{\text{vect}} = \mathcal{A}^{(m_a; m_k, n_k; k_\rho)} = \left\langle A; K \middle| \nu_A^{(m_a)}; \nu_k^{(m_k)}, \mu_k^{(n_k)}; \rho_A^{(k_\rho)} \right\rangle \tag{5.3.2}$$

is a polyadic vector space with the m_a-ary *vector addition* $\nu_A^{(m_a)}: A^{n_a} \to A$ and the *k_ρ-place multiaction* $\rho_A^{(k_\rho)}: K^{k_\rho} \times A \to A$;

(iii) The map $\mu_A^{(n_a)}: A^{n_a} \to A$ is a $\Bbbk$-linear map ('*vector multiplication*') satisfying total associativity

$$\mu_A^{(n_a)}\left[\hat{a}, \mu_A^{(n_a)}[\hat{b}], \hat{c}\right] = \text{invariant}, \tag{5.3.3}$$

where the second product $\mu_A^{(n_a)}$ can be on any place in brackets and $\hat{a}$, $\hat{b}$, $\hat{c}$ are polyads;

(iv) The multiaction $\rho_A^{(k_\rho)}$ is compatible with vector and field operations $\left(\nu_A^{(m_a)}, \mu_A^{(n_a)}; \nu_k^{(m_k)}, \mu_k^{(n_k)}\right)$.

Definition 5.3.2. We call the tuple $(m_a, n_a; m_k, n_k; k_\rho)$ an *arity shape* of the polyadic algebra $\mathcal{A}$ which is called a $(m_a, n_a; m_k, n_k; k_\rho)$-ary algebraic structure.

The compatibility of the multiaction $\rho_A^{(k_\rho)}$ ('linearity') consists of Duplij (2012, 2019):

(1) Distributivity with respect to the m_a-ary vector addition $\nu_A^{(m_a)}$ ('$\lambda(a+b)=\lambda a+\lambda b$')

$$\rho_A^{(k_\rho)}\left\{\lambda_1, \ldots\ldots, \lambda_{k_\rho}\middle|\nu_A^{(m_a)}[a_1, \ldots, a_{m_a}]\right\} \\ = \nu_A^{(m_a)}\left[\rho_A^{(k_\rho)}\left\{\lambda_1, \ldots\ldots, \lambda_{k_\rho}\middle|a_1\right\}, \ldots, \rho_A^{(k_\rho)}\left\{\lambda_1, \ldots\ldots, \lambda_{k_\rho}\middle|a_{m_a}\right\}\right]. \tag{5.3.4}$$

(2) Compatibility with n_a-ary 'vector multiplication' $\mu_A^{(n_a)}$ ('$(\lambda a)\cdot(\mu b)=(\lambda\mu)(a\cdot b)$')

$$\mu_A^{(n_a)}\left[\rho_A^{(k_\rho)}\left\{\lambda_1, \ldots\ldots, \lambda_{k_\rho}\middle|a_1\right\}, \ldots, \rho_A^{(k_\rho)}\left\{\lambda_{k_\rho(n_a-1)}, \ldots\ldots, \lambda_{k_\rho n_a}\middle|a_{n_a}\right\}\right] \\ = \rho_A^{(k_\rho)}\left\{\overbrace{\mu_k^{(n_k)}[\lambda_1, \ldots\ldots, \lambda_{m_k}], \ldots, \mu_k^{(n_k)}\left[\lambda_{m_k(\ell_\mu-1)}, \ldots\ldots, \lambda_{m_k\ell_\mu''}\right]}^{\ell_\mu''}, \right. \\ \left. \overbrace{\lambda_{m_k\ell_\mu''+1}, \ldots\ldots, \lambda_{k_\rho n_a}}^{\ell_{\mathrm{id}}''}\middle|\mu_A^{(n_a)}[a_1, \ldots, a_{n_a}]\right\}, \tag{5.3.5}$$

$$\ell_\mu''(n_k-1)=k_\rho(n_a-1), \tag{5.3.6}$$

where ℓ_μ'' and ℓ_{id}'' are the numbers of multiplications and intact elements in the resulting multiaction, respectively, and

$$\ell_\mu'' \leqslant k_\rho \leqslant \ell_\mu''(n_k-1),\ 2\leqslant n_a\leqslant n_k. \tag{5.3.7}$$

(3) Distributivity with respect to the m_k-ary field addition $\nu_k^{(m_k)}$ ('$(\lambda+\mu)a=\lambda a+\mu a$'), see (5.1.16),

$$\rho^{(k_\rho)}\left\{\overbrace{\nu_k^{(m_k)}[\lambda_1, \ldots\ldots, \lambda_{m_k}], \ldots, \nu_k^{(m_k)}\left[\lambda_{m_k(\ell_\mu'-1)}, \ldots\ldots, \lambda_{m_k\ell_\mu'}\right]}^{\ell_\mu'}, \overbrace{\lambda_{m_k\ell'+1}, \ldots\ldots, \lambda_{k_\rho m_a}}^{\ell_{\mathrm{id}}'}\middle|a\right\} \tag{5.3.8}$$

$$= \nu_A^{(m_a)}\left[\rho^{(k_\rho)}\left\{\lambda_1, \ldots\ldots, \lambda_{k_\rho}\middle|a\right\}, \ldots, \rho^{(k_\rho)}\left\{\lambda_{k_\rho(m_a-1)}, \ldots\ldots, \lambda_{k_\rho m_a}\middle|a\right\}\right],$$

$$\ell_\mu'(m_k-1)=k_\rho(m_a-1), \tag{5.3.9}$$

where ℓ_μ' and ℓ_{id}' are the numbers of multiplications and intact elements, and (see (5.1.20))

$$\ell'_\mu \leqslant k_\rho \leqslant \ell'_\mu(m_k - 1),\ 2 \leqslant m_a \leqslant m_k. \tag{5.3.10}$$

(4) Compatibility n_k-ary field multiplication $\mu_k^{(n_k)}$ ('$\lambda(\mu a) = (\lambda\mu)a$'), see (5.1.9),

$$\overbrace{\rho_A^{(k_\rho)}\left\{\lambda_1, \ldots\ldots, \lambda_{k_\rho}\middle| \ldots \rho_A^{(k_\rho)}\left\{\lambda_{k_\rho(n_\rho-1)}, \ldots\ldots, \lambda_{k_\rho n_\rho}\middle| a\right\}\ldots\right\}}^{n_\rho}$$
$$= \rho_A^{(k_\rho)}\left\{\overbrace{\mu_k^{(n_k)}[\lambda_1, \ldots\ldots, \lambda_{n_k}], \ldots, \mu_k^{(n_k)}\left[\lambda_{n_k(\ell_\mu-1)}, \ldots\ldots, \lambda_{n_k\ell_\mu}\right]}^{\ell_\mu}, \overbrace{\lambda_{n_k\ell_\mu+1}, \ldots\ldots, \lambda_{k_\rho n_\rho}}^{\ell_{\text{id}}}\middle| a\right\}, \tag{5.3.11}$$

$$\ell_\mu(n_k - 1) = k_\rho(n_\rho - 1), \tag{5.3.12}$$

where ℓ_μ and ℓ_{id} are the numbers of multiplications and intact elements, and (see (5.1.13))

$$\ell_\mu \leqslant k_\rho \leqslant \ell_\mu(n_k - 1),\ 2 \leqslant n_\rho \leqslant n_k. \tag{5.3.13}$$

Proposition 5.3.3. *The arities of the polyadic field $\Bbbk^{(m_k, n_k)}$, the arity n_a of the algebra multiplication $\mu_A^{(n_a)}$ and the ℓ-shape of the polyadic compatibility* (5.3.5) *satisfy*

$$k_\rho n_a = n_k \ell''_\mu + \ell''_{\text{id}}, \qquad k_\rho = \ell''_\mu + \ell''_{\text{id}}. \tag{5.3.14}$$

We can exclude from (5.3.14) ℓ''_ρ or ℓ''_{id} and obtain

$$n_a = n_k - \frac{n_k - 1}{k_\rho}\ell''_{\text{id}}, \quad n_a = \frac{n_k - 1}{k_\rho}\ell''_\mu + 1, \tag{5.3.15}$$

where $\frac{n_k - 1}{k_\rho}\ell''_{\text{id}} \geqslant 1$ and $\frac{n_k - 1}{k_\rho}\ell''_\mu \geqslant 1$ are integer, and the inequalities hold

$$1 \leqslant \ell''_\mu \leqslant k_\rho,\ \ 0 \leqslant \ell''_{\text{id}} \leqslant k_\rho - 1,\ \ \ell''_\mu \leqslant k_\rho \leqslant (n_k - 1)\ell''_\mu,\ \ 2 \leqslant n_a \leqslant n_k, \tag{5.3.16}$$

which gives (5.3.7). It follows from (5.3.14), that the ℓ-shape is determined by the arities and number of places k_ρ as

$$\ell''_\mu = \frac{k_\rho(n_a - 1)}{n_k - 1}, \quad \ell''_{\text{id}} = \frac{k_\rho(n_k - n_a)}{n_k - 1}. \tag{5.3.17}$$

Remark **5.3.4.** In the binary case, we have $m_a = n_a = m_k = n_k = n_\rho = 2$, $k_\rho = \ell_\mu = \ell'_\mu = \ell''_\mu = 1$. The n-ary algebras (de Azcarraga and Izquierdo 2010, Michor and Vinogradov 1996) have only one distinct arity $n_a = n$.

Definition 5.3.5. We call the triple $\left(\ell_\mu, \ell'_\mu, \ell''_\mu\right)$ a *ℓ-arity shape* of the polyadic algebra $\mathcal{A}$.

Proposition 5.3.6. *In the limiting ℓ-arity shapes the arity shape of $\mathcal{A}$ is determined by* three *integers (m, n, r), such that:*

(i) *For the maximal $\ell_\mu = \ell'_\mu = \ell''_\mu = k_\rho$, the arity shape of the algebra and underlying field coincide*

$$m_a = m_k = m, \quad n_a = n_k = n_\rho = n, \quad k_\rho = r.$$

(ii) *For the minimal ℓ-arities $\ell_\mu = \ell'_\mu = \ell''_\mu = 1$ it should be $k_\rho|(m_k - 1)$ and $k_\rho|(n_k - 1)$, and*

$$m_a = 1 + \frac{m-1}{r}, \quad n_a = n_\rho = 1 + \frac{n-1}{r}, \quad m_k = m, \quad n_k = n, \quad k_\rho = r.$$

Proof. This follows directly from the compatibility conditions (5.3.5)–(5.3.11). □

Proposition 5.3.7. *If the multiaction $\rho_A^{(k_\rho)}$ is an ordinary action $K \times A \to A$, then all ℓ-arities are minimal $\ell_\mu = \ell'_\mu = \ell''_\mu = 1$, and the arity shape of $\mathcal{A}$ is determined by* two *integers (m, n), such that the arities of the algebra and underlying field are equal, and the arity n_ρ of the action semigroup $\mathcal{S}_\rho$ is equal to the arity of multiplication in the underlying field*

$$m_a = m_k = m, \quad n_a = n_k = n_\rho = n. \tag{5.3.18}$$

In the case where the algebra multiplication $\mu_A^{(n_a)}$ is associative (5.3.70), then $\mathcal{A}^{(m_k, n_k; m_A, n_A; k_\rho)}$ is an *associative polyadic algebra* (for $k_\rho = 1$ see Carlsson 1980). If $\mu_A^{(n_a)}$ is commutative, $\mu_A^{(n_a)}[\mathbf{a}_A] = \mu_A^{(n_a)}[\sigma \circ \mathbf{a}_A]$, for any polyad in algebra $\mathbf{a}_A \in A^{\times n_A}$ for all permutations $\sigma \in S_n$, where S_n is the symmetry group, then $\mathcal{A}^{(m_k, n_k; m_A, n_A; k_\rho)}$ is called a (totally) *commutative polyadic algebra*. As in the n-ary (semi)group theory, for polyadic algebras one can introduce special kinds of associativity and partial commutativity. If the algebra multiplication $\mu_A^{(n_a)}$ contains the identity e_A (4.1.9) or a neutral polyad for any element, such a polyadic algebra is called *unital* or *neutral-unital*, respectively. It follows from (5.3.16) that:

Corollary 5.3.8. In a polyadic ('linear') algebra the arity of the algebra multiplication n_a is less than or equal to the arity of the field multiplication n_k.

Proposition 5.3.9. *It all the operation ℓ-shapes in* (5.1.9), (5.1.16) *and* (5.3.5) *coincide*

$$\ell''_\mu = \ell'_\mu = \ell_\mu, \qquad \ell''_{\mathrm{id}} = \ell'_{\mathrm{id}} = \ell_{\mathrm{id}}, \tag{5.3.19}$$

then, we obtain the conditions for the arities

$$n_k = m_k, \qquad n_\rho = n_A, \tag{5.3.20}$$

while m_A and k_ρ *are not connected.*

Proof. Use (5.1.14) and (5.3.17). □

Proposition 5.3.10. *In the case of equal ℓ-shapes the multiplication and addition of the polyadic ground field ('scalars') should coincide, while the arity n_ρ of the multiaction semigroup $\mathcal{S}_\rho$ should be the same as of the algebra multiplication n_A, while the arity of the algebra addition m_A and number of places k_ρ remain arbitrary.*

***Remark* 5.3.11.** The above ℓ-shapes (5.1.14), (5.1.21), and (5.3.17) are defined by a pair of integers, and therefore the arities in them are not arbitrary, but should be 'quantized' in the same manner as the arities of heteromorphisms in Duplij (2012).

Therefore, numerically the 'quantization' rules for the ℓ-shapes (5.1.14), (5.1.21), and (5.3.17) coincide and are given in table 5.1. Thus, we arrive at the following

Theorem 5.3.12. The arity partial freedom principle. The basic two-set polyadic algebraic structures have non-free underlying operation arities which are 'quantized' in such a way that their ℓ-shape is given by table 5.1.

The above definitions can be generalized, as in the binary case by considering a polyadic ring $\mathcal{R}^{(m_k,\,n_k)}$ instead of a polyadic field $\Bbbk^{(m_k,\,n_k)}$. In this way a polyadic

Table 5.1. 'Quantization' of arity ℓ-shapes.

k_ρ	$\ell_\mu \vert \ell'_\mu \vert \ell''_\mu$	$\ell_{\mathrm{id}} \vert \ell'_{\mathrm{id}} \vert \ell''_{\mathrm{id}}$	$n_K \vert m_K \vert n_K$ $n_\rho \vert n_\rho \vert n_A$
2	1	1	$3, 5, 7, \ldots$ $2, 3, 4, \ldots$
3	1	2	$4, 7, 10, \ldots$ $2, 3, 4, \ldots$
3	2	1	$4, 7, 10, \ldots$ $3, 5, 7, \ldots$
4	1	3	$5, 9, 13, \ldots$ $2, 3, 4, \ldots$
4	2	2	$3, 5, 7, \ldots$ $2, 3, 4, \ldots$
4	3	1	$5, 9, 13, \ldots$ $4, 7, 10, \ldots$

vector space becomes a *polyadic module over a ring* or *polyadic $\mathcal{R}$-module*, while a polyadic algebra over a polyadic field becomes a *polyadic algebra over a ring* or *polyadic $\mathcal{R}$-algebra*. All the axioms and relations between arities in the definition 5.1.6 and definition 5.3.1 remain the same. However, one should take into account that the ring multiplication $\mu_k^{(n_k)}$ can be noncommutative, and therefore for polyadic $\mathcal{R}$-modules and $\mathcal{R}$-algebras it is necessary to consider many different kinds of multiactions ρ_{k_p} (all of them are described in (5.1.9)). For instance, in the ternary case this corresponds to trimodules (Carlsson 1976) or ternary module structure (Michor and Vinogradov 1996).

As it was shown in Duplij (2017), there exist zeroless and nonunital polyadic fields and rings. Therefore, the main difference with the binary algebras is the possible absence of a zero and/or unit in the polyadic field $\Bbbk^{(m_k, n_k)}$ and/or in the polyadic ring

$$\mathcal{A}_{\text{ring}} = \mathcal{A}^{(m_a, n_a)} = \left\langle A \middle| \nu_A^{(m_a)}, \mu_A^{(n_a)} \right\rangle, \tag{5.3.21}$$

and so the additional axioms are needed iff such elements exist. This was the reason we have started from definition 5.3.1, where no existence of zeroes and units in $\Bbbk^{(m_k, n_k)}$ and $\mathcal{A}_{\text{ring}}$ is implied.

If they exist, denote possible units and zeroes by $e_k \in \Bbbk^{(m_k, n_k)}$, $z_k \in \Bbbk^{(m_k, n_k)}$ and $e_A \in \mathcal{A}^{(m_a, n_a)}$, $z_A \in \mathcal{A}^{(m_a, n_a)}$. In this way we have 4 choices for each $\Bbbk^{(m_k, n_k)}$ and $\mathcal{A}^{(m_a, n_a)}$, and these 16 possible kinds of polyadic algebras are presented in table 5.2. The most exotic case is at the bottom right, where both $\Bbbk^{(m_k, n_k)}$ and $\mathcal{A}^{(m_a, n_a)}$ are zeroless nonunital, which cannot exist in either binary algebras or n-ary algebras (de Azcarraga and Izquierdo 2010, Michor and Vinogradov 1996).

The standard case is that in the upper-left corner, when both $\Bbbk^{(m_k, n_k)}$ and $\mathcal{A}^{(m_a, n_a)}$ have a zero and unit.

***Example* 5.3.13.** Consider the ('$\Bbbk$-linear') associative polyadic algebra $\mathcal{A}^{(3,3;3,3;2)}$ over the zeroless nonunital (3, 3)-field $\Bbbk^{(3,3)}$ (from example 4.1.17). The elements of A are pairs $a = (\lambda, \lambda') \in \Bbbk^{(3,3)} \times \Bbbk^{(3,3)}$, and for them the ternary addition and ternary multiplication are defined by

$$\mu_A^{(3)}[(\lambda_1, \lambda_1')(\lambda_2, \lambda_2')(\lambda_3, \lambda_3')] = (\lambda_1\lambda_2'\lambda_3, \lambda_1'\lambda_2\lambda_3'), \tag{5.3.22}$$

$$\nu_A^{(3)}[(\lambda_1, \lambda_1')(\lambda_2, \lambda_2')(\lambda_3, \lambda_3')] = (\lambda_1 + \lambda_2 + \lambda_3, \lambda_1' + \lambda_2' + \lambda_3'), \quad \lambda_i, \lambda_i' \in \Bbbk^{(3,3)} \tag{5.3.23}$$

where operations on the rhs are in $\mathbb{C}$. If we introduce an element $0 \notin \Bbbk^{(3,3)}$ with the property $0 \cdot \lambda = \lambda \cdot 0 = 0$, then (5.3.22)–(5.3.23) can be presented as the ordinary multiplication and addition of three anti-diagonal 2×2 formal matrices $\begin{pmatrix} 0 & \lambda \\ \lambda' & 0 \end{pmatrix}$. There is no unit or zero in the ternary ring $\left\langle A \middle| \nu_A^{(3)}, \mu_A^{(3)} \right\rangle$, but both $\left\langle A \middle| \mu_A^{(3)} \right\rangle$ and $\langle A | \nu_A^{(3)} \rangle$ are ternary groups, because each $a = (ip/q, ip'/q') \in A$ has the unique additive querelement $\tilde{a} = (-ip/q, -ip'/q')$ and the unique multiplicative querelement $\bar{a} = (-iq'/p', -iq/p)$. The 2-place action ('2-scalar product') is defined by

Table 5.2. Kinds of polyadic algebras depending on zeroes and units.

$\mathbb{k}^{(m_k, n_k)} \mathbb{A}^{(m_a, n_a)}$	z_A e_A	z_A no e_A	no z_A e_A	no z_A no e_A
z_k	unital$\mathbb{A}$	nonunital$\mathbb{A}$	unital zeroless $\mathbb{A}$	nonunital zeroless$\mathbb{A}$
e_k	unital $\mathbb{k}$	unital $\mathbb{k}$	unital $\mathbb{k}$	unital $\mathbb{k}$
z_k	unital $\mathbb{A}$	nonunital $\mathbb{A}$	unital zeroless $\mathbb{A}$	nonunital zeroless $\mathbb{A}$
no e_k	nonunital $\mathbb{k}$	nonunital $\mathbb{k}$	nonunital $\mathbb{k}$	nonunital $\mathbb{k}$
no z_k	unital $\mathbb{A}$	nonunital $\mathbb{A}$	unital zeroless $\mathbb{A}$	nonunital zeroless$\mathbb{A}$
e_k	unital zeroless $\mathbb{k}$	unital zeroless $\mathbb{k}$	unital zeroless $\mathbb{k}$	unital zeroless$\mathbb{k}$
no z_k	unital$\mathbb{A}$	nonunital$\mathbb{A}$	unital zeroless $\mathbb{A}$	nonunital zeroless $\mathbb{A}$
no e_k	nonunital zeroless $\mathbb{k}$	nonunital zeroless $\mathbb{k}$	nonunital zeroless $\mathbb{k}$	nonunital zeroless $\mathbb{k}$

$\rho^{(2)}(\lambda_1, \lambda_2|(\lambda, \lambda')) = (\lambda_1\lambda_2\lambda, \lambda_1\lambda_2\lambda')$. The arity shape (see definition 5.3.5) of this zeroless nonunital polyadic algebra $\mathcal{A}^{(3,3;3,3;2)}$ is (2, 2, 2), and the compatibilities (5.3.4)–(5.3.12) hold.

5.3.2 Polyadic analog of the functions on group

In the search for a polyadic version of the algebra of $\Bbbk$-valued functions (which is isomorphic and dual to the corresponding group algebra) we cannot only have more complicated arity shapes than in the binary case, but also the exotic possibility that the arities of the field and group are different as can be possible for multiplace functions.

Let us consider an n_g-ary group $\mathcal{G} = \mathcal{G}^{(n_g)} = \left\langle G \middle| \mu_g^{(n_g)} \right\rangle$, which does not necessarily contain the identity e_g, and where each element is querable (see (1.2.13)). Now we introduce the set A_f of multiplace (s-place) functions $f_i(g_1, \ldots, g_s)$ (of finite support) which take value in the polyadic field $\Bbbk^{(m_k, n_k)}$ such that $f_i: G^s \to K$. To endow A_f with the structure of a polyadic associative algebra (5.3.1), we should consistently define the m_k-ary addition $\nu_f^{(m_k)}: A_f^{(m_k)} \to A_f$, n_k-ary multiplication ('convolution')$\mu_f^{(n_k)}: A_f^{(n_k)} \to A_f$ and the multiaction $\rho_f^{(r_f)}: K^{r_f} \times A_f \to A_f$ ('scalar multiplication'). Thus we write for the algebra of $\Bbbk$-valued functions

$$F_\Bbbk(\mathcal{G}) = \left\langle A_f \middle| \nu_f^{(m_k)}, \mu_f^{(n_k)}; \nu_k^{(m_k)}, \mu_k^{(n_k)}; \rho_f^{(r_f)} \right\rangle. \tag{5.3.24}$$

The simplest operation here is the addition of the $\Bbbk$-valued functions which, obviously, coincides with the field addition $\nu_f^{(m_k)} = \nu_k^{(m_k)}$.

Condition 5.3.14. Because all arguments of the multiaction $\rho_f^{(r_f)}$ are in the field, the only possibility for the rhs is its multiplication (similar to the regular representation)

$$\rho_f^{(r_f)}\left(\lambda_1, \ldots, \lambda_{r_f} \middle| f\right) = \mu_k^{(n_k)}\left[\lambda_1, \ldots, \lambda_{r_f}, f\right], \quad \lambda_i \in K, \; f \in A_f, \tag{5.3.25}$$

and in addition we have the arity shape relation

$$n_k = r_f + 1, \tag{5.3.26}$$

which is satisfied 'automatically' in the binary case.

The polyadic analog of $\Bbbk$-valued function convolution ('$(f_1 * f_2)(g) = \Sigma_{h_1 h_2 = g} f_1(h_1) f_2(h_2)$'), which is denoted by $\mu_f^{(n_k)}$ here, while the sum in the field is $\nu_k^{\ell_\nu(m_k - 1)+1}$, where ℓ_ν is the 'number of additions', can be constructed according to the arity rules from Duplij (2012, 2019).

Definition 5.3.15. The *polyadic convolution* of s-place $\Bbbk$-valued functions is defined as the admissible polyadic sum of $\ell_\nu(m_k - 1) + 1$ products

$$\mu_f^{(n_k)}[f_1(g_1, \ldots, g_s), \ldots, f_{n_k}(g_1, \ldots, g_s)] = \underset{\left\{\begin{array}{c} \mu_g^{(n_g)}[h_1,\ldots,h_{n_g}]=g_1, \\ \mu_g^{(n_g)}[h_{n_g+1},\ldots,h_{2n_g}]=g_2, \\ \vdots \\ \mu_g^{(n_g)}[h_{(s-\ell_{*id\%}-1)n_g},\ldots,h_{(s-\ell_{*id})n_g}]=g_{s-\ell_{*id}}, \\ h_{(s-\ell_{*id}+1)n_g}=g_{s-\ell_{*id}+1}, \\ \vdots \\ h_{sn_g}=g_s \end{array}\right.}{\left(\nu_k^{(m_k)}\right)^{\circ \ell_\nu}} \left[\mu_k^{(n_k)}[f_1(h_1, \ldots, h_s), \ldots, f_{n_k}(h_{s(n_k-1)}, \ldots, h_{sn_k})]\right], \tag{5.3.27}$$

where ℓ_{id} is the number of intact elements in the determining equations ('$h_1 h_2 = g$') under the field sum $\nu_k^{(m_k)}$. The arity shape is determined by

$$sn_k = (s - \ell_{\mathrm{id}})n_g + \ell_{\mathrm{id}}, \tag{5.3.28}$$

which gives the connection between the field and the group arities.

***Example* 5.3.16.** If $n_g = 3$, $n_k = 2$, $m_k = 3$, $s = 2$, $\ell_{\mathrm{id}} = 1$, then we obtain the arity changing polyadic convolution

$$\mu_f^{(2)}[f_1(g_1, g_2), f_2(g_1, g_2)] = \underset{\left\{\mu_g^{(3)}[h_1,h_2,h_3]=g_1, \mu_g^{(4)}=g_2\right.}{\left(\nu_k^{(3)}\right)^{\circ \ell_\nu}} \left[\mu_k^{(2)}[f_1(h_1, h_2), f_2(h_3, h_4)]\right], \tag{5.3.29}$$

where the ℓ_ν ternary additions are taken on the support. Now the multiaction (5.3.25) is one-place

$$\rho_f^{(1)}(\lambda | f) = \mu_k^{(2)}[\lambda, f], \quad \lambda \in K, \quad f \in A_f, \tag{5.3.30}$$

as it follows from (5.3.26).

***Remark* 5.3.17.** The general polyadic convolution (5.3.27) is inspired by the main heteromorphism equation (5.14) and the arity changing formula (5.15) of Duplij (2012). The graphical dependence of the field arity n_k on the number of places s is similar to that on figure 1, and the 'quantization' rules (following from the solutions of (5.3.28) in integers) are in table 1 there.

Proposition 5.3.18. *The multiplication (5.3.27) is associative.*

Proof. This follows from the associativity quiver technique of Duplij (2012) applied to the polyadic convolution. □

Corollary 5.3.19. The $\Bbbk$-valued multiplace functions $\{f_i\}$ form a polyadic associative algebra $F_{\Bbbk}(\mathcal{G})$.

5.3.3 'Diagrammatic' description

Here we formulate the polyadic algebra axioms in the more customary 'diagrammatic' form using the polyadic tensor products and mappings between them (denoted by bold corresponding letters). Informally, the $\Bbbk$-linearity is already 'automatically encoded' by the polyadic tensor algebra over $\Bbbk$, and therefore the axioms already contain the algebra multiplication (but not the scalar multiplication).

Let us denote the $\Bbbk$-*linear algebra multiplication map* by $\boldsymbol{\mu}^{(n)}$ ($\mu^{(n)} \equiv \mu_A^{(n_a)}$ from (5.3.1)) defined as

$$\boldsymbol{\mu}^{(n)} \circ (a_1 \otimes \ldots \otimes a_n) = \mu^{(n)}[a_1, \ldots, a_n], \quad a_1, \ldots, a_n \in A. \tag{5.3.31}$$

Definition 5.3.20. Algebra associativity axiom. A *polyadic (associative n-ary) algebra* (or $\Bbbk$-algebra) is a vector space $\mathcal{A}_{\text{vect}}$ over the polyadic field $\Bbbk$ (5.3.2) with the $\Bbbk$-linear algebra multiplication map

$$\mathcal{A}^{(n)} = \langle \mathcal{A}_{\text{vect}} | \boldsymbol{\mu}^{(n)} \rangle, \qquad \boldsymbol{\mu}^{(n)}: A^{\otimes n} \to A, \tag{5.3.32}$$

which is totally associative

$$\begin{gathered} \boldsymbol{\mu}^{(n)} \circ \left(\mathrm{id}_A^{\otimes(n-1-i)} \otimes \boldsymbol{\mu}^{(n)} \otimes \mathrm{id}_A^{\otimes i}\right) = \boldsymbol{\mu}^{(n)} \circ \left(\mathrm{id}_A^{\otimes(n-1-j)} \otimes \boldsymbol{\mu}^{(n)} \otimes \mathrm{id}_A^{\otimes j}\right), \\ \forall\, i, j = 0, \ldots n-1,\ i \neq j,\ \mathrm{id}_A: A \to A, \end{gathered} \tag{5.3.33}$$

such that the diagram

$$\begin{array}{ccc} A^{\otimes(2n-1)} & \xrightarrow{\ \mathrm{id}_A^{\otimes(n-1-i)} \otimes \boldsymbol{\mu}^{(n)} \otimes \mathrm{id}_A^{\otimes i}\ } & A^{\otimes n} \\ {\scriptstyle \mathrm{id}_A^{\otimes(n-1-j)} \otimes \boldsymbol{\mu}^{(n)} \otimes \mathrm{id}_A^{\otimes j}} \downarrow & & \downarrow {\scriptstyle \boldsymbol{\mu}^{(n)}} \\ A^{\otimes n} & \xrightarrow{\ \boldsymbol{\mu}^{(n)}\ } & A \end{array} \tag{5.3.34}$$

commutes.

Definition 5.3.21. A polyadic algebra $\mathcal{A}^{(n)}$ is called *totally commutative*, if

$$\boldsymbol{\mu}^{(n)} = \boldsymbol{\mu}^{(n)} \circ \boldsymbol{\tau}_n, \tag{5.3.35}$$

where $\boldsymbol{\tau}_n \in \mathrm{S}_n$, and S_n is the symmetry permutation group on n elements.

***Remark* 5.3.22.** Initially, there are no other axioms in the definition of a polyadic algebra, because polyadic fields and vector spaces do not necessarily contain zeroes and units (see table 5.2).

A special kind of polyadic algebra can appear, when the multiplication is 'iterated' from lower arity ones, which is one of three kinds of arity changing for polyadic systems (Duplij 2012).

Definition 5.3.23. A polyadic multiplication is called *derived*, if the map $\boldsymbol{\mu}_{\text{der}}^{(n)}$ is ℓ_μ-iterated from the maps $\boldsymbol{\mu}_0^{(n_0)}$ of lower arity $n_0 < n$

$$\boldsymbol{\mu}_{\text{der}}^{(n)} = \boldsymbol{\mu}_0^{(n_0)} \circ \overbrace{\left(\boldsymbol{\mu}_0^{(n_0)} \circ \ldots \left(\boldsymbol{\mu}_0^{(n_0)} \otimes \mathrm{id}^{\otimes(n_0-1)}\right) \otimes \ldots \otimes \mathrm{id}^{\otimes(n_0-1)}\right)}^{\ell_\mu}, \tag{5.3.36}$$

where

$$n = \ell_\mu(n_0 - 1) + 1, \quad \ell_\mu \geqslant 2, \tag{5.3.37}$$

and ℓ_μ is the 'number of iterations'.

***Example* 5.3.24.** In the ternary case $n = 3$ and $n_0 = 2$, we have $\boldsymbol{\mu}_{\text{der}}^{(3)} = \boldsymbol{\mu}_0^{(2)} \circ \left(\boldsymbol{\mu}_0^{(2)} \otimes \mathrm{id}\right)$, which in the 'elementwise' description is $[a_1, a_2, a_3]_{\text{der}} = a_1 \cdot (a_2 \cdot a_3)$, where $\mu_{\text{der}}^{(3)} = []_{\text{der}}$ and $\mu_0^{(2)} = (\cdot)$.

Introduce a $\Bbbk$-linear *multiaction map* $\boldsymbol{\rho}^{(r)}$ corresponding to the multiaction $\rho^{(r)} \equiv \rho_A^{(k_\rho = r)}$ (5.1.5) (by analogy with (5.3.31)) as

$$\boldsymbol{\rho}^{(r)} \circ (\lambda_1 \otimes \ldots \otimes \lambda_r \otimes a) = \rho_A^{(r)}(\lambda_1, \ldots, \lambda_r | a), \quad \lambda_1, \ldots, \lambda_r \in K, a \in A. \tag{5.3.38}$$

Let $\Bbbk$ and $\mathcal{A}^{(n)}$ both be unital, then we can construct a $\Bbbk$-linear *polyadic unit map* $\boldsymbol{\eta}$ by 'polyadizing' '$\mu \circ (\eta \otimes \mathrm{id}) = \mathrm{id}$' and the scalar product '$\lambda a = \rho(\lambda|a) = \eta(\lambda)a$' with '$\eta(e_k) = e_a$', using the normalization (5.1.6), and taking into account the standard identification $\Bbbk^{\otimes r} \otimes \mathcal{A} \cong \mathcal{A}$ (Yokonuma 1992).

Definition 5.3.25. Algebra unit axiom. The *unital polyadic algebra* $\mathcal{A}^{(n)}$ (5.3.32) contains in addition a $\Bbbk$-linear *polyadic (right) unit* map $\boldsymbol{\eta}^{(r,n)}: K^{\otimes r} \to A^{\otimes(n-1)}$ satisfying

$$\boldsymbol{\mu}^{(n)} \circ (\boldsymbol{\eta}^{(r,n)} \otimes \mathrm{id}_A) = \boldsymbol{\rho}^{(r)}, \tag{5.3.39}$$

such that the diagram

$$\begin{array}{ccc} K^{\otimes r} \otimes A & \xrightarrow{\boldsymbol{\eta}^{(r,n)} \otimes \mathrm{id}_A} & A^{\otimes n} \\ {\scriptstyle \boldsymbol{\rho}^{(r)}} \downarrow & \swarrow {\scriptstyle \boldsymbol{\mu}^{(n)}} & \\ A & & \end{array} \tag{5.3.40}$$

commutes.

The normalization of the multiaction (5.1.6) gives the corresponding normalization of the map $\boldsymbol{\eta}^{(r,n)}$ (instead of '$\eta(e_k) = e_a$')

$$\boldsymbol{\eta}^{(r,n)} \circ \left(\overbrace{e_k \otimes \ldots \otimes e_k}^{r} \right) = \overbrace{e_a \otimes \ldots \otimes e_a}^{n-1}. \tag{5.3.41}$$

Assertion 5.3.26. In the 'elementwise' description (see section 5.3.1) the polyadic unit $\eta^{(r,n)}$ of $\mathcal{A}^{(n)}$ is a $(n-1)$-valued function of r arguments.

Proposition 5.3.27. *The polyadic unit map $\boldsymbol{\eta}^{(r,n)}$ is (polyadically) multiplicative in the following sense*

$$\overbrace{\boldsymbol{\mu}^{(n)} \circ \ldots \circ \boldsymbol{\mu}^{(n)}}^{r} \circ \left(\overbrace{\boldsymbol{\eta}^{(r,n)} \otimes \ldots \otimes \boldsymbol{\eta}^{(r,n)}}^{r(n-1)+1} \right) = \boldsymbol{\eta}^{(r,n)} \circ \left(\mu_k^{(n_k)}\right)^{\circ \ell}. \tag{5.3.42}$$

Proof. This follows from the compatibility of the multiaction with the 'vector multiplication' (5.3.5) and the relation between corresponding arities (5.3.6), such that the number of arguments ('scalars' λ_i) in the rhs becomes $\ell(n_k - 1) + 1 = r(n-1) + 1$, where ℓ is an integer. □

Introduce a 'derived' version of the polyadic unit by analogy with the neutral sequence/polyad (1.1.26).

Definition 5.3.28. The $\Bbbk$-linear *derived polyadic unit* (*neutral unit sequence*) of n-ary algebra $\mathcal{A}^{(n)}$ is the set $\hat{\boldsymbol{\eta}}^{(r)} = \left\{\boldsymbol{\eta}_i^{(r)}\right\}$ of $n-1$ maps $\boldsymbol{\eta}_i^{(r)}: K^{\otimes r} \to A, i = 1, \ldots, n-1$, satisfying

$$\boldsymbol{\mu}^{(n)} \circ \left(\boldsymbol{\eta}_1^{(r)} \otimes \ldots \otimes \boldsymbol{\eta}_{n-1}^{(r)} \otimes \mathrm{id}_A\right) = \boldsymbol{\rho}^{(r)}, \tag{5.3.43}$$

where id_A can be on any place. If $\boldsymbol{\eta}_1^{(r)} = \ldots = \boldsymbol{\eta}_{n-1}^{(r)} = \boldsymbol{\eta}_0^{(r)}$, we call $\boldsymbol{\eta}_0^{(r)}$ the *strong derived polyadic unit*. Formally (comparing (5.3.39) and (5.3.43)), we can write

$$\boldsymbol{\eta}_{\mathrm{der}}^{(r,n)} = \boldsymbol{\eta}_1^{(r)} \otimes \ldots \otimes \boldsymbol{\eta}_{n-1}^{(r)}. \tag{5.3.44}$$

The normalization of the maps $\boldsymbol{\eta}_i^{(r)}$ is given by

$$\boldsymbol{\eta}_i^{(r)} \circ \left(\overbrace{e_k \otimes \ldots \otimes e_k}^{r} \right) = e_a, \quad i = 1, \ldots, n-1, \quad e_a \in A, \quad e_k \in K, \tag{5.3.45}$$

and in the 'elementwise' description $\eta_i^{(r)}$ is a function of r arguments, satisfying

$$\eta_i^{(r)}(\lambda_1, \ldots, \lambda_r) = \rho^{(r)}\{\lambda_1, \ldots, \lambda_r | e_a\}, \quad \lambda_i \in K, \tag{5.3.46}$$

where $\rho^{(r)}$ is the multiaction (5.1.5).

Definition 5.3.29. A polyadic associative algebra $\mathcal{A}_{\mathrm{der}}^{(n)} = \left\langle \mathcal{A}_{\mathrm{vect}} \middle| \boldsymbol{\mu}_{\mathrm{der}}^{(n)}, \boldsymbol{\eta}_{\mathrm{der}}^{(r,n)} \right\rangle$ is called *derived from* $\mathcal{A}_0^{(n_0)} = \left\langle \mathcal{A}_{\mathrm{vect}} \middle| \boldsymbol{\mu}_0^{(n_0)}, \boldsymbol{\eta}_0^{(r,n_0)} \right\rangle$, if (5.3.36) holds and further

$$\boldsymbol{\eta}_{\mathrm{der}}^{(r,n)} = \overbrace{\boldsymbol{\eta}_0^{(r,n_0)} \otimes \ldots \otimes \boldsymbol{\eta}_0^{(r,n_0)}}^{\ell_\mu} \tag{5.3.47}$$

is true, where $\boldsymbol{\eta}_0^{(r,n_0)} = \overbrace{\boldsymbol{\eta}_0^{(r)} \otimes \ldots \otimes \boldsymbol{\eta}_0^{(r)}}^{n_0\ \ 1}$ (formally, because id_A in (5.3.43) can be on any place).

The particular case $n = 3$ and $r = 1$ was considered in Duplij (2001) (with examples).

Invertibility in a polyadic algebra is not connected with the unit or zero (as in n-ary groups, Dörnte 1929), but is determined by the querelement (1.2.13). Introduce the corresponding mappings for the subsets of the *additively querable elements* $A_{\mathrm{quer}}^{(\mathrm{add})} \subseteq A$ and the *multiplicatively querable elements* $A_{\mathrm{quer}}^{(\mathrm{mult})} \subseteq A$.

Definition 5.3.30. In the polyadic algebra $\mathcal{A}^{(m,n)}$, the *additive quermap* $\mathbf{q}_{\mathrm{add}}: A_{\mathrm{quer}}^{(\mathrm{add})} \to A_{\mathrm{quer}}^{(\mathrm{add})}$ is defined by

$$\boldsymbol{\nu}^{(m)} \circ \left(\mathrm{id}_A^{\otimes(m-1)} \otimes \mathbf{q}_{\mathrm{add}}\right) \circ \mathbf{D}_a^{(m)} = \mathrm{id}_A, \tag{5.3.48}$$

and the *multiplicative quermap* $\mathbf{q}_{\mathrm{mult}}: A_{\mathrm{quer}}^{(\mathrm{mult})} \to A_{\mathrm{quer}}^{(\mathrm{mult})}$ is defined by

$$\boldsymbol{\mu}^{(n)} \circ \left(\mathrm{id}_A^{\otimes(n-1)} \otimes \mathbf{q}_{\mathrm{mult}}\right) \circ \mathbf{D}_a^{(n)} = \mathrm{id}_A, \tag{5.3.49}$$

where $\mathbf{D}_a^{(n)}: A \to A^{\otimes n}$ is the diagonal map given by $a \to \overbrace{a \otimes \ldots \otimes a}^{n}$, while $\mathbf{q}_{\mathrm{add}}$ and $\mathbf{q}_{\mathrm{mult}}$ can be on any place. They send an element to the additive querelement $a \overset{\mathbf{q}_{\mathrm{add}}}{\mapsto} \tilde{a}$, $a \in A_{\mathrm{quer}}^{(\mathrm{add})} \subseteq A$ and multiplicative querelement $a \overset{\mathbf{q}_{\mathrm{mult}}}{\mapsto} \bar{a}$, $a \in A_{\mathrm{quer}}^{(\mathrm{mult})} \subseteq A$ (see (1.2.13)), such that the diagrams

$$\begin{array}{ccc} A & \xrightarrow{\mathbf{D}_a^{(m)}} & A^{\otimes m} \\ \boldsymbol{\nu}^{(m)} \uparrow & \swarrow \mathrm{id}_A^{\otimes(m-1)} \otimes \mathbf{q}_{add} & \\ A^{\otimes m} & & \end{array} \qquad \begin{array}{ccc} A & \xrightarrow{\mathbf{D}_a^{(n)}} & A^{\otimes n} \\ \boldsymbol{\mu}^{(n)} \uparrow & \swarrow \mathrm{id}_A^{\otimes(n-1)} \otimes \mathbf{q}_{mult} & \\ A^{\otimes n} & & \end{array} \tag{5.3.50}$$

commute.

***Example* 5.3.31.** For the polyadic algebra $\mathcal{A}^{(3,3;3,3;2)}$ from example 5.3.13 all elements are additively and multiplicatively querable, and so the sets of querable elements coincide $A_{\mathrm{quer}}^{(\mathrm{add})} = A_{\mathrm{quer}}^{(\mathrm{mult})} = A$. The additive quermap $\mathbf{q}_{\mathrm{add}}$ and multiplicative quermap $\mathbf{q}_{\mathrm{mult}}$ act as follows (the operations are in $\mathbb{C}$)

$$\left(i\frac{p}{q}, i\frac{p'}{q'}\right) \overset{\mathbf{q}_{\mathrm{add}}}{\mapsto} \left(-i\frac{p}{q}, -i\frac{p'}{q'}\right), \tag{5.3.51}$$

$$\left(i\frac{p}{q}, i\frac{p'}{q'}\right) \overset{\mathfrak{q}_{\mathrm{mult}}}{\mapsto} \left(-i\frac{q'}{p'}, -i\frac{q}{p}\right), \quad i^2 = -1, \quad p, q \in \mathbb{Z}^{\mathrm{odd}}. \tag{5.3.52}$$

***Example* 5.3.32.** The polyadic field $\Bbbk^{(m_k, n_k)}$ is a polyadic algebra over itself. We identify $A = K$, $\mu_A^{(n)} = \mu_k^{(n_k)}$, and the multiplication is defined by the multiaction as

$$\boldsymbol{\mu}^{(n)} \circ (\lambda_1 \otimes \ldots \otimes \lambda_r \otimes \lambda) = \rho^{(r)}(\lambda_1, \ldots, \lambda_r | \lambda). \tag{5.3.53}$$

Therefore, we have the additional arity conditions

$$n = r + 1 = n_k, \tag{5.3.54}$$

which are trivially satisfied in the binary case. Now the polyadic unit map $\boldsymbol{\eta}^{(r, n)}$ (5.3.41) is the identity in each tensor component.

5.3.4 Medial map and polyadic permutations

Recall that the *binary medial map* for the tensor product of algebras (as vector spaces)

$$\boldsymbol{\tau}_{\mathrm{medial}}: (\mathcal{A}_1 \otimes \mathcal{A}_2) \otimes (\mathcal{A}_1 \otimes \mathcal{A}_2) \to (\mathcal{A}_1 \otimes \mathcal{A}_1) \otimes (\mathcal{A}_2 \otimes \mathcal{A}_2) \tag{5.3.55}$$

is defined by (evaluation)

$$\left(a_1^{(1)} \otimes a_1^{(2)}\right) \otimes \left(a_2^{(1)} \otimes a_2^{(2)}\right) \overset{\tau_{\mathrm{medial}}}{\mapsto} \left(a_1^{(1)} \otimes a_2^{(1)}\right) \otimes \left(a_1^{(2)} \otimes a_2^{(2)}\right). \tag{5.3.56}$$

It is obvious that

$$\boldsymbol{\tau}_{\mathrm{medial}} = \mathrm{id}_A \otimes \boldsymbol{\tau}_{\mathrm{op}} \otimes \mathrm{id}_A, \tag{5.3.57}$$

where $\boldsymbol{\tau}_{\mathrm{op}}: \mathcal{A}_1 \otimes \mathcal{A}_2 \to \mathcal{A}_1 \otimes \mathcal{A}_2$ is the permutation of 2 elements (twist/flip) of the tensor product, such that $a^{(1)} \otimes a^{(2)} \overset{\tau_{\mathrm{op}}}{\mapsto} a^{(2)} \otimes a^{(1)}$, $a^{(1)} \in \mathcal{A}_1$, $a^{(2)} \in \mathcal{A}_2$, $\tau_{\mathrm{op}} \in \mathrm{S}_2$. This may be presented (5.3.56) in the matrix form

$$\bigotimes(\mathbf{a})_{2\times 2} \overset{\tau_{\mathrm{medial}}}{\mapsto} \bigotimes(\mathbf{a}^T)_{2\times 2}, \quad \bigotimes(\mathbf{a})_{2\times 2} = \bigotimes\begin{pmatrix} a_1^{(1)} & a_1^{(2)} \\ a_2^{(1)} & a_2^{(2)} \end{pmatrix}, \tag{5.3.58}$$

where T is the ordinary matrix transposition.

Let us apply (5.3.55) to arbitrary tensor products. By analogy, if we have a tensor product of mn elements (of any nature) grouped by n elements (e.g., m elements from n different vector spaces), as in (5.3.56), (5.3.58), we can write the tensor product in the $(m \times n)$-matrix form (see (3.18)–(3.19) in Duplij 2012)

$$\bigotimes(\mathbf{a})_{m\times n} = \bigotimes\begin{pmatrix} a_1^{(1)} & a_1^{(2)} & \ldots & a_1^{(n)} \\ a_2^{(1)} & a_2^{(2)} & \ldots & a_2^{(n)} \\ \vdots & \vdots & \vdots & \vdots \\ a_m^{(1)} & a_m^{(2)} & \ldots & a_m^{(n)} \end{pmatrix} \tag{5.3.59}$$

Definition 5.3.33. The *polyadic medial map* $\boldsymbol{\tau}_{\text{medial}}^{(n,m)}\colon (A^{\otimes n})^{\otimes m} \to (A^{\otimes m})^{\otimes n}$ is defined as the transposition of the tensor product matrix (5.3.59) by the evaluation (see the binary case (5.3.56))

$$\bigotimes(\mathbf{a})_{m\times n} \overset{\tau_{\text{medial}}^{(n,m)}}{\mapsto} \bigotimes(\mathbf{a}^T)_{n\times m}. \tag{5.3.60}$$

We can extend the mediality concept (Evans 1963, Belousov 1972) to polyadic algebras using the medial map. If we have an algebra with n-ary multiplication (5.3.31), then the mediality relation follows from (5.3.59) with $m = n$ and contains $(n + 1)$ multiplications acting on n^2 elements.

Definition 5.3.34. A $\Bbbk$-linear polyadic algebra $\mathcal{A}^{(n)}$ (5.3.32) is called *medial*, if its n-ary multiplication map satisfies the relation

$$\boldsymbol{\mu}^{(n)} \circ ((\boldsymbol{\mu}^{(n)})^{\otimes n}) = \boldsymbol{\mu}^{(n)} \circ ((\boldsymbol{\mu}^{(n)})^{\otimes n}) \circ \boldsymbol{\tau}_{\text{medial}}^{(n,n)}, \tag{5.3.61}$$

where $\boldsymbol{\tau}_{\text{medial}}^{(n,n)}$ is given by (5.3.60), or in the manifest elementwise form (evaluation)

$$\begin{aligned}&\mu^{(n)}\Big[\mu^{(n)}\Big[a_1^{(1)}, a_1^{(2)}, \ldots, a_1^{(n)}\Big], \mu^{(n)}\Big[a_2^{(1)}, a_2^{(2)}, \ldots, a_2^{(n)}\Big], \ldots, \mu^{(n)}\Big[a_n^{(1)}, a_n^{(2)}, \ldots, a_n^{(n)}\Big]\Big]\\ &= \mu^{(n)}\Big[\mu^{(n)}\Big[a_1^{(1)}, a_2^{(1)}, \ldots, a_n^{(1)}\Big], \mu^{(n)}\Big[a_1^{(2)}, a_2^{(2)}, \ldots, a_n^{(2)}\Big], \ldots, \mu^{(n)}\Big[a_1^{(n)}, a_2^{(n)}, \ldots, a_n^{(n)}\Big]\Big].\end{aligned} \tag{5.3.62}$$

Let us 'polyadize' the binary twist map $\boldsymbol{\tau}_{\text{op}}$ from (5.3.57), which can be suitable for operations with polyadic tensor products. Informally, we can interpret (5.3.57), as 'omitting the fixed points' of the binary medial map $\boldsymbol{\tau}_{\text{medial}}$, and denote this procedure by '$\boldsymbol{\tau}_{\text{op}} = \boldsymbol{\tau}_{\text{medial}}\backslash\text{id}$'.

Definition 5.3.35. A (*medially allowed*) ℓ_τ*-place polyadic twist map* $\boldsymbol{\tau}_{\text{op}}^{(\ell_\tau)}$ is defined by

$$\text{‘}\boldsymbol{\tau}_{\text{op}}^{(\ell_\tau)} = \boldsymbol{\tau}_{\text{medial}}^{(n,m)}\backslash\text{id’}, \tag{5.3.63}$$

where $\ell_\tau = mn - k_{\text{fixed}}$, and k_{fixed} is the number of fixed points of the medial map $\boldsymbol{\tau}_{\text{medial}}^{(n,m)}$.

Assertion 5.3.36. If $m \neq n$, then $\ell_\tau = mn - 2$. If $m = n$, then the polyadic twist map $\boldsymbol{\tau}_{\text{op}}^{(\ell_\tau)}$ is the reflection

$$\boldsymbol{\tau}_{\text{op}}^{(\ell_\tau)} \circ \boldsymbol{\tau}_{\text{op}}^{(\ell_\tau)} = \text{id}_A \tag{5.3.64}$$

and $\ell_\tau = n(n - 1)$.

Table 5.3. Number of places ℓ_τ in the polyadic twist map $\boldsymbol{\tau}_{\text{op}}^{(\ell_\tau)}$.

mn	2	3	4	5	6	7
2	**2**	**4**	**6**	**8**	**10**	**12**
3	**4**	**6**	**10**	**13**	**16**	**19**
4	**6**	**10**	**12**	**18**	**22**	**26**
5	**8**	**13**	**18**	**20**	**28**	**33**
6	**10**	**16**	**22**	**28**	**30**	**40**
7	**12**	**19**	**26**	**33**	**40**	**42**

Proof. This follows from the matrix form (5.3.59) and (5.3.60). □

Therefore the number of places ℓ_τ is 'quantized' and for lowest m, n is presented in table 5.3.

This generalizes the binary twist in a more unique way, which gives polyadic commutativity.

Remark **5.3.37.** The polyadic twist map $\boldsymbol{\tau}_{\text{op}}^{(\ell_\tau)}$ is one element of the symmetry permutation group S_{ℓ_τ} which is fixed by the medial map $\boldsymbol{\tau}_{\text{medial}}^{(n,m)}$ and the special condition (5.3.63), and it therefore respects polyadic tensor product operations.

Example **5.3.38.** In the matrix representation we have

$$\tau_{\text{op}}^{(4)}|_{n=3,m=2} = \begin{pmatrix} 0 & 0 & 1 & 0 \\ 1 & 0 & 0 & 0 \\ 0 & 0 & 0 & 1 \\ 0 & 1 & 0 & 0 \end{pmatrix}, \tau_{\text{op}}^{(6)}|_{n=3,m=3} = \begin{pmatrix} 0 & 0 & 1 & 0 & 0 & 0 \\ 0 & 0 & 0 & 0 & 1 & 0 \\ 1 & 0 & 0 & 0 & 0 & 0 \\ 0 & 0 & 0 & 0 & 0 & 1 \\ 0 & 1 & 0 & 0 & 0 & 0 \\ 0 & 0 & 0 & 1 & 0 & 0 \end{pmatrix}, \tag{5.3.65}$$

$$\tau_{\text{op}}^{(6)}|_{n=4,m=2} = \begin{pmatrix} 0 & 0 & 0 & 1 & 0 & 0 \\ 1 & 0 & 0 & 0 & 0 & 0 \\ 0 & 0 & 0 & 0 & 1 & 0 \\ 0 & 1 & 0 & 0 & 0 & 0 \\ 0 & 0 & 0 & 0 & 0 & 1 \\ 0 & 0 & 1 & 0 & 0 & 0 \end{pmatrix}. \tag{5.3.66}$$

The introduction of the polyadic twist gives us the possibility to generalize (in a way consistent with the medial map) the notion of the opposite algebra.

Definition 5.3.39. For a polyadic algebra $\mathcal{A}^{(n)} = \langle A|\boldsymbol{\mu}^{(n)}\rangle$, an *opposite polyadic algebra*

$$\mathcal{A}_{\mathrm{op}}^{(n)} = \left\langle A \middle| \boldsymbol{\mu}^{(n)} \circ \boldsymbol{\tau}_{\mathrm{op}}^{(n)} \right\rangle \tag{5.3.67}$$

exists if the number of places for the polyadic twist map (which coincides in (5.3.67) with the arity of algebra multiplication $\ell_\tau = n$) is allowed (see table 5.3).

Definition 5.3.40. A polyadic algebra $\mathcal{A}^{(n)}$ is called *medially commutative*, if

$$\boldsymbol{\mu}_{\mathrm{op}}^{(n)} \equiv \boldsymbol{\mu}^{(n)} \circ \boldsymbol{\tau}_{\mathrm{op}}^{(n)} = \boldsymbol{\mu}^{(n)}, \tag{5.3.68}$$

where $\boldsymbol{\tau}_{\mathrm{op}}^{(n)}$ is the medially allowed polyadic twist map.

5.3.5 Tensor product of polyadic algebras

Let us consider a polyadic tensor product $\bigotimes_{i=1}^{n} \mathcal{A}_i^{(n)}$ of n polyadic associative n-ary algebras $\mathcal{A}_i^{(n)} = \left\langle A_i \middle| \boldsymbol{\mu}_i^{(n)} \right\rangle$, $i = 1, \ldots, n$, such that (see (5.3.31))

$$\begin{gathered} \boldsymbol{\mu}_i^{(n)} \circ \left(a_1^{(i)} \otimes \ldots \otimes a_n^{(i)}\right) = \mu_{A_i}^{(n)}\left[a_1^{(i)}, \ldots, a_n^{(i)}\right], \quad a_1^{(i)}, \ldots, a_n^{(i)} \in A_i, \\ \mu_{A_i}^{(n)}\colon A_i^{(n)} \to A_i. \end{gathered} \tag{5.3.69}$$

To endow $\bigotimes_{i=1}^{n} \mathcal{A}_i^{(n)}$ with the structure of an algebra, we will use the medial map $\boldsymbol{\tau}_{\mathrm{medial}}^{(n,m)}$ (5.3.60).

Proposition 5.3.41. *The tensor product of n associative n-ary algebras $\mathcal{A}_i^{(n)}$ has the structure of the polyadic algebra $\mathcal{A}_{\otimes}^{(n)} = \left\langle \bigotimes_{i=1}^{n} \mathcal{A}_i^{(n)} \middle| \boldsymbol{\mu}_{\otimes} \right\rangle$, which is associative (see (5.3.33))*

$$\begin{gathered} \boldsymbol{\mu}_{\otimes} \circ \left(\mathrm{id}_{A_\otimes}^{\otimes(n-1-i)} \otimes \boldsymbol{\mu}_{\otimes} \otimes \mathrm{id}_{A_\otimes}^{\otimes i}\right) = \boldsymbol{\mu}_{\otimes} \circ \left(\mathrm{id}_{A_\otimes}^{\otimes(n-1-j)} \otimes \boldsymbol{\mu}_{\otimes} \otimes \mathrm{id}_{A_\otimes}^{\otimes j}\right), \\ \forall\, i, j = 0, \ldots n-1,\ i \neq j, \quad \mathrm{id}_{A_\otimes}\colon A_1^{\otimes n} \otimes \ldots \otimes A_n^{\otimes n} \to A, \end{gathered} \tag{5.3.70}$$

if

$$\boldsymbol{\mu}_{\otimes} = \left(\boldsymbol{\mu}_1^{(n)} \otimes \ldots \otimes \boldsymbol{\mu}_n^{(n)}\right) \circ \boldsymbol{\tau}_{\mathrm{medial}}^{(n,n)}. \tag{5.3.71}$$

Proof. We act by the multiplication map $\boldsymbol{\mu}_{\otimes}$ on the element's tensor product matrix (5.3.59) and obtain

$$\begin{aligned}
&\boldsymbol{\mu}_\otimes \circ ((a_1^{(1)} \otimes a_1^{(2)} \otimes \ldots \otimes a_1^{(n)}) \otimes \ldots \otimes (a_n^{(1)} \otimes a_n^{(2)} \otimes \ldots \otimes a_n^{(n)})) \\
&= \left(\boldsymbol{\mu}_1^{(n)} \otimes \ldots \otimes \boldsymbol{\mu}_n^{(n)}\right) \circ \boldsymbol{\tau}_{\text{medial}}^{(n,n)} \circ ((a_1^{(1)} \otimes a_1^{(2)} \otimes \ldots \otimes a_1^{(n)}) \otimes \ldots \\
&\quad \otimes (a_n^{(1)} \otimes a_n^{(2)} \otimes \ldots \otimes a_n^{(n)})) \\
&= \left(\boldsymbol{\mu}_1^{(n)} \otimes \ldots \otimes \boldsymbol{\mu}_n^{(n)}\right) \circ ((a_1^{(1)} \otimes a_2^{(1)} \otimes \ldots \otimes a_n^{(1)}) \otimes \ldots \\
&\quad \otimes (a_1^{(n)} \otimes a_2^{(n)} \otimes \ldots \otimes a_n^{(n)})) \\
&= \mu_1^{(n)}[a_1^{(1)}, a_2^{(1)}, \ldots, a_n^{(1)}] \otimes \ldots \otimes \mu_n^{(n)}[a_1^{(n)}, a_2^{(n)}, \ldots, a_n^{(n)}],
\end{aligned} \tag{5.3.72}$$

which proves that $\boldsymbol{\mu}_\otimes$ is indeed a polyadic algebra multiplication. To prove the associativity (5.3.70) we repeat the same derivation (5.3.72) twice and show that the result is independent of i, j. □

If all$\mathcal{A}_i^{(n)}$ have their polyadic unit map $\boldsymbol{\eta}_i^{(r,n)}$ defined by (5.3.39) and acting as (5.3.41), then we have

Proposition 5.3.42. *The polyadic unit map of $\mathcal{A}_\otimes^{(n)}$ is $\boldsymbol{\eta}_\otimes$: $K^{\otimes nr} \to A_1^{\otimes(n-1)} \otimes \ldots \otimes A_n^{\otimes(n-1)}$ acting as*

$$\eta_\otimes \circ \left(\overbrace{e_k \otimes \ldots \otimes e_k}^{nr}\right) = \left(\overbrace{e_{a_1} \otimes \ldots \otimes e_{a_1}}^{n-1}\right) \otimes \ldots \otimes \left(\overbrace{e_{a_n} \otimes \ldots \otimes e_{a_n}}^{n-1}\right). \tag{5.3.73}$$

Assertion 5.3.43. *The polyadic unit of $\mathcal{A}_\otimes^{(n)}$ is a $(n^2 - n)$-valued function of* nr *arguments.*

Note that concepts of tensor product and derived polyadic algebras are different.

5.3.6 Heteromorphisms of polyadic associative algebras

The standard homomorphism between binary associative algebras is defined as a linear map φ which 'commutes' with the algebra multiplications ('$\varphi \circ \boldsymbol{\mu}_1 = \boldsymbol{\mu}_2 \circ (\varphi \otimes \varphi)$'). In the polyadic case there exists the possibility to change arity of the algebras, and for that, one needs to use the heteromorphisms (or multiplace maps) introduced in Duplij (2012). Let us consider two polyadic $\Bbbk$-algebras $\mathcal{A}_1^{(n_1)} = \left\langle A_1 \middle| \boldsymbol{\mu}_1^{(n_1)} \right\rangle$ and $\mathcal{A}_2^{(n_2)} = \left\langle A_2 \middle| \boldsymbol{\mu}_2^{(n_2)} \right\rangle$ (over the same polyadic field $\Bbbk$).

Definition 5.3.44. A *heteromorphism* between two polyadic $\Bbbk$-algebras $\mathcal{A}_1^{(n_1)}$ and $\mathcal{A}_2^{(n_2)}$ (of different arities n_1 and n_2) is a s-place $\Bbbk$-linear map $\boldsymbol{\Phi}_s^{(n_1,n_2)}$: $A_1^{\otimes s} \to A_2$, such that

$$\Phi_s^{(n_1,n_2)} \circ \left(\overbrace{\mu_1^{(n_1)} \otimes \ldots \otimes \mu_1^{(n_1)}}^{s-\ell_{\mathrm{id}}} \otimes \overbrace{\mathrm{id}_{A_1} \otimes \ldots \otimes \mathrm{id}_{A_1}}^{\ell_{\mathrm{id}}} \right)$$

$$= \mu_2^{(n_2)} \circ \left(\overbrace{\Phi_s^{(n_1,n_2)} \otimes \ldots \otimes \Phi_s^{(n_1,n_2)}}^{n_2} \right), \tag{5.3.74}$$

and the diagram

$$\begin{array}{ccc} A_1^{\otimes s n_2} & \xrightarrow{\left(\Phi_s^{(n_1,n_2)}\right)^{\otimes n_2}} & A_2^{\otimes n_2} \\ {\scriptstyle \left(\mu_1^{(n_1)}\right)^{\otimes(s-\ell_{\mathrm{id}})} \otimes \left(\mathrm{id}_{A_1}\right)^{\otimes \ell_{\mathrm{id}}}} \downarrow & & \downarrow {\scriptstyle \mu_2^{(n_2)}} \\ A_1^{\otimes s} & \xrightarrow{\Phi_s^{(n_1,n_2)}} & A_2 \end{array} \tag{5.3.75}$$

commutes. The arities satisfy

$$s n_2 = n_1 (s - \ell_{\mathrm{id}}) + \ell_{\mathrm{id}}, \tag{5.3.76}$$

where $0 \leqslant \ell_{\mathrm{id}} \leqslant s - 1$ is an integer (the number of 'intact elements' of $\mathcal{A}_1$), and therefore $2 \leqslant n_2 \leqslant n_1$.

Assertion 5.3.45. If $\ell_{\mathrm{id}} = 0$ (there are no 'intact elements'), then the (s-place) heteromorphism does not change the arity of the polyadic algebra.

Definition 5.3.46. A *homomorphism* between two polyadic $\Bbbk$-algebras $\mathcal{A}_1^{(n)}$ and $\mathcal{A}_2^{(n)}$ (of the same arity or *equiary*) is a 1-place $\Bbbk$-linear map $\Phi^{(n)} = \Phi_{s=1}^{(n,n)} \colon A_1 \to A_2$, such that

$$\Phi^{(n)} \circ \mu_1^{(n)} = \mu_2^{(n)} \circ \left(\overbrace{\Phi^{(n)} \otimes \ldots \otimes \Phi^{(n)}}^{n} \right), \tag{5.3.77}$$

and the diagram

$$\begin{array}{ccc} A_1^{\otimes n} & \xrightarrow{\left(\Phi^{(n)}\right)^{\otimes n}} & A_2^{\otimes n} \\ {\scriptstyle \mu_1^{(n)}} \downarrow & & \downarrow {\scriptstyle \mu_2^{(n)}} \\ A_1 & \xrightarrow{\Phi^{(n)}} & A_2 \end{array} \tag{5.3.78}$$

commutes.

The above definitions do not include the behavior of the polyadic unit under heteromorphism, because a polyadic associative algebra need not contain a unit. However, if both units exist, this will lead to strong arity restrictions.

Proposition 5.3.47. *If in* $\Bbbk$*-algebras* $\mathcal{A}_1^{(n_1)}$ *and* $\mathcal{A}_2^{(n_2)}$ *(of arities* n_1 *and* n_2*) there exist both polyadic units* (5.3.39) $\boldsymbol{\eta}_1^{(r,n_1)}\colon K^{\otimes r} \to A_1^{\otimes(n_1-1)}$ and $\boldsymbol{\eta}_2^{(r,n_2)}\colon K^{\otimes r} \to A_2^{\otimes(n_2-1)}$, *then*

(i) *The heteromorphism* (5.3.74) *connects them as*

$$\overbrace{\boldsymbol{\eta}_2^{(r,n_2)} \otimes \ldots \otimes \boldsymbol{\eta}_2^{(r,n_2)}}^{s} = \left(\overbrace{\boldsymbol{\Phi}_s^{(n_1,n_2)} \otimes \ldots \otimes \boldsymbol{\Phi}_s^{(n_1,n_2)}}^{n_1-1}\right) \circ \left(\overbrace{\boldsymbol{\eta}_1^{(r,n_1)} \otimes \ldots \otimes \boldsymbol{\eta}_1^{(r,n_1)}}^{s}\right), \quad (5.3.79)$$

and the diagram

$$\begin{array}{ccc} K^{rs} & \xrightarrow{\left(\boldsymbol{\eta}_1^{(r,n_1)}\right)^{\otimes s}} & A_1^{\otimes s(n_1-1)} \\ {\scriptstyle\left(\boldsymbol{\eta}_2^{(r,n_2)}\right)^{\otimes s}}\downarrow & \swarrow {\scriptstyle\left(\boldsymbol{\Phi}_s^{(n_1,n_2)}\right)^{\otimes(n_1-1)}} & \\ A_2^{\otimes s(n_2-1)} & & \end{array} \quad (5.3.80)$$

commutes.

(ii) *The number of 'intact elements' is fixed by its maximum value*

$$\ell_{\text{id}} = s - 1, \quad (5.3.81)$$

such that in the lhs of (5.3.74) *there is only one multiplication* $\boldsymbol{\mu}_1^{(n_1)}$.

(iii) *The number of places s in the heteromorphism* $\boldsymbol{\Phi}_s^{(n_1,n_2)}$ *is fixed by the arities of the polyadic algebras*

$$s(n_2 - 1) = n_1 - 1. \quad (5.3.82)$$

Proof. Using (5.3.76) we obtain $s(n_2 - 1) = (s - \ell_{\text{id}})(n_1 - 1)$, then the $(n_1 - 1)$ power of the heteromorphism $\boldsymbol{\Phi}_s^{(n_1,n_2)}$ maps $A_1^{\otimes s(n_1-1)} \to A_2^{\otimes(n_1-1)}$, and we have $s - \ell_{\text{id}} = 1$, which, together with (5.3.76), gives (5.3.81), and (5.3.82). □

5.3.7 Structure constants

Let $\mathcal{A}^{(n)}$ be a finite-dimensional polyadic algebra (5.3.1) having the basis $\mathrm{e}_i \in A$, $i = 1, \ldots, N$, where N is its dimension as a polyadic vector space

$$\mathcal{A}_{\text{vect}} = \left\langle A, K \middle| \nu^{(m)}; \nu_k^{(m_k)}, \mu_k^{(n_k)}; \rho^{(r)} \right\rangle \quad (5.3.83)$$

where we denote $\nu^{(m)} = \nu_A^{(m_a)}$ (see (5.1.22) and (5.3.2), here $N = d_v$). In the binary case '$a = \sum_i \lambda^{(i)} \mathrm{e}_i$', any element $a \in A$ is determined by the number N_λ of scalars $\lambda \in K$, which coincides with the algebra dimension $N_\lambda = N$, because $r = 1$. In the polyadic case, it can be that $r > 1$, and moreover with $m \geqslant 2$ the admissible number of 'words' (in the expansion of a by e_i) is 'quantized', such that $1, m, 2m - 1, 3m - 2, \ldots \ell_N(m - 1) + 1$, where $\ell_N \in \mathbb{N}_0$ is the 'number of additions'. So we have

Definition 5.3.48. In N-dimensional n-ary algebra $\mathcal{A}^{(n)}$ (with m-ary addition and r-place 'scalar' multiplication) the expansion of any element $a \in A$ by the basis $\{e_i | i = 1, \dots, N\}$ is

$$a = (\nu^{(m)})^{\circ \ell_N}\Big[\rho^{(r)}\Big\{\lambda_1^{(1)}, \dots, \lambda_r^{(1)}\Big|e_1\Big\}, \dots, \rho^{(r)}\Big\{\lambda_1^{(N)}, \dots, \lambda_r^{(N)}\Big|e_N\Big\}\Big], \tag{5.3.84}$$

and is determined by $N_\lambda \in \mathbb{N}$ 'scalars', where

$$N_\lambda = rN, \tag{5.3.85}$$

$$N = \ell_N(m-1) + 1, \quad \ell_N \in \mathbb{N}_0, \quad N \in \mathbb{N}, \quad m \geqslant 2. \tag{5.3.86}$$

In the binary case $m = 2$, the dimension N of an algebra is not restricted and is a natural number, because, $N = \ell_N + 1$.

Assertion 5.3.49. *The dimension of* n*-ary algebra* $\mathcal{A}^{(n)}$ *having* m*-ary addition is not arbitrary, but 'quantized' and can only have the following values for* $m \geqslant 3$

$$m = 3, \quad N = 1, 3, 5, \dots, 2\ell_N + 1, \tag{5.3.87}$$

$$m = 4, \quad N = 1, 4, 7, \dots, 3\ell_N + 1, \tag{5.3.88}$$

$$m = 5, \quad N = 1, 5, 9, \dots, 4\ell_N + 1, \tag{5.3.89}$$

$$\dots$$

Proof. It follows from (5.3.86) and demanding that the 'number of additions' ℓ_N is natural or zero. □

In a similar way, by considering a product of the basis elements, which can also be expanded in the basis '$e_i e_j = \sum_k \chi_{(i,j)}^{(k)} e_k$', we can define a polyadic analog of the structure constants $\chi_{(i,j)}^{(k)} \in K$.

Definition 5.3.50. The *polyadic structure constants* $\chi_{r,(i_1,\dots i_n)}^{(j)} \in K, i_1, \dots i_n, j = 1, \dots, n$ of the N-dimensional n-ary algebra $\mathcal{A}^{(n)}$ (with m-ary addition $\nu^{(m)}$ and r-place multiaction $\rho^{(r)}$) are defined by the expansion of the n-ary product of the basis elements $\{e_i | i = 1, \dots, N\}$ as

$$\begin{aligned} &\mu^{(n)}[e_{i_1}, \dots, e_{i_n}] \\ &= (\nu^{(m)})^{\circ \ell_N}\Big[\rho^{(r)}\Big\{\chi_{1,(i_1,\dots i_n)}^{(1)}, \dots, \chi_{r,(i_1,\dots i_n)}^{(1)}\Big|e_1\Big\}, \dots, \\ &\quad \rho^{(r)}\Big\{\chi_{1,(i_1,\dots i_n)}^{(N)}, \dots, \chi_{r,(i_1,\dots i_n)}^{(N)}\Big|e_N\Big\}\Big], \end{aligned} \tag{5.3.90}$$

where

$$N_\chi = rN^{n+1}, \ \ N, N_\chi \in \mathbb{N} \tag{5.3.91}$$

$$N = \ell_N(m-1) + 1, \ \ \ell_N \in \mathbb{N}_0, \ \ m \geqslant 2. \tag{5.3.92}$$

As in the binary case, we have

Corollary 5.3.51. The algebra multiplication $\mu^{(n)}$ of $\mathcal{A}^{(n)}$ is fully determined by the rN^{n+1} polyadic structure constants $\chi^{(j)}_{r,\,(i_1,\,\ldots i_n)} \in K$.

Contrary to the binary case $m = 2$, when N_χ can be any natural number, we now have

Assertion 5.3.52. *The number of the polyadic structure constants N_χ of the finite-dimensional* n*-ary algebra $\mathcal{A}^{(n)}$ with* m*-ary addition and* r*-place multiaction is not arbitrary, but 'quantized' according to*

$$N_\chi = r(\ell_N(m-1)+1)^{n+1}, \ \ r \in \mathbb{N}, \ \ \ell_N \in \mathbb{N}_0, \ \ m, n \geqslant 2. \tag{5.3.93}$$

Proof. This follows from (5.3.91) and 'quantization' of the algebra dimension N, see assertion 5.3.49. □

References

Belousov V D 1972 *N-ary Quasigroups* (Kishinev: Shtiintsa)

Carlsson R 1976 Cohomology of associative triple systems *Proc. Am. Math. Soc* 60 1–7

Carlsson R 1980 N-ary algebras *Nagoya Math. J.* **78** 45–56

de Azcarraga J and Izquierdo J M 2010 n-Ary algebras: a review with applications *J. Phys.* **A43** 293001

Dörnte W 1929 Unterschungen über einen verallgemeinerten Gruppenbegriff *Math. Z.* **29** 1–19

Duplij S 2001 Ternary Hopf algebras *Symmetry in Nonlinear Mathematical Physics* ed A G Nikitin and V M Boyko (Kyiv: Institute of Mathematics) pp 25–34

Duplij S 2012 Polyadic systems, representations and quantum groups *J. Kharkiv Natl Univ. Ser. Nuclei Part. Fields* **9** 28–59

Duplij S 2017 Polyadic integer numbers and finite (m, n)-fields *p-Adic Numbers Ultrametric Anal. Appl.* **9** 267–91

Duplij S 2019 Arity shape of polyadic algebraic structures *J. Math. Phys. Anal. Geom.* **15** 3–56

Duplij S and Werner W 2021 Structure of unital 3-fields *Math. Semesterber.* **68** 27–53

Evans T 1963 Abstract mean values *Duke Math. J.* **30** 331–47

Michor P W and Vinogradov A M 1996 n-Ary Lie and associative algebras *Rend. Sem. Mat. Univ. Pol. Torino* **54** 373–92

Yokonuma T 1992 *Tensor Spaces and Exterior Algebra* (Providence, RI: American Mathematical Society)

Chapter 6

Polyadic inner spaces and operators

Parts of this chapter have been reprinted with permission from Duplij (2019).

In this chapter we 'polyadize' dual vector spaces, direct sums and tensor products, and show that, as opposed to the binary case, they can be implemented in spaces of different arity signatures. In the resulting table 6.1 we present the arity signatures and shapes of the polyadic algebraic structures under consideration. Then, as application, we consider some starting points for polyadic operator theory by introducing multistars and polyadic analogs of adjoints, operator norms, isometries and projections. It is proved (theorem 6.2.7) that, if the polyadic inner pairing (the analog of the inner product) is symmetric, then all multistars coincide and all polyadic operators are self-adjoint (in contrast to the binary case). The polyadic analogs of C^*-algebras, Toeplitz algebras and Cuntz algebras are presented in terms of the polyadic operators introduced here, and a ternary example is given.

6.1 Polyadic inner pairing spaces and norms

Here we introduce the next important operation: a polyadic analog of the inner product for polyadic vector spaces—a polyadic inner pairing. However, this concept is different from the n-inner product spaces considered, e.g., in Misiak (1989).

Let $\mathcal{V}_{m_k,n_k,m_V,k_\rho} = \left\langle K;\ V \middle| \nu_k^{(m_k)},\ \mu_k^{(n_k)};\ \nu_V^{(m_V)} \middle| \rho_{k_\rho} \right\rangle$ be a polyadic vector space over the polyadic field $\Bbbk^{(m_k,\, n_k)}$ (5.1.22). By analogy with the binary inner product, we next introduce its polyadic counterpart and study its arity shape.

Definition 6.1.1. *A polyadic N*-place *inner pairing* (an analog of the inner product) is a mapping

$$\overbrace{\langle\langle \bullet|\bullet|\ldots|\bullet\rangle\rangle}^{N}\colon V^{\times N} \to K, \tag{6.1.1}$$

satisfying the following conditions:

Table 6.1. The arity signature and arity shape of polyadic algebraic structures.

	Sets		Operations and arities				
Structures	N	Name	N	Multiplications	Additions	Multiactions	Arity shape
Group-like polyadic algebraic structures							
n–ary magma (orgroupoid)	**1**	M	**1**	μ_n: $M^n \to M$			
n–ary semigroup (andmonoid)	**1**	S	**1**	μ_n: $S^n \to S$			
n–ary quasigroup (andloop)	**1**	Q	**1**	μ_n: $Q^n \to Q$			
n-ary group	**1**	G	**1**	μ_n: $G^n \to G$			
Ring-like polyadic algebraic structures							
(m, n)-ary ring	**1**	R	**2**	μ_n: $R^n \to R$	ν_m: $R^m \to R$		
(m, n)-ary field	**1**	K	**2**	μ_n: $K^n \to K$	ν_m: $K^m \to K$		
Module-like polyadic algebraic structures							
Module over (m, n)–ring	**2**	R, M	**4**	σ_n: $R^n \to R$	κ_m: $R^m \to R$; ν_{m_M}: $\mathrm{M}^{m_M} \to \mathrm{M}$	ρ_{k_p}: $R^{k_p} \times \mathrm{M} \to \mathrm{M}$	
Vector space over (m_K, n_K)–field	**2**	K, V	**4**	σ_{n_K}: $K^{n_K} \to K$	κ_{m_K}: $K^{m_K} \to K$; ν_{m_V}: $\mathrm{V}^{m_V} \to \mathrm{V}$	ρ_{k_p}: $K^{k_p} \times \mathrm{V} \to \mathrm{V}$	(5.1.14) (5.1.21)

Algebra-like polyadic algebraic structures

Inner pairing space over (m_K, n_K)–field	**2**	K, V	**5**	σ_{n_K}: $K^{n_K} \to K$	N–Form $\langle\langle \bullet..\bullet \rangle\rangle$: $\mathrm{V}^N \to K$	κ_{m_K}: $K^{m_K} \to K$	ν_{m_V}: $\mathrm{V}^{m_V} \to \mathrm{V}$	ρ_{k_ρ}: $K^{k_\rho} \times \mathrm{V} \to \mathrm{V}$	(6.1.6) (6.1.7) (6.1.8)
(m_A, n_A)–algebra over (m_K, n_K)–field	**2**	K, A	**5**	σ_{n_K}: $K^{n_K} \to K$	μ_{n_A}: $\mathrm{A}^n \to \mathrm{A}$	κ_{m_K}: $K^{m_K} \to K$	ν_{m_A}: $\mathrm{A}^{m_M} \to \mathrm{A}$	ρ_{k_ρ}: $K^{k_\rho} \times \mathrm{A} \to \mathrm{A}$	(5.3.17)

(1) Polyadic 'linearity' (5.1.9) (for first argument):

$$\left\langle\left\langle \rho_{k_\rho}\left\{\begin{array}{c}\lambda_1\\ \vdots \\ \lambda_{k_\rho}\end{array}\right| \mathsf{v}_1\right\}\Big|\mathsf{v}_2\Big|\ldots\Big|\mathsf{v}_N\right\rangle\right\rangle = \mu_k^{(n_k)}\left[\lambda_1, \ldots, \lambda_{k_\rho}, \langle\langle \mathsf{v}_1| \mathsf{v}_2|\ldots|\mathsf{v}_N\rangle\rangle\right]. \tag{6.1.2}$$

(2) Polyadic 'distributivity' (4.1.1)–(4.1.3) (on each place):

$$\begin{aligned}&\langle\langle \nu_V^{(m_V)}[\mathsf{v}_1, \mathsf{u}_1, \ldots \mathsf{u}_{m_V-1}]|\mathsf{v}_2|\ldots|\mathsf{v}_N\rangle\rangle\\ &= \nu_k^{(m_k)}\left[\langle\langle \mathsf{v}_1|\mathsf{v}_2|\ldots|\mathsf{v}_N\rangle\rangle, \langle\langle \mathsf{u}_1|\mathsf{v}_2|\ldots|\mathsf{v}_N\rangle\rangle\ldots\langle\langle \mathsf{u}_{m_V-1}|\mathsf{v}_2|\ldots|\mathsf{v}_N\rangle\rangle\right].\end{aligned} \tag{6.1.3}$$

If the polyadic field $\Bbbk^{(m_k, n_k)}$ contains the zero z_K and $\langle V|m_V\rangle$ has the zero 'vector' z_V (which is not always true in the polyadic case), we have the additional axiom:
(3) The polyadic inner pairing vanishes $\langle\langle \mathsf{v}_1|\mathsf{v}_2|\ldots|\mathsf{v}_N\rangle\rangle = z_K$, iff any of the 'vectors' vanishes, $\exists\, i \in 1, \ldots, N$, such that $\mathsf{v}_i = \mathsf{z}_V$.

If the standard binary ordering on $\Bbbk^{(m_k, n_k)}$ can be defined, then the polyadic inner pairing satisfies:
(4) The positivity condition

$$\overbrace{\langle\langle \mathsf{v}|\mathsf{v}|\ldots|\mathsf{v}\rangle\rangle}^{N} \geqslant z_K, \tag{6.1.4}$$

(5) The polyadic *Cauchy–Schwarz inequality ('triangle' inequality)*

$$\begin{aligned}&\mu_k^{(n_k)}\left[\overbrace{\overbrace{\langle\langle \mathsf{v}_1|\mathsf{v}_1|\ldots|\mathsf{v}_1\rangle\rangle}^{N}, \overbrace{\langle\langle \mathsf{v}_2|\mathsf{v}_2|\ldots|\mathsf{v}_2\rangle\rangle}^{N}\ldots\overbrace{\langle\langle \mathsf{v}_{n_k}|\mathsf{v}_{n_k}|\ldots|\mathsf{v}_{n_k}\rangle\rangle}^{N}}^{n_k}\right]\\ &\geqslant \mu_k^{(n_k)}\left[\overbrace{\langle\langle \mathsf{v}_1|\mathsf{v}_2|\ldots|\mathsf{v}_N\rangle\rangle, \langle\langle \mathsf{v}_1|\mathsf{v}_2|\ldots|\mathsf{v}_N\rangle\rangle\ldots, \langle\langle \mathsf{v}_1|\mathsf{v}_2|\ldots|\mathsf{v}_N\rangle\rangle}^{n_k}\right].\end{aligned} \tag{6.1.5}$$

To make the above relations consistent, the arity shapes should be fixed.

Definition 6.1.2. If the inner pairing is fully symmetric under permutations it is called a *polyadic inner product*.

Proposition 6.1.3. *The number of places in the multiaction* $\boldsymbol{\rho}_{k_\rho}$ *differs by 1 from the multiplication arity of the polyadic field*

$$n_k - k_\rho = 1. \tag{6.1.6}$$

Proof. It follows from the polyadic 'linearity' (6.1.2). □

Proposition 6.1.4. *The arities of 'vector addition' and 'field addition' coincide*

$$m_V = m_k. \tag{6.1.7}$$

Proof. Implied by the polyadic 'distributivity' (6.1.3). □

Proposition 6.1.5. *The arity of the 'field multiplication' is equal to the arity of the polyadic inner pairing space*

$$n_k = N. \tag{6.1.8}$$

Proof. This follows from the polyadic Cauchy–Schwarz inequality (6.1.5). □

Definition 6.1.6. The polyadic vector space $\mathcal{V}_{m_k,n_k,m_V,k_\rho}$ equipped with the polyadic inner pairing

$\overbrace{\langle\langle \bullet|\bullet|\ldots|\bullet\rangle\rangle}^{N} : V^{\times N} \to K$ is called a *polyadic inner pairing space* $\mathcal{H}_{m_k,n_k,m_V,k_\rho,N}$.

A polyadic analog of the binary norm $\|\bullet\|$: $V \to K$ can be induced by the inner pairing in a similar way to the binary case for the inner product (we use the form $\| \mathsf{v} \|^2 = \langle\langle \mathsf{v}|\mathsf{v}\rangle\rangle$).

Definition 6.1.7. A polyadic norm of a 'vector' v in the polyadic inner pairing space $\mathcal{H}_{m_k,n_k,m_V,k_\rho,N}$ is a mapping $\|\bullet\|_N$: $V \to K$, such that

$$\mu_k^{(n_k)}\left[\overbrace{\| \mathsf{v} \|_N, \| \mathsf{v} \|_N, \ldots, \| \mathsf{v} \|_N}^{n_k}\right] = \overbrace{\langle\langle \mathsf{v}|\mathsf{v}|\ldots|\mathsf{v}\rangle\rangle}^{N}, \tag{6.1.9}$$

$$n_k = N, \tag{6.1.10}$$

and the following axioms apply:

(1) The polyadic 'linearity'

$$\left\| \rho_{k_\rho}\left\{\begin{matrix}\lambda_1\\ \vdots\\ \lambda_{k_\rho}\end{matrix}\middle| \mathsf{v}\right\} \right\|_N = \mu_k^{(n_k)}\left[\lambda_1, \ldots, \lambda_{k_\rho}, \| \mathsf{v} \|_N\right], \tag{6.1.11}$$

$$n_k - k_\rho = 1. \tag{6.1.12}$$

If the polyadic field $\Bbbk^{(m_k, n_k)}$ contains the zero z_K and $\langle V|m_V\rangle$ has a zero 'vector' z_V, then:

(2) The polyadic norm vanishes $\| \mathsf{v} \|_N = z_K$, iff $\mathsf{v} = \mathsf{z}_V$.

If the binary ordering on $\langle V|m_V\rangle$ can be defined, then:

(3) The polyadic norm is positive $\| \mathsf{v} \|_N \geqslant \varkappa_K$.
(4) The polyadic 'triangle' inequality holds

$$\nu_k^{(m_k)}\left[\overbrace{\| \mathsf{v}_1 \|_N, \| \mathsf{v}_2 \|_N, \ldots, \| \mathsf{v}_N \|_N}^{m_k}\right] \geqslant \left\| \nu_k^{(m_k)}\left[\overbrace{\| \mathsf{v}_1 \|_N, \| \mathsf{v}_2 \|_N, \ldots, \| \mathsf{v}_N \|_N}^{m_V}\right] \right\|, \quad (6.1.13)$$

$$m_k = m_V = N. \quad (6.1.14)$$

Definition 6.1.8. The polyadic inner pairing space $\mathcal{H}_{m_k,n_k,m_V,k_p,N}$ equipped with the polyadic norm $\| \mathsf{v} \|_N$ is called a *polyadic normed space.*

Recall that in the binary vector space V over the field $\Bbbk$ equipped with the inner product $\langle\langle \bullet | \bullet \rangle\rangle$ and the norm $\|\bullet\|$, one can introduce the angle between vectors $\| \mathsf{v}_1 \| \cdot \| \mathsf{v}_2 \| \cdot \cos\theta = \langle\langle \mathsf{v}_1 | \mathsf{v}_2 \rangle\rangle$, where on the lhs there are two binary multiplications $(\cdot)$.

Definition 6.1.9. A *polyadic angle* between N vectors $\mathsf{v}_1, \mathsf{v}_2, \ldots, \mathsf{v}_{n_k}$ of the polyadic inner pairing space $\mathcal{H}_{m_k,n_k,m_V,k_p,N}$ is defined as a *set* of angles $\boldsymbol{\vartheta} = \{\{\theta_i\} | i = 1, 2, \ldots, n_k - 1\}$ satisfying

$$\begin{aligned}&\left(\mu_k^{(n_k)}\right)^{\circ 2}\left[\| \mathsf{v}_1 \|_N, \| \mathsf{v}_2 \|_N, \ldots, \| \mathsf{v}_{n_k} \|_N, \cos\theta_1, \cos\theta_2, \ldots, \cos\theta_{n_k-1}\right] \\ &= \langle\langle \mathsf{v}_1 | \mathsf{v}_2 | \ldots | \mathsf{v}_{n_k} \rangle\rangle,\end{aligned} \quad (6.1.15)$$

where $\left(\mu_k^{(n_k)}\right)^{\circ 2}$ is a long (iterated) product of two n_k-ary multiplications $\mu_k^{(n_k)}$, which consists of $2(n_k - 1) + 1$ terms.

We will not consider the completion with respect to the above norm (to obtain a polyadic analog of Hilbert space) and corresponding limits and boundedness questions, because this will not give us additional arity shapes, in which we are mostly interested here. Instead, we turn below to some applications and new general constructions which appear from the above polyadic structures.

To conclude, we present the resulting table 6.1 in which the polyadic algebraic structures are listed together with their arity shapes.

6.2 Elements of polyadic operator theory

Here we consider the 1-place polyadic operators $\mathbf{T} = \mathbf{F}_{k_F = 1}$ (the case $k_F = 1$ of the mapping $\mathbf{F}_{k_F}$ in definition 5.2.2) on polyadic inner pairing spaces and structurally generalize the concepts of adjointness and involution.

***Remark* 6.2.1.** A polyadic operator is a complicated mapping between polyadic vector spaces having nontrivial arity shapes (5.2.4) which is actually an action on a set of 'vectors'. However, only for $k_F = 1$ it can be written in a formal way multiplicatively, as is always done in the binary case.

Recall (to fix notations and observe analogies) the informal standard introduction of the operator algebra and the adjoint operator on a binary pre-Hilbert space $\mathcal{H}$ ($\equiv \mathcal{H}_{m_k = 2, n_k = 2, m_V = 2, k_\rho = 1, N = 2}$) over a binary field $\Bbbk$ ($\equiv \Bbbk^{m_k = 2,\, n_k = 2}$) (having the underlying set $\{K; V\}$). For the operator norm $\|\bullet\|_T : \{\mathbf{T}\} \to K$ we use (among many others) the following definition

$$\| \mathbf{T} \|_T = \inf \{ M \in K \| \| \mathbf{T}\mathsf{v} \| \leqslant M \| \mathsf{v} \|, \forall \mathsf{v} \in V \}, \tag{6.2.1}$$

which is convenient for further polyadic generalization. *Bounded* operators have $M < \infty$. If on the set of operators $\{\mathbf{T}\}$ (as 1-place mappings $V \to V$) one defines the addition ($+_T$), product ($\circ_T$) and scalar multiplication ($\cdot_T$) in the standard way

$$(\mathbf{T}_1 +_T \mathbf{T}_2)(\mathsf{v}) = \mathbf{T}_1\mathsf{v} + \mathbf{T}_2\mathsf{v}, \tag{6.2.2}$$

$$(\mathbf{T}_1 \circ_T \mathbf{T}_2)(\mathsf{v}) = \mathbf{T}_1(\mathbf{T}_2\mathsf{v}), \tag{6.2.3}$$

$$(\lambda \cdot_T \mathbf{T})(\mathsf{v}) = \lambda(\mathbf{T}\mathsf{v}), \quad \lambda \in K, \mathsf{v} \in V, \tag{6.2.4}$$

then $\langle \{\mathbf{T}\} | +_T, \circ_T | \cdot_T \rangle$ becomes an operator algebra $\mathcal{A}_T$ (associativity and distributivity are obvious). The unity $\mathbf{I}$ and zero $\mathbf{Z}$ of $\mathcal{A}_T$ (if they exist), satisfy

$$\mathbf{I}\mathsf{v} = \mathsf{v}, \tag{6.2.5}$$

$$\mathbf{Z}\mathsf{v} = \mathsf{z}_V, \qquad \forall \mathsf{v} \in V, \tag{6.2.6}$$

respectively, where $\mathsf{z}_V \in V$ is the polyadic 'zero-vector'.

The connection between operators, linear functionals and inner products is given by the Riesz representation theorem. Informally, it states that in a binary pre-Hilbert space $\mathcal{H} = \{K; V\}$ a (bounded) linear functional (sesquilinear form)$\mathbf{L}$: $V \times V \to K$ can be uniquely represented as

$$\mathbf{L}(\mathsf{v}_1, \mathsf{v}_2) = \langle\langle \mathbf{T}\mathsf{v}_1 | \mathsf{v}_2 \rangle\rangle_{\mathrm{sym}}, \qquad \forall \mathsf{v}_1, \mathsf{v}_2 \in V, \tag{6.2.7}$$

where $\langle\langle \bullet | \bullet \rangle\rangle_{\mathrm{sym}}$: $V \times V \to K$ is a (binary) inner product with standard properties and $\mathbf{T}$: $V \to V$ is a bounded linear operator, such that the norms of $\mathbf{L}$ and $\mathbf{T}$ coincide. Because the linear functionals form a dual space (see section 5.2.1), the relation (6.2.7) fixes the shape of its elements. The main consequence of the Riesz representation theorem is the existence of the adjoint: for any (bounded) linear operator $\mathbf{T}$: $V \to V$ there exists a (unique bounded) *adjoint operator* $\mathbf{T}^*$: $V \to V$ satisfying

$$\mathbf{L}(\mathsf{v}_1, \mathsf{v}_2) = \langle\langle \mathbf{T}\mathsf{v}_1 | \mathsf{v}_2 \rangle\rangle_{\mathrm{sym}} = \langle\langle \mathsf{v}_1 | \mathbf{T}^*\mathsf{v}_2 \rangle\rangle_{\mathrm{sym}}, \qquad \forall \mathsf{v}_1, \mathsf{v}_2 \in V, \tag{6.2.8}$$

and the norms of $\mathbf{T}$ and $\mathbf{T}^*$ are equal. It follows from the conjugation symmetry of the standard binary inner product, that (6.2.8) coincides with

$$\langle\langle \mathsf{v}_1 | \mathbf{T}\mathsf{v}_2 \rangle\rangle_{\mathrm{sym}} = \langle\langle \mathbf{T}^*\mathsf{v}_1 | \mathsf{v}_2 \rangle\rangle_{\mathrm{sym}}, \qquad \forall \mathsf{v}_1, \mathsf{v}_2 \in V. \tag{6.2.9}$$

However, when $\langle\langle \bullet | \bullet \rangle\rangle$ has no symmetry (permutation, conjugation, etc, see, e.g., Mignot 1976), it becomes the binary ($N = 2$) inner pairing (6.1.1), the binary adjoint consists of 2 operators $(\mathbf{T}^{\star_{12}}) \neq (\mathbf{T}^{\star_{21}})$, $\mathbf{T}^{\star_{ij}}\colon V \to V$, which should be defined by 2 equations

$$\langle\langle \mathbf{T}\mathsf{v}_1 | \mathsf{v}_2 \rangle\rangle = \langle\langle \mathsf{v}_1 | \mathbf{T}^{\star_{12}} \mathsf{v}_2 \rangle\rangle, \tag{6.2.10}$$

$$\langle\langle \mathsf{v}_1 | \mathbf{T}\mathsf{v}_2 \rangle\rangle = \langle\langle \mathbf{T}^{\star_{21}} \mathsf{v}_1 | \mathsf{v}_2 \rangle\rangle, \tag{6.2.11}$$

where $(\star_{12}) \neq (\star_{21})$ are 2 different star operations satisfying 2 relations

$$\mathbf{T}^{\star_{12}\star_{21}} = \mathbf{T}, \tag{6.2.12}$$

$$\mathbf{T}^{\star_{21}\star_{12}} = \mathbf{T}. \tag{6.2.13}$$

If $\langle\langle \bullet | \bullet \rangle\rangle = \langle\langle \bullet | \bullet \rangle\rangle_{\text{sym}}$ is symmetric, it becomes the inner product in the pre-Hilbert space $\mathcal{H}$ and the equations (6.2.12)–(6.2.13) coincide, while the operation $(*) = (\star_{12}) = (\star_{21})$ stands for the standard involution

$$\mathbf{T}^{**} = \mathbf{T}. \tag{6.2.14}$$

6.2.1 Multistars and polyadic adjoints

Consider now a special case of the polyadic inner pairing space (see definition 6.1.6)

$$\mathcal{H}_{m_k, n_k, m_V, k_\rho = 1, N} = \left\langle K;\, V \middle| \nu_k^{(m_k)},\, \mu_k^{(n_k)};\, \nu_k^{(m_k)} \middle| \rho_{k_\rho = 1} \middle| \overbrace{\langle\langle \bullet | \ldots | \bullet \rangle\rangle}^{N} \right\rangle \tag{6.2.15}$$

with 1-place multiaction $\rho_{k_\rho = 1}$.

Definition 6.2.2. The set of 1-place operators $\mathbf{T}\colon V \to V$ together with the set of 'scalars' K become a polyadic operator algebra $\mathcal{A}_T = \left\langle K;\, \{\mathbf{T}\} \middle| \nu_k^{(m_k)},\, \mu_k^{(n_k)};\, \nu_T^{(m_T)},\, \mu_T^{(m_T)} \middle| \boldsymbol{\theta}_{k_F = 1} \right\rangle$, if the operations $\nu_T^{(m_T)}$, $\mu_T^{(m_T)}$, $\boldsymbol{\theta}_{k_F = 1}$ to define by

$$\nu_T^{(m_T)}\left[\mathbf{T}_1, \mathbf{T}_2, \ldots, \mathbf{T}_{m_T}\right](\mathsf{v}) = \nu_k^{(m_k)}\left[\mathbf{T}_1\mathsf{v}, \mathbf{T}_2\mathsf{v}, \ldots, \mathbf{T}_{m_T}\mathsf{v}\right], \tag{6.2.16}$$

$$\mu_T^{(m_T)}\left[\mathbf{T}_1, \mathbf{T}_2, \ldots, \mathbf{T}_{n_T}\right](\mathsf{v}) = \mathbf{T}_1\left(\mathbf{T}_2 \ldots \left(\mathbf{T}_{n_T}\mathsf{v}\right)\right), \tag{6.2.17}$$

$$\boldsymbol{\theta}_{k_F = 1}\{\lambda | \mathbf{T}\}(\mathsf{v}) = \boldsymbol{\rho}_{k_\rho = 1}\{\lambda | \mathbf{T}\mathsf{v}\}, \quad \forall \lambda \in K, \forall \mathsf{v} \in V. \tag{6.2.18}$$

The arity shape is fixed by

Proposition 6.2.3. *In the polyadic algebra $\mathcal{A}_T$ the arity of the operator addition m_T coincides with the 'vector' addition of the inner pairing space m_V*, i.e.

$$m_T = m_V. \tag{6.2.19}$$

Proof. This follows from (6.2.16). □

To get relations between operators we assume (as in the binary case) uniqueness: for any $\mathbf{T}_1$, $\mathbf{T}_2$: $V \to V$ it follows from

$$\langle\langle \mathsf{v}_1|\mathsf{v}_2|\ldots|\mathbf{T}_1\mathsf{v}_i|\ldots\mathsf{v}_{N-1}|\mathsf{v}_N\rangle\rangle = \langle\langle \mathsf{v}_1|\mathsf{v}_2|\ldots|\mathbf{T}_2\mathsf{v}_i|\ldots\mathsf{v}_{N-1}|\mathsf{v}_N\rangle\rangle, \tag{6.2.20}$$

that $\mathbf{T}_1 = \mathbf{T}_2$ on any place $i = 1, \ldots, N$.

First, by analogy with the binary adjoint (6.2.8) we define N different adjoints for each operator $\mathbf{T}$.

Definition 6.2.4. Given a polyadic operator $\mathbf{T}$: $V \to V$ on the polyadic inner pairing space $\mathcal{H}_{m_k,n_k,m_V,k_\rho = 1,N}$ we define a *polyadic adjoint* as the set $\{\mathbf{T}^{\star_{ij}}\}$ of N operators $\mathbf{T}^{\star_{ij}}$ satisfying the following N equations

$$\begin{aligned}
&\langle\langle \mathbf{T}\mathsf{v}_1|\mathsf{v}_2|\mathsf{v}_3|\ldots|\mathsf{v}_N\rangle\rangle = \langle\langle \mathsf{v}_1|\mathbf{T}^{\star_{12}}\mathsf{v}_2|\mathsf{v}_3|\ldots|\mathsf{v}_N\rangle\rangle,\\
&\langle\langle \mathsf{v}_1|\mathbf{T}\mathsf{v}_2|\mathsf{v}_3|\ldots|\mathsf{v}_N\rangle\rangle = \langle\langle \mathsf{v}_1|\mathsf{v}_2|\mathbf{T}^{\star_{23}}\mathsf{v}_3|\ldots|\mathsf{v}_N\rangle\rangle,\\
&\vdots\\
&\langle\langle \mathsf{v}_1|\mathsf{v}_2|\mathsf{v}_3|\ldots\mathbf{T}\mathsf{v}_{N-1}|\mathsf{v}_N\rangle\rangle = \langle\langle \mathsf{v}_1|\mathsf{v}_2|\mathsf{v}_3|\ldots|\mathbf{T}^{\star_{N-1,N}}\mathsf{v}_N\rangle\rangle,\\
&\langle\langle \mathsf{v}_1|\mathsf{v}_2|\mathsf{v}_3|\ldots\mathsf{v}_{N-1}|\mathbf{T}\mathsf{v}_N\rangle\rangle = \langle\langle \mathbf{T}^{\star_{N,1}}\mathsf{v}_1|\mathsf{v}_2|\mathsf{v}_3|\ldots|\mathsf{v}_N\rangle\rangle, \quad \mathsf{v}_i \in V.
\end{aligned} \tag{6.2.21}$$

In what follows, for the composition we will use the notation $(\mathbf{T}^{\star_{ij}})^{\star_{kl}\cdots} \equiv \mathbf{T}^{\star_{ij}\star_{kl}\cdots}$. We have from (6.2.21) the N relations

$$\begin{aligned}
\mathbf{T}^{\star_{12}\star_{23}\star_{34}\cdots\star_{N-1,N}\star_{N,1}} &= \mathbf{T},\\
\mathbf{T}^{\star_{23}\star_{34}\cdots\star_{N-1,N}\star_{N,1}\star_{12}} &= \mathbf{T},\\
&\vdots\\
\mathbf{T}^{\star_{N,1}\star_{12}\star_{23}\star_{34}\cdots\star_{N-1,N}} &= \mathbf{T},
\end{aligned} \tag{6.2.22}$$

which are called *multistar cycles.*

Definition 6.2.5. We call the set of adjoint mappings $(\bullet^{\star_{ij}})$: $\mathbf{T} \to \mathbf{T}^{\star_{ij}}$ a *polyadic involution*, if they satisfy the multistar cycles (6.2.22).

If the inner pairing $\langle\langle\bullet|\ldots|\bullet\rangle\rangle$ has more than two places $N \geqslant 3$, we have some additional structural issues, which do not exist in the binary case.

First, we observe that the set of the adjointness relations (6.2.21) can be described in the framework of the associativity quiver approach introduced in Duplij (2012) for polyadic representations. That is, for general $N \geqslant 3$ in addition to (6.2.21) which corresponds to the so-called Post-like associativity quiver (they will be called the *Post-like adjointness relations*), there also exist other sets. It is cumbersome to write additional general formulas like (6.2.21) for other non-Post-like cases, so instead we give a clear example for $N = 4$.

Example 6.2.6. The polyadic adjointness relations for $N = 4$ consist of the sets corresponding to different associativity quivers

$$\begin{gathered}
\text{(1)Post-like adjointness relations}\\
\langle\langle \mathbf{T}\mathrm{v}_1|\mathrm{v}_2|\mathrm{v}_3|\mathrm{v}_4\rangle\rangle = \langle\langle \mathrm{v}_1|\mathbf{T}^{\star_{12}}\mathrm{v}_2|\mathrm{v}_3|\mathrm{v}_4\rangle\rangle,\\
\langle\langle \mathrm{v}_1|\mathbf{T}\mathrm{v}_2|\mathrm{v}_3|\mathrm{v}_4\rangle\rangle = \langle\langle \mathrm{v}_1|\mathrm{v}_2|\mathbf{T}^{\star_{23}}\mathrm{v}_3|\mathrm{v}_4\rangle\rangle,\\
\langle\langle \mathrm{v}_1|\mathrm{v}_2|\mathbf{T}\mathrm{v}_3|\mathrm{v}_4\rangle\rangle = \langle\langle \mathrm{v}_1|\mathrm{v}_2|\mathrm{v}_3|\mathbf{T}^{\star_{34}}\mathrm{v}_4\rangle\rangle,\\
\langle\langle \mathrm{v}_1|\mathrm{v}_2|\mathrm{v}_3|\mathbf{T}\mathrm{v}_4\rangle\rangle = \langle\langle \mathbf{T}^{\star_{41}}\mathrm{v}_1|\mathrm{v}_2|\mathrm{v}_3|\mathrm{v}_4\rangle\rangle,\\
\text{(2)Non-Post-like adjointness relations}\\
\langle\langle \mathbf{T}\mathrm{v}_1|\mathrm{v}_2|\mathrm{v}_3|\mathrm{v}_4\rangle\rangle = \langle\langle \mathrm{v}_1|\mathrm{v}_2|\mathrm{v}_3|\mathbf{T}^{\star_{14}}\mathrm{v}_4\rangle\rangle,\\
\langle\langle \mathrm{v}_1|\mathrm{v}_2|\mathrm{v}_3|\mathbf{T}\mathrm{v}_4\rangle\rangle = \langle\langle \mathrm{v}_1|\mathrm{v}_2|\mathbf{T}^{\star_{43}}\mathrm{v}_3|\mathrm{v}_4\rangle\rangle,\\
\langle\langle \mathrm{v}_1|\mathrm{v}_2|\mathbf{T}\mathrm{v}_3|\mathrm{v}_4\rangle\rangle = \langle\langle \mathrm{v}_1|\mathbf{T}^{\star_{32}}\mathrm{v}_2|\mathrm{v}_3|\mathrm{v}_4\rangle\rangle,\\
\langle\langle \mathrm{v}_1|\mathbf{T}\mathrm{v}_2|\mathrm{v}_3|\mathrm{v}_4\rangle\rangle = \langle\langle \mathbf{T}^{\star_{21}}\mathrm{v}_1|\mathrm{v}_2|\mathrm{v}_3|\mathrm{v}_4\rangle\rangle,
\end{gathered} \tag{6.2.23}$$

and the corresponding multistar cycles

$$\begin{array}{ll}
\text{(1)Post-like multistar cycles} & \text{(2)Non-Post-like multistar cycles}\\
\mathbf{T}^{\star_{12}\star_{23}\star_{34}\star_{41}} = \mathbf{T}, & \mathbf{T}^{\star_{14}\star_{43}\star_{32}\star_{21}} = \mathbf{T},\\
\mathbf{T}^{\star_{23}\star_{34}\star_{41}\star_{12}} = \mathbf{T}, & \mathbf{T}^{\star_{43}\star_{32}\star_{21}\star_{14}} = \mathbf{T},\\
\mathbf{T}^{\star_{34}\star_{41}\star_{12}\star_{23}} = \mathbf{T}, & \mathbf{T}^{\star_{32}\star_{21}\star_{14}\star_{43}} = \mathbf{T},\\
\mathbf{T}^{\star_{41}\star_{12}\star_{23}\star_{34}} = \mathbf{T}, & \mathbf{T}^{\star_{21}\star_{14}\star_{43}\star_{32}} = \mathbf{T}.
\end{array} \tag{6.2.24}$$

Thus, if the inner pairing has no symmetry, then both the Post-like and non-Post-like adjoints and corresponding multistar involutions are different.

Second, in the case $N \geqslant 3$ any symmetry of the multiplace inner pairing restricts the polyadic adjoint sets and multistar involutions considerably.

Theorem 6.2.7. *If the inner pairing with $N \geqslant 3$ has the full permutation symmetry*

$$\langle\langle \mathrm{v}_1|\mathrm{v}_2|\ldots|\mathrm{v}_N\rangle\rangle = \langle\langle \sigma\mathrm{v}_1|\sigma\mathrm{v}_2|\ldots|\sigma\mathrm{v}_N\rangle\rangle, \quad \forall\,\sigma \in S_N, \tag{6.2.25}$$

where S_N is the symmetric group of N *elements, then:*

(i) *All the multistars coincide* $(\star_{ij}) = (\star_{kl}) \coloneqq (*)$ *for any allowed* $i, j, k, l = 1, \ldots, N$;

(ii) *All the operators are self-adjoint* $\mathbf{T} = \mathbf{T}^*$.

Proof.

(i) In each adjointness relation from (6.2.21) we place the operator $\mathbf{T}$ on the lhs to the first position and its multistar adjoint $\mathbf{T}^{\star_{ij}}$ to the second position, using the full permutation symmetry, which together with (6.2.20) gives the equality of all multistar operations.

(ii) We place the operator $\mathbf{T}$ on the lhs to the first position and apply the derivation of the involution in the binary case to increasing cycles of size $i \leqslant N$ recursively, that is:

For $i = 2$

$$\begin{aligned}\langle\langle \mathbf{T}\mathsf{v}_1|\mathsf{v}_2|\mathsf{v}_3|\ldots|\mathsf{v}_N\rangle\rangle &= \langle\langle \mathsf{v}_1|\mathbf{T}^*\mathsf{v}_2|\mathsf{v}_3|\ldots|\mathsf{v}_N\rangle\rangle = \langle\langle \mathbf{T}^*\mathsf{v}_2|\mathsf{v}_1|\mathsf{v}_3|\ldots|\mathsf{v}_N\rangle\rangle \\ &= \langle\langle \mathsf{v}_2|\mathbf{T}^{**}\mathsf{v}_1|\mathsf{v}_3|\ldots|\mathsf{v}_N\rangle\rangle = \langle\langle \mathbf{T}^{**}\mathsf{v}_1|\mathsf{v}_2|\mathsf{v}_3|\ldots|\mathsf{v}_N\rangle\rangle,\end{aligned} \tag{6.2.26}$$

then, using (6.2.20) we get

$$\mathbf{T} = \mathbf{T}^{**}, \tag{6.2.27}$$

as in the standard binary case. However, for $N \geqslant 3$ we have N higher cycles in addition.

For $i = 3$

$$\begin{aligned}\langle\langle \mathbf{T}\mathsf{v}_1|\mathsf{v}_2|\mathsf{v}_3|\ldots|\mathsf{v}_N\rangle\rangle &= \langle\langle \mathsf{v}_1|\mathbf{T}^*\mathsf{v}_2|\mathsf{v}_3|\ldots|\mathsf{v}_N\rangle\rangle = \langle\langle \mathbf{T}^*\mathsf{v}_2|\mathsf{v}_3|\mathsf{v}_1|\ldots|\mathsf{v}_N\rangle\rangle \\ &= \langle\langle \mathsf{v}_2|\mathbf{T}^{**}\mathsf{v}_3|\mathsf{v}_1|\ldots|\mathsf{v}_N\rangle\rangle = \langle\langle \mathbf{T}^{**}\mathsf{v}_3|\mathsf{v}_1|\mathsf{v}_2|\ldots|\mathsf{v}_N\rangle\rangle \\ &= \langle\langle \mathsf{v}_3|\mathbf{T}^{***}\mathsf{v}_1|\mathsf{v}_2|\ldots|\mathsf{v}_N\rangle\rangle = \langle\langle \mathbf{T}^{***}\mathsf{v}_1|\mathsf{v}_2|\mathsf{v}_3|\ldots|\mathsf{v}_N\rangle\rangle,\end{aligned} \tag{6.2.28}$$

which together with (6.2.20) gives

$$\mathbf{T} = \mathbf{T}^{***}, \tag{6.2.29}$$

and after using (6.2.27)

$$\mathbf{T} = \mathbf{T}^{*}. \tag{6.2.30}$$

Similarly, for an arbitrary length of the cycle i we obtain $\mathbf{T} = \mathbf{T}^{\overbrace{**\ldots*}^{i}}$, which should be valid for each cycle recursively with $i = 2, 3, \ldots, N$. Therefore, for any $N \geqslant 3$ all the operators $\mathbf{T}$ are self-adjoint (6.2.30), while $N = 2$ is an exceptional case, when we have $\mathbf{T} = \mathbf{T}^{**}$ (6.2.27) only. □

Now we show that imposing a partial symmetry on the polyadic inner pairing will give more interesting properties to the adjoint operators. Recall that one of the possible binary commutativity generalizations of (semi)groups to the polyadic case is the semicommutativity concept, when in the multiplication only the first and last elements are exchanged. Similarly, we introduce

Definition 6.2.8. The polyadic inner pairing is called *semicommutative*, if

$$\langle\langle \mathsf{v}_1|\mathsf{v}_2|\mathsf{v}_3|\ldots|\mathsf{v}_N\rangle\rangle = \langle\langle \mathsf{v}_N|\mathsf{v}_2|\mathsf{v}_3|\ldots|\mathsf{v}_1\rangle\rangle, \quad \mathsf{v}_i \in V. \tag{6.2.31}$$

Proposition 6.2.9. *If the polyadic inner pairing is semicommutative, then for any operator* $\mathbf{T}$ *(satisfying Post-like adjointness* (6.2.21)*) the last multistar operation* ($\star_{N,1}$) *is a* binary involution *and is a composition of all the previous multistars*

$$\mathbf{T}^{\star_{N,1}} = \mathbf{T}^{\star_{12}\star_{23}\star_{34}\cdots\star_{N-1,N}}, \tag{6.2.32}$$

$$\mathbf{T}^{\star_{N,1}\star_{N,1}} = \mathbf{T}. \tag{6.2.33}$$

Proof. It follows from (6.2.21) and (6.2.31), that

$$\begin{aligned}
&\langle\langle \mathsf{v}_1|\mathsf{v}_2|\mathsf{v}_3|\ldots|\mathbf{T}\mathsf{v}_N\rangle\rangle \\
&= \langle\langle \mathbf{T}\mathsf{v}_N|\mathsf{v}_2|\mathsf{v}_3|\ldots|\mathsf{v}_1\rangle\rangle = \langle\langle \mathsf{v}_N|\mathsf{v}_2|\mathsf{v}_3|\ldots|\mathbf{T}^{\star_{12}\star_{23}\star_{34}\cdots\star_{N-1,N}}\mathsf{v}_1\rangle\rangle \\
&= \langle\langle \mathbf{T}^{\star_{12}\star_{23}\star_{34}\cdots\star_{N-1,N}}\mathsf{v}_1|\mathsf{v}_2|\mathsf{v}_3|\ldots|\mathsf{v}_N\rangle\rangle = \langle\langle \mathbf{T}^{\star_{N,1}}\mathsf{v}_1|\mathsf{v}_2|\mathsf{v}_3|\ldots|\mathsf{v}_N\rangle\rangle,
\end{aligned} \tag{6.2.34}$$

which using (6.2.20) gives (6.2.32), (6.2.33) follows from the first multistar cocycle in (6.2.22). □

The adjointness relations (6.2.21) (of all kinds) together with (6.2.18) and (6.2.19) allow us to fix the arity shape of the polyadic operator algebra $\mathcal{A}_T$. We will assume that the arity of the operator multiplication in $\mathcal{A}_T$ coincides with the number of places of the inner pairing N (6.1.1)

$$n_T = N, \tag{6.2.35}$$

because it is in agreement with (6.2.21). Thus, the arity shape of the polyadic operator algebra becomes

$$\mathcal{A}_T = \left\langle K;\ \{\mathbf{T}\} \middle| \nu_k^{(m_k)},\ \mu_k^{(n_k)};\ \nu_T^{(m_T = m_V)},\ \mu_T^{(m_T = N)} \middle| \boldsymbol{\theta}_{k_F = k_\rho = 1} \right\rangle, \tag{6.2.36}$$

Definition 6.2.10. We call the operator algebra $\mathcal{A}_T$ which has the arity $n_T = N$ a *nonderived polyadic operator algebra.*

Let us investigate some structural properties of $\mathcal{A}_T$ and types of polyadic operators.

***Remark* 6.2.11.** We can only define, but not derive, as in the binary case, the action of any multistar $(\star_{ij})$ on the product of operators, because in the nonderived n_T-ary algebra we have a fixed number of operators in a product and sum, that is $\ell'(n_T - 1) + 1$ and $\ell''(m_T - 1) + 1$, correspondingly, where ℓ' is the number of n_T-ary multiplications and ℓ' is the number of m_T-ary additions. Therefore, we cannot transfer (one at a time) all the polyadic operators from one place in the inner pairing to another place, as is done in the standard proof in the binary case.

Taking this into account, as well as consistency under the multistar cycles (6.2.22), we arrive at the following definition

Definition 6.2.12. The fixed multistar operation acts on the $\ell = 1$ product of n_T polyadic operators, depending on the *sequential number of the multistar* $(\star_{ij})$ (for the Post-like adjointness relations (6.2.21))

$$s_{ij} := \begin{cases} \frac{i+j-1}{2}, & \text{if } 3 \leqslant i+j \leqslant 2N-1 \\ N, & \text{if } i\,j = N, \end{cases} \qquad s_{ij} = 1, 2, \ldots, N-1, N, \quad (6.2.37)$$

in the following way

$$\begin{aligned} &\left(\mu_T^{(m_T)}\left[\mathbf{T}_1, \mathbf{T}_2, \ldots, \mathbf{T}_{n_T-1}, \mathbf{T}_{n_T}\right]\right)^{\star_{ij}} \\ &= \begin{cases} \mu_T^{(m_T)}\left[\mathbf{T}_{n_T}^{\star_{ij}}, \mathbf{T}_{n_T-1}^{\star_{ij}}, \ldots, \mathbf{T}_2^{\star_{ij}}, \mathbf{T}_1^{\star_{ij}}\right], & s_{ij} \text{ odd}, \\ \mu_T^{(m_T)}\left[\mathbf{T}_1^{\star_{ij}}, \mathbf{T}_2^{\star_{ij}}, \ldots, \mathbf{T}_{n_T-1}^{\star_{ij}}, \mathbf{T}_{n_T}^{\star_{ij}}\right], & s_{ij} \text{ even}. \end{cases} \end{aligned} \quad (6.2.38)$$

A rule similar to (6.2.38) holds also for non-Post-like adjointness relations, but their concrete form depends of the corresponding non-Post-like associative quiver.

Sometimes, to shorten notation, it will be more convenient to mark a multistar by the sequential number (6.2.37), such that $(\star_{ij}) \Rightarrow \left(\star_{s_{ij}}\right)$, e.g., $(\star_{23}) \Rightarrow (\star_2)$, $(\star_{N,1}) \Rightarrow (\star_N)$, etc. Also, in examples, for the ternary multiplication we will use the square brackets without the name of operation, if it is clear from the context, e.g., $\omega_3[\mathbf{T}_1, \mathbf{T}_2, \mathbf{T}_3] \Rightarrow [\mathbf{T}_1, \mathbf{T}_2, \mathbf{T}_3]$, etc.

Example 6.2.13. In the lowest ternary case $N = 3$ we have

$$\begin{aligned} &\text{(1)Post-like adjointness relations} \\ &\langle\langle \mathbf{T}\mathrm{v}_1 | \mathrm{v}_2 | \mathrm{v}_3 \rangle\rangle = \langle\langle \mathrm{v}_1 | \mathbf{T}^{\star_1}\mathrm{v}_2 | \mathrm{v}_3 \rangle\rangle, \\ &\langle\langle \mathrm{v}_1 | \mathbf{T}\mathrm{v}_2 | \mathrm{v}_3 \rangle\rangle = \langle\langle \mathrm{v}_1 | \mathrm{v}_2 | \mathbf{T}^{\star_2}\mathrm{v}_3 \rangle\rangle, \\ &\langle\langle \mathrm{v}_1 | \mathrm{v}_2 | \mathbf{T}\mathrm{v}_3 \rangle\rangle = \langle\langle \mathbf{T}^{\star_3}\mathrm{v}_1 | \mathrm{v}_2 | \mathrm{v}_3 \rangle\rangle, \\ &\text{(2)Non-Post-like adjointness relations} \\ &\langle\langle \mathbf{T}\mathrm{v}_1 | \mathrm{v}_2 | \mathrm{v}_3 \rangle\rangle = \langle\langle \mathrm{v}_1 | \mathrm{v}_2 | \mathbf{T}^{\star_3}\mathrm{v}_3 \rangle\rangle, \\ &\langle\langle \mathrm{v}_1 | \mathrm{v}_2 | \mathbf{T}\mathrm{v}_3 \rangle\rangle = \langle\langle \mathrm{v}_1 | \mathbf{T}^{\star_2}\mathrm{v}_2 | \mathrm{v}_3 \rangle\rangle, \\ &\langle\langle \mathrm{v}_1 | \mathbf{T}\mathrm{v}_2 | \mathrm{v}_3 \rangle\rangle = \langle\langle \mathbf{T}^{\star_1}\mathrm{v}_1 | \mathrm{v}_2 | \mathrm{v}_3 \rangle\rangle, \end{aligned} \quad (6.2.39)$$

and the corresponding multistar cycles

$$\begin{array}{ll} \text{(1)Post-like multistar cycles} & \text{(2)Non-Post-like multistar cycles} \\ \mathbf{T}^{\star_1\star_2\star_3} = \mathbf{T}, & \mathbf{T}^{\star_3\star_2\star_1} = \mathbf{T}, \\ \mathbf{T}^{\star_2\star_3\star_1} = \mathbf{T}, & \mathbf{T}^{\star_2\star_1\star_3} = \mathbf{T}, \\ \mathbf{T}^{\star_3\star_1\star_2} = \mathbf{T}, & \mathbf{T}^{\star_1\star_3\star_2} = \mathbf{T}. \end{array} \quad (6.2.40)$$

Using (6.2.38) we obtain the following ternary conjugation rules

$$([\mathbf{T}_1, \mathbf{T}_2, \mathbf{T}_3])^{\star_1} = [\mathbf{T}_3^{\star_1}, \mathbf{T}_2^{\star_1}, \mathbf{T}_1^{\star_1}], \tag{6.2.41}$$

$$([\mathbf{T}_1, \mathbf{T}_2, \mathbf{T}_3])^{\star_2} = [\mathbf{T}_1^{\star_2}, \mathbf{T}_2^{\star_2}, \mathbf{T}_3^{\star_2}], \tag{6.2.42}$$

$$([\mathbf{T}_1, \mathbf{T}_2, \mathbf{T}_3])^{\star_3} = [\mathbf{T}_3^{\star_3}, \mathbf{T}_2^{\star_3}, \mathbf{T}_1^{\star_3}], \tag{6.2.43}$$

which are common for both Post-like and non-Post-like adjointness relations (6.2.39).

Definition 6.2.14. A polyadic operator $\mathbf{T}$ is called *self-adjoint*, if all multistar operations are identities, i.e., $(\star_{ij}) = \mathrm{id}, \forall\, i, j$.

6.2.2 Polyadic isometry and projection

Now we introduce polyadic analogs for the following important types of operator: isometry, unitary, and (orthogonal) projection. Taking into account *remark* 6.2.11, we again cannot move operators singly, and instead of proving the operator relations, as is usually done in the binary case, we can only exploit some mnemonic rules to define the corresponding relations between polyadic operators.

If the polyadic operator algebra $\mathcal{A}_T$ contains a unit $\mathbf{I}$ and zero $\mathbf{Z}$ (see (6.2.5)–(6.2.6)) we define the conditions of polyadic isometry and orthogonality:

Definition 6.2.15. A polyadic operator $\mathbf{T}$ is called a *polyadic isometry*, if it preserves the polyadic inner pairing

$$\langle\langle \mathbf{T}\mathsf{v}_1 | \mathbf{T}\mathsf{v}_2 | \mathbf{T}\mathsf{v}_3 | \ldots | \mathbf{T}\mathsf{v}_N \rangle\rangle = \langle\langle \mathsf{v}_1 | \mathsf{v}_2 | \mathsf{v}_3 | \ldots | \mathsf{v}_N \rangle\rangle, \tag{6.2.44}$$

and satisfies

$$\begin{aligned} &\mu_T^{(m_T)}[\mathbf{T}^{\star_{N-1,N}}, \mathbf{T}^{\star_{N-2,N-1}\star_{N-1,N}}, \ldots \\ &\mathbf{T}^{\star_{23}\star_{34}\cdots\star_{N-2,N-1}\star_{N-1,N}}, \mathbf{T}^{\star_{12}\star_{23}\star_{34}\cdots\star_{N-2,N-1}\star_{N-1,N}}, \mathbf{T}] = \mathbf{I}, \\ &+ (N-1) \textit{ cycle permutations of multistars in the first } (N-1) \textit{ terms.} \end{aligned} \tag{6.2.45}$$

Remark **6.2.16.** If the multiplication in $\mathcal{A}_T$ is derived and all multistars are equal, then the polyadic isometry operators satisfy some kind of N-regularity (Duplij and Marcinek 2002) or regular N-cocycle condition (Duplij and Marcinek 2001).

Proposition 6.2.17. *The polyadic isometry operator* $\mathbf{T}$ *preserves the polyadic norm*

$$\| \mathbf{T}\mathsf{v} \|_N = \| \mathsf{v} \|_N, \qquad \forall \mathsf{v} \in V. \tag{6.2.46}$$

Proof. It follows from (6.1.9) and (6.2.44), that

$$\mu_k^{(n_k)}\Bigg[\overbrace{\| \mathbf{T}\mathsf{v} \|_N, \| \mathbf{T}\mathsf{v} \|_N, \ldots, \| \mathbf{T}\mathsf{v} \|_N}^{n_k}\Bigg] = \mu_k^{(n_k)}\Bigg[\overbrace{\| \mathsf{v} \|_N, \| \mathsf{v} \|_N, \ldots, \| \mathsf{v} \|_N}^{n_k}\Bigg], \tag{6.2.47}$$

which gives (6.2.46), when $n_k = N$. □

Definition 6.2.18. If for N polyadic operators $\mathbf{T}_i$ we have

$$\langle\langle \mathbf{T}_1\mathsf{v}_1|\mathbf{T}_2\mathsf{v}_2|\mathbf{T}_3\mathsf{v}_3|\ldots|\mathbf{T}_N\mathsf{v}_N\rangle\rangle = \mathsf{z}_K, \quad \forall \mathsf{v}_i \in V, \tag{6.2.48}$$

where $\mathsf{z}_K \in V$ is the zero of the underlying polyadic field $\Bbbk^{(m_k, n_k)}$, then we say that $\mathbf{T}_i$ are *(polyadically) orthogonal*, and they satisfy

$$\begin{aligned}&\mu_T^{(m_T)}\Big[\mathbf{T}_1^{\star N-1,N}, \mathbf{T}_2^{\star N-2,N-1\star N-1,N}, \ldots \\ &\mathbf{T}_3^{\star 23\star 34\cdots\star N-2,N-1\star N-1,N}, \mathbf{T}_{N-1}^{\star 12\star 23\star 34\cdots\star N-2,N-1\star N-1,N}, \mathbf{T}_N\Big] = \mathbf{Z}, \\ &+ (N-1) \text{ cycle permutations of multistars in the first } (N-1) \text{ terms.}\end{aligned} \tag{6.2.49}$$

The polyadic analog of projection is given by

Definition 6.2.19. If a polyadic operator $\mathbf{P} \in \mathcal{A}_T$ satisfies the polyadic idempotency condition

$$\mu_T^{(m_T)}\Bigg[\overbrace{\mathbf{P}, \mathbf{P}, \ldots\mathbf{P}}^{n_T}\Bigg] = \mathbf{P}, \tag{6.2.50}$$

then it is called a *polyadic projection.*

By analogy with the binary case, polyadic projections can be constructed from polyadic isometry operators in a natural way.

Proposition 6.2.20. *If* $\mathbf{T} \in \mathcal{A}_T$ *is a polyadic isometry, then*

$$\begin{aligned}&\mathbf{P}_\mathbf{T}^{(1)} = \mu_T^{(m_T)}\big[\mathbf{T}, \mathbf{T}^{\star N-1,N}, \mathbf{T}^{\star N-2,N-1\star N-1,N}, \ldots \\ &\mathbf{T}^{\star 23\star 34\cdots\star N-2,N-1\star N-1,N}, \mathbf{T}^{\star 12\star 23\star 34\cdots\star N-2,N-1\star N-1,N}\big], \\ &+ (N-1) \text{ cycle permutations of multistars in the last } (N-1) \text{ terms}\end{aligned} \tag{6.2.51}$$

are the corresponding polyadic projections $\mathbf{P}_\mathbf{T}^{(k)}$, $k = 1, \ldots, N$, *satisfying* (6.2.50).

Definition 6.2.21. A polyadic operator $\mathbf{T} \in \mathcal{A}_T$ is called *normal*, if

$$\begin{aligned}&\mu_T^{(m_T)}\big[\mathbf{T}^{\star N-1,N}, \mathbf{T}^{\star N-2,N-1\star N-1,N}, \ldots \\ &\mathbf{T}^{\star 23\star 34\cdots\star N-2,N-1\star N-1,N}, \mathbf{T}^{\star 12\star 23\star 34\cdots\star N-2,N-1\star N-1,N}, \mathbf{T}\big]= \\ &\mu_T^{(m_T)}\big[\mathbf{T}, \mathbf{T}^{\star N-1,N}, \mathbf{T}^{\star N-2,N-1\star N-1,N}, \ldots \\ &\mathbf{T}^{\star 23\star 34\cdots\star N-2,N-1\star N-1,N}, \mathbf{T}^{\star 12\star 23\star 34\cdots\star N-2,N-1\star N-1,N}\big], \\ &+ (N-1) \text{ cycle permutations of multistars in the } (N-1) \text{ terms.}\end{aligned} \tag{6.2.52}$$

Proof. Insert (6.2.51) into (6.2.50) and use (6.2.45) together with n_T-ary associativity. □

Definition 6.2.22. If all the polyadic projections (6.2.51) are equal to unity $\mathbf{P}_{\mathbf{T}}^{(k)} = \mathbf{I}$, then the corresponding polyadic isometry operator $\mathbf{T}$ is called a *polyadic unitary operator*.

It can be shown that each polyadic unitary operator is querable ('polyadically invertible'), such that it has a querelement in $\mathcal{A}_T$.

6.2.3 Towards a polyadic analog of C^*-algebras

Let us, first, generalize the operator binary norm (6.2.1) to the polyadic case. This can be done, provided that a binary ordering on the underlying polyadic field $\Bbbk^{(m_k,\, n_k)}$ can be introduced.

Definition 6.2.23. The polyadic operator norm $\|\bullet\|_T\colon \{\mathbf{T}\} \to K$ is defined by

$$\| \mathbf{T} \|_T = \inf\left\{ M \in K \,\middle|\, \| \mathbf{T}\mathsf{v} \|_N \leqslant \mu_k^{(n_k)}\left[\overbrace{M, \ldots, M}^{n_k - 1}, \| \mathsf{v} \|_N \right], \forall \mathsf{v} \in V \right\}, \tag{6.2.53}$$

where $\|\bullet\|_N$ is the polyadic norm in the inner pairing space $\mathcal{H}_{m_k, n_k, m_V, k_\rho = 1, N}$ and $\mu_k^{(n_k)}$ is the n_k-ary multiplication in $\Bbbk^{(m_k,\, n_k)}$.

Definition 6.2.24. The polyadic operator norm is called *submultiplicative*, if

$$\left\| \mu_T^{(m_T)}\left[\mathbf{T}_1, \mathbf{T}_2, \ldots, \mathbf{T}_{n_T} \right] \right\|_T \leqslant \mu_k^{(n_k)}\left[\| \mathbf{T}_1 \|_T, \| \mathbf{T}_2 \|_T, \ldots, \| \mathbf{T}_{n_k} \|_T \right], \tag{6.2.54}$$

$$n_T = n_k. \tag{6.2.55}$$

Definition 6.2.25. The polyadic operator norm is called *subadditive*, if

$$\left\| \nu_T^{(m_T)}\left[\mathbf{T}_1, \mathbf{T}_2, \ldots, \mathbf{T}_{n_T} \right] \right\|_T \leqslant \nu_{m_k}\left[\| \mathbf{T}_1 \|_T, \| \mathbf{T}_2 \|_T, \ldots, \| \mathbf{T}_{m_k} \|_T \right], \tag{6.2.56}$$

$$m_T = m_k. \tag{6.2.57}$$

By analogy with the binary case, we have

Definition 6.2.26. The polyadic operator algebra $\mathcal{A}_T$ equipped with the submultiplicative norm $\|\bullet\|_T$ is a *polyadic Banach algebra* of operators $\mathcal{B}_T$.

The connection between the polyadic norms of operators and their polyadic adjoints is given by

Proposition 6.2.27. *For polyadic operators in the inner pairing space* $\mathcal{H}_{m_k,n_k,m_V,k_\rho = 1,N}$

(i) *The following N multi-C^*-relations*

$$\Big\|\mu_T^{(m_T)}\Big[\mathbf{T}^{\star_{N-1,N}}, \mathbf{T}^{\star_{N-2,N-1}\star_{N-1,N}}, \ldots$$
$$\mathbf{T}^{\star_{23}\star_{34}\cdots\star_{N-2,N-1}\star_{N-1,N}}, \mathbf{T}^{\star_{12}\star_{23}\star_{34}\cdots\star_{N-2,N-1}\star_{N-1,N}}, \mathbf{T}_N\Big]\Big\|$$
$$= \mu_k^{(n_k)}\left[\overbrace{\| \mathbf{T} \|_T, \| \mathbf{T} \|_T, \ldots, \| \mathbf{T} \|_T}^{n_k}\right], \tag{6.2.58}$$

$+ (N - 1)$ cycle permutations of $(N - 1)$ terms with multistars,

take place, if $n_T = n_k$.

(ii) *The polyadic norms of operator and its all adjoints coincide*

$$\| \mathbf{T}^{\star_{i,j}} \|_T = \| \mathbf{T} \|_T, \forall i, j \in 1, \ldots, N. \tag{6.2.59}$$

Proof. Both statements follow from (6.2.21) and the definition of the polyadic operator norm (6.2.53). □

Therefore, we arrive at

Definition 6.2.28. The operator Banach algebra $\mathcal{B}_T$ satisfying the multi-C^*-relations is called a polyadic operator *multi-C^*-algebra.*

The first example of a multi-C^*-algebra (as in the binary case) can be constructed from one isometry operator (see definition 6.2.15).

Definition 6.2.29. A polyadic algebra generated by one isometry operator $\mathbf{T}$ satisfying (6.2.45) on the inner pairing space $\mathcal{H}_{m_k,n_k,m_V,k_\rho = 1,N}$ represents a *polyadic Toeplitz algebra* $\mathcal{T}_{m_T,n_T}$ and has the arity shape $m_T = m_V$, $n_T = N$.

Example 6.2.30. The ternary Toeplitz algebra $\mathcal{T}_{3,3}$ is represented by the operator $\mathbf{T}$ and relations

$$\begin{aligned}
[\mathbf{T}^{\star_1}, \mathbf{T}^{\star_3\star_1}, \mathbf{T}] &= \mathbf{I},\\
[\mathbf{T}^{\star_2}, \mathbf{T}^{\star_1\star_2}, \mathbf{T}] &= \mathbf{I},\\
[\mathbf{T}^{\star_3}, \mathbf{T}^{\star_2\star_3}, \mathbf{T}] &= \mathbf{I}.
\end{aligned} \tag{6.2.60}$$

Example 6.2.31. If the inner pairing is semicommutative (6.2.31), then $(\star_3)$ can be eliminated by

$$\mathbf{T}^{\star_3} = \mathbf{T}^{\star_1\star_2}, \tag{6.2.61}$$

$$\mathbf{T}^{\star_3\star_3} = \mathbf{T}, \tag{6.2.62}$$

and the corresponding relations representing $\mathcal{T}_{3,3}$ become

$$\begin{aligned} &[\mathbf{T}^{\star_1}, \mathbf{T}^{\star_1}, \mathbf{T}] = \mathbf{I}, \\ &[\mathbf{T}^{\star_2}, \mathbf{T}^{\star_1\star_2}, \mathbf{T}] = \mathbf{I}, \\ &[\mathbf{T}^{\star_1\star_2}, \mathbf{T}^{\star_2}, \mathbf{T}] = \mathbf{I}. \end{aligned} \tag{6.2.63}$$

Let us consider M polyadic operators $\mathbf{T}_1\mathbf{T}_2\ldots\mathbf{T}_M \in \mathcal{B}_T$ and the related partial (in the usual sense) isometries (6.2.50) which are mutually orthogonal (6.2.49). In the binary case, the algebra generated by M operators, such that the sum of the related orthogonal partial projections is unity, represents the Cuntz algebra $\mathcal{O}_M$ (Cuntz 1977).

Definition 6.2.32. A polyadic algebra generated by M polyadic isometric operators $\mathbf{T}_1\mathbf{T}_2\ldots\mathbf{T}_M \in \mathcal{B}_T$ satisfying

$$\left(\nu_T^{(m_T)}\right)^{\circ \ell_a}\left[\mathbf{P}_{\mathbf{T}_1}^{(k)}, \mathbf{P}_{\mathbf{T}_2}^{(k)}\ldots\mathbf{P}_{\mathbf{T}_M}^{(k)}\right] = \mathbf{I}, \quad k = 1, \ldots, N, \tag{6.2.64}$$

where $\mathbf{P}_{\mathbf{T}_i}^{(k)}$ are given by (6.2.51) and $(\nu_T^{(m_T)})^{\circ \ell_a}$ is a long (iterated) polyadic addition (6.2.16), represents a *polyadic Cuntz algebra* $p\mathcal{O}_{M|m_T,n_T}$, which has the arity shape

$$M = \ell_a(m_T - 1) + 1, \tag{6.2.65}$$

where ℓ_a is a number of 'm_T-ary additions'.

Below we use the same notations, as in example 6.2.13, also the ternary addition will be denoted by $(+_3)$ as follows: $\nu_T^{(3)}[\mathbf{T}_1, \mathbf{T}_2, \mathbf{T}_3] \equiv \mathbf{T}_1 +_3 \mathbf{T}_2 +_3 \mathbf{T}_3$.

Example 6.2.33. In the ternary case $m_T = n_T = 3$ and one ternary addition $\ell_a = 1$, we have $M = 3$ mutually orthogonal isometries $\mathbf{T}_1, \mathbf{T}_2, \mathbf{T}_3 \in \mathcal{B}_T$ and $N = 3$ multistars $(\star_i)$. In case of the Post-like multistar cocycles (6.2.40) they satisfy

$$\begin{array}{ll} \text{Isometry conditions} & \text{Orthogonality conditions} \\ \left[\mathbf{T}_i^{\star_1}, \mathbf{T}_i^{\star_3\star_1}, \mathbf{T}_i\right] = \mathbf{I}, & \left[\mathbf{T}_i^{\star_1}, \mathbf{T}_j^{\star_3\star_1}, \mathbf{T}_k\right] = \mathbf{Z}, \\ \left[\mathbf{T}_i^{\star_2}, \mathbf{T}_i^{\star_1\star_2}, \mathbf{T}_i\right] = \mathbf{I}, & \left[\mathbf{T}_i^{\star_2}, \mathbf{T}_j^{\star_1\star_2}, \mathbf{T}_k\right] = \mathbf{Z}, \\ \left[\mathbf{T}_i^{\star_3}, \mathbf{T}_i^{\star_2\star_3}, \mathbf{T}_i\right] = \mathbf{I}, & \left[\mathbf{T}_i^{\star_3}, \mathbf{T}_j^{\star_2\star_3}, \mathbf{T}_k\right] = \mathbf{Z}, \\ i = 1, 2, 3, & i, j, k = 1, 2, 3, \quad i \neq j \neq k, \end{array} \tag{6.2.66}$$

and the (sum of projections) relations

$$[\mathbf{T}_1, \mathbf{T}_1^{\star_1}, \mathbf{T}_1^{\star_3\star_1}] +_3 [\mathbf{T}_2, \mathbf{T}_2^{\star_1}, \mathbf{T}_2^{\star_3\star_1}] +_3 [\mathbf{T}_3, \mathbf{T}_3^{\star_1}, \mathbf{T}_3^{\star_3\star_1}] = \mathbf{I}, \tag{6.2.67}$$

$$[\mathbf{T}_1, \mathbf{T}_1^{\star_2}, \mathbf{T}_1^{\star_1\star_2}] +_3 [\mathbf{T}_2, \mathbf{T}_2^{\star_2}, \mathbf{T}_2^{\star_1\star_2}] +_3 [\mathbf{T}_3, \mathbf{T}_3^{\star_2}, \mathbf{T}_3^{\star_1\star_2}] = \mathbf{I}, \tag{6.2.68}$$

$$[\mathbf{T}_1, \mathbf{T}_1^{\star_3}, \mathbf{T}_1^{\star_2\star_3}] +_3 [\mathbf{T}_2, \mathbf{T}_2^{\star_3}, \mathbf{T}_2^{\star_2\star_3}] +_3 [\mathbf{T}_3, \mathbf{T}_3^{\star_3}, \mathbf{T}_3^{\star_2\star_3}] = \mathbf{I}, \tag{6.2.69}$$

which represent the ternary Cuntz algebra $p\mathcal{O}_{3|3,3}$.

Example 6.2.34. In the case where the inner pairing is semicommutative (6.2.31), we can eliminate the multistar ($\star_3$) by (6.2.61) and represent the *two-multistar ternary analog of the Cuntz algebra* $p\mathcal{O}_{3|3,3}$ by

$$\begin{array}{ll} \left[\mathbf{T}_i^{\star_1}, \mathbf{T}_i^{\star_2}, \mathbf{T}_i\right] = \mathbf{I}, & \left[\mathbf{T}_i^{\star_1}, \mathbf{T}_j^{\star_2}, \mathbf{T}_k\right] = \mathbf{Z}, \\ \left[\mathbf{T}_i^{\star_2}, \mathbf{T}_i^{\star_1\star_2}, \mathbf{T}_i\right] = \mathbf{I}, & \left[\mathbf{T}_i^{\star_1}, \mathbf{T}_j^{\star_1\star_2}, \mathbf{T}_k\right] = \mathbf{Z}, \\ \left[\mathbf{T}_i^{\star_1\star_2}, \mathbf{T}_i^{\star_2}, \mathbf{T}_i\right] = \mathbf{I}, & \left[\mathbf{T}_i^{\star_1\star_2}, \mathbf{T}_j^{\star_2}, \mathbf{T}_k\right] = \mathbf{Z}, \\ i = 1, 2, 3, & i, j, k = 1, 2, 3, \quad i \neq j \neq k, \end{array} \tag{6.2.70}$$

$$[\mathbf{T}_1, \mathbf{T}_1^{\star_1}, \mathbf{T}_1^{\star_1}] +_3 [\mathbf{T}_2, \mathbf{T}_2^{\star_1}, \mathbf{T}_2^{\star_1}] +_3 [\mathbf{T}_3, \mathbf{T}_3^{\star_1}, \mathbf{T}_3^{\star_1}] = \mathbf{I}, \tag{6.2.71}$$

$$[\mathbf{T}_1, \mathbf{T}_1^{\star_2}, \mathbf{T}_1^{\star_1\star_2}] +_3 [\mathbf{T}_2, \mathbf{T}_2^{\star_2}, \mathbf{T}_2^{\star_1\star_2}] +_3 [\mathbf{T}_3, \mathbf{T}_3^{\star_2}, \mathbf{T}_3^{\star_1\star_2}] = \mathbf{I}, \tag{6.2.72}$$

$$[\mathbf{T}_1, \mathbf{T}_1^{\star_1\star_2}, \mathbf{T}_1^{\star_2}] +_3 [\mathbf{T}_2, \mathbf{T}_2^{\star_1\star_2}, \mathbf{T}_2^{\star_2}] +_3 [\mathbf{T}_3, \mathbf{T}_3^{\star_1\star_2}, \mathbf{T}_3^{\star_2}] = \mathbf{I}. \tag{6.2.73}$$

References

Cuntz J 1977 Simple C^*-algebras generated by isometries *Commun. Math. Phys.* **57** 173–85

Duplij S 2012 Polyadic systems, representations and quantum groups *J. Kharkiv Natl Univ. Ser. Nuclei Part. Fields* **9** 28–59

Duplij S 2019 Arity shape of polyadic algebraic structures *J. Math. Phys. Anal. Geom.* **15** 3–56

Duplij S and Marcinek W 2001 Semisupermanifolds and regularization of categories, modules, algebras and Yang–Baxter equation (Int. Conf. on Supersymmetry and Quantum Field Theory: D.V. Volkov Memorial Conf., Kharkiv, Ukraine, July 2000) *Nucl. Phys. Proc. Suppl.* **102** 293–7

Duplij S and Marcinek W 2002 Regular obstructed categories and topological quantum field theory *J. Math. Phys* **43** 3329–41

Mignot F 1976 Contrôle dans les inéquations variationelles elliptiques *J. Funct. Anal.* **22** 130–85

Misiak A 1989 *n*-Inner product spaces *Math. Nachr.* **140** 299–319

IOP Publishing

Polyadic Algebraic Structures

Steven Duplij

Chapter 7

Medial deformation of n-ary algebras

Noncommutativity is the main mathematical idea of modern physics which, on the one hand, is grounded in quantum mechanics, in which generators of the algebra of observables do not commute, and, on the other hand, in supersymmetry and its variations based on the graded commutativity concept. Informally, 'deformation' of commutativity in algebras is mostly a special way to place a 'scalar' multiplier from the algebra field before the permuted product of two arbitrary elements. The general approach is based on projective representation theory and realized using almost commutative (ε-commutative) graded algebras, where the role of the multipliers is played by bicharacters of the grading group (as suitable 'scalar' objects taking values in the algebra field $\Bbbk$).

The commutativity property and its 'breaking' are quite obvious and unique for binary algebraic structures, because the permutation group S_2 has only one non-identity element. If the operation is n-ary, however, then one has $n! - 1$ non-identity permutations from S_n, and the uniqueness is lost. The standard way to bring uniqueness to an n-ary structure is by restricting to a particular n-ary commutation by fixing one chosen permutation using external (sometimes artificial) criteria (e.g., in Leibniz derivations or JB^*-triples). We introduce a different, canonical approach: to use another property which would be unique by definition but which can give commutativity in special cases. Murdoch (1939) (acting on n^2 elements) is such a property which can be substituted for commutativity (acting on n elements) in the generators/relations description of n-ary structures. For $n = 2$, mediality means $(ab)(cd) = (ac)(bd)$, and any medial magma is a commutative monoid; moreover, for binary groups commutativity immediately follows from mediality.

In this chapter we consider n-ary graded algebras and propose the following idea: instead of considering the non-unique commutativity property and its 'breaking', to investigate the unique property of mediality and its 'breaking' we exploit this 'commutativity-to-mediality' ansatz to introduce and study almost medial n-ary graded algebras by analogy with almost commutative algebras. By almost we refer to generalized or ε-commutative graded algebras (Rittenberg and Wyler 1978, Scheunert 1979) and β-commutative algebras (Bahturin *et al* 2003) (see, also Bongaarts and Pijls 1994, Covolo and Michel 2016, Morier-Genoud and Ovsienko 2010). For almost medial n-ary graded algebras, we prove an analog of Toyoda's theorem, which originally connected medial algebras with abelian algebras (Toyoda 1941). Note that almost co-mediality for polyadic bialgebras was introduced earlier in Duplij (2021). For other (binary) generalizations of grading, see, e.g., Dalzotto and Sbarra (2008), Elduque (2006), Nystedt (2020).

The proposed 'commutativity-to-mediality' ansatz can lead to medial n-ary superalgebras and Lie superalgebras as well as to a medial analog of noncommutative geometry.

7.1 Almost commutative graded algebra

The standard way to generalize the commutativity is using graded vector spaces and corresponding algebras together with the commutation factor defined on some abelian grading group (see, e.g., Rittenberg and Wyler 1978, Scheunert 1979, Bourbaki 1998, Nastasescu and Van Oystaeyen 2004).

Let $\mathcal{A} \equiv \mathcal{A}^{(2)} = \langle A | \mu^{(2)}, \nu^{(2)}; \lambda_1 \rangle$ be an associative (binary) algebra over a field $\Bbbk$ (having unit $1 \in \Bbbk$ and zero $0 \in \Bbbk$) with unit e (i.e., it is a unital $\Bbbk$-algebra) and zero $z \in A$. Here A is the underlying set and $\mu^{(2)}: A \otimes A \to A$ is the (bilinear) binary multiplication (which we write as $\mu^{(2)}[a, b]$, $a, b \in A$), usually in the binary case denoted by dot $\mu^{(2)} \equiv (\cdot)$, and $\nu^{(2)}: A \otimes A \to A$ is the (binary) addition denoted by $(+)$ a third (linear) operation λ_1 is the action $\lambda_1: K \otimes A \to A$ (widely called a 'scalar multiplication', but this is not always true, as can be seen from the polyadic case, Duplij 2019).

Informally, if $\mathcal{A}_2$ as a vector space can be decomposed into a direct sum, then one can introduce the *binary gradation concept*: each element $a \in A$ is endowed by an additional characteristic, its *gradation* denoted by a prime a' showing to which subspace it belongs, such that a' belongs to a discrete abelian group (initially $\mathbb{N}$ simply to 'enumerate' the subspaces, and this can be further generalized to a commutative semigroup). This group is called the *binary (abelian) grading group* $\mathcal{G}' = \langle G' | \nu_G^{(2)} \rangle$, and usually its operation is written as plus $\nu_G^{(2)} \equiv (+')$ and the neutral element (zero) by $0' \in G'$. Denote the subset of *homogeneous elements* of degree $a' \in G'$ by $A_{a'}$ (Scheunert 1979, Dade 1980).

Definition 7.1.1. An associative algebra, the algebra $\mathcal{A} \equiv \mathcal{A}^{\mathcal{G}}$ is called a *binary graded algebra* over $\Bbbk$ (or G-algebra $\mathcal{A}^{\mathcal{G}}$), if the algebra multiplication $\mu^{(2)}$ respects the gradation, i.e.,

$$\mu^{(2)} [A_{a'}, A_{b'}] \equiv A_{a'} \cdot A_{b'} \subseteq A_{a'+'b'}, \qquad \forall a', b' \in G', \tag{7.1.1}$$

where equality corresponds to *strong gradation*.

If there exist invertible elements of each degree $a' \in G'$, then $\mathcal{A}^{\mathcal{G}}$ is called a *cross product*, and if all non-zero homogeneous elements are invertible, $\mathcal{A}^{\mathcal{G}}$ is a *graded division algebra* (Dade 1980). Homogeneous (binary) morphisms $\varphi: \mathcal{A}^{\mathcal{G}} \to \mathcal{B}^{\mathcal{G}}$ preserve the grading $\varphi(A_{a'}) \subset B_{a'}, \forall a' \in G$, and the kernel of φ is an homogeneous ideal. The corresponding class of G-algebras and the homogeneous morphisms form a category of G-algebras G-`Alg` (for details, see, e.g., Bourbaki 1998, Dade 1980).

7.1.1 Almost commutativity

The graded algebras have a rich multiplicative structure, because of the possibility to deform (or twist) the algebra product $\mu^{(2)}$ by a function depending on the gradation. Let us consider the *twisting function* (*twist factor*)τ: $G' \times G' \to \Bbbk$.

Definition 7.1.2. A *twisted graded product* $\mu_\tau^{(2)}$ is defined for homogeneous elements by

$$\mu_\tau^{(2)}[a, b] = \tau(a', b')\mu^{(2)}[a, b], \quad a, b \in A; \; a', b' \in G'. \tag{7.1.2}$$

Proposition 7.1.3. *If the twisted algebra* $\left\langle A \middle| \mu_\tau^{(2)} \right\rangle$ *is associative, then the twisting function becomes a 2-cocycle* $\tau \mapsto \pi$: $G' \times G' \to \Bbbk^\times$ *on the abelian group* $\mathcal{G}$ *satisfying*

$$\pi(a', b')\pi(a' +' b', c') = \pi(a', b' +' c')\pi(b', c'), \quad a', b', c' \in G'. \tag{7.1.3}$$

Proof. The result follows from the binary associativity condition for $\mu_\pi^{(2)}$. □

Example **7.1.4.** An example of a solution to the functional equation (7.1.3) is $\pi(a', b') = (\exp(a'))^{b'}$.

The classes of π form the Schur multiplier group (Scheunert 1979), and for further properties of π and a connection with the cohomology classes $\mathsf{H}^2(\mathcal{G}', \Bbbk)$, see, e.g., Covolo and Michel (2016) and section 8.4.

In general, the twisted product (7.1.2) can be any polynomial in algebra elements. Nevertheless, the special cases where $\mu_{\varepsilon_0}^{(2)}[a, b]$ becomes a fixed expression for elements $a, b \in A$ are important.

Definition 7.1.5. If the twisted product coincides with the opposite product for all $a, b \in A$, we call the twisting function a0-*level commutation factor* $\tau \mapsto \varepsilon_0$: $G' \times G' \to \Bbbk^\times$, such that

$$\mu_{\varepsilon_0}^{(2)}[a, b] = \mu^{(2)}[b, a], \text{ or } \varepsilon_0(a', b')a \cdot b = b \cdot a, \quad \forall a, b \in A, \quad a', b' \in G'. \tag{7.1.4}$$

Definition 7.1.6. A binary algebra $\mathcal{A}_2^{(\varepsilon_0)}$ for which the twisted product coincides with the opposite product (7.1.4) is called 0-*level almost commutative* (ε_0-commutative).

Assertion 7.1.7. If the algebra for which (7.1.4) takes place is associative, the 0-level commutation factor[1] ε_0 satisfies the relations

$$\varepsilon_0(a', b')\varepsilon_0(b', a') = 1, \tag{7.1.5}$$

$$\varepsilon_0(a' +' b', c') = \varepsilon_0(a', c')\varepsilon_0(b', c'), \tag{7.1.6}$$

$$\varepsilon_0(a', b' +' c') = \varepsilon_0(a', b')\varepsilon_0(a', c'), \quad a', b', c', d' \in G', \ 1 \in \Bbbk. \tag{7.1.7}$$

Proof. The first relation (7.1.5) follows from permutation in (7.1.4) twice. The next ones follow from permutation in two ways: for (7.1.6) $a \cdot b \cdot c \mapsto a \cdot c \cdot b \mapsto c \cdot a \cdot b$ and $(a \cdot b) \cdot c \mapsto c \cdot (a \cdot b)$, and for (7.1.7) $a \cdot b \cdot c \mapsto b \cdot a \cdot c \mapsto b \cdot c \cdot a$ and $a \cdot (b \cdot c) \mapsto (b \cdot c) \cdot a$, using (7.1.4). □

In a more symmetric form this is

$$\varepsilon_0(a' +' b', c' +' d') = \varepsilon_0(a', c')\varepsilon_0(b', c')\varepsilon_0(a', d')\varepsilon_0(b', d'). \tag{7.1.8}$$

The following general expression

$$\varepsilon_0\left(\sum_{i_a=1}^{j_a} a'_{i_a}, \sum_{i_b=1}^{j_b} b'_{i_b}\right) = \prod_{i_a=1}^{j_a}\prod_{i_b=1}^{j_b} \varepsilon_0\left(a'_{i_a}, b'_{i_b}\right), \quad a'_{i_a}, b'_{i_b} \in G', \quad i_a, i_b, j_a, j_b \in \mathbb{N}, \tag{7.1.9}$$

can be written. In the case of equal elements we have

$$\varepsilon_0\left(j_a a', j_b b'\right) = \left(\varepsilon_0(a', b')\right)^{j_a j_b}. \tag{7.1.10}$$

***Remark* 7.1.8.** Recall that the standard commutation factor $\varepsilon\colon G' \times G' \to \Bbbk^\times$ of an almost commutative (ε-commutative or ε-symmetric) associative algebra is defined in a different way (Rittenberg and Wyler 1978, Scheunert 1979)

$$\varepsilon(a', b')b \cdot a = a \cdot b. \tag{7.1.11}$$

Comparing with (7.1.4) we have

$$\varepsilon_0(a', b') = \varepsilon(b', a'), \quad \forall a', b' \in G'. \tag{7.1.12}$$

[1] We denote levels of ε by lower indices which is 0 here.

7.1.2 Tower of higher level commutation brackets

Let us now construct the tower of higher level commutation factors and brackets using the following informal reasoning. We 'deform' the almost commutativity relation (7.1.4) by a function (a Lie bracket, Rittenberg and Wyler 1978, Scheunert 1979) $\mathcal{L}_0$: $A \times A \to A$ as

$$\varepsilon_0(a', b')a \cdot b = b \cdot a + \mathcal{L}_0^{(\varepsilon_0)}(a, b), \quad \forall\, a, b \in A, \quad a', b' \in G', \tag{7.1.13}$$

where $\varepsilon_0(a', b')$ is the 0-level commuting factor satisfying (7.1.5)–(7.1.7).

Consider the function (bracket)$\mathcal{L}_0^{(\varepsilon_0)}(a, b)$ as a multiplication $\mu^{(2)}_{\varepsilon_0, \mathcal{L}_0} = \mathcal{L}_0^{(\varepsilon_0)}(a, b)$ of a new algebra

$$\mathcal{A}_2^{\mathcal{L}_0} = \left\langle A \middle| \mu^{(2)}_{\varepsilon_0, \mathcal{L}_0} \right\rangle, \tag{7.1.14}$$

called a 0-*level bracket algebra*. Then (7.1.13) can be treated as its 'representation' by the associative algebra $\mathcal{A}$.

Proposition 7.1.9. *The algebra* $\mathcal{A}_2^{\mathcal{L}_0}$ *is almost commutative with the commutation factor* $(-\varepsilon_0^{-1})$.

Proof. Using (7.1.13) and (7.1.5)–(7.1.7) we get

$$\varepsilon_0(b', a')\mathcal{L}_0^{(\varepsilon_0)}(a, b) + \mathcal{L}_0^{(\varepsilon_0)}(b, a) = 0,$$

which can be rewritten in the almost commutativity form (7.1.4) as

$$(-\varepsilon_0(b', a'))\mathcal{L}_0^{(\varepsilon_0)}(a, b) = \mathcal{L}_0^{(\varepsilon_0)}(b, a).$$

It follows from (7.1.5) that

$$(-\varepsilon_0^{-1}(a', b'))\mathcal{L}_0^{(\varepsilon_0)}(a, b) = \mathcal{L}_0^{(\varepsilon_0)}(b, a). \tag{7.1.15}$$

□

The triple identity for $\mathcal{L}_0^{(\varepsilon_0)}(a, b)$ can be obtained using (7.1.5)–(7.1.7), (7.1.13) and (7.1.15)

$$\begin{aligned} &\varepsilon_0(c', a')\mathcal{L}_0^{(\varepsilon_0)}\Big(\mathcal{L}_0^{(\varepsilon_0)}(a, b), c\Big) + \varepsilon_0(a', b')\mathcal{L}_0^{(\varepsilon_0)}\Big(\mathcal{L}_0^{(\varepsilon_0)}(b, c), a\Big) \\ &+ \varepsilon_0(b', c')\mathcal{L}_0^{(\varepsilon_0)}\Big(\mathcal{L}_0^{(\varepsilon_0)}(c, a), b\Big) = 0, \quad \forall\, a, b, c \in A, \quad a', b', c' \in G'. \end{aligned} \tag{7.1.16}$$

In the more symmetric form using (7.1.8) we have

$$\begin{aligned} &\varepsilon_0(c', b')\varepsilon_0(d', a')\mathcal{L}_0^{(\varepsilon_0)}\Big(\mathcal{L}_0^{(\varepsilon_0)}(a, b), \mathcal{L}_0^{(\varepsilon_0)}(c, d)\Big) \\ &+ \varepsilon_0(d', c')\varepsilon_0(a', b')\mathcal{L}_0^{(\varepsilon_0)}\Big(\mathcal{L}_0^{(\varepsilon_0)}(b, c), \mathcal{L}_0^{(\varepsilon_0)}(d, a)\Big) \\ &+ \varepsilon_0(a', d')\varepsilon_0(b', c')\mathcal{L}_0^{(\varepsilon_0)}\Big(\mathcal{L}_0^{(\varepsilon_0)}(c, d), \mathcal{L}_0^{(\varepsilon_0)}(a, b)\Big) \\ &+ \varepsilon_0(b', a')\varepsilon_0(c', d')\mathcal{L}_0^{(\varepsilon_0)}\Big(\mathcal{L}_0^{(\varepsilon_0)}(d, a), \mathcal{L}_0^{(\varepsilon_0)}(b, c)\Big) = 0. \end{aligned} \tag{7.1.17}$$

By analogy with (7.1.13), we successively further 'deform' (7.1.15) then introduce 'deforming' functions and higher level commutation factors in the following way.

Definition 7.1.10. The *k-level commutation factors* $\varepsilon_k(a', b')$ and *k-level Lie brackets* $\mathscr{L}_k^{(\varepsilon_0,\varepsilon_1,\dots,\varepsilon_k)}(a, b)$ are defined by the following 'difference-like' equations

$$\varepsilon_1(a', b')\mathscr{L}_0^{(\varepsilon_0)}(a, b) = \mathscr{L}_0^{(\varepsilon_0)}(b, a) + \mathscr{L}_1^{(\varepsilon_0,\varepsilon_1)}(a, b), \tag{7.1.18}$$

$$\varepsilon_2(a', b')\mathscr{L}_1^{(\varepsilon_0,\varepsilon_1)}(a, b) = \mathscr{L}_1^{(\varepsilon_0,\varepsilon_1)}(b, a) + \mathscr{L}_2^{(\varepsilon_0,\varepsilon_1,\varepsilon_2)}(a, b), \tag{7.1.19}$$

$$\vdots$$

$$\varepsilon_k(a', b')\mathscr{L}_{k-1}^{(\varepsilon_0,\varepsilon_1,\dots,\varepsilon_{k-1})}(a, b) = \mathscr{L}_{k-1}^{(\varepsilon_0,\varepsilon_1,\dots,\varepsilon_{k-1})}(b, a) + \mathscr{L}_k^{(\varepsilon_0,\varepsilon_1,\dots,\varepsilon_k)}(a, b). \tag{7.1.20}$$

Definition 7.1.11. The *k-level almost commutativity* is defined by the vanishing of the last 'deforming' function

$$\mathscr{L}_k^{(\varepsilon_0,\varepsilon_1,\dots,\varepsilon_k)}(a, b) = 0, \quad \forall\, a, b \in A, \tag{7.1.21}$$

and can be expressed in a form analogous to (7.1.4)

$$\varepsilon_k(a', b')\mathscr{L}_{k-1}^{(\varepsilon_0,\varepsilon_1,\dots,\varepsilon_{k-1})}(a, b) = \mathscr{L}_{k-1}^{(\varepsilon_0,\varepsilon_1,\dots,\varepsilon_{k-1})}(b, a). \tag{7.1.22}$$

Proposition 7.1.12. *All higher level 'deforming' functions* $\mathscr{L}_i^{(\varepsilon_0,\varepsilon_1,\dots,\varepsilon_i)}$, $i = 1, \dots, k$ *can be expressed through* $\mathscr{L}_0^{(\varepsilon_0)}(a, b)$ *from* (7.1.13) *multiplied by a combination of the lower level commutation factors* $\varepsilon_i(a', b')$, $i = 1, \dots, k$.

Proof. This follows from equations (7.1.18)–(7.1.20). □

The first of such expressions are

$$\mathscr{L}_1^{(\varepsilon_0,\varepsilon_1)}(a, b) = [\varepsilon_1(a', b') + \varepsilon_0(b', a')]\mathscr{L}_0^{(\varepsilon_0)}(a, b), \tag{7.1.23}$$

$$\begin{aligned}\mathscr{L}_2^{(\varepsilon_0,\varepsilon_1,\varepsilon_2)}(a, b) = {} & [\varepsilon_2(a', b')(\varepsilon_1(a', b') + \varepsilon_0(b', a')) \\ & + \varepsilon_1(a', b')\varepsilon_0(b', a') + 1]\mathscr{L}_0^{(\varepsilon_0)}(a, b),\end{aligned} \tag{7.1.24}$$

$$\vdots$$

Recall the definition of the ε-Lie bracket (Scheunert 1979)

$$[a, b]_\varepsilon = a \cdot b - \varepsilon(a', b')b \cdot a, \quad \forall\, a, b \in A, \quad a', b' \in G'. \tag{7.1.25}$$

Assertion 7.1.13. The 0-level 'deforming' function $\mathscr{L}_0^{(\varepsilon_0)}(a, b)$ is the ε_0-twisted ε-Lie bracket

$$\mathscr{L}_0^{(\varepsilon_0)}(a, b) = \varepsilon_0(a', b')[a, b]_{\varepsilon=\varepsilon_0}. \tag{7.1.26}$$

Proof. This follows from (7.1.13) and (7.1.25). □

***Remark* 7.1.14.** The relations (7.1.16) and (7.1.17) are analogs of the ε-Jacobi identity of the ε-Lie algebra (Scheunert 1979).

Corollary 7.1.15. All higher level 'deforming' functions $\mathscr{L}_i^{(\varepsilon_0,\varepsilon_1,\ldots,\varepsilon_i)}$, $i = 1, \ldots, k$ can be expressed through the twisted ε-Lie bracket (7.1.25) with twisting coefficients.

7.2 Almost medial graded algebras

In search of a polyadic analog of almost commutativity, we will need some additional concepts, beyond the permutation of two elements (in the binary case), called commutativity, and various sums of permutations (of n elements, in n-ary case, which are usually non-unique). Instead we propose to consider a new concept, *polyadic mediality* (which gives a unique relation between n^2 elements in the n-ary case), as a polyadic inductive generalization of commutativity. We then twist the multiplication by a gradation (as in the binary case above) to obtain the polyadic version of almost commutativity as *almost mediality*. However, let us first recall the binary and polyadic versions of the mediality property.

7.2.1 Medial binary magmas and quasigroups

The mediality property was introduced as a generalization of the associative law for quasigroups, which are a direct generalization of abelian groups (Murdoch 1939). Other names for mediality are entropicity, bisymmetry, alternation and abelianness (see, e.g., Aczél 1948, Ježek and Kepka 1983, Evans 1963).

Let $\mathcal{M}^{(2)} = \left\langle M \middle| \mu_M^{(2)} \right\rangle$ be a binary *magma* (see definition 1.1.2).

Definition 7.2.1. A (binary) magma $\mathcal{M}^{(2)}$ is called *medial*, if

$$\mu_M^{(2)}\left[\mu_M^{(2)}[a, b], \mu_M^{(2)}[c, d]\right] = \mu_M^{(2)}\left[\mu_M^{(2)}[a, c], \mu_M^{(2)}[b, d]\right], \quad \forall a, b, c, d \in M. \tag{7.2.1}$$

Definition 7.2.2. We call the product of elements in the rhs of (7.2.1) *medially symmetric* to the lhs product.

Obviously, if a magma $\mathcal{M}^{(2)}$ contains a neutral element (identity) $e \in M$, such that $\mu_M^{(2)}[a, e] = \mu_M^{(2)}[e, a] = a$, $\forall a \in M$, then $\mathcal{M}^{(2)}$ is commutative $\mu_M^{(2)}[a, b] = \mu_M^{(2)}[b, a]$, $\forall a, b \in M$. Therefore, any commutative monoid is an example of a medial magma. Numerous different kinds of magma and their classifications are given in Ježek and Kepka (1983). If a magma $\mathcal{M}^{(2)}$ is cancellative ($\mu_M^{(2)}[a, b] = \mu_M^{(2)}[a, c] \Rightarrow b = c$, $\mu_M^{(2)}[a, c] = \mu_M^{(2)}[b, c] \Rightarrow a = b, \forall a, b, c \in M$), it is a binary quasigroup $\mathcal{Q}^{(2)} = \left\langle Q \middle| \mu_Q^{(2)} \right\rangle$

for which the equations $\mu_Q^{(2)}[a, x] = b$, $\mu_Q^{(2)}[y, a] = b$, $\forall a, b \in Q$, have a unique solution (Howroyd 1973). Moreover, as per (Sholander 1949), every medial cancellative magma can be embedded in a *medial quasigroup* (satisfying (7.2.1)), and the reverse statement is also true (Ježek and Kepka 1993). For a recent comprehensive review on quasigroups (including medial and n-ary ones), see, e.g., Shcherbacov (2017), and references therein.

The structure of medial quasigroups is determined by the Bruck–Murdoch–Toyoda theorem (Bruck 1944, Murdoch 1941, Toyoda 1941).

Theorem 7.2.3. Toyoda theorem. Any medial quasigroup $Q^{(2,\,\text{medial})} = \left\langle Q \middle| \mu_Q^{(2)} \right\rangle$ can be presented in the linear (functional) form

$$\mu_Q^{(2)}[a, b] = \nu_2[\nu_2[\varphi(a), \psi(b)], c] = \varphi(a) + \psi(b) + c, \quad \forall a, b, c \in Q, \tag{7.2.2}$$

where $\left\langle Q \middle| \nu_Q^{(2)} \equiv (+) \right\rangle$ is an abelian group and $\varphi, \psi \colon Q \to Q$ are commuting automorphisms $\varphi \circ \psi = \psi \circ \varphi$, and $c \in Q$ is fixed.

If $Q^{(2,\,\text{medial})}$ has an idempotent element (denoted here by 0), then

$$\mu_Q^{(2)}[a, b] = \nu_Q^{(2)}[\varphi(a), \psi(b)] = \varphi(a) + \psi(b), \quad \forall a, b \in Q. \tag{7.2.3}$$

It follows from the Toyoda theorem that medial quasigroups are isotopic to abelian groups, and their structure theories are very close (Bruck 1944, Murdoch 1941).

The mediality property (7.2.1) for semigroups leads to various consequences (Chrislock 1969, Nordahl 1974). Indeed, every medial semigroup $\mathcal{S}_{\text{medial}} = \left\langle S \middle| \mu_S^{(2)} \equiv (\cdot) \right\rangle$ is a *Putcha semigroup* ($b \in S^1aS^1 \Rightarrow b^m \in S^1a^2S^1$, $\forall a, b \in S$, $m \in \mathbb{N}$, $S^1 = S \cup \{1\}$), and therefore $\mathcal{S}_{\text{medial}}$ can be decomposed into the semilattice ($a^2 = a \wedge ab = ba$, $\forall a, b \in S$) of *Archimedean semigroups* ($\forall a, b \in S$, $\exists m, k \in \mathbb{N}$, $a^m = S^1bS^1 \wedge b^k = S^1aS^1 \wedge ab = ba$). If a medial semigroup $\mathcal{S}_{\text{medial}}$ is left (right) *cancellative*, $ab = ac \Rightarrow b = c$ ($ba = ca \Rightarrow b = c$), then it is left (right) *commutative* $abc = bac$ ($abc = acb$), $\forall a, b, c \in S$ and left (right) *separative*, $ab = a^2 \wedge ba = b^2$, $\forall a, b \in S$ ($ab = b^2 \wedge ba = a^2$) (for a review, see, Nagy 2001).

For a binary group $\left\langle G \middle| \mu_G^{(2)} \equiv (\cdot) \right\rangle$ mediality implies commutativity, because of cancellativity $abcd = acbd \Rightarrow bc = cb$, $\forall a, b, c, d \in G$. This is not the case for polyadic groups, where mediality implies semicommutativity only (see e.g., Głazek and Gleichgewicht 1982, Dog 2016).

7.2.2 Almost mediality

Let $\mathcal{A} = \mathcal{A}^{(2)} = \langle A | \mu^{(2)}, \nu^{(2)} \rangle$ be a binary $\Bbbk$-algebra which is not necessarily unital, cancellative and associative. Then mediality provides the corresponding behavior which depends on the properties of the 'vector multiplication' $\mu^{(2)} \equiv (\cdot)$.

For instance, for unital cancellative and associative algebras, mediality implies commutativity, as for groups (Głazek and Gleichgewicht 1982).

Consider an associative binary algebra $\mathcal{A}$ over a field $\Bbbk$. We introduce a weaker version of gradation than in (7.1.1).

Definition 7.2.4. An associative binary algebra $\mathcal{A}$ is called a *binary higher graded algebra* over $\Bbbk$, if the algebra multiplication of four (=2^2) elements respects the gradation

$$A_{a'} \cdot A_{b'} \cdot A_{c'} \cdot A_{d'} \subseteq A_{a' +' b' +' c' +' d'}, \forall a', b', c', d' \in G', \tag{7.2.4}$$

where equality corresponds to *strong higher gradation.*

Instead of (7.1.2) let us introduce the *higher twisting function* (*higher twist factor*) for four (=2^2) elements τ: $G'^{\times 4} \to \Bbbk$.

Definition 7.2.5. A *twisted (binary) higher graded product* $\mu_\tau^{(4)}$ is defined for homogeneous elements by

$$\mu_\tau^{(4)}[a, b, c, d] = \tau(a', b', c', d')a \cdot b \cdot c \cdot d, \quad a, b, c, d \in A; a', b', c', d' \in G'. \tag{7.2.5}$$

An analog of (total) associativity for the twisted binary higher graded product operation $\mu_\tau^{(4)}$ is the following condition on seven elements ($7 = 2 \cdot 2^2 - 1$) for all $a, b, c, d, t, u, v \in A$

$$\begin{aligned} &\mu_\tau^{(4)}\left[\mu_\tau^{(4)}[a, b, c, d], t, u, v\right] = \mu_\tau^{(4)}\left[a, \mu_\tau^{(4)}[b, c, d, t], u, v\right] \\ &= \mu_\tau^{(4)}\left[a, b, \mu_\tau^{(4)}[c, d, t, u], v\right] = \mu_\tau^{(4)}\left[a, b, c, \mu_\tau^{(4)}[d, t, u, v]\right]. \end{aligned} \tag{7.2.6}$$

Proposition 7.2.6. *If the twisted higher graded product satisfies the higher analog of associativity given by* (7.2.6), *then the twisting function becomes a higher analog of the cocycle* (7.1.3)$\tau \mapsto \pi$: $G'^{\times 4} \to \Bbbk^{\times}$ *on the abelian group* $\mathcal{G}$ *satisfying for all* $a', b', c', d', t', u', v' \in G'$

$$\begin{aligned} &\pi(a', b', c', d')\pi(a' +' b' +' c' +' d', t', u', v') \\ &= \pi(b', c', d', t')\pi(a', b' +' c' +' d' +' t', u', v') \\ &= \pi(c', d', t', u')\pi(a', b', c' +' d' +' t' +' u', v') \\ &= \pi(d', t', u', v')\pi(a', b', c', d' +' t' +' u' +' v'). \end{aligned} \tag{7.2.7}$$

Next we propose a medial analog of almost commutativity as follows. Instead of deforming commutativity by the grading twist factor ε_0 as in (7.1.4), we deform the mediality (7.2.1) by the higher twisting function τ (7.2.5).

Definition 7.2.7. If the higher twisted product coincides with the medially symmetric product (see (7.2.1)) for all $a, b \in A$, we call the twisting function a 0-*level mediality factor*[2] $\tau \mapsto \rho_0 : G'^{\times 4} \to \Bbbk^{\times}$, such that (see (7.1.4))

$$\mu_{\rho_0}^{(4)}[a, b, c, d] = a \cdot b \cdot c \cdot d, \text{ or} \tag{7.2.8}$$

$$\rho_0(a', b', c', d')a \cdot b \cdot c \cdot d = a \cdot c \cdot b \cdot d, \quad \forall a, b, c, d \in A, \quad a', b', c', d' \in G'. \tag{7.2.9}$$

From (7.2.9) follows the normalization condition for the mediality factor

$$\rho_0(a', a', a', a') = 1, \quad \forall a' \in G', \quad 1 \in \Bbbk. \tag{7.2.10}$$

Definition 7.2.8. A binary algebra $\mathcal{A}_{\rho_0}^{(2)} = \left\langle A \middle| \mu^{(2)}, \nu^{(2)}; \mu_{\rho_0}^{(4)} \right\rangle$ for which the higher twisted product coincides with the medially symmetric product $\mu_{\rho_0}^{(4)}[a, b, c, d] = a \cdot b \cdot c \cdot d$ (7.2.9) is called a 0-*level almost medial* (ρ_0-commutative) algebra.

Proposition 7.2.9. *If the algebra for which* (7.2.9) *holds is associative, the 0-level mediality factor* ρ_0 *satisfies the relations*

$$\rho_0(a', b', c', d')\rho_0(a', c', b', d') = 1, \quad a', b', c', d', f', g', h' \in G', \quad 1 \in \Bbbk, \tag{7.2.11}$$

$$\begin{aligned} &\rho_0(a', c' + 'd' + 'f' + 'g', b', h') = \rho_0(a', c', b', d') \\ &\times \rho_0(c', d', b', f')\rho_0(d', f', b', g')\rho_0(f', g', b', h'), \end{aligned} \tag{7.2.12}$$

$$\begin{aligned} &\rho_0(a', g', b' + 'c' + 'd' + 'f', h') = \rho_0(a', g', b'c') \\ &\times \rho_0(c', g', d', f')\rho_0(d', g', f', h')\rho_0(b', g', c', d'). \end{aligned} \tag{7.2.13}$$

Proof. As in (7.1.5), the relation (7.2.11) follows from applying (7.2.9) twice. The next ones follow from permutation in two ways using (7.2.8): for (7.2.12)

$$a \cdot b \cdot (c \cdot d \cdot f \cdot g) \cdot h \mapsto a \cdot (c \cdot d \cdot f \cdot g) \cdot b \cdot h, \quad a, c, d, f, g, b, h \in A, \tag{7.2.14}$$

[2] We denote the level by the lower index in ρ which is 0 here.

$$a\cdot b\cdot c\cdot d\cdot f\cdot g\cdot h\mapsto a\cdot c\cdot b\cdot d\cdot f\cdot g\cdot h\mapsto a\cdot c\cdot d\cdot b\cdot f\cdot g\cdot h \\ \mapsto a\cdot c\cdot d\cdot f\cdot b\cdot g\cdot h\mapsto a\cdot c\cdot d\cdot f\cdot g\cdot b\cdot h, \tag{7.2.15}$$

and for (7.2.13)

$$a\cdot(b\cdot c\cdot d\cdot f)\cdot g\cdot h\mapsto a\cdot g\cdot(b\cdot c\cdot d\cdot f)\cdot h, \tag{7.2.16}$$

$$a\cdot b\cdot c\cdot d\cdot f\cdot g\cdot h\mapsto a\cdot b\cdot c\cdot d\cdot g\cdot f\cdot h\mapsto a\cdot b\cdot c\cdot g\cdot d\cdot f\cdot h \\ \mapsto a\cdot b\cdot g\cdot c\cdot d\cdot f\cdot h\mapsto a\cdot g\cdot b\cdot c\cdot d\cdot f\cdot h. \tag{7.2.17}$$

Assertion 7.2.10. If the 0-level almost medial algebra $\mathcal{A}^{(2)}_{\rho_0}$ is cancellative, then it is isomorphic to an almost commutative algebra. □

Proof. After cancellation by a and d in (7.2.9), we obtain $\varepsilon_0(b', c')b\cdot c = c\cdot b$, where

$$\varepsilon_0(b', c') = \rho_0(a', b', c', d'), \qquad a', b', c', d' \in G'. \tag{7.2.18}$$

In case $\mathcal{A}^{(2)}_{\rho_0}$ is unital, one can choose $\varepsilon_0(b', c') = \rho_0(e', b', c', e') \equiv \rho_0(0', b', c', 0')$, $0' \in G'$, since the identity $e \in A$ is zero graded. □

7.2.3 Tower of higher binary mediality brackets

By analogy with (7.1.13), let us deform the medial twisted product $\mu^{(4)}_{\rho_0}$ (7.2.8) by the function (a medial analog of the Lie bracket (7.1.13))$M_0^{(\rho_0)}$: $A\times A\times A\times A\to A$ as follows

$$\rho_0(a', b', c', d')a\cdot b\cdot c\cdot d = a\cdot c\cdot b\cdot d + \mathcal{M}_0^{(\rho_0)}(a, b, c, d), \\ \forall a, b, c, d\in A, \quad a', b', c', d' \in G', \tag{7.2.19}$$

where ρ_0 is a 0-level mediality factor (7.2.8) which satisfies (7.2.11)–(7.2.13).

Let us next introduce a 4-ary multiplication $\mu^{(4)}_{\rho_0,\mathcal{M}_0}[a, b, c, d] = \mathcal{M}_0^{(\rho_0)}(a, b, c, d)$, $\forall a, b, c, d\in A$, see (7.2.19).

Definition 7.2.11. A 4-ary algebra

$$\mathcal{A}^{(4)}_{\rho_0,\mathcal{M}_0} = \left\langle A \middle| \mu^{(4)}_{\rho_0,\mathcal{M}_0}, \nu^{(2)} \right\rangle \tag{7.2.20}$$

is called a 0-*level medial bracket algebra*.

Proposition 7.2.12. *The 4-ary algebra* $\mathcal{A}^{(4)}_{\rho_0,\mathcal{M}_0}$ *is almost medial with the mediality factor* $\left(-\rho_0^{-1}\right)$.

Proof. Using (7.2.19) and (7.2.11)–(7.2.13) we get

$$\rho_0(a', c', b', d')\mathcal{M}_0^{(\rho_0)}(a, b, c, d) + \mathcal{M}_0^{(\rho_0)}(a, c, b, d) = 0,$$

which can be rewritten in the almost medial form (7.2.9) as

$$\left(-\rho_0(a', c', b', d')\right)\mathcal{M}_0^{(\rho_0)}(a, b, c, d) = \mathcal{M}_0^{(\rho_0)}(a, c, b, d).$$

From (7.2.11) we get

$$\left(-\rho_0^{-1}(a', b', c', d')\right)\mathcal{M}_0^{(\rho_0)}(a, b, c, d) = \mathcal{M}_0^{(\rho_0)}(a, c, b, d). \tag{7.2.21}$$

□

Let us 'deform' (7.2.9) again successively by introducing further 'deforming' functions $\mathcal{M}_k$ and higher level mediality factors ρ_k: $G' \times G' \times G' \times G' \to \Bbbk$ in the following way.

Definition 7.2.13. The *k-level mediality factors* $\rho_k(a', b', c', d')$ are defined by the following 'difference-like' recurrent equations

$$\rho_1(a', b', c', d')\mathcal{M}_0^{(\rho_0)}(a, b, c, d) = \mathcal{M}_0^{(\rho_0)}(a, c, b, d) + \mathcal{M}_1^{(\rho_0, \rho_1)}(a, b, c, d), \tag{7.2.22}$$

$$\rho_2(a', b', c', d')\mathcal{M}_1^{(\rho_0)}(a, b, c, d) = \mathcal{M}_1^{(\rho_0)}(a, c, b, d) + \mathcal{M}_2^{(\rho_0, \rho_1, \rho_2)}(a, b, c, d), \tag{7.2.23}$$

$$\vdots$$

$$\begin{aligned} \rho_k(a', b', c', d')\mathcal{M}_{k-1}^{(\rho_0, \rho_1, \ldots, \rho_{k-1})}(a, b, c, d) = \mathcal{M}_{k-1}^{(\rho_0, \rho_1, \ldots, \rho_{k-1})} \\ (a, c, b, d) + \mathcal{M}_k^{(\rho_0, \rho_1, \ldots, \rho_k)}(a, b, c, d), \end{aligned} \tag{7.2.24}$$

$$\forall a, b, c, d \in A, \quad a', b', c', d' \in G'.$$

Definition 7.2.14. The *k-level almost mediality* is defined by the vanishing of the last 'deforming' medial function

$$\mathcal{M}_k^{(\rho_0, \rho_1, \ldots, \rho_k)}(a, b, c, d) = 0, \quad \forall a, b, c, d \in A, \tag{7.2.25}$$

and can be expressed in a form analogous to (7.1.4) and (7.2.9)

$$\rho_k(a', b', c', d')\mathcal{M}_{k-1}^{(\rho_0, \rho_1, \ldots, \rho_{k-1})}(a, b, c, d) = \mathcal{M}_{k-1}^{(\rho_0, \rho_1, \ldots, \rho_{k-1})}(a, c, b, d). \tag{7.2.26}$$

Proposition 7.2.15. *The higher level 'deforming' functions* $\mathcal{M}_i^{(\rho_0, \rho_1, \ldots, \rho_i)}(a, b, c, d)$, $i = 1, \ldots, k$ *can be expressed through* $\mathcal{M}_0^{(\rho_0)}(a, b, c, d)$ *from* (7.2.19) *multiplied by a combination of the lower level mediality factors* $\rho_i(a', b', c', d')$, $i = 1, \ldots, k$.

Proof. It follows from equations (7.2.22)–(7.2.24). □

7.3 Medial n-ary algebras

We now extend the concept of almost mediality from binary to polyadic (n-ary) algebras in the unique way which uses the construction from the previous section.

Let $\mathcal{A}^{(n)} = \langle A | \mu^{(n)}, \nu^{(2)} \rangle$ be an associative n-ary algebra (with n-ary linear multiplication $A^{\otimes n} \to A$) over a field $\Bbbk$ with (possible) polyadic unit e (then $\mathcal{A}^{(n)}$ a unital $\Bbbk$-algebra) defined by $\mu^{(n)}[e^{n-1}, a] = a$, $\forall a \in A$ (where a can be on any place) and (binary) zero $z \in A$. We restrict ourselves (as in Michor and Vinogradov 1996, Goze *et al* 2010) by the binary addition $\nu^{(2)}: A \otimes A \to A$ which is denoted by $(+)$ (for more general cases, see Duplij 2019). To describe the mediality for arbitrary arity n we need the following matrix generalization of polyads (as was implicitly used in Duplij 2012, 2019).

Definition 7.3.1. A *matrix (n-ary) polyad* $\hat{A}_{(n^2)} \equiv \hat{A}_{(n\times n)}$ of size $n \times n$ is the sequence of n^2 elements $\hat{A}_{(n\times n)} = (a_{ij}) \in A^{\otimes n^2}$, $i, j = 1, \ldots, n$, and their product $\mathrm{A}_{n^2}^{(\mu)}: A^{\otimes n^2} \to A$ contains $n + 1$ of n-ary multiplications $\mu^{(n)}$, which can be written as (we use that for matrices of arguments, even informally)

$$\mathrm{A}_{n^2}^{(\mu)} \equiv \left(\mu^{(n)}\right)^{\circ(n+1)}\left[\hat{A}_{(n^2)}\right] = \mu^{(n)} \begin{bmatrix} \mu^{(n)}[a_{11}, a_{12}, \ldots, a_{1n}], \\ \mu^{(n)}[a_{21}, a_{22}, \ldots, a_{2n}], \\ \vdots \\ \mu^{(n)}[a_{n1}, a_{n2}, \ldots, a_{nn}] \end{bmatrix} \in A \tag{7.3.1}$$

due to the total associativity of $\mu^{(n)}$ (1.1.28) (by 'omitting brackets').

This construction is the stack reshape of a matrix or row-major order of an array.

***Example* 7.3.2.** In terms of matrix polyads the (binary) mediality property (7.2.1) becomes

$$\left(\mu^{(2)}\right)^{\circ(3)}\left[\hat{A}_{(4)}\right] = \left(\mu^{(2)}\right)^{\circ(3)}\left[\hat{A}_{(4)}^{T}\right], \quad \text{or} \quad \mathrm{A}_4^{(\mu)} = \mathrm{A}_4^{T(\mu)} \tag{7.3.2}$$

$$\hat{A}_{(4)} = \begin{pmatrix} a_{11} & a_{12} \\ a_{21} & a_{22} \end{pmatrix} \Rightarrow (a_{11}, a_{12}, a_{21}, a_{22}) \in A^{\otimes 4}, \tag{7.3.3}$$

where $\hat{A}_{(4)}^{T}$ is the transposed polyad matrix representing the sequence $(a_{11}, a_{21}, a_{12}, a_{22}) \in A^{\otimes 4}$, $\mathrm{A}_4^{(\mu)} = ((a_{11} \cdot a_{12}) \cdot (a_{21} \cdot a_{22})) \in A$ and $\mathrm{A}_4^{T(\mu)} = ((a_{11} \cdot a_{21}) \cdot (a_{12} \cdot a_{22})) \in A$ with $(\,\cdot\,) \equiv \mu_2$.

Definition 7.3.3. A *polyadic (n-ary) mediality* property is defined by the relation

$$\left(\mu^{(n)}\right)^{\circ(n+1)}\left[\hat{A}_{(n^2)}\right] = \left(\mu^{(n)}\right)^{\circ(n+1)}\left[\hat{A}_{(n^2)}^{T}\right], \quad \text{or} \quad \mathrm{A}_{n^2}^{(\mu)} = \mathrm{A}_{n^2}^{T(\mu)}, \tag{7.3.4}$$

$$\hat{A}_{(n^2)} = (a_{ij}) \in A^{\otimes n^2}. \tag{7.3.5}$$

Definition 7.3.4. A *polyadic medial twist map* $\chi^{(n^2)}_{\text{medial}}$ is defined on the matrix polyads as (Duplij 2021)

$$\hat{A}_{(n^2)} \overset{\chi^{(n^2)}_{\text{medial}}}{\mapsto} \hat{A}^{T}_{(n^2)}. \tag{7.3.6}$$

Definition 7.3.5. An n-ary algebra $\mathcal{A}^{(n)}$ is called *medial*, if it satisfies the n-ary mediality property (7.3.4) for all $a_{ij} \in A$.

It follows from (7.3.2) that not all medial binary algebras are abelian.

Corollary 7.3.6. If a binary medial algebra $\mathcal{A}^{(2)}$ is cancellative, it is abelian.

Assertion 7.3.7. *If an* n*-ary medial algebra* $\mathcal{A}^{(n)}$ *is cancellative, each matrix polyad* $\hat{A}_{(n^2)}$ *satisfies* $(n^2 - 2)$ *commutativity-like relations.*

7.4 Almost medial *n*-ary graded algebras

The gradation for associative n-ary algebras was considered in Michor and Vinogradov (1996), Gnedbaye (1995). Here we introduce a weaker version of gradation, because we need to define the grading twist not for n-ary multiplication, i.e., the polyads of the length n but only for the matrix polyads (7.3.5) of the length n^2 (for the binary case, see (7.2.4)).

Definition 7.4.1. An associative n-ary algebra $\mathcal{A}^{(n)}$ is called a *higher graded n-ary algebra* over $\Bbbk$, if the algebra multiplication of n^2 elements respects the gradation, i.e.,

$$\mu_n^{\circ(n+1)}\left[A_{\left(a'_{ij}\right)}\right] \equiv \mu_n\begin{bmatrix} \mu_n\left[A_{a'_{11}}, A_{a'_{12}}, \ldots, A_{a'_{1n}}\right], \\ \mu_n\left[A_{a'_{21}}, A_{a'_{22}}, \ldots, A_{a'_{21}}\right], \\ \vdots \\ \mu_n\left[A_{a'_{n1}}, A_{a'_{n2}}, \ldots, A_{a'_{nn}}\right] \end{bmatrix} \subseteq A_{a'_{11}+'\ldots+'a'_{nn}}, \tag{7.4.1}$$
$$\forall a'_{ij} \in G', \quad A_{a'_{ij}} \in A, \qquad i, j = 1, \ldots, n,$$

where equality corresponds to *strong higher gradation*.

Let us define the *higher twisting function* (*higher twist factor*) for n^2 elements τ_{n^2}: $G'^{\times n^2} \to \Bbbk$ by using matrix polyads (for $n = 2$ see (7.2.5)).

Definition 7.4.2. An *n-ary higher graded twisted product* $\mu_\tau^{(n^2)}$ is defined for homogeneous elements by

$$\mu_\tau^{(n^2)}\left[\hat{A}_{(n^2)}\right] = \tau_{n^2}\left(\hat{A}'_{(n^2)}\right)\mathrm{A}_{n^2}^{(\mu)}, \quad a_{ij} \in A; \;\; a'_{ij} \in G', \;\; i, j = 1, \ldots, n, \tag{7.4.2}$$

where $\hat{A}_{(n^2)} = (a_{ij}) \in A^{\otimes n^2}$ is the matrix polyad of elements (7.3.5), and $\hat{A}'_{(n^2)} = \left(a'_{ij}\right) \in G'^{\otimes n^2}$ is the matrix polyad of their gradings.

A medial analog of n-ary almost mediality can be introduced in a way analogous to the binary case (7.2.9).

Definition 7.4.3. If the higher twisted product coincides with the medially symmetric product (see (7.3.6)) for all $a_{ij} \in A$, we call the twisting function a 0-*level n-ary mediality factor* $\tau_{n^2} \mapsto \rho_0^{(n^2)}$: $G'^{\times n^2} \to \Bbbk^\times$, such that (see (7.1.4))

$$\mu_{\rho_0}^{(n^2)}\left[\hat{A}_{(n^2)}\right] = \mathrm{A}_{n^2}^{T(\mu)}, \quad \text{or} \tag{7.4.3}$$

$$\rho_0^{(n^2)}\left(\hat{A}'_{(n^2)}\right)\mathrm{A}_{n^2}^{(\mu)} = \mathrm{A}_{n^2}^{T(\mu)}, \quad a_{ij} \in A; \;\; a'_{ij} \in G', \;\; i, j = 1, \ldots, n. \tag{7.4.4}$$

It follows from (7.4.4) that the normalization condition for the n-ary mediality factor is

$$\rho_0^{(n^2)}\left(\overbrace{a', \ldots, a'}^{n^2}\right) = 1, \quad \forall a' \in G'. \tag{7.4.5}$$

Assertion 7.4.4. The 0-level n-ary mediality factor $\rho_0^{(n^2)}$ satisfies

$$\rho_0^{(n^2)}\left(\hat{A}'_{(n^2)}\right)\rho_0^{(n^2)}\left(\left(\hat{A}'_{(n^2)}\right)^T\right) = 1. \tag{7.4.6}$$

Proof. It follows from (7.4.4) and its transpose together with the relation $(B^T)^T = B$ for any matrix over $\Bbbk$. □

Definition 7.4.5. An n-ary algebra for which the higher twisted product coincides with the medially symmetric product (7.4.4), is called a 0-*level almost medial* (ρ_0-commutative) n-ary algebra $\mathcal{A}_n^{(\rho_0)}$.

Recall Bourbaki (1998), that a tensor product of binary algebras can be naturally endowed with a ε_0-graded structure in the following way (in our notation). Let $\mathcal{A}^{(2)}_{\varepsilon_0} = \left\langle A \middle| \mu^{(2)}_A, \nu^{(2)} \right\rangle$ and $\mathcal{B}^{(2)}_{\varepsilon_0} = \left\langle A \middle| \mu^{(2)}_B, \nu^{(2)} \right\rangle$ be binary graded algebras with the multiplications $\mu^{(2)}_A \equiv (\cdot_A)$ and $\mu^{(2)}_B \equiv (\cdot_B)$ and the same addition $\nu^{(2)}$ an the commutation factor ε_0 (see (7.1.4)), that is the same G-graded structure. Consider the tensor product $\mathcal{A}^{(2)}_{\varepsilon_0} \otimes \mathcal{B}^{(2)}_{\varepsilon_0}$ and introduce the *total ε_0-graded multiplication* $\left(\mathcal{A}^{(2)}_{\varepsilon_0} \otimes \mathcal{B}^{(2)}_{\varepsilon_0}\right) \star^{(\varepsilon_0)} \left(\mathcal{A}^{(2)}_{\varepsilon_0} \otimes \mathcal{B}^{(2)}_{\varepsilon_0}\right) \longrightarrow \mathcal{A}^{(2)}_{\varepsilon_0} \otimes \mathcal{B}^{(2)}_{\varepsilon_0}$ defined by the deformation (see (7.1.2))

$$\varepsilon_0(b_1', a_2')(a_1 \otimes b_1) \star^{(\varepsilon_0)} (a_2 \otimes b_2), \quad b_1', a_2' \in G', \, a_i \in A, \, b_i \in B, \, i = 1, 2. \quad (7.4.7)$$

Proposition 7.4.6. *If the ε_0-graded multiplication* (7.4.7) *satisfies (see* (7.1.4))

$$\varepsilon_0(b_1', a_2')(a_1 \otimes b_1) \star^{(\varepsilon_0)} (a_2 \otimes b_2) = (a_1 \cdot_A a_2) \otimes (b_1 \cdot_B b_2), \quad (7.4.8)$$

then $\mathcal{A}^{(2)}_{\varepsilon_0} \otimes \mathcal{B}^{(2)}_{\varepsilon_0}$ is a ε_0-graded commutative algebra.

Proposition 7.4.7. *If $\mathcal{A}^{(2)}_{\varepsilon_0}$ and $\mathcal{B}^{(2)}_{\varepsilon_0}$ are associative, then $\mathcal{A}^{(2)}_{\varepsilon_0} \otimes \mathcal{B}^{(2)}_{\varepsilon_0}$ is also associative.*

Proof. This follows from (7.4.7), (7.4.8) and the properties of the commutation factor ε_0 (7.1.6)–(7.1.7). □

In the matrix form (7.4.8) becomes (with $\star^{(\varepsilon_0)} \equiv \boldsymbol{\mu}^{(2)}_{\star\varepsilon_0}$)

$$\varepsilon_0(b_1', a_2')\boldsymbol{\mu}^{(2)}_{\star\varepsilon_0} \begin{bmatrix} \mu^{(2)}_{\otimes}[a_1, b_1] \\ \mu^{(2)}_{\otimes}[a_2, b_2] \end{bmatrix} = \mu^{(2)}_{\otimes} \begin{bmatrix} \mu^{(2)}_A[a_1, a_2] \\ \mu^{(2)}_B[b_1, b_2] \end{bmatrix}, \quad (7.4.9)$$

where $\mu^{(2)}_{\otimes}$ is the standard binary tensor product. For numerous generalizations (including braidings), see, e.g., López Peña *et al* (2007), and references therein.

Now we can extend (7.4.9) to almost medial algebras.

Definition 7.4.8. Let $\mathcal{A}^{(2)}_{\rho_0}$ and $\mathcal{B}^{(2)}_{\rho_0}$ be two binary medial algebras with the same addition and the mediality factor ρ_0. The total ρ_0-mediality graded multiplication $\boldsymbol{\mu}^{(2)}_{\star\rho_0}$: $\left(\mathcal{A}^{(2)}_{\rho_0} \otimes \mathcal{B}^{(2)}_{\rho_0}\right) \star^{(\rho_0)} \left(\mathcal{A}^{(2)}_{\rho_0} \otimes \mathcal{B}^{(2)}_{\rho_0}\right) \longrightarrow \mathcal{A}^{(2)}_{\rho_0} \otimes \mathcal{B}^{(2)}_{\rho_0}$ is defined by the mediality deformation (see (7.1.2))

$$\rho_0 \begin{pmatrix} a_1' & b_1' \\ a_2' & b_2' \end{pmatrix} (a_1 \otimes b_1) \star^{(\rho_0)} (a_2 \otimes b_2), \quad b_1', a_2' \in G', \, a_i \in A, \, b_i \in B, \, i = 1, 2. \quad (7.4.10)$$

Proposition 7.4.9. *If the ρ_0-graded multiplication* (7.4.10) *satisfies (see* (7.1.4))

$$\rho_0\begin{pmatrix} a_1' & b_1' \\ a_2' & b_2' \end{pmatrix}\boldsymbol{\mu}_{\star\rho_0}^{(2)}\begin{bmatrix} \mu_{\otimes}^{(2)}[a_1, b_1] \\ \mu_{\otimes}^{(2)}[a_2, b_2] \end{bmatrix} = \mu_{\otimes}^{(2)}\begin{bmatrix} \mu_A^{(2)}[a_1, a_2] \\ \mu_B^{(2)}[b_1, b_2] \end{bmatrix}, \tag{7.4.11}$$

then $\mathcal{A}_{\rho_0}^{(2)} \otimes \mathcal{B}_{\rho_0}^{(2)}$ is a ρ_0*-graded binary (almost medial) algebra.*

Using the matrix form (7.4.11) one can generalize the ρ_0-graded medial algebras to arbitrary arity.

Let $\mathcal{B}_{\rho_0,\,1}^{(n)}, \ldots, \mathcal{B}_{\rho_0,\,n}^{(n)}$ be n ρ_0-graded (almost medial) n-ary algebras ($\mathcal{B}_{\rho_0,\,i}^{(n)}= \left\langle B_i \middle| \mu_i^{(n)}, \nu^{(2)} \right\rangle$) with the same addition $\nu^{(2)}$, the mediality factor ρ_0 and the same graded structure. Consider their tensor product $\mathcal{B}_{\rho_0,\,1}^{(n)} \otimes \ldots \otimes \mathcal{B}_{\rho_0,\,n}^{(n)}$ and the ρ_0-graded n-ary multiplication $\boldsymbol{\mu}_{\star\rho_0}^{(n)}$ on it.

Proposition 7.4.10. *If the* ρ_0*-graded n-ary multiplication* $\boldsymbol{\mu}_{\star\varepsilon_0}^{(n)}$ *satisfies (see* (7.1.4)*)*

$$\rho_0\begin{pmatrix} b_1^{1'} & \ldots & b_1^{n'} \\ \vdots & \ldots & \vdots \\ b_n^{1'} & \ldots & b_n^{n'} \end{pmatrix}\boldsymbol{\mu}_{\star\rho_0}^{(n)}\begin{bmatrix} \mu_{\otimes}^{(n)}[b_1^1, \ldots, b_1^n] \\ \vdots \\ \mu_{\otimes}^{(n)}[b_n^1, \ldots, b_n^n] \end{bmatrix} = \mu_{\otimes}^{(n)}\begin{bmatrix} \mu_1^{(n)}[b_1^1, \ldots, b_n^1] \\ \vdots \\ \mu_n^{(n)}[b_n^1, \ldots, b_n^n] \end{bmatrix}, \tag{7.4.12}$$

$$b_1^i, \ldots, b_n^i \in \mathcal{B}_{\rho_0,\,i}^{(n)}, b_n^{i'}, \ldots, b_n^{i'} \in G', i = 1, \ldots, n, \tag{7.4.13}$$

then the tensor product $\mathcal{B}_{\rho_0,\,1}^{(n)} \otimes \ldots \otimes \mathcal{B}_{\rho_0,\,n}^{(n)}$ *is a* ρ_0*-graded n-ary (almost medial) algebra.*

Symbolically, we can write this in the form, similar to the almost mediality condition (7.4.4)

$$\rho_0\left(\hat{B}_{(n^2)}'\right)\boldsymbol{\mu}_{\star\rho_0}^{(n)} \circ \mu_{\otimes}^{(n^2)}\left[\hat{B}_{(n^2)}\right] = \mu_{\otimes}^{(n)} \circ \mu_1^{(n)} \circ \ldots \circ \mu_n^{(n)}\left[\hat{B}_{(n^2)}^T\right], \tag{7.4.14}$$

where

$$\hat{B}_{(n^2)}' = \begin{pmatrix} b_1^{1'} & \ldots & b_1^{n'} \\ \vdots & \ldots & \vdots \\ b_n^{1'} & \ldots & b_n^{n'} \end{pmatrix}, \quad \hat{B}_{(n^2)} = \begin{pmatrix} b_1^1 & \ldots & b_1^n \\ \vdots & \ldots & \vdots \\ b_n^1 & \ldots & b_n^n \end{pmatrix} \tag{7.4.15}$$

and $\hat{B}_{(n^2)}^T$ is its transpose.

***Example* 7.4.11.** In the lowest nonbinary example, for three ternary ρ_0-graded algebras $\mathcal{A}_{\rho_0}^{(3)} = \left\langle A \middle| \mu_A^{(3)}, \nu^{(2)} \right\rangle$, $\mathcal{B}_{\rho_0}^{(3)} = \left\langle B \middle| \mu_B^{(3)}, \nu^{(2)} \right\rangle$, $\mathcal{C}_{\rho_0}^{(3)} = \left\langle C \middle| \mu_C^{(3)}, \nu^{(2)} \right\rangle$, from

(7.4.12) we have the ternary multiplication $\boldsymbol{\mu}_{\star\rho_0}^{(3)}$ for their ternary tensor product $\mathcal{A}_{\rho_0}^{(3)} \otimes \mathcal{B}_{\rho_0}^{(3)} \otimes \mathcal{C}_{\rho_0}^{(3)}$ given by

$$\boldsymbol{\rho}_0 \begin{pmatrix} a_1' & b_1' & c_1' \\ a_2' & b_2' & c_2' \\ a_3' & b_3' & c_3' \end{pmatrix} \boldsymbol{\mu}_{\star\rho_0}^{(3)} \begin{pmatrix} (a_1 \otimes b_1 \otimes c_1) \\ (a_2 \otimes b_2 \otimes c_2) \\ (a_3 \otimes b_2 \otimes c_3) \end{pmatrix} \\ = \left(\mu_A^{(3)}[a_1, a_2, a_3] \otimes \mu_B^{(3)}[b_1, b_2, b_3] \otimes \mu_C^{(3)}[c_1, c_2, c_3]\right), \tag{7.4.16}$$

where $a_i \in A$, $b_i \in B$, $c_i \in C$, $a_i', b_i', c_i' \in G'$, $i = 1, 2, 3$.

7.4.1 Higher level mediality n^2-ary brackets

The almost mediality algebras were considered in (7.2.9) for the particular case $n = 2$, together with the tower of mediality factors (7.2.19), (7.2.22)–(7.2.24). Here we generalize this concept to any arity n which can be done using the matrix polyad construction.

First, we deform the almost mediality condition (7.4.4)

$$\rho_0^{(n^2)}\left(\hat{A}'_{(n^2)}\right)\mathrm{A}_{n^2}^{(\mu)} = \mathrm{A}_{n^2}^{T(\mu)} + \mathscr{M}_0^{(\rho_0)}\left(\hat{A}_{(n^2)}\right), \quad a_{ij} \in A; \;\; a_{ij}' \in G', \\ i, j = 1, \dots, n, \tag{7.4.17}$$

where $\mathscr{M}_0^{(\rho_0)}: A^{\otimes n^2} \to A$ is the *higher mediality n^2-ary bracket of 0-level*. Consider $\mathscr{M}_0^{(\rho_0)}$ as a new n^2-ary (bracket) multiplication

$$\mu_{\rho_0, \mathscr{M}_0}^{(n^2)}\left[\hat{A}_{(n^2)}\right] \equiv \mathscr{M}_0^{(\rho_0)}\left(\hat{A}_{(n^2)}\right). \tag{7.4.18}$$

Definition 7.4.12. An n^2-ary algebra

$$\mathcal{A}_{\rho_0, \mathscr{M}_0}^{(n^2)} = \left\langle A \middle| \mu_{\rho_0, \mathscr{M}_0}^{(n^2)}, \nu^{(2)} \right\rangle \tag{7.4.19}$$

is called a 0-*level mediality bracket* n^2*-ary algebra.*

Proposition 7.4.13. *The n^2-ary algebra $\mathcal{A}_{\rho_0, \mathscr{M}_0}^{(n^2)}$ is almost medial with the mediality factor $\left(-\left(\rho_0^{(n^2)}\right)^{-1}\right)$.*

Proof. We multiply the definition (7.4.17) by $\boldsymbol{\rho}_0^{(n^2)}\left(\left(\hat{A}'_{(n^2)}\right)^T\right)$ and use (7.4.6) to obtain

$$\boldsymbol{\rho}_0^{(n^2)}\left(\left(\hat{A}'_{(n^2)}\right)^T\right)\mathscr{M}_0^{(\rho_0)}\left(\hat{A}_{(n^2)}\right) = \mathrm{A}_{n^2}^{(\mu)} - \boldsymbol{\rho}_0^{(n^2)}\left(\left(\hat{A}'_{(n^2)}\right)^T\right)\mathrm{A}_{n^2}^{T(\mu)}. \tag{7.4.20}$$

Taking into account that the rhs here is exactly $-\mathcal{M}_0^{(\rho_0)}\left(\hat{A}_{(n^2)}^T\right)$, we have

$$-\rho_0^{(n^2)}\left(\left(\hat{A}'_{(n^2)}\right)^T\right)\mathcal{M}_0^{(\rho_0)}\left(\hat{A}_{(n^2)}\right) = \mathcal{M}_0^{(\rho_0)}\left(\hat{A}_{(n^2)}^T\right), \tag{7.4.21}$$

and using (7.4.6) again, we get

$$-\rho_0^{(n^2)}\left(\hat{A}'_{(n^2)}\right)^{-1}\mathcal{M}_0^{(\rho_0)}\left(\hat{A}_{(n^2)}\right) = \mathcal{M}_0^{(\rho_0)}\left(\hat{A}_{(n^2)}^T\right), \tag{7.4.22}$$

which should be compared with (7.4.4). □

Now we 'deform' (7.4.17) successively by defining further n^2-ary brackets $\mathcal{M}_k$ and higher level mediality factors $\rho_k^{(n^2)}: G'^{\times n^2} \to \Bbbk$ as follows.

Definition 7.4.14. The *k-level mediality* n^2-ary *brackets and factors are defined by*

$$\rho_1^{(n^2)}\left(\hat{A}'_{(n^2)}\right)\mathcal{M}_0^{(\rho_0)}\left(\hat{A}_{(n^2)}\right) = \mathcal{M}_0^{(\rho_0)}\left(\hat{A}_{(n^2)}^T\right) + \mathcal{M}_1^{(\rho_0,\rho_1)}\left(\hat{A}_{(n^2)}\right), \tag{7.4.23}$$

$$\rho_2^{(n^2)}\left(\hat{A}'_{(n^2)}\right)\mathcal{M}_1^{(\rho_0,\rho_1)}\left(\hat{A}_{(n^2)}\right) = \mathcal{M}_1^{(\rho_0,\rho_1)}\left(\hat{A}_{(n^2)}^T\right) + \mathcal{M}_2^{(\rho_0,\rho_1,\rho_2)}\left(\hat{A}_{(n^2)}\right), \tag{7.4.24}$$

$$\vdots$$

$$\rho_k^{(n^2)}\left(\hat{A}'_{(n^2)}\right)\mathcal{M}_{k-1}^{(\rho_0,\rho_1,\dots,\rho_{k-1})}\left(\hat{A}_{(n^2)}\right) = \mathcal{M}_{k-1}^{(\rho_0,\rho_1,\dots,\rho_{k-1})}\left(\hat{A}_{(n^2)}^T\right) + \mathcal{M}_k^{(\rho_0,\rho_1,\dots,\rho_k)}\left(\hat{A}_{(n^2)}\right)$$
$$\forall a_{ij} \in A;\ a'_{ij} \in G',\ i, j = 1, \dots, n. \tag{7.4.25}$$

Definition 7.4.15. *k-level* n^2-ary *almost mediality* is given by the vanishing of the last 'deforming' medial n^2-ary bracket

$$\mathcal{M}_k^{(\rho_0,\rho_1,\dots,\rho_k)}\left(\hat{A}_{(n^2)}\right) = 0, \quad \forall a_{ij} \in A, \tag{7.4.26}$$

and has the form

$$\rho_k^{(n^2)}\left(\hat{A}'_{(n^2)}\right)\mathcal{M}_{k-1}^{(\rho_0,\rho_1,\dots,\rho_{k-1})}\left(\hat{A}_{(n^2)}\right) = \mathcal{M}_{k-1}^{(\rho_0,\rho_1,\dots,\rho_{k-1})}\left(\hat{A}_{(n^2)}^T\right). \tag{7.4.27}$$

Proposition 7.4.16. *The higher level 'deforming' functions (n^2-ary brackets) $\mathcal{M}_i^{(\rho_0,\rho_1,\dots,\rho_k)}\left(\hat{A}_{(n^2)}\right)$, $i = 1, \dots, k$ can be expressed through $\mathcal{M}_0^{(\rho_0)}\left(\hat{A}_{(n^2)}\right)$ from* (7.4.17) *using a combination of the lower level n^2-ary mediality factors $\rho_k^{(n^2)}\left(\hat{A}'_{(n^2)}\right)$, $i = 1, \dots, k$.*

Proof. This follows from equations (7.4.23)–(7.4.25). □

7.5 Toyoda's theorem for almost medial algebras

The structure of the almost medial graded algebras (binary and n-ary) can be established by searching for possible analogs of Toyoda's theorem (7.2.2) (see, Bruck 1944, Murdoch 1941, Toyoda 1941) which is the main statement for medial groupoids (Ježek and Kepka 1983) and quasigroups (Shcherbacov 2017). As Toyoda's theorem connects medial algebras with abelian algebras, we can foresee that in the same way the almost medial algebras can be connected with almost commutative algebras.

First, let us consider almost medial graded binary algebras, as defined in (7.2.9)–(7.2.13).

Theorem 7.5.1. *Let* $\mathcal{A}_\rho^{(2)} = \left\langle A \middle| \mu_A^{(2)}, \nu^{(2)} \right\rangle$ *be an almost medial (ρ-commutative) G-graded binary algebra, then there exists an almost commutative (ε-commutative) G-graded binary algebra* $\mathcal{B}_\varepsilon^{(2)} = \left\langle B \middle| \mu_B^{(2)}, \nu^{(2)} \right\rangle$, *two grading preserving automorphisms* $\varphi_{1,2}$ *and a fixed element* $h \in A$, *such that (see* (7.2.2))

$$\mu_A^{(2)}[a, b] = \mu_B^{(2)}\left[\mu_B^{(2)}[\varphi_1(a), \varphi_2(b)], h\right] \text{ or } a \cdot b = \varphi_1(a)\varphi_2(b)h, \tag{7.5.1}$$

$$\rho(a', b', c', d') = \varepsilon(b', c'), \forall a, b, c, d \in A, \quad a', b', c', d' \in G', \tag{7.5.2}$$

where we denote $\mu_A^{(2)} \equiv (\cdot)$ *and* $\mu_B^{(2)}[a, b] \equiv ab$. *We use the multiplicative notation for the algebra* $\mathcal{B}_\varepsilon^{(2)}$, *because it is noncommutative, in general.*

Proof. We use the 'linear' presentation (7.5.1) for the product in $\mathcal{A}_\rho^{(2)}$ and insert it into the condition of almost mediality (7.2.9) to obtain

$$\begin{aligned}
&\rho(a', b', c', d')(a \cdot b) \cdot (c \cdot d) = (a \cdot c) \cdot (b \cdot d) \Rightarrow \\
&\rho(a', b', c', d')\varphi_1(\varphi_1(a)\varphi_2(b)h)\varphi_2((\varphi_1(c)\varphi_2(d)h))h \\
&\qquad = \varphi_1(\varphi_1(a)\varphi_2(c)h)\varphi_2((\varphi_1(b)\varphi_2(d)h)) \Rightarrow \\
&\rho(a', b', c', d')\varphi_1 \circ \varphi_1(a)\varphi_1 \circ \varphi_2(b)\varphi_1(h)\varphi_2 \circ \varphi_1(c)\varphi_2 \circ \varphi_2(d)\varphi_2(h)h \\
&\qquad = \varphi_1 \circ \varphi_1(a)\varphi_1 \circ \varphi_2(c)\varphi_1(h)\varphi_2 \circ \varphi_1(b)\varphi_2 \circ \varphi_2(d)\varphi_2(h)h,
\end{aligned} \tag{7.5.3}$$

where ($\circ$) is the composition of automorphisms. Using the cancellativity of $\mathcal{B}_\varepsilon^{(2)}$, we get

$$\rho(a', b', c', d')\varphi_1 \circ \varphi_2(b)\varphi_2 \circ \varphi_1(c) = \varphi_1 \circ \varphi_2(c)\varphi_2 \circ \varphi_1(b). \tag{7.5.4}$$

Because the automorphisms $\varphi_{1,2}$ preserve grading, after implementing almost (ε-) commutativity (7.1.4), the rhs of (7.5.4) becomes $\varepsilon(b', c')\varphi_2 \circ \varphi_1(b)\varphi_1 \circ \varphi_2(c)$ which gives (7.5.2) for commuting automorphisms. □

The higher arity cases are more nontrivial and very cumbersome. Therefore, we restrict ourselves by the case $n = 3$ only.

Theorem 7.5.2. *Let* $\mathcal{A}_{\rho}^{(3)} = \left\langle A \middle| \mu_A^{(3)}, \nu^{(2)} \right\rangle$ *be an almost medial* (ρ-commutative) *G-graded ternary algebra over a field* $\Bbbk$. *Then there exists an almost commutative* (ε-commutative) *G-graded binary algebra* $\mathcal{B}_{\varepsilon}^{(2)} = \left\langle A \middle| \mu_B^{(2)}, \nu^{(2)} \right\rangle$, *three commuting grading preserving automorphisms* $\varphi_{1,2,3}$ *and a fixed element* $h \in A$, *such that (see* (7.2.2))

$$\begin{aligned} \mu_A^{(3)}[a, b, c] = \mu_B^{(2)}\left[\mu_B^{(2)}\left[\mu_B^{(2)}[\varphi_1(a), \varphi_2(b)], h\right], h\right] \\ \equiv \varphi_1(a)\varphi_2(b)\varphi_3(c)h, \ \forall a, b, c, h \in A \end{aligned} \tag{7.5.5}$$

$$\begin{aligned} \rho^{(3^2)}\left(\hat{A}'_{(3^2)}\right) = \varepsilon(a'_{12}, a'_{31})\varepsilon(a'_{12}, a'_{21})\varepsilon(a'_{13}, a'_{31})\varepsilon(a'_{13}, a'_{32})\varepsilon(a'_{23}, a'_{32})\varepsilon(a'_{23}, a'_{31}), \\ \hat{A}'_{(3^2)} = \left(a'_{ij}\right), \ \forall a'_{ij} \in G', \quad i, j = 1, \dots, 3, \end{aligned} \tag{7.5.6}$$

where we denote $\mu_B^{(2)}[a, b] \equiv ab$.

Proof. Using the matrix form of ternary ($n = 3$) almost regularity (7.4.4) and inserting there the ternary 'linear' presentation (7.5.5) we get (in matrix form), $\forall a_{ij} \in A, \quad i, j = 1, \dots, 3$,

$$\rho^{(3^2)}\left(\hat{A}'_{(3^2)}\right)\begin{pmatrix} \varphi_1 \circ \varphi_1(a_{11}) & \varphi_1 \circ \varphi_2(a_{12}) & \varphi_1 \circ \varphi_3(a_{13}) \\ \varphi_2 \circ \varphi_1(a_{21}) & \varphi_2 \circ \varphi_2(a_{22}) & \varphi_2 \circ \varphi_3(a_{23}) \\ \varphi_3 \circ \varphi_1(a_{31}) & \varphi_3 \circ \varphi_2(a_{32}) & \varphi_3 \circ \varphi_3(a_{33}) \end{pmatrix} \tag{7.5.7}$$

$$= \begin{pmatrix} \varphi_1 \circ \varphi_1(a_{11}) & \varphi_1 \circ \varphi_2(a_{21}) & \varphi_1 \circ \varphi_3(a_{31}) \\ \varphi_2 \circ \varphi_1(a_{12}) & \varphi_2 \circ \varphi_2(a_{22}) & \varphi_2 \circ \varphi_3(a_{32}) \\ \varphi_3 \circ \varphi_1(a_{13}) & \varphi_3 \circ \varphi_2(a_{23}) & \varphi_3 \circ \varphi_3(a_{33}) \end{pmatrix}. \tag{7.5.8}$$

Applying the cancellativity of the binary algebra $\mathcal{B}_{\varepsilon}^{(2)}$, we have

$$\begin{aligned} \rho^{(3^2)}\left(\hat{A}'_{(3^2)}\right)\varphi_1 \circ \varphi_2(a_{12})\varphi_1 \circ \varphi_3(a_{13})\varphi_2 \circ \varphi_1(a_{21})\varphi_2 \circ \varphi_3(a_{23})\varphi_3 \circ \varphi_1(a_{31})\varphi_3 \circ \varphi_2(a_{32}) \\ = \varphi_1 \circ \varphi_2(a_{21})\varphi_1 \circ \varphi_3(a_{31})\varphi_2 \circ \varphi_1(a_{12})\varphi_2 \circ \varphi_3(a_{32})\varphi_3 \circ \varphi_1(a_{13})\varphi_3 \circ \varphi_2(a_{23}). \end{aligned} \tag{7.5.9}$$

Implementing almost (ε-) commutativity (7.1.4) on the rhs of (7.5.9), we arrive (for pairwise commuting grading preserving automorphisms $\varphi_i \circ \varphi_j = \varphi_j \circ \varphi_i$, $i, j = 1, 2, 3$) at (7.5.6). □

References

Aczél J 1948 On mean values *Bull. Am. Math. Soc.* **54** 392–400

Bahturin Y, Montgomery S and Zaicev M 2003 Generalized Lie solvability of associative algebras *Groups, Rings, Lie and Hopf Algebras* ed Y Bahturin (Dordrecht: Kluwer) pp 1–23

Bongaarts P J M and Pijls H G J 1994 Almost commutative algebra and differential calculus on the quantum hyperplane *J. Math. Phys.* **35** 959–70

Bourbaki N 1998 *Elements of Mathematics: Algebra* I (Berlin: Springer)

Bruck R H 1944 Some results in the theory of quasigroups *Trans. Am. Math. Soc.* **55** 19–52

Chrislock J L 1969 *On medial semigroups J. Alg* **12** 1–9

Covolo T and Michel J-P 2016 Determinants over graded-commutative algebras, a categorical viewpoint *L'Enseignement Math.* **62** 361–420

Dade E C 1980 Group-graded rings and modules *Math. Z.* **174** 241–62

Dalzotto G and Sbarra E 2008 On non-standard graded algebras *Toyama Math. J.* **31** 33–57

Dog S 2016 On entropicity in n-ary semigroups *Quasigr. Relat. Syst.* **24** 197–204

Duplij S 2012 Polyadic systems, representations and quantum groups *J. Kharkiv Natl Univ. Ser. Nuclei Part. Fields* **1017** 28–9

Duplij S 2019 Arity shape of polyadic algebraic structures *J. Math. Phys. Anal. Geom.* **15** 3–56

Duplij S 2021 Polyadic Hopf algebras and quantum groups East *Eur. J. Phys.* **2** 5–50

Elduque A 2006 A Lie grading which is not a semigroup grading *Linear Alg. Appl.* **418** 312–4

Evans T 1963 Abstract mean values *Duke Math. J.* **30** 331–47

Głazek K and Gleichgewicht B 1982 Abelian n-groups Universal algebra (Esztergom, 1977) *Colloq. Math. Soc. János Bolyai* vol 29 (Amsterdam: North-Holland) pp 321–9

Gnedbaye A V 1995 Les algèbres k-aires et leurs opérades *C. R. Acad. Sci. Paris Sér. I* **321** 147–52

Goze M, Goze N and Remm E 2010 n-Lie algebras *Afr. J. Math. Phys.* **8** 17–28

Howroyd T 1973 Cancellative medial groupoids and arithmetic means *Bull. Am. Math. Soc.* **8** 17–21

Ježek J and Kepka T 1983 *Medial Groupoids* (Praha: Academia)

Ježek J and Kepka T 1993 A note on medial division groupoids *Proc. Am. Math. Soc.* 119 423–6

López Peña J, Panaite F and Van Oystaeyen F 2007 General twisting of algebras *Adv. Math.* **212** 315–37

Michor P W and Vinogradov A M 1996 n-Ary Lie and associative algebras *Rend. Sem. Mat. Univ. Pol. Torino* **54** 373–92

Morier-Genoud S and Ovsienko V 2010 Simple graded commutative algebras *J. Algebra* **323** 1649–64

Murdoch D C 1939 Quasi-groups which satisfy certain generalized associative laws *Am. J. Math.* **61** 509–22

Murdoch D C 1941 Structure of abelian quasi-groups *Trans. Am. Math. Soc.* **49** 392–409

Nagy A 2001 *Special Classes of Semigroups* (Kluwer: Dordrecht)

Nastasescu C and Van Oystaeyen F 2004 *Methods of Graded Rings* (Berlin: Springer)

Nordahl T E 1974 Medial semigroups *PhD Thesis* University of California, Davis

Nystedt P 2020 Noncommutatively graded algebras *Algebraic Structures and Applications* ed S Silvestrov, A Malyarenko and M Rančić (Cham: Springer) pp 371–383

Rittenberg V and Wyler D 1978 Generalized superalgebras *Nucl. Phys.* B **139** 189–202

Scheunert M 1979 Generalized Lie algebras *J. Math. Phys.* **20** 712–20

Shcherbacov V 2017 *Elements of Quasigroup Theory and Applications* (Boca Raton, FL: CRC Press)

Sholander M 1949 On the existence of the inverse operation in alternation groupoids *Bull. Am. Math. Soc.* **55** 746–57

Toyoda K 1941 On axioms of linear functions *Proc. Imperial Acad. Tokyo.* **17** 221–7

Chapter 8

Membership deformations and obscure n-ary algebras

In this chapter we propose, principally, another new mechanism for 'deformation' of commutativity which comes from incorporating the ideas of vagueness in logic (Smith 2008) to algebra. First, take the underlying set of the algebra not as a crisp set, but as an obscure/fuzzy set (Dubois and Prade 2000). Second, consider an algebra (over $\Bbbk = \mathbb{C}$), such that each element can be endowed with a membership function (representing the degree of truth), a scalar function that takes values in the unit interval and describes the containment of a given element in the underlying set (Bělohlávek 2002). Third, introduce a special 'membership deformation' of the commutation relations, so to speak the difference of the degree of truth, which determines a 'measure of noncommutativity', whereby the elements having equal membership functions commute. Such procedure can be also interpreted as the 'continuous noncommutativity', because the membership function is usually continuous. Likewise we 'deform' ε-commutativity relations and ε-Lie algebras (Scheunert 1979, Rittenberg and Wyler 1978). Then we universalize and apply the above constructions to n-ary algebras (Michor and Vinogradov 1996) for which we also study projective representations generalizing the binary ones (Žmuď 1972).

8.1 Graded algebras and Shur factors

First recall the main features of the standard gradation concept and of generalized (almost) commutativity (or ε-commutativity) (see Scheunert 1979, Rittenberg and Wyler 1978 and section 7.1) in the slightly different notations which will be convenient to generalize them to the obscure and polyadic cases simultaneously.

Let $\Bbbk$ be a unital ground field (with unit $1 \in \Bbbk$ and zero $0 \in \Bbbk$) and $\mathcal{A} = \langle A | \cdot, + \rangle$ be an associative algebra over $\Bbbk$ having a zero $z \in A$ and unit $e \in A$ (for unital algebras). A *graded algebra* $\mathcal{A}^{\mathcal{G}}$ (G-graded $\Bbbk$-algebra) is a direct sum of subalgebras $\mathcal{A}^{\mathcal{G}} = \oplus_{a' \in G'} \mathcal{A}_{a'}$, where $\mathcal{G}' = \langle G' | +' \rangle$ is a *grading group* (an abelian (finite) group with 'unit' $0' \in G'$) and the set multiplication 'respects gradation' (7.1.1) $\mathcal{A}_{a'} \cdot \mathcal{A}_{b'} \subseteq \mathcal{A}_{a'+'b'}$, $a', b' \in G'$. Elements of the subsets $a = a_{(a')} \in \mathcal{A}_{a'}$ (with 'full membership') are *G-homogeneous of degree* $a' \in \mathcal{A}_{a'}$.

The graded algebra $\mathcal{A}^{\mathcal{G}}$ is called a *cross product,* if in each subalgebra $\mathcal{A}_{a'}$ there exists at least one invertible element. If all nonzero homogeneous elements are invertible, then $\mathcal{A}^{\mathcal{G}}$ is called a *graded division algebra* (Dade 1980). The morphisms $\varphi: \mathcal{A}^{\mathcal{G}} \to \mathcal{B}_{\mathcal{G}}$ acting on homogeneous elements (from $\mathcal{A}_{a'}$) should be compatible with the grading $\varphi(\mathcal{A}_{a'}) \subset \mathcal{B}_{a'}$, $\forall a' \in G'$, while $\ker \varphi$ is an homogeneous ideal. The category of binary G-graded algebras G-`Alg` consists of the corresponding class of algebras and the homogeneous morphisms (see for details, e.g., Bourbaki 1998, Dade 1980).

Recall, that in binary graded algebras there exists a way to generalize noncommutativity such that it can be dependent on the gradings ('coloring') (Scheunert 1979, Rittenberg and Wyler 1978). Indeed, some (two-place) function on grading degrees (bicharacter), a *(binary*[1]*) commutation factor* $\varepsilon^{(2)}: G' \times G' \to \Bbbk^{\times}$ (where $\Bbbk^{\times} = \Bbbk \backslash 0$) can be introduced (Scheunert 1979, Rittenberg and Wyler 1978)

$$a \cdot b = \varepsilon^{(2)}(a', b') b \cdot a, \quad \forall a, b \in \mathcal{A}, \quad \forall a', b' \in G'. \tag{8.1.1}$$

The properties of the commutation factor $\varepsilon^{(2)}$ under double permutation and associativity (7.1.5)–(7.1.7) make $\varepsilon^{(2)}$ a special 2-cocycle on the group $\mathcal{G}$ (Bourbaki 1998). The conditions (7.1.5)–(7.1.7) also imply that $\varepsilon^{(2)}(a', b') \neq 0$, $(\varepsilon^{(2)}(a', a'))^2 = 1$, $\varepsilon^{(2)}(a', 0') = \varepsilon^{(2)}(0', a') = 1$, where $0' \in G'$. A graded algebra $\mathcal{A}^{\mathcal{G}}$ endowed with the commutation factor $\varepsilon^{(2)}$ satisfying (7.1.5)–(7.1.7) is called an *almost-commutative* (*$\varepsilon^{(2)}$-commutative, color*) *binary algebra* (Scheunert 1979, Rittenberg and Wyler 1978) (for a review, see de Goursac *et al* 2012) and section 7.1.

The simplest example of a commutation factor is a *sign rule*

$$\varepsilon^{(2)}(a', b') = (-1)^{<a'', b''>}, \tag{8.1.2}$$

where $<,>: G' \times G' \to \mathbb{Z}_2$ is a bilinear form ('scalar product'), and for $G' = \mathbb{Z}_2$ the form is a product i.e., $<a', b'> = a'b' \in \mathbb{Z}_2$. This gives the standard supercommutative algebra (Kac 1977, Nahm *et al* 1976, Berezin 1987).

In the case $\mathcal{G}' = \mathbb{Z}_2^n$ the 'scalar product' $<,>: G' \times G' \to \mathbb{Z}$ is defined by

$$<(a_1' \ldots a_n'), (b_1' \ldots b_n')> = a_1'b_1' + \cdots + a_n'b_n' \in \mathbb{Z}, \tag{8.1.3}$$

which leads to $\mathbb{Z}_2^n$-commutative associative algebras (Covolo *et al* 2016).

A classification of the commutation factors $\varepsilon^{(2)}$ can be made in terms of the *factors (binary Schur multipliers)* $\pi^{(2)}: G' \times G' \to \Bbbk^{\times}$ such that (Žmuď 1960, Scheunert 1979)

$$\varepsilon_{\pi}^{(2)}(a', b') = \frac{\pi^{(2)}(a', b')}{\pi^{(2)}(b', a')}, \quad \forall a', b' \in G'. \tag{8.1.4}$$

[1] We denote arity (or number of places) by the upper index which is 2 here.

The Schur factors $\pi^{(2)}$ naturally appear in projective representation theory (Žmuď 1972) and satisfy the relation

$$\pi^{(2)}(a', b' + 'c')\pi^{(2)}(b', c') = \pi^{(2)}(a', b')\pi^{(2)}(a' + 'b', c'), \quad \forall a', b' \in G', \quad (8.1.5)$$

which follows from associativity. Using (8.1.4) we can rewrite the $\varepsilon^{(2)}$-commutation relation (8.1.1) of the graded algebra $\mathcal{A}^{\mathcal{G}}$ in the form

$$\pi^{(2)}(b', a')a \cdot b = \pi^{(2)}(a', b')b \cdot a, \quad \forall a, b \in A, \quad \forall a', b' \in G'. \quad (8.1.6)$$

We call (8.1.6) a *π-commutativity*. A factor set $\left\{\pi^{(2)}_{\text{sym}}\right\}$ is *symmetric*, if

$$\pi^{(2)}_{\text{sym}}(b', a') = \pi^{(2)}_{\text{sym}}(a', b'), \quad (8.1.7)$$

$$\varepsilon^{(2)}_{\pi_{\text{sym}}}(a', b') = 1, \quad \forall a', b' \in G', \quad (8.1.8)$$

and so the G-graded algebra $\mathcal{A}^{\mathcal{G}}$ becomes commutative in the ordinary sense (abelian). For further details, see section 8.4.

8.2 Membership function and obscure algebras

Now let us consider a generalization of associative algebras and graded algebras to the case where the degree of truth (containment of an element in the underlying set) is not full or distinct (for a review, see, e.g., Dubois and Prade 2000, Zimmermann 2011). In this case, an element $a \in A$ of the algebra $\mathcal{A} = \langle A | \cdot, + \rangle$ (over $\Bbbk$) can be endowed with an additional function, the *membership function* $\mathsf{f}\colon A \to [0, 1]$, $0, 1 \in \Bbbk$, which 'measures' the degree of truth as a 'grade of membership' (Bělohlávek 2002). If A is a *crisp set* then $\mathsf{f} \in \{0, 1\}$ and $\mathsf{f}^{\text{crisp}}(a) = 1$, if $a \in A$ and $\mathsf{f}^{\text{crisp}}(a) = 0$, if $a \notin A$. Also, we assume that the zero has full membership $\mathsf{f}(z) = 1$, $z \in A$ (for details, see, e.g., Dubois and Prade 2000, Zimmermann 2011). If the membership function f is positive, we can identify an *obscure (fuzzy) set* $A^{(\mathsf{f})}$ with support the universal set A consisting of pairs

$$A^{(\mathsf{f})} = \{(a|\mathsf{f}(a))\}, \; a \in A, \; \mathsf{f}(a) > 0. \quad (8.2.1)$$

Sometimes, instead of operations with obscure sets it is convenient to consider the corresponding operations only in terms of the membership function f itself. Denote $a \wedge b = \min\{a, b\}$, $a \vee b = \max\{a, b\}$. Then for inclusion $\subseteq$, union $\cup$, intersection $\cap$ and negation of obscure sets one can write $\mathsf{f}(a) \leqslant \mathsf{f}(b)$, $\mathsf{f}(a) \vee \mathsf{f}(b)$, $\mathsf{f}(a) \wedge \mathsf{f}(b)$, $1 - \mathsf{f}(a)$, $a, b \in A^{(\mathsf{f})}$, respectively.

Definition 8.2.1. An *obscure algebra* $\mathcal{A}(\mathsf{f}) = \langle A^{(\mathsf{f})} | \cdot, + \rangle$ is an algebra over $\Bbbk$ having an obscure set $A^{(\mathsf{f})}$ as its underlying set, where the following conditions hold

$$\mathsf{f}(a + b) \geqslant \mathsf{f}(a) \wedge \mathsf{f}(b), \quad (8.2.2)$$

$$\mathsf{f}(a \cdot b) \geqslant \mathsf{f}(a) \wedge \mathsf{f}(b), \quad (8.2.3)$$

$$\mathsf{f}(ka) \geqslant \mathsf{f}(a), \ \forall a, b \in A, \ k \in \Bbbk. \tag{8.2.4}$$

Definition 8.2.2. A *obscure unity* η is given by $\eta(a) = 1 \in \Bbbk$, if $a = 0$ and $\eta(a) = 0 \in \Bbbk$, if $a \neq 0$.

For two obscure sets their direct sum $A^{(\mathsf{f})} = A^{(\mathsf{f}_{a'})} \oplus A^{(\mathsf{f}_{b'})}$ can be defined, if $\mathsf{f}_{a'} \wedge \mathsf{f}_{b'} = \eta$, and the joint membership function has two 'nonintersecting' components

$$\mathsf{f}(a) = \mathsf{f}_{a,}(a_{(a')}) \vee \mathsf{f}_{b'}(a_{(b')}), \ \mathsf{f}_{a'} \wedge \mathsf{f}_{b'} = \eta, \tag{8.2.5}$$

$$a = a_{(a')} + a_{(b')}, \ a \in A^{(\mathsf{f})}, \ a_{(a')} \in A^{(\mathsf{f}_{a'})}, \ a_{(b')} \in A^{(\mathsf{f}_{b'})}, \ a', \ b' \in G', \tag{8.2.6}$$

satisfying

$$\mathsf{f}(a_{(a')} \cdot a_{(b')}) \geqslant \mathsf{f}_{a'}(a_{(a')}) \wedge \mathsf{f}_{b'}(a_{(b')}). \tag{8.2.7}$$

Definition 8.2.3. An *obscure G-graded algebra* is a direct sum decomposition

$$\mathcal{A}^{\mathcal{G}}(\mathsf{f}) = \oplus_{a' \in G'} \mathcal{A}(\mathsf{f}_{a'}), \tag{8.2.8}$$

such that the relation (7.1.1) holds and the joint membership function is

$$\mathsf{f}(a) = \bigvee_{a' \in G'} \mathsf{f}_{a'}(a_{(a')}), \qquad a = \sum_{a' \in G'} a_{(a')}, \ \forall a \in A^{(\mathsf{f})}, \ a_{(a')} \in A^{(\mathsf{f}_{a'})}, \ a' \in G'. \tag{8.2.9}$$

8.3 Membership deformation of commutativity

The membership concept leads to the question whether is it possible to generalize commutativity and $\varepsilon^{(2)}$-commutativity (8.1.1) for the obscure algebras? A positive answer to this question can be based on a consistent usage of the membership f as ordinary functions which are predefined for each element and satisfy some conditions (see, e.g., Dubois and Prade 2000, Zimmermann 2011). In this case, the commutation factor (and the Schur factors) may depend not only on the element grading, but also on the element membership function f, and therefore it becomes 'individual' for each pair of elements, and moreover it can be continuous. We call this procedure a *membership deformation* of commutativity.

8.3.1 Deformation of commutative algebras

Let us consider an obscure commutative algebra $\mathcal{A}(\mathsf{f}) = \langle A^{(\mathsf{f})} | \cdot, + \rangle$ (see definition 8.2.1), in which $a \cdot b = b \cdot a, \forall a, b \in A^{(\mathsf{f})}$. Now we 'deform' this commutativity by the membership function f and introduce a new algebra product $(*)$ for the elements in A to get a noncommutative algebra.

Definition 8.3.1. An *obscure membership deformed algebra* is $\mathcal{A}_*(\mathsf{f}) = \langle A^{(\mathsf{f})} | *, + \rangle$ in which the noncommutativity relation is

$$\mathsf{f}(b) a * b = \mathsf{f}(a) b * a, \quad \forall\, a, b \in A^{(\mathsf{f})}. \tag{8.3.1}$$

***Remark* 8.3.2.** The relation (8.3.1) is reminiscent of (8.1.6), where the role of the Schur factors is played by the membership function f which depends on the element itself, but not on the element's grading. Therefore, the membership deformation of commutativity (8.3.1) is highly nonlinear as opposed to gradation.

Because in our consideration the membership function f cannot be zero, it follows from (8.3.1) that

$$a * b = \epsilon_{\mathsf{f}}^{(2)}(a, b) b * a, \tag{8.3.2}$$

$$\epsilon_{\mathsf{f}}^{(2)}(a, b) = \frac{\mathsf{f}(a)}{\mathsf{f}(b)}, \quad \forall\, a, b \in A^{(\mathsf{f})}, \tag{8.3.3}$$

where $\epsilon_{\mathsf{f}}^{(2)}$ is the *membership commutation factor*.

Corollary 8.3.3. An obscure membership deformed algebra $\mathcal{A}_*(\mathsf{f})$ is a (kind of) ϵ_{f}-commutative algebra with a membership commutation factor (8.3.3) which now depends not on the element gradings (as in (8.1.4)), but on the membership function f.

***Remark* 8.3.4.** As can be seen from (8.3.3), the noncommutativity 'measures' the difference between the element degree of truth and $A^{(\mathsf{f})}$. So if two elements have the same membership function, they commute.

Proposition 8.3.5. *In* $\mathcal{A}_*(\mathsf{f})$ *the membership function satisfies the equality* (*see* (8.2.4))

$$\mathsf{f}(ka) = \mathsf{f}(a), \quad \forall\, a \in A^{(\mathsf{f})}, \quad k \in \Bbbk. \tag{8.3.4}$$

Proof. From the distributivity of the scalar multiplication $k(a * b) = ka * b = a * kb$ and the membership noncommutativity (8.3.1), we get

$$\mathsf{f}(b) ka * b = \mathsf{f}(ka) b * ka, \tag{8.3.5}$$

$$\mathsf{f}(kb) a * kb = \mathsf{f}(a) kb * a, \tag{8.3.6}$$

and so $\mathsf{f}(ka)\mathsf{f}(kb) = \mathsf{f}(a)\mathsf{f}(b)$, $\forall\, a \in A^{(\mathsf{f})}$, $k \in \Bbbk$, and thus (8.3.4) follows. □

In general, the algebra $\mathcal{A}_*(\mathsf{f})$ is not associative without further conditions (for instance, similar to (8.1.5)) on the membership function f which is assumed predefined and satisfies the properties (8.2.2)–(8.2.3) and (8.3.4) only.

Proposition 8.3.6. *The obscure membership deformed algebra* $\mathcal{A}_*(\mathsf{f})$ *cannot be associative with any additional conditions on the membership function* f.

Proof. Since the ε-commutativity (8.1.1) and the ϵ-commutativity (8.3.2) have the same form, the derivations of associativity coincide and give a cocycle relation similar to (7.1.6)–(7.1.7) also for $\epsilon_{\mathsf{f}}^{(2)}$, e.g.

$$\epsilon_{\mathsf{f}}^{(2)}(a, b * c) = \epsilon_{\mathsf{f}}^{(2)}(a, b)\epsilon_{\mathsf{f}}^{(2)}(a, c). \tag{8.3.7}$$

In terms of the membership function (8.3.3) this becomes $\mathsf{f}(a) = \mathsf{f}(b)\mathsf{f}(c)/\mathsf{f}(b * c)$, but this is impossible for all $a, b, c \in A^{(\mathsf{f})}$. On the other side, after double commutation in $(a * b) * c \to c * (b * a)$ and $a * (b * c) \to (c * b) * a$ we obtain (if the associativity of $(*)$ is implied)

$$\epsilon_{\mathsf{f}}^{(2)}(a, b * c)\epsilon_{\mathsf{f}}^{(2)}(b, c) = \epsilon_{\mathsf{f}}^{(2)}(a * b, c)\epsilon_{\mathsf{f}}^{(2)}(a, b), \tag{8.3.8}$$

which in terms of the membership function becomes $\mathsf{f}(b)^2 = \mathsf{f}(a * b)\mathsf{f}(b * c)$, and this is also impossible for arbitrary $a, b, c \in A^{(\mathsf{f})}$.

Nevertheless, distributivity of the algebra multiplication and algebra addition in $\mathcal{A}_*(\mathsf{f})$ is possible, but can only be one-sided.

Proposition 8.3.7. *The algebra* $\mathcal{A}_*(\mathsf{f})$ *is right-distributive, but has the membership deformed left distributivity*

$$\mathsf{f}(b + c)a * (b + c) = \mathsf{f}(b)a * b + \mathsf{f}(c)a * c, \tag{8.3.9}$$

$$(b + c) * a = b * a + c * a, \quad \forall a, b, c \in A^{(\mathsf{f})}. \tag{8.3.10}$$

Proof. Applying the membership noncommutativity (8.3.1) to (8.3.9), we obtain $\mathsf{f}(a)(b + c) * a = \mathsf{f}(a)b * a + \mathsf{f}(a)c * a$, and then (8.3.10), because $\mathsf{f} > 0$. □

***Theorem* 8.3.8.** *The binary obscure algebra* $\mathcal{A}_*(\mathsf{f}) = \langle A^{(\mathsf{f})} | *, + \rangle$ *is necessarily nonassociative and* $\epsilon_{\mathsf{f}}^{(2)}$*-commutative* (8.3.2), right-distributive (8.3.10) and membership deformed left-distributive (8.3.9).

***Remark* 8.3.9.** If membership noncommutativity is valid for generators of the algebra only, then the form of (8.3.2) coincides with that of the quantum polynomial algebra (Artamonov 1997), but the latter is two-sided distributive, in distinction to the obscure algebra $\mathcal{A}_*(\mathsf{f})$.

***Example* 8.3.10.** Deformed Weyl algebra. Consider an obscure algebra $\mathcal{A}_{\odot}^{\mathrm{Weyl}}(\mathsf{f}) = \langle A^{(\mathsf{f})} | \odot, + \rangle$ generated by two generators $x, y \in A^{(\mathsf{f})}$ satisfying the Weyl-like relation (see (8.3.1))

$$\mathsf{f}(y) x \odot y = \mathsf{f}(x) y \odot x + \mathsf{f}(x \odot y) e, \quad x, y, e \in A^{(\mathsf{f})}. \tag{8.3.11}$$

We call $\mathcal{A}_{\odot}^{\mathrm{Weyl}}(\mathsf{f})$ a *membership deformed Weyl algebra.* Because the membership function is predefined, and $\mathsf{f}(x \odot y)$ is not a symplectic form, the algebra $\mathcal{A}_{\odot}^{\mathrm{Weyl}}(\mathsf{f})$ is not isomorphic to the ordinary Weyl algebra (see, e.g., Lam 2001). In the same way, the graded Weyl algebra (Tilgner 1977) can be membership deformed in a similar way.

***Remark* 8.3.11.** The special property of the membership noncommutativity is the fact that each pair of elements has their own 'individual' commutation factor ϵ which depends on the membership function (8.3.3) that can also be continuous.

8.3.2 Deformation of almost-commutative algebras

Here we apply the membership deformation procedure (8.3.1) to the obscure G-graded algebras (8.2.8) which are $\varepsilon^{(2)}$-commutative (8.1.1). We now 'deform' (8.1.6) by analogy with (8.3.1).

Let $\mathcal{A}^{\mathcal{G}}(\mathsf{f})$ be a binary obscure G-graded algebra (8.2.8) which is $\varepsilon^{(2)}$-commutative with the Schur factor $\pi^{(2)}$ (8.1.4).

Definition 8.3.12. An *obscure membership deformed $\varepsilon^{(2)}$-commutative G-graded algebra* is $\mathcal{A}_{\mathcal{G}\star}(\mathsf{f}) = \langle A^{(\mathsf{f})} | \star, + \rangle$ in which the noncommutativity relation is given by

$$\mathsf{f}(b) \pi^{(2)}(b', a') a \star b = \mathsf{f}(a) \pi^{(2)}(a', b') b \star a, \quad \forall a, b \in A^{(\mathsf{f})}, \forall a', b' \in G'. \tag{8.3.12}$$

Because both functions $\pi^{(2)}$ and f are nonvanishing, we can combine (8.1.4) and (8.3.3).

Definition 8.3.13. An algebra $\mathcal{A}_{\mathcal{G}\star}(\mathsf{f})$ is called a *double $\varepsilon_{\pi}^{(2)} / \epsilon_{\mathsf{f}}^{(2)}$-commutative algebra,* when

$$a \star b = \varepsilon_{\pi}^{(2)}(a', b') \epsilon_{\mathsf{f}}^{(2)}(a, b) b \star a, \quad \forall a, b \in A^{(\mathsf{f})}, \forall a', b' \in G' \tag{8.3.13}$$

$$\epsilon_{\mathsf{f}}^{(2)}(a, b) = \frac{\mathsf{f}(a)}{\mathsf{f}(b)}, \tag{8.3.14}$$

$$\varepsilon_{\pi}^{(2)}(a', b') = \frac{\pi^{(2)}(a', b')}{\pi^{(2)}(b', a')}, \tag{8.3.15}$$

where $\varepsilon_{\pi}^{(2)}$ is the grading commutation factor and $\boldsymbol{\epsilon}_{\mathsf{f}}^{(2)}$ is the membership commutation factor.

In the first version, we assume that $\varepsilon_{\pi}^{(2)}$ is still a cocycle and satisfies (7.1.5)–(7.1.7), then in $\mathcal{A}_{\mathcal{G}\star}(\mathsf{f})$ the relation (8.3.4) is satisfied as well, because the Schur factors cancel in the derivation from (8.3.5)–(8.3.6). For the same reason assertion 8.3.6 holds, and therefore the double $\varepsilon_{\pi}^{(2)}/\boldsymbol{\epsilon}_{\mathsf{f}}^{(2)}$-commutative algebra $\mathcal{A}_{\mathcal{G}\star}(\mathsf{f})$ with the fixed grading commutation factor $\varepsilon_{\pi}^{(2)}$ is necessarily nonassociative.

As the second version, we consider the case when the grading commutation factor does not satisfy (7.1.5)–(7.1.7), but the double commutation factor does satisfy them (the membership function is fixed, being predefined for each element), which can lead to an associative algebra.

Proposition 8.3.14. *A double $\varepsilon_{\pi\mathsf{f}}^{(2)}/\boldsymbol{\epsilon}_{\mathsf{f}}^{(2)}$-commutative algebra $\mathcal{A}_{\circledast}^{\mathcal{G}}(\mathsf{f}) = \langle A^{(\mathsf{f})} | \circledast, + \rangle$ with*

$$a \circledast b = \varepsilon_{\pi\mathsf{f}}^{(2)}(a', b')\boldsymbol{\epsilon}_{\mathsf{f}}^{(2)}(a, b) b \circledast a, \quad \boldsymbol{\epsilon}_{\mathsf{f}}^{(2)}(a, b) = \frac{\mathsf{f}(a)}{\mathsf{f}(b)} \tag{8.3.16}$$

is associative, if the noncocycle commutation factor $\varepsilon_{\pi\mathsf{f}}^{(2)}$ satisfies the 'membership deformed cocycle-like' conditions

$$\varepsilon_{\pi\mathsf{f}}^{(2)}(a', b')\varepsilon_{\pi\mathsf{f}}^{(2)}(b', a') = \frac{1}{\boldsymbol{\epsilon}_{\mathsf{f}}^{(2)}(a, b)\boldsymbol{\epsilon}_{\mathsf{f}}^{(2)}(b, a)}, \tag{8.3.17}$$

$$\varepsilon_{\pi\mathsf{f}}^{(2)}(a', b' +' c') = \varepsilon_{\pi\mathsf{f}}^{(2)}(a', b')\varepsilon_{\pi\mathsf{f}}^{(2)}(a', c')\frac{\boldsymbol{\epsilon}_{\mathsf{f}}^{(2)}(a, b)\boldsymbol{\epsilon}_{\mathsf{f}}^{(2)}(a, c)}{\boldsymbol{\epsilon}_{\mathsf{f}}^{(2)}(a, b \circledast c)}, \tag{8.3.18}$$

$$\varepsilon_{\pi\mathsf{f}}^{(2)}(a' +' b', c') = \varepsilon_{\pi\mathsf{f}}^{(2)}(a', c')\varepsilon_{\pi\mathsf{f}}^{(2)}(b', c')\frac{\boldsymbol{\epsilon}_{\mathsf{f}}^{(2)}(a, c)\boldsymbol{\epsilon}_{\mathsf{f}}^{(2)}(b, c)}{\boldsymbol{\epsilon}_{\mathsf{f}}^{(2)}(a \circledast b, c)}, \tag{8.3.19}$$

$$\forall a, b \in A^{(\mathsf{f})}, \forall a', b' \in G'.$$

Proof. Now indeed the double commutation factor (the product of the grading and membership factors)$\varepsilon_{\pi\mathsf{f}}^{(2)}\boldsymbol{\epsilon}_{\mathsf{f}}^{(2)}$ satisfies (7.1.5)–(7.1.7). Then (8.3.17)–(8.3.19) immediately follow. □

We can find the deformed equation for the Schur-like factors (similar to (8.3.17)–(8.3.19))

$$\varepsilon_{\pi f}^{(2)}(a', b') = \frac{\pi_f^{(2)}(a', b')}{\pi_f^{(2)}(b', a')}, \tag{8.3.20}$$

such that the following 'membership deformed' commutation takes place (see (8.3.16))

$$\pi_f^{(2)}(b', a')\mathsf{f}(b)a \circledast b = \pi_f^{(2)}(a', b')\mathsf{f}(a)b \circledast a, \quad \forall a, b \in \boldsymbol{A}^{(\mathsf{f})}, \forall a', b' \in G', \tag{8.3.21}$$

and the algebra $\mathcal{A}_{\circledast}^{\mathcal{G}}(\mathsf{f})$ becomes associative (in distinct with (8.3.12), where the grading commutation factor $\varepsilon_{\pi}^{(2)}$, satisfies (7.1.5)–(7.1.7), and the algebra multiplications ($\star$) are different).

Thus the deformed binary Schur-like factors $\pi_f^{(2)}$ of the obscure membership deformed associative double commutative algebra $\mathcal{A}_{\circledast}^{\mathcal{G}}(\mathsf{f})$ satisfy

$$\pi_f^{(2)}(a', b' + 'c')\pi_f^{(2)}(b', c') = \pi_f^{(2)}(a', b')\pi_f^{(2)}(a' + 'b', c')\frac{\mathsf{f}(a \circledast b)}{\mathsf{f}(b)}, \tag{8.3.22}$$
$$\forall a, b \in \boldsymbol{A}^{(\mathsf{f})}, \ \forall a', b' \in G',$$

which should be compared with the corresponding nondeformed relation (8.1.5).

8.3.3 Double almost-Lie algebras

Consider the second version of an obscure double $\varepsilon_{\pi f}^{(2)}/\epsilon_f^{(2)}$-commutative algebra $\mathcal{A}_{\circledast}^{\mathcal{G}}(\mathsf{f})$ defined in (8.3.16)–(8.3.19) and construct a corresponding analog of the Lie algebra, following the same procedure as for associative ε-commutative algebras (Scheunert 1979, Rittenberg and Wyler 1978, Montgomery 1997).

Take $\mathcal{A}_{\circledast}^{\mathcal{G}}(\mathsf{f})$ and define a *double $\varepsilon\epsilon$-Lie bracket* $\mathscr{L}_{\varepsilon\epsilon}$: $\boldsymbol{A}^{(\mathsf{f})} \otimes \boldsymbol{A}^{(\mathsf{f})} \to \boldsymbol{A}^{(\mathsf{f})}$ by

$$\mathscr{L}_{\varepsilon\epsilon}[a, b] = a \circledast b - \varepsilon_{\pi f}^{(2)}(a', b')\epsilon_f^{(2)}(a, b)b \circledast a, \quad \forall a, b \in \boldsymbol{A}^{(\mathsf{f})}, \ \forall a', b' \in G', \tag{8.3.23}$$

where $\varepsilon_{\pi f}^{(2)}$ and $\epsilon_f^{(2)}$ are given in (8.3.16) and (8.3.20), respectively.

Proposition 8.3.15. *The double $\varepsilon\epsilon$-Lie bracket is $\varepsilon\epsilon$-skew commutative, i.e., it satisfies double commutativity with the commutation factor* $\left(-\varepsilon_{\pi f}^{(2)}\epsilon_f^{(2)}\right)$.

Proof. Multiply (8.3.23) by $\varepsilon_{\pi f}^{(2)}(b', a')\epsilon_f^{(2)}(b, a)$ and use (8.3.17) to obtain

$$\varepsilon_{\pi f}^{(2)}(b', a')\epsilon_f^{(2)}(b, a)\mathscr{L}_{\varepsilon\epsilon}[a, b] = \varepsilon_{\pi f}^{(2)}(b', a')\epsilon_f^{(2)}(b, a)a \circledast b - b \circledast a = -\mathscr{L}_{\varepsilon\epsilon}[b, a]. \tag{8.3.24}$$

Therefore,

$$\mathscr{L}_{\varepsilon\epsilon}[a, b] = -\varepsilon_{\pi f}^{(2)}(a', b')\epsilon_f^{(2)}(a, b)\mathscr{L}_{\varepsilon\epsilon}[b, a], \tag{8.3.25}$$

which should be compared with (8.3.16). □

Proposition 8.3.16. *The double $\varepsilon\epsilon$-Lie bracket satisfies the membership deformed $\varepsilon\epsilon$-Jacobi identity*

$$\begin{aligned}&\varepsilon_{\pi\mathsf{f}}^{(2)}(a', b')\boldsymbol{\epsilon}_{\mathsf{f}}^{(2)}(a, b)\mathscr{L}_{\varepsilon\epsilon}[a, \mathscr{L}_{\varepsilon\epsilon}[b, c]] + \text{cyclic permutations} = 0,\\ &\forall a, b, c \in A^{(\mathsf{f})}, \quad \forall a', b', c' \in G.\end{aligned} \tag{8.3.26}$$

Definition 8.3.17. A *double $\varepsilon\epsilon$-Lie algebra* is an obscure G-graded algebra with the double $\varepsilon\epsilon$-Lie bracket (satisfying $\varepsilon\epsilon$-skew commutativity (8.3.25) and the membership deformed Jacobi identity (8.3.26)) as a multiplication, that is $\mathcal{A}_{\mathscr{L}}^{\mathcal{G}}(\mathsf{f}) = \langle A^{(\mathsf{f})} | \mathscr{L}_{\varepsilon\epsilon}, + \rangle$.

8.4 Projective representations

To generalize the ε-commutative algebras to the n-ary case, we need to introduce n-ary projective representations and study them in brief.

8.4.1 Binary projective representations

First, briefly recall some general properties of the Schur factors and corresponding commutation factors in the projective representation theory of abelian groups (Žmud́ 1960, 1972) (see, also, Frucht 1955, and the unitary ray representations, Bargmann 1954). We show some known details in our notation which can be useful in further extensions of the well-known binary constructions to the n-ary case.

Let $\mathcal{H}^{(2)} = \langle H | \dotplus \rangle$ be a binary abelian group and $\boldsymbol{\Pi}\colon \mathcal{H}^{(2)} \to \mathcal{E}^{(2)}$, where $\mathcal{E}^{(2)} = \langle \operatorname{End} V | \circ \rangle$, and V is a vector space over a field $\Bbbk$. A map $\boldsymbol{\Pi}$ is a (binary) *projective representation* (*σ-representation,* Žmud́ 1960), if

$$\boldsymbol{\Pi}(x_1) \circ \boldsymbol{\Pi}(x_2) = \pi_0^{(2)}(x_1, x_2)\boldsymbol{\Pi}(x_1 \dotplus x_2), \; x_1, x_2 \in H, \tag{8.4.1}$$

and $\pi_0^{(2)}\colon \mathcal{H}^{(2)} \times \mathcal{H}^{(2)} \to \Bbbk^{\times}$ is a (Schur) factor, while $(\circ)$ is a (noncommutative binary) product in $\operatorname{End} V$. The 'associativity relation' of factors follows immediately from the associativity of $(\circ)$ such that (see (8.1.5))

$$\pi_0^{(2)}(x_1, x_2 \dotplus x_3)\pi_0^{(2)}(x_2, x_3) = \pi_0^{(2)}(x_1, x_2)\pi_0^{(2)}(x_1 \dotplus x_2, x_3), \quad \forall x_1, x_2, x_3 \in H. \tag{8.4.2}$$

Two factor systems are equivalent $\{\pi_0^{(2)}\} \overset{\lambda}{\sim} \{\tilde{\pi}_0^{(2)}\}$ (or associated, Žmud́ 1972), if there exists $\lambda\colon \mathcal{H}^{(2)} \to \Bbbk^{\times}$ such that

$$\tilde{\pi}_0^{(2)}(x_1, x_2) = \frac{\lambda(x_1)\lambda(x_2)}{\lambda(x_1 \dotplus x_2)}\pi_0^{(2)}(x_1, x_2), \qquad \forall x_1, x_2 \in H, \tag{8.4.3}$$

and the $\tilde{\pi}_0^{(2)}$-representation is given by $\tilde{\boldsymbol{\Pi}}(x) = \lambda(x)\boldsymbol{\Pi}(x)$, $x \in H$. The cocycle condition (8.4.2) means that $\pi_0^{(2)}$ belongs to the group $Z^2(\mathcal{H}^{(2)}, \Bbbk^{\times})$ of 2-cocycles of $\mathcal{H}^{(2)}$ over $\Bbbk^{\times}$, the quotient $\{\pi_0^{(2)}\}/\overset{\lambda}{\sim}$ gives the corresponding multiplier group, if $\Bbbk = \mathbb{C}$, and, in the general case, coincides with the exponents of the 2-cohomology classes $H^2(\mathcal{H}^{(2)}, \Bbbk)$ (for details, see, e.g., Žmud́ 1960, 1972, de Goursac *et al* 2012).

The group $\mathcal{H}^{(2)}$ is abelian, and therefore we have (see (8.1.6))

$$\begin{aligned}\pi_0^{(2)}(x_2, x_1)\mathbf{\Pi}(x_1) \circ \mathbf{\Pi}(x_2) &= \pi_0^{(2)}(x_2, x_1)\pi_0^{(2)}(x_1, x_2)\mathbf{\Pi}(x_1 \dotplus x_2) \\ &= \pi_0^{(2)}(x_1, x_2)\mathbf{\Pi}(x_2) \circ \mathbf{\Pi}(x_1),\end{aligned} \tag{8.4.4}$$

which allows us to introduce a (binary) *commutation factor* $\varepsilon_{\pi_0}^{(2)}: \mathcal{H}^{(2)} \times \mathcal{H}^{(2)} \to \Bbbk^{\times}$ by (see (8.1.1), (8.1.4))

$$\varepsilon_{\pi_0}^{(2)}(x_1, x_2) = \frac{\pi_0^{(2)}(x_1, x_2)}{\pi_0^{(2)}(x_2, x_1)}, \tag{8.4.5}$$

$$\mathbf{\Pi}(x_1) \circ \mathbf{\Pi}(x_2) = \varepsilon_{\pi_0}^{(2)}(x_1, x_2)\mathbf{\Pi}(x_2) \circ \mathbf{\Pi}(x_1),\ \forall\, x_1, x_2 \in H, \tag{8.4.6}$$

with the obvious 'normalization' (see (7.1.5))

$$\varepsilon_{\pi_0}^{(2)}(x_1, x_2)\varepsilon_{\pi_0}^{(2)}(x_2, x_1) = 1, \quad \varepsilon_{\pi_0}^{(2)}(x, x) = 1,\ \forall\, x_1, x_2, x \in H, \tag{8.4.7}$$

which will be important in the n-ary case below.

The multiplication of the commutation factors follows from different permutations of the three terms

$$\begin{aligned}&\mathbf{\Pi}(y) \circ \mathbf{\Pi}(x_1) \circ \mathbf{\Pi}(x_2) = \varepsilon_{\pi_0}^{(2)}(y, x_1)\mathbf{\Pi}(x_1) \circ \mathbf{\Pi}(y) \circ \mathbf{\Pi}(x_2) \\ &= \varepsilon_{\pi_0}^{(2)}(y, x_1)\varepsilon_{\pi_0}^{(2)}(y, x_2)\mathbf{\Pi}(x_1) \circ \mathbf{\Pi}(x_2) \circ \mathbf{\Pi}(y) = \pi_0^{(2)}(x_1, x_2)\mathbf{\Pi}(y) \circ \\ &\quad \mathbf{\Pi}(x_1 \dotplus x_2) \\ &= \varepsilon_{\pi_0}^{(2)}(y, x_1 \dotplus x_2)(\pi_0^{(2)}(x_1, x_2)\mathbf{\Pi}(x_1 \dotplus x_2)) \circ \mathbf{\Pi}(y) \\ &= \varepsilon_{\pi_0}^{(2)}(y, x_1 \dotplus x_2)\mathbf{\Pi}(x_1) \circ \mathbf{\Pi}(x_2) \circ \mathbf{\Pi}(y), \quad \forall\, x_1, x_2, y \in H.\end{aligned} \tag{8.4.8}$$

Thus, it follows that the commutation factor multiplication is (and similarly for the second place, see (7.1.6)–(7.1.7))

$$\varepsilon_{\pi_0}^{(2)}(y, x_1 \dotplus x_2) = \varepsilon_{\pi_0}^{(2)}(y, x_1)\varepsilon_{\pi_0}^{(2)}(y, x_2), \quad \forall\, x_1, x_2, y \in H, \tag{8.4.9}$$

which means that $\varepsilon_{\pi_0}^{(2)}$ is a (binary) bicharacter on $\mathcal{H}^{(2)}$, because for $\chi_y^{(2)}(x) \equiv \varepsilon_{\pi_0}^{(2)}(y, x)$ we have

$$\chi_y^{(2)}(x_1)\chi_y^{(2)}(x_2) = \chi_y^{(2)}(x_1 \dotplus x_2), \quad \forall\, x_1, x_2 \in H. \tag{8.4.10}$$

Denoting the group of bicharacters $\chi_y^{(2)}$ on $\mathcal{H}^{(2)}$ with the multiplication (8.4.10) by $\mathcal{B}^{(2)}(\mathcal{H}^{(2)}, \Bbbk)$, we observe that the mapping $\pi_0^{(2)} \to \varepsilon_{\pi_0}^{(2)}$ is a homomorphism of $\mathsf{Z}^2(\mathcal{H}^{(2)}, \Bbbk^{\times})$ to $\mathcal{B}^{(2)}(\mathcal{H}^{(2)}, \Bbbk)$ with the kernel a subgroup of the 2-coboundaries of $\mathcal{H}^{(2)}$ over $\Bbbk^{\times}$ (for more details, see Žmud́ 1972).

8.4.2 n-Ary projective representations

Here we consider some features of n-ary projective representations and corresponding particular generalizations of binary ε-commutativity.

Let $\mathcal{H}^{(n)} = \langle H | [\,]_n^{+} \rangle$ be an n-ary abelian group with the totally commutative multiplication[2] $\nu_H^{(n)} \equiv [\,]_n^{+}$, and the mapping $\mathbf{\Pi}: \mathcal{H}^{(n)} \to \mathcal{E}^{(n)}$, where $\mathcal{E}^{(n)} = \langle \operatorname{End} V | [\,]_n^{V} \rangle$ (for general polyadic representations, see Duplij 2012 and refs therein). Here we suppose that V is a vector space over a field $\Bbbk$, and $\nu_n^V \equiv [\ \ ]_n^V$ is an n-ary associative product in $\operatorname{End} V$, which means that

$$\begin{aligned}
&\mathbf{\Pi}\Big[\big[\mathbf{\Pi}(x_1), \ldots, \mathbf{\Pi}(x_n)\big]_n^V, \mathbf{\Pi}(x_{n+1}), \ldots, \mathbf{\Pi}(x_{2n-1})\Big]_n^V \\
&= \mathbf{\Pi}\Big[\mathbf{\Pi}(x_1), \big[\mathbf{\Pi}(x_2), \ldots, \mathbf{\Pi}(x_{n+1})\big]_n^V, \mathbf{\Pi}(x_{n+2}), \ldots, \mathbf{\Pi}(x_{2n-1})\Big]_n^V \\
&\ \vdots \\
&= \mathbf{\Pi}\Big[\mathbf{\Pi}(x_1), \mathbf{\Pi}(x_2), \ldots, \mathbf{\Pi}(x_{n-1}), \big[\mathbf{\Pi}(x_n), \ldots, \mathbf{\Pi}(x_{2n-1})\big]_n^V\Big]_n^V.
\end{aligned} \tag{8.4.11}$$

Definition 8.4.1. A map $\mathbf{\Pi}$ is an *n-ary projective representation*, if

$$\big[\mathbf{\Pi}(x_1), \ldots, \mathbf{\Pi}(x_n)\big]_n^V = \pi_0^{(n)}(x_1, \ldots, x_n)\mathbf{\Pi}\big([x_1, \ldots, x_n]_n^{+}\big), \quad x_1, \ldots, x_n \in H, \tag{8.4.12}$$

and $\pi_0^{(n)}: \overbrace{\mathcal{H}^{(n)} \times \ldots \times \mathcal{H}^{(n)}}^{n} \to \Bbbk^{\times}$ is an *n-ary (Schur-like) factor.*

Proposition 8.4.2. *The factors* $\pi_0^{(n)}$ *satisfy the n-ary 2-cocycle conditions (see* (8.4.2))

$$\begin{aligned}
&\pi_0^{(n)}(x_1, \ldots, x_n)\pi_0^{(n)}\big([x_1, \ldots, x_n]_n^{+}, x_{n+1}, \ldots, x_{2n-1}\big) \\
&= \pi_0^{(n)}(x_2, \ldots, x_{n+1})\pi_0^{(n)}\big(x_1, [x_2, \ldots, x_{n+1}]_n^{+}, x_{n+2}, \ldots, x_{2n-1}\big) \\
&\ \vdots \\
&= \pi_0^{(n)}(x_n, \ldots, x_{n+1})\pi_0^{(n)}\big(x_1, \ldots, x_{n-1}, [x_n, \ldots, x_{2n-1}]_n^{+}\big), \quad x_1, \\
&\quad \ldots, x_{2n-1} \in H.
\end{aligned} \tag{8.4.13}$$

Proof. These immediately follow from the n-ary associativity in $\operatorname{End} V$ (8.4.11) and (8.4.12). □

Two n-ary factor systems are equivalent $\{\pi_0^{(n)}\} \overset{\lambda}{\sim} \{\tilde{\pi}_0^{(n)}\}$, if there exists $\lambda: \mathcal{H}_n \to \Bbbk^{\times}$ such that

$$\tilde{\pi}_0^{(n)}(x_1, \ldots, x_n) = \frac{\lambda(x_1)\lambda(x_2)\ldots\lambda(x_n)}{\lambda([x_1, \ldots, x_n]_n^{+})}\pi_0^{(n)}(x_1, \ldots, x_n), \qquad \forall x_1, x_2 \in H, \tag{8.4.14}$$

and the $\tilde{\pi}_0^{(n)}$-representation is given by $\tilde{\mathbf{\Pi}}(x) = \lambda(x)\mathbf{\Pi}(x)$, $x \in H$.

[2] In this chapter polyadic operations in abelian groups and vector spaces (which are actually additions) are denoted by square brackets with various indices, just for clarity and convenience. The bracket notation is also convenient to consider the partial commutativity of $\mathcal{H}^{(n)}$.

To understand how properly and uniquely to introduce the commutation factors for n-ary projective representations, we need to consider an n-ary analog of (8.4.4) (see, also (8.1.6)).

Proposition 8.4.3. *The commutativity of a $\pi_0^{(n)}$-representation is given by* $(n! - 1)$ *relations of the form*

$$\pi_0^{(n)}(x_{\sigma(1)}, \ldots, x_{\sigma(n)})\left[\mathbf{\Pi}(x_1), \ldots, \mathbf{\Pi}(x_n)\right]_n^V = \pi_0^{(n)}(x_1, \ldots, x_n)\left[\mathbf{\Pi}(x_{\sigma(1)}), \ldots, \mathbf{\Pi}(x_{\sigma(n)})\right]_n^V, \tag{8.4.15}$$

where $\sigma \in S_n$, $\sigma \neq I$, S_n is the symmetry permutation group on n elements.

Proof. Using the definition of the n-ary projective representation (8.4.12), we obtain for the lhs and rhs of (8.4.15)

$$\pi_0^{(n)}(x_{\sigma(1)}, \ldots, x_{\sigma(n)})\pi_0^{(n)}(x_1, \ldots, x_n)\mathbf{\Pi}\left([x_1, \ldots, x_n]_n^+\right) \tag{8.4.16}$$

and

$$\pi_0^{(n)}(x_1, \ldots, x_n)\pi_0^{(n)}(x_{\sigma(1)}, \ldots, x_{\sigma(n)})\mathbf{\Pi}\left([x_{\sigma(1)}, \ldots, x_{\sigma(n)}]_n^+\right), \tag{8.4.17}$$

respectively.

Because $\mathcal{H}^{(n)}$ is totally commutative

$$[x_1, \ldots, x_n]_n^+ = [x_{\sigma(1)}, \ldots, x_{\sigma(n)}]_n^+, \quad \forall x_1, \ldots, x_n \in H, \forall \sigma \in S_n, \tag{8.4.18}$$

then $\mathbf{\Pi}([x_1, \ldots, x_n]_n^+) = \mathbf{\Pi}\left([x_{\sigma(1)}, \ldots, x_{\sigma(n)}]_n^+\right)$. Taking into account all nonidentical permutations, we get $(n! - 1)$ relations in (8.4.15). □

Corollary 8.4.4. To describe the noncommutativity of an n-ary projective representation, we need to have not one relation between Schur factors (as in the binary case (8.4.4) and (8.1.6)), but $(n! - 1)$ relations (8.4.15). This leads to the concept of the set of $(n! - 1)$ commutation factors.

Definition 8.4.5. The commutativity of the n-ary projective representation with the Schur-like factor $\pi_0^{(n)}(x_1, \ldots, x_n)$ is governed by the *set of* $(n! - 1)$ *commutation factors*

$$\varepsilon_{\sigma(1), \ldots, \sigma(n)}^{(n)} \equiv \varepsilon_{\sigma(1), \ldots, \sigma(n)}^{(n)}(x_1, \ldots, x_n) = \frac{\pi_0^{(n)}(x_1, \ldots, x_n)}{\pi_0^{(n)}(x_{\sigma(1)}, \ldots, x_{\sigma(n)})}, \tag{8.4.19}$$

$$\left[\mathbf{\Pi}(x_1), \ldots, \mathbf{\Pi}(x_n)\right]_n^V = \varepsilon_{\sigma(1), \ldots, \sigma(n)}^{(n)}\left[\mathbf{\Pi}(x_{\sigma(1)}), \ldots, \mathbf{\Pi}(x_{\sigma(n)})\right]_n^V, \tag{8.4.20}$$

where $\sigma \in S_n$. We call all $\varepsilon_{\sigma(1), \ldots, \sigma(n)}^{(n)}$ as a *commutation factor 'vector'* and denote it by $\vec{\varepsilon}$, where its components will be enumerated by lexicographic order.

Thus, each component of $\vec{\varepsilon}$ is responsible for the commutation of any two n-ary monomials with different permutations $\sigma, \sigma' \in S_n$, since from (8.4.20) it follows that

$$\begin{aligned}&\left[\mathbf{\Pi}(x_{\sigma'(1)}), \ldots, \mathbf{\Pi}(x_{\sigma'(n)})\right]_n^V \\ &= \varepsilon^{(n)}_{\sigma(1), \ldots, \sigma(n)}(x_{\sigma'(1)}, \ldots, x_{\sigma'(n)})\left[\mathbf{\Pi}(x_{\sigma(1)}), \ldots, \mathbf{\Pi}(x_{\sigma(n)})\right]_n^V.\end{aligned} \tag{8.4.21}$$

It follows from (8.4.19) that an n-ary analog of the normalization property (8.4.7) is

$$\varepsilon^{(n)}_{\sigma(1), \ldots, \sigma(n)}(x_{\sigma'(1)}, \ldots, x_{\sigma'(n)})\varepsilon^{(n)}_{\sigma'(1), \ldots, \sigma'(n)}(x_{\sigma(1)}, \ldots, x_{\sigma(n)}) = 1, \tag{8.4.22}$$

$$\varepsilon^{(n)}_{1, \ldots, n}(x, \ldots, x) = 1, \tag{8.4.23}$$

where $\sigma, \sigma' \in S_n$, $x_1, \ldots, x_n, x \in H$.

In this notation the binary commutation factor (8.4.5) is $\varepsilon^{(2)}_{\pi_0}(x_1, x_2) = \varepsilon^{(2)}_{21}(x_1, x_2)$.

***Example* 8.4.6.** Ternary projective representation. Consider the minimal nonbinary case $n = 3$. The ternary projective representation of the abelian ternary group $\mathcal{H}^{(3)}$ is given by

$$\left[\mathbf{\Pi}(x_1), \mathbf{\Pi}(x_2), \mathbf{\Pi}(x_3)\right]_3^V = \pi_0(x_1, x_2, x_3)\mathbf{\Pi}(x_1 \dotplus x_2 \dotplus x_3), \quad x_1, x_2, x_3 \in H, \tag{8.4.24}$$

where we denote $\pi_0^{(3)} \equiv \pi_0$, and $[x_1, x_2, x_3]_n^{\dotplus} \equiv x_1 \dotplus x_2 \dotplus x_3$.

The ternary 2-cocycle conditions for the ternary Schur-like factor π_0 now become

$$\begin{aligned}&\pi_0(x_1, x_2, x_3)\pi_0(x_1 \dotplus x_2 \dotplus x_3, x_4, x_5) \\ &= \pi_0(x_2, x_3, x_4)\pi_0(x_1, x_2 \dotplus x_3 \dotplus x_4, x_5) \\ &= \pi_0(x_3, x_4, x_5)\pi_0(x_1, x_2, x_3 \dotplus x_4 \dotplus x_5).\end{aligned} \tag{8.4.25}$$

Thus, we obtain $(3! - 1) = 5$ different ternary commutation relations and the corresponding 5-dimensional 'vector' $\vec{\varepsilon}$

$$\left[\mathbf{\Pi}(x_1), \mathbf{\Pi}(x_2), \mathbf{\Pi}(x_3)\right]_3^V = \varepsilon_{\sigma(1)\sigma(2)\sigma(3)}\left[\mathbf{\Pi}(x_{\sigma(1)}), \mathbf{\Pi}(x_{\sigma(2)}), \mathbf{\Pi}(x_{\sigma(3)})\right]_3^V, \tag{8.4.26}$$

$$\begin{aligned}&\varepsilon_{132} = \frac{\pi_0(x_1, x_2, x_3)}{\pi_0(x_1, x_3, x_2)}, \quad \varepsilon_{231} = \frac{\pi_0(x_1, x_2, x_3)}{\pi_0(x_2, x_3, x_1)}, \quad \varepsilon_{213} = \frac{\pi_0(x_1, x_2, x_3)}{\pi_0(x_2, x_1, x_3)}, \\ &\varepsilon_{312} = \frac{\pi_0(x_1, x_2, x_3)}{\pi_0(x_3, x_1, x_2)}, \quad \varepsilon_{321} = \frac{\pi_0(x_1, x_2, x_3)}{\pi_0(x_3, x_2, x_1)}, \quad x_1, x_2, x_3 \in H, \sigma \in S_3.\end{aligned} \tag{8.4.27}$$

8.5 n-ary double commutative algebras

8.5.1 n-ary almost-commutative algebras

Here we introduce grading noncommutativity for n-ary algebras ('n-ary coloring'), which is closest to the binary ('coloring') case (8.1.1). We are exploiting an n-ary analog of the (Schur) factor (8.1.4) and its relation (8.1.6) by means of the n-ary projective representation theory from section 8.4.

Let $\mathcal{A}^{(n)} = \langle A \mid \mu^{(n)}, + \rangle$ be an associative n-ary algebra (Carlsson 1980, Michor and Vinogradov 1996, de Oliveira 1960) over a binary field $\mathbb{k}$ (with the binary addition $(+)$) having zero $z \in A$ and unit $e \in A$, if $\mathcal{A}^{(n)}$ is unital (for polyadic algebras with all nonbinary operations, see Duplij 2019). A n-ary *graded algebra* $\mathcal{A}_{\mathcal{G}}^{(n)}$ (an n-ary G-graded $\mathbb{k}$-algebra) is a direct sum of subalgebras $\mathcal{A}_{\mathcal{G}}^{(n)} = \oplus_{a' \in G'} \mathcal{A}_{a'}$, where $\mathcal{G} = \langle G' \mid +' \rangle$ is a (binary abelian) *grading group* and the set n-ary multiplication 'respects the gradation'

$$\mu^{(n)}\left[A_{a_1'}, \ldots, A_{a_n'}\right] \subseteq A_{a_1' +' \ldots +' a_n'}, \qquad a_1', \ldots, a_n' \in G'. \tag{8.5.1}$$

As in the binary case the elements from $A_{a'} \subset A$ are homogeneous of degree $a' \in G'$.

It is natural to start our n-ary consideration from the Schur-like factors (8.4.13) which generalize (8.1.5) and (8.4.2).

Definition 8.5.1. In an n-ary graded algebra $\mathcal{A}_{\mathcal{G}}^{(n)}$ the *n-ary Schur-like factor* is an n-place function on gradings $\pi^{(n)}: \overbrace{G' \times \ldots \times G'}^{n} \to A$ satisfying the n-ary cocycle condition (see (8.1.5) and (8.4.13))

$$\begin{aligned}
&\pi^{(n)}(a_1', \ldots, a_n')\pi^{(n)}(\mu^{(n)}[a_1', \ldots, a_n'], a_{n+1}', \ldots, a_{2n-1}') \\
&= \pi^{(n)}(a_2', \ldots, a_{n+1}')\pi^{(n)}(a_1', \mu^{(n)}[a_2', \ldots, a_{n+1}'], a_{n+2}', \ldots, a_{2n-1}') \\
&\vdots \\
&= \pi^{(n)}(a_n', \ldots, a_{n+1}')\pi^{(n)}(a_1', \ldots, a_{n-1}', \mu^{(n)}[a_n', \ldots, a_{2n-1}']), \\
&\quad a_1', \ldots, a_{2n-1}' \in G'.
\end{aligned} \tag{8.5.2}$$

There are many possible ways to introduce noncommutativity for n-ary algebras (de Azcarraga and Izquierdo 2010, Michor and Vinogradov 1996). We propose a 'projective version' of noncommutativity in $\mathcal{A}_{\mathcal{G}}^{(n)}$ which naturally follows from the n-ary projective representations (8.4.19) and can be formulated in terms of the Schur-like factors as in (8.1.6) and (8.4.15).

Definition 8.5.2. An n-ary graded algebra $\mathcal{A}_{\mathcal{G}}^{(n)}$ is called *π-commutative*, if the $(n! - 1)$ relations (see (8.4.15))

$$\pi^{(n)}\left(a_{\sigma(1)}', \ldots, a_{\sigma(n)}'\right)\mu^{(n)}[a_1, \ldots, a_n] = \pi^{(n)}(a_1', \ldots, a_n')\mu^{(n)}[a_{\sigma(1)}, \ldots, a_{\sigma(n)}], \tag{8.5.3}$$

hold for all $a_1, \ldots, a_n \in A$, and $\sigma \in S_n, \sigma \neq I$, where $\pi^{(n)}$ are n-ary Schur-like factors satisfying (8.5.2).

Two n-ary Schur-like factor systems are equivalent $\{\pi^{(n)}\} \overset{\tilde{\lambda}}{\approx} \{\tilde{\pi}^{(n)}\}$, if there exists $\tilde{\lambda}: \mathcal{A}_{\mathcal{G}}^{(n)} \to \Bbbk^{\times}$ such that (see (8.4.14))

$$\tilde{\pi}^{(n)}(a_1', \ldots, a_n') = \frac{\tilde{\lambda}(a_1') \ldots \tilde{\lambda}(a_n')}{\tilde{\lambda}(\mu^{(n)}[a_1', \ldots, a_n'])} \pi^{(n)}(a_1', \ldots, a_n'), \qquad \forall a_1', \ldots, a_2' \in G'. \tag{8.5.4}$$

The quotient by this equivalence relation $\{\pi^{(n)}\}/\overset{\tilde{\lambda}}{\approx}$ is the corresponding multiplier group, as in the binary case (Scheunert 1979).

Let $\varphi \in \mathrm{Aut}\mathcal{G}'$ and $\{\pi^{(n)}\}$ be a factor system, then its pullback

$$\pi_*^{(n)}(a_1', \ldots, a_n') = \pi^{(n)}(\varphi(a_1'), \ldots, \varphi(a_n')) \tag{8.5.5}$$

is also a factor system $\{\pi_*^{(n)}\}$. In the n-ary case the 'homotopic' analog of (8.5.5) is possible

$$\pi_{**}^{(n)}(a_1', \ldots, a_n') = \pi^{(n)}(\varphi_1(a_1'), \ldots, \varphi_n(a_n')), \tag{8.5.6}$$

where $\varphi_1, \ldots, \varphi_n \in \mathrm{Aut}G$, such that $\{\pi_{**}^{(n)}\}$ is also a factor system.

Comparing (8.5.3) with binary π-commutativity (8.1.6) we observe that the most general description of n-ary graded algebras $\mathcal{A}_{\mathcal{G}}^{(n)}$ can be achieved by using at least $(n! - 1)$ commutation factors.

Definition 8.5.3. An n-ary graded algebra $\mathcal{A}_{\mathcal{G}}^{(n)}$ is $\varepsilon^{(n)}$*-commutative* if there are $(n! - 1)$ commutation factors

$$\varepsilon_{\sigma(1), \ldots, \sigma(n)}^{(n)} \equiv \varepsilon_{\sigma(1), \ldots, \sigma(n)}^{(n)}(a_1', \ldots, a_n') = \frac{\pi^{(n)}(a_1', \ldots, a_n')}{\pi^{(n)}\left(a_{\sigma(1)}', \ldots, a_{\sigma(n)}'\right)}, \tag{8.5.7}$$

$$\mu^{(n)}[a_1, \ldots, a_n] = \varepsilon_{\sigma(1), \ldots, \sigma(n)}^{(n)} \mu^{(n)}[a_{\sigma(1)}, \ldots, a_{\sigma(n)}], \tag{8.5.8}$$

where $\sigma \in S_n$, $\sigma \neq I$, and the factors $\pi^{(n)}$ satisfy (8.5.2). The set of $\varepsilon_{\sigma(1), \ldots, \sigma(n)}^{(n)}$ is a *commutation factor 'vector'* $\vec{\varepsilon}$ of the algebra $\mathcal{A}_{\mathcal{G}}^{(n)}$ having $(n! - 1)$ components.

Each component of the 'vector' $\vec{\varepsilon}$ is responsible for the commutation of two n-ary monomials in $\mathcal{A}_{\mathcal{G}}^{(n)}$ such that from (8.5.8) we have

$$\mu^{(n)}[a_{\sigma'(1)}, \ldots, a_{\sigma'(n)}] = \varepsilon_{\sigma(1), \ldots, \sigma(n)}^{(n)}\left(a_{\sigma'(1)}', \ldots, a_{\sigma'(n)}'\right) \mu^{(n)}[a_{\sigma(1)}, \ldots, a_{\sigma(n)}], \tag{8.5.9}$$

with permutations $\sigma, \sigma' \in S_n$.

It follows from (8.4.19) that an n-ary analog of the normalization property (8.4.7) is

$$\varepsilon_{\sigma(1), \ldots, \sigma(n)}^{(n)}\left(a_{\sigma'(1)}', \ldots, a_{\sigma'(n)}'\right) \varepsilon_{\sigma'(1), \ldots, \sigma'(n)}^{(n)}\left(a_{\sigma(1)}', \ldots, a_{\sigma(n)}'\right) = 1, \tag{8.5.10}$$

$$\varepsilon_{1, \ldots, n}^{(n)}(a', \ldots, a') = 1, \tag{8.5.11}$$

where $\sigma, \sigma' \in S_n, a_1', \dots, a_n', a' \in G', 1 \in \Bbbk$.

If $\varphi \in \mathrm{Aut}\, \mathcal{G}'$ and $\{\pi_*^{(n)}\}$ is a pullback of $\{\pi^{(n)}\}$ (8.5.5), then the corresponding

$$\varepsilon_{*\sigma(1), \dots, \sigma(n)}^{(n)}(a_1', \dots, a_n') = \frac{\pi_*^{(n)}(a_1', \dots, a_n')}{\pi_*^{(n)}\left(a_{\sigma(1)}', \dots, a_{\sigma(n)}'\right)} \tag{8.5.12}$$

are commutation factors as well (if $\Bbbk$ is algebraically closed, by analogy with Scheunert 1979). The same is true for their 'homotopic' analog $\{\pi_{**}^{(n)}\}$ (8.5.6).

Definition 8.5.4. A factor set $\{\pi^{(n)}\}$ is called *totally symmetric*, if $\vec{\varepsilon}$ is the unit commutation factor 'vector', such that all of its $(n! - 1)$ components are identities (in $\Bbbk$)

$$\varepsilon_{\sigma(1), \dots, \sigma(n)}^{(n)}(a_1', \dots, a_n') = 1, \quad \sigma \in S_n, \sigma \neq I, a_1', \dots, a_n' \in G', 1 \in \Bbbk. \tag{8.5.13}$$

Suppose we have two factor sets $\{\pi_1^{(n)}\}$ and $\{\pi_2^{(n)}\}$ which correspond to the same commutation factor $\varepsilon^{(n)}$ in $\mathcal{A}_{\mathcal{G}}^{(n)}$ such that

$$\varepsilon_{\sigma(1), \dots, \sigma(n)}^{(n)}(a_1', \dots, a_n') = \frac{\pi_1^{(n)}(a_1', \dots, a_n')}{\pi_1^{(n)}\left(a_{\sigma(1)}', \dots, a_{\sigma(n)}'\right)} = \frac{\pi_2^{(n)}(a_1', \dots, a_n')}{\pi_2^{(n)}\left(a_{\sigma(1)}', \dots, a_{\sigma(n)}'\right)}. \tag{8.5.14}$$

We can always choose the same order for the commutation factor 'vector' components σ. Define a new factor set $\{\pi_{12}^{(n)}\}$ by

$$\pi_{12}^{(n)}(a_1', \dots, a_n') = \frac{\pi_1^{(n)}(a_1', \dots, a_n')}{\pi_2^{(n)}(a_1', \dots, a_n')}, \quad a_1', \dots, a_n' \in G'. \tag{8.5.15}$$

Then $\{\pi_{12}^{(n)}\}$ becomes totally symmetric, because the corresponding communication factor is

$$\varepsilon_{12, \sigma(1), \dots, \sigma(n)}^{(n)}(a_1', \dots, a_n') = \frac{\pi_{12}^{(n)}(a_1', \dots, a_n')}{\pi_{12}^{(n)}\left(a_{\sigma(1)}', \dots, a_{\sigma(n)}'\right)} = 1, \tag{8.5.16}$$

as follows from (8.5.14)–(8.5.15). Therefore, as in the binary case, if the grading group $\mathcal{G}$ is finitely generated and $\Bbbk$ is algebraically closed, the communication factor $\varepsilon^{(n)}$ is constructed from the unique multiplier $\{\pi^{(n)}\}$ (Scheunert 1979).

The are two possible differences from the simple binary case (8.1.8), where for identity commutation factor the symmetry condition (8.1.7) is sufficient: (1) Not all $\varepsilon^{(n)}$ need be equal to the identity. (2) Some arguments of the Schur-like factors $\pi^{(n)}$ can be intact.

Definition 8.5.5. An $\varepsilon^{(n)}$-commutative n-ary graded algebra $\mathcal{A}_{\mathcal{G}}^{(n)}$ is called *m-partially* (or *partially*) *commutative* if exactly m commutation factors (with permutations $\tilde{\sigma}$) from the $(n! - 1)$ in total are equal to $1 \in \Bbbk$

$$\varepsilon^{(n)}_{\tilde{\sigma}(1),\ldots,\tilde{\sigma}(n)}(a'_1,\ldots,a'_n)=1,\quad \#\tilde{\sigma}\leqslant\#\sigma,\ \sigma\in S_n,\ \ a'_1,\ldots,a'_n\in G', \qquad (8.5.17)$$

where we denote $\#\tilde{\sigma}=m$, $\#\sigma=n!-1$. If $\#\tilde{\sigma}=\#\sigma$, then $\mathcal{A}^{(n)}_{\mathcal{G}}$ is totally commutative, see (8.5.13).

In an m-partially commutative n-ary graded algebra $\mathcal{A}^{(n)}_{\mathcal{G}}$ the Schur-like factors $\pi^{(n)}$ satisfy m additional symmetry conditions (see (8.1.7))

$$\pi^{(n)}(a'_1,\ldots,a'_n)=\pi^{(n)}\left(a'_{\tilde{\sigma}(1)},\ldots,a'_{\tilde{\sigma}(n)}\right),\quad \#\tilde{\sigma}\leqslant\#\sigma,\quad \sigma\in S_n. \qquad (8.5.18)$$

***Example* 8.5.6.** Ternary ε-commutative algebra. Let $\mathcal{A}^{\mathcal{G}}_3=\langle A\mid\mu^{(3)},+\rangle$ be a ternary associative G-graded algebra over $\Bbbk$. The ternary Schur-like factor $\pi_0(a'_1,a'_2,a'_3)$ satisfies the ternary 2-cocycle conditions (see (8.4.25))

$$\begin{aligned}&\pi^{(3)}(a'_1,a'_2,a'_3)\pi^{(3)}(a'_1+{}'a'_2+{}'a'_3,a'_4,a'_5)\\&=\pi^{(3)}(a'_2,a'_3,a'_4)\pi^{(3)}(a'_1,a'_2+{}'a'_3+{}'a'_4,a'_5)\\&=\pi^{(3)}(a'_3,a'_4,a'_5)\pi^{(3)}(a'_1,a'_2,a'_3+{}'a'_4+{}'a'_5),\quad a'_1,\ldots,a'_5\in G'.\end{aligned} \qquad (8.5.19)$$

We can introduce $(3!-1)=5$ different ternary commutation relations, explicitly

$$\mu^{(3)}[a_1,a_2,a_3]=\varepsilon^{(3)}_{\sigma(1)\sigma(2)\sigma(3)}\mu^{(3)}[a_{\sigma(1)},a_{\sigma(2)},a_{\sigma(3)}],\quad a_1,a_2,a_3\in A,,\sigma\in S_3, \qquad (8.5.20)$$

$$\begin{aligned}&\varepsilon^{(3)}_{132}=\frac{\pi^{(3)}(a'_1,a'_2,a'_3)}{\pi^{(3)}(a'_1,a'_3,a'_2)},\quad \varepsilon^{(3)}_{231}=\frac{\pi^{(3)}(a'_1,a'_2,a'_3)}{\pi^{(3)}(a'_2,a'_3,a'_1)},\quad \varepsilon^{(3)}_{213}=\frac{\pi^{(3)}(a'_1,a'_2,a'_3)}{\pi^{(3)}(a'_2,a'_1,a'_3)},\\&\varepsilon^{(3)}_{312}=\frac{\pi^{(3)}(a'_1,a'_2,a'_3)}{\pi^{(3)}(a'_3,a'_1,a'_2)},\quad \varepsilon^{(3)}_{321}=\frac{\pi^{(3)}(a'_1,a'_2,a'_3)}{\pi^{(3)}(a'_3,a'_2,a'_1)},\quad a'_1,a'_2,a'_3\in G'.\end{aligned} \qquad (8.5.21)$$

The 1-partial commutativity in the algebra $\mathcal{A}^{\mathcal{G}}_3$ can be realized, if, for instance,

$$\pi^{(3)}(a'_1,a'_2,a'_3)=\pi^{(3)}(a'_1,a'_3,a'_2),\quad a'_1,a'_2,a'_3\in G', \qquad (8.5.22)$$

and in this case $\varepsilon^{(3)}_{132}=1$ so that we obtain one commutativity relation

$$\mu^{(3)}[a_1,a_2,a_3]=\mu^{(3)}[a_1,a_3,a_2],\quad a_1,a_2,a_3\in A, \qquad (8.5.23)$$

while the other 4 relations in (8.5.20) will be ε-commutative.

8.5.2 Membership deformed n-ary algebras

Now we consider n-ary algebras over obscure sets as their underlying sets, where each element of them is endowed with the membership function f as a degree of truth (see section 8.2).

Definition 8.5.7. An *obscure n-ary algebra* $\mathcal{A}^{(n)}(\mathsf{f}) = \langle A^{(\mathsf{f})} | \mu^{(n)}, + \rangle$ is an n-ary algebra $\mathcal{A}^{(n)}$ over $\Bbbk$ having an obscure set $A^{(\mathsf{f})} = \{(a|\mathsf{f}(a)), a \in A, \mathsf{f} > 0\}$ as its underlying set (see (8.2.1)), and where the membership function $\mathsf{f}(a)$ satisfies

$$\mathsf{f}(a_1 + a_2) \geqslant \min \{\mathsf{f}(a_1), \mathsf{f}(a_2)\}, \tag{8.5.24}$$

$$\mathsf{f}([a_1, \ldots, a_n]_n) \geqslant \min \{\mathsf{f}(a_1), \ldots , \mathsf{f}(a_n)\}, \tag{8.5.25}$$

$$\mathsf{f}(ka) \geqslant \mathsf{f}(a), \ \ \forall a, a_i \in A, \ \ k \in \Bbbk, \ \ i = 1, \ldots , n. \tag{8.5.26}$$

Definition 8.5.8. An *obscure G-graded* n-ary *algebra* $\mathcal{A}_{\mathcal{G}}^{(n)}(\mathsf{f})$ is a direct sum decomposition (see (8.2.8))

$$\mathcal{A}_{\mathcal{G}}^{(n)}(\mathsf{f}) = \oplus_{a' \in G'} \mathcal{A}^{(n)}(\mathsf{f}_{a'}), \tag{8.5.27}$$

where $A^{(\mathsf{f})} = \bigcup_{a' \in G'} A_{a'}^{(\mathsf{f}_{a'})}$, $A^{(\mathsf{f}_{a'})} = \{(a|\mathsf{f}_{a'}(a)), a \in A_{a'}, \mathsf{f}_{a'} = (0, 1]\}$, and the joint membership function f is given by (8.2.9).

In the obscure totally commutative n-ary algebra $\mathcal{A}_{\mathcal{G}}^{(n)}(\mathsf{f})$ (for homogeneous elements) we have $(n! - 1)$ commutativity relations (see (8.4.18))

$$\mu^{(n)}[a_1, \ldots , a_n] = \mu^{(n)}[a_{\sigma(1)}, \ldots , a_{\sigma(n)}], \forall a_1, \ldots , a_n \in A^{(\mathsf{f})}, \forall \sigma \in S_n, \sigma \neq I. \tag{8.5.28}$$

Let us consider a 'linear' (in f) deformation of (8.5.28) analogous to the binary case (8.3.1).

Definition 8.5.9. An *obscure membership deformed n-ary algebra* is $\mathcal{A}_n^{*\mathcal{G}}(\mathsf{f}) = \left\langle A^{(\mathsf{f})} \middle| \mu_*^{(n)}, + \right\rangle$ in which there are $(n! - 1)$ possible noncommutativity relations

$$\begin{gathered} \mathsf{f}_{a_n'}(a_n)\mu_*^{(n)}[a_1, \ldots , a_n] = \mathsf{f}_{a_{\sigma(n)}'}(a_{\sigma(n)})\mu_*^{(n)}[a_{\sigma(1)}, \ldots , a_{\sigma(n)}], \\ \forall a_1, \ldots , a_n \in A^{(\mathsf{f})}, \ \ a_n', a_{\sigma(n)}' \in G', \ \ \forall \sigma \in S_n, \sigma \neq I. \end{gathered} \tag{8.5.29}$$

Since $\mathsf{f} > 0$, we can have

Definition 8.5.10. The $(n! - 1)$ membership commutation factors in $\mathcal{A}_{\mathcal{G}*}^{(n)}(\mathsf{f})$ are defined by

$$\epsilon_{\sigma(n)}^{(n)}\left(a_n', a_{\sigma(n)}', a_n, a_{\sigma(n)}\right) = \frac{\mathsf{f}_{a_{\sigma(n)}'}(a_{\sigma(n)})}{\mathsf{f}_{a_n'}(a_n)}, \ \ a_n', a_{\sigma(n)}' \in G', \ \ \forall \sigma \in S_n, \sigma \neq I, \tag{8.5.30}$$

and the relations (8.5.29) become

$$\mu_*^{(n)}[a_1, \ldots, a_n] = \epsilon_{\sigma(n)}^{(n)}\left(a_n', a_{\sigma(n)}', a_n, a_{\sigma(n)}\right)\mu_*^{(n)}[a_{\sigma(1)}, \ldots, a_{\sigma(n)}]. \tag{8.5.31}$$

These definitions are unique, if we require: (1) for connection only two monomials with different permutations; (2) membership 'linearity' (in f); (3) compatibility with the binary case (8.3.2)–(8.3.3).

***Example* 8.5.11.** or the obscure membership deformed ternary algebra $\mathcal{A}_{\mathcal{G}*}^{(n)}(\mathsf{f}) = \left\langle A^{(\mathsf{f})} \middle| \mu_*^{(3)}, + \right\rangle$, we obtain

$$\mu_*^{(3)}[a_1, a_2, a_3] = \epsilon_{\sigma(1)\sigma(2)\sigma(3)}^{(3)}\mu_*^{(3)}[a_{\sigma(1)}, a_{\sigma(2)}, a_{\sigma(3)}]_*^{(3)}, \quad a_1, a_2, a_3 \in A\,,, \sigma \in S_3,$$
$$\epsilon_{132}^{(3)} = \epsilon_{312}^{(3)} = \frac{\mathsf{f}_{a_3'}(a_3)}{\mathsf{f}_{a_2'}(a_2)}, \quad \epsilon_{231}^{(3)} = \epsilon_{321}^{(3)} = \frac{\mathsf{f}_{a_3'}(a_3)}{\mathsf{f}_{a_1'}(a_1)}, \quad \epsilon_{213}^{(3)} = 1\,,, \quad a_1', a_2', a_3' \in G', \tag{8.5.32}$$

which means that it is 1-partially commutative (see (8.5.17)) and has only 2 independent membership commutation factors (see the binary case (8.3.3)).

We now provide a sketch construction of the n-ary ε-commutative algebras (8.5.8) membership deformation (see section 8.3.2 for the binary case).

Definition 8.5.12. An *obscure membership deformed n-ary π-commutative algebra* over $\Bbbk$ is $\mathcal{A}_n^{\mathcal{G}\star}(\mathsf{f}) = \left\langle A^{(\mathsf{f})} \middle| \mu_\star^{(n)}, + \right\rangle$ in which the following $(n! - 1)$ noncommutativity relations are valid

$$\begin{aligned}
&\pi^{(n)}\left(a_{\sigma(1)}', \ldots, a_{\sigma(n)}'\right)\mathsf{f}_{a_n'}(a_n)\mu_\star^{(n)}[a_1, \ldots, a_n] \\
&= \pi^{(n)}(a_1', \ldots, a_n')\mathsf{f}_{a_{\sigma(n)}'}(a_{\sigma(n)})\mu_\star^{(n)}[a_{\sigma(1)}, \ldots, a_{\sigma(n)}], \\
&\forall\, a_i \in A^{(\mathsf{f})}, \ a_i' \in G', \ \forall\, \sigma \in S_n, \ \sigma \neq I,
\end{aligned} \tag{8.5.33}$$

where $\pi^{(n)}$ are n-ary Schur-like factors satisfying the 2-cocycle conditions (8.5.2).

Definition 8.5.13. An algebra $\mathcal{A}_n^{G\star}(\mathsf{f})$ is called a *double $\varepsilon_\pi^{(n)}/\epsilon_\mathsf{f}^{(n)}$-commutative algebra*, if

$$\begin{aligned}
&\mu_\star^{(n)}[a_1, \ldots, a_n] \\
&= \varepsilon_{\sigma(1), \ldots, \sigma(n)}^{(n)}(a_1', \ldots, a_n')\epsilon_{\sigma(n)}^{(n)}\left(a_n', a_{\sigma(n)}', a_n, a_{\sigma(n)}\right)\mu_\star^{(n)}[a_{\sigma(1)}, \ldots, a_{\sigma(n)}],
\end{aligned} \tag{8.5.34}$$

$$\varepsilon_{\sigma(1), \ldots, \sigma(n)}^{(n)}(a_1', \ldots, a_n') = \frac{\pi^{(n)}(a_1', \ldots, a_n')}{\pi^{(n)}\left(a_{\sigma(1)}', \ldots, a_{\sigma(n)}'\right)}, \tag{8.5.35}$$

$$\begin{aligned}&\epsilon^{(n)}_{\sigma(n)}\left(a'_n, a'_{\sigma(n)}, a_n, a_{\sigma(n)}\right) = \frac{\mathsf{f}_{a'_{\sigma(n)}}(a_{\sigma(n)})}{\mathsf{f}_{a'_n}(a_n)},\\ &a_1, \ldots, a_n, a_{\sigma(1)}, \ldots, a_{\sigma(n)} \in A^{(\mathsf{f})},\\ &\quad a'_1, \ldots, a'_n, a'_{\sigma(1)}, \ldots, a'_{\sigma(n)} \in G', \ \forall \sigma \in S_n, \sigma \neq I,\end{aligned} \tag{8.5.36}$$

where $\varepsilon^{(n)}_{\pi}$ is the n-ary grading commutation factor (8.5.7) and $\epsilon^{(n)}_{\mathsf{f}}$ is the n-ary membership commutation factor (in our definition (8.5.30)).

8.6 Conclusions

This procedure can be considered to be the membership deformation of the given ε-commutative algebra (8.5.8), which corresponds to the first version of the binary commutative algebra deformation as in (8.3.12) and leads to a nonassociative algebra, in general. To achieve associativity one should consider the second version of the algebra deformation as in the binary case (8.3.16), and to introduce a different n-ary multiplication $\mu^{\circledast}_n$ such that the product of commutation factors $\varepsilon^{(n)}\epsilon^{(n)}$ satisfies the n-ary 2-cocycle-like conditions. In this way the n-ary noncocycle commutation factor satisfies the 'membership deformed cocycle-like' conditions, analogous to (8.3.17)–(8.3.19).

References

Artamonov V A 1997 Quantum polynomial algebras *J. Math. Sci* **87** 3441–62

Bargmann V 1954 On unitary ray representations of continuous groups *Ann. Math. Second Ser.* **59** 1–46

Berezin F A 1987 *Introduction to Superanalysis* (Dordrecht: Reidel)

Bourbaki N 1998 *Elements of Mathematics: Algebra I* (Berlin: Springer)

Bělohlávek R 2002 *Fuzzy Relational Systems: Foundations and Principles* (New York: Springer)

Carlsson R 1980 N-ary algebras *Nagoya Math. J.* **78** 45–56

Covolo T, Grabowski J and Poncin N 2016 The category of $\mathbb{Z}_2^n$-supermanifolds *J. Math. Phys.* **57** 16

Dade E C 1980 Group-graded rings and modules *Math. Z.* **174** 241–62

de Azcarraga J and Izquierdo J M 2010 n-Ary algebras: a review with applications *J. Phys.* **A43** 293001

de Goursac A, Masson T and Wallet J C 2012 Noncommutative ε-graded connections *J. Noncommut. Geom.* **6** 343–87

de Oliveira T 1960 Modular systems *Univ. Lisboa Rev. Fac. Ci. A (2)* **8** 155–67

Dubois D and Prade H (ed) 2000 Fundamentals of fuzzy sets *The Handbooks of Fuzzy Sets Series* vol 7 (Boston, MA: Kluwer)

Duplij S 2012 Polyadic systems, representations and quantum groups *J. Kharkiv Natl Univ. Ser. Nuclei Part. Fields* **1017** 28–59 Expanded version arXiv: math.RT/1308.4060

Duplij S 2019 Arity shape of polyadic algebraic structures *J. Math. Phys. Anal. Geom.* **15** 3–56

Frucht R 1955 Zur Darstellung endlicher Abelscher Gruppen durch Kollineationen *Math. Z.* **63** 145–55

Kac V G 1977 Lie superalgebras *Adv. Math.* **26** 8–96

Lam T Y 2001 A first course in noncommutative rings *Graduate Texts in Mathematics* vol 131 2nd edn (New York: Springer)

Michor P W and Vinogradov A M 1996 *n*-Ary Lie and associative algebras *Rend. Sem. Mat. Univ. Pol. Torino.* **54** 373–92

Montgomery S 1997 Constructing simple Lie superalgebras from associative graded algebras *J. Alg.* **195** 558–79

Nahm W, Rittenberg V and Scheunert M 1976 The classification of graded Lie algebras *Phys. Lett.* **61B** 383–91

Rittenberg V and Wyler D 1978 Generalized superalgebras *Nucl. Phys.* B **139** 189–202

Scheunert M 1979 Generalized Lie algebras *J. Math. Phys.* **20** 712–20

Smith N J J 2008 *Vagueness and Degrees of Truth* (Oxford: Oxford University Press)

Tilgner H 1977 Graded generalizations of Weyl- and Clifford algebras *J. Pure Appl. Alg.* **10** 163–8

Zimmermann H-J 2011 *Fuzzy Set Theory and Its Applications* 2nd edn (New York: Springer)

Žmuď È M 1960 Isomorphisms of irreducible protective representations of finite groups *Zap. Meh.-Mat. Fak. i Hařkov Mat. Obšc.* **26** 333–72 (Russian, MR 0244409)

Žmuď È M 1972 Symplectic geometric and projective representations of finite abelian groups *Math. USSR Sbornik* **16** 1–16 (Mat. Sbornik, Tom 87 (129), (1972), no 1, pp 3–17, MR 0292963)

Part III

Polyadic quantum groups

Chapter 9

Polyadic Hopf algebras

Since Hopf algebras were introduced in connection with algebraic topology (Sweedler 1969, Abe 1980), their fundamental role has increased significantly (see, e.g., Radford 2012), with numerous applications in diverse areas, especially in relation to quantum groups (Drinfeld 1987, Shnider and Sternberg 1993, Chari and Pressley 1996, Kassel 1995, Majid 1995). There have been many generalizations of Hopf algebras (for a brief review, see, e.g., Karaali 2008). For instance, their Von Neumann generalization was introduced in Duplij and Li (2001), Duplij and Sinel'shchikov (2009), and Li and Duplij (2002); their actions on the quantum plane were classified in Duplij and Sinel'shchikov (2010); and ternary Hopf algebras were defined and studied in Duplij (2001) and Borowiec *et al* (2001).

From another perspective, the concepts of polyadic vector space, polyadic algebras and polyadic tensor product over general polyadic fields were introduced in Duplij (2019). They differ from the standard definitions of n-ary algebras (de Azcarraga and Izquierdo 2010, Michor and Vinogradov 1996, Goze and Rausch de Traubenberg 2009) in considering an arbitrary arity shape for all operations, and not the algebra multiplication alone. This means that the arities of addition in the algebra, the multiplication and addition in the underlying field can all be different from binary and the number of places in the multiaction (polyadic module) can be more than one (Duplij 2012). The connection between arities is determined by their arity shapes (Duplij 2019) ('arity freedom principle'). Note that our approach is somewhat different from the operad approach (see, e.g., Markl *et al* 2002, Loday and Vallette 2012).

In this chapter we propose a similar and consequent polyadic generalization of Hopf algebras (Sweedler 1969, Abe 1980). First, we define polyadic coalgebras and study their homomorphisms and tensor products. In the construction of the polyadic convolution product and bialgebras we propose considering different arities for the algebra and coalgebra, which is a crucial difference from the binary case. Instead of the antipode, we introduce its polyadic version, the querantipode, by analogy with the querelement in n-ary groups (Dörnte 1929).

doi:10.1088/978-0-7503-2648-3ch9

Here we use for polyadic fields, vector spaces and algebras the notations of chapter 5. Recall, in brief, the ground field $\Bbbk = \Bbbk^{(m_k, n_k)}$ is a (m_k, n_k)-ary field having m_k-ary addition $\nu_k^{(m_k)}$ and n_k-ary multiplication $\mu_k^{(n_k)}$. A polyadic vector space over the polyadic field $\Bbbk$ is $\mathcal{V} = \mathcal{V}^{(m_k, n_k; m_V; r_V)} = \left\langle V \middle| \nu_k^{(m_k)}, \mu_k^{(n_k)}; \nu_V^{(m_V)}; \rho_V^{(r_V)} \right\rangle$ with m_V-ary 'vector addition' $\nu_V^{(m_V)}$ and r_V-place action $\rho_V^{(r_V)}$ ('scalar multiplication'). The *dimension* D_V of a polyadic vector space is the number of elements in its polyadic basis. The polyadic direct sum and polyadic tensor product of polyadic vector spaces were constructed in section 5.2.2. The polyadic tensor product is '$\Bbbk$-linear' in the usual sense, only instead of 'multiplication by scalar' one uses the multiaction $\rho_V^{(r_V)}$. Because of associativity, we will use the binary notation for polyadic tensor products (implying that $\otimes = \otimes_{\Bbbk}$) and powers of them (for instance, $\mathcal{V}^{\otimes n}$) to be clearer in computations and as customary in diagrams.

9.1 Polyadic coalgebras

The standard motivation for introducing the comultiplication is from representation theory (Curtis and Reiner 1962, Kirillov 1976). The first examples come from so-called addition formulas for special functions ('anciently' started from sin/cos), which actually arise from representations of groups (Shnider and Sternberg 1993, Hazewinkel *et al* 2010).

In brief, informally, let $\mathbf{\Pi}$ be an ordinary finite-dimensional representation of a (binary) group $\mathcal{G} = \langle G | \mu^{(2)} \equiv (\cdot) \rangle$ in a vector space $\mathcal{V}$ over a ground (binary) field $\Bbbk = \Bbbk^{(2,2)}$, such that

$$\mathbf{\Pi}(g \cdot h) = \mathbf{\Pi}(g) \circ \mathbf{\Pi}(h), \quad \mathbf{\Pi} : G \to \operatorname{End} V, \quad g, h \in G. \tag{9.1.1}$$

In some basis of $\mathcal{V}$ the matrix elements $\varphi_{ij}(g)$ satisfy $\varphi_{ij}(g \cdot h) = \sum_k \varphi_{ik}(g) \varphi_{kj}(h)$ (from (9.1.1)) and span a finite-dimensional vector space $\mathcal{V}_f = \left\langle V_f \mid \mu_f^{(2)} \right\rangle$ of functions with a basis e_{φ_m} as $f_\varphi = \sum_m \alpha_m e_{\varphi_m}$, $f_\varphi \in V_f$, $\alpha_m \in \mathbb{C}$. Now (9.1.1) gives $f_\varphi(g \cdot h) = \sum_{m,n} \beta_{mn} e_{\varphi_m}(g) e_{\varphi_n}(h)$, $f_\varphi \in V_f$, $\beta_{mn} \in \mathbb{C}$. If we omit the evaluation, it can be written in the vector space $\mathcal{V}_f$ using an additional linear map Δ_φ: $V_f \to V_f \otimes V_f$, in the following way

$$\Delta_\varphi\left(f_\varphi\right) = \sum_{m,n} \beta_{mn} e_{\varphi_m} \otimes e_{\varphi_n} \in V_f \otimes V_f. \tag{9.1.2}$$

Thus, to any finite-dimensional representation $\mathbf{\Pi}$ one can define the map Δ_π of vector spaces $\mathcal{V}_f$ to functions on a group, called a *comultiplication*. It is important that all the above operations are binary, and the defining formula for comultiplication (9.1.2) is fully determined by the definition of a representation (9.1.1).

The polyadic analog of a representation is introduced in chapter 2. In the case of multiplace representations, arities of the initial group and its representation can be different. Indeed, let $\mathcal{G}^{(n)} = \left\langle G \middle| \mu_G^{(n)} \right\rangle$, $\mu_G^{(n)}$: $G^{\times n} \to G$, be an n-ary group and

$\mathcal{G}^{(n')}_{\mathrm{End}V} = \left\langle \mathrm{End}V \middle| \mu_E^{(n')} \right\rangle$, $\mu_E^{(n')}\colon (\mathrm{End}V)^{\times n'} \to \mathrm{End}V$, is an n'-ary group of endomorphisms of a polyadic vector space $\mathcal{V}$ (see (5.1.22)).

Definition 9.1.1. A *polyadic (multiplace) representation* of $\mathcal{G}^{(n)}$ in $\mathcal{V}$ is a s-place mapping

$$\mathbf{\Pi}_s^{(n,n')}\colon G^{\times s} \to \mathrm{End}V, \tag{9.1.3}$$

satisfying the associativity preserving heteromorphism equation (Duplij 2012)

$$\mathbf{\Pi}_s^{(n,n')}\left(\overbrace{\mu_G^{(n)}[g_1, \dots g_n], \dots, \mu_G^{(n)}\left[g_{n\left(s-\ell'_{\mathrm{id}}-1\right)}, \dots g_{n\left(s-\ell'_{\mathrm{id}}\right)}\right]}^{s-\ell'_{\mathrm{id}}}, \dots, \overbrace{g_{n\left(s-\ell'_{\mathrm{id}}\right)+1}, \dots, g_{n\left(s-\ell'_{\mathrm{id}}\right)+\ell'_{\mathrm{id}}}}^{\ell'_{\mathrm{id}}}\right) \tag{9.1.4}$$
$$= \mu_E^{(n')}\left[\mathbf{\Pi}_s^{(n,n')}(g_1, \dots g_s), \dots, \mathbf{\Pi}_s^{(n,n')}(g_{s(n'-1)}, \dots g_{sn'})\right],$$

such that the diagram

$$\begin{array}{ccc} G^{\times sn'} & \xrightarrow{\left(\mathbf{\Pi}_s^{(n,n')}\right)^{\times n'}} & (\mathrm{End}\,V)^{\times n'} \\ \Big\downarrow {\scriptstyle \left(\mu_G^{(n)}\right)^{\times(s-\ell'_{\mathrm{id}})} \times (\mathrm{id}_G)^{\times \ell'_{\mathrm{id}}}} & & \Big\downarrow {\scriptstyle \mu_E^{(n')}} \\ G^{\times n'} & \xrightarrow{\mathbf{\Pi}_s^{(n,n')}} & \mathrm{End}\,V \end{array} \tag{9.1.5}$$

commutes, and the arity changing formula

$$sn' = n(s - \ell'_{\mathrm{id}}) + \ell'_{\mathrm{id}}, \tag{9.1.6}$$

where ℓ'_{id} is the number of 'intact elements' in lhs of (9.1.4), $0 \leqslant \ell'_{\mathrm{id}} \leqslant s - 1$, $2 \leqslant n' \leqslant n$.

For various particular examples, see chapter 2.

9.1.1 Polyadic comultiplication

Our motivations say that in constructing a polyadic analog of the comultiplication, one should not only 'reverse arrows', but also pay thorough attention to the arities of all operations.

Assertion 9.1.2. *The arity of polyadic comultiplication coincides with the arity of the representation and can differ from the arity of the polyadic algebra.*

Proof. It follows from (9.1.1), (9.1.2) and (9.1.3). □

Let us consider a polyadic vector space over the polyadic field $\Bbbk^{(m_k, n_k)}$ as

$$\mathcal{C}_{\text{vect}} = \left\langle C; K \mid \nu_C^{(m_c)}; \nu_k^{(m_k)}, \mu_k^{(n_k)}; \rho_C^{(r_c)} \right\rangle, \tag{9.1.7}$$

where $\nu_C^{(m_c)}\colon C^{\times m_c} \to C$ is m_c-ary addition and $\rho_C^{(r_c)}\colon K^{\times r_c} \times C \to C$ is r_c-place action ('scalar multiplication' of coalgebra).

Definition 9.1.3. A *polyadic (n'-ary) comultiplication* is a $\Bbbk$-linear map $\mathbf{\Delta}^{(n')}\colon C \to C^{\otimes n'}$.

Definition 9.1.4. A *polyadic (coassociative) coalgebra $\mathcal{C}$ (or* $\Bbbk$-coalgebra) is the polyadic vector space $\mathcal{C}_{\text{vect}}$ equipped with the polyadic comultiplication

$$\mathcal{C} = \mathcal{C}^{(n')} = \langle \mathcal{C}_{\text{vect}} | \mathbf{\Delta}^{(n')} \rangle, \tag{9.1.8}$$

which is (totally) *coassociative*

$$\begin{gathered}\left(\mathrm{id}_C^{\otimes(n'-1-i)} \otimes \mathbf{\Delta}^{(n')} \otimes \mathrm{id}_C^{\otimes i}\right) \circ \mathbf{\Delta}^{(n')} = \left(\mathrm{id}_C^{\otimes(n-1-j)} \otimes \mathbf{\Delta}^{(n')} \otimes \mathrm{id}_C^{\otimes j}\right) \circ \mathbf{\Delta}^{(n')}, \\ \forall i, j = 0, \ldots n-1,\ i \neq j,\ \mathrm{id}_C\colon C \to C,\end{gathered} \tag{9.1.9}$$

and such that the diagram

$$\begin{array}{ccc} C^{\otimes(2n'-1)} & \xleftarrow{\mathrm{id}_C^{\otimes(n'-1-i)} \otimes \mathbf{\Delta}^{(n')} \otimes \mathrm{id}_C^{\otimes i}} & C^{\otimes n'} \\ {\scriptstyle \mathrm{id}_C^{\otimes(n'-1-j)} \otimes \mathbf{\Delta}^{(n')} \otimes \mathrm{id}_C^{\otimes j}} \Big\uparrow & & \Big\uparrow {\scriptstyle \mathbf{\Delta}^{(n')}} \\ C^{\otimes n'} & \xleftarrow{\mathbf{\Delta}^{(n')}} & C \end{array} \tag{9.1.10}$$

commutes (see (5.3.34)).

Definition 9.1.5. A polyadic coalgebra $\mathcal{C}^{(n')}$ is called *totally co-commutative*, if

$$\mathbf{\Delta}^{(n')} = \tau_{n'} \circ \mathbf{\Delta}^{(n')}, \tag{9.1.11}$$

where $\tau_{n'} \in S_{n'}$, and $S_{n'}$ is the permutation symmetry group on n' elements.

Definition 9.1.6. A polyadic coalgebra $\mathcal{C}^{(n')}$ is called *medially co-commutative*, if

$$\mathbf{\Delta}_{\text{cop}}^{(n')} \equiv \boldsymbol{\tau}_{\text{op}}^{(n')} \circ \mathbf{\Delta}^{(n')} = \mathbf{\Delta}^{(n')}, \tag{9.1.12}$$

where $\boldsymbol{\tau}_{\text{op}}^{(n')}$ is the medially allowed polyadic twist map (5.3.63).

There are no other axioms in the definition of a polyadic coalgebra, following the same reasoning as for a polyadic algebra: the possible absence of zeroes and units (see remark 5.3.22 and table 5.2). Obviously, in a polyadic coalgebra $\mathcal{C}^{(n')}$, there is no 'unit element', because there is no multiplication, and a polyadic analog of counit can be only defined, when the underlying field $\Bbbk^{(m_k, n_k)}$ is unital (which is not always the case, Duplij 2017).

By analogy with the iterated product (1.1.3), introduce the *ℓ'-coiterated n'-ary comultiplication* by

$$\left(\mathbf{\Delta}^{(n')}\right)^{\circ\ell'} = \overbrace{\left(\mathrm{id}_C^{\otimes(n'-1)} \otimes \ldots \left(\mathrm{id}_C^{\otimes(n'-1)}\mathbf{\Delta}^{(n')}\right) \ldots \circ \mathbf{\Delta}^{(n')}\right) \circ \mathbf{\Delta}^{(n')}}^{\ell'}, \quad \ell' \in \mathbb{N}. \tag{9.1.13}$$

Therefore, the *admissible length* of any co-word is fixed ('quantized') as $\ell'(n' - 1) + 1$, but not arbitrary, as opposite to the binary case.

Let us introduce a coanalog of the derived n-ary multiplication (5.3.36) by

Definition 9.1.7. A polyadic comultiplication $\mathbf{\Delta}_{\mathrm{der}}^{(n')}$ is called *derived,* if it is ℓ_d-coiterated from the comultiplication $\mathbf{\Delta}_0^{\left(n_0'\right)}$ of lower arity $n_0' < n'$

$$\mathbf{\Delta}_{\mathrm{der}}^{(n')} = \overbrace{\left(\mathrm{id}_C^{\otimes\left(n_0'-1\right)} \otimes \ldots \left(\mathrm{id}_C^{\otimes\left(n_0'-1\right)}\mathbf{\Delta}_0^{\left(n_0'\right)}\right) \ldots \circ \mathbf{\Delta}_0^{\left(n_0'\right)}\right) \circ \mathbf{\Delta}_0^{\left(n_0'\right)}}^{\ell_d}, \tag{9.1.14}$$

or

$$\mathbf{\Delta}_{\mathrm{der}}^{(n')} = \left(\mathbf{\Delta}_0^{\left(n_0'\right)}\right)^{\circ\ell'}, \tag{9.1.15}$$

where

$$n' = \ell_d(n_0' - 1) + 1, \tag{9.1.16}$$

and $\ell_d \geqslant 2$ is the 'number of coiterations'.

The standard coiterations of $\mathbf{\Delta}$ are binary and restricted by $n_0' = 2$ (Sweedler 1969).

***Example* 9.1.8.** The matrix coalgebra generated by the basis e_{ij}, $i, j = 1, \ldots, N$ of $\mathrm{Mat}_N(\mathbb{C})$ with the binary coproduct $\Delta_0^{(2)}(e_{ij}) = \sum_k e_{ik} \otimes e_{kj}$ (see, e.g., Abe 1980) can be extended to the derived ternary coalgebra by $\Delta_{\mathrm{der}}^{(3)}(e_{ij}) = \sum_{k,l} e_{ik} \otimes e_{kl} \otimes e_{lj}$, such that (9.1.14) becomes $\mathbf{\Delta}_{\mathrm{der}}^{(3)} = (\mathrm{id}_C \otimes \mathbf{\Delta}_0^{(2)})\mathbf{\Delta}_0^{(2)} = (\mathbf{\Delta}_0^{(2)} \otimes \mathrm{id}_C)\mathbf{\Delta}_0^{(2)}$.

***Example* 9.1.9.** Let us consider the ternary coalgebra $\langle C | \mathbf{\Delta}^{(3)} \rangle$ generated by two elements $\{a, b\} \in C$ with the von Neumann regular looking comultiplication

$$\Delta^{(3)}(a) = a \otimes b \otimes a, \quad \Delta^{(3)}(b) = b \otimes a \otimes b. \tag{9.1.17}$$

It is easy to check that $\mathbf{\Delta}^{(3)}$ is coassociative and nonderived.

Definition 9.1.10. A polyadic coalgebra $\mathcal{C}^{(n')}$ (9.1.8) is called *co-medial*, if its n'-ary multiplication map satisfies the relation

$$\left(\left(\boldsymbol{\Delta}^{(n')}\right)^{\otimes n'}\right) \circ \boldsymbol{\Delta}^{(n')} = \boldsymbol{\tau}_{\text{medial}}^{(n',n')} \circ \left(\left(\boldsymbol{\Delta}^{(n')}\right)^{\otimes n'}\right) \circ \boldsymbol{\Delta}^{(n')}, \tag{9.1.18}$$

where $\boldsymbol{\tau}_{\text{medial}}^{(n',n')}$ is the polyadic medial map given by (5.3.59)–(5.3.60).

Introduce a $\Bbbk$-linear *r'-place action map* $\bar{\boldsymbol{\rho}}^{(r')}\colon K^{\otimes r'} \otimes C \to C$ corresponding to $\rho_C^{(r_c)}$ by (see (5.2.36))

$$\bar{\boldsymbol{\rho}}^{(r')} \circ (\lambda_1 \otimes \ldots \otimes \lambda_{r'} \otimes c) = \rho_C^{(r')}(\lambda_1, \ldots, \lambda_{r'}|c), \; \lambda_1, \ldots, \lambda_{r'} \in K, \, c \in C. \tag{9.1.19}$$

Let $\Bbbk^{(m_k,\, n_k)}$ be unital with unit e_k.

Definition 9.1.11. A $\Bbbk$-linear *r'-place coaction map* $\boldsymbol{\sigma}^{(r')}\colon C \to K^{\otimes r'} \otimes C$ is defined by

$$c \mapsto \overbrace{e_k \otimes \ldots \otimes e_k}^{r'} \otimes c. \tag{9.1.20}$$

Assertion 9.1.12. The coaction map $\boldsymbol{\sigma}^{(r')}$ is a 'right inverse' for the multiaction map $\bar{\boldsymbol{\rho}}^{(r')}$

$$\bar{\boldsymbol{\rho}}^{(r')} \circ \boldsymbol{\sigma}^{(r')} = \mathrm{id}_C. \tag{9.1.21}$$

Proof. This follows from the normalization (5.1.6) and (9.1.20). □

***Remark* 9.1.13.** The maps (9.1.19) and (9.1.20) establish the isomorphism $\overbrace{\Bbbk \otimes \ldots \otimes \Bbbk}^{r'} \otimes \mathcal{C} \cong \mathcal{C}$, which is well-known in the binary case (see, e.g., Yokonuma 1992).

We can provide the definition of counit only in the case where the underlying field $\Bbbk$ has a unit.

Definition 9.1.14. Counit axiom. The polyadic coalgebra $\mathcal{C}^{(n)}$ (5.3.32) over the unital polyadic field $\Bbbk^{(m_k,\, n_k)}$ contains a $\Bbbk$-linear *polyadic (right) counit map* $\boldsymbol{\varepsilon}^{(n',\, r')}\colon C^{\otimes(n'-1)} \to K^{\otimes r'}$ satisfying

$$\left(\boldsymbol{\varepsilon}^{(n',\, r')} \otimes \mathrm{id}_C\right) \circ \boldsymbol{\Delta}^{(n')} = \boldsymbol{\sigma}^{(r')}, \tag{9.1.22}$$

such that the diagram

$$\begin{array}{ccc} K^{\otimes r'} \otimes C & \xleftarrow{\boldsymbol{\varepsilon}^{(n',r')}\otimes \mathrm{id}_C} & C^{\otimes n'} \\ {\scriptstyle \boldsymbol{\sigma}^{(r')}}\uparrow & \nearrow {\scriptstyle \boldsymbol{\Delta}^{(n')}} & \\ C & & \end{array} \tag{9.1.23}$$

commutes (see (5.3.40)).

***Remark* 9.1.15.** We cannot write the 'elementwise' normalization action for the counit analogous to (5.3.41) (and state the assertion 5.3.26), because a unit element in a (polyadic) coalgebra is not defined.

By analogy with the derived polyadic unit (see (5.3.43) and definition 5.3.28), consider a 'derived' version of the polyadic counit.

Definition 9.1.16. The $\Bbbk$-linear *derived polyadic counit* (*neutral counit sequence*) of the polyadic coalgebra $\mathcal{C}^{(n')}$ is the set $\hat{\boldsymbol{\varepsilon}}^{(r)} = \{\boldsymbol{\varepsilon}_i^{(r')}\}$ of $n' - 1$ maps $\boldsymbol{\varepsilon}_i^{(r')}\colon C \to K^{\otimes r'}$, $i = 1, \dots, n' - 1$, satisfying

$$\left(\boldsymbol{\varepsilon}_1^{(r')} \otimes \dots \otimes \boldsymbol{\varepsilon}_{n'-1}^{(r')} \otimes \mathrm{id}_C\right) \circ \boldsymbol{\Delta}^{(n')} = \boldsymbol{\sigma}^{(r')}, \tag{9.1.24}$$

where id_C can be on any place. If $\boldsymbol{\varepsilon}_1^{(r')} = \dots = \boldsymbol{\varepsilon}_{n'-1}^{(r')} = \boldsymbol{\varepsilon}_0^{(r')}$, we call it the *strong derived polyadic counit*. In general, we can define formally, see (5.3.44),

$$\boldsymbol{\varepsilon}_{\mathrm{der}}^{(n',r')} = \boldsymbol{\varepsilon}_1^{(r')} \otimes \dots \otimes \boldsymbol{\varepsilon}_{n'-1}^{(r')}. \tag{9.1.25}$$

Definition 9.1.17. A polyadic coassociative coalgebra $\mathcal{C}_{\mathrm{der}}^{(n')} = \left\langle \mathcal{C}_{\mathrm{vect}} \mid \boldsymbol{\Delta}_{\mathrm{der}}^{(n')}, \boldsymbol{\varepsilon}_{\mathrm{der}}^{(n',r')} \right\rangle$ is called *derived* from $\mathcal{C}_0^{(n')} = \left\langle \mathcal{C}_{\mathrm{vect}} \mid \boldsymbol{\Delta}_0^{\left(n_0'\right)}, \boldsymbol{\varepsilon}_0^{\left(n_0',r'\right)} \right\rangle$, if (9.1.14) and

$$\boldsymbol{\varepsilon}_{\mathrm{der}}^{(n',r')} = \overbrace{\boldsymbol{\varepsilon}_0^{\left(n_0',\,r'\right)} \otimes \dots \otimes \boldsymbol{\varepsilon}_0^{\left(n_0',r'\right)}}^{\ell_d} \tag{9.1.26}$$

hold, where $\boldsymbol{\varepsilon}_0^{\left(n_0',r'\right)} = \overbrace{\boldsymbol{\varepsilon}_0^{(r')} \otimes \dots \otimes \boldsymbol{\varepsilon}_0^{(r')}}^{n_0'-1}$ (formally, because id_C in (9.1.24) can be on any place).

In Duplij (2001) the particular case for $n' = 3$ and $r' = 1$ was considered.

9.1.2 Homomorphisms of polyadic coalgebras

In the binary case, a morphism of coalgebras is a linear map ψ: $C_1 \to C_2$ which 'commutes' with comultiplications ('$(\psi \otimes \psi) \circ \mathbf{\Delta}_1 = \mathbf{\Delta}_2 \circ \varphi$'). It seems that for the polyadic coalgebras, one could formally change the direction of all arrows in (5.3.75). However, we observed that arity changing is possible for multivalued morphisms only. Therefore, here we confine ourselves to homomorphisms (1-place heteromorphisms, Duplij 2012).

Let us consider two polyadic (equiary) $\Bbbk$-coalgebras $\mathcal{C}_1^{(n')} = \langle C_1 | \mathbf{\Delta}_1^{(n')} \rangle$ and $\mathcal{C}_2^{(n')} = \langle C_2 | \mathbf{\Delta}_2^{(n')} \rangle$ over the same polyadic field $\Bbbk^{(m_k, n_k)}$.

Definition 9.1.18. A (*coalgebra*) *homomorphism* between polyadic (equiary) coalgebras $\mathcal{C}_1^{(n')}$ and $\mathcal{C}_2^{(n')}$ is a $\Bbbk$-linear map $\mathbf{\Psi}^{(n')} : C_1 \to C_2$, such that

$$\left(\overbrace{\mathbf{\Psi}^{(n')} \otimes \ldots \otimes \mathbf{\Psi}^{(n')}}^{n'} \right) \circ \mathbf{\Delta}_1^{(n')} = \mathbf{\Delta}_2^{(n')} \circ \mathbf{\Psi}^{(n')}, \tag{9.1.27}$$

and the diagram

$$\begin{array}{ccc} C_2^{\otimes n'} & \xleftarrow{\left(\mathbf{\Psi}^{(n')}\right)^{\otimes n'}} & C_1^{\otimes n'} \\ {\scriptstyle \mathbf{\Delta}_2^{(n')}} \uparrow & & \uparrow {\scriptstyle \mathbf{\Delta}_1^{(n')}} \\ C_2 & \xleftarrow{\mathbf{\Psi}^{(n')}} & C_1 \end{array} \tag{9.1.28}$$

commutes (see (5.3.75)).

Only when the underlying field $\Bbbk$ is unital, we can also define a morphism for counits.

Definition 9.1.19. The *counit homomorphism* for $\boldsymbol{\varepsilon}_{1,2}^{(n',r')} : C_{1,2}^{\otimes(n'-1)} \to K^{\otimes r'}$ is given by

$$\boldsymbol{\varepsilon}_2^{(n',r')} = \boldsymbol{\varepsilon}_1^{(n',r')} \circ \left(\overbrace{\mathbf{\Psi}^{(n')} \otimes \ldots \otimes \mathbf{\Psi}^{(n')}}^{n'-1} \right), \tag{9.1.29}$$

and the diagram

$$\begin{array}{ccc} K^{r'} & \xleftarrow{\boldsymbol{\varepsilon}_2^{(n',r')}} & C_2^{\otimes(n'-1)} \\ {\scriptstyle \boldsymbol{\varepsilon}_1^{(n',r')}} \uparrow & \nearrow {\scriptstyle \left(\mathbf{\Psi}^{(n')}\right)^{\otimes(n'-1)}} & \\ C_1^{\otimes(n'-1)} & & \end{array} \tag{9.1.30}$$

commutes (see (5.3.80)).

9.1.3 Tensor product of polyadic coalgebras

Let us consider n' polyadic equiary coalgebras $\mathcal{C}_i^{(n')} = \langle C_i | \mathbf{\Delta}_i^{(n')} \rangle$, $i = 1, \ldots, n'$.

Proposition 9.1.20. *The tensor product of the coalgebras has a structure of the polyadic coassociative coalgebra* $\mathcal{C}_{\otimes}^{(n')} = \left\langle C_{\otimes} \mid \mathbf{\Delta}_{\otimes}^{(n')} \right\rangle$, $C_{\otimes} = \otimes_{i=1}^{n'} C_i$, *if*

$$\mathbf{\Delta}_{\otimes}^{(n')} = \boldsymbol{\tau}_{\mathrm{medial}}^{(n',n')} \circ \left(\mathbf{\Delta}_i^{(n')} \otimes \ldots \otimes \mathbf{\Delta}_i^{(n')} \right), \tag{9.1.31}$$

where $\boldsymbol{\tau}_{\mathrm{medial}}^{(n',n')}$ *is defined in (5.3.60) and* $\mathbf{\Delta}_{\otimes}^{(n')} : C_{\otimes} \to \overbrace{C_{\otimes} \otimes \ldots \otimes C_{\otimes}}^{n'}$.

The proof is in full analogy with that of proposition 5.3.41. If all of the coalgebras $\mathcal{C}_i^{(n')}$ have counits, we denote them $\boldsymbol{\varepsilon}_i^{(n',r')} : C_i^{\otimes(n'-1)} \to K^{\otimes r'}$, $i = 1, \ldots, n'$, and the counit map of $\mathcal{C}_{\otimes}^{(n')}$ will be denoted by $\boldsymbol{\varepsilon}_{\otimes}^{(n',r')} : C_{\otimes}^{\otimes(n'-1)} \to K^{\otimes r'}$. We have (in analogy to ' $\varepsilon_{C_1 \otimes C_2}(c_1 \otimes c_2) = \varepsilon_{C_1}(c_1)\varepsilon_{C_2}(c_2)$')

Proposition 9.1.21. *The tensor product coalgebra* $\mathcal{C}_{\otimes}^{(n')}$ *has a counit which is defined by*

$$\begin{aligned} &\boldsymbol{\varepsilon}_{\otimes}^{(n',r')} \circ \left(c_1 \otimes \ldots \otimes c_{n'(n'-1)} \right) \\ &= \boldsymbol{\mu}_k^{n_k} \circ \left(\boldsymbol{\varepsilon}_1^{(n',r')} \circ \left(c_1 \otimes \ldots \otimes c_{(n'-1)} \right) \otimes \ldots \otimes \boldsymbol{\varepsilon}_{n'}^{(n',r')} \circ \right. \\ &\left. \left(c_{(n'-1)(n'-1)} \otimes \ldots \otimes c_{n'(n'-1)} \right) \right), \\ &c_i \in C_i,\ i = 1, \ldots, n'(n'-1), \end{aligned} \tag{9.1.32}$$

and the arity of the comultiplication coincides with the arity of the underlying field

$$n' = n_k. \tag{9.1.33}$$

9.1.4 Polyadic coalgebras in the Sweedler notation

The $\Bbbk$-linear coalgebra comultiplication map $\mathbf{\Delta}^{(n')}$ defined in definition 9.1.3 is useful for a 'diagrammatic' description of polyadic coalgebras, and it corresponds to the algebra multiplication map $\boldsymbol{\mu}^{(n)}$, which both manipulate with sets. However, for concrete computations (with elements) we need an analog of the polyadic algebra multiplication $\mu^{(n)} \equiv \mu_A^{(n_a)}$ from (5.3.3). The connection of $\boldsymbol{\mu}^{(n)}$ and $\mu^{(n)}$ is given by (5.3.31), which can be treated as a 'bridge' between the 'diagrammatic' and 'elementwise' descriptions. The coanalog of (5.3.31) was not considered, because the comultiplication has only one argument. To be consistent, we introduce the 'elementwise' comultiplication $\Delta^{(n')}$ as the coanalog of $\mu^{(n)}$ by the evaluation

$$\mathbf{\Delta}^{(n')} \circ (c) = \Delta^{(n')}(c), \quad c \in C. \tag{9.1.34}$$

In general, one does not distinguish $\mathbf{\Delta}^{(n')}$ and $\Delta^{(n')}$ and may use one symbol in both descriptions (see, e.g., Radford 2012), but here (9.1.34) is important.

In real 'elementwise' coalgebra computations with many variables and comultiplications acting on them, the indices and various letters reproduce themselves in such a way that it is impossible to observe the structure of the expressions.

Therefore, instead of different letters in the binary decomposition ('$\Delta(c) = \sum_i a_i \otimes b_i$' and (9.1.2)) it was proposed (Sweedler 1968) to use the same letter ('$\Delta(c) = \sum_i c_{[1],i} \otimes c_{[2],i}$'), and then go from the real sum $\sum_i$ to the formal sum $\sum_{[c]}$ as ('$\Delta(c) = \sum_{[c]} c_{[1]} \otimes c_{[2]}$ ' remembering the place of the components $c_{[1]}$, $c_{[2]}$ only), because the real indices pullulate in complicated formulas enormously. In simple cases, the sum sign was also omitted ('$\Delta(c) = c_{[1]} \otimes c_{[2]}$'), which recalls the Einstein index summation rule in physics. This trick abbreviated tedious coalgebra computations and was called the (*sumless*) *Sweedler* (*sigma*) *notation* (sometimes it is called the *Heyneman–Sweedler notation*, Heyneman and Sweedler 1969).

Now we can write $\Delta^{(n')}$ as an n'-ary decomposition in the manifest 'elementwise' form

$$\begin{aligned} &\Delta^{(n')}(c) = (\nu^{(m)})^{\circ \ell_\Delta} \\ &[c_{[1],1} \otimes c_{[2],1} \otimes \ldots \otimes c_{[n'],1}, \ \ldots \ , c_{[1],N_\Delta} \otimes c_{[2],N_\Delta} \otimes \ldots \otimes c_{[n'],N_\Delta}], \ c_{[j],i} \in C, \end{aligned} \tag{9.1.35}$$

where $\ell_\Delta \in \mathbb{N}_0$ is a 'number of additions', and $N_\Delta \in \mathbb{N}$ is the 'number of summands'. In the binary case, the number of summands in the decomposition is not 'algebraically' restricted, because $N_\Delta = \ell_\Delta + 1$. In the polyadic case, we have

Assertion 9.1.22. The admissible 'number of summands' N_Δ in the polyadic comultiplication is

$$N_\Delta = \ell_\Delta(m - 1) + 1, \ \ \ell_\Delta \in \mathbb{N}_0, \ \ m \geqslant 2. \tag{9.1.36}$$

Therefore, the 'quantization' of N_Δ coincides with that of the N-dimensional polyadic algebra (see assertion 5.3.49).

Introduce the *polyadic Sweedler notation* by exchanging in (9.1.35) the real m-ary addition $\nu^{(m)}$ by the formal addition $\nu_{[c]}$ and writing

$$\Delta^{(n')}(c) = \nu_{[c]}[c_{[1]} \otimes c_{[2]} \otimes \ldots \otimes c_{[n']}] \Rightarrow c_{[1]} \otimes c_{[2]} \otimes \ldots \otimes c_{[n']}. \tag{9.1.37}$$

Remember here that we can formally add only N_Δ summands, because of the 'quantization rule' (9.1.36).

The polyadic Sweedler notation power can be seen in the following

***Example* 9.1.23.** We apply (9.1.37) to the coassociativity (9.1.9) with $n' = 3$, to obtain

$$\begin{aligned} (\mathrm{id} \otimes \mathrm{id} \otimes \Delta^{(3)}) \circ \Delta^{(3)}(c) &= (\mathrm{id} \otimes \Delta^{(3)} \otimes \mathrm{id}) \circ \Delta^{(3)}(c) \\ &= (\Delta^{(3)} \otimes \mathrm{id} \otimes \mathrm{id}) \circ \Delta^{(3)}(c) \Rightarrow \end{aligned} \tag{9.1.38}$$

$$\begin{aligned} &= \nu_{[c]}\Big[c_{[1]} \otimes c_{[2]} \otimes \nu_{[c_2]}\big[(c_{[3]})_{[1]} \otimes (c_{[3]})_{[2]} \otimes (c_{[3]})_{[3]}\big]\Big] \\ &= \nu_{[c]}\Big[c_{[1]} \otimes \nu_{[c_2]}\big[(c_{[2]})_{[1]} \otimes (c_{[2]})_{[2]} \otimes (c_{[2]})_{[3]}\big] \otimes c_{[3]}\Big] \\ &= \nu_{[c]}\Big[\nu_{[c_2]}\big[(c_{[1]})_{[1]} \otimes (c_{[1]})_{[2]} \otimes (c_{[1]})_{[3]}\big] \otimes c_{[2]} \otimes c_{[3]}\Big]. \end{aligned} \tag{9.1.39}$$

After dropping the brackets and applying the Sweedler trick for the second time, we get the same formal expression in all three cases

$$(\nu_{[c]})^{\circ 2}[c_{[1]} \otimes c_{[2]} \otimes c_{[3]} \otimes c_{[4]} \otimes c_{[5]}]. \tag{9.1.40}$$

Unfortunately, in the polyadic case the Sweedler notation loses too much information to be useful.

Assertion 9.1.24. The polyadic Sweedler notation can be applied to only the derived polyadic coalgebras (see definition 9.1.17).

Nevertheless, if in an expression there are no coiterations, one can formally use it (e.g., in the polyadic analog (9.1.22) of the counting axiom '$\sum \varepsilon(c_{[1]})c_{[2]} = c$').

9.1.5 Polyadic group-like and primitive elements

Let us consider some special kinds of elements in a polyadic coalgebra $C^{(n')}$. We should take into account that in the polyadic case, as in (9.1.35), there can only be the admissible 'number of summands' N_Δ (9.1.36).

Definition 9.1.25. An element g of $C^{(n')}$ is called *polyadic semigroup-like*, if

$$\Delta^{(n')}(g) = \overbrace{g \otimes \ldots \otimes g}^{n'}, g \in C. \tag{9.1.41}$$

When $C^{(n')}$ has the counit $\boldsymbol{\varepsilon}^{(n', r')}$ (9.1.22), then g is called *polyadic group-like*, if (' $\varepsilon(g) = 1$')

$$\boldsymbol{\varepsilon}^{(n', r')} \circ \left(\overbrace{g \otimes \ldots \otimes g}^{n' - 1} \right) = \overbrace{e_k \otimes \ldots \otimes e_k}^{r'}, \tag{9.1.42}$$

where e_k is the unit of the underlying polyadic field $\Bbbk$.

Definition 9.1.26. An element x of $C^{(n')}$ is called *polyadic skew k_p-primitive*, if ('$\Delta(x) = g_1 \otimes x + x \otimes g_2$')

$$\Delta^{(n')}(x) = (\nu^{(m)})^{\circ \ell_\Delta} \left[\left(\overbrace{g_1 \otimes \ldots \otimes g_{k_p}}^{k_p} \otimes \overbrace{x \otimes \ldots \otimes x}^{n' - k_p} \right), \ldots, \right.$$
$$\left. \left(\overbrace{x \otimes \ldots \otimes x}^{n' - k_p} \otimes \overbrace{g_{(N_\Delta - 1)k_p + 1} \otimes \ldots \otimes g_{N_\Delta k_p}}^{k_p} \right) \right], \tag{9.1.43}$$

where $1 \leqslant k_p \leqslant n' - 1$, $N_\Delta = \ell_\Delta(m - 1) + 1$ is the total 'number of summands', here $\ell_\Delta \in \mathbb{N}$ is the 'number of m-ary additions', and $g_i \in C$, $i = 1, \dots, N_\Delta k_p$ are polyadic (semi-)group-like (9.1.41). In (9.1.43) the $n' - k_p$ elements x move from the right to the left by one.

Assertion 9.1.27. If $k_p = n' - 1$, then $\Delta^{(n')}(x)$ is 'linear' in x, and $n' = \ell_\Delta(m - 1) + 1$.

In this case, we call x a *polyadic primitive element*.

***Example* 9.1.28.** Let $n' = 3$ and $k_p = 2$, then $m = 3$, and we have only one ternary addition $\ell_\Delta = 1$

$$\Delta^{(3)}(x) = \nu^{(3)}[g_1 \otimes g_2 \otimes x, g_3 \otimes x \otimes g_4, x \otimes g_5 \otimes g_6], \tag{9.1.44}$$

$$\Delta^{(3)}(g_i) = g_i \otimes g_i \otimes g_i, \quad i = 1, \dots, 6. \tag{9.1.45}$$

The ternary coassociativity gives $g_1 = g_2 = g_3$ and $g_4 = g_5 = g_6$. Therefore, the general form of the ternary primitive element is

$$\Delta^{(3)}(x) = \nu^{(3)}[g_1 \otimes g_1 \otimes x, g_1 \otimes x \otimes g_2, x \otimes g_2 \otimes g_2]. \tag{9.1.46}$$

Note that coassociativity leads to the derived comultiplication (9.1.14), because

$$\Delta^{(3)}(x) = (\mathrm{id} \otimes \Delta^{(2)})\Delta^{(2)}(x) = (\Delta^{(2)} \otimes \mathrm{id})\Delta^{(2)}(x), \tag{9.1.47}$$

$$\Delta^{(2)}(x) = g_1 \otimes x + x \otimes g_2. \tag{9.1.48}$$

The same situation occurs with the 'linear' comultiplication of any arity n', i.e., when $k_p = n' - 1$.

The most important difference with the binary case is the 'intermediate' possibility $k_p < n' - 1$, when the rhs is 'nonlinear' in x.

***Example* 9.1.29.** In the case where $n' = 3$ and $k_p = 1$, we have $m = 3$, and $\ell_\Delta = 1$

$$\Delta^{(3)}(x) = \nu^{(3)}[g_1 \otimes x \otimes x, x \otimes g_2 \otimes x, x \otimes x \otimes g_3], \tag{9.1.49}$$

$$\Delta^{(3)}(g_i) = g_i \otimes g_i \otimes g_i, \quad i = 1, \dots, 3. \tag{9.1.50}$$

Now ternary coassociativity cannot be achieved with any values of g_i. This is true for any arity n' and any 'nonlinear' comultiplication.

Therefore, we arrive at the general structure

Assertion 9.1.30. In a polyadic coassociative coalgebra $\mathcal{C}^{(n')}$ polyadic primitive elements exist, if and only if the n'-ary comultiplication $\Delta^{(n')}$ is derived (9.1.14) from the binary comultiplication $\Delta^{(2)}$.

9.1.6 Polyadic analog of duality

The connection between binary associative algebras and coassociative coalgebras (formally named as 'reversing arrows') is given in terms of the dual vector space (dual module) concept (see, e.g., Radford 2012). Informally, for a binary coalgebra $\mathcal{C}^{(2)} = \langle C | \Delta, \varepsilon \rangle$ considered as a vector space over a binary field $\Bbbk$ (a $\Bbbk$-vector space), its *dual* is $C_2^* = \mathrm{Hom}_{\Bbbk}(C, K)$ with the natural *pairing* $C^* \times C \to K$ given by $f(c)$, $f \in C^*$, $c \in C$. The *canonical injection* $\theta : C^* \otimes C^* \to (C \otimes C)^*$ is defined by

$$\theta(f_1 \otimes f_2) \circ (c_1 \otimes c_2) = f_1(c_1) f_2(c_2), \quad c_{1,2} \in C, \; f_{1,2} \in C^*, \tag{9.1.51}$$

which is an isomorphism in the finite-dimensional case. The transpose of $\Delta : C \to C \otimes C$ is a $\Bbbk$-linear map $\Delta_* : (C \otimes C)^* \to C^*$ acting as $\Delta_*(\xi)(c) = \xi \circ (\Delta(c))$, where $\xi \in (C \otimes C)^*$, $c \in C$. The multiplication μ_* on the set C^* is the map $C^* \otimes C^* \to C^*$, and therefore we have to use the canonical injection θ as follows

$$\mu_* : C^* \otimes C^* \xrightarrow{\theta} (C \otimes C)^* \xrightarrow{\Delta_*} C^*, \tag{9.1.52}$$

$$\mu_* = \Delta_* \circ \theta. \tag{9.1.53}$$

The associativity of μ_* follows from the coassociativity of Δ. Since $\Bbbk^* \simeq \Bbbk$, the dual of the counit is the unit $\eta_* : K \xrightarrow{\varepsilon^*} C^*$. Therefore, $\mathcal{C}^{(2)*} = \langle C^* | \mu_*, \eta_* \rangle$ is a binary associative algebra, which is called the *dual algebra* of the binary coalgebra $\mathcal{C}^{(2)} = \langle C | \Delta, \varepsilon \rangle$ (see, e.g., Radford 2012).

In the polyadic case, arities of the comultiplication, its dual multiplication and the underlying field can be different, but connected by (9.1.51). Let us consider a polyadic coassociative coalgebra $\mathcal{C}^{(n')}$ with n'-ary comultiplication $\Delta^{(n')}$ (9.1.34) over $\Bbbk^{(m_k, n_k)}$. In search of the most general polyadic analog of the injection (9.1.51), we arrive at the possibility of multiplace morphisms.

Definition 9.1.31. For the polyadic coalgebra $\mathcal{C}^{(n')}$ considered as a polyadic vector space over $\Bbbk^{(m_k, n_k)}$, a *polyadic dual* is $C^* = \mathrm{Hom}_{\Bbbk}(C^{\otimes s}, K)$ with *s-place pairing* $C^* \times \overbrace{C \times \ldots \times C}^{s} \to K$ given by $f^{(s)}(c_1, \ldots, c_s)$, $f \in C^*$, $c_i \in C$, $s \in \mathbb{N}$.

While constructing a polyadic analog of (9.1.51), recall that for any n'-ary operation the admissible length of a co-word is $\ell'(n' - 1) + 1$, where ℓ' is the number of the iterated operation (9.1.13).

Definition 9.1.32. A *polyadic canonical injection map* $\boldsymbol{\theta}^{(n^*, n', s)}$ of $\mathcal{C}^{(n')}$ is defined by

$$\boldsymbol{\theta}^{(n^*, n', s)} \circ \left(f_1^{(s)} \otimes \ldots \otimes f_{n^*}^{(s)}\right) \circ \left(c_1 \otimes \ldots \otimes c_{\ell'(n'-1)+1}\right) = \\ \left(\mu_k^{(n_k)}\right)^{\circ \ell_k}\left[f_1^{(s)}(c_1, \ldots, c_s), \ldots, f_{\ell_k(n_k-1)+1}^{(s)}(c_{(n^*-1)s+1}, \ldots, c_{n^*s})\right], \tag{9.1.54}$$

where

$$n^*s = \ell'(n'-1)+1, \;\; \ell' \in \mathbb{N}, n' \geqslant 2, \tag{9.1.55}$$

$$n^* = \ell_k\,(n_k-1)+1, \;\; \ell_k \in \mathbb{N}, n_k \geqslant 2. \tag{9.1.56}$$

It is obvious that $\theta^{(2,2,1)} = \theta$ from (9.1.51). Then, the *polyadic transpose map* of the n'-ary comultiplication $\Delta^{(n')} : C \to \overbrace{C \otimes \ldots \otimes C}^{n'}$ is a $\Bbbk$-linear map $\Delta_*^{(n'')} : \left(\overbrace{C \otimes \ldots \otimes C}^{n''}\right)^* \to C^*$ such that

$$\Delta_*^{(n'')} \circ \left(\xi^{(n'')}\right)(c) = \xi^{(n'')} \circ \left(\left(\Delta^{(n')}\right)^{\circ \ell'}(c)\right), \\ \xi^{(n'')} \in \left(\overbrace{C \otimes \ldots \otimes C}^{n''}\right)^*, \; n'' = \ell'(n'-1)+1, \, c \in C \tag{9.1.57}$$

where ℓ' is the 'number of comultiplications' (see (1.1.3) for multiplications and (9.1.13)).

Definition 9.1.33. A n^*-ary multiplication map $\boldsymbol{\mu}_*^{(n^*)}$ which is (*one way*) *dual* to the n'-ary comultiplication map $\boldsymbol{\Delta}^{(n')}$ is given by the composition of the polyadic canonical injection $\boldsymbol{\theta}^{(n^*, n', s)}$ (9.1.54) and the polyadic transpose $\boldsymbol{\Delta}_*^{(n'')}$ (9.1.57) by

$$\boldsymbol{\mu}_*^{(n^*)} = \boldsymbol{\Delta}_*^{(n'')} \circ \boldsymbol{\theta}^{(n^*, n', s)}. \tag{9.1.58}$$

Indeed, using (9.1.54) and (9.1.57) we obtain (in Sweedler notation)

$$\boldsymbol{\mu}_*^{(n^*)} \circ \left(f_1^{(s)} \otimes \ldots \otimes f_{n^*}^{(s)}\right) \circ (c) = \boldsymbol{\Delta}_*^{(n'')} \circ \boldsymbol{\theta}^{(n^*, n', s)} \circ \\ \left(f_1^{(s)} \otimes \ldots \otimes f_{n^*}^{(s)}\right) \circ (c) \\ = \boldsymbol{\theta}^{(n^*, n', s)} \circ \left(f_1^{(s)} \otimes \ldots \otimes f_{n^*}^{(s)}\right) \circ \left((\Delta^{(n')})^{\circ \ell'}(c)\right) \\ = \left(\mu_k^{(n_k)}\right)^{\circ \ell_k}\left[f_1^{(s)}(c_{[1]}, \ldots, c_{[s]}), \ldots, f_{\ell_k(n_k-1)+1}^{(s)}(c_{[(n^*-1)s+1]}, \ldots, c_{[n^*s]})\right], \tag{9.1.59}$$

and (9.1.55)–(9.1.56) are valid, from which we arrive at

Assertion 9.1.34. In the polyadic case the arity n^* of the multiplication $\boldsymbol{\mu}_*^{(n^*)}$ can be different from the arity n' of the initial coalgebra $\mathcal{C}^{(n')}$.

Remark **9.1.35.** If $n^* \neq n'$ and $s \geqslant 2$, the word 'duality' can only be used conditionally.

9.1.7 Polyadic convolution product

If $\mathcal{A}^{(2)} = \langle A | \mu, \eta \rangle$ is a binary algebra and $\mathcal{C}^{(2)} = \langle C | \Delta, \varepsilon \rangle$ is a binary coalgebra over a binary field $\Bbbk$, then a more general set of $\Bbbk$-linear maps $\mathrm{Hom}_{\Bbbk}(C, A)$ can be considered, while its particular case where $\mathcal{A}^{(2)} = \Bbbk$ corresponds to the above duality. The multiplication on $\mathrm{Hom}_{\Bbbk}(C, A)$ is the *convolution product* $(\star)$ which can be uniquely constructed in the natural way: by applying first comultiplication Δ and then multiplication $\mu \equiv (\cdot)$ to an element of C, as $C \xrightarrow{\Delta} C \otimes C \xrightarrow{f \otimes g} A \otimes A \xrightarrow{\mu} A$ or $f \star g = \mu \circ (f \otimes g) \circ \Delta$, where $f, g \in \mathrm{Hom}_{\Bbbk}(C, A)$. The associativity of the convolution product follows from the associativity of μ and coassociativity of Δ, and the role of the identity (neutral element) in $\mathrm{Hom}_{\Bbbk}(C, A)$ is played by the composition of the unit map $\eta : K \to A$ and the counit map $\varepsilon : C \to K$, such that $e_\star = \eta \circ \varepsilon \in \mathrm{Hom}_{\Bbbk}(C, A)$, because $e_\star \star f = f \star e_\star = f$. Indeed, from the obvious relation $\mathrm{id}_A \circ f \circ \mathrm{id}_C = f$ and the unit and counit axioms it follows that

$$\begin{gathered} \mu \circ (\eta \otimes \mathrm{id}_A) \circ (\mathrm{id}_K \otimes f) \circ (\varepsilon \otimes \mathrm{id}_C) \circ \Delta = \mu \circ \\ (\eta \circ \mathrm{id}_K \circ \varepsilon) \otimes (\mathrm{id}_A \circ f \circ \mathrm{id}_C) \circ \Delta = e_\star \star f = f, \end{gathered} \tag{9.1.60}$$

or in Sweedler notation $\varepsilon(c_{[1]}) \cdot f(c_{[2]}) = f(c_{[1]}) \cdot \varepsilon(c_{[2]}) = f(c)$.

The polyadic analog of duality and (9.1.59) offer an idea of how to generalize the binary convolution product to the most exotic case, when the algebra and coalgebra have different arities $n \neq n'$.

Let $\mathcal{A}^{(n)}$ and $\mathcal{C}^{(n')}$ be, respectively, a polyadic associative algebra and a coassociative coalgebra over the same polyadic field $\Bbbk^{(m_k, n_k)}$. If they are both unital and counital respectively, then we can consider a polyadic analog of the composition $\eta \circ \varepsilon$. The crucial difference from the binary case is that now $\boldsymbol{\eta}^{(r,n)}$ and $\boldsymbol{\varepsilon}^{(n',r')}$ are multiplace multivalued maps (5.3.39) and (9.1.22). Their composition is

$$\mathbf{e}_\star^{(n',n)} = \boldsymbol{\eta}^{(r,n)} \circ \boldsymbol{\gamma}^{(r',r)} \circ \boldsymbol{\varepsilon}^{(n',r')} \in \mathrm{Hom}_{\Bbbk}(C^{\otimes(n'-1)}, A^{\otimes(n-1)}), \tag{9.1.61}$$

where the multiplace multivalued map $\boldsymbol{\gamma}^{(r',r)} \in \mathrm{Hom}_{\Bbbk}(K^{\otimes r'}, K^{\otimes r})$ is, obviously, $(\simeq)$, and the diagram

$$\begin{array}{ccc} C^{\otimes(n'-1)} & \xrightarrow{\varepsilon^{(n',r')}} & K^{\otimes r'}, \\ \mathbf{e}_{\star}^{(n,n')}\Big\downarrow & & \Big\downarrow \gamma^{(r',r)}(\simeq) \\ A^{\otimes(n-1)} & \xleftarrow{\eta^{(r,n)}} & K^{\otimes r} \end{array} \tag{9.1.62}$$

commutes.

The formula (9.1.61) leads us to propose

Conjecture 9.1.36. A polyadic analog of the convolution should be considered for multiplace multivalued $\Bbbk$-linear maps in $\mathrm{Hom}_{\Bbbk}(C^{\otimes(n'-1)}, A^{\otimes(n-1)})$.

In this way, we arrive at the following

Condition 9.1.37. Introduce the $\Bbbk$-linear maps $f^{(i)} : C^{\otimes(n'-1)} \to A^{\otimes(n-1)}$, $i = 1, \ldots, n_\star$, where $n_\star \geqslant 2$. To create a closed $n_\star$-ary operation for them, we use the ℓ-iterated multiplication map $(\boldsymbol{\mu}^{(n)})^{\circ\ell} : A^{\otimes\ell(n-1)+1} \to A$ and ℓ'-iterated comultiplication map $(\boldsymbol{\Delta}^{(n')})^{\circ\ell'} : C \to C^{\otimes\ell'(n'-1)+1}$. Then we compose the above $\Bbbk$-linear maps in the same way as is done above for the binary case

$$\begin{array}{l} C^{\otimes(n'-1)} \xrightarrow{\left(\left(\boldsymbol{\Delta}^{(n')}\right)^{\circ\ell'}\right)^{\otimes(n'-1)}} C^{\otimes(n'-1)(\ell'(n'-1)+1)} \xrightarrow{\tau_{medial}^{(n_\star,n'-1)}} C^{\otimes(n'-1)(\ell'(n'-1)+1)} \\ \xrightarrow{f^{(1)}\otimes\ldots\otimes f^{(n_\star)}} A^{\otimes(n-1)(\ell'(n'-1)+1)} \xrightarrow{\tau_{medial}^{(n-1,n_\star)}} A^{\otimes(n-1)(\ell(n-1)+1)} \xrightarrow{\left(\left(\boldsymbol{\mu}^{(n)}\right)^{\circ\ell}\right)^{\otimes(n-1)}} A^{\otimes(n-1)}. \end{array} \tag{9.1.63}$$

where $\boldsymbol{\tau}_{\text{medial}}^{(n_\star,n'-1)}$ and $\boldsymbol{\tau}_{\text{medial}}^{(n-1,n_\star)}$ are the medial maps (5.3.60) acting on the Sweedler components of c and $f^{(i)}$, respectively. To make the sequence of maps (9.1.63) consistent, the arity $n_\star$ is connected with the iteration numbers ℓ, ℓ' by $n_\star = \ell(n-1) + 1 = \ell'(n'-1) + 1$, $\ell, \ell' \in \mathbb{N}$.

Definition 9.1.38. Let $\mathcal{A}^{(n)}$ and $\mathcal{C}^{(n')}$ be an n-ary associative algebra and n'-ary coassociative coalgebra over a polyadic field $\Bbbk$ (the existence of the unit and counit here is mandatory), then the set $\mathrm{Hom}_{\Bbbk}(C^{\otimes(n'-1)}, A^{\otimes(n-1)})$ is closed under the *$n_\star$-ary convolution product* map $\boldsymbol{\mu}_\star^{(n_\star)}$ defined by

$$\begin{gathered} \boldsymbol{\mu}_\star^{(n_\star)} \circ \left(\boldsymbol{f}^{(1)} \otimes \ldots \otimes \boldsymbol{f}^{(n_\star)}\right) = \\ \left(\left(\boldsymbol{\mu}^{(n)}\right)^{\circ\ell}\right)^{\otimes(n-1)} \circ \boldsymbol{\tau}_{\text{medial}}^{(n-1,n_\star)} \circ \left(\boldsymbol{f}^{(1)} \otimes \ldots \otimes \boldsymbol{f}^{(n_\star)}\right) \circ \boldsymbol{\tau}_{\text{medial}}^{(n_\star,n'-1)} \circ \left(\left(\boldsymbol{\Delta}^{(n')}\right)^{\circ\ell'}\right)^{\otimes(n'-1)}, \end{gathered} \tag{9.1.64}$$

and its arity is given by the following *$n_\star$-consistency condition*

$$n_\star - 1 = \ell(n-1) = \ell'(n'-1). \tag{9.1.65}$$

Definition 9.1.39. The set of $\Bbbk$-linear maps $f^{(i)} \in \mathrm{Hom}_{\Bbbk}(C^{\otimes(n'-1)}, A^{\otimes(n-1)})$ endowed with the convolution product (9.1.64) is called a *polyadic convolution algebra*

$$\mathcal{C}_{\star}^{(n',n)} = \left\langle \mathrm{Hom}_{\Bbbk}\left(C^{\otimes(n'-1)}, A^{\otimes(n-1)}\right) \middle| \boldsymbol{\mu}_{\star}^{(n_{\star})} \right\rangle. \tag{9.1.66}$$

***Example* 9.1.40.** An important case is given by the binary algebra $\mathcal{A}^{(2)}$ and coalgebra $\mathcal{C}^{(2)}$ $(n = n' = 2)$, when the number of iterations are equal $\ell = \ell'$, and the arity $n_{\star}$ becomes

$$n_{\star} = \ell + 1 = \ell' + 1, \quad \ell, \ell' \in \mathbb{N}, \tag{9.1.67}$$

while the $n_{\star}$-ary convolution product in $\mathrm{Hom}_{\Bbbk}(C, A)$ takes the form

$$\begin{aligned} \boldsymbol{\mu}_{\star}^{(n_{\star})} \circ (f^{(1)} \otimes \dots \otimes f^{(n_{\star})}) = \boldsymbol{\mu}^{\circ(n_{\star}-1)} \\ \circ (f^{(1)} \otimes \dots \otimes f^{(n_{\star})}) \circ \boldsymbol{\Delta}^{\circ(n_{\star}-1)}, f^{(i)} \in \mathrm{Hom}_{\Bbbk}(C, A), \end{aligned} \tag{9.1.68}$$

where $\boldsymbol{\mu} = \boldsymbol{\mu}^{(2)}$ and $\boldsymbol{\Delta} = \boldsymbol{\Delta}^{(2)}$ are the binary multiplication and comultiplication maps respectively.

Definition 9.1.41. The polyadic convolution algebra $\mathcal{C}_{\star}^{(2,\,2)}$ determined by the binary algebra and binary coalgebra (9.1.68) is called *derived.*

Corollary 9.1.42. The arity $n_{\star}$ of the derived polyadic convolution algebra is unrestricted and can take any integer value $n_{\star} \geqslant 2$.

***Remark* 9.1.43.** If the polyadic tensor product and the underlying polyadic field $\Bbbk$ are derived (see discussion in chapter 5 and Duplij 2019), while all maps coincide $f^{(i)} = f$, the convolution product (9.1.68) is called the *Sweedler power* of f (Kashina *et al* 2012) or the *Adams operator* (Aguiar and Lauve 2015). In the binary case they denoted it by $(f)^{n_{\star}}$, but for the $n_{\star}$-ary product this is the first polyadic power of f (see (1.1.3)).

Obviously, some interesting algebraic objects are nonderived, and here they are determined by $n + n' \geqslant 5$, and also the arities of the algebra and coalgebra can be different $n \neq n'$, which is a more exotic and exciting possibility. Generally, the arity $n_{\star}$ of the convolution product (9.1.64) is not arbitrary and is 'quantized' by solving (9.1.65) in integers. The values $n_{\star}$ for minimal arities n, n' are presented in table 9.1.

The most unusual possibility is the existence of nondiagonal entries, which correspond to unequal arities of multiplication and comultiplication $n \neq n'$. The table is symmetric, which means that the arity $n_{\star}$ is invariant under the exchange $(n, \ell) \longleftrightarrow (n', \ell')$ following from (9.1.65).

***Example* 9.1.44.** $\mathrm{Hom}_{\Bbbk}(C, A)$. In the simplest derived case (9.1.68), when both algebra $\mathcal{A}^{(2)} = \langle A | \boldsymbol{\mu} \rangle$ and coalgebra $\mathcal{C}^{(2)} = \langle C | \boldsymbol{\Delta} \rangle$ are binary with $n = 2, \ell = 2, n' = 2, \ell' = 2$, it is possible to obtain the ternary convolution product

Table 9.1. Arity values $n_\star$ of the polyadic convolution product (9.1.64), allowed by (9.1.65). The framed box corresponds to the binary convolution product.

n' \ n	μ	$n=2$			$n=3$			$n=4$			$n=5$		
Δ	ℓ' \ ℓ	$\ell=1$	$\ell=2$	$\ell=3$	$\ell=1$	$\ell=2$	$\ell=3$	$\ell=1$	$\ell=2$	$\ell=3$	$\ell=1$	$\ell=2$	$\ell=3$
$n'=2$	$\ell'=1$	**2**											
	$\ell'=2$		**3**		**3**								
	$\ell'=3$			**4**				**4**					
$n'=3$	$\ell'=1$		**3**		**3**								
	$\ell'=2$					**5**					**5**		
	$\ell'=3$						**7**		**7**				
$n'=4$	$\ell'=1$			**4**				**4**					
	$\ell'=2$						**7**		**7**				
	$\ell'=3$									**10**			
$n'=5$	$\ell'=1$					**5**					**5**		
	$\ell'=2$											**9**	
	$\ell'=3$												**13**

$\boldsymbol{\mu}_\star^{(3)}$ of the maps $\boldsymbol{f}^{(i)} : C \to A$, $i = 1, 2, 3$, using Sweedler notation for $\boldsymbol{\Delta}^{\circ 2} \equiv (\mathrm{id}_C \otimes \boldsymbol{\Delta}) \circ \boldsymbol{\Delta}$ as $\Delta^{\circ 2}(c) = c_{[1]} \otimes c_{[2]} \otimes c_{[3]}$, $\boldsymbol{\mu}^{\circ 2} \equiv \boldsymbol{\mu} \circ (\mathrm{id}_A \otimes \boldsymbol{\mu}) : A^{\otimes 3} \to A$, and the elementwise description using the evaluation

$$\boldsymbol{\mu}_{\star,\, der}^{(3)} \circ (\boldsymbol{f}^{(1)} \otimes \boldsymbol{f}^{(2)} \otimes \boldsymbol{f}^{(3)}) \circ (c) = \mu^{\circ 2}\left[\mathrm{f}^{(1)}(c_{[1]}),\, \mathrm{f}^{(2)}(c_{[2]}),\, \mathrm{f}^{(3)}(c_{[3]})\right]. \tag{9.1.69}$$

***Example* 9.1.45.** $\mathrm{Hom}_{\mathbb{k}}(C, A^{\otimes 2})$, $\mathrm{Hom}_{\mathbb{k}}(C^{\otimes 2}, A)$ .. Nonbinary, nonderived and nonsymmetric cases:

(i) The ternary algebra $\mathcal{A}^{(3)} = \langle A | \boldsymbol{\mu}^{(3)} \rangle$ and the binary coalgebra $\mathcal{C}^{(2)} = \langle C | \boldsymbol{\Delta} \rangle$, such that $n = 3$, $\ell = 1$, $n' = 2$, $\ell' = 2$ giving a ternary convolution product of the maps $\boldsymbol{f}^{(i)} : C \to A^{\otimes 2}$, $i = 1, 2, 3$. In the elementwise description $\boldsymbol{f}^{(i)} \circ (c) = \mathrm{f}_{[1]}^{(i)}(c) \otimes \mathrm{f}_{[2]}^{(i)}(c)$, $c \in C$. Using (9.1.64), we obtain the manifest form of the nonderived ternary convolution product by evaluation

$$\begin{gathered}\boldsymbol{\mu}_\star^{(3)} \circ (\boldsymbol{f}^{(1)} \otimes \boldsymbol{f}^{(2)} \otimes \boldsymbol{f}^{(3)}) \circ (c) \\ = \mu^{(3)}\left[\mathrm{f}_{[1]}^{(1)}(c_{[1]}),\, \mathrm{f}_{[1]}^{(2)}(c_{[2]}),\, \mathrm{f}_{[1]}^{(3)}(c_{[3]})\right] \otimes \mu^{(3)}\left[\mathrm{f}_{[2]}^{(1)}(c_{[1]}),\, \mathrm{f}_{[2]}^{(2)}(c_{[2]}),\, \mathrm{f}_{[2]}^{(3)}(c_{[3]})\right],\end{gathered} \tag{9.1.70}$$

where $\Delta^{\circ 2}(c) = c_{[1]} \otimes c_{[2]} \otimes c_{[3]}$.

(ii) The algebra is binary $\mathcal{A}^{(2)} = \langle A | \boldsymbol{\mu} \rangle$, and the coalgebra is ternary $\mathcal{C}^{(3)} = \langle C | \boldsymbol{\Delta}^{(3)} \rangle$, which corresponds to $n = 2$, $\ell = 2$, $n' = 3$, $\ell' = 1$, the maps $\boldsymbol{f}^{(i)} : C^{\otimes 2} \to A$, $i = 1, 2, 3$ in the elementwise description are two place, $\boldsymbol{f}^{(i)} \circ (c_1 \otimes c_2) = \mathrm{f}^{(i)}(c_1, c_2)$, $c_{1,2} \in C$, and $(\Delta^{(3)})^{\otimes 2}(c^{(1)} \otimes c^{(2)}) = \left(c_{[1]}^{(1)} \otimes c_{[2]}^{(1)} \otimes c_{[3]}^{(1)}\right) \otimes \left(c_{[1]}^{(2)} \otimes c_{[2]}^{(2)} \otimes c_{[3]}^{(2)}\right)$. The ternary convolution product is

$$\begin{gathered}\boldsymbol{\mu}_\star^{(3)} \circ (\boldsymbol{f}^{(1)} \otimes \boldsymbol{f}^{(2)} \otimes \boldsymbol{f}^{(3)}) \circ \\ (c^{(1)} \otimes c^{(2)}) = \mu^{\circ 2}\left[\mathrm{f}^{(1)}\left(c_{[1]}^{(1)}, c_{[1]}^{(2)}\right),\, \mathrm{f}^{(2)}\left(c_{[2]}^{(1)}, c_{[2]}^{(2)}\right),\, \mathrm{f}^{(3)}\left(c_{[3]}^{(1)}, c_{[3]}^{(2)}\right)\right].\end{gathered} \tag{9.1.71}$$

***Example* 9.1.46.** $\mathrm{Hom}_{\Bbbk}(C^{\otimes 2}, A^{\otimes 2})$.. The last (fourth) possibility for the ternary convolution product (see table 9.1) is nonderived and symmetric $n = 3$, $\ell = 1$, $n' = 3$, $\ell' = 1$, with both a ternary algebra $\mathcal{A}^{(3)} = \langle A | \boldsymbol{\mu}^{(3)} \rangle$ and coalgebra $\mathcal{C}^{(3)} = \langle C | \boldsymbol{\Delta}^{(3)} \rangle$. In the elementwise description the maps $\boldsymbol{f}^{(i)} : C^{\otimes 2} \to A^{\otimes 2}$, $i = 1, 2, 3$ are $\boldsymbol{f}^{(i)} \circ (c_1 \otimes c_2) = \mathrm{f}_{[1]}^{(i)}(c_1, c_2) \otimes \mathrm{f}_{[2]}^{(i)}(c_1, c_2)$, $c_{1,2} \in C$. Then

$$\mu_\star^{(3)} \circ \left(f^{(1)} \otimes f^{(2)} \otimes f^{(3)}\right) \circ \left(c^{(1)} \otimes c^{(2)}\right) = \mu^{(3)}\left[\mathrm{f}_{[1]}^{(1)}\left(c_{[1]}^{(1)}, c_{[1]}^{(2)}\right), \mathrm{f}_{[1]}^{(2)}\left(c_{[2]}^{(1)}, c_{[2]}^{(2)}\right), \mathrm{f}_{[1]}^{(3)}\left(c_{[3]}^{(1)}, c_{[3]}^{(2)}\right)\right]$$
$$\otimes \mu^{(3)}\left[\mathrm{f}_{[2]}^{(1)}\left(c_{[1]}^{(1)}, c_{[1]}^{(2)}\right), \mathrm{f}_{[2]}^{(2)}\left(c_{[2]}^{(1)}, c_{[2]}^{(2)}\right), \mathrm{f}_{[2]}^{(3)}\left(c_{[3]}^{(1)}, c_{[3]}^{(2)}\right)\right]. \tag{9.1.72}$$

The above examples present clearly the possible forms of the $n_\star$-ary convolution product, which can be convenient for lowest arity computations.

The *general polyadic convolution product* (9.1.64) in Sweedler notation can be presented as

$$\boldsymbol{\mu}_\star^{(n_\star)} \circ (\boldsymbol{f}^{(1)} \otimes \boldsymbol{f}^{(2)} \otimes \ldots \otimes \boldsymbol{f}^{(n_\star)}) = \boldsymbol{g},\ \boldsymbol{f}^{(i)}, \boldsymbol{g} \in \mathrm{Hom}_{\Bbbk}\left(C^{\otimes(n'-1)}, A^{\otimes(n-1)}\right),$$
$$g_{[j]} \circ \left(c^{(1)} \otimes \ldots \otimes c^{(n'-1)}\right)$$
$$= \left(\mu^{(n)}\right)^{\circ \ell}\left[\mathrm{f}_{[j]}^{(1)}\left(\overbrace{c_{[1]}^{(1)}, \ldots, c_{[1]}^{(n'-1)}}^{n'-1}\right), \mathrm{f}_{[j]}^{(2)}\left(\overbrace{c_{[2]}^{(1)}, \ldots, c_{[2]}^{(n'-1)}}^{n'-1}\right), \ldots, \mathrm{f}_{[j]}^{(n_\star)}\left(\overbrace{c_{[n_\star]}^{(1)}, \ldots, c_{[n_\star]}^{(n'-1)}}^{n'-1}\right)\right], \tag{9.1.73}$$
$$\mathrm{f}_{[j]}^{(i)} \in \mathrm{Hom}_{\Bbbk}\left(C^{\otimes(n'-1)}, A\right),\ i \in 1, \ldots, n_\star,\ j \in 1, \ldots, n-1,\ c \in C,$$

where $g_{[j]}$ are the Sweedler components of $\boldsymbol{g}$.

Recall that the associativity of the binary convolution product $(\star)$ is transparent in the Sweedler notation. Indeed, if $(f \star g) \circ (c) = f(c_{[1]}) \cdot g(c_{[2]})$, $f, g, h \in \mathrm{Hom}_{\Bbbk}(C, A)$, $c \in C$, $(\cdot) \equiv \mu_A^{(2)}$, then $((f \star g) \star h) \circ (c) = \left(f(c_{[1]}) \cdot g(c_{[2]})\right) \cdot h(c_{[3]}) = f(c_{[1]}) \cdot \left(g(c_{[2]}) \cdot h(c_{[3]})\right) = (f \star (g \star h)) \circ (c)$.

Lemma 9.1.47. The polyadic convolution algebra $\mathcal{C}_\star^{(n',n)}$ (9.1.66) is associative.

Proof. To prove the claimed associativity of polyadic convolution $\boldsymbol{\mu}_\star^{(n_\star)}$ we express (5.2.23) in Sweedler notation. Starting from

$$\boldsymbol{h} = \boldsymbol{\mu}_\star^{(n_\star)} \circ (\boldsymbol{g} \otimes \boldsymbol{f}^{(n_\star+1)} \otimes \boldsymbol{f}^{(n_\star+2)} \otimes \ldots \otimes \boldsymbol{f}^{(2n_\star-1)}),\ \boldsymbol{h} \in \mathrm{Hom}_{\Bbbk}\left(C^{\otimes(n'-1)}, A^{\otimes(n-1)}\right), \tag{9.1.74}$$

where $\boldsymbol{g}$ is given by (9.1.73), it follows that $\boldsymbol{h}$ should not depend on the place of $\boldsymbol{g}$ in (9.1.74). Applying $\boldsymbol{h}$ to $c \in C$ twice, we obtain for its Sweedler components $\boldsymbol{h}_{[j]}$, $j \in 1, \ldots, n-1$,

$$\boldsymbol{h}_{[j]} \circ \left(c^{(1)} \otimes \ldots \otimes c^{(n'-1)}\right) = \left(\mu^{(n)}\right)^{\circ\ell}\left[\left(\mu^{(n)}\right)^{\circ\ell}\left[\mathrm{f}_{[j]}^{(1)}\left(\overbrace{c_{[1]}^{(1)}, \ldots, c_{[1]}^{(n'-1)}}^{n'-1}\right), \mathrm{f}_{[j]}^{(2)}\left(\overbrace{c_{[2]}^{(1)}, \ldots, c_{[2]}^{(n'-1)}}^{n'-1}\right),\right.\right.$$
$$\left.\left.\ldots, \mathrm{f}_{[j]}^{(n_\star)}\left(\overbrace{c_{[2n_\star-1]}^{(1)}, \ldots, c_{[2n_\star-1]}^{(n'-1)}}^{n'-1}\right)\right], \mathrm{f}_{[j]}^{(n_\star+1)}\left(\overbrace{c_{[1]}^{(1)}, \ldots, c_{[1]}^{(n'-1)}}^{n'-1}\right), \mathrm{f}_{[j]}^{(n_\star+2)}\left(\overbrace{c_{[2]}^{(1)}, \ldots, c_{[2]}^{(n'-1)}}^{n'-1}\right),\right. \tag{9.1.75}$$
$$\left.\ldots, \mathrm{f}_{[j]}^{(2n_\star-1)}\left(\overbrace{c_{[2n_\star-1]}^{(1)}, \ldots, c_{[2n_\star-1]}^{(n'-1)}}^{n'-1}\right)\right],$$

and here coassociativity and (9.1.65) gives $\left(\Delta^{(n')}\right)^{\circ 2l'}(c) = c_{[1]} \otimes c_{[2]} \otimes \ldots \otimes c_{[2n_\star-1]}$. Since n-ary algebra multiplication $\mu^{(n)}$ is associative, the internal $\left(\mu^{(n)}\right)^{\circ\ell}$ in (9.1.75) can be on any place, and $\boldsymbol{g}$ in (9.1.74) can be on any place as well. This means that the polyadic convolution product $\boldsymbol{\mu}_\star^{(n_\star)}$ is associative. □

Observe the polyadic version of the identity used in (9.1.60): for any $\boldsymbol{f} \in \mathrm{Hom}_{\Bbbk}(C^{\otimes(n'-1)}, A^{\otimes(n-1)})$

$$\mathrm{id}_A^{\otimes(n-1)} \circ \boldsymbol{f} \circ \mathrm{id}_C^{\otimes(n'-1)} = \boldsymbol{f}. \tag{9.1.76}$$

Proposition 9.1.48. *If the polyadic associative algebra $\mathcal{A}^{(n)}$ is unital with $\boldsymbol{\eta}^{(r,\,n)} : K^r \to A^{\otimes(n-1)}$, and the polyadic coassociative coalgebra $\mathcal{C}^{(n')}$ is counital with $\boldsymbol{\varepsilon}^{(n',\,r')} : C^{\otimes(n'-1)} \to K^{r'}$, both over the same polyadic field $\Bbbk$, then the polyadic convolution algebra $\mathcal{C}_\star^{(n',n)}$ (9.1.66) is unital, and its unit is given by $\mathbf{e}_\star^{(n',n)}$ (9.1.65).*

Proof. In analogy with (9.1.60) we compose

$$\begin{aligned}
\boldsymbol{f} = &\left(\left(\boldsymbol{\mu}^{(n)}\right)^{\circ\ell}\right)^{\otimes(n-1)} \circ \boldsymbol{\tau}_{\mathrm{medial}}^{(n-1,n_\star)} \circ \left(\left(\boldsymbol{\eta}^{(r,\,n)}\right)^{\otimes(n_\star-1)} \otimes \mathrm{id}_A^{\otimes(n-1)}\right)\\
&\circ \left(\left(\boldsymbol{\gamma}^{(r',\,r)}\right)^{\otimes(n_\star-1)} \otimes \mathrm{id}_A^{\otimes(n-1)}\right)\\
&\circ \left(\left(\mathrm{id}_K^{\otimes r'}\right)^{\otimes(n_\star-1)} \otimes \boldsymbol{f}\right) \circ \left(\left(\boldsymbol{\varepsilon}^{(n',\,r')}\right)^{\otimes(n_\star-1)} \otimes \mathrm{id}_C^{\otimes(n'-1)}\right)\\
&\circ \boldsymbol{\tau}_{\mathrm{medial}}^{(n_\star,n'-1)} \circ \left(\left(\boldsymbol{\Delta}^{(n')}\right)^{\circ\ell'}\right)^{\otimes(n'-1)}\\
= &\left(\left(\boldsymbol{\mu}^{(n)}\right)^{\circ\ell}\right)^{\otimes(n-1)} \circ \boldsymbol{\tau}_{\mathrm{medial}}^{(n-1,n_\star)} \circ \left(\boldsymbol{\eta}^{(r,\,n)} \circ \boldsymbol{\gamma}^{(r',\,r)} \circ \boldsymbol{\varepsilon}^{(n',\,r')}\right)^{\otimes(n_\star-1)}\\
&\circ \left(\mathrm{id}_A^{\otimes(n-1)} \circ \boldsymbol{f} \circ \mathrm{id}_C^{\otimes(n'-1)}\right) \circ \boldsymbol{\tau}_{\mathrm{medial}}^{(n_\star,n'-1)} \circ \left(\left(\boldsymbol{\Delta}^{(n')}\right)^{\circ\ell'}\right)^{\otimes(n'-1)}\\
= &\boldsymbol{\mu}_\star^{(n_\star)} \circ \left(\left(\mathbf{e}_\star^{(n',n)}\right)^{\otimes(n_\star-1)} \otimes \boldsymbol{f}\right),
\end{aligned} \tag{9.1.77}$$

which coincides with the polyadic unit definition (5.3.43). We use the identity (9.1.76) and the axioms for a polyadic unit (5.3.39) and counit (9.1.22). The same derivation can be made for any place of $\boldsymbol{\mu}_\star^{(n_\star)}$. □

As in the general theory of n-ary groups (Dörnte 1929), the invertibility of maps in $\mathcal{C}_\star^{(n',n)}$ should be defined not by using the unit, but by using the querelement (1.2.13).

Definition 9.1.49. For a fixed $\boldsymbol{f} \in \mathcal{C}_\star^{(n',n)}$ its *coquerelement* $\mathbf{q}_\star(\boldsymbol{f}) \in \mathcal{C}_\star^{(n',n)}$ is the querelement in the $n_\star$-ary convolution product

$$\boldsymbol{\mu}_\star^{(n_\star)} \circ \left(\boldsymbol{f}^{\otimes(n_\star-1)} \otimes \mathbf{q}_\star(\boldsymbol{f})\right) = \boldsymbol{f}, \tag{9.1.78}$$

where $\mathbf{q}_\star(\boldsymbol{f})$ can be on any place and $n_\star \geqslant 3$. The maps in a polyadic convolution algebra which have a coquerelement are called *coquerable*.

Define the *positive convolution power* $\ell_\star$ of an element $\boldsymbol{f} \in \mathcal{C}_\star^{(n',n)}$ not recursively as in Post (1940), but through the $\ell_\star$-iterated multiplication (1.1.3)

$$\boldsymbol{f}^{\langle \ell_\star \rangle} = \left(\boldsymbol{\mu}_\star^{(n_\star)}\right)^{\circ \ell_\star} \circ \left(\boldsymbol{f}^{\otimes(\ell_\star(n_\star-1)+1)}\right), \tag{9.1.79}$$

and an element in the *negative convolution power* $\boldsymbol{f}^{\langle -\ell_\star \rangle}$ satisfies the equation

$$\begin{aligned}\left(\boldsymbol{\mu}_\star^{(n_\star)}\right)^{\circ \ell_\star} \circ \left(\boldsymbol{f}^{\langle \ell_\star-1 \rangle} \otimes \boldsymbol{f}^{\otimes(n_\star-2)} \otimes \boldsymbol{f}^{\langle -\ell_\star \rangle}\right) = \left(\boldsymbol{\mu}_\star^{(n_\star)}\right)^{\circ \ell_\star} \circ \\ \left(\boldsymbol{f}^{\otimes \ell_\star(n_\star-1)} \otimes \boldsymbol{f}^{\langle -\ell_\star \rangle}\right) = \boldsymbol{f}.\end{aligned} \tag{9.1.80}$$

It follows from (9.1.79) that the polyadic analogs of the exponent laws hold

$$\boldsymbol{\mu}_\star^{(n_\star)} \circ \left(\boldsymbol{f}^{\langle \ell_\star^{(1)} \rangle} \otimes \boldsymbol{f}^{\langle \ell_\star^{(2)} \rangle} \otimes \ldots \otimes \boldsymbol{f}^{\langle \ell_\star^{(n_\star)} \rangle}\right) = \boldsymbol{f}^{\langle \ell_\star^{(1)}+\ell_\star^{(2)} \ldots +\ell_\star^{(n_\star)}+1 \rangle}, \tag{9.1.81}$$

$$\left(\boldsymbol{f}^{\langle \ell_\star^{(1)} \rangle}\right)^{\langle \ell_\star^{(2)} \rangle} = \boldsymbol{f}^{\langle \ell_\star^{(1)}\ell_\star^{(2)}(n_\star-1)+\ell_\star^{(1)}+\ell_\star^{(2)} \rangle}. \tag{9.1.82}$$

Comparing (9.1.78) and (9.1.80), we have

$$\mathbf{q}_\star(\boldsymbol{f}) = \boldsymbol{f}^{\langle -1 \rangle}. \tag{9.1.83}$$

An arbitrary polyadic power ℓ_Q of the coquerelement $\mathbf{q}_\star^{\circ \ell_Q}(\boldsymbol{f})$ is defined by (9.1.78) recursively and can be expressed through the negative polyadic power of $\boldsymbol{f}$ (see, e.g., Dudek 2007 for n-ary groups). In terms of the Heine numbers (Heine 1878) (or q-deformed numbers, Kac and Cheung 2002)

$$[[l]]_q = \frac{q^l - 1}{q - 1}, \quad l \in \mathbb{N}_0, \, q \in \mathbb{Z}, \tag{9.1.84}$$

we obtain Duplij (2012)

$$\mathbf{q}_\star^{\circ \ell_Q}(\boldsymbol{f}) = \boldsymbol{f}^{\langle -[[\ell_Q]]_{2-n_\star} \rangle}. \tag{9.1.85}$$

9.2 Polyadic bialgebras

The next step is to combine algebras and coalgebras into a common algebraic structure in some 'natural' way. Informally, a bialgebra is defined as a vector space which is 'simultaneously' an algebra and a coalgebra with some compatibility conditions (e.g., Sweedler 1969, Abe 1980).

In search of a polyadic analog of bialgebras, we observe two structural differences with the binary case: (1) since the unit and counit do not necessarily exist, we obtain 4 different kinds of bialgebras (similar to the unit and zero in table 5.2); (2) where the most exotic is the possibility of unequal arities of multiplication and comultiplication $n \neq n'$ (see assertion 9.1.34). Initially, we take them as arbitrary and then try to find restrictions arising from some 'natural' relations.

Let $\mathcal{B}_{\text{vect}}$ be a polyadic vector space over the polyadic field $\Bbbk^{(m_k,\, n_k)}$ as (see (5.3.2) and (9.1.7))

$$\mathcal{B}_{\text{vect}} = \left\langle B;\, K \middle| \nu^{(m)};\, \nu_k^{(m_k)},\, \mu_k^{(n_k)};\, \rho^{(r)} \right\rangle, \tag{9.2.1}$$

where $\nu^{(m)} : B^{\times m} \to B$ is m-ary addition and $\rho^{(r)} : K^{\times r} \times B \to B$ is r-place action (see (5.1.5)).

Definition 9.2.1. A *polyadic bialgebra* $\mathcal{B}^{(n',\, n)}$ is $\mathcal{B}_{\text{vect}}$ equipped with a $\Bbbk$-linear n-ary multiplication map $\boldsymbol{\mu}^{(n)} : B^{\otimes n} \to B$ and a $\Bbbk$-linear n'-ary comultiplication map $\boldsymbol{\Delta}^{(n')} : B \to B^{\otimes n'}$ such that:

(i) (a) $\mathcal{B}_A^{(n)} = \langle \mathcal{B}_{\text{vect}} | \boldsymbol{\mu}^{(n)} \rangle$ is a n-ary algebra;
(b) The map $\boldsymbol{\mu}^{(n)}$ is a coalgebra (homo)morphism (9.1.27).

(ii) (a) $\mathcal{B}_C^{(n')} = \langle \mathcal{B}_{\text{vect}} | \boldsymbol{\Delta}^{(n')} \rangle$ is a n'-ary coalgebra;
(b) The map $\boldsymbol{\Delta}^{(n')}$ is an algebra (homo)morphism (9.4.11).

The equivalence of the compatibility conditions (i)(b) and (ii)(b) can be expressed in the form (polyadic analog of '$\Delta \circ \mu = (\mu \otimes \mu)(\mathrm{id} \otimes \tau \otimes \mathrm{id})(\Delta \otimes \Delta)$')

$$\boldsymbol{\Delta}^{(n')} \circ \boldsymbol{\mu}^{(n)} = \left(\overbrace{\boldsymbol{\mu}^{(n)} \otimes \ldots \otimes \boldsymbol{\mu}^{(n)}}^{n'} \right) \circ \boldsymbol{\tau}_{\text{medial}}^{(n,n')} \circ \left(\overbrace{\boldsymbol{\Delta}^{(n')} \otimes \ldots \otimes \boldsymbol{\Delta}^{(n')}}^{n} \right), \tag{9.2.2}$$

where $\boldsymbol{\tau}_{\text{medial}}^{(n,n')}$ is the medial map (5.3.60) acting on B, while the diagram

$$\begin{array}{ccccc} B^{\otimes n} & \xrightarrow{\left(\boldsymbol{\Delta}^{(n')}\right)^{\otimes n}} & \left(B^{\otimes n'}\right)^{\otimes n} & \xrightarrow{\boldsymbol{\tau}_{medial}^{(n,n')}} & \left(B^{\otimes n}\right)^{\otimes n'} \\ \Big\downarrow{\scriptstyle \boldsymbol{\mu}^{(n)}} & & & & \Big\downarrow{\scriptstyle \left(\boldsymbol{\mu}^{(n)}\right)^{\otimes n'}} \\ B & & \xrightarrow{\quad\boldsymbol{\Delta}^{(n')}\quad} & & B^{\otimes n'} \end{array} \tag{9.2.3}$$

commutes.

In an elementwise description it is the commutation of n-ary multiplication and n'-ary comultiplication

$$\Delta^{(n')}(\mu^{(n)}[b_1, \dots, b_n]) = \mu^{(n)}[\Delta^{(n')}(b_1), \dots, \Delta^{(n')}(b_n)], \tag{9.2.4}$$

which in the Sweedler notation becomes

$$\begin{aligned}&\mu^{(n)}[b_1, \dots, b_n]_{[1]} \otimes \mu^{(n)}[b_1, \dots, b_n]_{[2]} \otimes \dots \otimes \mu^{(n)}[b_1, \dots, b_n]_{[n']} \\ &= \mu^{(n)}\left[b_{[1]}^{(1)}, \dots, b_{[1]}^{(n)}\right] \otimes \mu^{(n)}\left[b_{[2]}^{(1)}, \dots, b_{[2]}^{(n)}\right] \otimes \dots \otimes \mu^{(n)}\left[b_{[n']}^{(1)}, \dots, b_{[n']}^{(n)}\right].\end{aligned} \tag{9.2.5}$$

Consider the example of a nonderived bialgebra $\mathcal{B}^{(n,n)}$ which reminds the von Neumann higher n-regularity relations (Duplij 1998, Duplij and Marcinek 2001, 2002, 2000).

Example* 9.2.2. *von Neumann n-regular bialgebra. Let $\mathcal{B}^{(n,n)} = \langle B | \boldsymbol{\mu}^{(n)}, \boldsymbol{\Delta}^{(n)} \rangle$ be a polyadic bialgebra generated by the elements $b_i \in B$, $i = 1, \dots, n-1$ subject to the nonderived n-ary multiplication

$$\mu^{(n)}(b_1, b_2, b_3 \dots, b_{n-2}, b_{n-1}, b_1) = b_1, \tag{9.2.6}$$

$$\mu^{(n)}(b_2, b_3, b_4 \dots b_{n-1}, b_1, b_2) = b_2, \tag{9.2.7}$$

$$\vdots$$

$$\mu^{(n)}(b_{n-1}, b_1, b_2 \dots b_{n-3}, b_{n-2}, b_{n-1}) = b_{n-1}, \tag{9.2.8}$$

and the nonderived n-ary comultiplication (see 9.1.17)

$$\Delta^{(n)}(b_1) = b_1 \otimes b_2 \otimes b_3 \dots, b_{n-2} \otimes b_{n-1} \otimes b_1, \tag{9.2.9}$$

$$\Delta^{(n)}(b_2) = b_2 \otimes b_3 \otimes b_4 \dots, b_{n-1} \otimes b_1 \otimes b_2, \tag{9.2.10}$$

$$\vdots$$

$$\Delta^{(n)}(b_{n-1}) = b_{n-1} \otimes b_1 \otimes b_2 \otimes \dots \otimes b_{n-3} \otimes b_{n-2} \otimes b_{n-1}. \tag{9.2.11}$$

It is straightforward to check that the compatibility condition (9.2.4) holds. Many possibilities exist for choosing other operations—algebra addition, field addition and multiplication, action—so to demonstrate the compatibility we have confined ourselves to only the algebra multiplication and comultiplication.

If the n-ary algebra $\mathcal{B}_A^{(n)}$ has unit and/or n'-ary coalgebra $\mathcal{B}_C^{(n')}$ has counit $\boldsymbol{\varepsilon}^{(n', r')}$, we should add the following additional axioms.

Definition 9.2.3. *Unit axiom.* If $\mathcal{B}_A^{(n)} = \langle \mathcal{B}_{\text{vect}} | \boldsymbol{\mu}^{(n)} \rangle$ is unital, then the unit $\boldsymbol{\eta}^{(r,n)}$ is a (homo)morphism of the coalgebra $\mathcal{B}_C^{(n')} = \langle \mathcal{B}_{\text{vect}} | \boldsymbol{\Delta}^{(n')} \rangle$ (see (9.1.27))

$$\left(\overbrace{\mathbf{\Delta}^{(n')} \otimes \ldots \otimes \mathbf{\Delta}^{(n')}}^{n-1} \right) \circ \boldsymbol{\eta}^{(r,\, n)} = \left(\overbrace{\boldsymbol{\eta}^{(r,\, n)} \otimes \ldots \otimes \boldsymbol{\eta}^{(r,\, n)}}^{n'} \right) \tag{9.2.12}$$

such that the diagram

$$\begin{array}{ccc} K^{\otimes r} & \xrightarrow{\boldsymbol{\eta}^{(r,n)}} & B^{\otimes(n-1)} \\ \simeq \big\downarrow & & \big\downarrow \left(\mathbf{\Delta}^{(n')}\right)^{\otimes(n-1)} \\ K^{\otimes rn'} & \xrightarrow{\left(\boldsymbol{\eta}^{(r,n)}\right)^{n'}} & B^{\otimes(n-1)n'} \end{array} \tag{9.2.13}$$

commutes.

Definition 9.2.4. ***Counit axiom.*** If $\mathcal{B}_C^{(n')} = \langle \mathcal{B}_{\text{vect}} | \mathbf{\Delta}^{(n')} \rangle$ is counital, then the counit $\boldsymbol{\varepsilon}^{(n',\, r')}$ is a (homo)morphism of the algebra $\mathcal{B}_A^{(n)} = \langle \mathcal{B}_{\text{vect}} | \boldsymbol{\mu}^{(n)} \rangle$ (see (9.4.11))

$$\boldsymbol{\varepsilon}^{(n',\, r')} \circ \left(\overbrace{\boldsymbol{\mu}^{(n)} \otimes \ldots \otimes \boldsymbol{\mu}^{(n)}}^{n'-1} \right) = \left(\overbrace{\boldsymbol{\varepsilon}^{(n',\, r')} \otimes \ldots \otimes \boldsymbol{\varepsilon}^{(n',\, r')}}^{n} \right) \tag{9.2.14}$$

such that the diagram

$$\begin{array}{ccc} K^{\otimes r} & \xleftarrow{\boldsymbol{\varepsilon}^{(n',r')}} & B^{\otimes(n'-1)} \\ \simeq \big\uparrow & & \big\uparrow \left(\boldsymbol{\mu}^{(n)}\right)^{\otimes(n'-1)} \\ K^{\otimes rn} & \longleftarrow & B^{\otimes(n'-1)n} \end{array} \tag{9.2.15}$$

commutes.

If both the polyadic unit and polyadic counit exist, then we include their compatibility condition

$$\left(\boldsymbol{\varepsilon}^{(n',\, r')}\right)^{\otimes(n-1)} \circ \left(\boldsymbol{\eta}^{(r,\, n)}\right)^{\otimes(n'-1)} \simeq \mathrm{id}_K, \tag{9.2.16}$$

such that the diagram

$$\begin{array}{ccccc} & & B^{\otimes(n'-1)(n-1)} & & \\ & \left(\boldsymbol{\eta}^{(r,n)}\right)^{\otimes(n'-1)} \nearrow & & \searrow \left(\boldsymbol{\varepsilon}^{(n',r')}\right)^{\otimes(n-1)} & \\ K^{\otimes r(n'-1)} & & \xrightarrow{\simeq} & & K^{\otimes r(n-1)} \end{array} \tag{9.2.17}$$

commutes.

Assertion 9.2.5. There are four kinds of polyadic bialgebras depending on whether the unit $\boldsymbol{\eta}^{(r,\, n)}$ and counit $\boldsymbol{\varepsilon}^{(n',\, r')}$ exist:

(1) nonunital–noncounital; (2) unital–noncounital; (3) nonunital–counital; (4) unital–counital.

Definition 9.2.6. A polyadic bialgebra $\mathcal{B}^{(n', n)}$ is called *totally co-commutative*, if

$$\boldsymbol{\mu}^{(n)} = \boldsymbol{\mu}^{(n)} \circ \boldsymbol{\tau}_n, \tag{9.2.18}$$

$$\boldsymbol{\Delta}^{(n')} = \boldsymbol{\tau}_{n'} \circ \boldsymbol{\Delta}^{(n')}, \tag{9.2.19}$$

where $\boldsymbol{\tau}_n \in \mathrm{S}_n$, $\boldsymbol{\tau}_{n'} \in \mathrm{S}_{n'}$, and S_n, $\mathrm{S}_{n'}$ are the symmetry permutation groups on n and n' elements respectively.

Definition 9.2.7. A polyadic bialgebra $\mathcal{B}^{(n', n)}$ is called *medially co-commutative*, if

$$\boldsymbol{\mu}^{(n)}_{\mathrm{op}} \equiv \boldsymbol{\mu}^{(n)} \circ \boldsymbol{\tau}^{(n)}_{\mathrm{op}} = \boldsymbol{\mu}^{(n)}, \tag{9.2.20}$$

$$\boldsymbol{\Delta}^{(n')}_{\mathrm{cop}} \equiv \boldsymbol{\tau}^{(n')}_{\mathrm{op}} \circ \boldsymbol{\Delta}^{(n')} = \boldsymbol{\Delta}^{(n')}, \tag{9.2.21}$$

where $\boldsymbol{\tau}^{(n)}_{\mathrm{op}}$ and $\boldsymbol{\tau}^{(n')}_{\mathrm{op}}$ are the medially allowed polyadic twist maps (5.3.63).

9.3 Polyadic Hopf algebras

Here we introduce the most general approach to 'polyadization' of the Hopf algebra concept (Abe 1980, Sweedler 1969, Radford 2012). Informally, the transition from bialgebra to Hopf algebra is, in some sense, 'dualizing' the passage from semigroup (containing noninvertible elements) to group (in which all elements are invertible). Schematically, if multiplication $\mu = (\cdot)$ in a semigroup G is binary, the invertibility of all elements demands two extra and necessary set-ups: (1) *An additional element* (identity $e \in G$ or the corresponding map from a one point set to group ϵ); (2) *An additional map* (inverse $\imath : G \to G$), such that $g \cdot \imath(g) = e$ in diagrammatic form is $\mu \circ (\mathrm{id}_G \times \imath) \circ D_2 = \epsilon$ ($D_2 : G \to G \times G$ is the diagonal map). When 'dualizing', in a (binary) bialgebra B (with multiplication μ and comultiplication Δ) again two set-ups should be considered in order to get a (binary) Hopf algebra: (1) *An analog of identity* $e_\star = \eta\varepsilon$ (where $\eta : k \to B$ is unit and $\varepsilon : B \to k$ is counit); (2) *An analog of inverse* $S : B \to B$ called the *antipode*, such that $\mu \circ (\mathrm{id}_B \otimes S) \circ \Delta = e_\star$ or in terms of the (binary) convolution product $\mathrm{id}_B \star S = e_\star$. By multiplying both sides by S from the left and by id_B from the right, we obtain weaker (von Neumann regularity) conditions $S \star \mathrm{id}_B \star S = S$, $\mathrm{id}_B \star S \star \mathrm{id}_B = \mathrm{id}_B$, which do not contain an identity $e_\star$ and lead to the concept of weak Hopf algebras (Duplij and Li 2001, Li and Duplij 2002, Szlachányi 1996).

The crucial peculiarity of the polyadic generalization is the possible absence of an identity or (1) in both cases. The role and necessity of the polyadic identity (1.1.25) is not so important: there polyadic groups without identity exist (see, e.g., Gal'mak 2003). Invertibility is determined by the querelement (1.2.13) in n-ary group or the quermap (5.3.48) in polyadic algebra. So there are two ways forward: 'dualize' the quermap (5.3.48) directly (as in the binary case) or use the most general version of the polyadic convolution product (9.1.64) and apply possible restrictions, if any. We will choose the second method, because the first one is a particular case of it. Thus, if the standard (binary) antipode is the convolution inverse (coinverse) to the identity in a bialgebra, then its polyadic counterpart should be a coquerelement (9.1.78) of some polyadic analog for the identity map in the polyadic bialgebra. We consider two possibilities to define a polyadic analog of identity: (1) *Singular case.* The comultiplication is binary $n' = 2$; (2) *Symmetric case.* The arities of multiplication and comultiplication need not be binary, but should coincide $n = n'$.

In the *singular case* a polyadic multivalued map in $\mathrm{End}_{\Bbbk}(B, B^{\otimes(n-1)})$ is a reminder of how an identity can be defined: its components are to be functions of one variable. That is, with more than one argument it is not possible to determine its value when these are unequal.

Definition 9.3.1. We take for a *singular polyadic identity* $\boldsymbol{Id}_0$ the diagonal map $\boldsymbol{Id}_0 = \boldsymbol{D} \in \mathrm{End}_{\Bbbk}(B, B^{\otimes(n-1)})$, such that $b \mapsto b^{\otimes(n-1)}$, for any $b \in B$.

We call the polyadic convolution product (9.1.64) with the binary comultiplication $n' = 2$ *reduced* and denote it by $\bar{\boldsymbol{\mu}}_{\star}^{(n_\star)}$ which in Sweedler notation can be obtained from (9.1.73)

$$\begin{aligned}
&\bar{\boldsymbol{\mu}}_{\star}^{(n_\star)} \circ (\boldsymbol{f}^{(1)} \otimes \boldsymbol{f}^{(2)} \otimes \ldots \otimes \boldsymbol{f}^{(n_\star)}) = \boldsymbol{g},\ \boldsymbol{f}^{(i)}, \boldsymbol{g} \in \mathrm{End}_{\Bbbk}(B, B^{\otimes(n-1)}),\\
&\boldsymbol{g}_{[j]} \circ (b) = (\mu^{(n)})^{\circ\ell}\left[\mathrm{f}_{[j]}^{(1)}(b_{[1]}), \mathrm{f}_{[j]}^{(2)}(b_{[2]}), \ldots, \mathrm{f}_{[j]}^{(n_\star)}(b_{[n_\star]})\right],\\
&\mathrm{f}_{[j]}^{(i)} \in \mathrm{End}_{\Bbbk}(B, B),\ i \in 1, \ldots, n_\star, j \in 1, \ldots, n-1,\ \ b \in B.
\end{aligned} \tag{9.3.1}$$

The consistency condition (9.1.65) becomes *reduced*

$$n_\star = \ell(n-1) + 1 = \ell' + 1. \tag{9.3.2}$$

Definition 9.3.2. The set of the multivalued maps $\boldsymbol{f}^{(i)} \in \mathrm{End}_{\Bbbk}(B, B^{\otimes(n-1)})$ (together with the polyadic identity $\boldsymbol{Id}_0$) endowed with the reduced convolution product $\bar{\boldsymbol{\mu}}_{\star}^{(n_\star)}$ is called a *reduced $n_\star$-ary convolution algebra*

$$\mathcal{C}_{\star}^{(2,n)} = \left\langle \mathrm{End}_{\Bbbk}(B, B^{\otimes(n-1)}) \middle| \bar{\boldsymbol{\mu}}_{\star}^{(n_\star)} \right\rangle. \tag{9.3.3}$$

***Remark* 9.3.3.** The reduced convolution algebra $\mathcal{C}_{\star}^{(2,n)}$ having $n \neq 2$ is not derived (definition 9.1.41).

Having the distinguished element $\boldsymbol{Id}_0 \in \mathcal{C}_{\star}^{(2,n)}$ as an analog of id_B and the querelement (9.1.78) for any $\boldsymbol{f} \in \mathcal{C}_{\star}^{(2,n)}$ (the polyadic version of inverse in the convolution algebra), we are now in a position to 'polyadize' the concept of the (binary) antipode.

Definition 9.3.4. A multivalued map $\mathbf{Q}_0 : B \to B^{\otimes(n-1)}$ in the polyadic bialgebra $\mathcal{B}^{(2,n)}$ is called a *singular querantipode*, if it is the coquerelement of the polyadic identity $\mathbf{Q}_0 = \mathbf{q}_\star(\boldsymbol{Id}_0)$ in the reduced $n_\star$-ary convolution algebra

$$\bar{\boldsymbol{\mu}}_\star^{(n_\star)} \circ \left(\boldsymbol{Id}_0^{\otimes(n_\star-1)} \otimes \mathbf{Q}_0\right) = \boldsymbol{Id}_0, \tag{9.3.4}$$

where $\mathbf{Q}_0$ can be on any place, such that the diagram

$$\begin{array}{ccccc} B & \xrightarrow{\quad\boldsymbol{\Delta}^{\circ(n_\star-1)}\quad} & & & B^{\otimes n_\star} \\ \Big\downarrow{\scriptstyle \boldsymbol{Id}_0} & & & & \Big\downarrow{\scriptstyle \boldsymbol{Id}_0^{\otimes(n_\star-1)}\otimes\mathbf{Q}_0} \\ B^{\otimes(n-1)} & \xleftarrow{\left(\left(\boldsymbol{\mu}^{(n)}\right)^{\circ\ell}\right)^{\otimes(n-1)}} & \left(B^{\otimes n_\star}\right)^{\otimes(n-1)} & \xleftarrow{\boldsymbol{\tau}_{medial}^{(n-1,n_\star)}} & \left(B^{\otimes(n-1)}\right)^{\otimes n_\star} \end{array} \tag{9.3.5}$$

commutes.

In Sweedler notation

$$\begin{gathered}\left(\mu^{(n)}\right)^{\circ\ell}\left[b_{[1]}, b_{[2]}, \ldots, b_{[n_\star-1]}, \mathrm{Q}_{0\,[j]}\left(b_{[n_\star]}\right)\right] = b, \\ j \in 1, \ldots, n-1, \quad i \in 1, \ldots, n_\star,\end{gathered} \tag{9.3.6}$$

where $\boldsymbol{\Delta}^{\circ(n_\star-1)}(b) = b_{[1]} \otimes b_{[2]} \otimes \ldots \otimes b_{[n_\star]}$, $\mathbf{Q}_0 \circ (b) = \mathrm{Q}_{0\,[1]}(b) \otimes \ldots \otimes \mathrm{Q}_{0\,[n-1]}(b)$, $b, b_{[i]} \in B$.

Definition 9.3.5. A polyadic bialgebra $\mathcal{B}^{(2,n)}$ equipped with the reduced $n_\star$-ary convolution product $\bar{\boldsymbol{\mu}}_\star^{(n_\star)}$ and the singular querantipode $\mathbf{Q}_0$ (9.3.4) is called a *singular polyadic Hopf algebra* and is denoted by $\mathcal{H}_{\mathrm{sing}}^{(n)} = \left\langle \mathcal{B}^{(n,n)} \middle| \bar{\boldsymbol{\mu}}_\star^{(n_\star)}, \mathbf{Q}_0 \right\rangle$.

Due to their exotic properties we will not consider singular polyadic Hopf algebras $\mathcal{H}_{\mathrm{sing}}^{(n)}$ in detail.

In the *symmetric case* a polyadic identity-like map in $\mathrm{End}_{\Bbbk}(B^{\otimes(n-1)}, B^{\otimes(n-1)})$ can be defined in a more natural way.

Definition 9.3.6. A *symmetric polyadic identity* $\boldsymbol{Id} : B^{\otimes(n-1)} \to B^{\otimes(n-1)}$ is a polyadic tensor product of ordinary identities in $\mathcal{B}^{(n,n)}$

$$\boldsymbol{Id} = \overbrace{\mathrm{id}_B \otimes \ldots \otimes \mathrm{id}_B}^{n-1}, \quad \mathrm{id}_B : B \to B. \tag{9.3.7}$$

Indeed, for any map $f \in \mathrm{End}_{\Bbbk}(B^{\otimes(n-1)}, B^{\otimes(n-1)})$, obviously $\boldsymbol{Id} \circ f = f \circ \boldsymbol{Id} = f$.

The numbers of iterations are now equal $\ell = \ell'$, and the consistency condition (9.1.65) becomes

$$n_\star - 1 = \ell(n - 1). \tag{9.3.8}$$

Definition 9.3.7. The set of the multiplace multivalued maps $f^{(i)} \in \mathrm{End}_{\Bbbk}(B^{\otimes(n-1)}, B^{\otimes(n-1)})$ (together with the polyadic identity $\boldsymbol{Id}$) endowed with the symmetric convolution product $\hat{\boldsymbol{\mu}}_\star^{(n_\star)} = \boldsymbol{\mu}_\star^{(n_\star)}|_{n=n'}$ (9.1.64) is called a *symmetric $n_\star$-ary convolution algebra*

$$\mathcal{C}_\star^{(n,n)} = \left\langle \mathrm{End}_{\Bbbk}(B^{\otimes(n-1)}, B^{\otimes(n-1)}) \middle| \hat{\boldsymbol{\mu}}_\star^{(n_\star)} \right\rangle. \tag{9.3.9}$$

For a polyadic analog of antipode in the symmetric case we have

Definition 9.3.8. A multiplace multivalued map $\mathbf{Q}_{\mathrm{id}} : B^{\otimes(n-1)} \to B^{\otimes(n-1)}$ in the polyadic bialgebra $\mathcal{B}^{(n,\,n)}$ is called a *symmetric querantipode*, if it is the coquerelement (see (9.1.78)) of the polyadic identity $\mathbf{Q}_{\mathrm{id}} = \mathbf{q}_\star(\boldsymbol{Id})$ in the symmetric $n_\star$-ary convolution algebra

$$\hat{\boldsymbol{\mu}}_\star^{(n_\star)} \circ (\boldsymbol{Id}^{\otimes(n_\star-1)} \otimes \mathbf{Q}_{\mathrm{id}}) = \boldsymbol{Id}, \tag{9.3.10}$$

where $\mathbf{Q}_{\mathrm{id}}$ can be on any place, such that the diagram

$$\begin{array}{ccccc}
B^{\otimes(n-1)} & \xrightarrow{\left(\left(\Delta^{(n)}\right)^{\circ\ell}\right)^{\otimes(n-1)}} & \left(B^{\otimes n_\star}\right)^{\otimes(n-1)} & \xrightarrow{\tau_{medial}^{(n_\star,n-1)}} & \left(B^{\otimes(n-1)}\right)^{\otimes n_\star} \\
\Big\downarrow \boldsymbol{Id} & & & & \Big\downarrow \boldsymbol{Id}^{\otimes(n_\star-1)}\otimes\mathbf{Q}_{\mathrm{id}} \\
B^{\otimes(n-1)} & \xleftarrow{\left(\left(\mu^{(n)}\right)^{\circ\ell}\right)^{\otimes(n-1)}} & \left(B^{\otimes n_\star}\right)^{\otimes(n-1)} & \xleftarrow{\tau_{medial}^{(n-1,n_\star)}} & \left(B^{\otimes(n-1)}\right)^{\otimes n_\star}
\end{array} \tag{9.3.11}$$

commutes.

In Sweedler notation we obtain (see (9.1.64) and (9.1.73))

$$\begin{gathered}
(\mu^{(n)})^{\circ\ell}\left[b_{[1]}^{(j)}, b_{[2]}^{(j)}, b_{[n_\star-1]}^{(j)}, \mathrm{Q}_{[j]}\left(b_{[n_\star]}^{(1)}, b_{[n_\star]}^{(2)}, \ldots, b_{[n_\star]}^{(n-1)}\right)\right] \\
= b^{(j)}, j \in 1, \ldots, n-1, i \in 1, \ldots, n_\star,
\end{gathered} \tag{9.3.12}$$

where $(\Delta^{(n)})^{\circ\ell}(b^{(j)}) = b_{[1]}^{(j)} \otimes b_{[2]}^{(j)} \otimes \ldots \otimes b_{[n_\star]}^{(j)}$, $b_{[i]}^{(j)} \in B, \ell \in \mathbb{N}, \mathrm{Q}_{[j]} \in \mathrm{End}_{\Bbbk}(B^{\otimes(n-1)}, B)$ are components of $\mathbf{Q}_{\mathrm{id}}$, and the convolution product arity is $n_\star = \ell(n-1) + 1$ (9.3.8).

Definition 9.3.9. A polyadic bialgebra $\mathcal{B}^{(n,\,n)}$ equipped with the symmetric $n_\star$-ary convolution product $\hat{\boldsymbol{\mu}}_\star^{(n_\star)}$ and the symmetric querantipode $\mathbf{Q}_{\mathrm{id}}$ (9.3.10) is called a *symmetric polyadic Hopf algebra* and is denoted by $\mathcal{H}_{\mathrm{sym}}^{(n)} = \left\langle \mathcal{B}^{(n,\,n)} \middle| \hat{\boldsymbol{\mu}}_\star^{(n_\star)}, \mathbf{Q}_{\mathrm{id}} \right\rangle$.

Example **9.3.10.** In the case where $n = n' = 3$ and $\ell = 1$ we have $\Delta^{(3)}(b^{(j)}) = b_{[1]}^{(j)} \otimes b_{[2]}^{(j)} \otimes b_{[3]}^{(j)}$, $j = 1, 2, 3$,

$$\begin{aligned}
&\mu^{(3)}\Big[b_{[1]}^{(1)}, b_{[2]}^{(1)}, \mathrm{Q}_{[1]}\Big(b_{[3]}^{(1)}, b_{[3]}^{(2)}\Big)\Big] = b^{(1)},\\
&\qquad\qquad\mu^{(3)}\Big[b_{[1]}^{(2)}, b_{[2]}^{(2)}, \mathrm{Q}_{[2]}\Big(b_{[3]}^{(1)}, b_{[3]}^{(2)}\Big)\Big] = b^{(2)},\\
&\mu^{(3)}\Big[b_{[1]}^{(1)}, \mathrm{Q}_{[1]}\Big(b_{[2]}^{(1)}, b_{[2]}^{(2)}\Big), b_{[3]}^{(1)}\Big] = b^{(1)},\\
&\qquad\qquad\mu^{(3)}\Big[b_{[1]}^{(2)}, \mathrm{Q}_{[2]}\Big(b_{[2]}^{(1)}, b_{[2]}^{(2)}\Big), b_{[3]}^{(2)}\Big] = b^{(2)},\\
&\mu^{(3)}\Big[\mathrm{Q}_{[1]}\Big(b_{[1]}^{(1)}, b_{[1]}^{(2)}\Big), b_{[2]}^{(1)}, b_{[3]}^{(1)}\Big] = b^{(1)},\\
&\qquad\qquad\mu^{(3)}\Big[\mathrm{Q}_{[2]}\Big(b_{[1]}^{(1)}, b_{[1]}^{(2)}\Big), b_{[2]}^{(2)}, b_{[3]}^{(2)}\Big] = b^{(2)},
\end{aligned} \tag{9.3.13}$$

which can be compared with the binary case $(b_{[1]}S\big(b_{[2]}\big) = S\big(b_{[1]}\big)b_{[2]} = \eta(\varepsilon(b)))$ and (1.2.13), (9.1.78).

Recall that the main property of the antipode S of a binary bialgebra B is its 'anticommutation' with the multiplication μ and comultiplication Δ (e.g., Sweedler 1969)

$$S \circ \mu = \mu \circ \tau_{\mathrm{op}} \circ (S \otimes S), \quad S \circ \eta = \eta, \tag{9.3.14}$$

$$\Delta \circ S = \tau_{\mathrm{op}} \circ (S \otimes S) \circ \Delta, \quad \varepsilon \circ S = \varepsilon, \tag{9.3.15}$$

where τ_{op} is the binary twist (see (5.3.57)). The first relation means that S is an algebra anti-endomorphism, because in the elementwise description $S(a \cdot b) = S(b) \cdot S(a)$, $a, b \in B$, $(\,\cdot\,) \equiv \mu$.

We propose the polyadic analogs of (9.3.14)–(9.3.15) without proofs, which are too cumbersome, but their derivations almost coincide with those for the binary case.

Proposition 9.3.11. *The querantipode $\mathbf{Q}_{\mathrm{id}} : B^{\otimes(n-1)} \to B^{\otimes(n-1)}$ of the polyadic bialgebra $\mathcal{B}^{(n,n)} = \langle B | \boldsymbol{\mu}^{(n)}, \boldsymbol{\Delta}^{(n)} \rangle$ satisfies the polyadic version of 'antimultiplicativity' ('antialgebra map')*

$$\mathbf{Q}_{\mathrm{id}} \circ \left(\left(\boldsymbol{\mu}^{(n)}\right)^{\circ\ell}\right)^{\otimes(n-1)} = \left(\left(\boldsymbol{\mu}^{(n)}\right)^{\circ\ell}\right)^{\otimes(n-1)} \circ \boldsymbol{\tau}_{\mathrm{op}}^{(\ell_\tau)} \circ \mathbf{Q}_{\mathrm{id}}^{\otimes n_\star} \circ \boldsymbol{\tau}_{\mathrm{medial}}^{(n_\star, n-1)}, \tag{9.3.16}$$

and 'anticomultiplicativity' ('anticoalgebra map')

$$\left(\left(\boldsymbol{\Delta}^{(n)}\right)^{\circ\ell}\right)^{\otimes(n-1)} \circ \mathbf{Q}_{\mathrm{id}} = \boldsymbol{\tau}_{\mathrm{op}}^{(\ell_\tau)} \circ \mathbf{Q}_{\mathrm{id}}^{\otimes n_\star} \circ \boldsymbol{\tau}_{\mathrm{medial}}^{(n-1, n_\star)} \circ \left(\left(\boldsymbol{\Delta}^{(n)}\right)^{\circ\ell}\right)^{\otimes(n-1)}, \tag{9.3.17}$$

where $\boldsymbol{\tau}_{\text{medial}}^{(n,m)}$ *is the medial map (7.3.6),* $\boldsymbol{\tau}_{\text{op}}^{(\ell_\tau)}$ *is the polyadic twist (5.3.63) and* $\ell_\tau = (n-1)n_\star$ *should be allowed (see table* 5.3).

Proposition 9.3.12. *If the polyadic unit* $\boldsymbol{\eta}^{(r,\,n)}$ *(5.3.39) and counit* $\boldsymbol{\varepsilon}^{(n,\,r)}$ (9.1.22) in $\mathcal{B}^{(n,\,n)}$ *exist, then*

$$\mathbf{Q}_{\text{id}} \circ \boldsymbol{\eta}^{(r,\,n)} = \boldsymbol{\eta}^{(r,\,n)}, \tag{9.3.18}$$

$$\boldsymbol{\varepsilon}^{(n,\,r)} \circ \mathbf{Q}_{\text{id}} = \boldsymbol{\varepsilon}^{(n,\,r)}. \tag{9.3.19}$$

Example **9.3.13.** If $n = 3$, $\ell = 1$, $n_\star = 3$, $\ell_\tau = 6$ (and (5.3.66)), then using Sweedler notation, for (9.3.16) we have

$$\begin{aligned} &\mathrm{Q}_{[1]}\left(\mu^{(3)}[a_1, a_2, a_3], \mu^{(3)}[b_1, b_2, b_3]\right) \\ &\quad= \mu^{(3)}\left[\mathrm{Q}_{[2]}(a_2, b_2), \mathrm{Q}_{[1]}(a_1, b_1), \mathrm{Q}_{[1]}(a_3, b_3)\right], \end{aligned} \tag{9.3.20}$$

$$\begin{aligned} &\mathrm{Q}_{[2]}\left(\mu^{(3)}[a_1, a_2, a_3], \mu^{(3)}[b_1, b_2, b_3]\right) \\ &\quad= \mu^{(3)}\left[\mathrm{Q}_{[2]}(a_1, b_1), \mathrm{Q}_{[2]}(a_3, b_3), \mathrm{Q}_{[1]}(a_2, b_2)\right], \end{aligned} \tag{9.3.21}$$

where $\mathbf{Q}_{\text{id}} \circ (a, b) = \mathrm{Q}_{[1]}(a, b) \otimes \mathrm{Q}_{[2]}(a, b) \in \mathrm{End}_{\Bbbk}(B \otimes B, B \otimes B)$, $a, b, a_i, b_i \in B$ (see (9.3.14)).

The key property of the binary antipode S is its involutivity $S^{\circ 2} = \mathrm{id}_B$ for either commutative ($\mu = \mu \circ \tau_{\text{op}}$) or co-commutative ($\Delta = \tau_{\text{op}} \circ \Delta$) Hopf algebras, which follows from (9.3.14) or (9.3.15) applied to $S \star S^{\circ 2}$ giving $\eta\varepsilon$ $(=S \star \mathrm{id}_B)$.

Proposition 9.3.14. *If in a symmetric Hopf algebra* $\mathcal{H}_{\text{sym}}^{(n)}$ *either multiplication or comultiplication is invariant under polyadic twist map* $\boldsymbol{\tau}_{\text{op}}^{(\ell_\tau)}$ *(5.3.63), then the querantipode* $\mathbf{Q}_{\text{id}}$ (9.3.10) *satisfies*

$$\hat{\boldsymbol{\mu}}_\star^{(n_\star)}\left[\overbrace{\mathbf{Q}_{\text{id}}, \ldots, \mathbf{Q}_{\text{id}}}^{(n_\star - 1)}, \mathbf{Q}_{\text{id}} \circ \mathbf{Q}_{\text{id}}\right] = \mathbf{Q}_{\text{id}}, \tag{9.3.22}$$

where $\mathbf{Q}_{\text{id}}^{\circ 2}$ *can be on any place, or the convolution querelement* (9.1.78) *of the querantipode* $\mathbf{Q}_{\text{id}}$ *is*

$$\mathbf{q}_\star(\mathbf{Q}_{\text{id}}) = \mathbf{Q}_{\text{id}}^{\circ 2}. \tag{9.3.23}$$

Proof. The proposition follows from applying either (9.3.16) or (9.3.17) to the lhs of (9.3.22), to use (9.1.78). □

9.4 Ternary examples

9.4.1 Ternary Sweedler Hopf algebra

One of the most important examples of noncocommutative Hopf algebras is the well-known Sweedler Hopf algebra (Sweedler 1969). In the binary case $\mathcal{H}_{\mathrm{Sw}}^{(2)} = \langle \{a, b, 0, e\} | (+), \mu^{(2)}, \Delta^{(2)}, \varepsilon^{(2)}, S^{(2)} \rangle$ has two generators a and b satisfying

$$\mu^{(2)}[a, a] = e, \quad \mu^{(2)}[b, b] = 0, \tag{9.4.1}$$

$$\sigma_{+}^{(2)}(\mu^{(2)}[a, b]) = -\sigma_{-}^{(2)}(\mu^{(2)}[a, b]), \tag{9.4.2}$$

where $\sigma_{\pm}^{(2)} \in S_2$ denote even and odd permutations, respectively, and the following comultiplication

$$\Delta^{(2)}(a) = a \otimes a, \tag{9.4.3}$$

$$\Delta^{(2)}(b) = b \otimes a + e \otimes b, \tag{9.4.4}$$

the counit $\varepsilon^{(2)}(a) = 1$, $\varepsilon^{(2)}(b) = 0$, and antipode $S^{(2)}(a) = a$, $S^{(2)}(b) = -b$, which respect the algebra structure (Sweedler 1969).

In the derived case a *ternary Sweedler Hopf algebra* $\mathcal{H}_{\mathrm{Sw}}^{(3)} = \langle \{a, b, 0, e\} | (+), \mu^{(3)}, \Delta^{(3)}, \varepsilon^{(3)}, S^{(3)} \rangle$ is generated also by two generators a and b obeying (Duplij 2001)

$$\mu^{(3)}[a, e, a] = \mu^{(3)}[e, a, a] = \mu^{(3)}[a, a, e] = e, \tag{9.4.5}$$

$$\sigma_{+}^{(3)}\left(\mu^{(3)}[b, e, b]\right) = 0, \quad \sigma_{+}^{(3)}\left(\mu^{(3)}[a, e, b]\right) = -\sigma_{-}^{(3)}\left(\mu^{(3)}[a, e, b]\right), \tag{9.4.6}$$

where $\sigma_{\pm}^{(3)} \in S_3$ denote a sum of terms with even and odd permutations, and the derived Hopf algebra structure is given by

$$\Delta^{(3)}(a) = a \otimes a \otimes a, \tag{9.4.7}$$

$$\Delta^{(3)}(b) = b \otimes a \otimes a + e \otimes b \otimes a + e \otimes e \otimes b, \tag{9.4.8}$$

$$\varepsilon^{(3)}(a) = \varepsilon^{(2)}(a) = 1, \; \varepsilon^{(3)}(b) = \varepsilon^{(2)}(b) = 0, \tag{9.4.9}$$

$$S^{(3)}(a) = S^{(2)}(a) = a, \; S^{(3)}(b) = S^{(2)}(b) = -b, \tag{9.4.10}$$

and it can be checked that (9.4.7)–(9.4.9) are algebra maps, while (9.4.10) are antialgebra maps. To obtain a nonderived ternary Sweedler example we have the following possibilities: (1) one 'even' generator a, two 'odd' generators $b_{1,2}$ and one ternary unit e; (2) two 'even' generators $a_{1,2}$, one 'odd' generator b and two ternary units $e_{1,2}$. In the first case the ternary algebra structure is (no summation, $i = 1, 2$)

$$\mu^{(3)}[a, a, a] = e, \quad \mu^{(3)}[b_i, b_i, b_i] = 0, \tag{9.4.11}$$

$$\mu^{(3)}[a, e, b_i] = -\mu^{(3)}[a, b_i, e], \mu^{(3)}[e, a, b_i] = -\mu^{(3)}[b_i, a, e], \mu^{(3)}[e, b_i, a] = -[b_i, e, a], \tag{9.4.12}$$

$$\sigma_+^{(3)}\left(\mu^{(3)}\left[b_i, a, b_i\right]\right) = \sigma_+^{(3)}\left(\mu^{(3)}\left[a, b_i, a\right]\right) = 0, \sigma_+^{(3)}\left(\mu^{(3)}[b_1, a, b_2]\right) = -\sigma_-^{(3)}\left(\mu^{(3)}[b_1, a, b_2]\right), \tag{9.4.13}$$

and the corresponding ternary Hopf algebra structure is

$$\Delta^{(3)}(a) = a \otimes a \otimes a, \quad \Delta^{(3)}(b_{1,2}) = b_{1,2} \otimes a \otimes a + e_{1,2} \otimes b_{2,1} \otimes a + e_{1,2} \otimes e_{2,1} \otimes b_{2,1}, \tag{9.4.14}$$

$$\varepsilon^{(3)}(a) = 1, \quad \varepsilon^{(3)}(b_i) = 0, \tag{9.4.15}$$

$$S^{(3)}(a) = a, \qquad S^{(3)}(b_i) = -b_i. \tag{9.4.16}$$

In the second case we have for the algebra structure

$$\mu^{(3)}[a_i, a_j, a_k] = \delta_{ij}\delta_{ik}\delta_{jk}e_i, \quad \mu^{(3)}[b, b, b] = 0, \tag{9.4.17}$$

$$\sigma_+^{(3)}\left(\mu^{(3)}\left[b, a_i, b\right]\right) = 0, \quad \sigma_+^{(3)}\left(\mu^{(3)}\left[a_i, b, a_i\right]\right) = 0, \tag{9.4.18}$$

$$\sigma_+^{(3)}\left(\mu^{(3)}\left[b_1, a, b_2\right]\right) = 0, \quad \sigma_-^{(3)}\left(\mu^{(3)}\left[b_1, a, b_2\right]\right) = 0, \tag{9.4.19}$$

and the ternary Hopf algebra structure is (Duplij 2001)

$$\Delta^{(3)}(a_i) = a_i \otimes a_i \otimes a_i, \quad \Delta^{(3)}(b) = b \otimes a_1 \otimes a_1 + e_1 \otimes b \otimes a_2 + e_1 \otimes e_2 \otimes b, \tag{9.4.20}$$

$$\varepsilon^{(3)}(a_i) = 1, \quad \varepsilon^{(3)}(b) = 0, \tag{9.4.21}$$

$$S^{(3)}(a_i) = a_i, \quad S^{(3)}(b) = -b. \tag{9.4.22}$$

9.4.2 Ternary quantum group examples

A ternary commutator can be obtained in different ways (Bremner and Hentzel 2000). We will consider the simplest version called the Nambu bracket (see, e.g., Takhtajan 1994, de Azcarraga and Izquierdo 2010). Let us introduce two maps $\omega_\pm^{(3)} : A \otimes A \otimes A \to A \otimes A \otimes A$ by

$$\omega_+^{(3)}(a \otimes b \otimes c) = a \otimes b \otimes c + b \otimes c \otimes a + c \otimes a \otimes b, \tag{9.4.23}$$

$$\omega_-^{(3)}(a \otimes b \otimes c) = b \otimes a \otimes c + c \otimes b \otimes a + a \otimes c \otimes b. \tag{9.4.24}$$

Thus, obviously $\mu^{(3)} \circ \omega_\pm^{(3)} = \sigma_\pm^{(3)} \circ \mu^{(3)}$, where $\sigma_\pm^{(3)} \in S_3$ denotes a sum of terms having even and odd permutations respectively. In the binary case $\omega_+^{(2)} = \mathrm{id} \otimes \mathrm{id}$

and $\omega_-^{(2)} = \tau$ is the twist operator $\tau : a \otimes b \to b \otimes a$, while $\mu^{(2)} \circ \omega_-^{(2)}$ is permutation $\sigma_-^{(2)}(\mu^{(2)}[a, b]) = \mu^{(2)}[b, a]$. So the Nambu product is $\omega_N^{(3)} = \omega_+^{(3)} - \omega_-^{(3)}$, and the ternary commutator is $\mu^{(3)}[,,]_N = \sigma_N^{(3)} = \sigma_+^{(3)} - \sigma_-^{(3)}$, or (Takhtajan 1994)

$$\begin{aligned}\mu^{(3)}[a, b, c]_N = \mu^{(3)}[a, b, c] + \mu^{(3)}[b, c, a]\\ + \mu^{(3)}[c, a, b] - \mu^{(3)}[c, b, a] - \mu^{(3)}[a, c, b]\\ -\mu^{(3)}[b, a, c].\end{aligned} \tag{9.4.25}$$

An *abelian ternary algebra* is defined by the vanishing of the Nambu bracket $[a, b, c]_N = 0$ or *ternary commutation* relation $\sigma_+^{(3)} = \sigma_-^{(3)}$. By analogy with the binary case a *deformed ternary algebra* can be defined by

$$\begin{aligned}\sigma_+^{(3)} = q\sigma_-^{(3)} \text{ or } \mu^{(3)}[a, b, c] + \mu^{(3)}[b, c, a] + \mu^{(3)}[c, a, b]\\ = q(\mu^{(3)}[c, b, a] + \mu^{(3)}[a, c, b] + \mu^{(3)}[b, a, c]),\end{aligned} \tag{9.4.26}$$

where multiplication by q is treated as an external operation.

Let us consider a ternary analog of the *Woronowicz example* of a bialgebra construction, which has two generators satisfying $\mu^{(2)}[a, b] = q\mu^{(2)}[b, a]$ (or $\sigma_+^{(2)}(\mu^{(2)}[a, b]) = q\sigma_-^{(2)}(\mu^{(2)}[a, b])$), then the following coproducts

$$\Delta^{(2)}(a) = a \otimes a \tag{9.4.27}$$

$$\Delta^{(2)}(b) = b \otimes a + e \otimes b \tag{9.4.28}$$

are the algebra maps. In the derived ternary case using (9.4.26) we have

$$\sigma_+^{(3)}(\mu^{(3)}[a, e, b]) = q\sigma_-^{(3)}(\mu^{(3)}[a, e, b]), \tag{9.4.29}$$

where e is the ternary unit and ternary coproducts are

$$\Delta^{(3)}(e) = e \otimes e \otimes e, \tag{9.4.30}$$

$$\Delta^{(3)}(a) = a \otimes a \otimes a, \tag{9.4.31}$$

$$\Delta^{(3)}(b) = b \otimes a \otimes a + e \otimes b \otimes a + e \otimes e \otimes b, \tag{9.4.32}$$

which are *ternary algebra maps*, i.e., they satisfy

$$\sigma_+^{(3)}\left(\mu^{(3)}\left[\Delta^{(3)}(a)\Delta^{(3)}(e)\Delta^{(3)}(b)\right]\right) = q\sigma_-^{(3)}\left(\mu^{(3)}\left[\Delta^{(3)}(a)\Delta^{(3)}(e)\Delta^{(3)}(b)\right]\right). \tag{9.4.33}$$

Let us consider the group $G = SL(n, \Bbbk)$. Then the algebra generated by $a_j^i \in SL(n, \Bbbk)$ can be endowed with the structure of a ternary Hopf algebra (see, e.g., Madore 1995 for the binary case) by choosing the ternary coproduct, counit and antipode as (here summation is implied)

$$\Delta^{(3)}\left(a_j^i\right) = a_k^i \otimes a_l^k \otimes a_j^l, \quad \varepsilon\left(a_j^i\right) = \delta_j^i, \quad S^{(3)}\left(a_j^i\right) = \left(a^{-1}\right)_j^i. \tag{9.4.34}$$

This antipode is a skew one since from (5.3.33) it follows that

$$\mu^{(3)} \circ \left(S^{(3)} \otimes \mathrm{id} \otimes \mathrm{id}\right) \circ \Delta^{(3)}\left(a_j^i\right) = S^{(3)}\left(a_k^i\right) a_l^k a_j^l = \left(a^{-1}\right)_k^i a_l^k a_j^l = \delta_l^i a_j^l$$
$$= a_j^i. \tag{9.4.35}$$

This ternary Hopf algebra is derived since for $\Delta^{(2)} = a_j^i \otimes a_k^j$ we have

$$\Delta^{(3)} = \left(\mathrm{id} \otimes \Delta^{(2)}\right) \otimes \Delta^{(2)}\left(a_j^i\right) = \left(\mathrm{id} \otimes \Delta^{(2)}\right)\left(a_k^i \otimes a_j^k\right) = a_k^i \otimes \Delta^{(2)}\left(a_j^k\right)$$
$$= a_k^i \otimes a_l^k \otimes a_j^l. \tag{9.4.36}$$

In the most important case $n = 2$ we can obtain the manifest action of the ternary coproduct $\Delta^{(3)}$ on components. Possible nonderived matrix representations of the ternary product can be done only by four-rank $n \times n \times n \times n$ twice covariant and twice contravariant tensors $\{a_{kl}^{ij}\}$. Among all products the nonderived ones are only the following: $a_{jk}^{oi} b_{oo}^{jl} c_{il}^{ko}$ and $a_{ok}^{ij} b_{io}^{ol} c_{il}^{ko}$ (where o is any index). So using, e.g., the first choice we can define the nonderived Hopf algebra structure by

$$\Delta^{(3)}\left(a_{kl}^{ij}\right) = a_{\nu\rho}^{i\mu} \otimes a_{kl}^{\nu\sigma} \otimes a_{\mu\sigma}^{\rho j}, \tag{9.4.37}$$

$$\varepsilon\left(a_{kl}^{ij}\right) = \frac{1}{2}\left(\delta_k^i \delta_l^j + \delta_l^i \delta_k^j\right), \tag{9.4.38}$$

and the skew antipode $s_{kl}^{ij} = S^{(3)}(a_{kl}^{ij})$ which is a solution of the equation $s_{\nu\rho}^{i\mu} a_{kl}^{\nu\sigma} = \delta_\rho^i \delta_k^\mu \delta_l^\sigma$.

9.5 Polyadic almost co-commutativity and co-mediality

Bialgebras with a special relaxation of co-commutativity, almost co-commutativity, are the ground objects in the construction of quantum groups identified with the non-commutative and non-co-commutative quasitriangular Hopf algebras (Drinfeld 1987, 1989a). Here we consider polyadic analogs of braidings, almost co-commutativity and the R-matrix, together with the quasitriangularity equations. This description is not unique, as with the polyadic analog of the twist map, while the medial map is unique for all arities. A new (unique) concept of deformation is proposed: almost co-mediality with the corresponding M-matrix. The medial analogs of braidings and quasitriangularity are introduced, and the equations for M-matrix are obtained.

9.5.1 Quantum Yang–Baxter equation

Here we recall the binary case (informally) in a notation that will allow us to provide the 'polyadization' in a clearer way.

Let us consider a (binary) bialgebra $\mathcal{B}^{(2,2)} = \langle B | \mu, \Delta \rangle$, where $\mu = \mu^{(2)}$ is the binary multiplication, $\Delta = \Delta^{(2)}$ (see definition 9.2.1), and the opposite comultiplication is given by $\Delta_{\mathrm{cop}} \equiv \tau_{\mathrm{op}} \circ \Delta$, where τ_{op} is the binary twist (5.3.57). To relax the co-commutativity condition ($\Delta_{\mathrm{cop}} = \Delta$), the following construction inspired by conjugation in groups was proposed (Drinfeld 1987, 1989a). A bialgebra $\mathcal{B}^{(2,2)}$ is *almost co-commutative*, if there exists $\mathcal{R} \in B \otimes B$ such that (in the elementwise notation)

$$\mu\left[\Delta_{cop}(b), \mathcal{R}\right] = \mu[\mathcal{R}, \Delta(b)], \quad \forall b \in B. \tag{9.5.1}$$

A fixed element $\mathcal{R}$ of a bialgebra satisfying (9.5.1) is called a *universal R-matrix*. For a co-commutative bialgebra we have $\mathcal{R} = e_B \otimes e_B$, where $e_B \in B$ is the unit (element) of the algebra $\langle B|\mu\rangle$.

If we demand that $\left\langle B|\Delta_{cop}\right\rangle$ is the opposite coalgebra of $\langle B|\Delta\rangle$, and therefore Δ_{cop} be coassociative, then $\mathcal{R}$ cannot be arbitrary, but has to satisfy some additional conditions, which we will call the *almost co-commutativity equations* for the R-matrix. Indeed, using (9.5.1) we can write

$$\begin{aligned} &\mu\left[\left(\Delta_{cop} \otimes \mathrm{id}_B\right) \circ \Delta_{cop}(b), \mu[(\mathcal{R} \otimes e_B), (\Delta \otimes \mathrm{id}_B)(\mathcal{R})]\right] \\ &= \mu[\mu[(\mathcal{R} \otimes e_B), (\Delta \otimes \mathrm{id}_B)(\mathcal{R})], (\Delta \otimes \mathrm{id}_B) \circ \Delta(b)], \end{aligned} \tag{9.5.2}$$

$$\begin{aligned} &\mu\left[\left(\mathrm{id}_B \otimes \Delta_{cop}\right) \circ \Delta_{cop}(b), \mu[(e_B \otimes \mathcal{R}), (\mathrm{id}_B \otimes \Delta)(\mathcal{R})]\right] \\ &= \mu[\mu[(e_B \otimes \mathcal{R}), (\mathrm{id}_B \otimes \Delta)(\mathcal{R})], (\mathrm{id}_B \otimes \Delta) \circ \Delta(b)]. \end{aligned} \tag{9.5.3}$$

Therefore, the coassociativity of Δ_{cop} leads to the first almost co-commutativity equation

$$\mu[(\mathcal{R} \otimes e_B), (\Delta \otimes \mathrm{id}_B)(\mathcal{R})] = \mu[(e_B \otimes \mathcal{R}), (\mathrm{id}_B \otimes \Delta)(\mathcal{R})]. \tag{9.5.4}$$

On the other hand, directly from (9.5.1), we have relations which can be treated as the next two almost co-commutativity equations (unconnected to the coassociativity of Δ_{cop})

$$\begin{aligned} \mu[(\mathcal{R} \otimes e_B), (\Delta \otimes \mathrm{id}_B)(\mathcal{R})] &= \mu\left[\left(\Delta_{cop} \otimes \mathrm{id}_B\right)(\mathcal{R}), (\mathcal{R} \otimes e_B)\right] \\ &= \mu\left[\left(\tau_{op} \otimes \mathrm{id}_B\right) \circ (\Delta \otimes \mathrm{id}_B)(\mathcal{R}), (\mathcal{R} \otimes e_B)\right], \end{aligned} \tag{9.5.5}$$

$$\begin{aligned} \mu[(e_B \otimes \mathcal{R}), (\mathrm{id}_B \otimes \Delta)(\mathcal{R})] &= \mu\left[\left(\mathrm{id}_B \otimes \Delta_{cop}\right)(\mathcal{R}), (e_B \otimes \mathcal{R})\right] \\ &= \mu\left[\left(\mathrm{id}_B \otimes \tau_{op}\right) \circ (\mathrm{id}_B \otimes \Delta)(\mathcal{R}), (e_B \otimes \mathcal{R})\right]. \end{aligned} \tag{9.5.6}$$

The equations (9.5.4)–(9.5.6) for the components of

$$\mathcal{R} = \sum_{\alpha} \mathrm{r}_{\alpha}^{(1)} \otimes \mathrm{r}_{\alpha}^{(2)} \in B \otimes B \tag{9.5.7}$$

are on $B \otimes B \otimes B$. In components the almost co-commutativity (9.5.1) can be expressed as follows

$$\sum_{[b]} \sum_{\alpha} \mu\left[b_{[2]}, \mathrm{r}_{\alpha}^{(1)}\right] \otimes \mu\left[b_{[1]}, \mathrm{r}_{\alpha}^{(2)}\right] = \sum_{[b]'} \sum_{\alpha'} \mu\left[\mathrm{r}_{\alpha'}^{(1)}, b_{[1]'}\right] \otimes \mu\left[\mathrm{r}_{\alpha'}^{(2)}, b_{[2]'}\right]. \tag{9.5.8}$$

Now introduce the 'extended' form of the R-matrix $\mathcal{R}_{ij} \in B \otimes B \otimes B, i, j = 1, 2, 3$, by

$$\mathcal{R}_{12} = \sum_{\alpha} \mathsf{r}_{\alpha}^{(1)} \otimes \mathsf{r}_{\alpha}^{(2)} \otimes e_B \equiv \mathscr{R} \otimes e_B, \tag{9.5.9}$$

$$\mathcal{R}_{13} = \sum_{\alpha} \mathsf{r}_{\alpha}^{(1)} \otimes e_B \otimes \mathsf{r}_{\alpha}^{(2)} = \left(\mathrm{id}_B \otimes \tau_{\mathrm{op}}\right) \circ (\mathscr{R} \otimes e_B), \tag{9.5.10}$$

$$\mathcal{R}_{23} = \sum_{\alpha} e_B \otimes \mathsf{r}_{\alpha}^{(1)} \otimes \mathsf{r}_{\alpha}^{(2)} \equiv e_B \otimes \mathscr{R}. \tag{9.5.11}$$

Obviously, one can try to solve (9.5.4)–(9.5.6) with respect to the $\mathsf{r}_{\alpha}^{(1)}, \mathsf{r}_{\alpha}^{(2)}$ directly, but then we are confronted with a difficulty arising from the Sweedler components, because now (see (9.1.35)–(9.1.37))

$$(\Delta \otimes \mathrm{id}_B)(\mathscr{R}) = \sum_{[\mathsf{r}_{\alpha}^{(1)}]} \sum_{\alpha} \mathsf{r}_{\alpha,[1]}^{(1)} \otimes \mathsf{r}_{\alpha,[2]}^{(1)} \otimes \mathsf{r}_{\alpha}^{(2)}, \tag{9.5.12}$$

$$(\mathrm{id}_B \otimes \Delta)(\mathscr{R}) = \sum_{[\mathsf{r}_{\alpha}^{(2)}]} \sum_{\alpha} \mathsf{r}_{\alpha}^{(1)} \otimes \mathsf{r}_{\alpha,[1]}^{(2)} \otimes \mathsf{r}_{\alpha,[2]}^{(2)}. \tag{9.5.13}$$

To avoid computations in the Sweedler components, one can substitute them by the components of $\mathscr{R}$ directly as $\mathsf{r}_{[j]}^{(i)} \longrightarrow \mathsf{r}^{(i)}$ (schematically). This allows us to express (9.5.12)–(9.5.13) solely through elements of the 'extended' R-matrix $\mathcal{R}_{ij}$ by

$$(\Delta \otimes \mathrm{id}_B)(\mathscr{R}) = \mu[\mathcal{R}_{13}, \mathcal{R}_{23}] \equiv \sum_{\alpha,\beta} \mathsf{r}_{\alpha}^{(1)} \otimes \mathsf{r}_{\beta}^{(1)} \otimes \mu\left[\mathsf{r}_{\alpha}^{(2)}, \mathsf{r}_{\beta}^{(2)}\right], \tag{9.5.14}$$

$$(\mathrm{id}_B \otimes \Delta)(\mathscr{R}) = \mu[\mathcal{R}_{13}, \mathcal{R}_{12}] \equiv \sum_{\alpha,\beta} \mu\left[\mathsf{r}_{\alpha}^{(1)}, \mathsf{r}_{\beta}^{(1)}\right] \otimes \mathsf{r}_{\beta}^{(2)} \otimes \mathsf{r}_{\alpha}^{(2)}, \tag{9.5.15}$$

which do not contain Sweedler components of $\mathscr{R}$ at all. The equations (9.5.14)–(9.5.15) define a *quasitriangular* R-matrix (Drinfeld 1987). The corresponding almost co-commutative (binary) bialgebra $\mathcal{B}_{\mathrm{braid}}^{(2)} = \langle \mathcal{B}^{(2,2)}, \mathscr{R} \rangle$ is called a *quasitriangular almost co-commutative bialgebra* (or *braided bialgebra*, Kassel 1995). Only for them can the almost co-commutativity equations (9.5.4)–(9.5.6) be expressed solely in terms of R-matrix components or through the 'extended' R-matrix $\mathcal{R}_{ij}$, using (9.5.14)–(9.5.15).

Theorem 9.5.1. In the binary case, three almost co-commutativity equations for the R-matrix coincide with

$$\mu^{\circ 2}[\mathcal{R}_{12}, \mathcal{R}_{13}, \mathcal{R}_{23}] = \mu^{\circ 2}[\mathcal{R}_{23}, \mathcal{R}_{13}, \mathcal{R}_{12}]. \tag{9.5.16}$$

Conversely, any quasitriangular R-matrix is a solution of (9.5.16) by the above construction. The equation for the 'extended' R-matrix $\mathcal{R}_{ij}$ (9.5.16) is called the *quantum Yang–Baxter equation* (Lambe and Radford 1997, Majid 1995) (or the triangle relation, Drinfeld 1989a). In terms of the R-matrix components (9.5.7) the quantum Yang–Baxter equation (9.5.16) takes the form

$$\begin{aligned}&\sum_{\alpha,\beta,\gamma}\mu\left[\mathrm{r}_\alpha^{(1)},\mathrm{r}_\beta^{(1)}\right]\otimes\mu\left[\mathrm{r}_\alpha^{(2)},\mathrm{r}_\gamma^{(1)}\right]\otimes\mu\left[\mathrm{r}_\beta^{(2)},\mathrm{r}_\gamma^{(2)}\right]\\ &=\sum_{\alpha',\beta',\gamma'}\mu\left[\mathrm{r}_{\beta'}^{(1)},\mathrm{r}_{\alpha'}^{(1)}\right]\otimes\mu\left[\mathrm{r}_{\gamma'}^{(1)},\mathrm{r}_{\alpha'}^{(2)}\right]\otimes\mu\left[\mathrm{r}_{\gamma'}^{(2)},\mathrm{r}_{\beta'}^{(2)}\right].\end{aligned}\tag{9.5.17}$$

Let us consider modules over the braided bialgebra $\mathcal{B}_{\text{braid}}^{(2,2)}$ and recall (Drinfeld 1989b) how the universal R-matrix generalizes the standard flip $\tau_{V_1V_2}:V_1\otimes V_2\to V_2\otimes V_1$. Define the isomorphism of modules (which in our notation correspond to a 1-place action ρ (5.1.5)) $\mathbf{c}_{V_1V_2}:V_1\otimes V_2\to V_2\otimes V_1$ by

$$\begin{aligned}&\mathbf{c}_{V_1V_2}\circ(v_1\otimes v_2)=\tau_{V_1V_2}\circ\mathcal{R}\circ\\ &(v_1\otimes v_2)=\sum_\alpha\rho\left(\mathrm{r}_\alpha^{(2)}\middle|v_2\right)\otimes\rho\left(\mathrm{r}_\alpha^{(1)}\middle|v_1\right),\ v_i\in V_i,\ \mathrm{r}_\alpha^{(i)}\in B,\ i=1,2.\end{aligned}\tag{9.5.18}$$

The quasitriangularity (9.5.14)–(9.5.15) and (9.5.18) on $V_1\otimes V_2\otimes V_3$ leads to (see, e.g., Kassel 1995)

$$(\mathbf{c}_{V_1V_3}\otimes\mathrm{id}_{V_2})\circ(\mathrm{id}_{V_1}\otimes\mathbf{c}_{V_2V_3})=\mathbf{c}_{V_1\otimes V_2,V_3},\tag{9.5.19}$$

$$(\mathrm{id}_{V_2}\otimes\mathbf{c}_{V_1V_3})\circ(\mathbf{c}_{V_1V_2}\otimes\mathrm{id}_{V_3})=\mathbf{c}_{V_1,V_2\otimes V_3}.\tag{9.5.20}$$

Similarly, the quantum Yang–Baxter equation (9.5.16) gives the *braid equation* (Drinfeld 1989b) mapping $V_1\otimes V_2\otimes V_3\to V_3\otimes V_2\otimes V_1$:

$$\begin{aligned}&(\mathbf{c}_{V_2V_3}\otimes\mathrm{id}_{V_1})\circ(\mathrm{id}_{V_2}\otimes\mathbf{c}_{V_1V_3})\circ(\mathbf{c}_{V_1V_2}\otimes\mathrm{id}_{V_3})\\ &=(\mathrm{id}_{V_3}\otimes\mathbf{c}_{V_1V_2})\circ(\mathbf{c}_{V_1V_3}\otimes\mathrm{id}_{V_2})\circ(\mathrm{id}_{V_1}\otimes\mathbf{c}_{V_2V_3}).\end{aligned}\tag{9.5.21}$$

Putting $V_1=V_2=V_3=V$ shows that $\mathbf{c}_{VV}$ is a solution of the braid equation (9.5.21) for any module V, if the $\mathcal{R}$ is a solution of the Yang–Baxter equation (Drinfeld 1989b, Kassel 1995).

9.5.2 n'-Ary braid equation

Let us consider possible higher arity generalizations of the braid equation (9.5.21), informally. Introduce the modules V_i over the polyadic bialgebra $\mathcal{B}^{(n',n)}$ (definition 9.2.1) by the r-place actions $\rho_{V_i}^{(r)}(b_1,\ldots b_r|v_i)$, $b_j\in B$, $v_i\in V_i$, $i=1,\ldots s$, $j=1,\ldots,r$ (see (5.2.36)). Define the following morphisms of modules

$$\mathbf{c}_{V_1\ldots V_{n'}}:V_1\otimes\ldots\otimes V_{n'}\to V_{n'}\otimes\ldots\otimes V_1.\tag{9.5.22}$$

We use the shorthand notation $\mathbf{c}_{V^{n'}}\equiv\mathbf{c}_{V_1\ldots V_{n'}}$, $\mathrm{id}_V\equiv\mathrm{id}_{V_i}$ and introduce indices manifestly only when it will be needed.

Proposition 9.5.2. *The n'-ary braid equation has the form*

$$\left(\mathbf{c}_{V^{n'}} \otimes \overbrace{\mathrm{id}_V \otimes \ldots \otimes \mathrm{id}_V}^{n'-1}\right) \circ \left(\mathrm{id}_V \otimes \mathbf{c}_{V^{n'}} \otimes \overbrace{\mathrm{id}_V \otimes \ldots \otimes \mathrm{id}_V}^{n'-2}\right) \circ \ldots$$
$$\circ \left(\overbrace{\mathrm{id}_V \otimes \ldots \otimes \mathrm{id}_V}^{n'-1} \otimes \mathbf{c}_{V^{n'}}\right) \circ \left(\mathbf{c}_{V^{n'}} \otimes \overbrace{\mathrm{id}_V \otimes \ldots \otimes \mathrm{id}_V}^{n'-1}\right)$$
$$= \left(\overbrace{\mathrm{id}_V \otimes \ldots \otimes \mathrm{id}_V}^{n'-1} \otimes \mathbf{c}_{V^{n'}}\right) \circ \left(\mathbf{c}_{V^{n'}} \otimes \overbrace{\mathrm{id}_V \otimes \ldots \otimes \mathrm{id}_V}^{n'-1}\right) \circ \ldots \tag{9.5.23}$$
$$\circ \left(\overbrace{\mathrm{id}_V \otimes \ldots \otimes \mathrm{id}_V}^{n'-2} \otimes \mathbf{c}_{V^{n'}} \otimes \mathrm{id}_V\right) \circ \left(\overbrace{\mathrm{id}_V \otimes \ldots \otimes \mathrm{id}_V}^{n'-1} \otimes \mathbf{c}_{V^{n'}}\right),$$

where each side consists of $(n'+1)$ brackets with $(2n'-1)$ multipliers.

Proof. Use the associative quiver technique from Duplij (2012). □

***Remark* 9.5.3.** There can be additional equations depending on the concrete values of n' which can contain a different number of brackets determined by the corresponding diagram commutation.

***Example* 9.5.4.** In case $n' = 3$ we have the *ternary braided equation* for $\mathbf{c}_{V_1V_2V_3} : V_1 \otimes V_2 \otimes V_3 \to V_3 \otimes V_2 \otimes V_1$ on the tensor product of modules $V_1 \otimes V_2 \otimes V_3 \otimes V_4 \otimes V_5$, as

$$\begin{aligned}
&(\mathbf{c}_{V_3V_4V_5} \otimes \mathrm{id}_{V_2} \otimes \mathrm{id}_{V_1}) \circ (\mathrm{id}_{V_3} \otimes \mathbf{c}_{V_2V_5V_4} \otimes \mathrm{id}_{V_1}) \circ \\
&\quad (\mathrm{id}_{V_3} \otimes \mathrm{id}_{V_2} \otimes \mathbf{c}_{V_1V_4V_5}) \\
&\quad \circ (\mathbf{c}_{V_1V_2V_3} \otimes \mathrm{id}_{V_4} \otimes \mathrm{id}_{V_5}) = \\
&(\mathrm{id}_{V_5} \otimes \mathrm{id}_{V_4} \otimes \mathbf{c}_{V_1V_2V_3}) \circ (\mathbf{c}_{V_1V_4V_5} \otimes \mathrm{id}_{V_2} \otimes \mathrm{id}_{V_3}) \circ \\
&\quad (\mathrm{id}_{V_1} \otimes \mathbf{c}_{V_2V_5V_4} \otimes \mathrm{id}_{V_3}) \\
&\quad \circ (\mathrm{id}_{V_1} \otimes \mathrm{id}_{V_2} \otimes \mathbf{c}_{V_3V_4V_5}).
\end{aligned} \tag{9.5.24}$$

The ternary compatibility conditions for $\mathbf{c}_{V_1V_2V_3}$ (corresponding to (9.5.19)–(9.5.20)) are

$$\begin{aligned}
&(\mathbf{c}_{V_1V_4V_5} \otimes \mathrm{id}_{V_2} \otimes \mathrm{id}_{V_3}) \circ (\mathrm{id}_{V_1} \otimes \mathbf{c}_{V_2V_5V_4} \otimes \mathrm{id}_{V_3}) \\
&\quad \circ (\mathrm{id}_{V_1} \otimes \mathrm{id}_{V_2} \otimes \mathbf{c}_{V_3V_4V_5}) = \mathbf{c}_{V_1 \otimes V_2 \otimes V_3, V_4, V_5},
\end{aligned} \tag{9.5.25}$$

$$\begin{aligned}
&(\mathrm{id}_{V_3} \otimes \mathbf{c}_{V_2V_5V_4} \otimes \mathrm{id}_{V_1}) \circ (\mathrm{id}_{V_3} \otimes \mathrm{id}_{V_2} \otimes \mathbf{c}_{V_1V_4V_5}) \\
&\quad \circ (\mathbf{c}_{V_1V_2V_3} \otimes \mathrm{id}_{V_4} \otimes \mathrm{id}_{V_5}) = \mathbf{c}_{V_1, V_2, V_3 \otimes V_4 \otimes V_5}.
\end{aligned} \tag{9.5.26}$$

Now we follow the opposite (to the standard, Drinfeld 1989b), but consistent way: using the equations (9.5.24)–(9.5.26) we find polyadic analogs of the corresponding equations for the R-matrix and the quasitriangularity conditions (9.5.14)–(9.5.15), which will fix the comultiplication structure of a polyadic bialgebra $\mathcal{B}^{(n',n)}$.

9.5.3 Polyadic almost co-commutativity

We will see that the almost co-commutativity equations for the R-matrix are more complicated in the polyadic case, because the main condition (9.5.1) will have a different form coming from n-ary group theory (Gal'mak 2003). Indeed, let $\mathcal{G}^{(n)} = \langle G | \mu^{(n)} \rangle$ be an n-ary group and $\mathcal{H}' = \langle H' | \mu^{(n)} \rangle$, $\mathcal{H}'' = \langle H'' | \mu^{(n)} \rangle$ are its n-ary subgroups. Recall Gal'mak (2003) that $\mathcal{H}'$ and $\mathcal{H}''$ are *semiconjugated* in $\mathcal{G}^{(n)}$, if there exist $g \in G$, such that $\mu^{(n)}[g, h_1', \dots, h_{n-1}'] = \mu^{(n)}[h_1'', \dots, h_{n-1}'', g]$, $h_i' \in H'$, $h_i'' \in H''$, and if g can be on any place, then $\mathcal{H}'$ and $\mathcal{H}''$ are *conjugated* in $\mathcal{G}^{(n)}$. Based on this notion and on analogy with (1.1.26), we can 'polyadize' the almost co-commutativity condition (9.5.1) in the following way.

Let $\mathcal{B}^{(n',n)} = \langle B | \mu^{(n)}, \Delta^{(n')} \rangle$, be a polyadic bialgebra (see definition 9.2.1), and the opposite comultiplication $\Delta_{\text{cop}}^{(n')} = \tau_{\text{op}}^{(n')} \circ \Delta^{(n')}$, where $\tau_{\text{op}}^{(n')}$ is the polyadic twist (5.3.63).

Definition 9.5.5. A polyadic bialgebra $\mathcal{B}^{(n',n)}$ is called *polyadic sequenced almost co-commutative*, if there exist fixed $(n-1)$ elements $\mathcal{R}_i^{(n')} \in B^{\otimes n'}$, $i = 1, \dots, n-1$, called a *polyadic R-matrix sequence*, such that

$$\begin{aligned}
&\mu^{(n)}\left[\Delta_{\text{cop}}^{(n')}(b), \mathcal{R}_1^{(n')}, \mathcal{R}_2^{(n')}, \dots, \mathcal{R}_{n-1}^{(n')}\right] \\
&= \mu^{(n)}[\mathcal{R}_1^{(n')}, \Delta^{(n')}(b), \mathcal{R}_2^{(n')}, \dots, \mathcal{R}_{n-1}^{(n')}] \\
&\vdots \\
&= \mu^{(n)}[\mathcal{R}_1^{(n')}, \mathcal{R}_2^{(n')}, \dots, \mathcal{R}_{n-1}^{(n')}, \Delta^{(n')}(b)], \quad \forall b \in B.
\end{aligned} \tag{9.5.27}$$

Definition 9.5.6. A polyadic bialgebra $\mathcal{B}^{(n',n)}$ is called *polyadic sequenced almost semico-commutative*, if only the first and the last relations in (9.5.27) hold

$$\begin{aligned}
&\mu^{(n)}\left[\Delta_{\text{cop}}^{(n')}(b), \mathcal{R}_1^{(n')}, \mathcal{R}_2^{(n')}, \dots, \mathcal{R}_{n-1}^{(n')}\right] = \\
&\mu^{(n)}[\mathcal{R}_1^{(n')}, \mathcal{R}_2^{(n')}, \dots, \mathcal{R}_{n-1}^{(n')}, \Delta^{(n')}(b)], \quad \forall b \in B.
\end{aligned} \tag{9.5.28}$$

***Remark* 9.5.7.** Using $(n-1)$ polyadic R-matrices $\mathcal{R}_i^{(n')}$ is the only way to build a polyadic analog for the almost commutativity concept, since now there is no binary multiplication.

The definition (9.5.27) is too general and needs to consider $(n-1)$ different polyadic analogs of the R-matrix which might not be unique. Therefore, in a similar way to the correspondence of the neutral sequence (1.1.26) and the polyadic unit (1.1.25), we arrive at

Definition 9.5.8. A polyadic bialgebra $\mathcal{B}^{(n',n)} = \langle B | \mu^{(n)}, \Delta^{(n')} \rangle$ is called *polyadic almost (semi)co-commutative*, if there exists one fixed element $\mathcal{R}^{(n')} \in B^{\otimes n'}$ called an *n′-ary R-matrix*, such that

$$\mu^{(n)}\left[\Delta_{\mathrm{cop}}^{(n')}(b), \overbrace{\mathcal{R}^{(n')}, \ldots, \mathcal{R}^{(n')}}^{n-1}\right] = \mu^{(n)}\left[\overbrace{\mathcal{R}^{(n')}, \ldots, \mathcal{R}^{(n')}}^{n-1}, \Delta^{(n')}(b)\right], \quad \forall b \in B. \tag{9.5.29}$$

In components the n'-ary R-matrix $\mathcal{R}^{(n')}$ is

$$\mathcal{R}^{(n')} = \sum_{\alpha} \mathsf{r}_{\alpha}^{(1)} \otimes \ldots \otimes \mathsf{r}_{\alpha}^{(n')}, \quad \mathsf{r}_{\alpha}^{(i)} \in B. \tag{9.5.30}$$

***Remark* 9.5.9.** Polyadic almost co-commutativity (9.5.29) can be expressed in component form, as in the binary case (9.5.8), only if we know concretely the polyadic twist $\tau_{\mathrm{op}}^{(n')} \in \mathrm{S}_{n'}$ (where $\mathrm{S}_{n'}$ is the symmetry permutation group ON n' elements), which is not unique for arbitrary $n' > 2$.

***Example* 9.5.10.** For $\mathcal{B}^{(3,3)}$ the ternary almost (semi)co-commutativity (9.5.29) is given by

$$\mu^{(3)}\left[\Delta_{\mathrm{cop}}^{(3)}(b), \mathcal{R}^{(3)}, \mathcal{R}^{(3)}\right] = \mu^{(3)}[\mathcal{R}^{(3)}, \mathcal{R}^{(3)}, \Delta^{(3)}(b)], \quad \forall b \in B, \tag{9.5.31}$$

which with $\tau_{\mathrm{op}}^{(3)} = \left(\frac{123}{321}\right)$ becomes, in components,

$$\begin{aligned}&\sum_{[b]}\sum_{\alpha,\beta} \mu^{(3)}\left[b_{[3]}, \mathsf{r}_{\alpha}^{(1)}, \mathsf{r}_{\beta}^{(1)}\right] \otimes \mu^{(3)}\left[b_{[2]}, \mathsf{r}_{\alpha}^{(2)}, \mathsf{r}_{\beta}^{(2)}\right] \otimes \mu^{(3)}\left[b_{[1]}, \mathsf{r}_{\alpha}^{(3)}, \mathsf{r}_{\beta}^{(3)}\right] \\ &= \sum_{[b]'}\sum_{\alpha'\beta'} \mu^{(3)}\left[\mathsf{r}_{\alpha'}^{(1)}, \mathsf{r}_{\beta'}^{(1)}, b_{[1]'}\right] \otimes \mu^{(3)}\left[\mathsf{r}_{\alpha'}^{(2)}, \mathsf{r}_{\beta'}^{(2)}, b_{[2]'}\right] \otimes \mu^{(3)}\left[\mathsf{r}_{\alpha'}^{(3)}, \mathsf{r}_{\beta'}^{(3)}, b_{[3]'}\right].\end{aligned} \tag{9.5.32}$$

***Example* 9.5.11.** In the exotic mixed case $\mathcal{B}^{(4,3)}$ where the polyadic twist 'without fixed points' (5.3.63) is $\tau_{\mathrm{op}}^{(4)} = \left(\frac{1234}{3142}\right)$, the polyadic almost co-commutativity (9.5.29) becomes

$$\begin{aligned}&\sum_{[b]}\sum_{\alpha,\beta} \mu^{(3)}\left[b_{[3]}, \mathsf{r}_{\alpha}^{(1)}, \mathsf{r}_{\beta}^{(1)}\right] \otimes \mu^{(3)}\left[b_{[1]}, \mathsf{r}_{\alpha}^{(2)}, \mathsf{r}_{\beta}^{(2)}\right] \otimes \mu^{(3)}\left[b_{[4]}, \mathsf{r}_{\alpha}^{(3)}, \mathsf{r}_{\beta}^{(3)}\right] \otimes \mu^{(3)}\left[b_{[2]}, \mathsf{r}_{\alpha}^{(4)}, \mathsf{r}_{\beta}^{(4)}\right] \\ &= \sum_{[b]'}\sum_{\alpha'\beta'} \mu^{(3)}\left[\mathsf{r}_{\alpha'}^{(1)}, \mathsf{r}_{\beta'}^{(1)}, b_{[1]'}\right] \otimes \mu^{(3)}\left[\mathsf{r}_{\alpha'}^{(2)}, \mathsf{r}_{\beta'}^{(2)}, b_{[2]'}\right] \otimes \mu^{(3)}\left[\mathsf{r}_{\alpha'}^{(3)}, \mathsf{r}_{\beta'}^{(3)}, b_{[3]'}\right] \\ &\otimes \mu^{(3)}\left[\mathsf{r}_{\alpha'}^{(4)}, \mathsf{r}_{\beta'}^{(4)}, b_{[4]'}\right].\end{aligned} \tag{9.5.33}$$

9.5.4 Equations for the n'-ary R-matrix

Here we consider the most consistent way (from a categorical viewpoint) to derive equations for the polyadic R-matrix, in other words, through using the braided equation (9.5.21) (and n'-ary braided equation (9.5.23)) with the concrete choice of the braiding $\mathbf{c}_{V^{n'}}$.

Suppose that the n'-ary braiding $\mathbf{c}_{V^{n'}}$ is defined still by a 1-place action $\rho^{(1)}$, as in the binary case (9.5.18). At first glance, we could define the braiding (similar to (9.5.18))

$$\begin{aligned}&\mathbf{c}_{V_1 \ldots V_{n'}} \circ (v_1 \otimes \ldots \otimes v_{n'}) = \tau_{V_1 \ldots V_{n'}} \circ \mathcal{R}^{(n')} \circ (v_1 \otimes \ldots \otimes v_{n'}) \\ &= \tau_{V_1 \ldots V_{n'}} \circ \left(\sum_\alpha \rho^{(1)}(\mathsf{r}_\alpha^{(1)} | v_1) \otimes \ldots \otimes \rho^{(1)}(\mathsf{r}_\alpha^{(n')} | v_{n'}) \right), \; v_i \in V_i, \; \mathsf{r}_\alpha^{(i)} \in B,\end{aligned} \tag{9.5.34}$$

where $\rho^{(1)} : B \otimes V_i \to V_i$ is the 1-place action (see (5.1.5)). We recall that only the n-ary composition of 1-place actions (n is the arity of multiplication $\mu^{(n)}$) is defined here (see Duplij 2019)

$$\overbrace{\rho^{(1)}(b_1 | \rho^{(1)}(b_2 \ldots \rho^{(1)}(b_n | v)))}^{n} = \rho^{(1)}(\mu^{(n)}[b_1, \ldots b_n] | v), \; b_i \in B, \; v \in V. \tag{9.5.35}$$

As in the binary case (9.5.9)–(9.5.11), we need the 'extended' polyadic R-matrix.

***Remark* 9.5.12.** The standard definition of the 'extended' n'-ary R-matrix can be possible, if the algebra $\langle B | \mu^{(n)} \rangle$ contains one polyadic unit (element) e_B, because in the polyadic case there are new intriguing possibilities (which did not exist in the binary case) of having several units, or even where all elements are units (see the discussion after (1.1.26) and Duplij 2012).

Definition 9.5.13. The 'extended' form of the n'-ary R-matrix is defined by $\mathcal{R}_{i_1 \ldots i_{n'}}^{(2n'-1)} \in B^{\otimes (2n'-1)}$, such that

$$\begin{aligned}&\mathcal{R}_{i_1 \ldots i_{n'}}^{(2n'-1)} = \sum_\alpha e_B \otimes \ldots \otimes \mathsf{r}_\alpha^{(i_1)} \otimes \ldots \otimes \mathsf{r}_\alpha^{(i_{n'})} \otimes \ldots \otimes e_B, \\ &\qquad i_1, \ldots, i_{n'} \in \{1, \ldots 2n' - 1\},\end{aligned} \tag{9.5.36}$$

where $\mathsf{r}_\alpha^{(i_k)}$ are on the i_k-place.

In this way we can express in terms of the 'extended' n'-ary R-matrix (9.5.36) the n'-ary braided equation (9.5.23), in full analogy with the binary case (9.5.16).

***Example* 9.5.14.** For the ternary case

$$\begin{aligned}&\mathbf{c}_{V_1 V_2 V_3} \circ (v_1 \otimes v_2 \otimes v_3) = \tau_{V_1 V_2 V_3} \circ \mathcal{R}^{(3)} \circ (v_1 \otimes v_2 \otimes v_3) \\ &= \sum_\alpha \rho^{(1)}\left(\mathsf{r}_\alpha^{(1)} \middle| v_3\right) \otimes \rho^{(1)}\left(\mathsf{r}_\alpha^{(2)} \middle| v_2\right) \otimes \rho^{(1)}\left(\mathsf{r}_\alpha^{(3)} \middle| v_1\right), \; v_i \in V_i, \; \mathsf{r}_\alpha^{(i)} \in B,\end{aligned} \tag{9.5.37}$$

and we define $\mathcal{R}^{(5)}_{i_1i_2i_3}$ by (9.5.36), $i_1, i_2, i_3 \in \{1, \dots, 5\}$, $\tau_{V_1V_2V_3} = \binom{123}{321}$, and consider the ternary braid equation (9.5.24). Using (9.5.35) we obtain (informally)

$$\mathcal{R}^{(5)}_{123}\mathcal{R}^{(5)}_{145}\mathcal{R}^{(5)}_{254}\mathcal{R}^{(5)}_{345} = \mathcal{R}^{(5)}_{345}\mathcal{R}^{(5)}_{254}\mathcal{R}^{(5)}_{145}\mathcal{R}^{(5)}_{123}. \tag{9.5.38}$$

***Remark* 9.5.15.** Unfortunately, a 'linear' $\mathscr{R}^{(n')}$ n'-ary braiding $\mathbf{c}_{V_1 \dots V_{n'}}$ (as in (9.5.34) and (9.5.37)) is not consistent with the polyadic analog of the quasitriangularity equations (9.5.19)–(9.5.20), because the polyadic almost co-commutativity (9.5.29) contains $(n-1)$ copies of n'-ary R-matrix $\mathscr{R}^{(n')}$.

Therefore, in order to agree with (9.5.29), instead of (9.5.34), we have

Definition 9.5.16. The polyadic braiding $\mathbf{c}_{V_1 \dots V_{n'}}$ is defined by

$$\begin{gathered}\mathbf{c}_{V_1\dots V_{n'}} \circ (v_1 \otimes \dots \otimes v_{n'}) = \tau_{V_1\dots V_{n'}} \circ \rho^{(n-1)}\left(\overbrace{\mathscr{R}^{(n')}, \dots, \mathscr{R}^{(n')}}^{n-1} \middle| (v_1 \otimes \dots \otimes v_{n'}) \right) \\ = \tau_{V_1\dots V_{n'}} \circ \left(\sum_{\alpha_1,\dots \alpha_{n-1}} \rho^{(n-1)}\left(\mathsf{r}^{(1)}_{\alpha_1}, \dots \mathsf{r}^{(1)}_{\alpha_{n-1}} \middle| v_1\right) \otimes \dots \otimes \rho^{(n-1)}\left(\mathsf{r}^{(n')}_{\alpha_1}, \dots \mathsf{r}^{(n')}_{\alpha_{n-1}} \middle| v_{n'}\right)\right), \\ v_i \in V_i,\ \mathsf{r}^{(i)}_{\alpha} \in B,\end{gathered} \tag{9.5.39}$$

where $\rho^{(n-1)} : B^{n-1} \otimes V \to V$ the $(n-1)$-place action (see (5.1.5)).

***Remark* 9.5.17.** The twist of the modules $\tau_{V_1\dots V_{n'}}$ should be compatible with the polyadic twist $\tau^{(n')}_{\text{op}}$ in (9.5.27). In the binary case they are both the same flip $\binom{12}{21}$, but in the n'-ary case they can be different.

***Example* 9.5.18.** Consider the ternary braided equation (9.5.24), but now for the braiding $\mathbf{c}_{V_1V_2V_3}$, instead of (9.5.37), where we have

$$\begin{gathered}\mathbf{c}_{V_1V_2V_3} \circ (v_1 \otimes v_2 \otimes v_3) = \tau_{V_1V_2V_3} \circ \rho^{(2)}(\mathscr{R}^{(3)}, \mathscr{R}^{(3)} | (v_1 \otimes v_2 \otimes v_3)) \\ = \sum_{\alpha,\beta} \rho^{(2)}\left(\mathsf{r}^{(3)}_{\alpha}, \mathsf{r}^{(3)}_{\beta} \middle| v_3\right) \otimes \rho^{(2)}\left(\mathsf{r}^{(2)}_{\alpha}, \mathsf{r}^{(2)}_{\beta} \middle| v_2\right) \otimes \rho^{(2)}\left(\mathsf{r}^{(1)}_{\alpha}, \mathsf{r}^{(1)}_{\beta} \middle| v_1\right), \\ v_i \in V_i,\ \mathsf{r}^{(i)}_{\alpha,\beta} \in B,\end{gathered} \tag{9.5.40}$$

where $\rho^{(2)} : B \otimes B \otimes V \to V$ is a 2-place action (5.1.5). In this way (9.5.40) is consistent with (9.5.31). In each place of the 2-place action $\rho^{(2)}$ we then obtain the relation (9.5.38).

9.5.5 Polyadic triangularity

A polyadic analog of the *triangularity* (Drinfeld 1987) can be defined, if we rewrite (9.5.15) as

$$(\mathrm{id}_B \otimes \Delta)(\mathscr{R}) = \mu[\mathcal{R}_{13}, \mathcal{R}_{12}] \equiv \sum_{\alpha,\beta} \mu \circ \tau_{\mathrm{op}}\left[\mathrm{r}_\alpha^{(1)}, \mathrm{r}_\beta^{(1)}\right] \otimes \mathrm{r}_\alpha^{(2)} \otimes \mathrm{r}_\beta^{(2)}, \tag{9.5.41}$$

where τ_{op} is the binary twist. Instead of the R-matrix formulation (the left equality in (9.5.41)), we use the component approach by Radford (2012), and propose the following

Definition 9.5.19. A polyadic almost co-commutative bialgebra $\mathcal{B}^{(n',n)} = \langle B | \mu^{(n)}, \Delta^{(n')} \rangle$ with the polyadic R-matrix $\mathscr{R}^{(n')} = \sum_\alpha \mathrm{r}_\alpha^{(1)} \otimes \ldots \otimes \mathrm{r}_\alpha^{(n')}$, $\mathrm{r}_\alpha^{(i)} \in B$ is called *quasipolyangular*, if the following n' relations hold

$$\begin{aligned}\sum_\alpha \Delta^{(n')}\left(\mathrm{r}_\alpha^{(1)}\right) \otimes \mathrm{r}_\alpha^{(2)} \otimes \ldots \otimes \mathrm{r}_\alpha^{(n')} = \sum_{\alpha_1, \ldots \alpha_{n'}} \mathrm{r}_{\alpha_1}^{(1)} \otimes \mathrm{r}_{\alpha_2}^{(1)} \otimes \ldots \otimes \mathrm{r}_{\alpha_{n'}}^{(1)} \\ \otimes \left(\mu^{(n)}\right)^{\circ \ell}\left[\mathrm{r}_{\alpha_1}^{(2)} \otimes \mathrm{r}_{\alpha_2}^{(2)} \otimes \ldots \otimes \mathrm{r}_{\alpha_{n'}}^{(2)}\right] \otimes \ldots \\ \otimes \left(\mu^{(n)}\right)^{\circ \ell}\left[\mathrm{r}_{\alpha_1}^{(n')} \otimes \mathrm{r}_{\alpha_2}^{(n')} \otimes \ldots \otimes \mathrm{r}_{\alpha_{n'}}^{(n')}\right],\end{aligned} \tag{9.5.42}$$

$$\begin{aligned}\sum_\alpha \mathrm{r}_\alpha^{(1)} \otimes \Delta^{(n')}\left(\mathrm{r}_\alpha^{(2)}\right) \otimes \ldots \otimes \mathrm{r}_\alpha^{(n')} = \sum_{\alpha_1, \ldots \alpha_{n'}} \left(\mu^{(n)}\right)^{\circ \ell} \circ \\ \tau_{\mathrm{op}}^{(n')}\left[\mathrm{r}_{\alpha_1}^{(1)} \otimes \mathrm{r}_{\alpha_2}^{(1)} \otimes \ldots \otimes \mathrm{r}_{\alpha_{n'}}^{(1)}\right] \\ \otimes \mathrm{r}_{\alpha_1}^{(2)} \otimes \mathrm{r}_{\alpha_2}^{(2)} \otimes \ldots \otimes \mathrm{r}_{\alpha_{n'}}^{(2)} \otimes \ldots \otimes \left(\mu^{(n)}\right)^{\circ \ell}\left[\mathrm{r}_{\alpha_1}^{(n')} \otimes \mathrm{r}_{\alpha_2}^{(n')} \otimes \ldots \otimes \mathrm{r}_{\alpha_{n'}}^{(n')}\right],\end{aligned} \tag{9.5.43}$$

$$\vdots$$

$$\begin{aligned}\sum_\alpha \mathrm{r}_\alpha^{(1)} \otimes \mathrm{r}_\alpha^{(2)} \otimes \ldots \otimes \Delta^{(n')}\left(\mathrm{r}_\alpha^{(n')}\right) = \sum_{\alpha_1, \ldots \alpha_{n'}} \left(\mu^{(n)}\right)^{\circ \ell} \circ \\ \tau_{\mathrm{op}}^{(n')}\left[\mathrm{r}_{\alpha_1}^{(1)} \otimes \mathrm{r}_{\alpha_2}^{(1)} \otimes \ldots \otimes \mathrm{r}_{\alpha_{n'}}^{(1)}\right] \\ \otimes \left(\mu^{(n)}\right)^{\circ \ell} \circ \tau_{\mathrm{op}}^{(n')}\left[\mathrm{r}_{\alpha_1}^{(2)} \otimes \mathrm{r}_{\alpha_2}^{(2)} \otimes \ldots \otimes \mathrm{r}_{\alpha_{n'}}^{(2)}\right] \otimes \ldots \otimes \mathrm{r}_{\alpha_1}^{(n')} \otimes \mathrm{r}_{\alpha_2}^{(n')} \otimes \ldots \otimes \mathrm{r}_{\alpha_{n'}}^{(n')},\end{aligned} \tag{9.5.44}$$

where $\tau_{\mathrm{op}}^{(n')}$ is the polyadic twist map (5.3.63). The arity shape of a quasipolyangular $\mathcal{B}^{(n',n)}$ is fixed by

$$n' = \ell(n-1) + 1, \quad \ell \in \mathbb{N}. \tag{9.5.45}$$

***Remark* 9.5.20.** As opposed to the binary case (9.5.14)–(9.5.15), the right-hand sides here can be expressed in terms of the extended R-matrix in the first equation (9.5.42) and the last one (9.5.44) only, because in the intermediate equations the sequences of R-matrix elements are permuted. For instance, it is clear that the binary product $\sum_{\alpha,\beta}\left(\mathrm{r}_\beta^{(1)} \otimes \mathrm{r}_\alpha^{(2)} \otimes e_B\right) \cdot \left(\mathrm{r}_\alpha^{(1)} \otimes \mathrm{r}_\beta^{(2)} \otimes e_B\right) = \sum_{\alpha,\beta}\left(\mathrm{r}_\beta^{(1)} \cdot \mathrm{r}_\alpha^{(1)} \otimes \mathrm{r}_\alpha^{(2)} \cdot \mathrm{r}_\beta^{(2)} \otimes e_B\right)$ cannot be expressed in terms of the extended binary R-matrix (9.5.9).

9.5.6 Almost co-medial polyadic bialgebras

The previous considerations showed that co-commutativity and almost co-commutativity in the polyadic case are not unique and do not describe the bialgebras to the fullest extent. This happens because mediality is a more general and consequent property of polyadic algebraic structures, while commutativity can be treated as a particular case of it (see section 5.3.4 and (5.3.57)). Therefore, we propose here to deform co-mediality (rather than co-commutativity as in Drinfeld 1987, 1989a).

Let $\mathcal{B}^{(n',n)} = \langle B | \mu^{(n)}, \Delta^{(n')} \rangle$, be a polyadic bialgebra (see definition 9.2.1). Now we deform the co-mediality condition (9.1.18) in a similar way to the polyadic R-matrix (9.5.29).

Definition 9.5.21. A polyadic bialgebra $\mathcal{B}^{(n',n)}$ is called *polyadic sequenced almost co-medial*, if there exist $(n' - 1)$ fixed elements $\mathcal{M}_i^{(n'^2)} \in B^{\otimes n'^2}$, $i = 1, \ldots, n - 1$, called a *polyadic M-matrix sequence*, such that (see (9.1.18) and (9.5.29))

$$\begin{aligned} &\mu^{(n)}\left[\tau_{\text{medial}}^{(n',n')} \circ \left((\Delta^{(n')})^{\otimes n'}\right) \circ \Delta^{(n')}(b), \mathcal{M}_1^{(n'^2)}, \mathcal{M}_2^{(n'^2)}, \ldots, \mathcal{M}_{n-1}^{(n'^2)}\right] \\ &= \mu^{(n)}\left[\mathcal{M}_1^{(n'^2)}, \left(\left(\Delta^{(n')}\right)^{\otimes n'}\right) \circ \Delta^{(n')}(b), \mathcal{M}_2^{(n'^2)}, \ldots, \mathcal{M}_{n-1}^{(n'^2)}\right] \\ &= \mu^{(n)}\left[\mathcal{M}_1^{(n'^2)}, \mathcal{M}_2^{(n'^2)}, \ldots, \mathcal{M}_{n-1}^{(n'^2)}, ((\Delta^{(n')})^{\otimes n'}) \circ \Delta^{(n')}(b)\right], \quad \forall b \in B, \end{aligned} \tag{9.5.46}$$

where $\tau_{\text{medial}}^{(n',n')}$ is the polyadic medial map (7.3.6).

Definition 9.5.22. A polyadic bialgebra $\mathcal{B}^{(n',n)}$ is called *polyadic sequenced almost (semi)co-medial*, if only the first and the last relations in (9.5.46) hold

$$\begin{aligned} &\mu^{(n)}\left[\tau_{\text{medial}}^{(n',n')} \circ \left(\left(\Delta^{(n')}\right)^{\otimes n'}\right) \circ \Delta^{(n')}(b), \mathcal{M}_1^{(n'^2)}, \ldots, \mathcal{M}_{n-1}^{(n'^2)}\right] \\ &= \mu^{(n)}\left[\mathcal{M}_1^{(n'^2)}, \ldots, \mathcal{M}_{n-1}^{(n'^2)}, \left((\Delta^{(n')})^{\otimes n'}\right) \circ \Delta^{(n')}(b)\right], \quad \forall b \in B. \end{aligned} \tag{9.5.47}$$

If all the elements in the sequence (similar to the neutral sequence for n-ary groups (1.1.26)) are the same $\mathcal{M}_1^{(n'^2)} = \mathcal{M}_2^{(n'^2)} = \ldots = \mathcal{M}_{n-1}^{(n'^2)} \equiv \mathcal{M}^{(n'^2)}$, we have

Definition 9.5.23. A polyadic bialgebra $\mathcal{B}^{(n',n)}$ is called *polyadic almost (semi)co-medial*, if there exist one fixed element $\mathcal{M}^{(n'^2)} \in B^{\otimes n'^2}$ called a *polyadic M-matrix*, such that (see (9.1.18) and (9.5.29))

$$\begin{aligned} &\mu^{(n)}\left[\tau_{\text{medial}}^{(n',n')} \circ \left((\Delta^{(n')})^{\otimes n'}\right) \circ \Delta^{(n')}(b), \overbrace{\mathcal{M}^{(n'^2)}, \ldots, \mathcal{M}^{(n'^2)}}^{n-1}\right] \\ &= \mu^{(n)}\left[\overbrace{\mathcal{M}^{(n'^2)}, \ldots, \mathcal{M}^{(n'^2)}}^{n-1}, \left((\Delta^{(n')})^{\otimes n'}\right) \circ \Delta^{(n')}(b)\right], \quad \forall b \in B. \end{aligned} \tag{9.5.48}$$

***Remark* 9.5.24.** The main advantage of the polyadic almost co-mediality property over polyadic almost co-commutativity is the uniqueness of the medial map $\boldsymbol{\tau}_{\mathrm{medial}}^{(n,n)}$ and nonuniqueness of the polyadic twist map $\boldsymbol{\tau}_{\mathrm{op}}^{(\ell_\tau)}$ (5.3.63).

The polyadic M-matrix $\mathscr{M}^{(n')}$ in components is given by

$$\mathscr{M}^{(n'^2)} = \sum_{\alpha} \mathsf{m}_\alpha^{(1)} \otimes \ldots \otimes \mathsf{m}_\alpha^{(n'^2)}, \quad \mathsf{m}_\alpha^{(i)} \in B,\ i = 1, \ldots, n'^2. \tag{9.5.49}$$

***Example* 9.5.25.** In the binary case for $\mathcal{B}^{(2,2)} = \langle B | \mu, \Delta \rangle$ we have an almost co-mediality (9.5.48) as

$$\mu[\tau_{\mathrm{medial}} \circ (\Delta \otimes \Delta) \circ \Delta(b), \mathscr{M}^{(4)}] = \mu[\mathscr{M}^{(4)}, (\Delta \otimes \Delta) \circ \Delta(b)], \quad \forall b \in B, \tag{9.5.50}$$

which gives, in components (cf for R-matrix (9.5.8))

$$\begin{aligned} &\sum_{[b]_{[b]}} \sum_{\alpha} \mu\left[b_{[1]_{[1]}}, \mathsf{m}_\alpha^{(1)}\right] \otimes \mu\left[b_{[2]_{[1]}}, \mathsf{m}_\alpha^{(2)}\right] \otimes \mu\left[b_{[1]_{[2]}}, \mathsf{m}_\alpha^{(3)}\right] \otimes \mu\left[b_{[2]_{[2]}}, \mathsf{m}_\alpha^{(4)}\right] \\ &= \sum_{[b]'_{[b]}} \sum_{\alpha'} \mu\left[\mathsf{m}_{\alpha'}^{(1)}, b_{[1]'_{[1]}}\right] \otimes \mu\left[\mathsf{m}_{\alpha'}^{(2)}, b_{[1]'_{[2]}}\right] \otimes \mu\left[\mathsf{m}_{\alpha'}^{(3)}, b_{\left[2\right]'_{[1]}}\right] \otimes \mu\left[\mathsf{m}_{\alpha'}^{(4)}, b_{[2]'_{[2]}}\right]. \end{aligned} \tag{9.5.51}$$

Let us clarify the connection between the almost co-commutativity and almost co-mediality properties.

Theorem 9.5.26. If $\mathcal{B}^{(n',n)}$ is polyadic almost (semi)co-commutative with the polyadic twist map $\tau_{\mathrm{op}}^{(n')}$ (5.3.63) and the n'-ary R-matrix $\mathscr{R}^{(n')}$ (9.5.30), then (9.5.29) can be presented in the 'medial-like' form

$$\begin{aligned} &\mu^{(n)}\left[\tau_R^{(n',n')} \circ \left((\Delta^{(n')})^{\otimes n'}\right) \circ \Delta^{(n')}(b), \overbrace{\mathscr{M}_R^{(n'^2)}, \ldots, \mathscr{M}_R^{(n'^2)}}^{n-1}\right] \\ &= \mu^{(n)}\left[\overbrace{\mathscr{M}_R^{(n'^2)}, \ldots, \mathscr{M}_R^{(n'^2)}}^{n-1}, \left((\Delta^{(n')})^{\otimes n'}\right) \circ \Delta^{(n')}(b)\right], \quad \forall b \in B, \end{aligned} \tag{9.5.52}$$

where

$$\tau_R^{(n',n')} = \overbrace{\tau_{\mathrm{op}}^{(n')} \otimes \ldots \otimes \tau_{\mathrm{op}}^{(n')}}^{n'}, \tag{9.5.53}$$

$$\mathcal{M}_R^{(n'^2)} = \overbrace{\mathcal{R}^{(n')} \otimes \ldots \otimes \mathcal{R}^{(n')}}^{n'}. \tag{9.5.54}$$

Proof. Applying (9.5.29) to each Sweedler component $b_{[i]}$ of $\Delta^{(n')}(b)$, $i = 1, \ldots, n'$, we obtain n' relations for the polyadic almost (semi)co-commutativity. Then multiplying them tensorially, we obtain

$$\left(\overbrace{\tau_{\mathrm{op}}^{(n')} \otimes \ldots \otimes \tau_{\mathrm{op}}^{(n')}}^{n'}\right) \circ \left(\Delta^{(n')}\left(b_{[1]}\right) \otimes \ldots \otimes \Delta^{(n')}\left(b_{[n']}\right)\right) \circ \left(\overbrace{\mathcal{R}^{(n')} \otimes \ldots \otimes \mathcal{R}^{(n')}}^{n'}\right)$$

$$= \left(\overbrace{\mathcal{R}^{(n')} \otimes \ldots \otimes \mathcal{R}^{(n')}}^{n'}\right) \circ \left(\Delta^{(n')}\left(b_{[1]}\right) \otimes \ldots \otimes \Delta^{(n')}\left(b_{[n']}\right)\right),$$

which immediately gives (9.5.52). The converse statement is obvious. □

Corollary 9.5.27. Polyadic almost co-commutativity is a particular case of polyadic co-mediality with the special 'medial-like' twist map $\boldsymbol{\tau}_R^{(n',n')}$ (9.5.53) and the composite M-matrix (9.5.54) consisting of n' copies of the R-matrix (9.5.30).

***Example* 9.5.28.** In the binary case we compare the medial map (5.3.57) with the composed 'medial-like' twist map (9.5.53) as

$$\boldsymbol{\tau}_{\mathrm{medial}} = \mathrm{id}_B \otimes \boldsymbol{\tau}_{\mathrm{op}} \otimes \mathrm{id}_B, \tag{9.5.55}$$

$$\boldsymbol{\tau}_R = \boldsymbol{\tau}_{\mathrm{op}} \otimes \boldsymbol{\tau}_{\mathrm{op}}, \tag{9.5.56}$$

or in components

$$b_1 \otimes b_2 \otimes b_3 \otimes b_4 \overset{\tau_{\mathrm{medial}}}{\mapsto} b_1 \otimes b_3 \otimes b_2 \otimes b_4, \tag{9.5.57}$$

$$b_1 \otimes b_2 \otimes b_3 \otimes b_4 \overset{\tau_R}{\mapsto} b_2 \otimes b_1 \otimes b_4 \otimes b_3. \tag{9.5.58}$$

This shows manifestly the difference between (polyadic) almost co-commutativity and (polyadic) almost co-mediality.

9.5.7 Equations for the M-matrix

Let us find the equations for the M-matrix (9.5.48) using the medial analog of the n'-ary braid equation. Now the morphism of modules $\mathbf{c}_{V_1 \ldots V_{n'^2}}$ becomes (see for the R-matrix (9.5.34))

$$
\begin{aligned}
&\mathbf{c}_{V_1 \ldots V_{n'^2}} \circ (v_1 \otimes \ldots \otimes v_{n'^2}) \\
&= \tau^{(n',n')}_{\text{medial},\, V_1 \ldots V_{n'^2}} \circ \rho^{(n-1)} \left(\overbrace{\mathcal{M}^{(n'^2)}, \ldots, \mathcal{M}^{(n'^2)}}^{n-1} \middle| (v_1 \otimes \ldots \otimes v_{n'^2}) \right) \\
&= \tau^{(n',n')}_{\text{medial},\, V_1 \ldots V_{n'^2}} \circ \left(\sum_{\alpha_1, \ldots \alpha_{n-1}} \rho^{(n-1)} \left(\mathsf{m}^{(1)}_{\alpha_1}, \ldots \mathsf{m}^{(1)}_{\alpha_{n-1}} \middle| v_1 \right) \otimes \ldots \right. \\
&\qquad \left. \otimes \rho^{(n-1)} \left(\mathsf{m}^{(n'^2)}_{\alpha_1}, \ldots \mathsf{m}^{(n'^2)}_{\alpha_{n-1}} \middle| v_{n'^2} \right) \right), \\
&v_i \in V_i, \quad \mathsf{r}^{(i)}_{\alpha} \in B, \quad i = 1, \ldots, n'^2,
\end{aligned}
\tag{9.5.59}
$$

where $\tau^{(n',n')}_{\text{medial},\, V_1 \ldots V_{n'^2}}$ is the medial map (7.3.6) acting on n'^2 modules V_i, $\mathcal{M}^{(n'^2)}$ is the polyadic M-matrix (9.5.49), and $\rho^{(n-1)}$ is the $(n-1)$-place action (5.1.5). Now instead of the n'-ary braid equation (9.5.23) we can have (see remark 9.5.3)

Proposition 9.5.29. *The n'-ary medial braid equation is*

$$
\begin{aligned}
&\left(\mathbf{c}_{V^{n'^2}} \otimes \overbrace{\mathrm{id}_V \otimes \ldots \otimes \mathrm{id}_V}^{n'^2 - 1} \right) \circ \left(\mathrm{id}_V \otimes \mathbf{c}_{V^{n'^2}} \otimes \overbrace{\mathrm{id}_V \otimes \ldots \otimes \mathrm{id}_V}^{n'^2 - 2} \right) \circ \ldots \\
&\circ \left(\overbrace{\mathrm{id}_V \otimes \ldots \otimes \mathrm{id}_V}^{n'^2 - 1} \otimes \mathbf{c}_{V^{n'^2}} \right) \circ \left(\mathbf{c}_{V^{n'^2}} \otimes \overbrace{\mathrm{id}_V \otimes \ldots \otimes \mathrm{id}_V}^{n'^2 - 1} \right) \\
&= \left(\overbrace{\mathrm{id}_V \otimes \ldots \otimes \mathrm{id}_V}^{n'^2 - 1} \otimes \mathbf{c}_{V^{n'^2}} \right) \circ \left(\mathbf{c}_{V^{n'^2}} \otimes \overbrace{\mathrm{id}_V \otimes \ldots \otimes \mathrm{id}_V}^{n'^2 - 1} \right) \circ \ldots \\
&\circ \left(\overbrace{\mathrm{id}_V \otimes \ldots \otimes \mathrm{id}_V}^{n'^2 - 2} \otimes \mathbf{c}_{V^{n'^2}} \otimes \mathrm{id}_V \right) \circ \left(\overbrace{\mathrm{id}_V \otimes \ldots \otimes \mathrm{id}_V}^{n'^2 - 1} \otimes \mathbf{c}_{V^{n'^2}} \right),
\end{aligned}
\tag{9.5.60}
$$

where we use the notation $\mathbf{c}_{V^{n'^2}} \equiv \mathbf{c}_{V_1 \ldots V_{n'^2}}$, $\mathrm{id}_V \equiv \mathrm{id}_{V_i}$, *and each side consists of* $(n'^2 + 1)$ *brackets with* $(2n'^2 - 1)$ *multipliers.*

Proof. This follows from the associative quiver technique (Duplij 2012). □

We observe that even in the binary case the medial braid equations are cumbersome and nontrivial.

***Example* 9.5.30.** In the binary case $n' = 2$ we have the map $\mathbf{c}_{V^{n'2}}$ (see (9.5.55), (9.5.57))

$$\mathbf{c}_{V_1V_2V_3V_4} : V_1 \otimes V_2 \otimes V_3 \otimes V_4 \to V_1 \otimes V_3 \otimes V_2 \otimes V_4, \tag{9.5.61}$$

which acts on $V_1 \otimes V_2 \otimes V_3 \otimes V_4 \otimes V_5 \otimes V_6 \otimes V_7$. There are two medial braid equations which correspond to diagrams of different lengths (see the standard braid equation (9.5.21)).

(1) $V_1 \otimes V_2 \otimes V_3 \otimes V_4 \otimes V_5 \otimes V_6 \otimes V_7 \longrightarrow V_1 \otimes V_4 \otimes V_5 \otimes V_6 \otimes V_3 \otimes V_2 \otimes V_7$:

$$\begin{aligned}
&(\mathbf{c}_{V_1V_5V_4V_6} \otimes \mathrm{id}_{V_3} \otimes \mathrm{id}_{V_2} \otimes \mathrm{id}_{V_7}) \circ (\mathrm{id}_{V_1} \otimes \mathbf{c}_{V_5V_6V_4V_3} \otimes \mathrm{id}_{V_2} \otimes \mathrm{id}_{V_7})\\
&\circ (\mathrm{id}_{V_1} \otimes \mathrm{id}_{V_5} \otimes \mathbf{c}_{V_6V_3V_4V_2} \otimes \mathrm{id}_{V_7}) \circ (\mathrm{id}_{V_1} \otimes \mathrm{id}_{V_5} \otimes \mathrm{id}_{V_6} \otimes \mathbf{c}_{V_3V_2V_4V_7})\\
&\circ (\mathbf{c}_{V_1V_6V_5V_3} \otimes \mathrm{id}_{V_2} \otimes \mathrm{id}_{V_4} \otimes \mathrm{id}_{V_7}) \circ (\mathrm{id}_{V_1} \otimes \mathbf{c}_{V_6V_3V_5V_2} \otimes \mathrm{id}_{V_4} \otimes \mathrm{id}_{V_7})\\
&\circ (\mathrm{id}_{V_1} \otimes \mathrm{id}_{V_6} \otimes \mathbf{c}_{V_3V_2V_5V_4} \otimes \mathrm{id}_{V_7}) \circ (\mathrm{id}_{V_1} \otimes \mathrm{id}_{V_6} \otimes \mathrm{id}_{V_6} \otimes \mathbf{c}_{V_2V_4V_5V_7})\\
&\circ (\mathbf{c}_{V_1V_3V_6V_2} \otimes \mathrm{id}_{V_4} \otimes \mathrm{id}_{V_5} \otimes \mathrm{id}_{V_7}) \circ (\mathrm{id}_{V_1} \otimes \mathbf{c}_{V_3V_2V_6V_4} \otimes \mathrm{id}_{V_5} \otimes \mathrm{id}_{V_7})\\
&\circ (\mathrm{id}_{V_1} \otimes \mathrm{id}_{V_3} \otimes \mathbf{c}_{V_2V_4V_6V_5} \otimes \mathrm{id}_{V_7}) \circ (\mathrm{id}_{V_1} \otimes \mathrm{id}_{V_3} \otimes \mathrm{id}_{V_2} \otimes \mathbf{c}_{V_4V_5V_6V_7})\\
&\circ (\mathbf{c}_{V_1V_2V_3V_4} \otimes \mathrm{id}_{V_5} \otimes \mathrm{id}_{V_6} \otimes \mathrm{id}_{V_7}) = (\mathrm{id}_{V_1} \otimes \mathrm{id}_{V_4} \otimes \mathrm{id}_{V_5} \otimes \mathbf{c}_{V_6V_2V_3V_7})\\
&\circ (\mathbf{c}_{V_1V_5V_4V_6} \otimes \mathrm{id}_{V_2} \otimes \mathrm{id}_{V_3} \otimes \mathrm{id}_{V_7}) \circ (\mathrm{id}_{V_1} \otimes \mathbf{c}_{V_5V_6V_4V_2} \otimes \mathrm{id}_{V_3} \otimes \mathrm{id}_{V_7})\\
&\circ (\mathrm{id}_{V_1} \otimes \mathrm{id}_{V_5} \otimes \mathbf{c}_{V_6V_2V_4V_3} \otimes \mathrm{id}_{V_7}) \circ (\mathrm{id}_{V_1} \otimes \mathrm{id}_{V_5} \otimes \mathrm{id}_{V_6} \otimes \mathbf{c}_{V_2V_3V_4V_7})\\
&\circ (\mathbf{c}_{V_1V_6V_5V_2} \otimes \mathrm{id}_{V_3} \otimes \mathrm{id}_{V_4} \otimes \mathrm{id}_{V_7}) \circ (\mathrm{id}_{V_1} \otimes \mathbf{c}_{V_6V_2V_5V_3} \otimes \mathrm{id}_{V_4} \otimes \mathrm{id}_{V_7})\\
&\circ (\mathrm{id}_{V_1} \otimes \mathrm{id}_{V_6} \otimes \mathbf{c}_{V_2V_3V_5V_4} \otimes \mathrm{id}_{V_7}) \circ (\mathrm{id}_{V_1} \otimes \mathrm{id}_{V_6} \otimes \mathrm{id}_{V_2} \otimes \mathbf{c}_{V_3V_4V_5V_7})\\
&\circ (\mathbf{c}_{V_1V_2V_6V_3} \otimes \mathrm{id}_{V_4} \otimes \mathrm{id}_{V_5} \otimes \mathrm{id}_{V_7}) \circ (\mathrm{id}_{V_1} \otimes \mathbf{c}_{V_6V_3V_5V_2} \otimes \mathrm{id}_{V_4} \otimes \mathrm{id}_{V_7})\\
&\circ (\mathrm{id}_{V_1} \otimes \mathrm{id}_{V_2} \otimes \mathbf{c}_{V_3V_4V_6V_5} \otimes \mathrm{id}_{V_7}) \circ (\mathrm{id}_{V_1} \otimes \mathrm{id}_{V_2} \otimes \mathrm{id}_{V_3} \otimes \mathbf{c}_{V_4V_5V_6V_7}).
\end{aligned} \tag{9.5.62}$$

(2) $V_1 \otimes V_2 \otimes V_3 \otimes V_4 \otimes V_5 \otimes V_6 \otimes V_7 \longrightarrow V_1 \otimes V_6 \otimes V_5 \otimes V_4 \otimes V_2 \otimes V_3 \otimes V_7$:

$$\begin{aligned}
&(\mathrm{id}_{V_1} \otimes \mathrm{id}_{V_6} \otimes \mathbf{c}_{V_5V_2V_4V_3} \otimes \mathrm{id}_{V_7}) \circ (\mathrm{id}_{V_1} \otimes \mathrm{id}_{V_6} \otimes \mathrm{id}_{V_5} \otimes \mathbf{c}_{V_2V_3V_4V_7})\\
&\circ (\mathbf{c}_{V_1V_5V_6V_2} \otimes \mathrm{id}_{V_3} \otimes \mathrm{id}_{V_4} \otimes \mathrm{id}_{V_7}) \circ (\mathrm{id}_{V_1} \otimes \mathbf{c}_{V_5V_2V_6V_3} \otimes \mathrm{id}_{V_4} \otimes \mathrm{id}_{V_7})\\
&\circ (\mathrm{id}_{V_1} \otimes \mathrm{id}_{V_5} \otimes \mathbf{c}_{V_2V_3V_6V_4} \otimes \mathrm{id}_{V_7}) \circ (\mathrm{id}_{V_1} \otimes \mathrm{id}_{V_5} \otimes \mathrm{id}_{V_2} \otimes \mathbf{c}_{V_3V_4V_6V_7})\\
&\circ (\mathbf{c}_{V_1V_2V_5V_3} \otimes \mathrm{id}_{V_4} \otimes \mathrm{id}_{V_6} \otimes \mathrm{id}_{V_7}) \circ (\mathrm{id}_{V_1} \otimes \mathbf{c}_{V_2V_3V_5V_4} \otimes \mathrm{id}_{V_6} \otimes \mathrm{id}_{V_7})\\
&\circ (\mathrm{id}_{V_1} \otimes \mathrm{id}_{V_2} \otimes \mathbf{c}_{V_3V_4V_5V_6} \otimes \mathrm{id}_{V_7}) = (\mathrm{id}_{V_1} \otimes \mathbf{c}_{V_6V_4V_5V_2} \otimes \mathrm{id}_{V_3} \otimes \mathrm{id}_{V_7})\\
&\circ (\mathrm{id}_{V_1} \otimes \mathrm{id}_{V_6} \otimes \mathbf{c}_{V_4V_2V_5V_3} \otimes \mathrm{id}_{V_7}) \circ (\mathrm{id}_{V_1} \otimes \mathrm{id}_{V_6} \otimes \mathrm{id}_{V_4} \otimes \mathbf{c}_{V_2V_3V_5V_7})\\
&\circ (\mathbf{c}_{V_1V_4V_6V_2} \otimes \mathrm{id}_{V_3} \otimes \mathrm{id}_{V_5} \otimes \mathrm{id}_{V_7}) \circ (\mathrm{id}_{V_1} \otimes \mathbf{c}_{V_4V_2V_6V_3} \otimes \mathrm{id}_{V_5} \otimes \mathrm{id}_{V_7})\\
&\circ (\mathrm{id}_{V_1} \otimes \mathrm{id}_{V_4} \otimes \mathbf{c}_{V_2V_3V_6V_5} \otimes \mathrm{id}_{V_7}) \circ (\mathrm{id}_{V_1} \otimes \mathrm{id}_{V_4} \otimes \mathrm{id}_{V_2} \otimes \mathbf{c}_{V_3V_5V_6V_7})\\
&\circ (\mathbf{c}_{V_1V_2V_4V_3} \otimes \mathrm{id}_{V_5} \otimes \mathrm{id}_{V_6} \otimes \mathrm{id}_{V_7}) \circ (\mathrm{id}_{V_1} \otimes \mathbf{c}_{V_2V_3V_4V_5} \otimes \mathrm{id}_{V_6} \otimes \mathrm{id}_{V_7}).
\end{aligned} \tag{9.5.63}$$

The equations for the M-matrix can be obtained by introducing the 'extended' M-matrix, as in the case of the R-matrix (see remark 9.5.12), and this can also be possible if the n-ary algebra $\langle B | \mu^{(n)} \rangle$ has the unit (element) $e_B \in B$.

Definition 9.5.31. The 'extended' M-matrix is defined by $\mathcal{M}^{(2n'^2-1)}_{i_1 \ldots i_{n'^2}} \in B^{\otimes(2n'^2-1)}$, such that

$$\mathcal{M}^{(2n'^2-1)}_{i_1 \ldots\, i_{n'^2}} = \sum_\alpha e_B \otimes \ldots \otimes \mathsf{m}^{(i_1)}_\alpha \otimes \ldots \otimes \mathsf{m}^{(i_{n'^2})}_\alpha \otimes \ldots \otimes e_B, \qquad (9.5.64)$$

$$i_1, \ldots, i_{n'^2} \in \{1, \ldots, 2n'^2 - 1\},$$

where $\mathsf{m}^{(i_k)}_\alpha$ are on the i_k-place.

It is difficult to write the general compatibility equations for the 'extended' M-matrix (9.5.64).

***Example* 9.5.32.** In the binary case $n = n' = 2$ we have for the polyadic M-matrix $\mathscr{M}^{(4)}$ in components

$$\mathscr{M}^{(4)} = \sum_\alpha \mathsf{m}^{(1)}_\alpha \otimes \mathsf{m}^{(2)}_\alpha \otimes \mathsf{m}^{(3)}_\alpha \otimes \mathsf{m}^{(4)}_\alpha, \quad \mathsf{m}^{(i)}_\alpha \in B, \qquad (9.5.65)$$

and $\mathcal{M}^{(7)}_{i_1 \ldots\, i_4} \in B^{\otimes 7}$ with

$$\mathcal{M}^{(7)}_{i_1 \ldots\, i_4} = \sum_\alpha e_B \otimes \ldots \otimes \mathsf{m}^{(i_1)}_\alpha \otimes \ldots \otimes \mathsf{m}^{(i_4)}_\alpha \otimes \ldots \otimes e_B, \qquad (9.5.66)$$

$$i_1, \ldots, i_4 \in \{1, \ldots, 7\}.$$

The map of modules $\mathbf{c}_{V_1 \ldots V_4}$ (9.5.61) in the manifest form is

$$\mathbf{c}_{V_1 V_2 V_3 V_4} \circ (v_1 \otimes v_2 \otimes v_3 \otimes \otimes v_4) = \tau_{\mathrm{medial}} \circ \rho(\mathscr{M}^{(4)} | (v_1 \otimes v_2 \otimes v_3 \otimes v_4))$$
$$= \tau_{\mathrm{medial}} \circ \left(\sum_\alpha \rho(\mathsf{m}^{(1)}_\alpha | v_1) \otimes \rho(\mathsf{m}^{(2)}_\alpha | v_2) \otimes \rho(\mathsf{m}^{(3)}_\alpha | v_3) \otimes \rho(\mathsf{m}^{(4)}_\alpha | v_4) \right), \qquad (9.5.67)$$
$$v_i \in V_i, \quad \mathsf{m}^{(i)}_\alpha \in B, \quad i = 1, 2, 3, 4,$$

where τ_{medial} is the medial map (9.5.55), and $\rho : B \times V \to V$ is the ordinary 1-place action (5.1.5).

After inserting (9.5.67) into (9.5.62) and (9.5.63), using (9.5.66) we obtain the equations for M-matrix

$$\mathcal{M}^{(7)}_{1546} \mathcal{M}^{(7)}_{5643} \mathcal{M}^{(7)}_{6342} \mathcal{M}^{(7)}_{3247} \mathcal{M}^{(7)}_{1653} \mathcal{M}^{(7)}_{6352} \mathcal{M}^{(7)}_{3254} \mathcal{M}^{(7)}_{2457} \mathcal{M}^{(7)}_{1362} \mathcal{M}^{(7)}_{3264} \mathcal{M}^{(7)}_{2465} \mathcal{M}^{(7)}_{4567} \mathcal{M}^{(7)}_{1234}$$
$$= \mathcal{M}^{(7)}_{6237} \mathcal{M}^{(7)}_{1546} \mathcal{M}^{(7)}_{5642} \mathcal{M}^{(7)}_{6243} \mathcal{M}^{(7)}_{2347} \mathcal{M}^{(7)}_{1652} \mathcal{M}^{(7)}_{6253} \mathcal{M}^{(7)}_{2354} \mathcal{M}^{(7)}_{3457} \mathcal{M}^{(7)}_{1263} \mathcal{M}^{(7)}_{2364} \mathcal{M}^{(7)}_{3465} \mathcal{M}^{(7)}_{4567} \qquad (9.5.68)$$

and

$$\mathcal{M}^{(7)}_{5243} \mathcal{M}^{(7)}_{2347} \mathcal{M}^{(7)}_{1562} \mathcal{M}^{(7)}_{5263} \mathcal{M}^{(7)}_{2364} \mathcal{M}^{(7)}_{3467} \mathcal{M}^{(7)}_{1253} \mathcal{M}^{(7)}_{2354} \mathcal{M}^{(7)}_{3456}$$
$$= \mathcal{M}^{(7)}_{6452} \mathcal{M}^{(7)}_{4253} \mathcal{M}^{(7)}_{2357} \mathcal{M}^{(7)}_{1462} \mathcal{M}^{(7)}_{4263} \mathcal{M}^{(7)}_{2365} \mathcal{M}^{(7)}_{3567} \mathcal{M}^{(7)}_{1243} \mathcal{M}^{(7)}_{2345}, \qquad (9.5.69)$$

which respect the braid equations (9.5.62) and (9.5.63).

***Remark* 9.5.33.** The unequal number of terms in (9.5.68) and (9.5.69) is governed by different commutative diagrams of modules (9.5.62) and (9.5.63), respectively (see (9.5.16) and (9.5.19)–(9.5.21)).

9.5.8 Medial analog of triangularity

Now we consider the possible analogs of the quasitriangularity conditions (similar to (9.5.14)–(9.5.15) and quasipolyangularity (9.5.42–(9.5.44)) for a polyadic almost co-medial bialgebra $\mathcal{B}^{(n',n)}$ (see definition 9.5.23).

Definition 9.5.34. A polyadic almost co-medial bialgebra $\mathcal{B}^{(n',n)} = \langle B \mid \mu^{(n)}, \Delta^{(n')} \rangle$ with the polyadic M-matrix $\mathcal{M}^{(n'^2)} = \sum_\alpha \mathsf{m}_\alpha^{(1)} \otimes \ldots \otimes \mathsf{m}_\alpha^{(n'^2)}$, $\mathsf{m}_\alpha^{(i)} \in B$ is called *medial quasipolyangular*, if the following n'^2 relations hold

$$\sum_\alpha \left(\Delta^{(n')}\right)^{\otimes n'} \circ \Delta^{(n')}(\mathsf{m}_\alpha^{(1)}) \otimes \mathsf{m}_\alpha^{(2)} \otimes \ldots \otimes \mathsf{m}_\alpha^{(n'^2)} = \sum_{\alpha_1, \ldots \alpha_{n'^2}} \mathsf{m}_{\alpha_1}^{(1)} \otimes \mathsf{m}_{\alpha_2}^{(1)} \otimes \ldots \otimes \mathsf{m}_{\alpha_{n'^2}}^{(1)}$$
$$\otimes \left(\mu^{(n)}\right)^{\circ\ell}\left[\mathsf{m}_{\alpha_1}^{(2)} \otimes \mathsf{m}_{\alpha_2}^{(2)} \otimes \ldots \otimes \mathsf{m}_{\alpha_{n'^2}}^{(2)}\right] \otimes \ldots \otimes \left(\mu^{(n)}\right)^{\circ\ell}\left[\mathsf{m}_{\alpha_1}^{(n'^2)} \otimes \mathsf{m}_{\alpha_2}^{(n'^2)} \otimes \ldots \otimes \mathsf{m}_{\alpha_{n'^2}}^{(n'^2)}\right], \tag{9.5.70}$$

$$\sum_\alpha \mathsf{m}_\alpha^{(1)} \otimes \left(\Delta^{(n')}\right)^{\otimes n'} \circ \Delta^{(n')}(\mathsf{m}_\alpha^{(2)}) \otimes \ldots \otimes \mathsf{m}_\alpha^{(n'^2)}$$
$$= \sum_{\alpha_1, \ldots \alpha_{n'^2}} \left(\mu^{(n)}\right)^{\circ\ell} \circ \tau_{\text{medial}}^{(n'^2,n'^2)}\left[\mathsf{m}_{\alpha_1}^{(1)} \otimes \mathsf{m}_{\alpha_2}^{(1)} \otimes \ldots \otimes \mathsf{m}_{\alpha_{n'^2}}^{(1)}\right] \otimes \mathsf{m}_{\alpha_1}^{(2)} \otimes \mathsf{m}_{\alpha_2}^{(2)} \otimes \ldots$$
$$\otimes \mathsf{m}_{\alpha_{n'^2}}^{(2)} \otimes \ldots$$
$$\otimes \left(\mu^{(n)}\right)^{\circ\ell}\left[\mathsf{m}_{\alpha_1}^{(n'^2)} \otimes \mathsf{m}_{\alpha_2}^{(n'^2)} \otimes \ldots \otimes \mathsf{m}_{\alpha_{n'^2}}^{(n'^2)}\right], \tag{9.5.71}$$

$$\vdots$$

$$\sum_\alpha \mathsf{m}_\alpha^{(1)} \otimes \ldots \otimes \mathsf{m}_\alpha^{(n'^2-1)} \otimes \left(\Delta^{(n')}\left(\mathsf{m}_\alpha^{(n'^2)}\right)\right)^{\otimes n'} \circ \Delta^{(n')}\left(\mathsf{m}_\alpha^{(n'^2)}\right)$$
$$= \sum_{\alpha_1, \ldots \alpha_{n'^2}} \left(\mu^{(n)}\right)^{\circ\ell} \circ \tau_{\text{medial}}^{(n'^2,n'^2)}\left[\mathsf{m}_{\alpha_1}^{(1)} \otimes \mathsf{m}_{\alpha_2}^{(1)} \otimes \ldots \otimes \mathsf{m}_{\alpha_{n'^2}}^{(1)}\right] \otimes \ldots \tag{9.5.72}$$
$$\otimes \left(\mu^{(n)}\right)^{\circ\ell} \circ \tau_{\text{medial}}^{(n'^2,n'^2)}\left[\mathsf{m}_{\alpha_1}^{(n'^2-1)} \otimes \mathsf{m}_{\alpha_2}^{(n'^2-1)} \otimes \ldots \otimes \mathsf{m}_{\alpha_{n'^2}}^{(n'^2-1)}\right] \otimes \mathsf{m}_{\alpha_1}^{(n'^2)} \otimes \mathsf{m}_{\alpha_2}^{(n'^2)} \otimes \ldots \otimes \mathsf{m}_{\alpha_{n'^2}}^{(n'^2)},$$

where $\tau_{\text{medial}}^{(n'^2,n'^2)}$ is the unique medial twist map (7.3.6). The arity shape of a medial quasipolyangular bialgebra $\mathcal{B}^{(n',n)}$ is given by (see (9.5.45))

$$n'^2 = \ell(n-1) + 1, \quad \ell \in \mathbb{N}. \tag{9.5.73}$$

***Remark* 9.5.35.** Similar to remark 9.5.20, the medial quasipolyangularity equations (9.5.70)–(9.5.72) can be expressed in terms of the extended M-matrix for the first equation (9.5.70) and the last one (9.5.72) only, because in the intermediate equations the sequences of M-matrix elements are permuted.

***Example* 9.5.36.** In the case where $n' = n = 2$, $\ell = 3$, for the bialgebra $\mathcal{B}^{(2,2)} = \langle B | \mu = (\cdot), \Delta \rangle$ with the polyadic M-matrix

$$\mathcal{M}^{(4)} = \sum_{\alpha} \mathsf{m}_{\alpha}^{(1)} \otimes \mathsf{m}_{\alpha}^{(2)} \otimes \mathsf{m}_{\alpha}^{(3)} \otimes \mathsf{m}_{\alpha}^{(4)}, \quad \mathsf{m}_{\alpha}^{(i)} \in B \tag{9.5.74}$$

we have the binary medial quasipolyangularity equations

$$\sum_{\alpha} (\Delta \otimes \Delta) \circ \Delta(\mathsf{m}_{\alpha}^{(1)}) \otimes \mathsf{m}_{\alpha}^{(2)} \otimes \mathsf{m}_{\alpha}^{(3)} \otimes \mathsf{m}_{\alpha}^{(4)} = \sum_{\alpha_1,\alpha_2,\alpha_3,\alpha_4} \mathsf{m}_{\alpha_1}^{(1)} \otimes \mathsf{m}_{\alpha_2}^{(1)} \otimes \mathsf{m}_{\alpha_3}^{(1)} \otimes \mathsf{m}_{\alpha_4}^{(1)} \\ \otimes \mathsf{m}_{\alpha_1}^{(2)} \cdot \mathsf{m}_{\alpha_2}^{(2)} \cdot \mathsf{m}_{\alpha_3}^{(2)} \cdot \mathsf{m}_{\alpha_4}^{(2)} \otimes \mathsf{m}_{\alpha_1}^{(3)} \cdot \mathsf{m}_{\alpha_2}^{(3)} \cdot \mathsf{m}_{\alpha_3}^{(3)} \cdot \mathsf{m}_{\alpha_4}^{(3)} \otimes \mathsf{m}_{\alpha_1}^{(4)} \cdot \mathsf{m}_{\alpha_2}^{(4)} \cdot \mathsf{m}_{\alpha_3}^{(4)} \cdot \mathsf{m}_{\alpha_4}^{(4)}, \tag{9.5.75}$$

$$\sum_{\alpha} \mathsf{m}_{\alpha}^{(1)} \otimes (\Delta \otimes \Delta) \circ \Delta\left(\mathsf{m}_{\alpha}^{(2)}\right) \otimes \mathsf{m}_{\alpha}^{(3)} \otimes \mathsf{m}_{\alpha}^{(4)} = \sum_{\alpha_1,\alpha_2,\alpha_3,\alpha_4} \mathsf{m}_{\alpha_1}^{(1)} \cdot \mathsf{m}_{\alpha_3}^{(1)} \cdot \mathsf{m}_{\alpha_2}^{(1)} \cdot \mathsf{m}_{\alpha_4}^{(1)} \\ \otimes \mathsf{m}_{\alpha_1}^{(2)} \otimes \mathsf{m}_{\alpha_2}^{(2)} \otimes \mathsf{m}_{\alpha_3}^{(2)} \otimes \mathsf{m}_{\alpha_4}^{(2)} \otimes \mathsf{m}_{\alpha_1}^{(3)} \cdot \mathsf{m}_{\alpha_2}^{(3)} \cdot \mathsf{m}_{\alpha_3}^{(3)} \cdot \mathsf{m}_{\alpha_4}^{(3)} \otimes \mathsf{m}_{\alpha_1}^{(4)} \cdot \mathsf{m}_{\alpha_2}^{(4)} \cdot \mathsf{m}_{\alpha_3}^{(4)} \cdot \mathsf{m}_{\alpha_4}^{(4)}, \tag{9.5.76}$$

$$\sum_{\alpha} \mathsf{m}_{\alpha}^{(1)} \otimes \mathsf{m}_{\alpha}^{(2)} \otimes (\Delta \otimes \Delta) \circ \Delta\left(\mathsf{m}_{\alpha}^{(3)}\right) \otimes \mathsf{m}_{\alpha}^{(4)} = \sum_{\alpha_1,\alpha_2,\alpha_3,\alpha_4} \mathsf{m}_{\alpha_1}^{(1)} \cdot \mathsf{m}_{\alpha_3}^{(1)} \cdot \mathsf{m}_{\alpha_2}^{(1)} \cdot \mathsf{m}_{\alpha_4}^{(1)} \\ \otimes \mathsf{m}_{\alpha_1}^{(2)} \cdot \mathsf{m}_{\alpha_3}^{(2)} \cdot \mathsf{m}_{\alpha_2}^{(2)} \cdot \mathsf{m}_{\alpha_4}^{(2)} \otimes \mathsf{m}_{\alpha_1}^{(3)} \otimes \mathsf{m}_{\alpha_2}^{(3)} \otimes \mathsf{m}_{\alpha_3}^{(3)} \otimes \mathsf{m}_{\alpha_4}^{(3)} \otimes \mathsf{m}_{\alpha_1}^{(4)} \cdot \mathsf{m}_{\alpha_2}^{(4)} \cdot \mathsf{m}_{\alpha_3}^{(4)} \cdot \mathsf{m}_{\alpha_4}^{(4)}, \tag{9.5.77}$$

$$\sum_{\alpha} \mathsf{m}_{\alpha}^{(1)} \otimes \mathsf{m}_{\alpha}^{(2)} \otimes \mathsf{m}_{\alpha}^{(3)} \otimes (\Delta \otimes \Delta) \circ \Delta(\mathsf{m}_{\alpha}^{(4)}) = \sum_{\alpha_1,\alpha_2,\alpha_3,\alpha_4} \mathsf{m}_{\alpha_1}^{(1)} \cdot \mathsf{m}_{\alpha_3}^{(1)} \cdot \mathsf{m}_{\alpha_2}^{(1)} \cdot \mathsf{m}_{\alpha_4}^{(1)} \\ \otimes \mathsf{m}_{\alpha_1}^{(2)} \cdot \mathsf{m}_{\alpha_3}^{(2)} \cdot \mathsf{m}_{\alpha_2}^{(2)} \cdot \mathsf{m}_{\alpha_4}^{(2)} \otimes \mathsf{m}_{\alpha_1}^{(3)} \cdot \mathsf{m}_{\alpha_3}^{(3)} \cdot \mathsf{m}_{\alpha_2}^{(3)} \cdot \mathsf{m}_{\alpha_4}^{(3)} \otimes \mathsf{m}_{\alpha_1}^{(4)} \otimes \mathsf{m}_{\alpha_2}^{(4)} \otimes \mathsf{m}_{\alpha_3}^{(4)} \otimes \mathsf{m}_{\alpha_4}^{(4)}. \tag{9.5.78}$$

According to remark 9.5.35, we can express through the extended M-matrix (9.5.66) the first medial quasipolyangularity equation (9.5.75) and the last one (9.5.78) only, as follows

$$((\Delta \otimes \Delta) \circ \Delta \otimes \mathrm{id}_B \otimes \mathrm{id}_B \otimes \mathrm{id}_B)(\mathcal{M}^{(4)}) = \mathcal{M}_{1567}^{(7)} \cdot \mathcal{M}_{2567}^{(7)} \cdot \mathcal{M}_{3567}^{(7)} \cdot \mathcal{M}_{4567}^{(7)}, \tag{9.5.79}$$

$$(\mathrm{id}_B \otimes \mathrm{id}_B \otimes \mathrm{id}_B \otimes (\Delta \otimes \Delta) \circ \Delta)(\mathcal{M}^{(4)}) = \mathcal{M}_{1234}^{(7)} \cdot \mathcal{M}_{1236}^{(7)} \cdot \mathcal{M}_{1235}^{(7)} \cdot \mathcal{M}_{1237}^{(7)}. \tag{9.5.80}$$

The compatibility of (9.5.79)–(9.5.80) with the (binary) almost co-mediality (9.5.50) leads to

Proposition 9.5.37. *An extended binary M-matrix* (9.5.66) *of the binary almost co-medial bialgebra* $\mathcal{B}^{(2,2)} = \langle B | \mu = (\cdot), \Delta \rangle$ *satisfies the compatibility equations* (see (9.5.16))

$$\mathcal{M}_{1234}^{(7)} \cdot \mathcal{M}_{1567}^{(7)} \cdot \mathcal{M}_{2567}^{(7)} \cdot \mathcal{M}_{3567}^{(7)} \cdot \mathcal{M}_{4567}^{(7)} = \mathcal{M}_{1567}^{(7)} \cdot \mathcal{M}_{3567}^{(7)} \cdot \mathcal{M}_{2567}^{(7)} \cdot \mathcal{M}_{4567}^{(7)} \cdot \mathcal{M}_{1234}^{(7)}, \tag{9.5.81}$$

$$\mathcal{M}^{(7)}_{4567} \cdot \mathcal{M}^{(7)}_{1234} \cdot \mathcal{M}^{(7)}_{1236} \cdot \mathcal{M}^{(7)}_{1235} \cdot \mathcal{M}^{(7)}_{1237} = \mathcal{M}^{(7)}_{1234} \cdot \mathcal{M}^{(7)}_{1235} \cdot \mathcal{M}^{(7)}_{1236} \cdot \mathcal{M}^{(7)}_{1237} \cdot \mathcal{M}^{(7)}_{4567}. \tag{9.5.82}$$

Proof. The identities for the M-matrix

$$\begin{aligned}&\left(\mathscr{M}^{(4)} \otimes \mathrm{id}_B \otimes \mathrm{id}_B \otimes \mathrm{id}_B\right) \circ \left(\left(\Delta \otimes \Delta\right) \circ \Delta \otimes \mathrm{id}_B \otimes \mathrm{id}_B \otimes \mathrm{id}_B\right)\left(\mathscr{M}^{(4)}\right)\\ &= \left(\tau_{\mathrm{medial}} \circ (\Delta \otimes \Delta) \circ \Delta \otimes \mathrm{id}_B \otimes \mathrm{id}_B \otimes \mathrm{id}_B\right)\left(\mathscr{M}^{(4)}\right) \circ\\ &\quad \left(\mathscr{M}^{(4)} \otimes \mathrm{id}_B \otimes \mathrm{id}_B \otimes \mathrm{id}_B\right),\end{aligned} \tag{9.5.83}$$

$$\begin{aligned}&\left(\mathrm{id}_B \otimes \mathrm{id}_B \otimes \mathrm{id}_B \otimes \mathscr{M}^{(4)}\right) \circ \left(\mathrm{id}_B \otimes \mathrm{id}_B \otimes \mathrm{id}_B \otimes \left(\Delta \otimes \Delta\right) \circ \Delta\right)(\mathscr{M}^{(4)})\\ &= \left(\mathrm{id}_B \otimes \mathrm{id}_B \otimes \mathrm{id}_B \otimes \tau_{\mathrm{medial}} \circ \left(\Delta \otimes \Delta\right) \circ \Delta\right)\left(\mathscr{M}^{(4)}\right) \circ\\ &\quad \left(\mathrm{id}_B \otimes \mathrm{id}_B \otimes \mathrm{id}_B \otimes \mathscr{M}^{(4)}\right),\end{aligned} \tag{9.5.84}$$

follow from the almost co-mediality condition(9.5.50), and then we apply quasipolyangularity (9.5.79)–(9.5.80). □

***Remark* 9.5.38.** Two other compatibility equations corresponding to the intermediate quasipolyangularity equations (9.5.76)–(9.5.77) can be written in component form only (see remark 9.5.35).

The solutions to (9.5.81)–(9.5.82) can be found in matrix form by choosing an appropriate basis and using the standard methods (see, e.g., Lambe and Radford 1997; Kassel 1995).

References

Abe E 1980 *Hopf Algebras* (Cambridge: Cambridge University Press)

Aguiar M and Lauve A 2015 The characteristic polynomial of the Adams operators on graded connected Hopf algebras *Algebra Number Theory* **9** 547–83

Borowiec A, Dudek W and Duplij S 2001 Basic concepts of ternary Hopf algebras *J. Kharkiv Natl Univ. Ser. Nuclei Part. Fields* **529** 21–9 arXiv:math.QA/0306208

Bremner M and Hentzel I 2000 Identities for generalized Lie and Jordan products on totally associative triple systems *J. Algebra* **231** 387–405

Chari V and Pressley A 1996 *A Guide to Quantum Groups* (Cambridge: Cambridge University Press)

Curtis C W and Reiner I 1962 *Representation Theory of Finite Groups and Associative Algebras* (Providence, RI: American Mathematical Society)

de Azcarraga J and Izquierdo J M 2010 *n*-Ary algebras: a review with applications *J. Phys.* **A43** 293001

Dörnte W 1929 Unterschungen über einen verallgemeinerten Gruppenbegriff *Math. Z.* **29** 1–19

Drinfeld V G 1987 Quantum groups *Proceedings of the ICM, Berkeley* A Gleason (Providence, RI: American Mathematical Society) pp 798–820

Drinfeld V G 1989a On almost cocommutative Hopf algebras *Leningrad Math. J.* **1** 321–42

Drinfeld V G 1989b Quasi-Hopf algebras and Knizhnik–Zamolodchikov equations *Problems of Modern Quantum Field Theory* ed A A Belavin, A V Klimyk and A B Zamolodchikov (Heidelberg: Springer) pp 1–13

Dudek W A 2007 Remarks to Głazek's results on n-ary groups Discuss *Math. Gen. Algebra Appl.* **27** 199–233

Duplij S 1998 On semi-supermanifolds *Pure Math. Appl.* **9** 283–310

Duplij S 2001 Ternary Hopf algebras *Symmetry in Nonlinear Mathematical Physics* ed A G Nikitin and V M Boyko (Kyiv: Institute of Mathematics) pp 25–34

Duplij S 2012 Polyadic systems, representations and quantum groups *J. Kharkiv Natl Univ. Ser. Nuclei Part. Fields* **1017** 28–59 arXiv:math.RT/1308.4060

Duplij S 2017 Polyadic integer numbers and finite (m, n)-fields *p-Adic Numbers Ultrametric Anal. Appl.* **9** 267–91

Duplij S 2019 Arity shape of polyadic algebraic structures *J. Math. Phys. Anal. Geom.* **15** 3–56

Duplij S and Li F 2001 Regular solutions of quantum Yang-Baxter equation from weak Hopf algebras *Czech. J. Phys.* **51** 1306–11

Duplij S and Marcinek W 2000 On higher regularity and monoidal categories *Kharkiv State Univ. J. (Vestnik KSU), Ser. Nuclei, Particles and Fields* **481** 27–30

Duplij S and Marcinek W 2001 Semisupermanifolds and regularization of categories, modules, algebras and Yang–Baxter equation (Int. Conf. on Supersymmetry and Quantum Field Theory: D.V. Volkov Memorial Conference, Kharkiv, Ukraine, July 2000) *Nucl. Phys. Proc. Suppl.* **102** 293–7

Duplij S and Marcinek W 2002 Regular obstructed categories and topological quantum field theory *J. Math. Phys.* **43** 3329–41

Duplij S and Sinel'shchikov S 2009 Quantum enveloping algebras with von Neumann regular Cartan-like generators and the Pierce decomposition *Commun. Math. Phys.* **287** 769–85

Duplij S and Sinel'shchikov S 2010 Classification of $U_q(sl_2)$-module algebra structures on the quantum plane *J. Math. Phys. Anal. Geom.* **6** 21–46

Gal'mak A M 2003 n-*Ary Groups, Part 1* (Gomel: Gomel University)

Goze M and Rausch de Traubenberg M 2009 Hopf algebras for ternary algebras and groups *J. Math. Phys.* **50** 063508

Hazewinkel M, Gubareni N and Kirichenko V V 2010 *Algebras, Rings and Modules: Lie Algebras and Hopf Algebras* (Providence, RI: American Mathematical Society)

Heine E 1878 *Handbuch der Kugelfunktionan* (Berlin: Reimer)

Heyneman R G and Sweedler M E 1969 Affine Hopf algebras, I *J. Algebra* **13** 192–241

Kac V and Cheung P 2002 *Quantum Calculus* (New York: Springer)

Karaali G 2008 On Hopf algebras and their generalizations *Commun. Algebra* **36** 4341–67

Kashina Y, Montgomery S and Ng S-H 2012 On the trace of the antipode and higher indicators *Israel J. Math.* **188** 57–89

Kassel C 1995 *Quantum Groups* (New York: Springer)

Kirillov A A 1976 *Elements of the Theory of Representations* (Berlin: Springer)

Lambe L A and Radford D E 1997 *Introduction to the Quantum Yang-Baxter Equation and Quantum Groups: An Algebraic Approach* (Dordrecht: Kluwer)

Li F and Duplij S 2002 Weak Hopf algebras and singular solutions of quantum Yang-Baxter equation *Commun. Math. Phys.* **225** 191–217

Loday J-L and Vallette B 2012 *Algebraic Operads* (Heidelberg: Springer)

Madore J 1995 *Introduction to Noncommutative Geometry and Its Applications* (Cambridge: Cambridge University Press)

Majid S 1995 *Foundations of Quantum Group Theory* (Cambridge: Cambridge University Press)

Markl M, Shnider S and Stasheff J 2002 Operads in Algebra *Topology and Physics* (Providence, RI: American Mathematical Society)

Michor P W and Vinogradov A M 1996 *n*-Ary Lie and associative algebras *Rend. Sem. Mat. Univ. Pol. Torino* **54** 373–92

Post E L 1940 Polyadic groups *Trans. Am. Math. Soc.* **48** 208–350

Radford D E 2012 *Hopf Algebras* (Hackensack: World Scientific)

Shnider S and Sternberg S 1993 *Quantum Groups* (Boston, MA: International Press)

Sweedler M E 1968 The Hopf algebra of an algebra applied to field theory *J. Algebra* **8** 262–76

Sweedler M E 1969 *Hopf Algebras* (New York: Benjamin)

Szlachányi K 1996 Weak Hopf algebras *Operator Algebras and Quantum Field Theory* ed S Doplicher, R Longo, J E Roberts and L Zsidó (New York: International Press) pp 221–9

Takhtajan L 1994 On foundation of the generalized Nambu mechanics *Commun. Math. Phys.* **160** 295–315

Yokonuma T 1992 *Tensor Spaces and Exterior Algebra* (Providence, RI: American Mathematical Society)

Chapter 10

Solutions to higher braid equations

In this chapter we obtain and study the solutions to the higher arity (polyadic) braid equations introduced in Duplij (2021b) (see, also, Duplij 2021a), as a polyadic generalization of the constant Yang–Baxter equation (which is considerably different from the generalized Yang–Baxter equation of Rowell *et al* 2010, Kitaev and Wang 2012, Vasquez *et al* 2016, Padmanabhan *et al* 2020). We introduce special classes of matrices (star and circle types), to which most of the solutions belong, and find that the so-called magic matrices (Khaneja and Glaser 2001, Kraus and Cirac 2001, Ballard and Wu 2011) belong to the star class. We investigate their general non-trivial group properties and polyadic generalizations.

10.1 Yang–Baxter operators

Consider here the solutions to the *constant Yang–Baxter equation* (Lambe and Radford 1997) (called also *algebraic* Yang–Baxter equation, Dye 2003), or the (binary) *braid equation* (Duplij 2021b).

10.1.1 Yang–Baxter maps and braid group

First we introduce a general abstract construction of the (binary) braid equation (see section 9.5). Let $\mathcal{V} = \mathcal{V}^{(2,2,2,1)} = \left\langle K, V \middle| \nu_k^{(2)}, \mu_k^{(2)}, \mu_V^{(2)}; \rho^{(1)} \right\rangle$ be a (binary) vector space over a (binary) field $\Bbbk = \Bbbk^{(2,2)} = \left\langle K \middle| \nu_k^{(2)}, \mu_k^{(2)} \right\rangle$ (see definition 5.1.6) and the mapping $\mathbf{c}_{V^2}\colon V \otimes V \to V \otimes V$, where $\otimes = \otimes_{\Bbbk}$ is the tensor product over $\Bbbk$. A linear operator (*braid operator*) $\mathbf{c}_{V^2}$ is called a *Yang–Baxter operator* (denoted by R in Kauffman and Lomonaco 2004 and by B in Lambe and Radford 1997) or *Yang–Baxter map* (Veselov 2003) (denoted by F in Bukhshtaber 1998), if it satisfies the *braid equation* (Drinfeld 1989, 1992, Kassel 1995)

$$(\mathbf{c}_{V^2} \otimes \mathrm{id}_V) \circ (\mathrm{id}_V \otimes \mathbf{c}_{V^2}) \circ (\mathbf{c}_{V^2} \otimes \mathrm{id}_V) = (\mathrm{id}_V \otimes \mathbf{c}_{V^2}) \circ (\mathbf{c}_{V^2} \otimes \mathrm{id}_V) \circ (\mathrm{id}_V \otimes \mathbf{c}_{V^2}), \tag{10.1.1}$$

where id_V: $V \to V$, is the identity operator in $\mathcal{V}$ (see section 9.5).

Let us introduce the operators $\mathrm{A}_{1,2}$: $V \otimes V \otimes V \to V \otimes V \otimes V$ by

$$\mathrm{A}_1 = \mathbf{c}_{V^2} \otimes \mathrm{id}_V, \quad \mathrm{A}_2 = \mathrm{id}_V \otimes \mathbf{c}_{V^2}. \tag{10.1.2}$$

Using (10.1.2), it follows from (10.1.1) that

$$\mathrm{A}_1 \circ \mathrm{A}_2 \circ \mathrm{A}_1 = \mathrm{A}_2 \circ \mathrm{A}_1 \circ \mathrm{A}_2. \tag{10.1.3}$$

If $\mathbf{c}_{V^2}$ is invertible, then $\mathbf{c}_{V^2}^{-1}$ is also the Yang–Baxter map with A_1^{-1} and A_2^{-1}. Therefore, the operators A_i represent the braid group $\mathcal{B}_3 = \{e, \sigma_1, \sigma_2 | \sigma_1\sigma_2\sigma_1 = \sigma_2\sigma_1\sigma_2\}$ by the mapping π_3 as

$$\mathcal{B}_3 \xrightarrow{\pi_3} \mathrm{End}(V \otimes V \otimes V), \; \sigma_1 \overset{\pi_3}{\mapsto} \mathrm{A}_1, \; \sigma_2 \overset{\pi_3}{\mapsto} \mathrm{A}_2, \; e \overset{\pi_3}{\mapsto} \mathrm{id}_V. \tag{10.1.4}$$

The representation π_m of the braid group with m strands

$$\mathcal{B}_m = \left\{e, \sigma_1, \dots, \sigma_{m-1} \;\middle|\; \begin{array}{c} \sigma_i\sigma_{i+1}\sigma_i = \sigma_{i+1}\sigma_i\sigma_{i+1}, \; i = 1, \dots, m-1, \\ \sigma_i\sigma_j = \sigma_j\sigma_i, \; |\, i - j \,| \geqslant 2, \end{array}\right\} \tag{10.1.5}$$

can be obtained using operators $\mathrm{A}_i(m)$: $V^{\otimes m} \to V^{\otimes m}$ analogous to (10.1.2)

$$\begin{aligned} \mathrm{A}_i(m) &= \overbrace{\mathrm{id}_V \otimes \dots \otimes \mathrm{id}_V}^{i-1} \otimes \mathbf{c}_{V^2} \otimes \overbrace{\mathrm{id}_V \otimes \dots \mathrm{id}_V}^{m-i-1}, \\ \mathrm{A}_0(m) &= (\mathrm{id}_V)^{\otimes m}, \; i = 1, \dots, m-1, \end{aligned} \tag{10.1.6}$$

by the mapping π_m: $\mathcal{B}_m \to \mathrm{End}\ \mathrm{V}^{\otimes m}$ in the following way

$$\pi_m(\sigma_i) = \mathrm{A}_i(m), \quad \pi_m(e) = \mathrm{A}_0(m). \tag{10.1.7}$$

In this notation (10.1.2) is $\mathrm{A}_i = \mathrm{A}_i(2)$, $i = 1, 2$, and therefore (10.1.3) represents $\mathcal{B}_3$ by (10.1.4).

10.1.2 Constant matrix solutions to the Yang–Baxter equation

Consider next a concrete version of the vector space V which is used in the quantum computation, a d-dimensional Euclidean vector space V_d over complex numbers $\mathbb{C}$ with a basis $\{e_i\}$, $i = 1, \dots, d$. A linear operator $V_d \to V_d$ is given by a complex $d \times d$ matrix, the identity operator id_V becomes the identity $d \times d$ matrix I_d, and the Yang–Baxter map $\mathbf{c}_{V^2}$ is a $d^2 \times d^2$ matrix C_{d^2} (denoted by R in Dye 2003) satisfying the matrix algebraic Yang–Baxter equation

$$(C_{d^2} \otimes I_d)(I_d \otimes C_{d^2})(C_{d^2} \otimes I_d) = (I_d \otimes C_{d^2})(C_{d^2} \otimes I_d)(I_d \otimes C_{d^2}), \tag{10.1.8}$$

being an equality between two matrices of size $d^3 \times d^3$. We use the unified notations which can be straightforwardly generalized for higher braid operators. In components

$$C_{d^2} \circ (e_{i_1} \otimes e_{i_2}) = \sum_{j_1', j_2'=1}^{d} c_{i_1 i_2}{}^{j_1' j_2'} \cdot e_{j_1'} \otimes e_{j_2'}, \tag{10.1.9}$$

the Yang–Baxter equation (10.1.8) has the shape (where summing is by primed indices)

$$\sum_{j_1', j_2', j_3'=1}^{d} c_{i_1 i_2}{}^{j_1' j_2'} \cdot c_{j_2' i_3}{}^{j_3' k_3} \cdot c_{j_1' j_3'}{}^{k_1 k_2} = \sum_{l_1', l_2', l_3'=1}^{d} c_{i_2 i_3}{}^{l_2' l_3'} \cdot c_{i_1 l_2'}{}^{k_1 l_1'} \cdot c_{l_1' l_3'}{}^{k_2 k_3} \equiv q_{i_1 i_2 i_3}{}^{k_1 k_2 k_3}. \tag{10.1.10}$$

The system (10.1.10) is highly overdetermined, because the matrix C_{d^2} contains d^4 unknown entries, while there are d^6 cubic polynomial equations for them. So for $d = 2$ we have 64 equations for 16 unknowns, while for $d = 3$ there are 729 equations for the 81 unknown entries of C_{d^2}. The unitarity of C_{d^2} imposes a further d^2 quadratic equations, and so for $d = 2$ we have in total 68 equations for 16 unknowns. This makes the direct discovery of solutions for the matrix Yang–Baxter equation (10.1.10) very cumbersome. Nevertheless, using a conjugation classes method, the unitary solutions and their classification for $d = 2$ were presented in Dye (2003).

In the standard matrix form (10.1.9) can be presented by introducing the 4-dimensional vector space $\tilde{V}_4 = V \otimes V$ with the natural basis $\tilde{e}_{\tilde{k}} = \{e_1 \otimes e_1, e_1 \otimes e_2, e_2 \otimes e_1, e_2 \otimes e_2\}$, where $\tilde{k} = 1, \ldots, 8$ is a cumulative index. The linear operator $\tilde{C}_4$: $\tilde{V}_4 \to \tilde{V}_4$ corresponding to (10.1.9) is given by 4×4 matrix $\tilde{c}_{\tilde{i}\tilde{j}}$ as $\tilde{C}_4 \circ \tilde{e}_{\tilde{i}} = \sum_{\tilde{j}=1}^{4} \tilde{c}_{\tilde{i}\tilde{j}} \cdot \tilde{e}_{\tilde{j}}$. The operators (10.1.2) become two 8×8 matrices $\tilde{A}_{1,2}$ as

$$\tilde{A}_1 = \tilde{c} \otimes_K I_2, \; \tilde{A}_2 = I_2 \otimes_K \tilde{c}, \tag{10.1.11}$$

where $\otimes_K$ is the Kronecker product of matrices and I_2 is the 2×2 identity matrix. In this notation (which is universal and also used for higher braid equations) the operator binary braid equations (10.2.7) become a single matrix equation

$$\tilde{A}_1 \tilde{A}_2 \tilde{A}_1 = \tilde{A}_2 \tilde{A}_1 \tilde{A}_2, \tag{10.1.12}$$

which we call the *matrix binary braid equation* (and also the constant Yang–Baxter equation, Dye 2003). In component form (10.1.12) is a highly overdetermined system of 64 cubic equations for 16 unknowns, the entries of $\tilde{c}$.

The matrix equation (10.1.12) has the following 'gauge invariance', which allows a classification of Yang–Baxter maps (Hietarinta 1993). Introduce an invertible operator Q: $V \to V$ in the two-dimensional vector space $V \equiv V_{d=2}$. In the basis $\{e_1, e_2\}$ its 2×2 matrix q is given by $Q \circ e_i = \sum_{j=1}^{2} q_{ij} \cdot e_j$. In the natural 4-dimensional basis $\tilde{e}_{\tilde{k}}$ the tensor product of operators $Q \otimes Q$ is presented by the Kronecker product of matrices $\tilde{q}_4 = q \otimes_K q$. If the 4×4 matrix $\tilde{c}$ is a fixed solution to the Yang–Baxter equation (10.1.12), then the family of solutions $\tilde{c}(q)$ corresponding to the invertible 2×2 matrix q is the conjugation of $\tilde{c}$ by $\tilde{q}_4$ such that

$$\tilde{c}(q) = \tilde{q}_4 \tilde{c} \tilde{q}_4^{-1} = (q \otimes_K q) \tilde{c} (q^{-1} \otimes_K q^{-1}), \tag{10.1.13}$$

which follows from conjugating (10.1.12) by $q \otimes_K q$ and using (10.1.11). If we include the obvious invariance of (10.1.12) with respect to an overall factor $t \in \mathbb{C}$, the general family of solutions becomes (see the Yang–Baxter equation, Hietarinta 1993)

$$\tilde{c}(q, t) = t\tilde{q}_4\tilde{c}\tilde{q}_4^{-1} = t(q \otimes_K q)\tilde{c}(q^{-1} \otimes_K q^{-1}). \tag{10.1.14}$$

It follows from (10.1.13) that the matrix $q \in \mathrm{GL}(2, \mathbb{C})$ is defined up to a complex nonzero factor. In this case we can put

$$q = \begin{pmatrix} a & 1 \\ c & d \end{pmatrix}, \tag{10.1.15}$$

and the manifest form of $\tilde{q}_4$ is

$$\tilde{q}_4 = \begin{pmatrix} a^2 & a & a & 1 \\ ac & ad & c & d \\ ac & c & ad & d \\ c^2 & cd & cd & d^2 \end{pmatrix}. \tag{10.1.16}$$

The matrix $\tilde{q}_4^\star\tilde{q}_4$ (where $\star$ represents Hermitian conjugation) is diagonal (this case is important in a further classification similar to the binary one, Dye 2003), when the condition

$$c = -a/d^* \tag{10.1.17}$$

holds, and so the matrix q takes the special form (depending on 2 complex parameters)

$$q = \begin{pmatrix} a & 1 \\ -a/d^* & d \end{pmatrix}. \tag{10.1.18}$$

We call two solutions $\tilde{c}_1$ and $\tilde{c}_2$ of the constant Yang–Baxter equation (10.1.12) *q-conjugated*, if

$$\tilde{c}_1\tilde{q}_4 = \tilde{q}_4\tilde{c}_2, \tag{10.1.19}$$

and we will not distinguish between them. The q-conjugation in the form (10.1.19) does not require the invertibility of the matrix q, and therefore the solutions of different ranks (or invertible and not invertible) can be q-conjugated (for the invertible case, see Hietarinta 1993, Alagic *et al* 2014, Padmanabhan *et al* 2021).

The matrix equation (10.1.12) does not imply the invertibility of solutions, i.e., matrices $\tilde{c}$ being of full rank (in the binary Yang–Baxter case of rank 4 and $d = 2$). Below we introduce in a unified way invertible and non-invertible solutions to the matrix Yang–Baxter equation (10.1.10) for any rank of the corresponding matrices.

10.1.3 Partial identity and unitarity

To be as close as possible to the invertible case, we introduce 'non-invertible analogs' of identity and unitarity. Let M be a diagonal $n \times n$ matrix of rank $r \leqslant n$, and therefore with $n - r$ zeroes on the diagonal. If the other diagonal elements are units, such a diagonal M can be reduced by row operations to a block matrix, being a direct sum of the identity matrix $I_{r\times r}$ and the zero matrix $Z_{(n-r)\times(n-r)}$. We call such a diagonal matrix a *block r-partial identity* $I_n^{(\text{block})}(r) = \text{diag}\left\{\overbrace{1, \ldots, 1}^{r}, \overbrace{0, \ldots, 0}^{n-r}\right\}$, and without the block reduction—a *shuffle r-partial identity* $I_n^{(\text{shuffle})}(r)$ (these are connected by conjugation). We will use the term partial identity and $I_n(r)$ to denote any matrix of this form. Obviously, with the full rank $r = n$ we have $I_n(n) \equiv I_n$, where I_n is the identity $n \times n$ matrix. As with the invertible case and identities, the partial identities (of the corresponding form) are *trivial solutions* of the Yang–Baxter equation.

If a matrix $M = M(r)$ of size $n \times n$ and rank r satisfies the following *r-partial unitarity* condition

$$M(r)^\star M(r) = I_n^{(1)}(r), \tag{10.1.20}$$

$$M(r)M(r)^\star = I_n^{(2)}(r), \tag{10.1.21}$$

where $M(r)^\star$ is the conjugate-transposed matrix and $I_n^{(1)}(r)$, $I_n^{(2)}(r)$ are partial identities (of any kind, they can be different), then $M(r)$ is called a *r-partial unitary matrix*. In the case, when $I_n^{(1)}(r) = I_n^{(2)}(r)$, the matrix $M(r)$ is called *normal*. If $M(r)^\star = M(r)$, then it is called *r-partial self-adjoint*. In the case of full rank $r = n$, the conditions (10.1.20)–(10.1.21) become ordinary unitarity, and $M(n)$ becomes an unitary (and normal) matrix, while a r-partial self-adjoint matrix becomes a self-adjoint matrix or Hermitian matrix.

As an example, we consider a 4×4 matrix of rank 3

$$M(3) = \begin{pmatrix} 0 & 0 & 0 & 0 \\ 0 & e^{i\beta} & 0 & 0 \\ 0 & 0 & 0 & e^{i\gamma} \\ e^{i\alpha} & 0 & 0 & 0 \end{pmatrix}, \quad \alpha, \beta, \gamma \in \mathbb{R}, \tag{10.1.22}$$

which satisfies the 3-partial unitarity conditions (10.1.20)–(10.1.21) with two different 3-partial identities on the rhs

$$\begin{aligned} M(3)^\star M(3) &= \begin{pmatrix} 1 & 0 & 0 & 0 \\ 0 & 1 & 0 & 0 \\ 0 & 0 & 0 & 0 \\ 0 & 0 & 0 & 1 \end{pmatrix} = I_4^{(1)}(3) \\ &\neq I_4^{(2)}(3) = \begin{pmatrix} 0 & 0 & 0 & 0 \\ 0 & 1 & 0 & 0 \\ 0 & 0 & 1 & 0 \\ 0 & 0 & 0 & 1 \end{pmatrix} = M(3)M(3)^\star. \end{aligned} \tag{10.1.23}$$

For a non-invertible matrix $M(r)$ one can define a *pseudoinverse* $M(r)^+$ (or the *Moore–Penrose inverse*) (Nashed 1976) by

$$M(r)M(r)^+M(r) = M(r), \; M(r)^+M(r)M(r)^+ = M(r)^+, \tag{10.1.24}$$

and $M(r)M(r)^{+}$, $M(r)^{+}M(r)$ are Hermitian. In the case of (10.1.22) the partial unitary matrix $M(3)$ coincides with its pseudoinverse

$$M(3)^{\star} = M(3)^{+}, \tag{10.1.25}$$

which is similar to the standard unitarity $M_{\mathrm{inv}}^{\star} = M_{\mathrm{inv}}^{-1}$ for an invertible matrix M_{inv}. It is important that (10.1.22) is a solution of the matrix Yang–Baxter equation (10.1.12), and so is an example of a non-invertible Yang–Baxter map.

If only the first (second) of the conditions (10.1.20)–(10.1.21) holds, we call such $M(r)$ a *left (right) r-partial unitary matrix*. An example of such a non-invertible Yang–Baxter map of rank 2 is the left 2-partial unitary matrix

$$M(2) = \frac{1}{\sqrt{2}}\begin{pmatrix} 0 & 0 & 0 & e^{i\alpha} \\ 0 & e^{i\beta} & 0 & 0 \\ 0 & e^{i\beta} & 0 & 0 \\ 0 & 0 & 0 & e^{i\beta} \end{pmatrix}, \quad \alpha, \beta \in \mathbb{R}, \tag{10.1.26}$$

which satisfies (10.1.20), but not (10.1.21), and so $M(2)$ is not normal

$$M(2)^{\star}M(2) = \begin{pmatrix} 0 & 0 & 0 & 0 \\ 0 & 1 & 0 & 0 \\ 0 & 0 & 0 & 0 \\ 0 & 0 & 0 & 1 \end{pmatrix} \neq \begin{pmatrix} 1 & 0 & 0 & e^{i(\alpha-\beta)} \\ 0 & 1 & 1 & 0 \\ 0 & 1 & 1 & 0 \\ e^{i(\beta-\alpha)} & 0 & 0 & 1 \end{pmatrix} = M(2)M(2)^{\star}. \tag{10.1.27}$$

Nevertheless, the property (10.1.25) still holds and $M(2)^{\star} = M(2)^{+}$.

10.1.4 Permutation and parameter-permutation 4-vertex Yang–Baxter maps

The system (10.1.12) with respect to all 16 variables is too cumbersome for direct solution. The classification of all solutions can only be accomplished in special cases, e.g., for matrices over finite fields (Hietarinta 1993) or for fewer than 16 vertices. Here we will start from 4-vertex permutation and parameter-permutation matrix solutions and investigate their group structure. It was shown (Dye 2003, Kauffman and Lomonaco 2004) that the special 8-vertex solutions to the Yang–Baxter equation are most important for further applications including braiding gates. We will therefore study the 8-vertex solutions in the most general way: over $\mathbb{C}$ and in various configurations, invertible and not invertible, and also consider their group structure.

First, we introduce the *permutation Yang–Baxter maps* which are presented by the permutation matrices (binary matrices with a single 1 in each row and column), i.e., 4-vertex solutions. In total, there are 64 permutation matrices of size 4×4, while only 4 of them have the full rank 4 and simultaneously satisfy the Yang–Baxter equation (10.1.12). These are the following

$$\tilde{c}_{\mathrm{bisymm}}^{\mathrm{perm}} = \begin{pmatrix} 1 & 0 & 0 & 0 \\ 0 & 0 & 1 & 0 \\ 0 & 1 & 0 & 0 \\ 0 & 0 & 0 & 1 \end{pmatrix}, \begin{pmatrix} 0 & 0 & 0 & 1 \\ 0 & 1 & 0 & 0 \\ 0 & 0 & 1 & 0 \\ 1 & 0 & 0 & 0 \end{pmatrix}, \quad \begin{matrix} \mathrm{tr}\,\tilde{c} = 2, \ \det \tilde{c} = -1, \\ \text{eigenvalues: } \{1\}^{[2]}, \{-1\}^{[2]}, \end{matrix} \tag{10.1.28}$$

$$\tilde{c}^{\text{perm}}_{90\text{symm}} = \begin{pmatrix} 0 & 1 & 0 & 0 \\ 0 & 0 & 0 & 1 \\ 1 & 0 & 0 & 0 \\ 0 & 0 & 1 & 0 \end{pmatrix}, \begin{pmatrix} 0 & 0 & 1 & 0 \\ 1 & 0 & 0 & 0 \\ 0 & 0 & 0 & 1 \\ 0 & 1 & 0 & 0 \end{pmatrix}, \quad \begin{array}{c} \text{tr}\tilde{c} = 0, \det \tilde{c} = -1, \\ \text{eigenvalues: } 1, i, -1, -i. \end{array} \tag{10.1.29}$$

Here and next we list eigenvalues to understand which matrices are conjugated, and after that, if and only if the conjugation matrix is of the form (10.1.16), then such solutions to the Yang–Baxter equation (10.1.12) coincide. The traces are important in the construction of corresponding link invariants (Turaev 1988) and local invariants (Balakrishnan and Sankaranarayanan 2010, Sudbery 2001), and determinants are connected with the concurrence (Walter *et al* 2016, Jaffali and Oeding 2020). Note that the first matrix in (10.1.28) is the SWAP quantum gate (Nielsen and Chuang 2000).

To understand symmetry properties of (10.1.28)–(10.1.29), we introduce the so-called *reverse matrix* $J \equiv J_n$ of size $n \times n$ by $(J_n)_{ij} = \delta_{i,n+1-i}$. For $n = 4$ it is

$$J_4 = \begin{pmatrix} 0 & 0 & 0 & 1 \\ 0 & 0 & 1 & 0 \\ 0 & 1 & 0 & 0 \\ 1 & 0 & 0 & 0 \end{pmatrix}. \tag{10.1.30}$$

For any $n \times n$ matrix $M \equiv M_n$ the matrix JM is the matrix M reflected vertically, and the product MJ is M reflected horizontally. In addition to the standard *symmetric matrix* satisfying $M = M^T$ (T is the transposition), one can introduce

$$M \text{ is per symmetric, if } JM = (JM)^T, \tag{10.1.31}$$

$$M \text{ is } 90°\text{-symmetric, if } M^T = JM. \tag{10.1.32}$$

Thus, a persymmetric matrix is symmetric with respect to the minor diagonal, while a 90°-symmetric matrix is symmetric under 90°-rotations. A *bisymmetric matrix* is symmetric and persymmetric simultaneously. In this notation, the first family of the permutation solutions (10.1.28) are bisymmetric, but not 90°-symmetric, while the second family of the solutions (10.1.29) are, oppositely, 90°-symmetric, but not symmetric and not persymmetric (which explains their notation).

In the next step, we define the corresponding *parameter-permutation solutions* replacing the units in (10.1.28) by parameters. We found the following four 4-vertex solutions to the Yang–Baxter equation (10.1.12) over $\mathbb{C}$

$$\tilde{c}^{\text{perm,star}}_{\text{rank}=4}(x, y, z, t) = \begin{pmatrix} x & 0 & 0 & 0 \\ 0 & 0 & y & 0 \\ 0 & z & 0 & 0 \\ 0 & 0 & 0 & t \end{pmatrix}, \begin{pmatrix} 0 & 0 & 0 & y \\ 0 & x & 0 & 0 \\ 0 & 0 & t & 0 \\ z & 0 & 0 & 0 \end{pmatrix}, \quad \begin{array}{c} \text{tr}\tilde{c} = x + t, \\ \det \tilde{c} = -xyzt,\ x, y, z, t \neq 0, \\ \text{eigenvalues: } x, t, \sqrt{yz}, -\sqrt{yz}, \end{array} \tag{10.1.33}$$

$$\tilde{c}^{\text{perm,circ}}_{\text{rank}=4}(x, y) = \begin{pmatrix} 0 & 0 & x & 0 \\ y & 0 & 0 & 0 \\ 0 & 0 & 0 & x \\ 0 & y & 0 & 0 \end{pmatrix}, \begin{pmatrix} 0 & x & 0 & 0 \\ 0 & 0 & 0 & x \\ y & 0 & 0 & 0 \\ 0 & 0 & y & 0 \end{pmatrix}, \quad \begin{array}{c} \text{tr}\tilde{c} = 0, \\ \det \tilde{c} = -x^2y^2,\ x, y \neq 0, \\ \text{eigenvalues: } \sqrt{xy}, -\sqrt{xy}, i\sqrt{xy}, -i\sqrt{xy}. \end{array} \tag{10.1.34}$$

The first pair of solutions (10.1.33) correspond to the bisymmetric permutation matrices (10.1.28), and we call them *star-like solutions*, while the second two solutions (10.1.34) correspond to the 90°-symmetric matrices (10.1.28) which are called *circle-like solutions*.

The first (second) star-like solution in (10.1.33) with $y = z$ ($x = t$) becomes symmetric (persymmetric), while on the other hand with $x = t$ ($y = z$) it becomes persymmetric (symmetric). They become bisymmetric parameter-permutation solutions if all the parameters are equal $x = y = z = t$. The circle-like solutions (10.1.34) are 90°-symmetric when $x = y$.

Using q-conjugation (10.1.14) one can next get families of solutions depending from the entries of q, the additional complex parameters in (10.1.15).

10.1.5 Group structure of 4-vertex and 8-vertex matrices

Let us analyze the group structure of 4-vertex matrices (10.1.33)–(10.1.34) with respect to matrix multiplication, i.e., which kinds of subgroups in GL(4, $\mathbb{C}$) they can form. For this we introduce four 4-vertex 4×4 matrices over $\mathbb{C}$: two star-like matrices

$$N_{\text{star1}} = \begin{pmatrix} x & 0 & 0 & 0 \\ 0 & 0 & y & 0 \\ 0 & z & 0 & 0 \\ 0 & 0 & 0 & t \end{pmatrix}, \quad N_{\text{star2}} = \begin{pmatrix} 0 & 0 & 0 & y \\ 0 & x & 0 & 0 \\ 0 & 0 & t & 0 \\ z & 0 & 0 & 0 \end{pmatrix}, \quad \begin{array}{c} \text{tr} N = x + t, \\ \det N = -xyzt,\ x, y, z, t \neq 0, \\ \text{eigenvalues: } x,\ t,\ \sqrt{yz},\ -\sqrt{yz}, \end{array} \tag{10.1.35}$$

and two circle-like matrices

$$N_{\text{circ1}} = \begin{pmatrix} 0 & 0 & x & 0 \\ y & 0 & 0 & 0 \\ 0 & 0 & 0 & z \\ 0 & t & 0 & 0 \end{pmatrix}, \quad N_{\text{circ2}} = \begin{pmatrix} 0 & x & 0 & 0 \\ 0 & 0 & 0 & y \\ z & 0 & 0 & 0 \\ 0 & 0 & t & 0 \end{pmatrix}, \quad \begin{array}{c} \text{tr} N = 0, \\ \det N = -xyzt,\ x, y, z, t \neq 0, \\ \text{eigenvalues:} \\ \sqrt[4]{xyzt},\ -\sqrt[4]{xyzt},\ i\sqrt[4]{xyzt},\ -\sqrt[4]{xyzt}, \end{array}. \tag{10.1.36}$$

Denoting the corresponding sets by $\mathsf{N}_{\text{star1,2}} = \{N_{\text{star1,2}}\}$ and $\mathsf{N}_{\text{circ1,2}} = \{N_{\text{circ1,2}}\}$, these do not intersect and are closed with respect to the following multiplications

$$\mathsf{N}_{\text{star1}}\mathsf{N}_{\text{star1}}\mathsf{N}_{\text{star1}} = \mathsf{N}_{\text{star1}}, \tag{10.1.37}$$

$$\mathsf{N}_{\text{star2}}\mathsf{N}_{\text{star2}}\mathsf{N}_{\text{star2}} = \mathsf{N}_{\text{star2}}, \tag{10.1.38}$$

$$\mathsf{N}_{\text{circ1}}\mathsf{N}_{\text{circ1}}\mathsf{N}_{\text{circ1}}\mathsf{N}_{\text{circ1}}\mathsf{N}_{\text{circ1}} = \mathsf{N}_{\text{circ1}}, \tag{10.1.39}$$

$$\mathsf{N}_{\text{circ2}}\mathsf{N}_{\text{circ2}}\mathsf{N}_{\text{circ2}}\mathsf{N}_{\text{circ2}}\mathsf{N}_{\text{circ2}} = \mathsf{N}_{\text{circ2}}. \tag{10.1.40}$$

Note that there are no closed binary multiplications among the sets of 4-vertex matrices (10.1.35)–(10.1.36).

To give a proper group interpretation of (10.1.37)–(10.1.40), we introduce a *k-ary (polyadic) general linear semigroup* GLS $^{[k]}(n, \mathbb{C}) = \{\mathsf{M}_{\text{full}} | \mu^{\circ(k-1)}\}$, where $\mathsf{M}_{\text{full}} = \{M_{n \times n}\}$ is the set of $n \times n$ matrices over $\mathbb{C}$ and μ is an ordinary matrix multiplication. The full semigroup GLS$^{[k]}(n, \mathbb{C})$ is *derived* in the sense that its product can be obtained by repeating the binary products that are (binary) closed at each step. However, $n \times n$ matrices of special shape can form k-ary subsemigroups of $\mathrm{GLS}^{[k]}(n, \mathbb{C})$, which can be closed with respect to the product of at minimum k matrices, but not of 2 matrices, and we call such semigroups *k-non-derived.* Moreover, we have for the sets $\mathsf{N}_{\mathrm{star}1,2}$ and $\mathsf{N}_{\mathrm{circ}1,2}$

$$\begin{aligned} \mathsf{M}_{\mathrm{full}} &= \mathsf{N}_{\mathrm{star}1} \bigcup \mathsf{N}_{\mathrm{star}2} \bigcup \mathsf{N}_{\mathrm{circ}1} \bigcup \mathsf{N}_{\mathrm{circ}2}, \\ & \mathsf{N}_{\mathrm{star}1} \bigcap \mathsf{N}_{\mathrm{star}2} \bigcap \mathsf{N}_{\mathrm{circ}1} \bigcap \mathsf{N}_{\mathrm{circ}2} = \varnothing. \end{aligned} \tag{10.1.41}$$

A simple example of a 3-nonderived subsemigroup of the full semigroup $\mathrm{GLS}^{[k]}(n, \mathbb{C})$ is the set of antidiagonal matrices $\mathsf{M}_{\mathrm{adiag}} = \{M_{\mathrm{adiag}}\}$ (having nonzero elements on the minor diagonal only): the product of 3 matrices from $\mathsf{M}_{\mathrm{adiag}}$ is closed, and therefore $\mathsf{M}_{\mathrm{adiag}}$ is a subsemigroup $\mathcal{S}^{(3)}_{\mathrm{adiag}} = \{\mathsf{M}_{\mathrm{adiag}} | \mu^{\circ 2}\}$ of the full ternary general linear semigroup $\mathrm{GLS}^{(3)}(n, \mathbb{C})$ with the multiplication as the ordinary triple matrix product.

In the theory of polyadic groups (Dörnte 1929) an analog of the binary inverse M^{-1} is given by the *querelement*, which is denoted by $\bar{M}$ and in the matrix k-ary case is defined by

$$\overbrace{M \ldots M}^{k-1} \bar{M} = M, \tag{10.1.42}$$

where $\bar{M}$ can be on any place. If each element of the k-ary semigroup $\mathrm{GLS}^{[k]}(n, \mathbb{C})$ (or its subsemigroup) has its querelement $\bar{M}$, then this semigroup is a *k-ary general linear group* $\mathrm{GL}^{[k]}(n, \mathbb{C})$.

In the set of $n \times n$ matrices the binary (ordinary) product is defined (even it is not closed), and for invertible matrices we formally determine the standard inverse M^{-1}, but for arity $k \geqslant 4$ it does not coincide with the querelement $\bar{M}$, because, as follows from (10.1.42) and cancellativity in $\mathbb{C}$ that

$$\bar{M} = M^{2-k}. \tag{10.1.43}$$

The *k-ary (polyadic) identity* $I_n^{[k]}$ in $\mathrm{GLS}^{[k]}(n, \mathbb{C})$ is defined by

$$\overbrace{I_n^{[k]} \ldots I_n^{[k]}}^{k-1} M = M, \tag{10.1.44}$$

which holds when M in the lhs is on any place. If M is only on one or another side (but not in the middle places) in (10.1.44), $I_n^{[k]}$ is called *left (right) polyadic identity.* For instance, in the subsemigroup (in $\mathrm{GLS}^{[k]}(n, \mathbb{C})$) of antidiagonal matrices $\mathcal{S}^{(3)}_{\mathrm{adiag}}$ the ternary identity $I_n^{(3)}$ can be chosen as the $n \times n$ reverse matrix (10.1.30) having units on the minor diagonal, while the ordinary $n \times n$ unit matrix I_n is not in $\mathcal{S}^{(3)}_{\mathrm{adiag}}$. It follows from (10.1.44), that for matrices over $\mathbb{C}$ the (left, right) polyadic identity $I_n^{[k]}$ is

$$\left(I_n^{[k]}\right)^{k-1} = I_n, \tag{10.1.45}$$

which means that for the ordinary matrix product $I_n^{[k]}$ is a $(k-1)$-root of I_n (or $I_n^{[k]}$ is a reflection of $(k-1)$ degree), while both sides cannot belong to a subsemigroup $\mathcal{S}^{[k]}$ of $\mathrm{GLS}^{[k]}(n, \mathbb{C})$ under consideration (as in $\mathcal{S}_{\mathrm{adiag}}^{(3)}$). As the solutions of (10.1.45) are not unique, there can be many k-ary identities in a k-ary matrix semigroup. We denote the set of k-ary identities by $\mathsf{I}_n^{[k]} = \{I_n^{[k]}\}$. In the case of $\mathcal{S}_{\mathrm{adiag}}^{(3)}$ the ternary identity $I_n^{(3)}$ can be chosen as any of the $n \times n$ reverse matrices (10.1.30) with unit complex numbers $e^{i\alpha_j}$, $j = 1, \ldots, n$ on the minor diagonal, where α_j satisfy additional conditions depending on the semigroup. In the concrete case of $\mathcal{S}_{\mathrm{adiag}}^{(3)}$ the conditions, giving (10.1.45), are $(k-1)\alpha_j = 1 + 2\pi r_j$, $r_j \in \mathbb{Z}$, $j = 1, \ldots, n$.

In the framework of the above definitions, we can interpret the closed products (10.1.37)–(10.1.38) as the multiplications of the *ternary semigroups* $\mathcal{S}_{\mathrm{star1,\,2}}^{(3)}(4, \mathbb{C}) = \{\mathsf{N}_{\mathrm{star1,2}} | \mu^{\circ 2}\}$. The corresponding querelements are given by

$$\bar{N}_{\mathrm{star1}} = N_{\mathrm{star1}}^{-1} = \begin{pmatrix} \frac{1}{x} & 0 & 0 & 0 \\ 0 & 0 & \frac{1}{z} & 0 \\ 0 & \frac{1}{y} & 0 & 0 \\ 0 & 0 & 0 & \frac{1}{t} \end{pmatrix}, \quad \bar{N}_{\mathrm{star2}} = N_{\mathrm{star2}}^{-1} = \begin{pmatrix} 0 & 0 & 0 & \frac{1}{z} \\ 0 & \frac{1}{x} & 0 & 0 \\ 0 & 0 & \frac{1}{t} & 0 \\ \frac{1}{y} & 0 & 0 & 0 \end{pmatrix}, \quad x, y, z, t \neq 0. \quad (10.1.46)$$

The ternary semigroups having querelements for each element (i.e., the additional operation $\overline{(\ )}$ defined by (10.1.46)) are the *ternary groups* $\mathcal{G}_{\mathrm{star1,\,2}}^{(3)}(4, \mathbb{C}) = \{\mathsf{N}_{\mathrm{star1,2}} | \mu^{\circ 2}, \overline{(\ )}\}$ which are two (non-intersecting because $\mathsf{N}_{\mathrm{star1}} \cap \mathsf{N}_{\mathrm{star2}} = \varnothing$) subgroups of the ternary general linear group $\mathrm{GL}^{(3)}(4, \mathbb{C})$. The ternary identities in $\mathcal{G}_{\mathrm{star1,\,2}}^{(3)}(4, \mathbb{C})$ are the following different continuous sets $\mathsf{I}_{\mathrm{star1,\,2}}^{(3)} = \left\{I_{\mathrm{star1,\,2}}^{(3)}\right\}$, where

$$I_{\mathrm{star1}}^{(3)} = \begin{pmatrix} e^{i\alpha_1} & 0 & 0 & 0 \\ 0 & 0 & e^{i\alpha_2} & 0 \\ 0 & e^{i\alpha_3} & 0 & 0 \\ 0 & 0 & 0 & e^{i\alpha_4} \end{pmatrix}, \quad e^{2i\alpha_1} = e^{2i\alpha_4} = e^{i(\alpha_2 + \alpha_3)} = 1, \ \alpha_j \in \mathbb{R}, \quad (10.1.47)$$

$$I_{\mathrm{star2}}^{(3)} = \begin{pmatrix} 0 & 0 & 0 & e^{i\alpha_1} \\ 0 & e^{i\alpha_2} & 0 & 0 \\ 0 & 0 & e^{i\alpha_3} & 0 \\ e^{i\alpha_4} & 0 & 0 & 0 \end{pmatrix}, \quad e^{2i\alpha_2} = e^{2i\alpha_3} = e^{i(\alpha_1 + \alpha_4)} = 1, \ \alpha_j \in \mathbb{R}. \quad (10.1.48)$$

In the particular case $\alpha_j = 0$, $j = 1, 2, 3, 4$, the ternary identities (10.1.47)–(10.1.48) coincide with the bisymmetric permutation matrices (10.1.28).

Next we treat the closed set products (10.1.39)–(10.1.40) as the multiplications $\mu^{(5)}$ of the *5-ary semigroups* $\mathcal{S}_{\mathrm{circ1,\,2}}^{(5)}(4, \mathbb{C}) = \{\mathsf{N}_{\mathrm{circ1,2}} | \mu^{\circ 4}\}$. The querelements are

$$\bar{N}_{\text{circ1}} = N_{\text{circ1}}^{-3} = \begin{pmatrix} 0 & 0 & \frac{1}{yzt} & 0 \\ \frac{1}{xzt} & 0 & 0 & 0 \\ 0 & 0 & 0 & \frac{1}{xyt} \\ 0 & \frac{1}{xyz} & 0 & 0 \end{pmatrix}, \quad \bar{N}_{\text{circ2}} = N_{\text{circ2}}^{-3} = \begin{pmatrix} 0 & \frac{1}{yzt} & 0 & 0 \\ 0 & 0 & 0 & \frac{1}{xzt} \\ \frac{1}{xyt} & 0 & 0 & 0 \\ 0 & 0 & \frac{1}{xyz} & 0 \end{pmatrix}, \qquad (10.1.49)$$

where $x, y, z, t \neq 0$, and the corresponding *5-ary groups* $\mathcal{G}^{(5)}_{\text{circ1, 2}}(4, \mathbb{C}) = \{\mathsf{N}_{\text{circ1,2}} | \mu^{\circ 4}, \overline{(\,)}\}$ which are two (non-intersecting because $\mathsf{N}_{\text{circ1}} \cap \mathsf{N}_{\text{circ2}} = \emptyset$) subgroups of the 5-ary general linear group $\mathrm{GL}^{(5)}(n, \mathbb{C})$. We have the following continuous sets of 5-ary identities $\mathsf{I}^{(3)}_{\text{circ1, 2}} = \left\{I^{(3)}_{\text{circ1, 2}}\right\}$ in $\mathcal{G}^{(5)}_{\text{circ1, 2}}(4, \mathbb{C})$ satisfying

$$I^{(5)}_{\text{circ1}} = \begin{pmatrix} 0 & 0 & e^{i\alpha_1} & 0 \\ e^{i\alpha_2} & 0 & 0 & 0 \\ 0 & 0 & 0 & e^{i\alpha_3} \\ 0 & e^{i\alpha_4} & 0 & 0 \end{pmatrix}, \quad e^{i(\alpha_1 + \alpha_2 + \alpha_3 + \alpha_4)} = 1, \ \alpha_j \in \mathbb{R}, \qquad (10.1.50)$$

$$I^{(5)}_{\text{circ2}} = \begin{pmatrix} 0 & e^{i\alpha_1} & 0 & 0 \\ 0 & 0 & 0 & e^{i\alpha_2} \\ e^{i\alpha_3} & 0 & 0 & 0 \\ 0 & 0 & e^{i\alpha_4} & 0 \end{pmatrix}, \quad e^{i(\alpha_1 + \alpha_2 + \alpha_3 + \alpha_4)} = 1, \ \alpha_j \in \mathbb{R}. \qquad (10.1.51)$$

In the case $\alpha_j = 0$, $j = 1, 2, 3, 4$, the 5-ary identities (10.1.50)–(10.1.51) coincide with the 90°-symmetric permutation matrices (10.1.29).

Thus, it follows from (10.1.46)–(10.1.51) that the 4-vertex star-like (10.1.35) and circle-like (10.1.36) matrices form subgroups of the k-ary general linear group $\mathrm{GL}^{[k]}(4, \mathbb{C})$ with significantly different properties: they have different querelements and (sets of) polyadic identities, and even the arities of the subgroups $\mathcal{G}^{(3)}_{\text{star1, 2}}(4, \mathbb{C})$ and $\mathcal{G}^{(5)}_{\text{circ1, 2}}(4, \mathbb{C})$ do not coincide (10.1.37)–(10.1.40). If we take into account that 4-vertex star-like (10.1.35) and circle-like (10.1.36) matrices are (binary) additive and distributive, then they form (with respect to the binary matrix addition (+) and the multiplications $\mu^{\circ 2}$ and $\mu^{\circ 4}$) the (2, 3)-ring $\mathcal{R}^{(3)}_{\text{star1, 2}}(4, \mathbb{C}) = \{\mathsf{N}_{\text{star1,2}} | +, \mu^{\circ 2}\}$ and (2, 5)-ring $\mathcal{R}^{[5]}_{\text{circ1, 2}}(4, \mathbb{C}) = \{\mathsf{N}_{\text{star1,2}} | +, \mu^{\circ 4}\}$.

Next we consider the 'interaction' of the 4-vertex star-like (10.1.35) and circle-like (10.1.36) matrix sets, i.e., their exotic module structure. For this, let us recall the ternary (polyadic) module (Duplij 2001) and s-place action (Duplij 2012) definitions, which are suitable for our case. An abelian group $\mathcal{M}$ is a ternary left (middle, right) $\mathcal{R}$-module (or a module over $\mathcal{R}$), if there exists a ternary operation $\mathcal{R} \times \mathcal{R} \times \mathcal{M} \to \mathcal{M}$ ($\mathcal{R} \times \mathcal{M} \times \mathcal{R} \to \mathcal{M}$, $\mathcal{M} \times \mathcal{R} \times \mathcal{R} \to \mathcal{M}$) which satisfies some compatibility conditions (associativity and distributivity) which hold in the matrix case under consideration (and where the module operation is the triple ordinary matrix product) (Duplij 2001). A 5-ary left (right) module $\mathcal{M}$ over $\mathcal{R}$ is a 5-ary operation $\mathcal{R} \times \mathcal{R} \times \mathcal{R} \times \mathcal{R} \times \mathcal{M} \to \mathcal{M}$ ($\mathcal{M} \times \mathcal{R} \times \mathcal{R} \times \mathcal{R} \times \mathcal{R} \to \mathcal{M}$) with analogous conditions (and where the module operation is the pentuple matrix product) (Duplij 2012).

First, we have the triple relations 'inside' star and circle matrices

$$\mathsf{N}_{\text{star1}}(\mathsf{N}_{\text{star2}})\mathsf{N}_{\text{star1}} = (\mathsf{N}_{\text{star2}}), \quad \mathsf{N}_{\text{circ1}}\mathsf{N}_{\text{circ2}}\mathsf{N}_{\text{circ1}} = \mathsf{N}_{\text{circ1}}, \tag{10.1.52}$$

$$\mathsf{N}_{\text{star1}}\mathsf{N}_{\text{star1}}(\mathsf{N}_{\text{star2}}) = (\mathsf{N}_{\text{star2}}), \quad \mathsf{N}_{\text{circ1}}\mathsf{N}_{\text{circ1}}\mathsf{N}_{\text{circ2}} = \mathsf{N}_{\text{circ1}}, \tag{10.1.53}$$

$$(\mathsf{N}_{\text{star2}})\mathsf{N}_{\text{star1}}\mathsf{N}_{\text{star1}} = (\mathsf{N}_{\text{star2}}), \quad \mathsf{N}_{\text{circ2}}\mathsf{N}_{\text{circ1}}\mathsf{N}_{\text{circ1}} = \mathsf{N}_{\text{circ1}}, \tag{10.1.54}$$

$$\mathsf{N}_{\text{star2}}\mathsf{N}_{\text{star2}}(\mathsf{N}_{\text{star1}}) = (\mathsf{N}_{\text{star1}}), \quad \mathsf{N}_{\text{circ2}}\mathsf{N}_{\text{circ2}}\mathsf{N}_{\text{circ1}} = \mathsf{N}_{\text{circ2}}, \tag{10.1.55}$$

$$\mathsf{N}_{\text{star2}}(\mathsf{N}_{\text{star1}})\mathsf{N}_{\text{star2}} = (\mathsf{N}_{\text{star1}}), \quad \mathsf{N}_{\text{circ2}}\mathsf{N}_{\text{circ1}}\mathsf{N}_{\text{circ2}} = \mathsf{N}_{\text{circ2}}, \tag{10.1.56}$$

$$(\mathsf{N}_{\text{star1}})\mathsf{N}_{\text{star2}}\mathsf{N}_{\text{star2}} = (\mathsf{N}_{\text{star1}}), \quad \mathsf{N}_{\text{circ1}}\mathsf{N}_{\text{circ2}}\mathsf{N}_{\text{circ2}} = \mathsf{N}_{\text{circ2}}. \tag{10.1.57}$$

We observe the following module structures on the left column above (elements of the corresponding module are in brackets, and we informally denote modules by their sets):

(1) from (10.1.52)–(10.1.54), the set $\mathsf{N}_{\text{star2}}$ is a middle, right and left module over $\mathsf{N}_{\text{star1}}$,

(2) from (10.1.55)–(10.1.57), the set $\mathsf{N}_{\text{star1}}$ is a middle, right and left module over $\mathsf{N}_{\text{star2}}$;

$$\mathsf{N}_{\text{star1}}\mathsf{N}_{\text{circ1}}\mathsf{N}_{\text{star1}} = \mathsf{N}_{\text{circ2}}, \quad \mathsf{N}_{\text{star1}}\mathsf{N}_{\text{circ2}}\mathsf{N}_{\text{star1}} = \mathsf{N}_{\text{circ1}}, \tag{10.1.58}$$

$$\mathsf{N}_{\text{star2}}\mathsf{N}_{\text{circ1}}\mathsf{N}_{\text{star2}} = \mathsf{N}_{\text{circ2}}, \quad \mathsf{N}_{\text{star2}}\mathsf{N}_{\text{circ2}}\mathsf{N}_{\text{star2}} = \mathsf{N}_{\text{circ1}}, \tag{10.1.59}$$

$$\mathsf{N}_{\text{star1}}\mathsf{N}_{\text{star1}}(\mathsf{N}_{\text{circ1}}) = (\mathsf{N}_{\text{circ1}}), \quad (\mathsf{N}_{\text{circ1}})\mathsf{N}_{\text{star1}}\mathsf{N}_{\text{star1}} = (\mathsf{N}_{\text{circ1}}), \tag{10.1.60}$$

$$\mathsf{N}_{\text{star1}}\mathsf{N}_{\text{star1}}(\mathsf{N}_{\text{circ2}}) = (\mathsf{N}_{\text{circ2}}), \quad (\mathsf{N}_{\text{circ2}})\mathsf{N}_{\text{star1}}\mathsf{N}_{\text{star1}} = (\mathsf{N}_{\text{circ2}}), \tag{10.1.61}$$

$$\mathsf{N}_{\text{star2}}\mathsf{N}_{\text{star2}}(\mathsf{N}_{\text{circ1}}) = (\mathsf{N}_{\text{circ1}}), \quad (\mathsf{N}_{\text{circ1}})\mathsf{N}_{\text{star2}}\mathsf{N}_{\text{star2}} = (\mathsf{N}_{\text{circ1}}), \tag{10.1.62}$$

$$\mathsf{N}_{\text{star2}}\mathsf{N}_{\text{star2}}(\mathsf{N}_{\text{circ2}}) = (\mathsf{N}_{\text{circ2}}), \quad (\mathsf{N}_{\text{circ2}})\mathsf{N}_{\text{star2}}\mathsf{N}_{\text{star2}} = (\mathsf{N}_{\text{circ2}}), \tag{10.1.63}$$

(3) from (10.1.60)–(10.1.63), the sets $\mathsf{N}_{\text{circ1,2}}$ are a right and left module over $\mathsf{N}_{\text{star1,2}}$;

$$\mathsf{N}_{\text{circ1}}(\mathsf{N}_{\text{star1}})\mathsf{N}_{\text{circ1}} = (\mathsf{N}_{\text{star1}}), \quad \mathsf{N}_{\text{circ1}}(\mathsf{N}_{\text{star2}})\mathsf{N}_{\text{circ1}} = (\mathsf{N}_{\text{star2}}), \tag{10.1.64}$$

$$\mathsf{N}_{\text{circ2}}(\mathsf{N}_{\text{star1}})\mathsf{N}_{\text{circ2}} = (\mathsf{N}_{\text{star1}}), \quad \mathsf{N}_{\text{circ2}}(\mathsf{N}_{\text{star2}})\mathsf{N}_{\text{circ2}} = (\mathsf{N}_{\text{star2}}), \tag{10.1.65}$$

$$\mathsf{N}_{\text{circ1}}\mathsf{N}_{\text{circ1}}\mathsf{N}_{\text{star1}} = \mathsf{N}_{\text{star2}}, \quad \mathsf{N}_{\text{star1}}\mathsf{N}_{\text{circ1}}\mathsf{N}_{\text{circ1}} = \mathsf{N}_{\text{star2}}, \tag{10.1.66}$$

$$\mathsf{N}_{\text{circ1}}\mathsf{N}_{\text{circ1}}\mathsf{N}_{\text{star2}} = \mathsf{N}_{\text{star1}}, \quad \mathsf{N}_{\text{star2}}\mathsf{N}_{\text{circ1}}\mathsf{N}_{\text{circ1}} = \mathsf{N}_{\text{star1}}, \tag{10.1.67}$$

$$\mathsf{N}_{\text{circ2}}\mathsf{N}_{\text{circ2}}\mathsf{N}_{\text{star1}} = \mathsf{N}_{\text{star2}}, \quad \mathsf{N}_{\text{star1}}\mathsf{N}_{\text{circ2}}\mathsf{N}_{\text{circ2}} = \mathsf{N}_{\text{star2}}, \tag{10.1.68}$$

$$\mathsf{N}_{\text{circ2}}\mathsf{N}_{\text{circ2}}\mathsf{N}_{\text{star2}} = \mathsf{N}_{\text{star1}}, \quad \mathsf{N}_{\text{star2}}\mathsf{N}_{\text{circ2}}\mathsf{N}_{\text{circ2}} = \mathsf{N}_{\text{star1}}, \tag{10.1.69}$$

(4) from (10.1.64)–(10.1.65), the sets $\mathsf{N}_{\text{star1,2}}$ are a middle ternary module over $\mathsf{N}_{\text{circ1,2}}$;

$$\mathsf{N}_{\text{circ1}}\mathsf{N}_{\text{circ1}}\mathsf{N}_{\text{circ1}}\mathsf{N}_{\text{circ1}}(\mathsf{N}_{\text{star1}}) = (\mathsf{N}_{\text{star1}}), \ \mathsf{N}_{\text{circ1}}\mathsf{N}_{\text{circ1}}\mathsf{N}_{\text{circ1}}\mathsf{N}_{\text{circ1}} (\mathsf{N}_{\text{star2}}) = (\mathsf{N}_{\text{star2}}), \tag{10.1.70}$$

$$(\mathsf{N}_{\text{star1}})\mathsf{N}_{\text{circ1}}\mathsf{N}_{\text{circ1}}\mathsf{N}_{\text{circ1}}\mathsf{N}_{\text{circ1}} = (\mathsf{N}_{\text{star1}}), \ (\mathsf{N}_{\text{star2}})\mathsf{N}_{\text{circ1}}\mathsf{N}_{\text{circ1}}\mathsf{N}_{\text{circ1}} \mathsf{N}_{\text{circ1}} = (\mathsf{N}_{\text{star2}}), \tag{10.1.71}$$

$$\mathsf{N}_{\text{circ2}}\mathsf{N}_{\text{circ2}}\mathsf{N}_{\text{circ2}}\mathsf{N}_{\text{circ2}}(\mathsf{N}_{\text{star1}}) = (\mathsf{N}_{\text{star1}}), \ \mathsf{N}_{\text{circ2}}\mathsf{N}_{\text{circ2}}\mathsf{N}_{\text{circ2}}\mathsf{N}_{\text{circ2}} (\mathsf{N}_{\text{star2}}) = (\mathsf{N}_{\text{star2}}), \tag{10.1.72}$$

$$(\mathsf{N}_{\text{star1}})\mathsf{N}_{\text{circ2}}\mathsf{N}_{\text{circ2}}\mathsf{N}_{\text{circ2}}\mathsf{N}_{\text{circ2}} = (\mathsf{N}_{\text{star1}}), \ (\mathsf{N}_{\text{star2}})\mathsf{N}_{\text{circ2}}\mathsf{N}_{\text{circ2}}\mathsf{N}_{\text{circ2}} \mathsf{N}_{\text{circ2}} = (\mathsf{N}_{\text{star2}}), \tag{10.1.73}$$

$$\mathsf{N}_{\text{circ1}}\mathsf{N}_{\text{circ1}}\mathsf{N}_{\text{circ1}}\mathsf{N}_{\text{circ1}}(\mathsf{N}_{\text{circ2}}) = (\mathsf{N}_{\text{circ2}}), \ (\mathsf{N}_{\text{circ2}})\mathsf{N}_{\text{circ1}}\mathsf{N}_{\text{circ1}}\mathsf{N}_{\text{circ1}} \mathsf{N}_{\text{circ1}} = (\mathsf{N}_{\text{circ2}}), \tag{10.1.74}$$

$$\mathsf{N}_{\text{circ2}}\mathsf{N}_{\text{circ2}}\mathsf{N}_{\text{circ2}}\mathsf{N}_{\text{circ2}}(\mathsf{N}_{\text{circ1}}) = (\mathsf{N}_{\text{circ1}}), \ (\mathsf{N}_{\text{circ1}})\mathsf{N}_{\text{circ2}}\mathsf{N}_{\text{circ2}}\mathsf{N}_{\text{circ2}} \mathsf{N}_{\text{circ2}} = (\mathsf{N}_{\text{circ1}}), \tag{10.1.75}$$

(5) from (10.1.70)–(10.1.75), the sets $\mathsf{N}_{\text{circ1,2}}$ are right and left 5-ary modules over $\mathsf{N}_{\text{circ2,1}}$ and $\mathsf{N}_{\text{star1,2}}$.

Note that the sum of 4-vertex star solutions of the Yang–Baxter equations (10.1.33) (with different parameters) gives the shape of 8-vertex matrices, and the same with the 4-vertex circle solutions (10.1.34). Let us introduce two kinds of 8-vertex 4×4 matrices over $\mathbb{C}$: an 8-vertex *star matrix* M_{star} and an 8-vertex *circle matrix* M_{circ} as

$$M_{\text{star}} = \begin{pmatrix} x & 0 & 0 & y \\ 0 & z & s & 0 \\ 0 & t & u & 0 \\ v & 0 & 0 & w \end{pmatrix}, \ \det M_{\text{star}} = (xw - yv)(st - uz), \ \operatorname{tr} M_{\text{star}} = x + z + u + w, \tag{10.1.76}$$

$$M_{\text{circ}} = \begin{pmatrix} 0 & x & y & 0 \\ z & 0 & 0 & s \\ t & 0 & 0 & u \\ 0 & v & w & 0 \end{pmatrix}, \ \det M_{\text{circ}} = (xw - yv)(st - uz), \ \operatorname{tr} M_{\text{circ}} = 0. \tag{10.1.77}$$

If $M_{\rm star}$ and $M_{\rm circ}$ are invertible (the determinants in (10.1.76)–(10.1.77) are non-vanishing), then

$$M_{\rm star}^{-1}=\begin{pmatrix}\frac{w}{xw-yv} & 0 & 0 & -\frac{v}{xw-yv}\\ 0 & -\frac{u}{st-uz} & \frac{t}{st-uz} & 0\\ 0 & \frac{s}{st-uz} & -\frac{z}{st-uz} & 0\\ -\frac{y}{xw-yv} & 0 & 0 & \frac{x}{xw-yv}\end{pmatrix},$$
$$M_{\rm circ}^{-1}=\begin{pmatrix}0 & \frac{w}{xw-yv} & -\frac{v}{xw-yv} & 0\\ -\frac{u}{st-uz} & 0 & 0 & \frac{t}{st-uz}\\ \frac{s}{st-uz} & 0 & 0 & -\frac{z}{st-uz}\\ 0 & -\frac{y}{xw-yv} & \frac{x}{xw-yv} & 0\end{pmatrix}, \tag{10.1.78}$$

and therefore the parameter conditions for invertibility are the same in both $M_{\rm star}$ and $M_{\rm circ}$

$$xw-yv\neq 0,\qquad st-uz\neq 0. \tag{10.1.79}$$

The corresponding sets $\mathsf{M}_{\rm star}=\{M_{\rm star}\}$ and $\mathsf{M}_{\rm circ}=\{M_{\rm circ}\}$ are closed under the following multiplications

$$\mathsf{M}_{\rm star}\mathsf{M}_{\rm star}=\mathsf{M}_{\rm star}, \tag{10.1.80}$$

$$\mathsf{M}_{\rm star}\mathsf{M}_{\rm circ}=\mathsf{M}_{\rm circ},\qquad \mathsf{M}_{\rm circ}\mathsf{M}_{\rm star}=\mathsf{M}_{\rm circ}, \tag{10.1.81}$$

$$\mathsf{M}_{\rm circ}\mathsf{M}_{\rm circ}=\mathsf{M}_{\rm star}, \tag{10.1.82}$$

and in terms of sets we can write $\mathsf{M}_{\rm star}=\mathsf{N}_{\rm star1}\cup\mathsf{N}_{\rm star2}$ and $\mathsf{M}_{\rm circ}=\mathsf{N}_{\rm circ1}\cup\mathsf{N}_{\rm circ2}$, while $\mathsf{N}_{\rm star1}\cap\mathsf{N}_{\rm star2}=\varnothing$ and $\mathsf{N}_{\rm circ1}\cap\mathsf{N}_{\rm circ2}=\varnothing$ (see (10.1.41)). Note that, if $\mathsf{M}_{\rm star}$ and $\mathsf{M}_{\rm circ}$ are treated as elements of an algebra, then (10.1.80)–(10.1.82) are reminiscent of the Cartan decomposition (see, e.g., Helgason 1962), but we will consider them from a more general viewpoint, which will treat such structures as semigroups, ternary groups and modules.

Thus the set $\mathsf{M}_{\rm 8vertex}=\mathsf{M}_{\rm star}\cup\mathsf{M}_{\rm circ}$ is closed, and because of the associativity of matrix multiplication, $\mathsf{M}_{\rm 8vertex}$ forms a non-commutative semigroup which we call a 8-*vertex matrix semigroup* $\mathcal{S}_{\rm 8vertex}(4,\mathbb{C})$, which contains the zero matrix $Z\in\mathcal{S}_{\rm 8vertex}(4,\mathbb{C})$ and is a subsemigroup of the (binary) general linear semigroup $\mathrm{GLS}(4,\mathbb{C})$. It follows from (10.1.80), that $\mathsf{M}_{\rm star}$ is its subsemigroup $\mathcal{S}_{\rm 8vertex}^{\rm star}(4,\mathbb{C})$. Moreover, the invertible elements of $\mathcal{S}_{\rm 8vertex}(4,\mathbb{C})$ form a 8-*vertex matrix group* $\mathcal{G}_{\rm 8vertex}(4,\mathbb{C})$, because its identity is a unit 4×4 matrix $I_4\in\mathsf{M}_{\rm star}$, and so $\mathsf{M}_{\rm star}$ is a subgroup $\mathcal{G}_{\rm 8vertex}^{\rm star}(4,\mathbb{C})$ of $\mathcal{G}_{\rm 8vertex}(4,\mathbb{C})$ and a subgroup of the (binary) general linear group $\mathrm{GL}(4, \mathbb{C})$. The structure of $\mathcal{S}_{8\mathrm{vertex}}(4, \mathbb{C})$ (10.1.80) is similar to that of block-diagonal and block-antidiagonal matrices (of the necessary sizes). So the 8-vertex (binary) matrix semigroup $\mathcal{S}_{8\mathrm{vertex}}(4, \mathbb{C})$ in which the parameters satisfy (10.1.79) is a 8-vertex (binary) matrix group $\mathcal{G}_{8\mathrm{vertex}}(4, \mathbb{C})$, having a subgroup $\mathcal{G}_{8\mathrm{vertex}}^{\mathrm{star}}(4, \mathbb{C}) = \langle \mathsf{M}_{\mathrm{star}} | \mu = (\cdot), I_4 \rangle$, where $(\cdot)$ is an ordinary matrix product, and I_4 is its identity.

The group structure of the circle matrices $\mathsf{M}_{\mathrm{circ}}$ (10.1.77) follows from

$$\mathsf{M}_{\mathrm{circ}} \mathsf{M}_{\mathrm{circ}} \mathsf{M}_{\mathrm{circ}} = \mathsf{M}_{\mathrm{circ}}, \tag{10.1.83}$$

which means that $\mathsf{M}_{\mathrm{circ}}$ is closed with respect to the product of three matrices (the product of two matrices from $\mathsf{M}_{\mathrm{circ}}$ is outside the set (10.1.82)). We define a ternary multiplication $\nu^{(3)}$ as the ordinary triple product of matrices, then $\mathcal{S}_{8\mathrm{vertex}}^{\mathrm{circ}(3)}(4, \mathbb{C}) = \langle \mathsf{M}_{\mathrm{circ}} | \nu^{\circ 2} \rangle$ is a ternary (3-nonderived) semigroup with the zero $Z \in \mathsf{M}_{\mathrm{circ}}$ which is a subsemigroup of the ternary (derived) general linear semigroup $\mathrm{GLS}^{(3)}(4, \mathbb{C})$. Instead of the inverse, for each invertible element $M_{\mathrm{circ}} \in \mathsf{M}_{\mathrm{circ}} / Z$ we introduce the unique querelement $\bar{M}_{\mathrm{circ}}$ (Dörnte 1929) by (10.1.42), and since the ternary product is the triple ordinary product, we have $\bar{M}_{\mathrm{circ}} = M_{\mathrm{circ}}^{-1}$ from (10.1.43). Thus, if the conditions of invertibility (10.1.79) hold valid, then the ternary semigroup $\mathcal{S}_{8\mathrm{vertex}}^{\mathrm{circ}(3)}(4, \mathbb{C})$ becomes the ternary group $\mathcal{G}_{8\mathrm{vertex}}^{\mathrm{circ}(3)}(4, \mathbb{C}) = \langle \mathsf{M}_{\mathrm{circ}} | \nu^{\circ 2}, \overline{()} \rangle$ which does not contain the ordinary (binary) identity, since $I_4 \notin \mathsf{M}_{\mathrm{circ}}$. Nevertheless, the ternary group of circle matrices $\mathcal{G}_{8\mathrm{vertex}}^{\mathrm{circ}(3)}(4, \mathbb{C})$ has the following set $\mathsf{I}_{\mathrm{circ}}^{(3)} = \left\{ I_{\mathrm{circ}}^{(3)} \right\}$ of left-right 6-vertex and 8-vertex ternary identities (see (10.1.44)–(10.1.45))

$$I_{\mathrm{circ}}^{(3)} = \begin{pmatrix} 0 & \frac{1}{a} & b & 0 \\ a & 0 & 0 & -\frac{ab}{c} \\ 0 & 0 & 0 & \frac{1}{c} \\ 0 & 0 & c & 0 \end{pmatrix}, \begin{pmatrix} 0 & -\frac{ab}{c} & \frac{1}{c} & 0 \\ 0 & 0 & 0 & \frac{1}{b} \\ c & 0 & 0 & a \\ 0 & b & 0 & 0 \end{pmatrix}, \begin{pmatrix} 0 & -\frac{ab}{c} & \frac{1 - ad}{c} & 0 \\ -\frac{cd}{b} & 0 & 0 & \frac{1 - ad}{b} \\ c & 0 & 0 & a \\ 0 & b & d & 0 \end{pmatrix}, \tag{10.1.84}$$

which (without additional conditions) depend upon the free parameters $a, b, c, d \in \mathbb{C}$, $b, c \neq 0$, and $\left(I_{\mathrm{circ}}^{(3)} \right)^2 = I_4$, $I_{\mathrm{circ}}^{(3)} \in \mathsf{M}_{\mathrm{circ}}$. In the binary sense, the matrices from (10.1.84) are mutually similar, but as ternary identities they are different.

If we consider the second operation for matrices (as elements of a general matrix ring), the binary matrix addition $(+)$, the structure of $\mathsf{M}_{8\mathrm{vertex}} = \mathsf{M}_{\mathrm{star}} \cup \mathsf{M}_{\mathrm{circ}}$ becomes more exotic: the set $\mathsf{M}_{\mathrm{star}}$ is a $(2, 2)$-ring $\mathcal{R}_{8\mathrm{vertex}}^{\mathrm{star}[2, 2]} = \langle \mathsf{M}_{\mathrm{star}} | +, \cdot \rangle$ with the binary addition $(+)$ and binary multiplication $(\cdot)$ from the semigroup $\mathcal{S}_{8\mathrm{vertex}}^{\mathrm{star}}$, while $\mathsf{M}_{\mathrm{circ}}$ is a $(2, 3)$-ring $\mathcal{R}_{8\mathrm{vertex}}^{\mathrm{circ}[2, 3]} = \langle \mathsf{M}_{\mathrm{circ}} | +, \nu^{\circ 2} \rangle$ with the binary matrix addition $(+)$, the ternary matrix multiplication and the zero Z.

Moreover, because of the distributivity and associativity of binary matrix multiplication, the relations (10.1.81) mean that the set $\mathsf{M}_{\mathrm{circ}}$ (being an abelian group under binary addition) can be treated as a left and right binary module $\mathcal{M}_{8\mathrm{vertex}}^{\mathrm{circ}}$ over the ring $\mathcal{R}_{8\text{vertex}}^{\text{star}(2,2)}$ with an operation (*): the module action $M_{\text{circ}}{*}M_{\text{star}} = M_{\text{circ}}$, $M_{\text{star}}{*}M_{\text{circ}} = M_{\text{circ}}$ (coinciding with the ordinary matrix product (10.1.81)). The left and right modules are compatible, since the associativity of ordinary matrix multiplication gives the compatibility condition $(M_{\text{circ}}M_{\text{star}})M'_{\text{circ}} = M_{\text{circ}}(M_{\text{star}}M'_{\text{circ}})$, $M_{\text{star}} \in \mathcal{R}_{8\text{vertex}}^{\text{star}(2,2)}$, $M_{\text{circ}}, M'_{\text{circ}} \in \mathcal{R}_{8\text{vertex}}^{\text{circ}(2,3)}$, and therefore M_{circ} (as an abelian group under the binary addition (+) and the module action (*)) is a $\mathcal{R}_{8\text{vertex}}^{\text{star}(2,2)}$-bimodule $\mathcal{M}_{8\text{vertex}}^{\text{circ}}$. The last relation (10.1.82) shows another interpretation of M_{circ} as a formal 'square root' of M_{star} (as sets).

10.1.6 Star 8-vertex and circle 8-vertex Yang–Baxter maps

Let us consider the star 8-vertex solutions $\tilde{c}$ to the Yang–Baxter equation (10.1.12), having the shape (10.1.76), in the most general setting, over $\mathbb{C}$ and for different ranks (i.e., including non-invertible ones). In components they are determined by

$$\begin{aligned}
&y(s(x-z)+t(u-x))=0,\ \ vy(u-z)=0,\ \ y(t^2-wz-x^2+xz)=0,\\
&y(u(w-x)+x^2-s^2)=0,\ \ svy-tuz=0,\ \ vwy+xz(x-z)-stz=0,\\
&y(w^2-wz+xz-s^2)=0,\ \ uz(z-u)=0,\ \ suz-tvy=0,\\
&y(s(w-u)+t(z-w))=0,\ \ v(s^2-wz-x^2+xz)=0,\ \ tuz-svy=0,\\
&stz-vxy+wz(z-w)=0,\ \ stu+u^2x-ux^2-vwy=0,\ \ tvy-suz=0,\\
&v(s(z-w)+t(w-u))=0,\ \ uz(u-z)=0,\ \ v(s^2+u(w-x)-w^2)=0,\\
&y(t^2+u(w-x)-w^2)=0,\ \ v(s(u-x)+t(x-z))=0,\\
&v(u(w-x)+x^2-t^2)=0,\ \ uw^2+vxy-stu-u^2w=0,\\
&v(w^2-t^2-wz+xz)=0,\ \ vy(z-u)=0.
\end{aligned} \tag{10.1.85}$$

Solutions from, e.g., Dye (2003), Hietarinta (1993), etc, should satisfy this over-determined system of 24 cubic equations for 8 variables.

We search for the 8-vertex constant solutions to the Yang–Baxter equation over $\mathbb{C}$ without additional conditions, unitarity, etc (which will be considered in the next sections). We also will need the matrix functions tr and det which are related to link invariants, as well as eigenvalues which help to find similar matrices and q-conjugated solutions to braid equations. Take into account that the Yang–Baxter maps are determined up to a general complex factor $t \in \mathbb{C}$ (10.1.14). For eigenvalues (which are determined up to the same factor t) we use the notation: $\{\text{eigenvalue}\}^{[\text{algebraic multiplicity}]}$.

We found the following 8-vertex solutions, classified by rank and number of parameters.

• Rank = 4 (invertible star Yang–Baxter maps) are

(1) quadratic in two parameters

$$\tilde{c}_{\text{rank}=4}^{\text{par}=2}(x,y)=\begin{pmatrix} xy & 0 & 0 & y^2\\ 0 & xy & \pm xy & 0\\ 0 & \mp xy & xy & 0\\ -x^2 & 0 & 0 & xy \end{pmatrix},\qquad \begin{array}{c}\operatorname{tr}\tilde{c}=4xy,\\ \det\tilde{c}=4x^4y^4,\quad x\neq 0,\ y\neq 0,\\ \text{eigenvalues: } \{(1+i)xy\}^{[2]},\ \{(1-i)xy\}^{[2]},\end{array} \tag{10.1.86}$$

(2) quadratic in three parameters

$$\tilde{c}^{\,par=3}_{rank=4,\,1}(x,y,z)=\begin{pmatrix} xy & 0 & 0 & y^2 \\ 0 & zy & \pm xy & 0 \\ 0 & \pm xy & zy & 0 \\ z^2 & 0 & 0 & xy \end{pmatrix},\quad \begin{array}{c} \mathrm{tr}\tilde{c}=2y(x+z), \\ \det\tilde{c}=y^4\left(z^2-x^2\right)^2,\ z\neq\pm x,\ y\neq 0, \\ \text{eigenvalues:} y(x-z),\ -y(x-z),\ \{y(x+z)\}^{[2]}, \end{array} \tag{10.1.87}$$

(3) irrational in three parameters

$$\tilde{c}^{\,par=3}_{rank=4}(x,y,z)=\begin{pmatrix} xy & 0 & 0 & y^2 \\ 0 & \frac{x+z}{2}y & \pm y\sqrt{\frac{x^2+z^2}{2}} & 0 \\ 0 & \pm y\sqrt{\frac{x^2+z^2}{2}} & \frac{x+z}{2}y & 0 \\ \frac{(x+z)^2}{4} & 0 & 0 & yz \end{pmatrix},\quad \begin{array}{c} \mathrm{tr}\tilde{c}=2y(x+z), \\ \det\tilde{c}=\frac{1}{16}y^4(x-z)^4,\ y\neq 0,\ z\neq x, \end{array} \tag{10.1.88}$$

$$\text{eigenvalues}:\left\{\frac{1}{2}y\left(x+z-\sqrt{2}\sqrt{x^2+z^2}\right)\right\}^{[2]},\ \left\{\frac{1}{2}y\left(x+z+\sqrt{2}\sqrt{x^2+z^2}\right)\right\}^{[2]}.$$

Note that only the first and the last cases are genuine 8-vertex Yang–Baxter maps, because the three-parameter matrices (10.1.87) are q-conjugated with the 4-vertex parameter-permutation solutions (10.1.33). Indeed,

$$\begin{pmatrix} xy & 0 & 0 & y^2 \\ 0 & zy & xy & 0 \\ 0 & xy & zy & 0 \\ z^2 & 0 & 0 & xy \end{pmatrix}=(q\otimes_K q)\begin{pmatrix} y(x+z) & 0 & 0 & 0 \\ 0 & 0 & y(x-z) & 0 \\ 0 & y(x-z) & 0 & 0 \\ 0 & 0 & 0 & y(x+z) \end{pmatrix}\left(q^{-1}\otimes_K q^{-1}\right), \tag{10.1.89}$$

$$q=\begin{pmatrix} \pm\sqrt{\frac{y}{z}} & b \\ 1 & \mp b\sqrt{\frac{z}{y}} \end{pmatrix}, \tag{10.1.90}$$

where $b\in\mathbb{C}$ is a free parameter. If $b=\frac{y}{z}$ two matrices q in (10.1.90) are similar, and we have the unique q-conjugation (10.1.89). Another solution in (10.1.87) is q-conjugated to the second 4-vertex parameter-permutation solutions (10.1.33) such that

$$\begin{pmatrix} xy & 0 & 0 & y^2 \\ 0 & zy & -xy & 0 \\ 0 & -xy & zy & 0 \\ z^2 & 0 & 0 & xy \end{pmatrix}=(q\otimes_K q)\begin{pmatrix} 0 & 0 & 0 & y(x-z) \\ 0 & y(x+z) & 0 & 0 \\ 0 & 0 & y(x+z) & 0 \\ y(x-z) & 0 & 0 & 0 \end{pmatrix}\left(q^{-1}\otimes_K q^{-1}\right), \tag{10.1.91}$$

$$q=\begin{pmatrix} i\sqrt{\frac{y}{z}} & \pm i\sqrt{\frac{y}{z}} \\ \pm 1 & 1 \end{pmatrix},\ \begin{pmatrix} -i\sqrt{\frac{y}{z}} & \pm i\sqrt{\frac{y}{z}} \\ \pm 1 & 1 \end{pmatrix}, \tag{10.1.92}$$

where q are pairwise similar in (10.1.92), and therefore we have 2 different q-conjugations.

• Rank =2 (non-invertible star Yang–Baxter maps) are quadratic in parameters

$$\tilde{c}_{\text{rank}=2}^{\text{par}=2}(x, y) = \begin{pmatrix} xy & 0 & 0 & y^2 \\ 0 & xy & \pm xy & 0 \\ 0 & \pm xy & xy & 0 \\ x^2 & 0 & 0 & xy \end{pmatrix}, \quad \begin{array}{c} \text{tr}\tilde{c} = 4xy, \\ \text{eigenvalues:}\{2xy\}^{[2]}, \ \{0\}^{[2]}. \end{array} \tag{10.1.93}$$

There are no star 8-vertex solutions of rank 3. The above two solutions for $\tilde{c}_{\text{rank}=4}^{\text{par}=2}$ with different signs are q-conjugated (10.1.19) with the matrix q being one of the following

$$q = \begin{pmatrix} 0 & 1 \\ \pm i\dfrac{x}{y} & 0 \end{pmatrix}. \tag{10.1.94}$$

Further families of solutions can be obtained from (10.1.86)–(10.1.93) by applying the general q-conjugation (10.1.14).

Particular cases of the star solutions are called also X-type operators (Padmanabhan *et al* 2021) or magic matrices (Ballard and Wu 2011) connected with the Cartan decomposition of $SU(4)$ (Khaneja and Glaser 2001, Kraus and Cirac 2001, Bullock 2004, Bullock and Brennen 2004).

The circle 8-vertex solutions $\tilde{c}$ to the Yang–Baxter equation (10.1.12) of the shape (10.1.77) are determined by the following system of 32 cubic equations for 8 unknowns over $\mathbb{C}$

$$\begin{gathered} x(ty + z(u - y) - vx) = 0, \ tx^2 + y^2(v - z) - wx^2 = 0, \ y(-st + tx + wy - xz) = 0, \\ z(t(y - x) - sz + wx) = 0, \ v(sy + x^2) - z(s^2 + ux) = 0, \ swy - s^2v + xy(v - z) = 0, \\ su(x - y) - sxy + uxy = 0, \ swx - s^2w + yz(u - y) = 0, \ s(tu - uw + yz) - ty^2 = 0, \\ st^2 - t^2x + z^2(y - u) = 0, \ su(v - z) + x(xz - tu) = 0, \ su(w - v) + xy(z - t) = 0, \\ s(tv + z^2) - x(v^2 + wz) = 0, \ svw - vwx + z(xz - wy) = 0, \ sw(w - t) + yz(z - v) = 0, \\ t(tu - vy + z(y - x)) = 0, \ tx(x - s) + u^2v - uvy = 0, \ xy(t - w) + u^2w - uvx = 0, \\ tz(s - x) - sv^2 + tuv = 0, \ tz(x - y) - svw + uvw = 0, \ u(w^2 - tz) - swz + tyz = 0, \\ s(sz + u(v - w) - vy) = 0, \ t(sy + u^2) - w(ux + y^2) = 0, \ u(s(v - w) - tu + wx) = 0, \\ tx(w - t) + uv(z - v) = 0, \ tvx - t^2y - uvw + vwy = 0, \ tvy - t^2u + w(wy - uz) = 0, \\ twz - tv(w + z) + vwz = 0, \ v(s(w - t) - uw + vx) = 0, \ sw^2 - uv^2 + v^2y - w^2x = 0, \\ s^2(t - w) + u^2(v - z) = 0, \ w(sv + u(z - v) - wy) = 0. \end{gathered} \tag{10.1.95}$$

We found the 8-vertex solutions, classified by rank and number of parameters.

• Rank =4 (invertible circle Yang–Baxter map) are quadratic in parameters

$$\tilde{c}_{\text{rank}=4}^{\text{par}=3}(x, y, z) = \begin{pmatrix} 0 & xy & yz & 0 \\ z^2 & 0 & 0 & xy \\ xz & 0 & 0 & yz \\ 0 & z^2 & xz & 0 \end{pmatrix}, \quad \begin{array}{c} \text{tr}\tilde{c} = 0, \\ \det\tilde{c} = y^2z^2(z^2 - x^2), \ y \neq 0, \ z \neq 0, \ z \neq \pm x, \\ \text{eigenvalues:}\sqrt{-yz}(x - z), \ -\sqrt{-yz}(x - z), \\ \sqrt{yz}(x + z), \ -\sqrt{yz}(x + z). \end{array} \tag{10.1.96}$$

• Rank =2 (non-invertible circle Yang–Baxter map) are linear in parameters

$$\tilde{c}^{\,\text{par}\,=\,2}_{\text{rank}\,=\,2}(x,y)=\begin{pmatrix} 0 & -y & -y & 0 \\ -x & 0 & 0 & y \\ -x & 0 & 0 & y \\ 0 & x & x & 0 \end{pmatrix},\ \text{eigenvalues: } 2\sqrt{xy},\ -2\sqrt{xy},\ \{0\}^{[2]}. \quad (10.1.97)$$

There are no circle 8-vertex solutions of rank 3. The corresponding families of solutions can be derived from the above by using the q-conjugation (10.1.14).

A particular case of the 8-vertex circle solution (10.1.96) was considered in Asaulko and Korablev (2019).

10.1.7 Triangle invertible 9- and 10-vertex solutions

There are some higher vertex solutions to the Yang–Baxter equations which are not in the above star/circle classification. They are determined by the following system of 15 cubic equations for 9 unknowns over $\mathbb{C}$

$$\begin{aligned}
&y(-py-x(u+w-y)+v(y+z))+s(v-x)(v+x)\\
&\quad=0,\ x(-ty+vz+x(y-z))=0,\\
&sx(t-v)+ty(w-z)+vz(y-u)\\
&\quad=0,\ z(pz-t(y+z)+x(u+w-z))+s(x^2-t^2)=0,\\
&ps(-u+w-y+z)+s(-t(u+z)+x(u-w+y-z)+v(w+y))+uwy\\
&\quad-uwz-uyz+wyz=0,\\
&t(y(p-t)+u(x-v))=0,\ t(pz-t(u+z)+ux)=0,\\
&p^2s+pu(-u+y+z)-t(st+u(u+w))+u^2x\\
&\quad=0,\ t(z(p-v)-tw+wx)=0,\\
&ps(t-v)+tw(y-u)+uv(w-z)=0,\ v(-py+v(w+y)-wx)=0,\\
&v(z(v-p)+tw-wx)=0,\ v(y(t-p)+u(v-x))=0,\\
&p^2(-s)+pw(w-y-z)+sv^2+w(v(u+w)-wx)\\
&\quad=0,\ p(p(w-u)-tw+uv)=0,
\end{aligned} \quad (10.1.98)$$

We found the following 9-vertex Yang–Baxter maps

$$\tilde{c}^{\,9-\text{vert},\,1}_{\text{rank}\,=\,4}=\begin{pmatrix} x & y & z & s \\ 0 & 0 & x & y \\ 0 & x & 0 & z \\ 0 & 0 & 0 & x \end{pmatrix},\begin{pmatrix} x & y & y & z \\ 0 & 0 & -x & -y \\ 0 & -x & 0 & -y \\ 0 & 0 & 0 & x \end{pmatrix},\begin{pmatrix} x & y & -y & z \\ 0 & 0 & x & -\dfrac{zx}{y} \\ 0 & x & 0 & \dfrac{zx}{y} \\ 0 & 0 & 0 & x \end{pmatrix}, \quad (10.1.99)$$

$$\operatorname{tr}\tilde{c}=2x,\quad \det\tilde{c}=-x^4,\quad x\neq 0,\ \text{eigenvalues: } \{x\}^{[3]},\ -x. \quad (10.1.100)$$

The third matrix in (10.1.99) is conjugated with the 4-vertex parameter-permutation solutions (10.1.33) of the form (which has the same eigenvalues (10.1.100))

$$\tilde{c}^{4-\text{vert}}_{\text{rank} = 4}(x) = \begin{pmatrix} x & 0 & 0 & 0 \\ 0 & 0 & x & 0 \\ 0 & x & 0 & 0 \\ 0 & 0 & 0 & x \end{pmatrix} \tag{10.1.101}$$

by the conjugated matrix

$$U^{9\text{to}4} = \begin{pmatrix} 1 & -\frac{y}{2x} & \frac{y}{2x} & 0 \\ 0 & 1 & 0 & -\frac{z}{y} \\ 0 & 0 & 1 & 0 \\ 0 & 0 & 0 & 1 \end{pmatrix}. \tag{10.1.102}$$

The matrix (10.1.102) cannot be presented as the Kronecker product $q \otimes_K q$ (10.1.16), and so the third matrix in (10.1.99) and (10.1.101) are different solutions of the Yang–Baxter equation (10.1.12). Despite the first two matrices in (10.1.99) have the same eigenvalues (10.1.100), they are not similar, because they have different from (10.1.101) middle Jordan blocks.

Then we have another 3-parameter solutions with fractions

$$\tilde{c}^{9-\text{vert},\,2}_{\text{rank} = 4}(x, y, z) = \begin{pmatrix} x & y & y & z \\ 0 & 0 & -x & y - \frac{2xz}{y} \\ 0 & -x & 0 & y - \frac{2xz}{y} \\ 0 & 0 & 0 & x\left(\frac{4xz}{y^2} - 3\right) \end{pmatrix}, \quad \begin{array}{c} \operatorname{tr}\tilde{c} = 2x\frac{2xz - y^2}{y^2}, \\ \det \tilde{c} = x^4\left(3 - \frac{4xz}{y^2}\right), \; x \neq 0, \; y \neq 0, \; z \neq \frac{3y^2}{4x}, \\ \text{eigenvalues: } \{x\}^{[2]}, \; -x, \; x\left(\frac{4xz}{y^2} - 3\right), \end{array} \tag{10.1.103}$$

and

$$\tilde{c}^{9-\text{vert},\,3}_{\text{rank} = 4}(x, y, z) = \begin{pmatrix} x & y & -y & z \\ 0 & 0 & -x & \frac{2zx}{y} + y \\ 0 & 3x & 0 & \frac{2zx}{y} - y \\ 0 & 0 & 0 & \frac{4zx^2}{y^2} + x \end{pmatrix}, \quad \begin{array}{c} \operatorname{tr}\tilde{c} = 2x\left(1 + 2\frac{xz}{y^2}\right) \\ \det \tilde{c} = 3x^4\left(\frac{4zx}{y^2} + 1\right), \; x \neq 0, \; y \neq 0, \; z \neq \frac{y^2}{4x} \\ \text{eigenvalues: } x, \; i\sqrt{3}\,x, \; -i\sqrt{3}\,x, \; x\left(1 + \frac{4zx}{y^2}\right) \end{array} \tag{10.1.104}$$

The 4-parameter 9-vertex solution is

$$\tilde{c}^{9-\text{vert},\text{par} = 4}_{\text{rank} = 4}(x, y, z, s) = \begin{pmatrix} x & y & z & s \\ 0 & 0 & -x & y - \frac{2sx}{z} \\ 0 & x - \frac{2xy}{z} & 0 & z - \frac{2sx}{z} \\ 0 & 0 & 0 & \frac{x(4sx - z(2y + z))}{z^2} \end{pmatrix}, \quad \begin{array}{c} \operatorname{tr}\tilde{c} = 2x\frac{2sx - yz}{z^2} \\ \det \tilde{c} = \frac{x^4(2y - z)\left(z(2y + z) - 4sx\right)}{z^3}, \\ x \neq 0, \; y \neq \frac{z}{2}, \; z \neq 0, \end{array} \tag{10.1.105}$$

$$\text{eigenvalues}: \; x, \; x\sqrt{\frac{2y}{z}-1}, \; -x\sqrt{\frac{2y}{z}-1}, \; \frac{x(4sx - z(2y+z))}{z^2}. \tag{10.1.106}$$

We also found 5-parameter, 9-vertex solution of the form

$$\tilde{c}^{9-\text{vert,par}=5}_{\text{rank}=4}(x, y, z, s, t) = \begin{pmatrix} x & y & z & s \\ 0 & 0 & t & \frac{s(t-x)}{z} + y \\ 0 & \frac{y(t-x)}{z} + x & 0 & \frac{s(t-x)}{z} + z \\ 0 & 0 & 0 & \frac{s(t-x)^2 + tz(y+z) - xyz}{z^2} \end{pmatrix}, \tag{10.1.107}$$

$$\operatorname{tr}\tilde{c} = \frac{st^2 + sx^2 + tz^2 + xz^2 - 2stx + tyz - xyz}{z^2}, \tag{10.1.108}$$

$$\det\tilde{c} = \frac{xt(x(y-z) - ty)(s(t-x)^2 + tz(y+z) - xyz)}{z^3}, \; x \neq 0, \; z \neq 0, \; t \neq 0, \tag{10.1.109}$$

$$\begin{gathered} \text{eigenvalues}: \; x, \; \sqrt{\frac{t}{z}(ty - xy + xz)}, \; -\sqrt{\frac{t}{z}(ty - xy + xz)}, \\ \frac{st^2 - 2stx + tz^2 + ytz + sx^2 - yxz}{z^2}. \end{gathered} \tag{10.1.110}$$

Finally, we found the following 3-parameter 10-vertex solution

$$\tilde{c}^{10-\text{vert}}_{\text{rank}=4}(x, y, z) = \begin{pmatrix} x & y & y & \frac{y^2}{x} \\ 0 & 0 & -x & -y \\ 0 & -x & 0 & -y \\ z & 0 & 0 & x \end{pmatrix}, \quad \begin{gathered} \operatorname{tr}\tilde{c} = 2x, \\ \det\tilde{c} = -x(x^3 + zy^2), \; x \neq 0, \\ \text{eigenvalues: } \{x\}^{[2]}, \; \sqrt{x^2 + \frac{zy^2}{x}}, \; -\sqrt{x^2 + \frac{zy^2}{x}}. \end{gathered} \tag{10.1.111}$$

This solution is conjugated with the 4-vertex parameter-permutation solutions (10.1.33) of the form (which has the same the same eigenvalues as (10.1.111))

$$\tilde{c}^{4-\text{vert}}_{\text{rank}=4}(x, y, z) = \begin{pmatrix} x & 0 & 0 & 0 \\ 0 & 0 & x + \frac{y^2 z}{x^2} & 0 \\ 0 & x & 0 & 0 \\ 0 & 0 & 0 & x \end{pmatrix} \tag{10.1.112}$$

by the conjugated matrix

$$U^{10\text{to}4} = \begin{pmatrix} 0 & \frac{x}{z} & -\frac{x}{z} & 0 \\ -1 & -\frac{x^2}{yz} & \frac{x^2}{yz} & -\frac{y}{x} \\ 1 & -\frac{x^2}{yz} & \frac{x^2}{yz} & 0 \\ 0 & 0 & 1 & 1 \end{pmatrix}. \tag{10.1.113}$$

Because the matrix (10.1.113) cannot be presented as the Kronecker product $q \otimes_K q$ (10.1.16), therefore (10.1.111) and (10.1.112) are different solutions of the Yang–Baxter equation (10.1.12).

Further families of the higher vertex solutions to the constant Yang–Baxter equation (10.1.12) can be obtained from the above ones by using the q-conjugation (10.1.14).

10.2 Polyadic braid operators and higher braid equations

The polyadic version of the braid equation (10.1.1) was introduced in Duplij (2021b) (see section 9.5). Here we define higher analog of the Yang–Baxter operator and develop its connection with higher braid groups and quantum computations.

Let us consider a vector space $\mathcal{V}$ over a field $\Bbbk$. A *polyadic (n-ary) braid operator* $\mathbf{c}_{V^n}$ is defined as the mapping (Duplij 2021b)

$$\mathbf{c}_{V^n}\colon \overbrace{V \otimes \ldots \otimes V}^{n} \to \overbrace{V \otimes \ldots \otimes V}^{n}. \tag{10.2.1}$$

The polyadic analog of the braid equation (10.1.1) was introduced in Duplij (2021b) using the associative quiver technique (Duplij 2012).

Let us introduce n operators

$$\mathrm{A}_p\colon \overbrace{V \otimes \ldots \otimes V}^{2n-1} \to \overbrace{V \otimes \ldots \otimes V}^{2n-1}, \tag{10.2.2}$$

$$\mathrm{A}_p = \mathrm{id}_V^{\otimes(p-1)} \otimes \mathbf{c}_{V^n} \otimes \mathrm{id}_V^{\otimes(n-p)},\quad p = 1, \ldots, n, \tag{10.2.3}$$

i.e., p is a place of $\mathbf{c}_{V^n}$ instead of one id_V in $\mathrm{id}_V^{\otimes n}$. A system of $(n-1)$ *polyadic (n-ary) braid equations* is defined by

$$\mathrm{A}_1 \circ \mathrm{A}_2 \circ \mathrm{A}_3 \circ \mathrm{A}_4 \circ \ldots \circ \mathrm{A}_{n-2} \circ \mathrm{A}_{n-1} \circ \mathrm{A}_n \circ \mathrm{A}_1 \tag{10.2.4}$$

$$= \mathrm{A}_2 \circ \mathrm{A}_3 \circ \mathrm{A}_4 \circ \mathrm{A}_5 \circ \ldots \circ \mathrm{A}_{n-1} \circ \mathrm{A}_n \circ \mathrm{A}_1 \circ \mathrm{A}_2 \tag{10.2.5}$$

$$\vdots$$

$$= \mathrm{A}_n \circ \mathrm{A}_1 \circ \mathrm{A}_2 \circ \mathrm{A}_3 \circ \ldots \circ \mathrm{A}_{n-3} \circ \mathrm{A}_{n-2} \circ \mathrm{A}_{n-1} \circ \mathrm{A}_n. \tag{10.2.6}$$

Example 10.2.1. In the lowest non-binary case $n = 3$, we have the ternary braid operator $\mathbf{c}_{V^3}\colon V \otimes V \otimes V \to V \otimes V \otimes V$ and two ternary braid equations on $V^{\otimes 5}$ (see section 9.5)

$$\left(\mathbf{c}_{V^3}\otimes \mathrm{id}_V\otimes \mathrm{id}_V\right)\circ\left(\mathrm{id}_V\otimes \mathbf{c}_{V^3}\otimes \mathrm{id}_V\right)\circ\left(\mathrm{id}_V\otimes \mathrm{id}_V\otimes \mathbf{c}_{V^3}\right)\circ\left(\mathbf{c}_{V^3}\otimes \mathrm{id}_V\otimes \mathrm{id}_V\right)$$
$$=\left(\mathrm{id}_V\otimes \mathbf{c}_{V^3}\otimes \mathrm{id}_V\right)\circ\left(\mathrm{id}_V\otimes \mathrm{id}_V\otimes \mathbf{c}_{V^3}\right)\circ\left(\mathbf{c}_{V^3}\otimes \mathrm{id}_V\otimes \mathrm{id}_V\right)\circ\left(\mathrm{id}_V\otimes \mathbf{c}_{V^3}\otimes \mathrm{id}_V\right) \quad (10.2.7)$$
$$=\left(\mathrm{id}_V\otimes \mathrm{id}_V\otimes \mathbf{c}_{V^3}\right)\circ\left(\mathbf{c}_{V^3}\otimes \mathrm{id}_V\otimes \mathrm{id}_V\right)\circ\left(\mathrm{id}_V\otimes \mathbf{c}_{V^3}\otimes \mathrm{id}_V\right)\circ\left(\mathrm{id}_V\otimes \mathrm{id}_V\otimes \mathbf{c}_{V^3}\right).$$

Remark 10.2.2. The higher braid equations presented above differ from the generalized Yang–Baxter equations of Rowell *et al* (2010), Kitaev and Wang (2012), Chen (2012).

The higher braid operators (10.2.1) satisfying the higher braid equations (10.2.4)–(10.2.6) can represent the higher braid group (Duplij 2021a) using (10.1.6) and (10.2.3). By analogy with (10.1.6) we introduce m operators by

$$\mathrm{B}_i(m)\colon \overbrace{V\otimes \ldots\otimes V}^{m+n-2} \to \overbrace{V\otimes \ldots\otimes V}^{m+n-2},\quad \mathrm{B}_0(m)=(\mathrm{id}_V)^{\otimes(m+n-2)}, \quad (10.2.8)$$

$$\mathrm{B}_i(m)=\mathrm{id}_V^{\otimes(i-1)}\otimes \mathbf{c}_{V^n}\otimes \mathrm{id}_V^{\otimes(m-i-1)},\quad i=1,\ldots,m-1. \quad (10.2.9)$$

The representation $\pi_m^{[n]}$ of the higher braid group $\mathcal{B}_m^{[n+1]}$ (of $(n+1)$-degree in the notation of Duplij 2021a) (having $m-1$ generators $\boldsymbol{\sigma}_i$ and identity $\mathbf{e}$) is given by

$$\pi_m^{[n]}\colon \mathcal{B}_m^{[n+1]} \longrightarrow \mathrm{End}\, V^{\otimes(m+n-2)}, \quad (10.2.10)$$

$$\pi_m^{[n]}(\boldsymbol{\sigma}_i)=\mathrm{B}_i(m),\quad i=1,\ldots,m-1. \quad (10.2.11)$$

In this way, the generators $\boldsymbol{\sigma}_i$ of the higher braid group $\mathcal{B}_m^{[n+1]}$ satisfy the relations

- *n higher braid relations*

$$\overbrace{\boldsymbol{\sigma}_i\boldsymbol{\sigma}_{i+1}\ldots\boldsymbol{\sigma}_{i+n-2}\boldsymbol{\sigma}_{i+n-1}\boldsymbol{\sigma}_i}^{n+1} \quad (10.2.12)$$

$$=\boldsymbol{\sigma}_{i+1}\boldsymbol{\sigma}_{i+2}\ldots\boldsymbol{\sigma}_{i+n-1}\boldsymbol{\sigma}_i\boldsymbol{\sigma}_{i+1} \quad (10.2.13)$$
$$\vdots$$

$$=\boldsymbol{\sigma}_{i+n-1}\boldsymbol{\sigma}_i\boldsymbol{\sigma}_{i+1}\boldsymbol{\sigma}_{i+2}\ldots\boldsymbol{\sigma}_i\boldsymbol{\sigma}_{i+1}\boldsymbol{\sigma}_{i+n-1}, \quad (10.2.14)$$

$$i=1,\ldots,m-n, \quad (10.2.15)$$

- *n-ary far commutativity*

$$\overbrace{\boldsymbol{\sigma}_{i_1}\boldsymbol{\sigma}_{i_2}\ldots\boldsymbol{\sigma}_{i_{n-2}}\boldsymbol{\sigma}_{i_{n-1}}\boldsymbol{\sigma}_{i_n}}^{n} \quad (10.2.16)$$
$$\vdots$$

$$=\boldsymbol{\sigma}_{\tau(i_1)}\boldsymbol{\sigma}_{\tau(i_2)}\ldots\boldsymbol{\sigma}_{\tau(i_{n-2})}\boldsymbol{\sigma}_{\tau(i_{n-1})}\boldsymbol{\sigma}_{\tau(i_n)}, \quad (10.2.17)$$

$$\text{if all } \left|i_p-i_s\right|\geqslant n,\quad p,s=1,\ldots,n \quad (10.2.18)$$

where τ is an element of the permutation symmetry group $\tau \in S_n$. The relations (10.2.12)–(10.2.17) coincide with those from Duplij (2021a), obtained by another method, that is via the polyadic-binary correspondence.

In the case $m = 4$ and $n = 3$ the higher braid group $\mathcal{B}_4^{[4]}$ is represented by (10.2.7) and generated by 3 generators $\boldsymbol{\sigma}_1, \boldsymbol{\sigma}_2, \boldsymbol{\sigma}_3$, which satisfy 2 braid relations only (without far commutativity)

$$\boldsymbol{\sigma}_1\boldsymbol{\sigma}_2\boldsymbol{\sigma}_3\boldsymbol{\sigma}_1 = \boldsymbol{\sigma}_2\boldsymbol{\sigma}_3\boldsymbol{\sigma}_1\boldsymbol{\sigma}_2 = \boldsymbol{\sigma}_3\boldsymbol{\sigma}_1\boldsymbol{\sigma}_2\boldsymbol{\sigma}_3. \tag{10.2.19}$$

According to (10.2.16)–(10.2.17), the far commutativity relations appear when the number of elements of the higher braid groups satisfy

$$m \geqslant m_{\min} = n(n-1) + 2, \tag{10.2.20}$$

such that all conditions (10.2.18) should hold. Thus, to have the far commutativity relations in the ordinary (binary) braid group (10.1.5) we need 3 generators and $\mathcal{B}_4$, while for $n = 3$ we need at least 7 generators $\boldsymbol{\sigma}_i$ and $\mathcal{B}_8^{[4]}$ (see example 7.12 in Duplij 2021a).

In the concrete realization of $\mathcal{V}$ as a d-dimensional Euclidean vector space V_d over the complex numbers $\mathbb{C}$ and basis $\{e_i\}$, $i = 1, \dots, d$, the polyadic (n-ary) braid operator $\mathbf{c}_{V^n}$ becomes a matrix C_{d^n} of size $d^n \times d^n$, which satisfies the $n-1$ higher braid (10.2.4)–(10.2.6) in matrix form. In the components, the matrix braid operator is

$$C_{d^n} \circ (e_{i_1} \otimes e_{i_2} \otimes \dots \otimes e_{i_n}) = \sum_{j_1', j_2' \dots j_n' = 1}^{d} c_{i_1 i_2 \dots i_n}^{\ j_1' j_2' \dots j_n'} \cdot e_{j_1'} \otimes e_{j_2'} \otimes \dots \otimes e_{j_n'}. \tag{10.2.21}$$

Thus, we have d^{2n} unknowns in C_{d^n} satisfying $(n-1)d^{4n-2}$ equations (10.2.4)–(10.2.6) in components of polynomial power $n+1$. In the minimal non-binary case $n = 3$, we have $2d^{10}$ equations of power 4 for d^6 unknowns, e.g., even for $d = 2$ we have 2048 for 64 components, and for $d = 3$ there are 118 098 equations for 729 components. Therefore, solving the matrix higher braid equations directly is cumbersome, and only particular cases are possible to investigate, for instance by using permutation matrices (10.1.28), or the star and circle matrices (10.1.76)–(10.1.77).

10.3 Solutions to the ternary braid equations

Here we consider some special solutions to the minimal ternary version ($n = 3$) of the polyadic braid equation (10.2.4)–(10.2.6), the ternary braid equation (10.2.7).

10.3.1 Constant matrix solutions

Let us consider the following two-dimensional vector space $V \equiv V_{d=2}$ (which is important for quantum computations) and the component matrix realization (10.2.21) of the ternary braiding operator C_8: $V \otimes V \otimes V \to V \otimes V \otimes V$ as

$$C_8 \circ (e_{i_1} \otimes e_{i_2} \otimes e_{i_3}) = \sum_{j_1', j_2' j_3' = 1}^{2} c_{i_1 i_2 i_3}^{\ j_1' j_2' j_3'} \cdot e_{j_1'} \otimes e_{j_2'} \otimes e_{j_3'}, \quad i_{1,2,3}, j'_{1,2,3} = 1, 2. \tag{10.3.1}$$

We now turn (10.3.1) to the standard matrix form (just to fix notations) by introducing the 8-dimensional vector space $\tilde{V}_8 = V \otimes V \otimes V$ with the natural basis $\tilde{e}_{\tilde{k}} = \{e_1 \otimes e_1 \otimes e_1, e_1 \otimes e_1 \otimes e_2, \dots, e_2 \otimes e_2 \otimes e_2\}$, where $\tilde{k} = 1, \dots, 8$ is a cumulative index. The linear operator $\tilde{C}_8: \tilde{V}_8 \to \tilde{V}_8$ corresponding to (10.3.1) is given by the 8×8 matrix $\tilde{c}_{\tilde{i}\tilde{j}}$ as $\tilde{C}_8 \circ \tilde{e}_{\tilde{i}} = \sum_{\tilde{j}=1}^{8} \tilde{c}_{\tilde{i}\tilde{j}} \cdot \tilde{e}_{\tilde{j}}$. The operators (10.2.2)–(10.2.3) become three 32×32 matrices $\tilde{A}_{1,2,3}$ as

$$\tilde{A}_1 = \tilde{c} \otimes_K I_2 \otimes_K I_2, \quad \tilde{A}_2 = I_2 \otimes_K \tilde{c} \otimes_K I_2, \quad \tilde{A}_3 = I_2 \otimes_K I_2 \otimes_K \tilde{c}, \tag{10.3.2}$$

where $\otimes_K$ is the Kronecker product of matrices and I_2 is the 2×2 identity matrix. In this notation the operator ternary braid equations (10.2.7) become the matrix equations (see (10.2.4)–(10.2.6) with $n = 3$)

$$\tilde{A}_1\tilde{A}_2\tilde{A}_3\tilde{A}_1 = \tilde{A}_2\tilde{A}_3\tilde{A}_1\tilde{A}_2 = \tilde{A}_3\tilde{A}_1\tilde{A}_2\tilde{A}_3, \tag{10.3.3}$$

which we call the *total matrix ternary braid equations.* Some weaker versions of ternary braiding are described by the *partial braid equations*

$$\text{partial 12–braid equation } \tilde{A}_1\tilde{A}_2\tilde{A}_3\tilde{A}_1 = \tilde{A}_2\tilde{A}_3\tilde{A}_1\tilde{A}_2, \tag{10.3.4}$$

$$\text{partial 13–braid equation } \tilde{A}_1\tilde{A}_2\tilde{A}_3\tilde{A}_1 = \tilde{A}_3\tilde{A}_1\tilde{A}_2\tilde{A}_3, \tag{10.3.5}$$

$$\text{partial 23–braid equation } \tilde{A}_2\tilde{A}_3\tilde{A}_1\tilde{A}_2 = \tilde{A}_3\tilde{A}_1\tilde{A}_2\tilde{A}_3, \tag{10.3.6}$$

where, obviously, two of them are independent. It follows from (10.2.4)–(10.2.6) that the weaker versions of braiding are possible for $n \geqslant 3$, only, so for higher than binary braiding (the Yang–Baxter equation (10.1.8)).

Thus, comparing (10.3.3) and (10.2.19) we conclude that (for each invertible matrix $\tilde{c}$ in (10.3.2) satisfying (10.3.3)) the isomorphism $\tilde{\pi}_4^{[4]}: \boldsymbol{\sigma}_i \mapsto \tilde{A}_i$, $i = 1, 2, 3$ gives a representation of the braid group $\mathcal{B}_4^{[4]}$ by 32×32 matrices over $\mathbb{C}$.

Now we can generate families of solutions corresponding to (10.3.2)–(10.3.3) in the following way. Consider an invertible operator $Q: V \to V$ in the two-dimensional vector space $V \equiv V_{d=2}$. In the basis $\{e_1, e_2\}$ its 2×2 matrix q is given by $Q \circ e_i = \sum_{j=1}^{2} q_{ij} \cdot e_j$. In the natural 8-dimensional basis $\tilde{e}_{\tilde{k}}$ the tensor product of operators $Q \otimes Q \otimes Q$ is presented by the Kronecker product of matrices $\tilde{q}_8 = q \otimes_K q \otimes_K q$. Let the 8×8 matrix $\tilde{c}$ be a fixed solution to the ternary braid matrix equations (10.3.3). Then, the family of solutions $\tilde{c}(q)$ corresponding to the invertible 2×2 matrix q is the conjugation of $\tilde{c}$ by $\tilde{q}_8$ so that

$$\tilde{c}(q) = \tilde{q}_8 \tilde{c} \tilde{q}_8^{-1} = (q \otimes_K q \otimes_K q)\tilde{c}(q^{-1} \otimes_K q^{-1} \otimes_K q^{-1}). \tag{10.3.7}$$

This also follows directly from the conjugation of the braid equations (10.3.3)–(10.3.6) by $q \otimes_K q \otimes_K q \otimes_K q \otimes_K q$ and (10.3.2). If we include the obvious invariance of the braid equations with the respect of an overall factor $t \in \mathbb{C}$, the general family of solutions becomes (see the Yang–Baxter equation; Hietarinta 1993)

$$\tilde{c}(q, t) = t\tilde{q}_8\tilde{c}\tilde{q}_8^{-1} = t(q \otimes_K q \otimes_K q)\tilde{c}(q^{-1} \otimes_K q^{-1} \otimes_K q^{-1}). \tag{10.3.8}$$

Let

$$q = \begin{pmatrix} a & b \\ c & d \end{pmatrix} \in \mathrm{GL}(2, \mathbb{C}), \tag{10.3.9}$$

and then the manifest form of $\tilde{q}_8$ is

$$\tilde{q}_8 = \begin{pmatrix} a^3 & a^2b & a^2b & ab^2 & a^2b & ab^2 & ab^2 & b^3 \\ a^2c & a^2d & abc & abd & abc & abd & b^2c & b^2d \\ a^2c & abc & a^2d & abd & abc & b^2c & abd & b^2d \\ ac^2 & acd & acd & ad^2 & bc^2 & bcd & bcd & bd^2 \\ a^2c & abc & abc & b^2c & a^2d & abd & abd & b^2d \\ ac^2 & acd & bc^2 & bcd & acd & ad^2 & bcd & bd^2 \\ ac^2 & bc^2 & acd & bcd & acd & bcd & ad^2 & bd^2 \\ c^3 & c^2d & c^2d & cd^2 & c^2d & cd^2 & cd^2 & d^3 \end{pmatrix}. \tag{10.3.10}$$

It is important that not every conjugation matrix has this very special form (10.3.10), and that therefore, in general, conjugated matrices are different solutions of the ternary braid equations (10.3.3). The matrix $\tilde{q}_8^{\star}\tilde{q}_8$ ($\star$ being the Hermitian conjugation) is diagonal (this case is important for further classification similar to the binary one, Dye 2003), when the conditions

$$ab^* + cd^* = 0 \tag{10.3.11}$$

hold, and so the matrix q has the special form (depending of 3 complex parameters, for $d \neq 0$)

$$q = \begin{pmatrix} a & b \\ -a\dfrac{b^*}{d^*} & d \end{pmatrix}. \tag{10.3.12}$$

We can present the families (10.3.7) for different ranks, because the conjugation by an invertible matrix does not change rank. To avoid demanding (10.3.11), due to the cumbersome calculations involved, we restrict ourselves to a triangle matrix for q (10.3.9).

In general, there are $8 \times 8 = 64$ unknowns (elements of the matrix $\tilde{c}$), and each partial braid equation (10.3.4)–(10.3.6) gives $32 \times 32 = 1024$ conditions (of power 4) for the elements of $\tilde{c}$, while the total braid equations (10.3.3) give twice as many conditions $1024 \times 2 = 2048$ (see the binary case: 64 cubic equations for 16 unknowns (10.1.8)). This means that even in the ternary case the higher braid system of equations is hugely overdetermined, and finding even the simplest solutions is a non-trivial task.

10.3.2 Permutation and parameter-permutation 8-vertex solutions

First we consider the case when $\tilde{c}$ is a binary (or logical) matrix consisting of $\{0, 1\}$ only, and, moreover, it is a permutation matrix (see subsection 10.1.4). In the latter case $\tilde{c}$ can be considered as a matrix over the field $\mathbb{F}_2$ (Galois field $GF(2)$). In total, there are $8! = 40\,320$ permutation matrices of the size 8×8. All of them are invertible of full rank 8, because they are obtained from the identity matrix by permutation of rows and columns.

We have found the following four invertible 8-vertex permutation matrix solutions to the ternary braid equations (10.3.3)

$$\tilde{c}^{\,\mathrm{bisymm1}}_{\mathrm{rank}\,=\,8} = \begin{pmatrix} 1&0&0&0&0&0&0&0\\ 0&0&0&0&0&0&1&0\\ 0&0&1&0&0&0&0&0\\ 0&0&0&0&1&0&0&0\\ 0&0&0&1&0&0&0&0\\ 0&0&0&0&0&1&0&0\\ 0&1&0&0&0&0&0&0\\ 0&0&0&0&0&0&0&1 \end{pmatrix}, \quad \tilde{c}^{\,\mathrm{bisymm2}}_{\mathrm{rank}\,=\,8} = \begin{pmatrix} 0&0&0&0&0&0&0&1\\ 0&1&0&0&0&0&0&0\\ 0&0&0&0&0&1&0&0\\ 0&0&0&1&0&0&0&0\\ 0&0&0&0&1&0&0&0\\ 0&0&1&0&0&0&0&0\\ 0&0&0&0&0&0&1&0\\ 1&0&0&0&0&0&0&0 \end{pmatrix}, \tag{10.3.13}$$

$$\mathrm{tr}\tilde{c} = 4, \quad \det\tilde{c} = 1, \quad \text{eigenvalues:}\{1\}^{[4]},\ \{-1\}^{[4]},$$

$$\tilde{c}^{\,\mathrm{symm1}}_{\mathrm{rank}\,=\,8} = \begin{pmatrix} 1&0&0&0&0&0&0&0\\ 0&0&0&0&1&0&0&0\\ 0&0&0&0&0&0&0&1\\ 0&0&0&1&0&0&0&0\\ 0&1&0&0&0&0&0&0\\ 0&0&0&0&0&1&0&0\\ 0&0&0&0&0&0&1&0\\ 0&0&1&0&0&0&0&0 \end{pmatrix}, \quad \tilde{c}^{\,\mathrm{symm2}}_{\mathrm{rank}\,=\,8} = \begin{pmatrix} 0&0&0&0&0&1&0&0\\ 0&1&0&0&0&0&0&0\\ 0&0&1&0&0&0&0&0\\ 0&0&0&0&0&0&1&0\\ 0&0&0&0&1&0&0&0\\ 1&0&0&0&0&0&0&0\\ 0&0&0&1&0&0&0&0\\ 0&0&0&0&0&0&0&1 \end{pmatrix}, \tag{10.3.14}$$

$$\mathrm{tr}\tilde{c} = 4, \quad \det\tilde{c} = 1, \quad \text{eigenvalues:}\{1\}^{[4]},\ \{-1\}^{[4]}.$$

The first two solutions (10.3.13) are given by bisymmetric permutation matrices (see (10.1.31)), and we call them 8-vertex bisymm1 and bisymm2 respectively. The second two solutions (10.3.14) are symmetric matrices only (we call them 8-vertex symm1 and symm2), but one matrix is a reflection of the other with respect to the minor diagonal (making them mutually persymmetric). No 90°-symmetric (see (10.1.32)) solution for the ternary braid equations (10.3.3) was found. The bisymmetric and symmetric matrices have the same eigenvalues, and are therefore pairwise conjugate, but not q-conjugate, because the conjugation matrices do not have the form (10.3.10). Thus they are 4 different permutation solutions to the ternary braid equations (10.3.3). Note that the bisymm1 solution (10.3.13) coincides with the three-qubit swap operator introduced in Ballard and Wu (2011).

All the permutation solutions are reflections (or involutions) $\tilde{c}^2 = I_8$ having $\det\tilde{c} = +1$, eigenvalues $\{1, -1\}$, and are semi-magic squares (the sums in rows and columns are 1, but not the sums in both diagonals). The 8-vertex permutation matrix solutions do not form a binary or ternary group, because they are not closed with respect to multiplication.

By analogy with (10.1.33)–(10.1.34), we obtain the 8-vertex parameter-permutation solutions from (10.3.13)–(10.3.14) by replacing units with parameters and then solving the ternary braid equations (10.3.3). Each type of the permutation solutions bisymm1, 2 and symm1, 2 from (10.3.13)–(10.3.14) will give a corresponding series of parameter-permutation solutions over $\mathbb{C}$. The ternary braid maps are determined up to a general complex factor (see (10.1.14) for the Yang–Baxter maps and (10.3.8)), and therefore we can present all the parameter-permutation solutions in polynomial form.

• The bisymm1 series consists of 2 two-parameter matrices with and 2 two-parameter matrices

$$\tilde{c}^{\,\mathrm{bisymm1},\,1}_{\mathrm{rank}\,=\,8}(x,y)=\begin{pmatrix} xy & 0 & 0 & 0 & 0 & 0 & 0 & 0\\ 0 & 0 & 0 & 0 & 0 & 0 & \pm y^2 & 0\\ 0 & 0 & xy & 0 & 0 & 0 & 0 & 0\\ 0 & 0 & 0 & 0 & \pm x^2 & 0 & 0 & 0\\ 0 & 0 & 0 & \pm y^2 & 0 & 0 & 0 & 0\\ 0 & 0 & 0 & 0 & 0 & xy & 0 & 0\\ 0 & \pm x^2 & 0 & 0 & 0 & 0 & 0 & 0\\ 0 & 0 & 0 & 0 & 0 & 0 & 0 & xy \end{pmatrix}, \tag{10.3.15}$$

$\mathrm{tr}\tilde{c}=4xy$, $\det\tilde{c}=x^8y^8$, $x,y\neq 0$, eigenvalues: $\{xy\}^{[6]}$, $\{-xy\}^{[2]}$,

$$\tilde{c}^{\,\mathrm{bisymm1},\,2}_{\mathrm{rank}\,=\,8}(x,y)=\begin{pmatrix} xy & 0 & 0 & 0 & 0 & 0 & 0 & 0\\ 0 & 0 & 0 & 0 & 0 & 0 & \pm y^2 & 0\\ 0 & 0 & xy & 0 & 0 & 0 & 0 & 0\\ 0 & 0 & 0 & 0 & \pm x^2 & 0 & 0 & 0\\ 0 & 0 & 0 & \mp y^2 & 0 & 0 & 0 & 0\\ 0 & 0 & 0 & 0 & 0 & xy & 0 & 0\\ 0 & \mp x^2 & 0 & 0 & 0 & 0 & 0 & 0\\ 0 & 0 & 0 & 0 & 0 & 0 & 0 & xy \end{pmatrix}, \tag{10.3.16}$$

$\mathrm{tr}\tilde{c}=4xy$, $\det\tilde{c}=x^8y^8$, $x,y\neq 0$, eigenvalues: $\{xy\}^{[4]}$, $\{ixy\}^{[2]}$, $\{-ixy\}^{[2]}$

.

• The bisymm2 series consists of 4 two-parameter matrices

$$\tilde{c}^{\,\mathrm{bisymm2},\,1}_{\mathrm{rank}\,=\,8}(x,y)=\begin{pmatrix} 0 & 0 & 0 & 0 & 0 & 0 & 0 & x^6\\ 0 & \pm x^3y^3 & 0 & 0 & 0 & 0 & 0 & 0\\ 0 & 0 & 0 & 0 & 0 & x^4y^2 & 0 & 0\\ 0 & 0 & 0 & \pm x^3y^3 & 0 & 0 & 0 & 0\\ 0 & 0 & 0 & 0 & \pm x^3y^3 & 0 & 0 & 0\\ 0 & 0 & x^2y^4 & 0 & 0 & 0 & 0 & 0\\ 0 & 0 & 0 & 0 & 0 & 0 & \pm x^3y^3 & 0\\ y^6 & 0 & 0 & 0 & 0 & 0 & 0 & 0 \end{pmatrix}, \tag{10.3.17}$$

$\mathrm{tr}\tilde{c}=\pm 4x^3y^3$, $\det\tilde{c}^{\,\mathrm{bisymm2}}_{\mathrm{rank}\,=\,8}(x,y)=x^{24}y^{24}$, $x,y\neq 0$,

eigenvalues : $\{x^3y^3\}^{[2]}$, $\{-x^3y^3\}^{[2]}$, $\{\pm x^3y^3\}^{[4]}$,

$$\tilde{c}_{\mathrm{rank}\,=\,8}^{\mathrm{bisymm2},\,2}(x,\,y)=\begin{pmatrix} 0 & 0 & 0 & 0 & 0 & 0 & 0 & x^6 \\ 0 & \pm x^3y^3 & 0 & 0 & 0 & 0 & 0 & 0 \\ 0 & 0 & 0 & 0 & 0 & x^4y^2 & 0 & 0 \\ 0 & 0 & 0 & \pm x^3y^3 & 0 & 0 & 0 & 0 \\ 0 & 0 & 0 & 0 & \pm x^3y^3 & 0 & 0 & 0 \\ 0 & 0 & -x^2y^4 & 0 & 0 & 0 & 0 & 0 \\ 0 & 0 & 0 & 0 & 0 & 0 & \pm x^3y^3 & 0 \\ -y^6 & 0 & 0 & 0 & 0 & 0 & 0 & 0 \end{pmatrix}, \tag{10.3.18}$$

$$\mathrm{tr}\tilde{c} = \pm\, 4x^3y^3,\ \det \tilde{c}_{\mathrm{rank}\,=\,8}^{\mathrm{bisymm2}}(x,\,y) = x^{24}y^{24},\ x,\ y \neq 0,$$

$$\text{eigenvalues}: \quad \{ix^3y^3\}^{[2]},\ \{-ix^3y^3\}^{[2]},\ \{\pm x^3y^3\}^{[4]}.$$

- The symm1 series consists of 4 two-parameter matrices

$$\tilde{c}_{\mathrm{rank}\,=\,8}^{\mathrm{symm1},\,1}(x,\,y)=\begin{pmatrix} xy & 0 & 0 & 0 & 0 & 0 & 0 & 0 \\ 0 & 0 & 0 & 0 & \pm xy & 0 & 0 & 0 \\ 0 & 0 & 0 & 0 & 0 & 0 & 0 & y^2 \\ 0 & 0 & 0 & xy & 0 & 0 & 0 & 0 \\ 0 & \pm xy & 0 & 0 & 0 & 0 & 0 & 0 \\ 0 & 0 & 0 & 0 & 0 & xy & 0 & 0 \\ 0 & 0 & 0 & 0 & 0 & 0 & xy & 0 \\ 0 & 0 & x^2 & 0 & 0 & 0 & 0 & 0 \end{pmatrix}, \tag{10.3.19}$$

$$\mathrm{tr}\tilde{c} = 4xy,\ \det \tilde{c} = x^8y^8,\ x,\ y \neq 0,\ \text{eigenvalues: } \{xy\}^{[6]},\ \{-xy\}^{[2]},$$

$$\tilde{c}_{\mathrm{rank}\,=\,8}^{\mathrm{symm1},\,2}(x,\,y)=\begin{pmatrix} xy & 0 & 0 & 0 & 0 & 0 & 0 & 0 \\ 0 & 0 & 0 & 0 & \pm xy & 0 & 0 & 0 \\ 0 & 0 & 0 & 0 & 0 & 0 & 0 & y^2 \\ 0 & 0 & 0 & xy & 0 & 0 & 0 & 0 \\ 0 & \mp xy & 0 & 0 & 0 & 0 & 0 & 0 \\ 0 & 0 & 0 & 0 & 0 & xy & 0 & 0 \\ 0 & 0 & 0 & 0 & 0 & 0 & xy & 0 \\ 0 & 0 & -x^2 & 0 & 0 & 0 & 0 & 0 \end{pmatrix}, \tag{10.3.20}$$

$$\mathrm{tr}\tilde{c} = 4xy,\ \det \tilde{c} = x^8y^8,\ x,\ y \neq 0,\ \text{eigenvalues: } \{xy\}^{[6]},\ \{-xy\}^{[2]},$$

- The symm2 series consists of 4 two-parameter matrices

$$\tilde{c}_{\mathrm{rank}\,=\,8}^{\mathrm{symm2},\,1}(x,\,y)=\begin{pmatrix} 0 & 0 & 0 & 0 & 0 & y^2 & 0 & 0 \\ 0 & xy & 0 & 0 & 0 & 0 & 0 & 0 \\ 0 & 0 & xy & 0 & 0 & 0 & 0 & 0 \\ 0 & 0 & 0 & 0 & 0 & 0 & \pm xy & 0 \\ 0 & 0 & 0 & 0 & xy & 0 & 0 & 0 \\ x^2 & 0 & 0 & 0 & 0 & 0 & 0 & 0 \\ 0 & 0 & 0 & \pm xy & 0 & 0 & 0 & 0 \\ 0 & 0 & 0 & 0 & 0 & 0 & 0 & xy \end{pmatrix}, \tag{10.3.21}$$

$$\mathrm{tr}\tilde{c} = 4xy,\ \det \tilde{c} = x^8y^8,\ x,\ y \neq 0,\ \text{eigenvalues: } \{xy\}^{[4]},\ \{ixy\}^{[2]},\ \{-ixy\}^{[2]}.$$

$$\tilde{c}^{\text{symm}2,\,2}_{\text{rank}\,=\,8}(x,y) = \begin{pmatrix} 0 & 0 & 0 & 0 & 0 & y^2 & 0 & 0 \\ 0 & xy & 0 & 0 & 0 & 0 & 0 & 0 \\ 0 & 0 & xy & 0 & 0 & 0 & 0 & 0 \\ 0 & 0 & 0 & 0 & 0 & 0 & \pm xy & 0 \\ 0 & 0 & 0 & 0 & xy & 0 & 0 & 0 \\ -x^2 & 0 & 0 & 0 & 0 & 0 & 0 & 0 \\ 0 & 0 & 0 & \mp xy & 0 & 0 & 0 & 0 \\ 0 & 0 & 0 & 0 & 0 & 0 & 0 & xy \end{pmatrix}, \tag{10.3.22}$$

$$\text{tr}\tilde{c} = 4xy,\ \det\tilde{c} = x^8y^8,\ x, y \neq 0,\ \text{eigenvalues: } \{xy\}^{[4]},\ \{ixy\}^{[2]},\ \{-ixy\}^{[2]}.$$

The above matrices with the same eigenvalues are similar, but their conjugation matrices do not have the form of the triple Kronecker product (10.3.10), and therefore all of them together are 16 different two-parameter invertible solutions to the ternary braid equations (10.3.3). Further families of solutions can be obtained using ternary q-conjugation (10.3.8).

10.3.3 Group structure of the star and circle 8-vertex matrices

Here we investigate the group structure of 8×8 matrices by analogy with the star-like (10.1.35) and circle-like (10.1.36) 4×4 matrices which are connected with our 8-vertex constant solutions (10.3.15)–(10.3.22) to the ternary braid equations (10.3.3).

Let us introduce the *star-like* 8×8 *matrices* (see (10.1.35)) which correspond to the bisymm series (10.3.15)–(10.3.18)

$$N'_{\text{star}1} = \begin{pmatrix} x & 0 & 0 & 0 & 0 & 0 & 0 & 0 \\ 0 & 0 & 0 & 0 & 0 & 0 & y & 0 \\ 0 & 0 & z & 0 & 0 & 0 & 0 & 0 \\ 0 & 0 & 0 & 0 & s & 0 & 0 & 0 \\ 0 & 0 & 0 & t & 0 & 0 & 0 & 0 \\ 0 & 0 & 0 & 0 & 0 & u & 0 & 0 \\ 0 & v & 0 & 0 & 0 & 0 & 0 & 0 \\ 0 & 0 & 0 & 0 & 0 & 0 & 0 & w \end{pmatrix}, \quad N'_{\text{star}2} = \begin{pmatrix} 0 & 0 & 0 & 0 & 0 & 0 & 0 & y \\ 0 & x & 0 & 0 & 0 & 0 & 0 & 0 \\ 0 & 0 & 0 & 0 & 0 & s & 0 & 0 \\ 0 & 0 & 0 & z & 0 & 0 & 0 & 0 \\ 0 & 0 & 0 & 0 & u & 0 & 0 & 0 \\ 0 & 0 & t & 0 & 0 & 0 & 0 & 0 \\ 0 & 0 & 0 & 0 & 0 & 0 & w & 0 \\ v & 0 & 0 & 0 & 0 & 0 & 0 & 0 \end{pmatrix}, \tag{10.3.23}$$

$$\text{tr}N' = x + z + u + w,\ \det N' = stuvwxyz,\ s, t, u, v, w, x, y, z \neq 0,$$
$$\text{eigenvalues: } x, z, u, w, -\sqrt{yv}, \sqrt{yv}, -\sqrt{st}, \sqrt{st},$$

and the *circle-like* 8×8 *matrices* (see (10.1.36)) which correspond to the symm series (10.3.19)–(10.3.22)

$$N'_{\text{circ}1} = \begin{pmatrix} x & 0 & 0 & 0 & 0 & 0 & 0 & 0 \\ 0 & 0 & 0 & 0 & y & 0 & 0 & 0 \\ 0 & 0 & 0 & 0 & 0 & 0 & 0 & z \\ 0 & 0 & 0 & s & 0 & 0 & 0 & 0 \\ 0 & t & 0 & 0 & 0 & 0 & 0 & 0 \\ 0 & 0 & 0 & 0 & 0 & u & 0 & 0 \\ 0 & 0 & 0 & 0 & 0 & 0 & v & 0 \\ 0 & 0 & w & 0 & 0 & 0 & 0 & 0 \end{pmatrix}, \quad N'_{\text{circ}2} = \begin{pmatrix} 0 & 0 & 0 & 0 & 0 & y & 0 & 0 \\ 0 & x & 0 & 0 & 0 & 0 & 0 & 0 \\ 0 & 0 & s & 0 & 0 & 0 & 0 & 0 \\ 0 & 0 & 0 & 0 & 0 & 0 & z & 0 \\ 0 & 0 & 0 & 0 & u & 0 & 0 & 0 \\ t & 0 & 0 & 0 & 0 & 0 & 0 & 0 \\ 0 & 0 & 0 & w & 0 & 0 & 0 & 0 \\ 0 & 0 & 0 & 0 & 0 & 0 & 0 & v \end{pmatrix}, \tag{10.3.24}$$

$$\text{tr}N' = x + s + u + v,\ \det N' = stuvwxyz,\ s, t, u, v, w, x, y, z \neq 0,$$
$$\text{eigenvalues: } x, s, u, v, -\sqrt{ty}, \sqrt{ty}, -\sqrt{wz}, \sqrt{wz}. \tag{10.3.25}$$

Denote the corresponding sets by $\mathsf{N}'_{\text{star1},\,2} = \left\{N'_{\text{star1},\,2}\right\}$ and $\mathsf{N}'_{\text{circ1},\,2} = \left\{N'_{\text{circ1},\,2}\right\}$, then we have for them (which differs from 4×4 matrix sets (10.1.41))

$$\begin{aligned} \mathsf{M}'_{\text{full}} &= \mathsf{N}'_{\text{star1}} \bigcup \mathsf{N}'_{\text{star2}} \bigcup \mathsf{N}'_{\text{circ1}} \bigcup \mathsf{N}'_{\text{circ2}}, \\ \mathsf{N}'_{\text{star1}} &\bigcap \mathsf{N}'_{\text{star2}} \bigcap \mathsf{N}'_{\text{circ1}} \bigcap \mathsf{N}'_{\text{circ2}} = \mathsf{D}, \end{aligned} \tag{10.3.26}$$

where D is the set of diagonal 8×8 matrices. Again, as for 4×4 star-like and circle-like matrices, there are no closed binary multiplications for the sets of 8-vertex matrices (10.3.23)–(10.3.25). Nevertheless, we have the following triple set products

$$\mathsf{N}'_{\text{star1}}\mathsf{N}'_{\text{star1}}\mathsf{N}'_{\text{star1}} = \mathsf{N}'_{\text{star1}}, \tag{10.3.27}$$

$$\mathsf{N}'_{\text{star2}}\mathsf{N}'_{\text{star2}}\mathsf{N}'_{\text{star2}} = \mathsf{N}'_{\text{star2}}, \tag{10.3.28}$$

$$\mathsf{N}'_{\text{circ1}}\mathsf{N}'_{\text{circ1}}\mathsf{N}'_{\text{circ1}} = \mathsf{N}'_{\text{circ1}}, \tag{10.3.29}$$

$$\mathsf{N}'_{\text{circ2}}\mathsf{N}'_{\text{circ2}}\mathsf{N}'_{\text{circ2}} = \mathsf{N}'_{\text{circ2}}, \tag{10.3.30}$$

which should be compared with the analogous 4×4 matrices (10.1.39)–(10.1.40): note that now we do not have pentuple products.

Using the definitions (10.1.42)–(10.1.45), we interpret the closed products (10.3.27)–(10.3.28) and (10.3.29)–(10.3.30) as the multiplication $\mu^{\circ 2}$ (being the ordinary triple matrix product) of the *ternary semigroups* $\mathcal{S}^{(3)}_{\text{star1},\,2}(8, \mathbb{C}) = \left\{\mathsf{N}'_{\text{star1},\,2} \middle| \mu^{\circ 2}\right\}$ and $\mathcal{S}^{(3)}_{\text{circ1},\,2}(8, \mathbb{C}) = \left\{\mathsf{N}'_{\text{circ1},\,2} \middle| \mu^{\circ 2}\right\}$, respectively. The corresponding querelements (10.1.42) are given by

$$\bar{N}'_{\text{star1}} = N'^{-1}_{\text{star1}} = \begin{pmatrix} \frac{1}{x} & 0 & 0 & 0 & 0 & 0 & 0 & 0 \\ 0 & 0 & 0 & 0 & 0 & 0 & \frac{1}{v} & 0 \\ 0 & 0 & \frac{1}{z} & 0 & 0 & 0 & 0 & 0 \\ 0 & 0 & 0 & 0 & \frac{1}{t} & 0 & 0 & 0 \\ 0 & 0 & 0 & \frac{1}{s} & 0 & 0 & 0 & 0 \\ 0 & 0 & 0 & 0 & 0 & \frac{1}{u} & 0 & 0 \\ 0 & \frac{1}{y} & 0 & 0 & 0 & 0 & 0 & 0 \\ 0 & 0 & 0 & 0 & 0 & 0 & 0 & \frac{1}{w} \end{pmatrix}, \tag{10.3.31}$$

$$\bar{N}'_{\mathrm{star2}} = N'^{-1}_{\mathrm{star2}} = \begin{pmatrix} 0 & 0 & 0 & 0 & 0 & 0 & 0 & \frac{1}{v} \\ 0 & \frac{1}{x} & 0 & 0 & 0 & 0 & 0 & 0 \\ 0 & 0 & 0 & 0 & 0 & \frac{1}{t} & 0 & 0 \\ 0 & 0 & 0 & \frac{1}{z} & 0 & 0 & 0 & 0 \\ 0 & 0 & 0 & 0 & \frac{1}{u} & 0 & 0 & 0 \\ 0 & 0 & \frac{1}{s} & 0 & 0 & 0 & 0 & 0 \\ 0 & 0 & 0 & 0 & 0 & 0 & \frac{1}{w} & 0 \\ \frac{1}{y} & 0 & 0 & 0 & 0 & 0 & 0 & 0 \end{pmatrix}, s, t, u, v, w, x, y, z \neq 0, \tag{10.3.32}$$

and

$$\bar{N}'_{\mathrm{circ1}} = N'^{-1}_{\mathrm{circ1}} = \begin{pmatrix} \frac{1}{x} & 0 & 0 & 0 & 0 & 0 & 0 & 0 \\ 0 & 0 & 0 & 0 & \frac{1}{t} & 0 & 0 & 0 \\ 0 & 0 & 0 & 0 & 0 & 0 & 0 & \frac{1}{w} \\ 0 & 0 & 0 & \frac{1}{s} & 0 & 0 & 0 & 0 \\ 0 & \frac{1}{y} & 0 & 0 & 0 & 0 & 0 & 0 \\ 0 & 0 & 0 & 0 & 0 & \frac{1}{u} & 0 & 0 \\ 0 & 0 & 0 & 0 & 0 & 0 & \frac{1}{v} & 0 \\ 0 & 0 & \frac{1}{z} & 0 & 0 & 0 & 0 & 0 \end{pmatrix}, \tag{10.3.33}$$

$$\bar{N}'_{\mathrm{circ2}} = N'^{-1}_{\mathrm{circ2}} = \begin{pmatrix} 0 & 0 & 0 & 0 & 0 & \frac{1}{t} & 0 & 0 \\ 0 & \frac{1}{x} & 0 & 0 & 0 & 0 & 0 & 0 \\ 0 & 0 & \frac{1}{s} & 0 & 0 & 0 & 0 & 0 \\ 0 & 0 & 0 & 0 & 0 & 0 & \frac{1}{w} & 0 \\ 0 & 0 & 0 & 0 & \frac{1}{u} & 0 & 0 & 0 \\ \frac{1}{y} & 0 & 0 & 0 & 0 & 0 & 0 & 0 \\ 0 & 0 & 0 & \frac{1}{z} & 0 & 0 & 0 & 0 \\ 0 & 0 & 0 & 0 & 0 & 0 & 0 & \frac{1}{v} \end{pmatrix}, s, t, u, v, w, x, y, z \neq 0. \tag{10.3.34}$$

The ternary semigroups $\mathcal{S}^{(3)}_{\mathrm{star1},\,2}(8,\,\mathbb{C}) = \left\{\mathsf{N}'_{\mathrm{star1},\,2}\middle|\mu^{\circ 2}\right\}$ and $\mathcal{S}^{(3)}_{\mathrm{circ1},\,2}(8,\,\mathbb{C}) = \left\{\mathsf{N}'_{\mathrm{circ1},\,2}\middle|\mu^{\circ 2}\right\}$ in which every element has its querelement given by (10.3.31)–(10.3.33) become the *ternary groups* $\mathcal{G}^{(3)}_{\mathrm{star1},\,2}(8,\,\mathbb{C}) = \left\{\mathsf{N}'_{\mathrm{star1},\,2}\middle|\mu^{\circ 2},\,\overline{(\,)}\right\}$ and $\mathcal{G}^{(3)}_{\mathrm{circ1},\,2}(8,\,\mathbb{C}) = \left\{\mathsf{N}'_{\mathrm{circ1},\,2}\middle|\mu^{\circ 2},\,\overline{(\,)}\right\}$, which are four different (3-nonderived) ternary subgroups of the derived ternary general linear group $\mathrm{GL}^{(3)}(8,\,\mathbb{C})$. The ternary identities in $\mathcal{G}^{(3)}_{\mathrm{star1},\,2}(8,\,\mathbb{C})$ and $\mathcal{G}^{(3)}_{\mathrm{circ1},\,2}(8,\,\mathbb{C})$ are the following different continuous sets $\mathsf{I}'^{(3)}_{\mathrm{star1},\,2} = \left\{I'^{(3)}_{\mathrm{star1},\,2}\right\}$ and $\mathsf{I}'^{(3)}_{\mathrm{circ1},\,2} = \left\{I'^{(3)}_{\mathrm{circ1},\,2}\right\}$, where

$$I'^{(3)}_{\mathrm{star1}} = \begin{pmatrix} e^{i\alpha_1} & 0 & 0 & 0 & 0 & 0 & 0 & 0 \\ 0 & 0 & 0 & 0 & 0 & 0 & e^{i\alpha_2} & 0 \\ 0 & 0 & e^{i\alpha_3} & 0 & 0 & 0 & 0 & 0 \\ 0 & 0 & 0 & 0 & e^{i\alpha_4} & 0 & 0 & 0 \\ 0 & 0 & 0 & e^{i\alpha_5} & 0 & 0 & 0 & 0 \\ 0 & 0 & 0 & 0 & 0 & e^{i\alpha_6} & 0 & 0 \\ 0 & e^{i\alpha_7} & 0 & 0 & 0 & 0 & 0 & 0 \\ 0 & 0 & 0 & 0 & 0 & 0 & 0 & e^{i\alpha_8} \end{pmatrix}, \tag{10.3.35}$$

$$I'^{(3)}_{\mathrm{star2}} = \begin{pmatrix} 0 & 0 & 0 & 0 & 0 & 0 & 0 & e^{i\alpha_2} \\ 0 & e^{i\alpha_1} & 0 & 0 & 0 & 0 & 0 & 0 \\ 0 & 0 & 0 & 0 & 0 & e^{i\alpha_4} & 0 & 0 \\ 0 & 0 & 0 & e^{i\alpha_3} & 0 & 0 & 0 & 0 \\ 0 & 0 & 0 & 0 & e^{i\alpha_6} & 0 & 0 & 0 \\ 0 & 0 & e^{i\alpha_5} & 0 & 0 & 0 & 0 & 0 \\ 0 & 0 & 0 & 0 & 0 & 0 & e^{i\alpha_8} & 0 \\ e^{i\alpha_7} & 0 & 0 & 0 & 0 & 0 & 0 & 0 \end{pmatrix}, \tag{10.3.36}$$

$$e^{2i\alpha_1} = e^{2i\alpha_3} = e^{2i\alpha_6} = e^{2i\alpha_8} = e^{i(\alpha_2+\alpha_7)} = e^{i(\alpha_4+\alpha_5)}$$
$$= 1,\ \alpha_1,\,\ldots,\,\alpha_8 \in \mathbb{R},$$

and

$$I'^{(3)}_{\mathrm{circ1}} = \begin{pmatrix} e^{i\alpha_1} & 0 & 0 & 0 & 0 & 0 & 0 & 0 \\ 0 & 0 & 0 & 0 & e^{i\alpha_2} & 0 & 0 & 0 \\ 0 & 0 & 0 & 0 & 0 & 0 & 0 & e^{i\alpha_3} \\ 0 & 0 & 0 & e^{i\alpha_4} & 0 & 0 & 0 & 0 \\ 0 & e^{i\alpha_5} & 0 & 0 & 0 & 0 & 0 & 0 \\ 0 & 0 & 0 & 0 & 0 & e^{i\alpha_6} & 0 & 0 \\ 0 & 0 & 0 & 0 & 0 & 0 & e^{i\alpha_7} & 0 \\ 0 & 0 & e^{i\alpha_8} & 0 & 0 & 0 & 0 & 0 \end{pmatrix}, \tag{10.3.37}$$

$$I'^{(3)}_{\text{circ2}} = \begin{pmatrix} 0 & 0 & 0 & 0 & 0 & e^{i\alpha_2} & 0 & 0 \\ 0 & e^{i\alpha_1} & 0 & 0 & 0 & 0 & 0 & 0 \\ 0 & 0 & e^{i\alpha_4} & 0 & 0 & 0 & 0 & 0 \\ 0 & 0 & 0 & 0 & 0 & 0 & e^{i\alpha_3} & 0 \\ 0 & 0 & 0 & 0 & e^{i\alpha_6} & 0 & 0 & 0 \\ e^{i\alpha_5} & 0 & 0 & 0 & 0 & 0 & 0 & 0 \\ 0 & 0 & 0 & e^{i\alpha_8} & 0 & 0 & 0 & 0 \\ 0 & 0 & 0 & 0 & 0 & 0 & 0 & e^{i\alpha_7} \end{pmatrix}, \tag{10.3.38}$$

$$e^{2i\alpha_1} = e^{2i\alpha_4} = e^{2i\alpha_6} = e^{2i\alpha_7} = e^{i(\alpha_3+\alpha_8)} = e^{i(\alpha_2+\alpha_5)} = 1,\ \alpha_1, \ldots, \alpha_8 \in \mathbb{R},$$

such that all the identities are the 8×8 matrix reflections $\left(I'^{(3)}\right)^2 = I_8$ (see (10.1.45)). If $\alpha_j = 0$, $j = 1, \ldots, 8$, the ternary identities (10.3.36)–(10.3.38) coincide with the 8×8 permutation matrices (10.3.13)–(10.3.14), which are solutions to the ternary braid equations (10.3.3).

The module structure of the 8-vertex star-like (10.3.23) and circle-like (10.3.25) 8×8 matrix sets differs from the 4×4 matrix sets (10.1.52)–(10.1.75). Firstly, because of the absence of pentuple matrix products (10.1.70)–(10.1.75), and secondly through some differences in the ternary closed products of sets.

We have the following triple relations between star and circle matrices separately (the sets corresponding to modules are in brackets, and we informally denote modules by their sets)

$$\mathsf{N}'_{\text{star1}}(\mathsf{N}'_{\text{star2}})\mathsf{N}'_{\text{star1}} = (\mathsf{N}'_{\text{star2}}),\quad \mathsf{N}'_{\text{circ1}}(\mathsf{N}'_{\text{circ2}})\mathsf{N}'_{\text{circ1}} = (\mathsf{N}'_{\text{circ2}}), \tag{10.3.39}$$

$$\mathsf{N}'_{\text{star1}}\mathsf{N}'_{\text{star1}}(\mathsf{N}'_{\text{star2}}) = (\mathsf{N}'_{\text{star2}}),\quad \mathsf{N}'_{\text{circ1}}\mathsf{N}'_{\text{circ1}}(\mathsf{N}'_{\text{circ2}}) = \mathsf{N}'_{\text{circ2}}, \tag{10.3.40}$$

$$(\mathsf{N}'_{\text{star2}})\mathsf{N}'_{\text{star1}}\mathsf{N}'_{\text{star1}} = (\mathsf{N}'_{\text{star2}}),\quad (\mathsf{N}'_{\text{circ2}})\mathsf{N}'_{\text{circ1}}\mathsf{N}'_{\text{circ1}} = (\mathsf{N}'_{\text{circ2}}), \tag{10.3.41}$$

$$\mathsf{N}'_{\text{star2}}\mathsf{N}'_{\text{star2}}(\mathsf{N}'_{\text{star1}}) = (\mathsf{N}'_{\text{star1}}),\quad \mathsf{N}'_{\text{circ2}}\mathsf{N}'_{\text{circ2}}(\mathsf{N}'_{\text{circ1}}) = (\mathsf{N}'_{\text{circ1}}), \tag{10.3.42}$$

$$\mathsf{N}'_{\text{star2}}(\mathsf{N}'_{\text{star1}})\mathsf{N}'_{\text{star2}} = (\mathsf{N}'_{\text{star1}}),\quad \mathsf{N}'_{\text{circ2}}(\mathsf{N}'_{\text{circ1}})\mathsf{N}'_{\text{circ2}} = (\mathsf{N}'_{\text{circ1}}), \tag{10.3.43}$$

$$(\mathsf{N}'_{\text{star1}})\mathsf{N}'_{\text{star2}}\mathsf{N}'_{\text{star2}} = (\mathsf{N}'_{\text{star1}}),\quad (\mathsf{N}'_{\text{circ1}})\mathsf{N}'_{\text{circ2}}\mathsf{N}'_{\text{circ2}} = (\mathsf{N}'_{\text{circ1}}). \tag{10.3.44}$$

So we may observe the following module structures:

(1) from (10.3.39)–(10.3.41), the sets $\mathsf{N}'_{\text{star2}}$ ($\mathsf{N}'_{\text{circ2}}$) are the middle, right and left ternary modules over $\mathsf{N}'_{\text{star1}}$ ($\mathsf{N}'_{\text{circ1}}$);

(2) from (10.3.42)–(10.3.44), the set $\mathsf{N}'_{\text{star1}}$ ($\mathsf{N}'_{\text{circ1}}$) are middle, right and left ternary modules over $\mathsf{N}'_{\text{star2}}$ ($\mathsf{N}'_{\text{circ2}}$);

$$\mathsf{N}'_{\text{star1}}\mathsf{N}'_{\text{star1}}(\mathsf{N}'_{\text{circ1}}) = (\mathsf{N}'_{\text{circ1}}),\quad (\mathsf{N}'_{\text{circ1}})\mathsf{N}'_{\text{star1}}\mathsf{N}'_{\text{star1}} = (\mathsf{N}'_{\text{circ1}}), \tag{10.3.45}$$

$$\mathsf{N}'_{\text{star1}}\mathsf{N}'_{\text{star1}}(\mathsf{N}'_{\text{circ2}}) = (\mathsf{N}'_{\text{circ2}}),\quad (\mathsf{N}'_{\text{circ2}})\mathsf{N}'_{\text{star1}}\mathsf{N}'_{\text{star1}} = (\mathsf{N}'_{\text{circ2}}), \tag{10.3.46}$$

$$\mathsf{N}'_{\text{star2}}\mathsf{N}'_{\text{star2}}(\mathsf{N}'_{\text{circ1}}) = (\mathsf{N}'_{\text{circ1}}),\quad (\mathsf{N}'_{\text{circ1}})\mathsf{N}'_{\text{star2}}\mathsf{N}'_{\text{star2}} = (\mathsf{N}'_{\text{circ1}}), \tag{10.3.47}$$

$$\mathsf{N}'_{\text{star2}}\mathsf{N}'_{\text{star2}}(\mathsf{N}'_{\text{circ2}}) = (\mathsf{N}'_{\text{circ2}}),\quad (\mathsf{N}'_{\text{circ2}})\mathsf{N}'_{\text{star2}}\mathsf{N}'_{\text{star2}} = (\mathsf{N}'_{\text{circ2}}), \tag{10.3.48}$$

(3) from (10.3.45)–(10.3.48), the sets $\mathsf{N}'_{\mathrm{circ}1,\,2}$ are right and left ternary modules over $\mathsf{N}'_{\mathrm{star}1,\,2}$;

$$\mathsf{N}'_{\mathrm{circ}1}\mathsf{N}'_{\mathrm{circ}1}(\mathsf{N}'_{\mathrm{star}1}) = (\mathsf{N}'_{\mathrm{star}1}),\quad (\mathsf{N}'_{\mathrm{star}1})\mathsf{N}'_{\mathrm{circ}1}\mathsf{N}'_{\mathrm{circ}1} = (\mathsf{N}'_{\mathrm{star}1}), \tag{10.3.49}$$

$$\mathsf{N}'_{\mathrm{circ}1}\mathsf{N}'_{\mathrm{circ}1}(\mathsf{N}'_{\mathrm{star}2}) = (\mathsf{N}'_{\mathrm{star}2}),\quad (\mathsf{N}'_{\mathrm{star}2})\mathsf{N}'_{\mathrm{circ}1}\mathsf{N}'_{\mathrm{circ}1} = (\mathsf{N}'_{\mathrm{star}2}), \tag{10.3.50}$$

$$\mathsf{N}'_{\mathrm{circ}2}\mathsf{N}'_{\mathrm{circ}2}(\mathsf{N}'_{\mathrm{star}1}) = (\mathsf{N}'_{\mathrm{star}1}),\quad (\mathsf{N}'_{\mathrm{star}1})\mathsf{N}'_{\mathrm{circ}2}\mathsf{N}'_{\mathrm{circ}2} = (\mathsf{N}'_{\mathrm{star}1}), \tag{10.3.51}$$

$$\mathsf{N}'_{\mathrm{circ}2}\mathsf{N}'_{\mathrm{circ}2}(\mathsf{N}'_{\mathrm{star}2}) = (\mathsf{N}'_{\mathrm{star}2}),\quad (\mathsf{N}'_{\mathrm{star}2})\mathsf{N}'_{\mathrm{circ}2}\mathsf{N}'_{\mathrm{circ}2} = (\mathsf{N}'_{\mathrm{star}2}), \tag{10.3.52}$$

(4) from (10.3.49)–(10.3.52), the sets $\mathsf{N}'_{\mathrm{star}1,\,2}$ are right and left ternary modules over $\mathsf{N}'_{\mathrm{circ}1,\,2}$.

10.3.4 Group structure of the star and circle 16-vertex matrices

Next we will introduce 8×8 matrices of a special form similar to the star 8-vertex matrices (10.1.76) and the circle 8-vertex matrices (10.1.77), analyze their group structure and establish which ones could be 16-vertex solutions to the ternary braid equations (10.3.3). We will derive the solutions in the opposite way to that for the 8-vertex Yang–Baxter maps, following the note after (10.1.34). Indeed, the sum of the permutation bisymm solutions (10.3.13) gives the shape of the 8×8 star matrix M'_{star} (as in (10.1.76)), while the sum of symm solutions (10.3.14) gives the 8×8 circle matrix M'_{circ} (as in (10.1.77))

$$M'_{\mathrm{star}} = \begin{pmatrix} x & 0 & 0 & 0 & 0 & 0 & 0 & y \\ 0 & z & 0 & 0 & 0 & 0 & s & 0 \\ 0 & 0 & t & 0 & 0 & u & 0 & 0 \\ 0 & 0 & 0 & v & w & 0 & 0 & 0 \\ 0 & 0 & 0 & a & b & 0 & 0 & 0 \\ 0 & 0 & c & 0 & 0 & d & 0 & 0 \\ 0 & f & 0 & 0 & 0 & 0 & g & 0 \\ h & 0 & 0 & 0 & 0 & 0 & 0 & p \end{pmatrix}, \tag{10.3.53}$$

$$M'_{\mathrm{circ}} = \begin{pmatrix} x & 0 & 0 & 0 & 0 & y & 0 & 0 \\ 0 & z & 0 & 0 & s & 0 & 0 & 0 \\ 0 & 0 & t & 0 & 0 & 0 & 0 & u \\ 0 & 0 & 0 & v & 0 & 0 & w & 0 \\ 0 & f & 0 & 0 & g & 0 & 0 & 0 \\ h & 0 & 0 & 0 & 0 & p & 0 & 0 \\ 0 & 0 & 0 & a & 0 & 0 & b & 0 \\ 0 & 0 & c & 0 & 0 & 0 & 0 & d \end{pmatrix}, \tag{10.3.54}$$

$$\begin{aligned} \mathrm{tr} M' &= x + z + t + v + b + d + g + p, \\ \det M' &= (bv - aw)(cu - dt)(fs - gz)(px - hy), \end{aligned} \tag{10.3.55}$$

$$\begin{gathered} \text{eigenvalues}: \frac{1}{2}\left(d + t - \sqrt{4cu + (d - t)^2}\right), \frac{1}{2}\left(d + t + \sqrt{4cu + (d - t)^2}\right), \\ \frac{1}{2}\left(b + v - \sqrt{4aw + (b - v)^2}\right), \frac{1}{2}\left(b + v + \sqrt{4aw + (b - v)^2}\right), \frac{1}{2}\left(p + x - \sqrt{4hy + (p - x)^2}\right), \\ \frac{1}{2}\left(p + x + \sqrt{4hy + (p - x)^2}\right), \frac{1}{2}\left(g + z - \sqrt{4fs + (g - z)^2}\right), \frac{1}{2}\left(g + z + \sqrt{4fs + (g - z)^2}\right). \end{gathered} \tag{10.3.56}$$

The 16-vertex matrices are invertible, if $\det M'_{\mathrm{star}} \neq 0$ and $\det M'_{\mathrm{circ}} \neq 0$, which give the following joint conditions on the parameters (see (10.1.79))

$$bv - aw \neq 0,\ cu - dt \neq 0,\ fs - gz \neq 0,\ px - hy \neq 0. \tag{10.3.57}$$

Only in this concrete parametrization (10.3.53) and (10.3.54) do the matrices M'_{star} and M'_{circ} have the same trace, determinant and eigenvalues, and they are diagonalizable and conjugate via

$$U' = \begin{pmatrix} 1 & 0 & 0 & 0 & 0 & 0 & 0 & 0 \\ 0 & 1 & 0 & 0 & 0 & 0 & 0 & 0 \\ 0 & 0 & 1 & 0 & 0 & 0 & 0 & 0 \\ 0 & 0 & 0 & 1 & 0 & 0 & 0 & 0 \\ 0 & 0 & 0 & 0 & 1 & 0 & 1 & 0 \\ 0 & 0 & 0 & 0 & 0 & 0 & 0 & 1 \\ 0 & 0 & 0 & 0 & 1 & 0 & 0 & 0 \\ 0 & 0 & 0 & 0 & 0 & 1 & 0 & 0 \end{pmatrix}. \tag{10.3.58}$$

The matrix U' cannot be presented in the form of a triple Kronecker product (10.3.10), and so two matrices M'_{star} and M'_{circ} are not q-conjugate in the parametrization (10.3.53) and (10.3.54), and can lead to different solutions to the ternary braid equations (10.3.3). It follows from (10.3.57) that 16-vertex matrices with all nonzero entries equal to 1 are non-invertible, having vanishing determinant and rank 4 (despite each one being a sum of two permutation matrices). In the case all the conditions (10.3.57) holding, the inverse matrices become

$$(M'_{\mathrm{star}})^{-1} = \begin{pmatrix} \frac{p}{px - hy} & 0 & 0 & 0 & 0 & 0 & 0 & -\frac{y}{px - hy} \\ 0 & -\frac{g}{fs - gz} & 0 & 0 & 0 & 0 & \frac{s}{fs - gz} & 0 \\ 0 & 0 & -\frac{d}{cu - dt} & 0 & 0 & \frac{u}{cu - dt} & 0 & 0 \\ 0 & 0 & 0 & \frac{b}{bv - aw} & -\frac{w}{bv - aw} & 0 & 0 & 0 \\ 0 & 0 & 0 & -\frac{a}{bv - aw} & \frac{v}{bv - aw} & 0 & 0 & 0 \\ 0 & 0 & \frac{c}{cu - dt} & 0 & 0 & -\frac{t}{cu - dt} & 0 & 0 \\ 0 & \frac{f}{fs - gz} & 0 & 0 & 0 & 0 & -\frac{z}{fs - gz} & 0 \\ -\frac{h}{px - hy} & 0 & 0 & 0 & 0 & 0 & 0 & \frac{x}{px - hy} \end{pmatrix}, \tag{10.3.59}$$

$$\left(M'_{\text{circ}}\right)^{-1}=\begin{pmatrix} \frac{p}{px-hy} & 0 & 0 & 0 & 0 & -\frac{y}{px-hy} & 0 & 0 \\ 0 & -\frac{g}{fs-gz} & 0 & 0 & \frac{s}{fs-gz} & 0 & 0 & 0 \\ 0 & 0 & -\frac{d}{cu-dt} & 0 & 0 & 0 & 0 & \frac{u}{cu-dt} \\ 0 & 0 & 0 & \frac{b}{bv-aw} & 0 & 0 & -\frac{w}{bv-aw} & 0 \\ 0 & \frac{f}{fs-gz} & 0 & 0 & -\frac{z}{fs-gz} & 0 & 0 & 0 \\ -\frac{h}{px-hy} & 0 & 0 & 0 & 0 & \frac{x}{px-hy} & 0 & 0 \\ 0 & 0 & 0 & -\frac{a}{bv-aw} & 0 & 0 & \frac{v}{bv-aw} & 0 \\ 0 & 0 & \frac{c}{cu-dt} & 0 & 0 & 0 & 0 & -\frac{t}{cu-dt} \end{pmatrix}. \quad (10.3.60)$$

Denoting the sets of matrices corresponding to (10.3.53) and (10.3.54) by $\mathsf{M}'_{\text{star}}$ and $\mathsf{M}'_{\text{circ}}$, their multiplications are

$$\mathsf{M}'_{\text{star}}\mathsf{M}'_{\text{star}}=\mathsf{M}'_{\text{star}}, \qquad \mathsf{M}'_{\text{circ}}\mathsf{M}'_{\text{circ}}=\mathsf{M}'_{\text{circ}}, \quad (10.3.61)$$

and in term of sets $\mathsf{M}'_{\text{star}}=\mathsf{N}'_{\text{star1}}\cup\mathsf{N}'_{\text{star2}}$ and $\mathsf{M}'_{\text{circ}}=\mathsf{N}'_{\text{circ1}}\cup\mathsf{N}'_{\text{circ2}}$, and $\mathsf{N}'_{\text{star1}}\cap\mathsf{N}'_{\text{star2}}=\mathsf{D}$ and $\mathsf{N}'_{\text{circ1}}\cap\mathsf{N}'_{\text{circ2}}=\mathsf{D}$ (see (10.3.26)). Note that the structure (10.3.61) is considerably different from the binary case (10.1.80)–(10.1.82), and therefore it may not necessarily be related to the Cartan decomposition.

The products (10.3.61) mean that both $\mathsf{M}'_{\text{star}}$ and $\mathsf{M}'_{\text{circ}}$ are separately closed with respect to binary matrix multiplication $(\cdot)$, and therefore $\mathcal{S}^{\text{star}}_{16\text{vert}}=\langle\mathsf{M}'_{\text{star}}\mid\cdot\rangle$ and $\mathcal{S}^{\text{circ}}_{16\text{vert}}=\langle\mathsf{M}'_{\text{circ}}\mid\cdot\rangle$ are semigroups. We denote their intersection by $\mathcal{S}^{\text{diag}}_{8\text{vert}}=\mathcal{S}^{\text{star}}_{16\text{vert}}\cap\mathcal{S}^{\text{circ}}_{16\text{vert}}$ which is a semigroup of diagonal 8-vertex matrices. In case, the invertibility conditions (10.3.57) are fulfilled, the sets $\mathsf{M}'_{\text{star}}$ and $\mathsf{M}'_{\text{circ}}$ form subgroups $\mathcal{G}^{\text{star}}_{16\text{vert}}=\langle\mathsf{M}'_{\text{star}}\mid\cdot,(\_\_)^{-1},I_8\rangle$ and $\mathcal{G}^{\text{circ}}_{16\text{vert}}=\langle\mathsf{M}'_{\text{circ}}\mid\cdot,(\_\_)^{-1},I_8\rangle$ (where I_8 is the 8×8 identity matrix) of $GL(8,\mathbb{C})$ with the inverse elements given explicitly by (10.3.59)–(10.3.60). Because the elements M'_{star} and M'_{circ} in (10.3.53) and (10.3.54) are conjugates by the invertible matrix U' (10.3.58), the subgroups $\mathcal{G}^{\text{star}}_{16\text{vert}}$ and $\mathcal{G}^{\text{circ}}_{16\text{vert}}$ (as well as the semigroups $\mathcal{S}^{\text{star}}_{16\text{vert}}$ and $\mathcal{S}^{\text{circ}}_{16\text{vert}}$) are isomorphic by the obvious isomorphism

$$M'_{\text{star}}\mapsto U'M'_{\text{circ}}U'^{-1}, \quad (10.3.62)$$

where U' is in (10.3.58).

The 'interaction' between $\mathsf{M}'_{\text{star}}$ and $\mathsf{M}'_{\text{circ}}$ also differs from the binary case (10.1.81), because

$$\mathsf{M}'_{\text{star}}\mathsf{M}'_{\text{circ}}=\mathsf{M}'_{\text{quad}}, \quad \mathsf{M}'_{\text{circ}}\mathsf{M}'_{\text{star}}=\mathsf{M}'_{\text{quad}}, \quad (10.3.63)$$

$$\mathsf{M}'_{\text{quad}}\mathsf{M}'_{\text{quad}}=\mathsf{M}'_{\text{quad}}, \quad (10.3.64)$$

where $\mathsf{M}'_{\text{quad}}$ is a set of 32-vertex so-called *quad-matrices* of the form

$$M'_{\text{quad}} = \begin{pmatrix} x_1 & 0 & y_1 & 0 & 0 & z_1 & 0 & s_1 \\ 0 & t_1 & 0 & u_1 & v_1 & 0 & w_1 & 0 \\ a_1 & 0 & b_1 & 0 & 0 & c_1 & 0 & d_1 \\ 0 & f_1 & 0 & g_1 & h_1 & 0 & p_1 & 0 \\ 0 & x_2 & 0 & y_2 & z_2 & 0 & s_2 & 0 \\ t_2 & 0 & u_2 & 0 & 0 & v_2 & 0 & w_2 \\ 0 & a_2 & 0 & b_2 & c_2 & 0 & d_2 & 0 \\ f_2 & 0 & g_2 & 0 & 0 & h_2 & 0 & p_2 \end{pmatrix}. \tag{10.3.65}$$

Because of (10.3.64), the set $\mathsf{M}'_{\text{quad}}$ is closed with respect to matrix multiplication, and therefore (for invertible matrices M'_{quad}) the group $\mathcal{G}^{\text{quad}}_{32\text{vert}} = \left\langle \mathsf{M}'_{\text{quad}} \middle| \cdot, (\text{—})^{-1}, I_8 \right\rangle$ is a subgroup of GL(8, $\mathbb{C}$). So, in trying to find higher 32-vertex solutions (having at most half as many unknown variables as a general 8×8 matrix) to the ternary braid equations (10.3.3) it is worthwhile to search within the class of quad-matrices (10.3.65).

Thus, the group structure of the above 16-vertex 8×8 matrices (10.3.61)–(10.3.64) is considerably different to that of 8-vertex 4×4 matrices (10.1.76)–(10.1.77) as the former contains two isomorphic binary subgroups $\mathcal{G}^{\text{star}}_{16\text{vert}}$ and $\mathcal{G}^{\text{circ}}_{16\text{vert}}$ of GL(8, $\mathbb{C}$) (see (10.1.80)–(10.1.82) and (10.3.61)).

The sets $\mathsf{M}'_{\text{star}}$, $\mathsf{M}'_{\text{circ}}$ and $\mathsf{M}'_{\text{quad}}$ are closed with respect to matrix addition as well, and therefore (because of the distributivity of $\mathbb{C}$) they are the matrix rings $\mathcal{R}^{\text{star}}_{16\text{vert}}$, $\mathcal{R}^{\text{circ}}_{16\text{vert}}$ and $\mathcal{R}^{\text{quad}}_{32\text{vert}}$, respectively. In the invertible case (10.3.57) and $\det M'_{\text{quad}} \neq 0$, these become matrix fields.

10.3.5 Pauli matrix presentation of the star and circle 16-vertex constant matrices

The main peculiarity of the 16-vertex 8×8 matrices (10.3.61)–(10.3.64) is the fact that they can be expressed as special tensor (Kronecker) products of the Pauli matrices (see, also, Khaneja and Glaser 2001, Ballard and Wu 2011). Indeed, let

$$\Sigma_{ijk} = \rho_i \otimes_K \rho_j \otimes_K \rho_k, \quad i, j, k = 1, 2, 3, 4, \tag{10.3.66}$$

where ρ_i are Pauli matrices (we have already used the letter 'σ' for the braid group generators (10.1.5))

$$\rho_1 = \begin{pmatrix} 0 & 1 \\ 1 & 0 \end{pmatrix}, \rho_2 = \begin{pmatrix} 0 & -i \\ i & 0 \end{pmatrix}, \rho_3 = \begin{pmatrix} 1 & 0 \\ 0 & -1 \end{pmatrix}, \rho_4 = I_2 = \begin{pmatrix} 1 & 0 \\ 0 & 1 \end{pmatrix}. \tag{10.3.67}$$

Among the total of 64 8×8 matrices Σ_{ijk} (10.3.66) there are 24 which generate M'_{star} (10.3.53) and M'_{circ} (10.3.54):

- 8 diagonal matrices: $\Sigma_{\text{diag}} = \{\Sigma_{333}, \Sigma_{334}, \Sigma_{343}, \Sigma_{344}, \Sigma_{433}, \Sigma_{434}, \Sigma_{443}, \Sigma_{444}\}$;
- 8 antidiagonal matrices: $\Sigma_{\text{adiag}} = \{\Sigma_{111}, \Sigma_{112}, \Sigma_{121}, \Sigma_{122}, \Sigma_{211}, \Sigma_{212}, \Sigma_{221}, \Sigma_{222}\}$;
- 8 circle-like matrices (M'_{circ} with 0's on diagonal):

$\Sigma_{\text{circ}} = \{\Sigma_{131}, \Sigma_{132}, \Sigma_{141}, \Sigma_{142}, \Sigma_{231}, \Sigma_{232}, \Sigma_{241}, \Sigma_{242}\}$.

Thus, in general we have the following set structure for the star and circle 16-vertex matrices (10.3.53) and (10.3.54)

$$\mathsf{M}'_{\text{star}} = \Sigma_{\text{diag}} \bigcup \Sigma_{\text{adiag}}, \tag{10.3.68}$$

$$\mathsf{M}'_{\text{circ}} = \Sigma_{\text{diag}} \bigcup \Sigma_{\text{circ}}, \tag{10.3.69}$$

$$\mathsf{M}'_{\text{star}} \bigcap \mathsf{M}'_{\text{circ}} = \Sigma_{\text{diag}}. \tag{10.3.70}$$

In particular, for the 8-vertex permutation solutions (10.3.13)–(10.3.14) of the ternary braid equations (10.3.3) we have

$$\tilde{c}^{\,\text{bisymm1, 2}}_{\text{rank} = 8} = \frac{1}{2}(\Sigma_{111} + \Sigma_{444} \pm \Sigma_{212} \pm \Sigma_{343}), \tag{10.3.71}$$

$$\tilde{c}^{\,\text{symm1, 2}}_{\text{rank} = 8} = \frac{1}{2}(\Sigma_{141} + \Sigma_{444} \pm \Sigma_{232} \pm \Sigma_{333}). \tag{10.3.72}$$

The non-invertible 16-vertex solutions M'_{star} (10.3.53) and M'_{circ} (10.3.54) having 1's on nonzero places are of rank = 4 and can be presented by (10.3.66) as follows

$$M'_{\text{star}}(1) = \begin{pmatrix} 1&0&0&0&0&0&0&1\\ 0&1&0&0&0&0&1&0\\ 0&0&1&0&0&1&0&0\\ 0&0&0&1&1&0&0&0\\ 0&0&0&1&1&0&0&0\\ 0&0&1&0&0&1&0&0\\ 0&1&0&0&0&0&1&0\\ 1&0&0&0&0&0&0&1 \end{pmatrix} = \Sigma_{111} + \Sigma_{444}, \tag{10.3.73}$$

$$M'_{\text{circ}}(1) = \begin{pmatrix} 1&0&0&0&0&1&0&0\\ 0&1&0&0&1&0&0&0\\ 0&0&1&0&0&0&0&1\\ 0&0&0&1&0&0&1&0\\ 0&1&0&0&1&0&0&0\\ 1&0&0&0&0&1&0&0\\ 0&0&0&1&0&0&1&0\\ 0&0&1&0&0&0&0&1 \end{pmatrix} = \Sigma_{141} + \Sigma_{444}. \tag{10.3.74}$$

Similarly, one can obtain the Pauli matrix presentation for the general star and circle 16-vertex matrices (10.3.53) and (10.3.54) which will contain linear combinations of the 16 parameters as coefficients before the Σ.

10.3.6 Invertible and non-invertible 16-vertex solutions to the ternary braid equations

First, consider the 16-vertex solutions to (10.3.3) having the star matrix shape (10.3.53). We found the following 2 one-parameter invertible solutions

$$\tilde{c}_{\text{rank}=8}^{16-\text{vert,star}}(x) = \begin{pmatrix} x^3 & 0 & 0 & 0 & 0 & 0 & 0 & -1 \\ 0 & x^3 & 0 & 0 & 0 & 0 & \mp x^2 & 0 \\ 0 & 0 & x^3 & 0 & 0 & -x^2 & 0 & 0 \\ 0 & 0 & 0 & x^3 & \mp x^4 & 0 & 0 & 0 \\ 0 & 0 & 0 & \pm x^2 & x^3 & 0 & 0 & 0 \\ 0 & 0 & x^4 & 0 & 0 & x^3 & 0 & 0 \\ 0 & \pm x^4 & 0 & 0 & 0 & 0 & x^3 & 0 \\ x^6 & 0 & 0 & 0 & 0 & 0 & 0 & x^3 \end{pmatrix}, \tag{10.3.75}$$

$\operatorname{tr}\tilde{c} = 8x^3$, $\det\tilde{c} = 16x^{24}$, $x \neq 0$, eigenvalues: $\{(1+i)x^3\}^{[4]}$, $\{(1-i)x^3\}^{[4]}$.

Both matrices in (10.3.75) are diagonalizable and are conjugates via

$$U_{\text{star}} = \begin{pmatrix} 1 & 0 & 0 & 0 & 0 & 0 & 0 & 0 \\ 0 & -1 & 0 & 0 & 0 & 0 & 0 & 0 \\ 0 & 0 & 1 & 0 & 0 & 0 & 0 & 0 \\ 0 & 0 & 0 & -1 & 0 & 0 & 0 & 0 \\ 0 & 0 & 0 & 0 & 1 & 0 & 0 & 0 \\ 0 & 0 & 0 & 0 & 0 & 1 & 0 & 0 \\ 0 & 0 & 0 & 0 & 0 & 0 & 1 & 0 \\ 0 & 0 & 0 & 0 & 0 & 0 & 0 & 1 \end{pmatrix}, \tag{10.3.76}$$

which cannot be presented in the form of a triple Kronecker product (10.3.10). Therefore, the two solutions in (10.3.75) are not q-conjugate and become different 16-vertex one-parameter invertible solutions of the braid equations (10.3.3).

In search of 16-vertex solutions to the total braid equations (10.3.3) of the circle matrix shape (10.3.54) we found that only non-invertible ones exist. They are the following two 2-parameter solutions of rank 4

$$\tilde{c}_{\text{rank}=4}^{16-\text{vert,circ}}(x,y) = \begin{pmatrix} \pm xy & 0 & 0 & 0 & 0 & y^2 & 0 & 0 \\ 0 & \pm xy & 0 & 0 & xy & 0 & 0 & 0 \\ 0 & 0 & \pm xy & 0 & 0 & 0 & 0 & y^2 \\ 0 & 0 & 0 & \pm xy & 0 & 0 & xy & 0 \\ 0 & xy & 0 & 0 & \pm xy & 0 & 0 & 0 \\ x^2 & 0 & 0 & 0 & 0 & \pm xy & 0 & 0 \\ 0 & 0 & 0 & xy & 0 & 0 & \pm xy & 0 \\ 0 & 0 & x^2 & 0 & 0 & 0 & 0 & \pm xy \end{pmatrix}, \tag{10.3.77}$$

$$\operatorname{tr}\tilde{c} = \pm 8xy, \text{ eigenvalues: } \{2xy\}^{[4]}, \{0\}^{[4]}. \tag{10.3.78}$$

Two matrices in (10.3.78) are not even conjugates in the standard way, and so they are different 16-vertex two-parameter non-invertible solutions to the braid equations (10.3.3).

For the only partial 13-braid equation (10.3.5), there are 4 polynomial 16-vertex two-parameter invertible solutions

$$\tilde{c}_{\text{rank}=8}^{16-\text{vert},\ 13\text{circ1}}(x,y)=\begin{pmatrix} x & 0 & 0 & 0 & 0 & y^2 & 0 & 0 \\ 0 & xy & 0 & 0 & x & 0 & 0 & 0 \\ 0 & 0 & x & 0 & 0 & 0 & 0 & \pm y^2 \\ 0 & 0 & 0 & xy & 0 & 0 & \pm x & 0 \\ 0 & x & 0 & 0 & xy & 0 & 0 & 0 \\ x^2 & 0 & 0 & 0 & 0 & x & 0 & 0 \\ 0 & 0 & 0 & \pm x & 0 & 0 & xy & 0 \\ 0 & 0 & \pm x^2 & 0 & 0 & 0 & 0 & x \end{pmatrix}, \tag{10.3.79}$$

$$\tilde{c}_{\text{rank}=8}^{16-\text{vert},\ 13\text{circ2}}(x,y)=\begin{pmatrix} x & 0 & 0 & 0 & 0 & y^2 & 0 & 0 \\ 0 & xy & 0 & 0 & -x & 0 & 0 & 0 \\ 0 & 0 & x & 0 & 0 & 0 & 0 & \pm y^2 \\ 0 & 0 & 0 & xy & 0 & 0 & \mp x & 0 \\ 0 & -x & 0 & 0 & xy & 0 & 0 & 0 \\ x^2 & 0 & 0 & 0 & 0 & x & 0 & 0 \\ 0 & 0 & 0 & \mp x & 0 & 0 & xy & 0 \\ 0 & 0 & \pm x^2 & 0 & 0 & 0 & 0 & x \end{pmatrix}, \tag{10.3.80}$$

$$\begin{aligned} &\operatorname{tr}\tilde{c}=4x(y+1),\ \det\tilde{c}=x^8(y^2-1)^4,\ x\neq 0,\ y\neq 1,\\ &\text{eigenvalues}:\ \{x(y+1)\}^{[4]},\ \{x(y-1)\}^{[2]},\ \{-x(y-1)\}^{[2]}. \end{aligned} \tag{10.3.81}$$

Also, for the partial 13-braid equation (10.3.5), we found 4 exotic irrational (an analog of (10.1.88) for the Yang–Baxter equation (10.1.12)) 16-vertex, two-parameter invertible solutions of rank 8 of the form

$$\begin{aligned}&\tilde{c}_{\text{rank}=8}^{16-\text{vert},\ 13\text{circ,irr1}}(x,y)\\ &=\begin{pmatrix} x(2y-1) & 0 & 0 & 0 & 0 & y^2 & 0 & 0 \\ 0 & xy & 0 & 0 & x\sqrt{2(y-1)y+1} & 0 & 0 & 0 \\ 0 & 0 & x(2y-1) & 0 & 0 & 0 & 0 & \pm y^2 \\ 0 & 0 & 0 & xy & 0 & 0 & \pm x\sqrt{2(y-1)y+1} & 0 \\ 0 & x\sqrt{2(y-1)y+1} & 0 & 0 & xy & 0 & 0 & 0 \\ x^2 & 0 & 0 & 0 & 0 & x & 0 & 0 \\ 0 & 0 & 0 & \pm x\sqrt{2(y-1)y+1} & 0 & 0 & xy & 0 \\ 0 & 0 & \pm x^2 & 0 & 0 & 0 & 0 & x \end{pmatrix},\end{aligned} \tag{10.3.82}$$

$$\tilde{c}^{16-\text{vert},\ 13\text{circ},\text{irr}2}_{\text{rank}\,=\,8}(x,y)$$
$$=\begin{pmatrix} x(2y-1) & 0 & 0 & 0 & 0 & y^2 & 0 & 0 \\ 0 & xy & 0 & 0 & -x\sqrt{2(y-1)y+1} & 0 & 0 & 0 \\ 0 & 0 & x(2y-1) & 0 & 0 & 0 & 0 & \pm y^2 \\ 0 & 0 & 0 & xy & 0 & 0 & \mp x\sqrt{2(y-1)y+1} & 0 \\ 0 & -x\sqrt{2(y-1)y+1} & 0 & 0 & xy & 0 & 0 & 0 \\ x^2 & 0 & 0 & 0 & 0 & x & 0 & 0 \\ 0 & 0 & 0 & \mp x\sqrt{2(y-1)y+1} & 0 & 0 & xy & 0 \\ 0 & 0 & \pm x^2 & 0 & 0 & 0 & 0 & x \end{pmatrix}, \qquad (10.3.83)$$

$$\text{tr}\tilde{c} = 8xy,\ \det\tilde{c} = x^8(y-1)^8, \qquad x \neq 0,\ y \neq 1, \qquad (10.3.84)$$

$$\text{eigenvalues}: \left\{x\left(y+\sqrt{2(y-1)y+1}\right)\right\}^{[4]}, \left\{x\left(y-\sqrt{2(y-1)y+1}\right)\right\}^{[4]}. \qquad (10.3.85)$$

The matrices in (10.3.80)–(10.3.83) are diagonalizable, have the same eigenvalues (10.3.85) and are pairwise conjugate by

$$U_{\text{circ}} = \begin{pmatrix} 1 & 0 & 0 & 0 & 0 & 0 & 0 & 0 \\ 0 & 1 & 0 & 0 & 0 & 0 & 0 & 0 \\ 0 & 0 & -1 & 0 & 0 & 0 & 0 & 0 \\ 0 & 0 & 0 & -1 & 0 & 0 & 0 & 0 \\ 0 & 0 & 0 & 0 & 1 & 0 & 0 & 0 \\ 0 & 0 & 0 & 0 & 0 & 1 & 0 & 0 \\ 0 & 0 & 0 & 0 & 0 & 0 & 1 & 0 \\ 0 & 0 & 0 & 0 & 0 & 0 & 0 & 1 \end{pmatrix}. \qquad (10.3.86)$$

Because U_{circ} cannot be presented in the form (10.3.10), all solutions in (10.3.80)–(10.3.83) are not mutually q-conjugate and become 8 different 16-vertex two-parameter invertible solutions to the partial 13-braid equation (10.3.5). If $y = 1$, then the matrices (10.3.80)–(10.3.83) become of rank 4 with vanishing determinants (10.3.81), (10.3.84), and therefore in this case they are a 16-vertex one-parameter circle of non-invertible solutions to the total braid equations (10.3.3).

Further families of solutions could be constructed using additional parameters: the scaling parameter t in (10.3.8) and the complex elements of the matrix q (10.3.9).

10.3.7 Higher 2^n-vertex constant solutions to n-ary braid equations

Next we considered the 4-ary constant braid equations (10.2.4)–(10.2.6) and found the following 32-vertex star solution

$$\tilde{c}_{16} = \begin{pmatrix} 1 & 0 & 0 & 0 & 0 & 0 & 0 & 0 & 0 & 0 & 0 & 0 & 0 & 0 & 0 & -1 \\ 0 & 1 & 0 & 0 & 0 & 0 & 0 & 0 & 0 & 0 & 0 & 0 & 0 & 0 & -1 & 0 \\ 0 & 0 & 1 & 0 & 0 & 0 & 0 & 0 & 0 & 0 & 0 & 0 & 0 & -1 & 0 & 0 \\ 0 & 0 & 0 & 1 & 0 & 0 & 0 & 0 & 0 & 0 & 0 & 0 & -1 & 0 & 0 & 0 \\ 0 & 0 & 0 & 0 & 1 & 0 & 0 & 0 & 0 & 0 & 0 & -1 & 0 & 0 & 0 & 0 \\ 0 & 0 & 0 & 0 & 0 & 1 & 0 & 0 & 0 & 0 & -1 & 0 & 0 & 0 & 0 & 0 \\ 0 & 0 & 0 & 0 & 0 & 0 & 1 & 0 & 0 & -1 & 0 & 0 & 0 & 0 & 0 & 0 \\ 0 & 0 & 0 & 0 & 0 & 0 & 0 & 1 & -1 & 0 & 0 & 0 & 0 & 0 & 0 & 0 \\ 0 & 0 & 0 & 0 & 0 & 0 & 0 & 1 & 1 & 0 & 0 & 0 & 0 & 0 & 0 & 0 \\ 0 & 0 & 0 & 0 & 0 & 0 & 1 & 0 & 0 & 1 & 0 & 0 & 0 & 0 & 0 & 0 \\ 0 & 0 & 0 & 0 & 0 & 1 & 0 & 0 & 0 & 0 & 1 & 0 & 0 & 0 & 0 & 0 \\ 0 & 0 & 0 & 0 & 1 & 0 & 0 & 0 & 0 & 0 & 0 & 1 & 0 & 0 & 0 & 0 \\ 0 & 0 & 0 & 1 & 0 & 0 & 0 & 0 & 0 & 0 & 0 & 0 & 1 & 0 & 0 & 0 \\ 0 & 0 & 1 & 0 & 0 & 0 & 0 & 0 & 0 & 0 & 0 & 0 & 0 & 1 & 0 & 0 \\ 0 & 1 & 0 & 0 & 0 & 0 & 0 & 0 & 0 & 0 & 0 & 0 & 0 & 0 & 1 & 0 \\ 1 & 0 & 0 & 0 & 0 & 0 & 0 & 0 & 0 & 0 & 0 & 0 & 0 & 0 & 0 & 1 \end{pmatrix}. \tag{10.3.87}$$

We may compare (10.3.87) with particular cases of the star solutions to the Yang–Baxter equation (10.1.86) and the ternary braid equation (10.3.75)

$$\tilde{c}_4 = \begin{pmatrix} 1 & 0 & 0 & -1 \\ 0 & 1 & -1 & 0 \\ 0 & 1 & 1 & 0 \\ 1 & 0 & 0 & 1 \end{pmatrix}, \ \tilde{c}_8 = \begin{pmatrix} 1 & 0 & 0 & 0 & 0 & 0 & 0 & -1 \\ 0 & 1 & 0 & 0 & 0 & 0 & -1 & 0 \\ 0 & 0 & 1 & 0 & 0 & -1 & 0 & 0 \\ 0 & 0 & 0 & 1 & -1 & 0 & 0 & 0 \\ 0 & 0 & 0 & 1 & 1 & 0 & 0 & 0 \\ 0 & 0 & 1 & 0 & 0 & 1 & 0 & 0 \\ 0 & 1 & 0 & 0 & 0 & 0 & 1 & 0 \\ 1 & 0 & 0 & 0 & 0 & 0 & 0 & 1 \end{pmatrix}. \tag{10.3.88}$$

Informally we call such solutions the 'Minkowski' star solutions, since their legs have the 'Minkowski signature'. Thus, we assume that in the general case for the n-ary braid equation there exist 2^{n+1}-vertex $2^n \times 2^n$ matrix 'Minkowski' star invertible solutions of the above form

$$\tilde{c}_{2^n} = \begin{pmatrix} 1 & 0 & 0 & 0 & 0 & -1 \\ 0 & \ddots & 0 & 0 & \ddots & 0 \\ 0 & 0 & 1 & -1 & 0 & 0 \\ 0 & 0 & 1 & 1 & 0 & 0 \\ 0 & \ddots & 0 & 0 & \ddots & 0 \\ 1 & 0 & 0 & 0 & 0 & 1 \end{pmatrix}. \tag{10.3.89}$$

This allows us to use the general solution (10.3.89) as n-ary braiding quantum gates with an arbitrary number of qubits (Duplij and Vogl 2021).

References

Alagic G, Bapat A and Jordan S 2014 *Classical simulation of Yang–Baxter gates 9th Conference on the Theory of Quantum Computation, Communication and Cryptography LIPIcs. Leibniz Int. Proc. Inform.* vol 27, ed S T Flammia and A W Harrow *(Schloss Dagstuhl. Leibniz-Zent. Inform., Wadern)* pp 161–75

Asaulko K S and Korablev P G 2019 On a Yang-Baxter operator and the corresponding knots invariant *Chelyabinsk Phys. Math. J.* **4** 255–64

Balakrishnan S and Sankaranarayanan R 2010 Entangling power and local invariants of two-qubit gates *Phys. Rev.* A **82** 034301

Ballard A D and Wu Y-S 2011 Cartan decomposition and entangling power of braiding quantum gates *Cross Disciplinary Advances in Quantum Computing Contemporary Mathematics* vol 536 ed K Mahdavi, D Koslover and L L Brown (Providence, RI: American Mathematical Society) pp 1–15

Bukhshtaber V M 1998 The Yang-Baxter transformation *Russ. Math. Surv.* **53** 1343–4

Bullock S S 2004 Note on the Khaneja Glaser decomposition *Quantum Inf. Comp.* **4** 396–400

Bullock S S and Brennen G K 2004 Canonical decompositions of n-qubit quantum computations and concurrence *J. Math. Phys.* **45** 2447–67

Chen R S 2012 Generalized Yang–Baxter equations and braiding quantum gates *J. Knot Theory Ramif.* **21** 18

Dörnte W 1929 Unterschungen über einen verallgemeinerten Gruppenbegriff *Math. Z.* **29** 1–19

Drinfeld V G 1989 Quasi-Hopf algebras and Knizhnik– equations Zamolodchikov *Problems of Modern Quantum Field Theory* ed A A Belavin, A V Klimyk and A B Zamolodchikov (Heidelberg: Springer) pp 1–13

Drinfeld V G 1992 On some unsolved problems in quantum group theory *Quantum Groups* ed P P Kulish (Berlin: Springer) pp 1–8

Duplij S 2001 Ternary Hopf algebras *Symmetry in Nonlinear Mathematical Physics* ed A G Nikitin and V M Boyko (Kyiv: Institute of Mathematics) pp 25–34

Duplij S 2012 Polyadic systems, representations and quantum groups *J. Kharkiv Natl Univ. Ser. Nuclei Part. Fields* **1017** 28–59 Expanded version arXiv:math.RT/1308.4060

Duplij S 2021a Higher braid groups and regular semigroups from polyadic-binary correspondence *Mathematics* **9** 972

Duplij S 2021b Polyadic Hopf algebras and quantum groups *East Eur. J. Phys.* **2** 5–50

Duplij S and Vogl R 2021 Polyadic braid operators and higher braiding gates *Universe* **7** 301

Dye H A 2003 Unitary solutions to the Yang-Baxter equation in dimension four *Quantum Inf. Process.* **2** 117–50

Helgason S 1962 *Differential Geometry and Symmetric Spaces* (New York: Academic)

Hietarinta J 1993 Solving the two-dimensional constant quantum Yang-Baxter equation *J. Math. Phys.* **34** 1725–56

Jaffali H and Oeding L 2020 Learning algebraic models of quantum entanglement *Quantum Inf. Process.* **19** 279

Kassel C 1995 *Quantum Groups* (New York: Springer)

Kauffman L H and Lomonaco S J 2004 Braiding operators are universal quantum gates *New J. Phys.* **6** 134–9

Khaneja N and Glaser S J 2001 Cartan decomposition of $SU(2n)$ and control of spin systems *Chem. Phys.* **267** 11–23

Kitaev A and Wang Z 2012 Solutions to generalized Yang–Baxter equations via ribbon fusion categories *Proceedings of the Freedman Fest Geometry & Topology Monographs* vol 18 (Coventry: Geometry & Topology Publications) pp 191–7

Kraus B and Cirac J I 2001 Optimal creation of entanglement using a two-qubit gate *Phys. Rev.* **A63** 8

Lambe L A and Radford D E 1997 *Introduction to the Quantum Yang-Baxter Equation and Quantum Groups: An Algebraic Approach* (Dordrecht: Kluwer)

Nashed M Z 1976 *Generalized Inverses and Applications* (New York: Academic)

Nielsen M A and Chuang I L 2000 *Quantum Computation and Quantum Information* (Cambridge: Cambridge University Press)

Padmanabhan P, Sugino F and Trancanelli D 2020 Generating W states with braiding operators *Quantum Inf. Comput.* **13 & 14** 1154–62

Padmanabhan P, Sugino F and Trancanelli D 2021 Local invariants of braiding quantum gates—associated link polynomials and entangling power *J. Phys. A: Math. Theor.* **54** 135301

Rowell E C, Zhang Y, Wu Y-S and Ge M-L 2010 Extraspecial two-groups, generalized Yang-Baxter equations and braiding quantum gates *Quantum Inf. Comput.* **10** 685–702

Sudbery A 2001 On local invariants of pure three-qubit states *J. Phys. A: Math. Gen.* **34** 643–52

Turaev V G 1988 The Yang-Baxter equation and invariants of links *Invent. Math.* **92** 527–53

Vasquez J F, Wang Z and Wong H M 2016 Qubit representations of the braid groups from generalized Yang-Baxter matrices *Quantum Inf. Process.* **15** 3035–42

Veselov A P 2003 Yang-Baxter maps and integrable dynamics *Phys. Lett.* **A314** 214–21

Walter M, Gross D and Eisert J 2016 *Multipartite entanglement Quantum Information* ed D Bruss and G Leuchs (New York: Wiley) pp 293–330

Part IV

Polyadic categories

IOP Publishing

Polyadic Algebraic Structures

Steven Duplij

Chapter 11

Polyadic tensor categories

This chapter is devoted to a similar to medial algebras (see chapter 7) consideration of tensor categories (Mac Lane 1971, Etingof *et al* 2015). We define polyadic tensor categories by considering an n-ary tensor product (which may not be iterated from binary tensor products) and n-ary coherence conditions for the corresponding associators. The peculiarities of polyadic semigroupal and monoidal categories are studied and the differences from the corresponding binary tensor categories are outlined. We introduce a new kind of tensor categories, polyadic nonunital 'groupal' categories, which contain a 'querfunctor' and 'quertors' (similar to querelements in n-ary groups, Dörnte 1929, Post 1940). We introduce arity-nonreducible n-ary braidings and find the equation for them that in the binary case turns into the Yang–Baxter equation in the tensor product form.

11.1 Binary tensor categories

We now apply the above ideas to construct a special kind of *categories with multiplication* (Bénabou 1963, Mac Lane 1963) which appeared already in Tannaka (1939) and later on they were called tensor categories and monoidal categories (as they 'remind' us of the structure of a monoid) (Mac Lane 1971). For reviews, see, e.g., Calaque and Etingof (2008), Müger (2010). The monoidal categories can be considered as the *categorification* (Baez and Dolan 1998a) of a monoid object and can be treated as an instance of the *microcosm principle*: 'certain algebraic structures can be defined in any category equipped with a categorified version of the same structure' (Baez and Dolan 1998b). We start from the definitions of categories (Adámek *et al* 1990, Borceux 1994) and binary tensor categories (Mac Lane 1971) (in our notation).

Let $\mathscr{C}$ be a category with the class of objects Ob $\mathscr{C}$ and morphisms Mor $\mathscr{C}$, such that the arrow from the source X_1 to the target X_2 is defined by Mor $\mathscr{C} \ni \mathsf{f}_{12}$: $X_1 \to X_2$, $X_{1,2} \in$ Ob $\mathscr{C}$, and usually $\mathrm{Hom}_C(X_1, X_2)$ denotes all arrows which do not intersect. If Ob $\mathscr{C}$ and Mor $\mathscr{C}$ are sets, the category is *small*. The composition ($\circ$) of three morphisms, their associativity and the identity morphism (id_X) are defined in the standard way (Mac Lane 1971).

If $\mathscr{C}$ and $\mathscr{C}'$ are two categories, then a mapping between them is called a *covariant functor* $\mathscr{F}: \mathscr{C} \to \mathscr{C}'$ which consists of two different components: (1) the X-component is a mapping of objects $\mathscr{F}_{\mathrm{Ob}}$: Ob $\mathscr{C} \to$ Ob $\mathscr{C}'$; (2) the f-component is a mapping of morphisms $\mathscr{F}_{\mathrm{Mor}}$: Mor $\mathscr{C} \to$ Mor $\mathscr{C}'$ such that $\mathscr{F} = \{\mathscr{F}_{\mathrm{Ob}}, \mathscr{F}_{\mathrm{Mor}}\}$. A functor preserves the identity morphism $\mathscr{F}_{\mathrm{Mor}}(\mathrm{id}_X) = \mathrm{id}_{\mathscr{F}_{\mathrm{Ob}}(X)}$ and the composition of morphisms $\mathscr{F}_{\mathrm{Mor}}(\mathsf{f}_{23} \circ \mathsf{f}_{12}) = \mathscr{F}_{\mathrm{Mor}}(\mathsf{f}_{23}) \circ' \mathscr{F}_{\mathrm{Mor}}(\mathsf{f}_{12})$ $(= \mathscr{F}_{\mathrm{Mor}}(\mathsf{f}_{12}) \circ' \mathscr{F}_{\mathrm{Mor}}(\mathsf{f}_{23})$ for a *contravariant functor*), where ($\circ'$) is the composition in $\mathscr{C}'$.

The (binary) *product category* $\mathscr{C} \times \mathscr{C}'$ consists of all pairs of objects (Ob $\mathscr{C}$, Ob $\mathscr{C}'$), morphisms (Mor $\mathscr{C}$, Mor $\mathscr{C}'$) and identities (id_X, $\mathrm{id}_{X'}$), while the composition ($\circ''$) is made component-wise

$$\begin{gathered}(\mathsf{f}_{23}, \mathsf{f}'_{23}) \circ'' (\mathsf{f}_{12}, \mathsf{f}'_{12}) = (\mathsf{f}_{23} \circ \mathsf{f}_{12}, \mathsf{f}'_{23} \circ' \mathsf{f}'_{12}),\\ \mathsf{f}_{ij}: X_i \to X_j,\ \forall X_i \in \mathrm{Ob}\,\mathscr{C},\ \mathsf{f}'_{ij}: X'_i \to X'_j,\ \forall X'_i \in \mathrm{Ob}\,\mathscr{C}',\ i, j = 1, 2, 3,\end{gathered} \tag{11.1.1}$$

and by analogy this may be extended for more multipliers. A functor on a binary product category is called a *bifunctor* (*multifunctor*). A functor consists of two components $\{\mathscr{F}_{\mathrm{Ob}}, \mathscr{F}_{\mathrm{Mor}}\}$, and therefore a mapping between two functors $\mathscr{F}$ and $\mathscr{G}$ should also be two-component $\boldsymbol{T}^{FG} = \{\boldsymbol{T}^{FG}_{\mathrm{Ob}}, \boldsymbol{T}^{FG}_{\mathrm{Mor}}\}$. Usually (Mac Lane 1971), the components are denoted by the same letter, but for clarity we will distinguish them, because their action, arguments and corresponding commutative diagrams are different. Without other conditions $\boldsymbol{T}^{FG}$ is called an *infranatural transformation* from $\mathscr{F}$ to $\mathscr{G}$. A *natural transformation* (denoted by the double arrow $\boldsymbol{T}^{FG}: \mathscr{F} \Rightarrow \mathscr{G}$) is defined by the consistency condition of the above mappings in $\mathscr{C}'$

$$\boldsymbol{T}^{FG}_{\mathrm{Ob}} \circ' \mathscr{F}_{\mathrm{Mor}} = \mathscr{G}_{\mathrm{Mor}} \circ' \boldsymbol{T}^{FG}_{\mathrm{Ob}}. \tag{11.1.2}$$

Application to objects gives the following commutative diagram for the natural transformations (bifunctoriality)

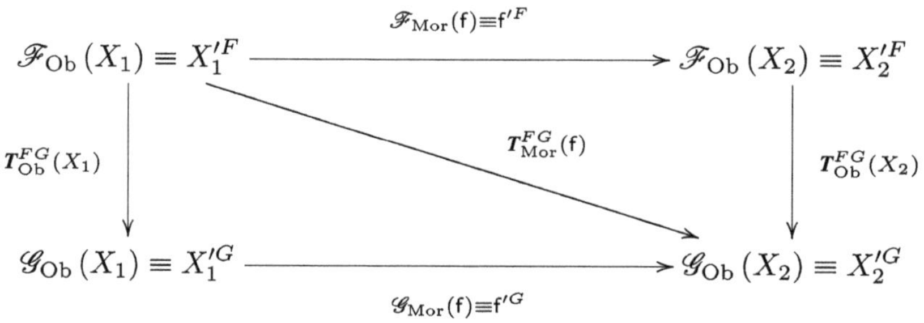

(11.1.3)

which is the consistency of the objects in $\mathscr{C}'$ transformed by $\mathscr{F}$ and $\mathscr{G}$. The diagonal in (11.1.3) may also be interpreted as the action of the natural transformation on a morphism $\boldsymbol{T}^{FG}_{\mathrm{Mor}}(\mathsf{f})$: $\mathscr{F}_{\mathrm{Ob}}(X_1) \to \mathscr{G}_{\mathrm{Ob}}(X_2)$, $\mathsf{f}: X_1 \to X_2$, $\mathsf{f} \in$ Mor $\mathscr{C}$, $X_1, X_2 \in$ Ob $\mathscr{C}$, such that

$$\boldsymbol{T}^{FG}_{\mathrm{Mor}}(\mathsf{f}) = \boldsymbol{T}^{FG}_{\mathrm{Ob}}(X_2) \circ' \mathscr{F}_{\mathrm{Mor}}(\mathsf{f}) = \mathscr{G}_{\mathrm{Mor}}(\mathsf{f}) \circ' \boldsymbol{T}^{FG}_{\mathrm{Ob}}(X_1), \tag{11.1.4}$$

where the second equality holds valid due to the naturality (11.1.2).

In a concise form the natural transformations are described by the commutative diagram

$$\mathscr{C} \underset{\mathscr{G}}{\overset{\mathscr{F}}{\genfrac{}{}{0pt}{}{\longrightarrow}{\Downarrow T^{FG}} \atop \longrightarrow}} \mathscr{C}' \tag{11.1.5}$$

For a category $\mathscr{C}$, the *identity functor* $\mathscr{I}d_{\mathscr{C}} = (\mathscr{I}d_{\mathscr{C},\,\mathrm{Ob}}, \mathscr{I}d_{\mathscr{C},\,\mathrm{Mor}})$ is defined by $\mathscr{I}d_{\mathscr{C},\,\mathrm{Ob}}(X) = X$, $\mathscr{I}d_{\mathscr{C},\,\mathrm{Mor}}(\mathsf{f}) = \mathsf{f}$, $\forall X \in \mathrm{Ob}\,\mathscr{C}$, $\forall \mathsf{f} \in \mathrm{Mor}\,\mathscr{C}$. Two categories $\mathscr{C}$ and $\mathscr{C}'$ are *equivalent*, if there exist two functors $\mathscr{F}$ and $\mathscr{G}$ and two natural transformations $\boldsymbol{T}^{FG}\colon \mathcal{I}d_{\mathscr{C}'} \Rightarrow \mathscr{F} \circ' \mathscr{G}$ and $\boldsymbol{T}^{GF}\colon \mathscr{G} \circ \mathscr{F} \Rightarrow \mathcal{I}d_{\mathscr{C}}$.

For more details and standard properties of categories, see, e.g., Mac Lane (1971), Adámek *et al* (1990), Borceux (1994) and references therein.

The categorification (Baez and Dolan 1998a, Crane and Yetter 1998) of most algebraic structures can be provided by endowing categories with an additional operation (Bénabou 1963, Mac Lane 1963) 'reminding' us of the tensor product (Mac Lane 1971).

A binary '*magmatic*' *tensor category* is $(\mathscr{C}, \mathscr{M}^{(2\otimes)})$, where $\mathscr{M}^{(2\otimes)} \equiv \otimes\colon \mathscr{C} \times \mathscr{C} \to \mathscr{C}$ is a bifunctor. We use this notation with brackets $\mathscr{M}^{(2\otimes)}$ (Aguiar and Mahajan 2010), because they are convenient for further consideration of the n-ary case (Duplij 2019). In component form the bifunctor is $\mathscr{M}^{(2\otimes)} = \left\{\mathscr{M}^{(2\otimes)}_{\mathrm{Ob}}, \mathscr{M}^{(2\otimes)}_{\mathrm{Mor}}\right\}$, where $\mathscr{M}^{(2\otimes)}_{\mathrm{Mor}}$ is

$$\begin{gathered}\mathscr{M}^{(2\otimes)}_{\mathrm{Mor}}\left[\mathsf{f}_{11'}, \mathsf{f}_{22'}\right] = \mathscr{M}^{(2\otimes)}_{\mathrm{Ob}}\left[X_1, X_2\right] \to \mathscr{M}^{(2\otimes)}_{\mathrm{Ob}}\left[X_1', X_2'\right], \\ \mathsf{f}_{ii'}\colon X_i \to X_i',\ \mathsf{f}_{ii'} \in \mathrm{Mor}\,\mathscr{C},\ \forall X_i, X_{i'} \in \mathrm{Ob}\,\mathscr{C}, \quad i = 1, 2.\end{gathered} \tag{11.1.6}$$

The composition of the f-components is determined by the binary mediality property (see (5.3.71))

$$\begin{gathered}\mathscr{M}^{(2\otimes)}_{\mathrm{Mor}}\left[\mathsf{f}_{23}, \mathsf{g}_{23}\right] \circ \mathscr{M}^{(2\otimes)}_{\mathrm{Mor}}\left[\mathsf{f}_{12}, \mathsf{g}_{12}\right] = \mathscr{M}^{(2\otimes)}_{\mathrm{Mor}}\left[\mathsf{f}_{23} \circ \mathsf{f}_{12}, \mathsf{g}_{23} \circ \mathsf{g}_{12}\right], \\ \mathsf{f}_{ij}\colon X_i \to X_j,\ \mathsf{g}_{ij}\colon Y_i \to Y_j,\ \mathsf{f}_{ij}, \mathsf{g}_{ij} \in \mathrm{Mor}\,\mathscr{C},\ \forall X_i, \\ Y_i \in \mathrm{Ob}\,\mathscr{C},\ i = 1, 2, 3.\end{gathered} \tag{11.1.7}$$

The identity of the tensor product satisfies

$$\mathscr{M}^{(2\otimes)}_{\mathrm{Mor}}[\mathrm{id}_{X_1}, \mathrm{id}_{X_2}] = \mathrm{id}_{\mathscr{M}^{(2\otimes)}_{\mathrm{Ob}}[X_1, X_2]}. \tag{11.1.8}$$

We call a category $\mathscr{C}$ a *strict* (*binary*) *semigroupal* (Yetter 2001, Lu *et al* 2019) (or *strictly associative semigroupal category*, Boyarchenko 2007, also, *semi-monoidal*, Kock 2008), if the bifunctor $\mathscr{M}^{(2\otimes)}$ satisfies only (without unit objects and unitors) the binary associativity condition $(X_1 \otimes X_2) \otimes X_3 = X_1 \otimes (X_2 \otimes X_3)$ and $(\mathsf{f}_1 \otimes \mathsf{f}_2) \otimes \mathsf{f}_3 = \mathsf{f}_1 \otimes (\mathsf{f}_2 \otimes \mathsf{f}_3)$, where $X_i \in \mathrm{Ob}\,\mathscr{C}$, $\mathsf{f}_i \in \mathrm{Mor}\,\mathscr{C}$, $i = 1, 2, 3$ (also denoted by `sSGCat`). Strict associativity is the equivalence

$$\mathscr{M}^{(2\otimes)}_{\mathrm{Ob}}[\mathscr{M}^{(2\otimes)}_{\mathrm{Ob}}[X_1, X_2], X_3] = \mathscr{M}^{(2\otimes)}_{\mathrm{Ob}}[X_1, \mathscr{M}^{(2\otimes)}_{\mathrm{Ob}}[X_2, X_3]], \tag{11.1.9}$$

$$\mathscr{M}^{(2\otimes)}_{\mathrm{Mor}}[\mathscr{M}^{(2\otimes)}_{\mathrm{Mor}}[\mathsf{f}_1, \mathsf{f}_2], \mathsf{f}_3] = \mathscr{M}^{(2\otimes)}_{\mathrm{Mor}}[\mathsf{f}_1, \mathscr{M}^{(2\otimes)}_{\mathrm{Mor}}[\mathsf{f}_2, \mathsf{f}_3]]. \tag{11.1.10}$$

***Remark* 11.1.1.** Usually, only the first equation for the X-components is presented in the definition of associativity (and other properties), while the equation for the f-components is assumed to be satisfied 'automatically' having the same form (Mac Lane 1971, Stasheff 1970). In some cases, the diagrams for $\mathcal{M}_{\mathrm{Ob}}^{(2\otimes)}$ and $\mathcal{M}_{\mathrm{Mor}}^{(2\otimes)}$ can fail to coincide and have different shapes, for instance, in the case of the *dagger categories* dealing with the 'reverse' morphisms (Abramsky and Coecke 2008).

The associativity relations guarantee that in any product of objects or morphisms different ways of inserting parentheses lead to equivalent results (as for semigroups).

In the case of a *non-strict semigroupal category* `SGCat` (with no unit objects and unitors) (Yetter 2001, Boyarchenko 2007) (see, also, Lu *et al* 2019, Elgueta 2004, Davydov 2007) a collection of mappings can be introduced which are just the isomorphisms (*associators*)$\mathrm{A}^{(3\otimes)} = \left\{\mathrm{A}_{\mathrm{Ob}}^{(3\otimes)}, \mathrm{A}_{\mathrm{Mor}}^{(3\otimes)}\right\}$ from the left side functor to the right side functor of (11.1.9) and (11.1.10) as

$$\mathrm{A}_{\mathrm{Ob}}^{(3\otimes)}(X_1, X_2, X_3): \mathcal{M}_{\mathrm{Ob}}^{(2\otimes)}[\mathcal{M}_{\mathrm{Ob}}^{(2\otimes)}[X_1, X_2], X_3] \overset{\cong}{\Rightarrow} \mathcal{M}_{\mathrm{Ob}}^{(2\otimes)}[X_1, \mathcal{M}_{\mathrm{Ob}}^{(2\otimes)}[X_2, X_3]], \quad (11.1.11)$$

where $\mathrm{A}_{\mathrm{Mor}}^{(3\otimes)}$ may be interpreted similarly to the diagonal in (11.1.3), because the associators are natural transformations (Mac Lane 1971) or *trifunctorial isomorphisms* (in the terminology of Boyarchenko 2007). Now different ways of inserting parentheses in a product of N objects give different results in the absence of conditions on the associator $\mathrm{A}^{(3\otimes)}$. However, if the associator $\mathrm{A}^{(3\otimes)}$ satisfies some consistency relations, they can give isomorphic results, such that the corresponding diagrams commute, which is the statement of the *coherence theorem* (Mac Lane 1963, Kelly 1964). This can also be applied to `SGCat`, because it can be proved independently of existence of units (Yetter 2001, Boyarchenko 2007, Lu *et al* 2019). It was shown (Mac Lane 1963) that it is sufficient to consider one commutative diagram using the associator (the *associativity constraint*) for two different rearrangements of parentheses for three tensor multiplications of four objects, giving the following isomorphism

$$\mathcal{M}_{\mathrm{Ob}}^{(2\otimes)}\left[\mathcal{M}_{\mathrm{Ob}}^{(2\otimes)}\left[\mathcal{M}_{\mathrm{Ob}}^{(2\otimes)}[X_1, X_2], X_3\right], X_4\right] \overset{\cong}{\Rightarrow} \mathcal{M}_{\mathrm{Ob}}^{(2\otimes)}\left[X_1, \mathcal{M}_{\mathrm{Ob}}^{(2\otimes)}\left[X_2, \mathcal{M}_{\mathrm{Ob}}^{(2\otimes)}[X_3, X_4]\right]\right]. \quad (11.1.12)$$

The constraint of associativity is called a *pentagon axiom* (Mac Lane 1971), such that the diagram

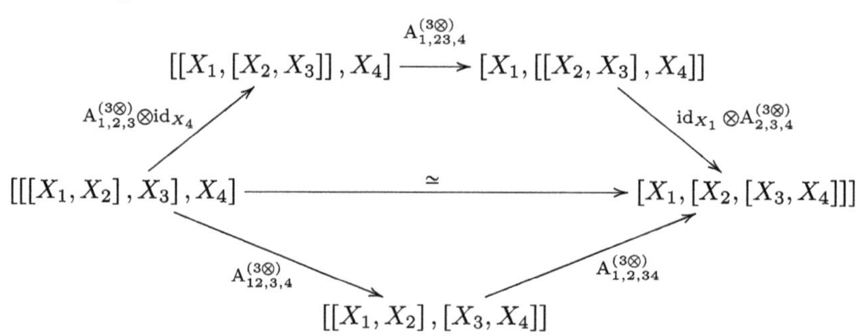

(11.1.13)

commutes.

We omit $\mathscr{M}_{\mathrm{Ob}}^{(2\otimes)}$ in diagrams by leaving the square brackets only and using the obvious subscripts in $\mathrm{A}^{(3\otimes)}$.

A similar condition for morphisms, but in another context (for H-spaces), was presented in Stasheff (1963, 1970). Note that there exists a different (but not alternative) approach to natural associativity without the use of the pentagon axiom (Joyce 2001).

The transition from the semigroupal non-strict category `SGCat` to the monoidal non-strict category `MonCat` can be done in a way similar to passing from a semigroup to a monoid: by adding the *unit object* $E \in \mathrm{Ob}\,\mathscr{C}$ and the (*right and left*) *unitors* $\boldsymbol{U}_{(1)}^{(2\otimes)} = \left\{\boldsymbol{U}_{(1)\mathrm{Ob}}^{(2\otimes)}, \boldsymbol{U}_{(1)\mathrm{Mor}}^{(2\otimes)}\right\}$ and $\boldsymbol{U}_{(2)}^{(2\otimes)} = \left\{\boldsymbol{U}_{(2)\mathrm{Ob}}^{(2\otimes)}, \boldsymbol{U}_{(2)\mathrm{Mor}}^{(2\otimes)}\right\}$ ('unit morphisms' which are functorial isomorphisms, natural transformations) (Mac Lane 1971)

$$\boldsymbol{U}_{(1)\mathrm{Ob}}^{(2\otimes)}\colon \mathscr{M}_{\mathrm{Ob}}^{(2\otimes)}[X, E] \overset{\simeq}{\Rightarrow} X, \tag{11.1.14}$$

$$\boldsymbol{U}_{(2)\mathrm{Ob}}^{(2\otimes)}\colon \mathscr{M}_{\mathrm{Ob}}^{(2\otimes)}[E, X] \overset{\simeq}{\Rightarrow} X, \quad \forall X \in \mathrm{Ob}\,\mathscr{C}, \tag{11.1.15}$$

and $\boldsymbol{U}_{(1,\,2)\mathrm{Mor}}^{(2\otimes)}$ can be viewed as the diagonal in the diagram of naturality similar to (11.1.3). The unitors are connected with the associator $\mathrm{A}^{(3\otimes)}$, such that the diagram (*triangle axiom*)

$$\begin{array}{ccc} [[X_1, E], X_2] & \xrightarrow{\mathrm{A}_{\mathrm{Ob}}^{(3\otimes)}} & [X_1, [E, X_2]] \\ & {\scriptstyle \boldsymbol{U}_{(1)\,\mathrm{Ob}}^{(2\otimes)}\otimes \mathrm{id}_{X_2}} \searrow \quad \swarrow {\scriptstyle \mathrm{id}_{X_1}\otimes \boldsymbol{U}_{(2)\,\mathrm{Ob}}^{(2\otimes)}} & \\ & [X_1, X_2] & \end{array} \tag{11.1.16}$$

commutes.

Using the above, the definition of a *binary non-strict monoidal category* `MonCat` can be given as the 6-tuple $(\mathscr{C}, \mathscr{M}^{(2\otimes)}, \mathrm{A}^{(3\otimes)}, E, \boldsymbol{U}^{(2\otimes)})$ such that the pentagon axiom (11.1.13) and the triangle axiom (11.1.16) are satisfied (Mac Lane 1963, 1971) (see, also, Kelly 1964, 1965).

The following 'normalizing' relations for the unitors of a monoidal non-strict category

$$\boldsymbol{U}_{(1)\mathrm{Ob}}^{(2\otimes)}(E) = \boldsymbol{U}_{(2)\mathrm{Ob}}^{(2\otimes)}(E), \tag{11.1.17}$$

can be proven (Joyal and Street 1993), as well as that the diagrams

$$\begin{array}{ccc} [[X_1, X_2], E] & \xrightarrow{\mathrm{A}_{\mathrm{Ob}}^{(3\otimes)}} & [X_1, [X_2, E]] \\ & {\scriptstyle \boldsymbol{U}_{(1)\,\mathrm{Ob}}^{(2\otimes)}} \searrow \quad \swarrow {\scriptstyle \mathrm{id}_{X_1}\otimes \boldsymbol{U}_{(1)\,\mathrm{Ob}}^{(2\otimes)}} & \\ & [X_1, X_2] & \end{array} \qquad \begin{array}{ccc} [[E, X_1], X_2] & \xrightarrow{\mathrm{A}_{\mathrm{Ob}}^{(3\otimes)}} & [E, [X_1, X_2]] \\ & {\scriptstyle \boldsymbol{U}_{(2)\,\mathrm{Ob}}^{(2\otimes)}\otimes \mathrm{id}_{X_2}} \searrow \quad \swarrow {\scriptstyle \boldsymbol{U}_{(2)\,\mathrm{Ob}}^{(2\otimes)}} & \\ & [X_1, X_2] & \end{array} \tag{11.1.18}$$

commute.

The *coherence theorem* (Bénabou 1963, Mac Lane 1963) proves that any diagram in a non-strict monoidal category, which can be built from an associator satisfying the pentagon axiom (11.1.13) and unitors satisfying the triangle axiom (11.1.16), commutes. Another formulation (Mac Lane 1971) states that every monoidal non-strict category is (monoidally) equivalent to a monoidal strict one (see also Kassel 1995).

Thus, it is important to prove analogs of the coherence theorem for various existing generalizations of categories (having weak modification of units, Kock 2008, Joyal and Kock 2013, Andrianopoulos 2017, and from the 'periodic table' of higher categories, Baez and Dolan 1995) as well as for further generalizations (e.g., n-ary ones below).

11.2 Polyadic tensor categories

The arity of the additional multiplication in a category (the tensor product) was previously taken to be binary. Here we introduce categories with tensor multiplication which 'remind' n-ary semigroups, n-ary monoids and n-ary groups (Dörnte 1929, Post 1940) (see, also Gal'mak 2003), i.e., we provide the categorification (Crane and Frenkel 1994, Crane and Yetter 1998) of 'higher-arity' structures according to the Baez–Dolan microcosm principle (Baez and Dolan 1998b). In our considerations we use the term 'tensor category' in a wider context, because it can include not only binary monoid-like structures and their combinations, but also n-ary-like algebraic structures. It is important to note that our construction is different from other higher generalizations of categories, such as 2-categories (Kelly and Street 1974) and bicategories (Bénabou 1967), n-categories (Baez 1997, Leinster 2002) and n-categories of n-groups (Aldrovandi and Noohi 2009), multicategories (Lambek 1969, Leinster 1998, Cruttwell and Shulman 2010), n-tuple categories and multiple categories (Grandis 2020), iterated (n-fold) monoidal categories (Balteanu *et al* 2003), iterated icons (Cheng and Gurski 2014), and obstructed categories (Duplij and Marcinek 2000, 2002). We introduce the categorification of 'higher-arity' structures like (Duplij 2019) and consider their properties, some of them are different from the binary case (as in n-ary (semi)groups and n-ary monoids). Also, the terms 'k-ary algebraic category' and 'k-ary category' appeared in Herrlich (1971) and Shulman (2012), respectively, but they describe different constructions.

Let $\mathscr{C}$ be a category (Mac Lane 1971), we introduce an additional multiplication as an n-ary tensor product similar to Duplij (2012, 2019).

Definition 11.2.1. An n-ary tensor product in a category $\mathscr{C}$ is an n-ary functor

$$\mathscr{M}^{(n\otimes)}:\overbrace{\mathscr{C}\times\ldots\times\mathscr{C}}^{n}\rightarrow\mathscr{C} \tag{11.2.1}$$

having the component form $\mathscr{M}^{(n\otimes)}=\left\{\mathscr{M}^{(n\otimes)}_{\mathrm{Ob}},\mathscr{M}^{(n\otimes)}_{\mathrm{Mor}}\right\}$ where the f-component $\mathscr{M}^{(n\otimes)}_{\mathrm{Mor}}$ is

$$\begin{gathered}\mathscr{M}^{(n\otimes)}_{\mathrm{Mor}}\left[\mathsf{f}_{11'},\mathsf{f}_{22'},\ldots\mathsf{f}_{nn'}\right]=\mathscr{M}^{(n\otimes)}_{\mathrm{Ob}}\left[X_1,X_2,\ldots X_n\right]\rightarrow\mathscr{M}^{(n\otimes)}_{\mathrm{Ob}}\left[X_1',X_2',\ldots X_n'\right],\\ \mathsf{f}_{ii'}:X_i\rightarrow X_i',\ \mathsf{f}_{ii'}\in\mathrm{Mor}\,\mathscr{C},\ \forall X_i,X_i'\in\mathrm{Ob}\,\mathscr{C},\ i=1,\ldots,n.\end{gathered} \tag{11.2.2}$$

The n-ary composition of the f-components (morphism products of length n) is determined by the n-ary mediality property (see (7.3.4))

$$\mathcal{M}^{(n\otimes)}_{\mathrm{Mor}}\Big[\mathsf{f}^{(1,1)}, \mathsf{f}^{(1,2)}, \dots, \mathsf{f}^{(1,n)}\Big] \circ \dots \circ \mathcal{M}^{(n\otimes)}_{\mathrm{Mor}}\Big[\mathsf{f}^{(n,1)}, \mathsf{f}^{(n,2)}, \dots, \mathsf{f}^{(n,n)}\Big],$$
$$= \mathcal{M}^{(n\otimes)}_{\mathrm{Mor}}\Big[\mathsf{f}^{(1,1)} \circ \mathsf{f}^{(2,1)} \circ \dots \circ \mathsf{f}^{(n,1)}, \dots, \mathsf{f}^{(1,n)} \circ \dots \circ \mathsf{f}^{(n,n)}\Big], \tag{11.2.3}$$
$$\mathsf{f}^{(i,j)} \in \mathrm{Mor}\,\mathcal{C}, \quad i, j = 1, 2 \dots, n.$$

The identity morphism of the n-ary tensor product satisfies

$$\mathcal{M}^{(n\otimes)}_{\mathrm{Mor}}\Big[\mathrm{id}_{X_1}, \mathrm{id}_{X_2}, \dots, \mathrm{id}_{X_n}\Big] = \mathrm{id}_{\mathcal{M}^{(n\otimes)}_{\mathrm{Ob}}[X_1, X_2 \dots, X_n]}. \tag{11.2.4}$$

Definition 11.2.2. An n-ary tensor product $\mathcal{M}^{(n\otimes)}$ which can be constructed from a binary tensor product $\mathcal{M}'^{(2\otimes)}$ by successive (iterative) repetitions is called an *arity-reduced tensor product* (by analogy with the 'derived n-ary group') (Dörnte 1929, Post 1940), and otherwise it is called an *arity-nonreduced tensor product.*

Categories containing iterations of the binary tensor product were considered in Balteanu *et al* (2003), Cheng and Gurski (2014). We will mostly be interested in the arity-nonreducible tensor products and their corresponding categories.

Definition 11.2.3. A polyadic (n-ary) '*magmatic*' *tensor category* is $(\mathcal{C}, \mathcal{M}^{(n\otimes)})$, where $\mathcal{M}^{(n\otimes)}$ is an n-ary tensor product (functor (11.2.1)), and it is called an *arity-reduced category* or *arity-nonreduced category* depending on its tensor product.

11.2.1 Polyadic semigroupal categories

We call sequences of objects and morphisms X-polyads and f-polyads (Post 1940) and denote them $\mathbf{X}$ and $\mathbf{f}$, respectively (as in (1.1.28)).

Definition 11.2.4. The n-ary functor $\mathcal{M}^{(n\otimes)}$ is *totally* (n-ary) *associative*, if it satisfies the following $(n-1)$ pairs of X equivalences

$$\mathcal{M}^{(n\otimes)}_{\mathrm{Ob}}[\mathbf{X}, \mathcal{M}^{(n\otimes)}_{\mathrm{Ob}}[\mathbf{Y}], \mathbf{Z}] = \text{equivalent}, \tag{11.2.5}$$

where $\mathbf{X}, \mathbf{Y}, \mathbf{Z}$ are X-polyads of the necessary length, and the total length of each $(\mathbf{X}, \mathbf{Y}, \mathbf{Z})$-polyad is $2n-1$, while the internal tensor products in (11.2.5) can be on any of the n places.

Example 11.2.5. In the ternary case ($n = 3$) the total associativity for the X-polyads of the length $5 = 2 \cdot 3 - 1$ gives $2 = 3 - 1$ pairs of equivalences

$$\mathcal{M}^{(3\otimes)}_{\mathrm{Ob}}\Big[\mathcal{M}^{(3\otimes)}_{\mathrm{Ob}}[X_1, X_2, X_3], X_4, X_5\Big] = \mathcal{M}^{(3\otimes)}_{\mathrm{Ob}}\Big[X_1, \mathcal{M}^{(3\otimes)}_{\mathrm{Ob}}[X_2, X_3, X_4], X_5\Big]$$
$$= \mathcal{M}^{(3\otimes)}_{\mathrm{Ob}}\Big[X_1, X_2, \mathcal{M}^{(3\otimes)}_{\mathrm{Ob}}[X_3, X_4, X_5]\Big], \tag{11.2.6}$$
$$\forall X_i \in \mathrm{Ob}\,\mathcal{C}, \quad \forall \mathsf{f}_i \in \mathrm{Mor}\,\mathcal{C},$$
$$i = 1, \dots, 5.$$

Definition 11.2.6. A category $(\mathscr{C}, \mathscr{M}^{(n\otimes)})$ is called a *polyadic (n-ary) strict semigroupal category* $\mathtt{sSGCat}_n$, if the bifunctor $\mathscr{M}^{(n\otimes)}$ satisfies objects and unitors) the n-ary associativity condition (11.2.5).

Thus, in a polyadic strict semigroupal category for any (allowed, i.e., having the size $k(n-1)+1$, $\forall k \in \mathbb{N}$, where k is the number of n-ary tensor multiplications) product of objects (or morphisms), all different ways of inserting parentheses give equivalent results (as for n-ary semigroups).

11.2.2 *n*-Ary coherence

As in the binary case (11.1.11), the transition to non-strict categories results in the consideration of independent isomorphisms instead of the equivalence (11.2.5).

Definition 11.2.7. The $(n-1)$ pairs of X and f isomorphisms $\mathrm{A}^{(2n-1)\otimes} = \left\{\mathrm{A}_{\mathrm{Ob}}^{(2n-1)\otimes}, \mathrm{A}_{\mathrm{Mor}}^{(2n-1)\otimes}\right\}$ such that

$$\mathrm{A}_{i,\,\mathrm{Ob}}^{(2n-1)\otimes}\colon \mathscr{M}_{\mathrm{Ob}}^{(n\otimes)}\left[\mathbf{X}, \mathscr{M}_{i,\,\mathrm{Ob}}^{(n\otimes)}[\mathbf{Y}], \mathbf{Z}\right] \overset{\cong}{\rightrightarrows} \mathscr{M}_{\mathrm{Ob}}^{(n\otimes)}\left[\mathbf{X}, \mathscr{M}_{i+1,\,\mathrm{Ob}}^{(n\otimes)}[\mathbf{Y}], \mathbf{Z}\right], \tag{11.2.7}$$

are called *n-ary associators* being $(2n-1)$-place natural transformations, where $\mathrm{A}_{\mathrm{Mor}}^{(2n-1)\otimes}$ may be viewed as corresponding diagonals as in (11.1.3). Here $i = 1, \ldots, n-1$ is the place of the internal brackets.

In the ternary case ($n = 3$) we have $2 = 3 - 1$ pairs of the ternary associators

$$\begin{aligned} \mathrm{A}_{1,\,\mathrm{Ob}}^{(5\otimes)}\colon &\mathscr{M}_{\mathrm{Ob}}^{(3\otimes)}\left[\mathscr{M}_{\mathrm{Ob}}^{(3\otimes)}\left[X_1, X_2, X_3\right], X_4, X_5\right] \\ &\overset{\cong}{\rightrightarrows} \mathscr{M}_{\mathrm{Ob}}^{(3\otimes)}\left[X_1, \mathscr{M}_{\mathrm{Ob}}^{(3\otimes)}\left[X_2, X_3, X_4\right], X_5\right], \end{aligned} \tag{11.2.8}$$

and

$$\begin{aligned} \mathrm{A}_{2,\,\mathrm{Ob}}^{(5\otimes)}\colon &\mathscr{M}_{\mathrm{Ob}}^{(3\otimes)}\left[X_1, \mathscr{M}_{\mathrm{Ob}}^{(3\otimes)}\left[X_2, X_3, X_4\right], X_5\right] \\ &\overset{\cong}{\rightrightarrows} \mathscr{M}_{\mathrm{Ob}}^{(3\otimes)}\left[X_1, X_2, \mathscr{M}_{\mathrm{Ob}}^{(3\otimes)}\left[X_3, X_4, X_5\right]\right]. \end{aligned} \tag{11.2.9}$$

It is now definite that different ways of inserting parentheses in a product of N objects will give different results (the same will be true for morphisms as well), if we do not impose constraints on the associators. We anticipate that we will need (as in the binary case (11.1.12)) only one more (i.e., three) tensor multiplication than appears in the associativity conditions (11.2.5) to make a commutative diagram for the following isomorphism of $3 \cdot (n-1) + 1 = 3n - 2$ objects

$$\begin{aligned} &\mathscr{M}_{\mathrm{Ob}}^{(n\otimes)}\left[\mathscr{M}_{\mathrm{Ob}}^{(n\otimes)}\left[\mathscr{M}_{\mathrm{Ob}}^{(n\otimes)}\left[X_1, \ldots, X_n\right], X_{n+1}, \ldots, X_{2n-1}\right], X_{2n}, \ldots, X_{3n-2}\right] \\ &\overset{\cong}{\rightrightarrows} \mathscr{M}_{\mathrm{Ob}}^{(n\otimes)}\left[X_1, \ldots, X_{n-1}, \mathscr{M}_{\mathrm{Ob}}^{(n\otimes)}\left[X_n, \ldots, X_{2n-2}, \mathscr{M}_{\mathrm{Ob}}^{(n\otimes)}\left[X_{2n-1}, \ldots, X_{3n-2}\right]\right]\right]. \end{aligned} \tag{11.2.10}$$

Conjecture 11.2.8. *N*-ary coherence. If the n-ary associator $\mathrm{A}^{(2n-1)\otimes}$ satisfies such *n-ary coherence conditions* that the isomorphism (11.2.10) takes place, then any diagram containing $\mathrm{A}^{(2n-1)\otimes}$ together with the identities (11.2.4) commutes.

The n-ary coherence conditions are described by a '(n^2+1)-gon', which is the pentagon (11.1.13) for $n = 2$ (for classification of 'N-gons' see, e.g., Wenninger 1974).

Definition 11.2.9. A category $(\mathscr{C}, \mathscr{M}^{(n\otimes)})$ is called a *polyadic* (*n-ary*) *non-strict semigroupal category* $\mathtt{sSGCat}_n$, if the bifunctor $\mathscr{M}^{(n\otimes)}$ satisfies the n-ary coherence.

Example 11.2.10. In the ternary case $n = 3$ we have 2 pairs of 5-place associators (11.2.8)–(11.2.9)$\mathrm{A}_1^{(5\otimes)}$ and $\mathrm{A}_2^{(5\otimes)}$ which act on $7 = 3 \cdot 3 - 2$ objects (11.2.10). We consider the diagram for objects only, then the associativity constraint for the associators $\mathrm{A}_{1,\,\mathrm{Ob}}^{(5\otimes)}$ and $\mathrm{A}_{2,\,\mathrm{Ob}}^{(5\otimes)}$ will be a *decagon axiom* requiring that the diagram

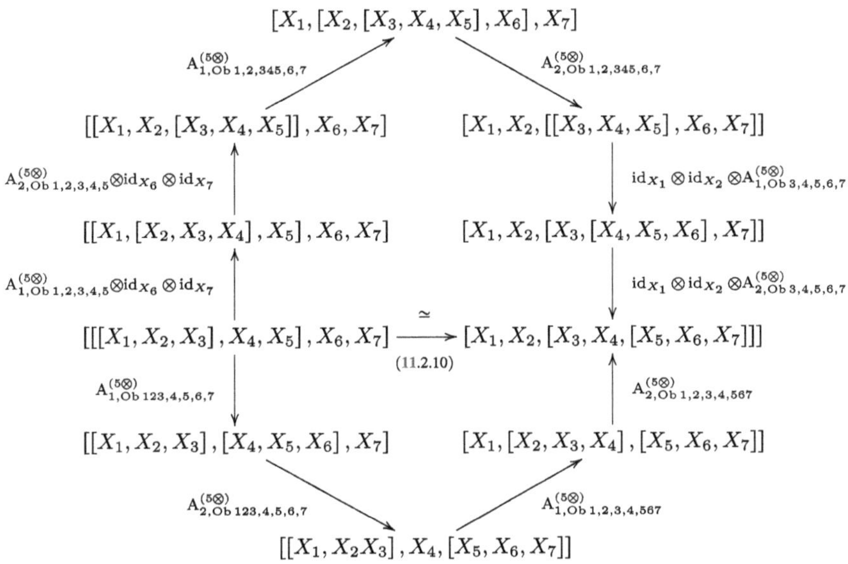

(11.2.11)

commutes (see (11.2.11) and the pentagon axiom (11.1.13) for binary non-strict tensor categories).

11.3 Polyadic units, unitors and quertors

Introducing n-ary (polyadic) analogs of units and unitors is nontrivial, because in n-ary (polyadic) structures there are various possibilities: one unit, many units, all elements are units or there are no units at all (see, e.g., for n-ary groups, Dörnte 1929,

Post 1940, Gal'mak 2003, and for n-ary monoids, Pop and Pop 2010). A similar situation is expected in the category theory after proper categorification (Crane and Frenkel 1994, Crane and Yetter 1998, Baez and Dolan 1998a) of n-ary (polyadic) structures.

11.3.1 Polyadic monoidal categories

Let $(\mathscr{C}, \mathscr{M}^{(n\otimes)}, \mathrm{A}^{(2n-1)\otimes})$ be an n-ary non-strict semigroupal category $\mathtt{SGCat}_n$ (see definition 11.2.6) with n-ary tensor product $\mathscr{M}^{(n\otimes)}$ and the associator $\mathrm{A}^{(2n-1)\otimes}$ satisfying n-ary coherence. If a category has a *unit neutral sequence* of objects $\mathbf{E}_{(n-1)} = (E_1, \dots, E_i)$, $E_i \in \mathrm{Ob}\,\mathscr{C}$, $i = 1, \dots, n-1$, we call it a *unital* category. Note that the unit neutral sequence may not be unique. If all E_i coincide $E_i = E \in \mathrm{Ob}\,\mathscr{C}$, then E is called a *unit object* of $\mathscr{C}$. The n-ary *unitors* $\boldsymbol{U}_{(i)}^{(n\otimes)}$, $i = 1, \dots, n$ (n-ary 'unit morphisms' being natural transformations) are defined by

$$\boldsymbol{U}_{(i)\mathrm{Ob}}^{(n\otimes)}: \mathscr{M}_{\mathrm{Ob}}^{(n\otimes)}\left[E_1, \dots E_{i-1}, X, E_{i+1}, \dots E_n\right] \overset{\cong}{\Rrightarrow} X, \qquad \forall X, E_i \in \mathrm{Ob}\,\mathscr{C}, i = 1, \dots, n-1. \tag{11.3.1}$$

The n-ary unitors $\boldsymbol{U}_{(i)}^{(n\otimes)}$ are compatible with the n-ary associators $\mathrm{A}^{(2n-1)\otimes}$ by the analog of the triangle axiom (11.1.16). In the binary case (11.1.14)–(11.1.15), we have $\boldsymbol{U}_{(1)}^{(2\otimes)} = \mathrm{R}^{(2\otimes)}$, $\boldsymbol{U}_{(2)}^{(2\otimes)} = \mathrm{L}^{(2\otimes)}$.

Definition 11.3.1. A polyadic (n*-ary*) *non-strict monoidal category* $\mathtt{MonCat}_n$ is a polyadic (n-ary) non-strict semigroupal category $\mathtt{SGCat}_n$ endowed with a unit neutral sequence $\mathbf{E}_{(n-1)}$ and n unitors $\boldsymbol{U}_{(i)}^{(n\otimes)}$, $i = 1, \dots, n$, that is a 5-tuple $\left(\mathscr{C}, \mathscr{M}^{(n\otimes)}, \mathrm{A}^{(n\otimes)}, \mathbf{E}_{(n-1)}, \boldsymbol{U}^{(n\otimes)}\right)$ satisfying the '(n^2+1)-gon' axiom for the $(n-1)$ associators $\mathrm{A}_{(i)}^{(2n-1)\otimes}$ and the triangle axiom (the analog of (11.1.16)) for the unitors and associators compatibility condition.

Example 11.3.2. If we consider the ternary non-strict monoidal category $\mathtt{MonCat}_3$ with one unit object $E \in \mathrm{Ob}\,\mathscr{C}$, then we have 2 associators $\mathrm{A}_1^{(5\otimes)}$ and $\mathrm{A}_2^{(5\otimes)}$ satisfying the decagon axiom (11.2.11) and 3 unitors

$$\boldsymbol{U}_{(1)\mathrm{Ob}}^{(3\otimes)}: \mathscr{M}_{\mathrm{Ob}}^{(3\otimes)}[X, E, E] \overset{\cong}{\Rrightarrow} X, \tag{11.3.2}$$

$$\boldsymbol{U}_{(2)\mathrm{Ob}}^{(3\otimes)}: \mathscr{M}_{\mathrm{Ob}}^{(3\otimes)}\left[E, X, E\right] \overset{\cong}{\Rrightarrow} X, \tag{11.3.3}$$

$$\boldsymbol{U}_{(3)\mathrm{Ob}}^{(3\otimes)}: \mathscr{M}_{\mathrm{Ob}}^{(3\otimes)}\left[E, E, X\right] \overset{\cong}{\Rrightarrow} X, \quad \forall X \in \mathrm{Ob}\,\mathscr{C}, \tag{11.3.4}$$

which satisfy the 'normalizing' conditions $\boldsymbol{U}_{(i)\mathrm{Ob}}^{(3\otimes)}(E) = E$, $i = 1, 2, 3$ and the ternary analog of the triangle axiom (11.1.16), such that the diagram

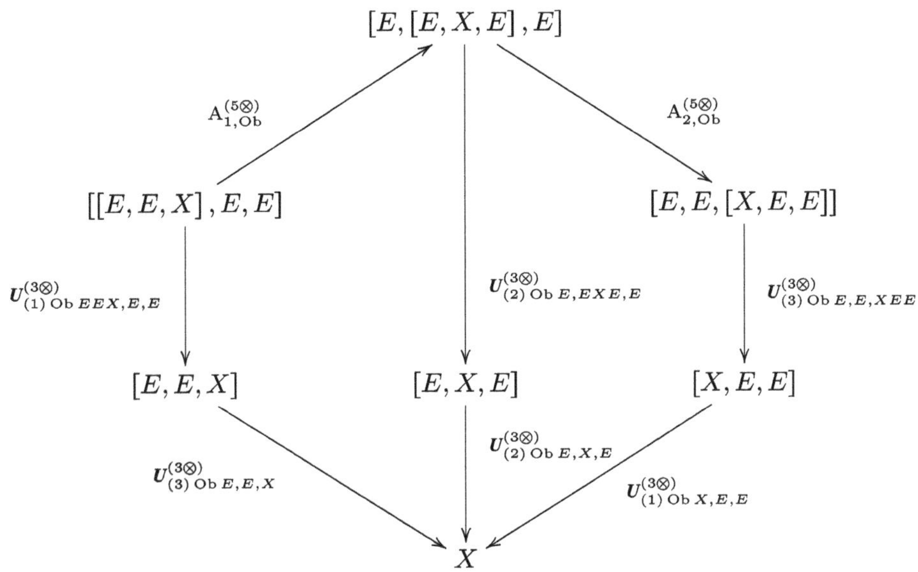

(11.3.5)

commutes.

11.3.2 Polyadic nonunital groupal categories

The main result of n-ary group theory (Dörnte 1929, Post 1940) is connected with units and neutral polyads: if they exist, then such n-ary group is reducible to a binary group. A similar statement can be true in some sense for categories.

Conjecture 11.3.3. If a polyadic (n-ary) tensor category has unit object and unitors, it can be arity-reducible to a binary category, such that the n-ary product can be obtained by iterations of the binary tensor product.

Therefore, it would be worthwhile to introduce and study nonreducible polyadic tensor categories which do not possess unit objects and unitors at all. This can be done by 'categorification' of the *querelement* concept (Dörnte 1929). Recall that, for instance, in a ternary group $\langle G | \mu_3 \rangle$ for an element $g \in G$ a querelement $\bar{g}$ is uniquely defined by $\mu_3[g, g, \bar{g}] = g$, which can be treated as a generalization of the inverse element concept to the n-ary case. The mapping $g \to \bar{g}$ can be considered as an additional unary operation (*queroperation*) in the ternary (and n-ary) group, while viewing it as an abstract algebra (Gleichgewicht and Głazek 1967) such that the notion of the identity is not used. The (binary) category of n-ary groups and corresponding functors were considered in Michalski (1979, 1984), Iancu (1991).

Let $(\mathcal{C}, \mathcal{M}^{(n\otimes)}, \mathrm{A}^{(2n-1)\otimes})$ be a polyadic (n-ary) non-strict semigroupal category, where $\mathcal{M}^{(n\otimes)}$ is the n-ary tensor product, and $\mathrm{A}^{(2n-1)\otimes}$ is the associator making the '(n^2+1)-gon' diagram of n-ary coherence commutative. We propose a 'categorification' analog of the queroperation to be a covariant endofunctor of $\mathcal{C}$.

Definition 11.3.4. A *querfunctor* $\mathcal{Q}\colon \mathscr{C} \to \mathscr{C}$ is an endofunctor of $\mathscr{C}$ sending $\mathcal{Q}_{\mathrm{Ob}}(X) = \bar{X}$ and $\mathcal{Q}_{\mathrm{Mor}}(\mathsf{f}) = \bar{\mathsf{f}}$, where $\bar{X}$ and $\bar{\mathsf{f}}$ are the *querobject* and the *quermorphism* of X and f, respectively, such that the i diagrams ($i = 1, \dots, n$)

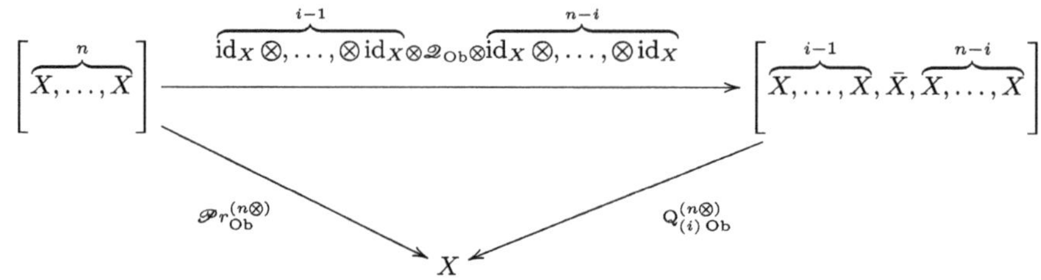

(11.3.6)

commute (and analogously for morphisms), where $\mathrm{Q}^{(n\otimes)}_{(i)}$ are *quertors*

$$\mathrm{Q}^{(n\otimes)}_{(i)\mathrm{Ob}}\colon \mathscr{M}^{(n\otimes)}_{\mathrm{Ob}}\Big[\overbrace{X, \dots, X}^{i-1}, \bar{X}, \overbrace{X, \dots, X}^{n-i}\Big] \xrightarrow{\cong} X, \quad \forall X \in \mathrm{Ob}\,\mathscr{C},\ i = 1, \dots, n, \tag{11.3.7}$$

and $\mathscr{P}r^{(n\otimes)}\colon \mathscr{C}^{n\otimes} \to \mathscr{C}$ is the projection. The action on morphisms $\mathrm{Q}^{(n\otimes)}_{(i)\mathrm{Mor}}$ can be found using the diagonal arrow in the corresponding natural transformation, as in (11.1.3).

Example 11.3.5. In the ternary case we have (for objects) the querfunctor $\mathcal{Q}_{\mathrm{Ob}}(X) = \bar{X}$ and three quertor isomorphisms

$$\mathrm{Q}^{(3\otimes)}_{(1)\mathrm{Ob}}\colon \mathscr{M}^{(3\otimes)}_{\mathrm{Ob}}\big[\bar{X}, X, X\big] \xrightarrow{\cong} X, \tag{11.3.8}$$

$$\mathrm{Q}^{(3\otimes)}_{(2)\mathrm{Ob}}\colon \mathscr{M}^{(3\otimes)}_{\mathrm{Ob}}\big[X, \bar{X}, X\big] \xrightarrow{\cong} X, \tag{11.3.9}$$

$$\mathrm{Q}^{(3\otimes)}_{(3)\mathrm{Ob}}\colon \mathscr{M}^{(3\otimes)}_{\mathrm{Ob}}\big[X, X, \bar{X}\big] \xrightarrow{\cong} X, \quad \forall X \in \mathrm{Ob}\,\mathscr{C}. \tag{11.3.10}$$

The three quertors $\mathrm{Q}^{(3\otimes)}_{(i)\mathrm{Ob}}$ and the querfunctor $\mathcal{Q}$ are connected with two ternary associators $\mathrm{A}^{(5\otimes)}_{1,\,\mathrm{Ob}}$, $\mathrm{A}^{(5\otimes)}_{2,\,\mathrm{Ob}}$ (11.2.8)–(11.2.9) such that the following diagram

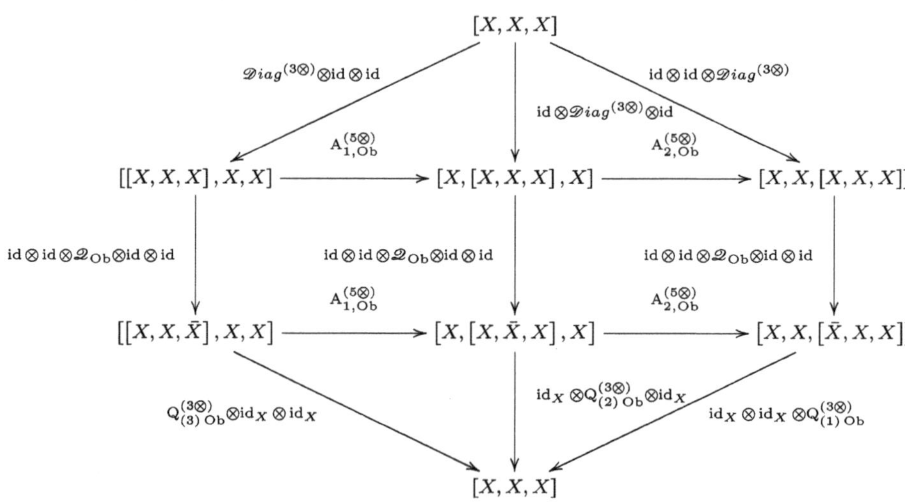

(11.3.11)

commutes, where $\mathscr{D}iag^{(n\otimes)}\colon \mathscr{C} \to \mathscr{C}^{n\otimes}$ is the diagonal.

Definition 11.3.6. A *nonunital non-strict groupal category* is

$$\mathtt{GCat}_n = (\mathscr{C}, \mathscr{M}^{(n\otimes)}, \mathrm{A}^{(2n-1)\otimes}, \mathscr{Q}, \mathrm{Q}^{(n\otimes)}),$$

i.e., a polyadic non-strict semigroupal category $\mathtt{SGCat}_n$ equipped with the querfunctor $\mathscr{Q}$ and the quertors $\mathrm{Q}^{(n\otimes)}$ satisfying (11.3.6).

Conjecture 11.3.7. There exist polyadic nonunital non-strict groupal categories which are arity-nonreducible (see definition 11.2.3), and so their n-ary tensor product cannot be presented in the form of binary tensor product iterations.

11.4 Braided tensor categories

The next step in the investigation of binary tensor categories is consideration of the tensor product 'commutativity' property. The tensor product can be 'commutative' such that for a tensor category $\mathcal{C}$ there exists the equivalence $X \otimes Y = Y \otimes X$, $\forall X, Y, \in \mathrm{Ob}\mathcal{C}$, and such tensor categories are called *symmetric* (Mac Lane 1971). By analogy with associativity, one can introduce non-strict 'commutativity', which leads to the notion of a braided (binary) tensor category and the corresponding coherence theorems (Joyal and Street 1993). Various generalizations of braiding were considered in Garner and Franco (2016), Duplij and Marcinek (2002, 2003), and their higher versions are found, e.g., in Kapranov and Voevodsky (1994), Batanin (2010), Weber (2005).

11.4.1 Braided binary tensor categories

Let $(\mathcal{C}, \mathscr{M}^{(2\otimes)}, \mathrm{A}^{(3\otimes)})$ be a non-strict semigroupal category with the bifunctor $\mathscr{M}^{(2\otimes)}$ and the associator $\mathrm{A}^{(3\otimes)}$ (11.1.11) satisfying the pentagon axiom (11.1.13) (Yetter 2001, Boyarchenko 2007).

Definition 11.4.1. A *(binary) braiding* $\mathrm{B}^{(2\otimes)}=\left\{\mathrm{B}^{(2\otimes)}_{\mathrm{Ob}}, \mathrm{B}^{(2\otimes)}_{\mathrm{Mor}}\right\}$ of a semigroupal category $\mathtt{SGCat}_2$ is a natural transformation of the bifunctor $\mathscr{M}^{(2\otimes)}$ (bifunctorial isomorphism) such that

$$\mathrm{B}^{(2\otimes)}_{\mathrm{Ob}}: \mathscr{M}^{(2\otimes)}_{\mathrm{Ob}}\left[X_1, X_2\right] \overset{\cong}{\Rightarrow} \mathscr{M}^{(2\otimes)}_{\mathrm{Ob}}\left[X_2, X_1\right], \quad \forall X_i \in \mathrm{Ob}\mathcal{C},\ i = 1, 2, \qquad (11.4.1)$$

and the action on morphisms $\mathrm{B}^{(2\otimes)}_{\mathrm{Mor}}$ may be interpreted as a diagonal, similarly to (11.1.3).

Definition 11.4.2. A non-strict semigroupal category endowed with a binary braiding is called a *(binary) braided semigroupal category* $\mathtt{bSGCat}_2$ $(\mathcal{C}, \mathscr{M}^{(2\otimes)}, \mathrm{A}^{(3\otimes)}, \mathrm{B}^{(2\otimes)})$.

The braiding $\mathrm{B}^{(2\otimes)}$ is connected with the associator $\mathrm{A}^{(3\otimes)}$ by the *hexagon identity*

$$\begin{array}{ccccc}
 & & [[X_1,X_2],X_3] & & \\
 & \swarrow^{\mathrm{B}^{(2\otimes)}_{\mathrm{Ob}\,1,2}\otimes \mathrm{id}_{X_3}} & & \searrow^{\mathrm{A}^{(3\otimes)}_{\mathrm{Ob}\,1,2,3}} & \\
[[X_2,X_1],X_3] & & & & [X_1,[X_2,X_3]] \\
\downarrow \mathrm{A}^{(3\otimes)}_{\mathrm{Ob}\,2,1,3} & & & & \downarrow \mathrm{B}^{(2\otimes)}_{(3)\,\mathrm{Ob}\,1,23} \\
[X_2,[X_1,X_3]] & & & & [[X_2,X_3],X_1] \\
 & \searrow_{\mathrm{id}_{X_2}\otimes \mathrm{B}^{(2\otimes)}_{\mathrm{Ob}\,1,3}} & & \swarrow_{\mathrm{A}^{(3\otimes)}_{\mathrm{Ob}\,2,3,1}} & \\
 & & [X_2,[X_3,X_1]] & &
\end{array} \tag{11.4.2}$$

for objects, and similarly for the inverse associator.

Definition 11.4.3. A *symmetric braided semigroupal category* $\mathtt{sbSGCat}_2$ has the 'invertible' braiding

$$\mathrm{B}^{(2\otimes)}_{\mathrm{Ob}X_1,\,X_2} \circ \mathrm{B}^{(2\otimes)}_{\mathrm{Ob}X_2,\,X_1} = \mathrm{id}_{X_1\otimes X_2} \qquad \text{or} \tag{11.4.3}$$

$$\mathrm{B}^{(2\otimes)}_{\mathrm{Ob}X_2,\,X_1} = \mathrm{B}^{(2\otimes),-1}_{\mathrm{Ob}X_1,\,X_2}, \quad \forall X_i \in \mathrm{Ob}\mathcal{C}. \tag{11.4.4}$$

A von Neumann regular generalization (von Neumann 1936) (weakening) of (11.4.3) leads to

Definition 11.4.4. A (*von Neumann*) *regular braided semigroupal category* is defined by a braiding which satisfies (Duplij and Marcinek 2001, 2003)

$$\mathrm{B}^{(2\otimes)}_{\mathrm{Ob}X_1,\,X_2} \circ \mathrm{B}^{*(2\otimes)}_{\mathrm{Ob}X_1,\,X_2} \circ \mathrm{B}^{(2\otimes)}_{\mathrm{Ob}X_1,\,X_2} = \mathrm{B}^{(2\otimes)}_{\mathrm{Ob}X_1,\,X_2}, \tag{11.4.5}$$

where $\mathrm{B}^{*(2\otimes)}_{\mathrm{Ob}X_1,\,X_2}$ is a generalized inverse (Penrose 1955, Nashed 1976) of $\mathrm{B}^{(2\otimes)}_{\mathrm{Ob}X_1,\,X_2}$, and such that $\mathrm{B}^{*(2\otimes)}_{\mathrm{Ob}X_1,\,X_2} \neq \mathrm{B}^{(2\otimes),-1}_{\mathrm{Ob}X_1,\,X_2}$ (see (11.4.4)).

Proposition 11.4.5. If the (binary) braided semigroupal category is strict (the associator becomes the equivalence (11.1.9) and (11.1.10), and we can omit internal brackets), then the diagram

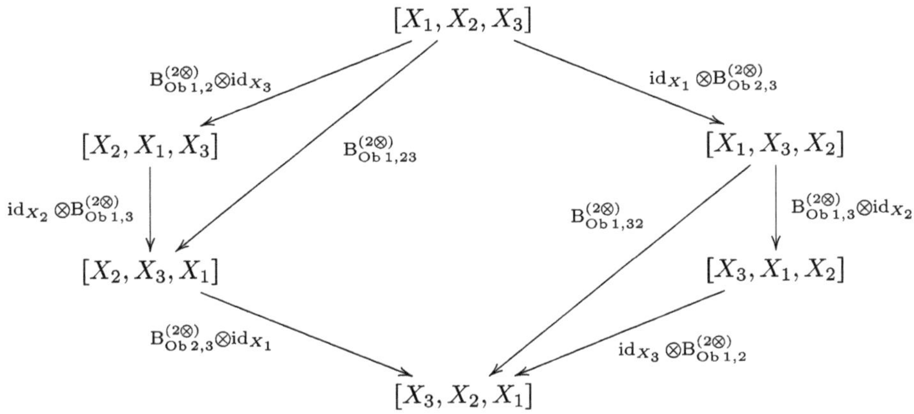

(11.4.6)

commutes (Stasheff 1963, 1970).

Proof. The triangles commute due to the hexagon identity (11.4.2) and the internal rectangle commutes, because the binary braiding $B^{(2\otimes)}$ is a natural transformation (bifunctorial isomorphism).

Omitting indices (11.4.6) becomes the *Yang–Baxter equation* in terms of tensor products (Drinfeld 1989) (or the *binary braid group* relation—for their difference, see Street 1995)

$$\left(B^{(2\otimes)}_{Ob} \otimes id\right) \circ \left(id \otimes B^{(2\otimes)}_{Ob}\right) \circ \left(B^{(2\otimes)}_{Ob} \otimes id\right) = \left(id \otimes B^{(2\otimes)}_{Ob}\right) \circ \left(B^{(2\otimes)}_{Ob} \otimes id\right) \circ \left(id \otimes B^{(2\otimes)}_{Ob}\right). \quad (11.4.7)$$

If the braided semigroupal category $\mathtt{bSGCat}_2$ contains a unit object, then we have

Definition 11.4.6. A (*binary*) *braided monoidal category*

$\mathtt{MonCat}_2(\mathcal{C}, \mathcal{M}^{(2\otimes)}, A^{(3\otimes)}, E, U^{(2\otimes)}, B^{(2\otimes)})$

is the category $\mathtt{bSGCat}_2$ together with a unit object $E \in \mathrm{Ob}\mathcal{C}$ satisfying the triangle axiom (11.1.16) and a unitor $U^{(2\otimes)}$ (11.1.14) and (11.1.15) the compatibility condition with the braiding $B^{(2\otimes)}$ such that the diagram (for objects)

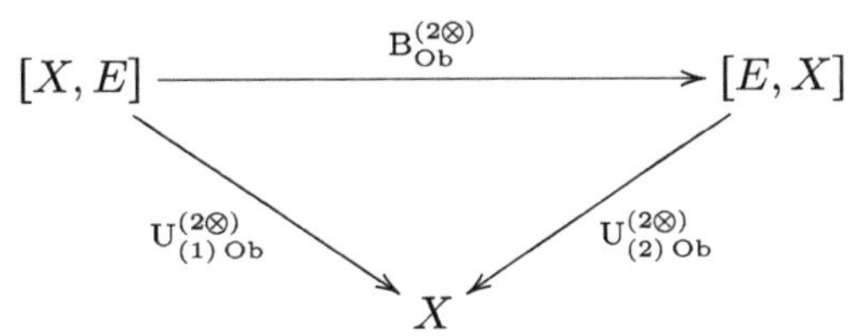

(11.4.8)

commutes.

For more details on binary braided monoidal categories, see Freyd and Yetter (1989), Joyal and Street (1993) and for review, see, e.g., Chari and Pressley (1996), Etingof *et al* (2015), Bulacu *et al* (2019).

11.4.2 Braided polyadic tensor categories

Higher braidings for binary tensor categories were considered (from an n-category viewpoint) in Manin and Schechtman (1989), Kapranov and Voevodsky (1994). We will discuss them for polyadic categories, defined above in section 11.2. The difference will be clearer if a polyadic category is not arity-reduced (see definition 11.2.3) and for nonunital groupal categories (section 11.3.2).

Let $(\mathcal{C}, \mathscr{M}^{(n\otimes)}, \mathrm{A}^{(2n-1)\otimes})$ be a polyadic non-strict semigroupal category, where $\mathscr{M}^{(n\otimes)}$ is an (not arity reduced) n-ary tensor product (n-ary functor) and $\mathrm{A}^{(2n-1)\otimes}$ is an associator, i.e., $n-1$ different $(2n-1)$-ary natural transformations (see definition 11.2.6). Now the braiding becomes an n-ary natural transformation, which leads to any of n permutations from the symmetry (permutation) group S_n rather than one possibility only, as for the binary braiding (11.4.1). Note that in the consideration of higher braidings (Manin and Schechtman 1989, Kapranov and Voevodsky 1994) one ('order reversing') element of S_n was used $\sigma_n^{(rev)} \equiv \begin{pmatrix} 1 & 2 & \dots & n \\ n & n-1 & \dots & 1 \end{pmatrix} \in S_n$. Thus, we arrive at the most general

Definition 11.4.7. An *n-ary braiding* $\mathrm{B}^{n\otimes}=\left\{\mathrm{B}_{\mathrm{Ob}}^{(n\otimes)}, \mathrm{B}_{\mathrm{Mor}}^{(n\otimes)}\right\}$ of a polyadic non-strict semigroupal category is an n-ary natural (or infranatural) transformation

$$\mathrm{B}_{\mathrm{Ob}}^{(n\otimes)}: \mathscr{M}_{\mathrm{Ob}}^{(n\otimes)}\left[\mathbf{X}\right] \overset{\cong}{\Rightarrow} \mathscr{M}_{\mathrm{Ob}}^{(n\otimes)}\left[\sigma_n \circ \mathbf{X}\right], \tag{11.4.9}$$

where $\mathbf{X}$ is an X-polyad (see definition 11.2.4) of the necessary length (which is n here), and $\sigma_n \in S_n$ are permutations that may satisfy some consistency conditions. The action on morphisms $\mathrm{B}_{\mathrm{Mor}}^{(n\otimes)}$ may be found from the corresponding diagonal of the natural transformation square (see (11.1.3)).

The binary non-mixed (standard) braiding (11.4.1) has $\sigma_2 = \sigma_2^{(rev)} = \begin{pmatrix} 1 & 2 \\ 2 & 1 \end{pmatrix} \in S_2$.

Definition 11.4.8. A polyadic (non-strict) semigroupal category endowed with the n-ary braiding $(\mathcal{C}, \mathscr{M}^{(n\otimes)}, \mathrm{A}^{(2n-1)\otimes}, \mathrm{B}^{(n\otimes)})$ is called a *braided semigroupal polyadic category* $\mathtt{bSGCat}_n$.

The n-ary braiding $\mathrm{B}^{(n\otimes)}$ is connected with the associator $\mathrm{A}^{(2n-1)\otimes}$ by a polyadic analog of the hexagon identity (11.4.2).

Example 11.4.9. In the case $n = 3$, the braided non-strict semigroupal ternary category $\mathtt{bSGCat}_3$ contains two associators $\mathrm{A}_1^{(5\otimes)}$ and $\mathrm{A}_2^{(5\otimes)}$ (see example 11.2.10) satisfying the decagon axiom (11.2.11). Let us take for the ternary braiding $\mathrm{B}^{(3\otimes)}$ its 'order reversing' version

$$\mathrm{B}_{\mathrm{Ob}}^{(3\otimes)}: \mathcal{M}_{\mathrm{Ob}}^{(2\otimes)}\left[X_1, X_2, X_3\right] \overset{\simeq}{\rightrightarrows} \mathcal{M}_{\mathrm{Ob}}^{(2\otimes)}\left[X_3, X_2, X_1\right], \quad \forall X_i \in \mathrm{Ob}\mathcal{C}, \quad i = 1, 2, 3. \tag{11.4.10}$$

Then the ternary analog of the hexagon identity is the *decagon identity* such that the diagram

$$\begin{aligned}
&[[X_1,X_2,X_3],X_4,X_5] \xrightarrow{\mathrm{B}^{(3\otimes)}_{\mathrm{Ob}\,1,2,3}\otimes \mathrm{id}_{X_4}\otimes \mathrm{id}_{X_5}} [[X_3,X_2,X_1],X_4,X_5] \xrightarrow{\mathrm{A}^{(5\otimes)}_{1,\mathrm{Ob}\,3,2,1,4,5}} [X_3,[X_2,X_1,X_4],X_5] \\
&\xrightarrow{\mathrm{id}_{X_3}\otimes \mathrm{B}^{(3\otimes)}_{\mathrm{Ob}\,2,1,4}\otimes \mathrm{id}_{X_5}} [X_3,[X_4,X_1,X_2],X_5] \xrightarrow{\mathrm{A}^{(5\otimes)}_{2,\mathrm{Ob}\,3,4,1,2,5}} [X_3,X_4,[X_1,X_2,X_5]] \xrightarrow{\mathrm{id}_{X_3}\otimes \mathrm{id}_{X_4}\otimes \mathrm{B}^{(3\otimes)}_{\mathrm{Ob}\,1,2,5}} [X_3,X_4,[X_5,X_2,X_1]], \\
&[[X_1,X_2,X_3],X_4,X_5] \xrightarrow{\mathrm{A}^{(5\otimes)}_{1,\mathrm{Ob}\,1,2,3,4,5}} [X_1,[X_2,X_3,X_4],X_5] \xrightarrow{\mathrm{A}^{(5\otimes)}_{2,\mathrm{Ob}\,1,2,3,4,5}} [X_1,X_2,[X_3,X_4,X_5]] \xrightarrow{\mathrm{B}^{(3\otimes)}_{\mathrm{Ob}\,1,2,345}} [[X_3,X_4,X_5],X_2,X_1] \\
&\xrightarrow{\mathrm{A}^{(5\otimes)}_{1,\mathrm{Ob}\,3,4,5,2,1}} [X_3,[X_4,X_5,X_2],X_1] \xrightarrow{\mathrm{A}^{(5\otimes)}_{2,\mathrm{Ob}\,3,4,5,2,1}} [X_3,X_4,[X_5,X_2,X_1]]
\end{aligned} \tag{11.4.11}$$

commutes.

Conjecture 11.4.10. Braided *n*-ary coherence. If the n-ary associator $\mathrm{A}^{(2n-1)\otimes}$ satisfies such *n-ary coherence conditions* that the isomorphism (11.2.10) takes place, and the n-ary braiding $\mathrm{B}^{(n\otimes)}$ satisfies the polyadic analog of the hexagon identity, then any diagram containing $\mathrm{A}^{(2n-1)\otimes}$ and $\mathrm{B}^{(n\otimes)}$ commutes.

Proposition 11.4.11. If the braided semigroupal ternary category $\mathtt{bSGCat}_3$ is strict (the associators becomes equivalences, and we can omit internal brackets), then the diagram containing only the ternary braidings $\mathrm{B}^{(3\otimes)}$

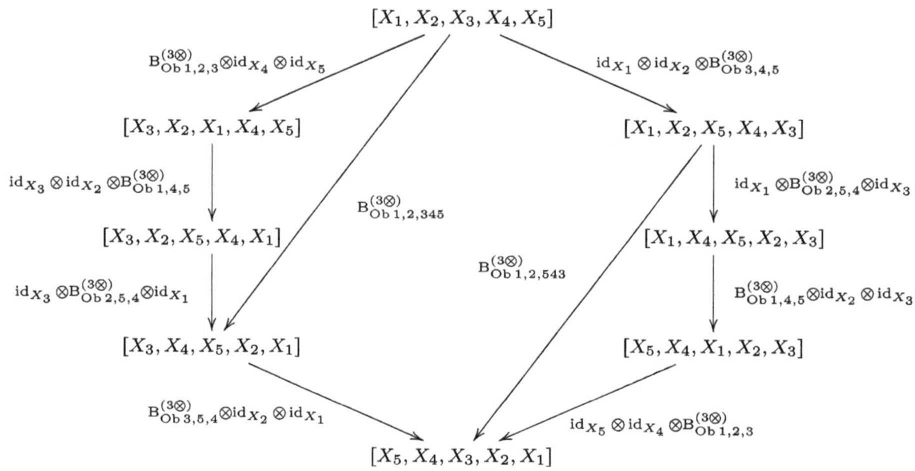

(11.4.12)

commutes (see the binary braiding (11.4.6)).

Proof. This is analogous to (11.4.6).

There follows from (11.4.12), omitting indices, the *ternary braid group relation* in terms of tensor products (see the tetrahedron equation, Bazhanov and Stroganov 1982, Kapranov and Voevodsky 1994, Baez and Neuchl 1996)

$$\begin{aligned}&(\mathrm{B}^{(3\otimes)} \otimes \mathrm{id} \otimes \mathrm{id}) \circ (\mathrm{id} \otimes \mathrm{B}^{(3\otimes)} \otimes \mathrm{id}) \circ (\mathrm{id} \otimes \mathrm{id} \otimes \mathrm{B}^{(3\otimes)}) \circ (\mathrm{B}^{(3\otimes)} \otimes \mathrm{id} \otimes \mathrm{id})\\ &= (\mathrm{id} \otimes \mathrm{id} \otimes \mathrm{B}^{(3\otimes)}) \circ (\mathrm{B}^{(3\otimes)} \otimes \mathrm{id} \otimes \mathrm{id}) \circ (\mathrm{id} \otimes \mathrm{B}^{(3\otimes)} \otimes \mathrm{id}) \circ (\mathrm{id} \otimes \mathrm{id} \otimes \mathrm{B}^{(3\otimes)}),\end{aligned} \tag{11.4.13}$$

which was obtained in Duplij (2021) using another approach: by the associative quiver technique from Duplij (2012). For instance, the 4-*ary braid group* relation for 4-ary braiding $\mathrm{B}^{(4\otimes)}$ has the form

$$\begin{aligned}&(\mathrm{B}^{(4\otimes)} \otimes \mathrm{id} \otimes \mathrm{id} \otimes \mathrm{id}) \circ (\mathrm{id} \otimes \mathrm{B}^{(4\otimes)} \otimes \mathrm{id} \otimes \mathrm{id}) \circ (\mathrm{id} \otimes \mathrm{id} \otimes \mathrm{B}^{(4\otimes)} \otimes \mathrm{id})\\ &\circ (\mathrm{id} \otimes \mathrm{id} \otimes \mathrm{id} \otimes \mathrm{B}^{(4\otimes)}) \circ (\mathrm{B}^{(4\otimes)} \otimes \mathrm{id} \otimes \mathrm{id} \otimes \mathrm{id}) = (\mathrm{id} \otimes \mathrm{id} \otimes \mathrm{id} \otimes \mathrm{B}^{(4\otimes)})\\ &\circ (\mathrm{B}^{(4\otimes)} \otimes \mathrm{id} \otimes \mathrm{id} \otimes \mathrm{id}) \circ (\mathrm{id} \otimes \mathrm{id} \otimes \mathrm{B}^{(4\otimes)} \otimes \mathrm{id}) \circ (\mathrm{id} \otimes \mathrm{B}^{(4\otimes)} \otimes \mathrm{id} \otimes \mathrm{id})\\ &\circ (\mathrm{id} \otimes \mathrm{id} \otimes \mathrm{id} \otimes \mathrm{B}^{(4\otimes)}).\end{aligned} \tag{11.4.14}$$

For the non-mixed 'order reversing' n-ary braiding (see definition 11.4.7) we have (Duplij 2021)

Proposition 11.4.12. The n-ary braid equation contains $(n + 1)$ multipliers, and each one acts on $(2n - 1)$ tensor products as

$$\begin{aligned}
&\Big(\mathrm{B}^{(n\otimes)}\otimes\overbrace{\mathrm{id}\otimes\ldots\otimes\mathrm{id}}^{n-1}\Big)\circ\Big(\mathrm{id}\otimes\mathrm{B}^{(n\otimes)}\otimes\overbrace{\mathrm{id}\otimes\ldots\otimes\mathrm{id}}^{n-2}\Big)\circ\\
&\quad\Big(\mathrm{id}\otimes\mathrm{id}\otimes\mathrm{B}^{(n\otimes)}\otimes\overbrace{\mathrm{id}\otimes\ldots\otimes\mathrm{id}}^{n-3}\Big)\ldots\\
&\circ\Big(\overbrace{\mathrm{id}\otimes\ldots\otimes\mathrm{id}}^{n-2}\otimes\mathrm{B}^{(n\otimes)}\otimes\mathrm{id}\Big)\circ\Big(\overbrace{\mathrm{id}\otimes\ldots\otimes\mathrm{id}}^{n-1}\otimes\mathrm{B}^{(n\otimes)}\Big)\circ\Big(\mathrm{B}^{(n\otimes)}\otimes\overbrace{\mathrm{id}\otimes\ldots\otimes\mathrm{id}}^{n-1}\Big)\\
&=\Big(\overbrace{\mathrm{id}\otimes\ldots\otimes\mathrm{id}}^{n-1}\otimes\mathrm{B}^{(n\otimes)}\Big)\circ\Big(\mathrm{B}^{(n\otimes)}\otimes\overbrace{\mathrm{id}\otimes\ldots\otimes\mathrm{id}}^{n-1}\Big)\circ\Big(\mathrm{id}\otimes\mathrm{B}^{(n\otimes)}\otimes\overbrace{\mathrm{id}\otimes\ldots\otimes\mathrm{id}}^{n-2}\Big)\ldots\\
&\circ\Big(\overbrace{\mathrm{id}\otimes\ldots\otimes\mathrm{id}}^{n-3}\otimes\mathrm{B}^{(n\otimes)}\otimes\mathrm{id}\otimes\mathrm{id}\Big)\circ\Big(\overbrace{\mathrm{id}\otimes\ldots\otimes\mathrm{id}}^{n-2}\otimes\mathrm{B}^{(n\otimes)}\otimes\mathrm{id}\Big)\\
&\circ\Big(\overbrace{\mathrm{id}\otimes\ldots\otimes\mathrm{id}}^{n-1}\otimes\mathrm{B}^{(n\otimes)}\Big).
\end{aligned}\tag{11.4.15}$$

***Remark* 11.4.13.** If a polyadic category is arity-nonreducible, then the higher n-ary braid relations cannot be 'iterated', i.e., obtained from the lower n ones.

Consider a polyadic monoidal category $\mathtt{MonCat}_n$ with one unit object E (see definition 11.3.1). Then the n-ary braiding $\mathrm{B}^{(n\otimes)}$ satisfies the triangle identity connecting it with the unitors $\mathrm{U}^{(n\otimes)}$.

Example 11.4.14. In the case of the ternary monoidal category $\mathtt{MonCat}_3$ (see example 11.3.2) the 'order reversing' braiding $\mathrm{B}^{(3\otimes)}$ (11.4.10) satisfies an additional triangle identity analogous to (11.4.8) such that the diagram

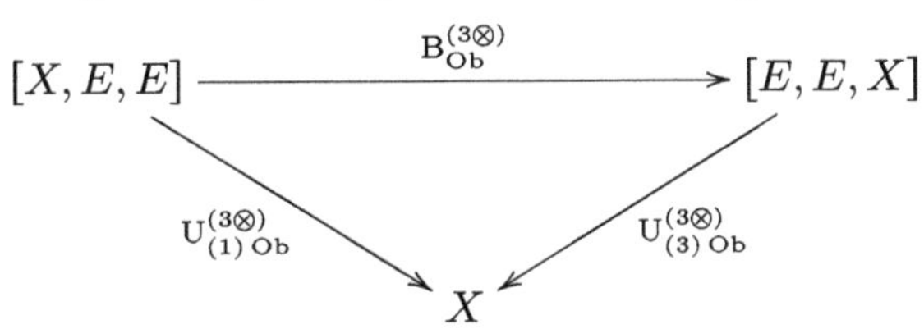

(11.4.16)

commutes.

For the polyadic nonunital groupal category $\mathtt{GCat}_n$ (see definition 11.3.6) the n-ary braiding $\mathrm{B}^{(n\otimes)}$ should be consistent with the quertors $\mathrm{U}^{(n\otimes)}$ and the querfunctor $\mathcal{Q}$ (see definition 11.3.4).

Definition 11.4.15. A *braided polyadic groupal category* $\mathtt{bGCat}_n$ is a polyadic groupal category $\mathtt{GCat}_n$ endowed with the n-ary braiding $(\mathcal{C}, \mathcal{M}^{(n\otimes)}, \mathrm{A}^{(2n-1)\otimes}, \mathcal{Q}, \mathrm{Q}^{(n\otimes)}, \mathrm{B}^{(n\otimes)})$.

Example 11.4.16. In the ternary groupal category $\mathtt{GCat}_3$ (see example 11.3.5) the 'order reversing' braiding $\mathrm{B}^{(3\otimes)}$ (11.4.10) satisfies the additional identity of consistency with the querfunctor $\mathcal{Q}$ and quertor $\mathrm{Q}^{(3\otimes)}$ such that the diagram

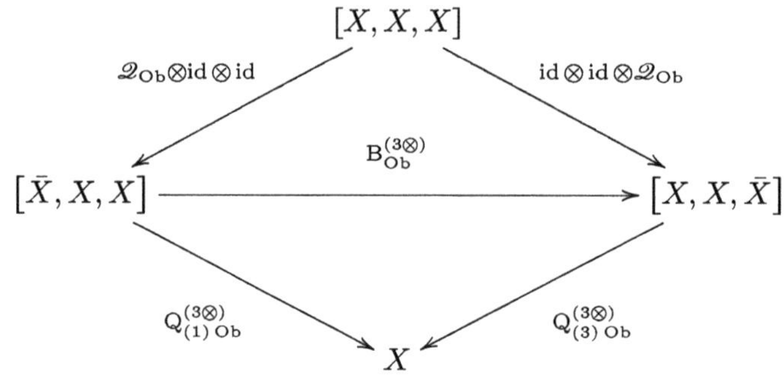

(11.4.17)

commutes.

The above diagrams ensure that some version of coherence can also be proven for braided polyadic categories.

11.5 Medialed polyadic tensor categories

Here we consider a medial approach to braiding inspired by the first part of our paper. As opposed to binary braiding which is defined by one unique permutation (11.4.1), the n-ary braiding can be defined by the enormous number of possible allowed permutations (11.4.9). Therefore, in most cases only one permutation, that is the 'order reversing', is usually (and artificially) used (see, e.g., Manin and Schechtman 1989) ignoring other possible cases. On the other side, for n-ary structures it is natural to use the mediality property (7.3.1) which is unique in the n-ary case and for binary groups reduces to commutativity. So we introduce a *medialing* instead of braiding for the tensor product in categories, and (by analogy with braided categories) we call them *medialed categories*.

Let $(\mathcal{C}, \mathcal{M}^{(n\otimes)}, \mathrm{A}^{(2n-1)\otimes})$ be a polyadic non-strict semigroupal category $\mathtt{SGCat}_n$ (see definition 11.2.6).

Definition 11.5.1. An *n-ary medialing* $\mathrm{M}^{(n^2\otimes)}$ (or 'medial braiding') is a *mediality constraint* which is a natural (or infranatural) transformation of two composed n-ary tensor product functors $\mathcal{M}^{(n\otimes)}$ (or functorial n^2-isomorphism)

$$\mathrm{M}_{\mathrm{Ob}}^{(n^2\otimes)}: \mathcal{M}_{\mathrm{Ob}}^{(n\otimes)}\begin{bmatrix} \mathcal{M}_{\mathrm{Ob}}^{(n\otimes)}\left[X_{11}, X_{12}, \dots, X_{1n}\right], \\ \mathcal{M}_{\mathrm{Ob}}^{(n\otimes)}\left[X_{21}, X_{22}, \dots, X_{2n}\right], \\ \vdots \\ \mathcal{M}_{\mathrm{Ob}}^{(n\otimes)}\left[X_{n1}, X_{n2}, \dots, X_{nn}\right] \end{bmatrix} \overset{\cong}{\Rightarrow} \mathcal{M}_{\mathrm{Ob}}^{(n\otimes)}\begin{bmatrix} \mathcal{M}_{\mathrm{Ob}}^{(n\otimes)}\left[X_{11}, X_{21}, \dots, X_{n1}\right], \\ \mathcal{M}_{\mathrm{Ob}}^{(n\otimes)}\left[X_{12}, X_{22}, \dots, X_{n2}\right], \\ \vdots \\ \mathcal{M}_{\mathrm{Ob}}^{(n\otimes)}[X_{1n}, X_{2n}, \dots, X_{nn}] \end{bmatrix}, \quad (11.5.1)$$

where the action on morphisms $\mathcal{M}_{\mathrm{Mor}}^{(n\otimes)}$ can be viewed as the corresponding diagonal in the natural transformation diagram as in (11.1.3).

Remark **11.5.2.** The advantage of n-ary medialing is its *uniqueness*, because it does not contain a huge number of possible permutations $\sigma_n \in S_n$ as does the n-ary braiding (11.4.9).

Example 11.5.3. In the binary case $n = 2$ we have (using the standard notation $\mathcal{M}^{(2\otimes)} \longrightarrow \otimes$)

$$\mathrm{M}_{\mathrm{Ob}}^{(4\otimes)}: (X_1 \otimes X_2) \otimes (X_3 \otimes X_4) \overset{\cong}{\Rightarrow} (X_1 \otimes X_3) \otimes (X_2 \otimes X_4), \qquad \forall X_i \in \mathrm{Ob}\mathcal{C},\ i = 1, \dots, 4, \quad (11.5.2)$$

which is called a *binary medialing* by analogy with binary braiding (11.4.1).

In the compact matrix notation (see definition 7.3.1) instead of (11.5.1) we have (symbolically)

$$\mathrm{M}_{\mathrm{Ob}}^{(n^2\otimes)}: \left(\mathcal{M}_{\mathrm{Ob}}^{(n\otimes)}\right)^2\left[\hat{X}_{(n^2)}\right] \overset{\cong}{\Rightarrow} \left(\mathcal{M}_{\mathrm{Ob}}^{(n\otimes)}\right)^2\left[\hat{X}_{(n^2)}^T\right], \quad (11.5.3)$$

where the matrix polyads of objects is (see (7.3.4))

$$\hat{X}_{(n^2)} = \left(X_{ij}\right) \in (\mathrm{Ob}\mathcal{C})^{\otimes n^2}, \qquad X_{ij} \in \mathrm{Ob}\mathcal{C}, \quad (11.5.4)$$

and $(\ )^T$ is matrix transposition.

Definition 11.5.4. A *medialed polyadic semigroupal category*

$$\mathtt{mSGCat}_n = \left(\mathcal{C}, \mathcal{M}^{(n\otimes)}, \mathrm{A}^{(2n-1)\otimes}, \mathrm{M}^{(n^2\otimes)}\right) \quad (11.5.5)$$

is a polyadic non-strict semigroupal category $\mathtt{SGCat}_n$ (see definition 11.2.9) endowed with the n-ary medialing $\mathrm{M}^{(n^2\otimes)}$ satisfying the *n-ary medial coherence condition* (a medial analog of the hexagon identity (11.4.2)).

Definition 11.5.5. A *medialed polyadic monoidal category*

$$\texttt{mMonCat}_n = \left(\mathcal{C}, \mathscr{M}^{(n\otimes)}, \mathrm{A}^{(2n-1)\otimes}, E, \mathrm{U}^{(n\otimes)}, \mathrm{M}^{(n^2\otimes)}\right)$$

is a medialed polyadic semigroupal category $\texttt{mSGCat}_n$ with the unit object $E \in \mathrm{Ob}\mathcal{C}$ and the unitor $\mathrm{U}^{(n\otimes)}$ satisfying some compatibility condition.

Let us consider the polyadic nonunital groupal category $\texttt{GCat}_n$ (see definition 11.3.6), then the n-ary medialing $\mathrm{M}^{(n^2\otimes)}$ should be consistent with the quertors $\mathrm{U}^{(n\otimes)}$ and the querfunctor $\mathscr{Q}$ (see definition 11.3.4 and also the consistency condition for the ternary braiding (11.4.17)).

Definition 11.5.6. A *braided polyadic groupal category*

$$\texttt{mGCat}_n = \left(\mathcal{C}, \mathscr{M}^{(n\otimes)}, \mathrm{A}^{(2n-1)\otimes}, \mathscr{Q}, \mathrm{Q}^{(n\otimes)}, \mathrm{M}^{(n^2\otimes)}\right)$$

is a polyadic groupal category $\texttt{GCat}_n$ endowed with the n-ary medialing $\mathrm{M}^{(n^2\otimes)}$.

11.5.1 Medialed binary and ternary categories

Due to the complexity of the relevant polyadic diagrams, it is not possible to draw them in a general case for arbitrary arity n. Therefore, it would be worthwhile to consider first the binary case, and then some of the diagrams for the ternary case.

Example 11.5.7. Let

$$\texttt{mSGCat}_2 = \left(\mathcal{C}, \mathscr{M}^{(2\otimes)}, \mathrm{A}^{(3\otimes)}, \mathrm{M}^{(4\otimes)}\right)$$

be a binary medialed semigroupal category and the binary medialing be as in (11.5.2). Then the medial analog of the hexagon identity (11.4.2) is given by the binary medial coherence condition such that the diagram

$$
\begin{array}{ccc}
[[[[X_1,X_2],X_3],X_4],X_5] & \xrightarrow{\mathrm{A}^{(3\otimes)}_{\mathrm{Ob}\,1,2,3}\otimes \mathrm{id}_{X_4}\otimes \mathrm{id}_{X_5}} & [[[X_1,[X_2,X_3]],X_4],X_5] \\
\downarrow{\scriptstyle \mathrm{A}^{(3\otimes)}_{\mathrm{Ob}\,12,3,4}\otimes \mathrm{id}_{X_5}} & & \downarrow{\scriptstyle \mathrm{A}^{(3\otimes)}_{\mathrm{Ob}\,123,4,5}} \\
[[[X_1,X_2],[X_3,X_4]],X_5] & & [[X_1,[X_2,X_3]],[X_4,X_5]] \\
\downarrow{\scriptstyle \mathrm{M}^{(4\otimes)}_{\mathrm{Ob}\,1,2,3,4}\otimes \mathrm{id}_{X_5}} & & \downarrow{\scriptstyle \mathrm{M}^{(4\otimes)}_{\mathrm{Ob}\,1,23,4,5}} \\
[[[X_1,X_3],[X_2,X_4]],X_5] & & [[X_1,X_4],[[X_2,X_3],X_5]] \\
\downarrow{\scriptstyle \mathrm{A}^{(3\otimes)}_{\mathrm{Ob}\,13,24,5}} & & \downarrow{\scriptstyle \mathrm{A}^{(3\otimes)}_{\mathrm{Ob}\,1,4,235}} \\
[[X_1,X_3],[[X_2,X_4],X_5]] & & [X_1,[X_4,[[X_2,X_3],X_5]]] \\
\downarrow{\scriptstyle \mathrm{M}^{(4\otimes)}_{\mathrm{Ob}\,1,3,24,5}} & & \downarrow{\scriptstyle \mathrm{id}_{X_1}\otimes \mathrm{id}_{X_4}\otimes \mathrm{A}^{(3\otimes)}_{\mathrm{Ob}\,2,3,5}} \\
[[X_1,[X_2,X_4]],[X_3,X_5]] & & [X_1,[X_4,[X_2,[X_3,X_5]]]] \\
\downarrow{\scriptstyle \mathrm{A}^{(3\otimes)-1}_{\mathrm{Ob}\,1,2,4}\otimes \mathrm{id}_{X_3}\otimes \mathrm{id}_{X_5}} & & \uparrow{\scriptstyle \mathrm{A}^{(3\otimes)}_{\mathrm{Ob}\,1,4,235}} \\
[[[X_1,X_2],X_4],[X_3,X_5]] & & [[X_1,X_4],[X_2,[X_3,X_5]]] \\
\downarrow{\scriptstyle \mathrm{A}^{(3\otimes)-1}_{\mathrm{Ob}\,124,3,5}} & & \uparrow{\scriptstyle \mathrm{id}_{X_1}\otimes \mathrm{id}_{X_4}\otimes \mathrm{A}^{(3\otimes)}_{\mathrm{Ob}\,2,3,5}} \\
[[[[X_1,X_2],X_4],X_3],X_5] & & [[X_1,X_4],[[X_2,X_3],X_5]] \\
\downarrow{\scriptstyle \mathrm{A}^{(3\otimes)}_{\mathrm{Ob}\,12,4,3}\otimes \mathrm{id}_{X_5}} & & \uparrow{\scriptstyle \mathrm{A}^{(3\otimes)}_{\mathrm{Ob}\,14,23,5}} \\
[[[X_1,X_2],[X_4,X_3]],X_5] & \xrightarrow{\mathrm{M}^{(4\otimes)}_{\mathrm{Ob}\,1,2,4,3}\otimes \mathrm{id}_{X_5}} & [[[X_1,X_4],[X_2,X_3]],X_5]
\end{array}
\tag{11.5.6}
$$

commutes.

If a medialed semigroupal category $\mathtt{mSGCat}_2$ contains a unit object and the unitor, then we have

Definition 11.5.8. A *medialed monoidal category*

$\mathtt{mMonCat}_2 = \left(\mathcal{C}, \mathscr{M}^{(2\otimes)}, \mathrm{A}^{(3\otimes)}, E, \mathrm{U}^{(2\otimes)}, \mathrm{M}^{(4\otimes)}\right)$

is a (*binary*) medialed semigroupal category $\mathtt{mSGCat}_2$ together with a unit object $E \in \mathrm{Ob}\mathcal{C}$ and a unitor $\mathrm{U}^{(2\otimes)}$ (11.1.14) and (11.1.15) satisfying the triangle axiom (11.1.16).

For $\mathtt{mMonCat}_2$ the compatibility condition of the medialing $\mathrm{M}^{(4\otimes)}$ with E and $\mathrm{U}^{(2\otimes)}$ is given by the commutative diagram

$$
\begin{array}{ccc}
[[X_1,E],[X,X_2]] & \xrightarrow{\mathrm{M}^{(4\otimes)}_{\mathrm{Ob}}} & [[X_1,X],[E,X_2]] \\
\Big\downarrow {\scriptstyle \mathrm{A}^{(3\otimes)}_{\mathrm{Ob}\,X_1,E,XX_2}} & & \Big\downarrow {\scriptstyle \mathrm{A}^{(3\otimes)}_{\mathrm{Ob}\,X_1,X,EX_2}} \\
[X_1,[E,[X,X_2]]] & & [X_1,[X,[E,X_2]]] \\
\Big\downarrow {\scriptstyle \mathrm{id}_{X_1}\otimes \mathrm{A}^{(3\otimes)-1}_{\mathrm{Ob}}} & & \Big\downarrow {\scriptstyle \mathrm{id}_{X_1}\otimes \mathrm{A}^{(3\otimes)-1}_{\mathrm{Ob}}} \\
[X_1,[[E,X],X_2]] & & [X_1,[[X,E],X_2]] \\
& \searrow {\scriptstyle \mathrm{id}_{X_1}\otimes \mathrm{U}^{(2\otimes)}_{(2)\,\mathrm{Ob}}\otimes \mathrm{id}_{X_2}} \quad {\scriptstyle \mathrm{id}_{X_1}\otimes \mathrm{U}^{(2\otimes)}_{(1)\,\mathrm{Ob}}\otimes \mathrm{id}_{X_2}} \swarrow & \\
& [X_1,[X,X_2]] &
\end{array}
\tag{11.5.7}
$$

which is an analog of the triangle diagram for braiding (11.4.8).

Example 11.5.9. In the ternary nonunital groupal category $\mathtt{GCat}_3$ (see example 11.3.5) the medialing $\mathrm{M}^{(9\otimes)}$ satisfies the additional identity of consistency with the querfunctor $\mathcal{Q}$ and quertor $\mathrm{Q}^{(3\otimes)}$ such that the diagram

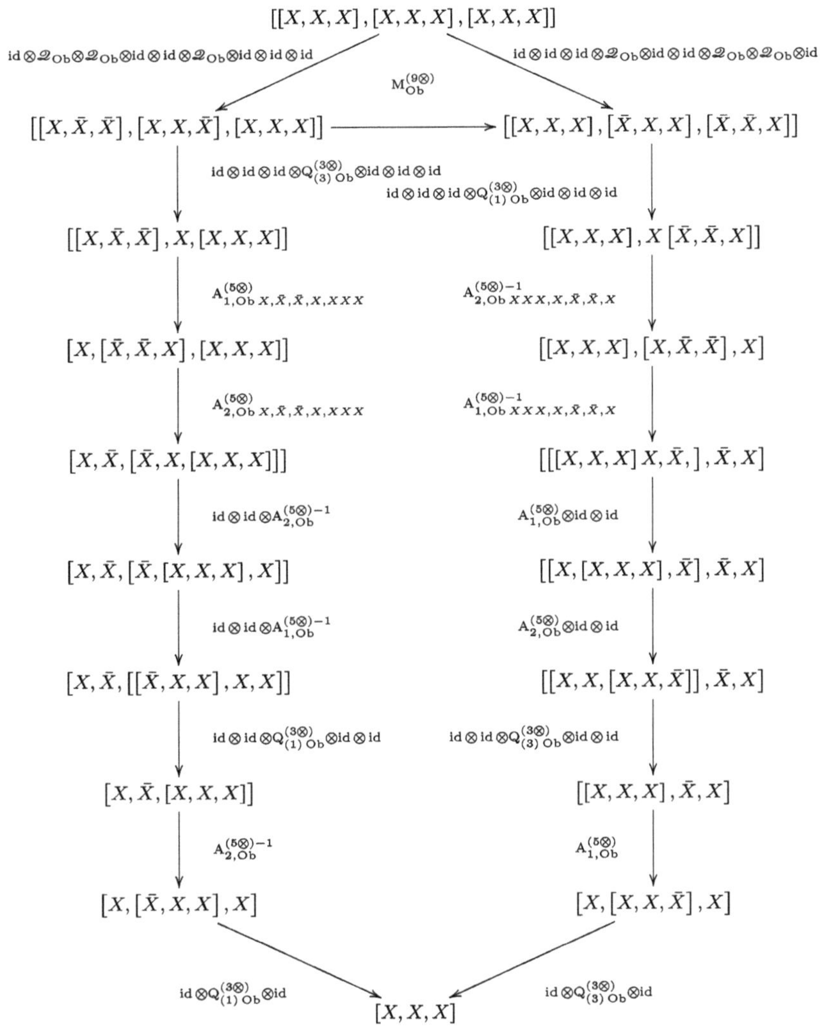

(11.5.8)

commutes. An analog of the hexagon identity in $\mathtt{GCat}_3$ can be expressed by a diagram which is similar to (11.5.6).

11.6 Conclusions

Commutativity in polyadic algebraic structures is defined non-uniquely, if one considers permutations and their combinations. We proposed a canonical way out: to substitute the commutativity property by mediality. Following this 'commutativity-to-mediality' ansatz, we first investigated mediality for graded linear n-ary algebras and arrived at the concept of almost mediality, which is an analog of almost commutativity. We constructed 'deforming' medial brackets, which could be treated as a medial analog of Lie brackets. We then proved Toyoda's theorem for almost medial n-ary algebras. Inspired by the above as examples, we proposed generalizing tensor and braided categories in a similar way. We defined polyadic tensor categories with an additional n-ary tensor multiplication for which a polyadic analog of the pentagon axiom was given. Instead of braiding we introduced n-ary 'medialing' which satisfies a medial analog of the hexagon identity and constructed the 'medialed' polyadic version of tensor categories.

References

Abramsky S and Coecke B 2008 Categorical quantum mechanics *Handbook of Quantum Logic and Quantum Structures* ed K Engesser, D M Gabbay and D Lehmann vol 2 (Amsterdam: Elsevier) pp 261–324

Adámek J, Herrlich H and Strecker G E 1990 *Abstract and Concrete Categories: The Joy of Cats* (New York: Wiley)

Aguiar M and Mahajan S A 2010 Monoidal functors, species and Hopf algebras *CRM Monograph Series* vol 29 (Providence, RI: American Mathematical Society)

Aldrovandi E and Noohi B 2009 Butterflies I: Morphisms of 2-group stacks *Adv. Math.* **221** 687–773

Andrianopoulos J 2017 Remarks on units of skew monoidal categories *Appl. Categ. Struct.* **25** 863–73

Baez J C 1997 An introduction to n-categories *7th Conference on Category Theory and Computer Science* (*Lecture Notes in Computer Science* vol 1290) ed E Moggi and G Rosolini (Berlin: Springer) pp 1–33

Baez J C and Dolan J 1995 Higher-dimensional algebra and topological quantum field theory *J. Math. Phys.* **36** 6073–105

Baez J C and Dolan J 1998a Categorification *Higher Category Theory Contemporary Mathematics* vol 230, ed E Getzler and M Kapranov (Providence, RI: American Mathematical Society) pp 1–36

Baez J C and Dolan J 1998b Higher-dimensional algebra III. n-Categories and the algebra of opetopes *Adv. Math.* **135** 145–206

Baez J C and Neuchl M 1996 Higher-dimensional algebra I: braided monoidal 2-categories *Adv. Math.* **121** 196–244

Balteanu C, Fiedorowicz Z, Schwänzl R and Vogt R 2003 Iterated monoidal categories *Adv. Math.* **176** 277–349

Batanin M 2010 Locally constant n-operads as higher braided operads *J. Noncommut. Geom.* **4** 237–63

Bazhanov V V and Stroganov Y G 1982 Conditions of commutativity of transfer matrices on a multidimensional lattice *Theor. Math. Phys.* **52** 685–91

Bénabou J 1963 Catégories avec multiplication *C. R. Acad. Sci. Paris* **256** 1887–90

Bénabou J 1967 Introduction to bicategories *Reports of the Midwest Category Seminar* (Berlin: Springer) pp 1–77

Borceux F 1994 Handbook of categorical algebra 1. Basic category theory *Encyclopedia of Mathematics and Its Applications* vol 50 (Cambridge: Cambridge University Press)

Boyarchenko M 2007 *Associativity Constraints in Monoidal Categories* (Chicago: University of Chicago Press) p 18 preprint

Bulacu D, Caenepeel S, Panaite F and Van Oystaeyen F 2019 *Quasi-Hopf Algebras: A Categorical Approach* (Cambridge: Cambridge University Press)

Calaque D and Etingof P 2008 Lectures on tensor categories *Quantum Groups IRMA Lectures in Mathematics and Theoretical Physics* vol 12, ed B Enriquez (Strasbourg: IRMA-EMS) pp 1–38

Chari V and Pressley A 1996 *A Guide to Quantum Groups* (Cambridge: Cambridge University Press)

Cheng E and Gurski N 2014 Iterated icons *Theory Appl. Categ.* **29** 929–77

Crane L and Frenkel I B 1994 Four-dimensional topological quantum field theory, Hopf categories, and the canonical bases *J. Math. Phys.* **35** 5136–54

Crane L and Yetter D N 1998 Examples of categorification *Cahiers Topol. Geom. Diff. Categ.* **39** 3–25

Cruttwell G S H and Shulman M A 2010 A unified framework for generalized multicategories *Theory Appl. Categ.* **24** 580–655

Davydov A 2007 Nuclei of categories with tensor products *Theory Appl. Categ.* **18** 440–72

Dörnte W 1929 Unterschungen über einen verallgemeinerten Gruppenbegriff *Math. Z.* **29** 1–19

Drinfeld V G 1989 Quasi-Hopf algebras and Knizhnik–Zamolodchikov equations *Problems of Modern Quantum Field Theory* ed A A Belavin, A V Klimyk and A B Zamolodchikov (Heidelberg: Springer) pp 1–13

Duplij S 2012 Polyadic systems, representations and quantum groups *J. Kharkiv Natl Univ. Ser. Nuclei Part. Fields* **1017** 28–59 Expanded version arXiv: math.RT/1308.4060

Duplij S 2019 Arity shape of polyadic algebraic structures *J. Math. Phys. Anal. Geom.* **15** 3–56

Duplij S 2021 Polyadic Hopf algebras and quantum groups *East Eur. J. Phys.* **2** 5–50

Duplij S and Marcinek W 2000 On higher regularity and monoidal categories *Kharkiv State Univ. J. (Vestnik KSU), Ser. Nuclei, Particles and Fields* **481** 27–30

Duplij S and Marcinek W 2001 Semisupermanifolds and regularization of categories, modules, algebras and Yang–Baxter equation *Supersymmetry and Quantum Field Theory* (Amsterdam: Elsevier) pp 110–5

Duplij S and Marcinek W 2002 Regular obstructed categories and topological quantum field theory *J. Math. Phys.* **43** 3329–41

Duplij S and Marcinek W 2003 Braid semistatistics and doubly regular R-matrix *J. Kharkiv Natl Univ. Ser. Nuclei Part. Fields* **601** 75–80

Elgueta J 2004 Cohomology and deformation theory of monoidal 2-categories *I Adv. Math.* **182** 204–77

Etingof P, Gelaki S, Nikshych D and Ostrik V 2015 *Tensor Categories* (Providence, RI: American Mathematical Society)

Freyd P J and Yetter D N 1989 Braided compact closed categories with applications to low dimensional topology *Adv. Math.* **77** 156–82

Gal'mak A M 2003 n-*Ary Groups, Part 1* (Gomel: Gomel University)

Garner R and Franco I L 2016 Commutativity *J. Pure Appl. Algebra* **220** 1707–51

Gleichgewicht B and Głazek K 1967 Remarks on n-groups as abstract algebras *Colloq. Math.* **17** 209–19

Grandis M 2020 *Higher Dimensional Categories: From Double to Multiple Categories* (Singapore: World Scientific)

Herrlich H 1971 A characterization of k-ary algebraic categories *Manuscr. Math.* **4** 277–84

Iancu L 1991 On the category of n-groups *Bul. Stiint. Univ. Baia Mare Ser.* B **7** 9–14

Joyal A and Kock J 2013 Coherence for weak units *Doc. Math.* **18** 71–110

Joyal A and Street R 1993 Braided tensor categories *Adv. Math.* **102** 20–78

Joyce W P 2001 Natural associativity without the pentagon condition *preprint Univ. Canterbury, Christchurch* p 64 (arXiv:math/0109088)

Kapranov M and Voevodsky V 1994 Braided monoidal 2-categories and Manin–Schechtman higher braid groups *J. Pure Appl. Algebra* **92** 241–67

Kassel C 1995 *Quantum Groups* (New York: Springer)

Kelly G M 1964 On Mac Lane's conditions for coherence of natural associativities, commutativities, etc *J. Algebra* **1** 397–402

Kelly G M 1965 Tensor products in categories *J. Algebra* **2** 15–37

Kelly G M and Street R 1974 Review of the elements of 2-categories *Category Seminar (Proc. Sem., Sydney, 1972/1973) Lecture Notes in Mathematics* vol 420 (Berlin: Springer) pp 75–103

Kock J 2008 Elementary remarks on units in monoidal categories *Math. Proc. Camb. Phil. Soc.* **144** 53–76

Lambek 1969 Deductive systems and categories III *Category Theory, Homology Theory and Their Applications, I Lecture Notes in Mathematics* vol 86 (Berlin: Springer) pp 76–122

Leinster T 1998 General operads and multicategories *preprint University of Cambridge, Cambridge* p 35 arXiv:math/9810053

Leinster T 2002 A survey of definitions of n-category *Theory Appl. Categ.* **10** 1–70

Lu X, Ye Y and Hu S 2019 A graphical calculus for semi-groupal categories *Appl. Categor. Struct.* **27** 163–97

Mac Lane S 1963 Natural associativity and commutativity *Rice Univ. Stud.* **49** 28–46

Mac Lane S 1971 *Categories for the Working Mathematician* (Berlin: Springer)

Manin Y I and Schechtman V 1989 Arrangements of hyperplanes, higher braid groups and higher Bruhat orders *Adv. Stud. Pure Math.* **17** 289–308

Michalski J 1979 On some functors from the category of n-groups *Bull. Acad. Polon. Sci. Ser. Sci. Math.* **27** 345–9

Michalski J 1984 On the category of n-groups *Fundam. Math.* **122** 187–97

Müger M 2010 Tensor categories: a selective guided tour *Rev. Union Mat. Argentina* **51** 95–163

Nashed M Z 1976 *Generalized Inverses and Applications* (New York: Academic)

Penrose R 1955 A generalized inverse for matrices *Math. Proc. Camb. Phil. Soc.* **51** 406–13

Pop A and Pop M S 2010 On generalized algebraic structures *Creat. Math. Inf.* **19** 184–90

Post E L 1940 Polyadic groups *Trans. Am. Math. Soc.* **48** 208–350

Shulman M 2012 Exact completions and small sheaves *Theory Appl. Categ.* **27** 97–173

Stasheff J 1970 H-*Spaces from a Homotopy Point of View* (Berlin: Springer)

Stasheff J D 1963 Homotopy associativity of H-spaces. I, II *Trans. Am. Math. Soc.* **108** 293–312

Street R 1995 Higher categories, strings, cubes and simplex equations *Appl. Categ. Struct.* **3** 29–77

Tannaka T 1939 Über den dualitätssatz der nichtkommutativen topologischen gruppen *Tôhoku Math. J.* **45** 1–12

von Neumann J 1936 On regular rings *Proc. Nat. Acad. Sci. USA* **22** 707–13

Weber M 2005 Operads within monoidal pseudo algebras *Appl. Categ. Struct.* **13** 389–420

Wenninger M J 1974 *Polyhedron Models* (Cambridge: Cambridge University Press)

Yetter D N 2001 Functorial knot theory. Categories of tangles, coherence, categorical deformations, and topological invariants *Series on Knots and Everything* vol 26 (River Edge: World Scientific)

www.ingramcontent.com/pod-product-compliance
Ingram Content Group UK Ltd.
Pitfield, Milton Keynes, MK11 3LW, UK
UKHW051908100926
8556IPUK00005B/813

9 780750 326469